The Formation and Evolution of Africa: A Synopsis of
3.8 Ga of Earth History

Geological Society books refereeing procedures

The Society makes every effort to ensure that the scientific and production quality of its books matches that of its journals. Since 1997, all book proposals have been refereed by specialist reviewers as well as by the Society's Books Editorial Committee. If the referees identify weaknesses in the proposal, these must be addressed before the proposal is accepted.

Once the book is accepted, the Society Book Editors ensure that the volume editors follow strict guidelines on refereeing and quality control. We insist that individual papers can only be accepted after satisfactory review by two independent referees. The questions on the review forms are similar to those for *Journal of the Geological Society*. The referees' forms and comments must be available to the Society's Book Editors on request.

Although many of the books result from meetings, the editors are expected to commission papers that were not presented at the meeting to ensure that the book provides a balanced coverage of the subject. Being accepted for presentation at the meeting does not guarantee inclusion in the book.

More information about submitting a proposal and producing a book for the Society can be found on its web site: www.geolsoc.org.uk.

GEOLOGICAL SOCIETY SPECIAL PUBLICATION NO. 357

The Formation and Evolution of Africa: A Synopsis of 3.8 Ga of Earth History

EDITED BY

D. J. J. VAN HINSBERGEN
University of Oslo, Norway

S. J. H. BUITER
Geological Survey of Norway

T. H. TORSVIK
University of Oslo

C. GAINA
Geological Survey of Norway

and

S. J. WEBB
University of the Witwatersrand, South Africa

2011
Published by
The Geological Society
London

THE GEOLOGICAL SOCIETY

The Geological Society of London (GSL) was founded in 1807. It is the oldest national geological society in the world and the largest in Europe. It was incorporated under Royal Charter in 1825 and is Registered Charity 210161.

The Society is the UK national learned and professional society for geology with a worldwide Fellowship (FGS) of over 10 000. The Society has the power to confer Chartered status on suitably qualified Fellows, and about 2000 of the Fellowship carry the title (CGeol). Chartered Geologists may also obtain the equivalent European title, European Geologist (EurGeol). One fifth of the Society's fellowship resides outside the UK. To find out more about the Society, log on to www.geolsoc.org.uk.

The Geological Society Publishing House (Bath, UK) produces the Society's international journals and books, and acts as European distributor for selected publications of the American Association of Petroleum Geologists (AAPG), the Indonesian Petroleum Association (IPA), the Geological Society of America (GSA), the Society for Sedimentary Geology (SEPM) and the Geologists' Association (GA). Joint marketing agreements ensure that GSL Fellows may purchase these societies' publications at a discount. The Society's online bookshop (accessible from www.geolsoc.org.uk) offers secure book purchasing with your credit or debit card.

To find out about joining the Society and benefiting from substantial discounts on publications of GSL and other societies worldwide, consult www.geolsoc.org.uk, or contact the Fellowship Department at: The Geological Society, Burlington House, Piccadilly, London W1J 0BG: Tel. +44 (0)20 7434 9944; Fax +44 (0)20 7439 8975; E-mail: enquiries@geolsoc.org.uk.

For information about the Society's meetings, consult *Events* on www.geolsoc.org.uk. To find out more about the Society's Corporate Affiliates Scheme, write to enquiries@geolsoc.org.uk.

Published by The Geological Society from:
The Geological Society Publishing House, Unit 7, Brassmill Enterprise Centre, Brassmill Lane, Bath BA1 3JN, UK

(*Orders*:   Tel. +44 (0)1225 445046, Fax +44 (0)1225 442836)
Online bookshop: www.geolsoc.org.uk/bookshop

The publishers make no representation, express or implied, with regard to the accuracy of the information contained in this book and cannot accept any legal responsibility for any errors or omissions that may be made.

British Library Cataloguing in Publication Data

A catalogue record for this book is available from the British Library.
ISBN 978-1-86239-335-6

Distributors
For details of international agents and distributors see:
www.geolsoc.org.uk/agentsdistributors

Typeset by Techset Composition Ltd, Salisbury, UK
Printed by MPG Books Ltd, Bodmin, UK

Contents

The formation and evolution of Africa from the Archaean to Present: introduction

DOUWE J. J. VAN HINSBERGEN[1,2]*, SUSANNE J. H. BUITER[1,2,3], TROND
H. TORSVIK[1,2,3,4], CARMEN GAINA[1,2,3] & SUSAN J. WEBB[4]

[1]*Physics of Geological Processes, University of Oslo, Sem Sælands vei 24,
NO-0316 Oslo, Norway*

[2]*Center for Advanced Study, Norwegian Academy of Science and Letters,
Drammensveien 78, 0271 Oslo, Norway*

[3]*Centre for Geodynamics, Geological Survey of Norway (NGU), Leiv Eirikssons vei 39,
7491 Trondheim, Norway*

[4]*School of Geosciences, University of the Witwatersrand, WITS 2050 Johannesburg,
South Africa*

*Corresponding author (e-mail: d.v.hinsbergen@fys.uio.no)

The African continent preserves a long geological record that covers almost 75% of Earth's history. The Pan-African orogeny (c. 600–500 Ma) brought together old continental kernels (or cratons such as West African, Congo, Kalahari and Tanzania) forming Gondwana and subsequently the supercontinent Pangea by the late Palaeozoic (Fig. 1).

The break-up of Pangea since the Jurassic and Cretaceous, primarily through the opening of the Central Atlantic (e.g. Torsvik et al. 2008; Labails et al. 2010), Indian (e.g. Gaina et al. 2007; Müller et al. 2008; Cande et al. 2010) and South Atlantic (e.g. Torsvik et al. 2009) oceans and the complicated subduction history to the north gradually shaped the African continent and its surrounding oceanic basins. Many first-order questions of African geology are still unanswered. How many accretion phases do the Proterozoic belts represent? What triggers extension and formation of the East African Rift on a continent that is largely surrounded by spreading centres and, therefore, expected to be mainly in compression? What is the role of shallow mantle and edge-driven convection (King & Ritsema 2000)? What are the sources of the volcanic centres of Northern Africa (e.g. Tibesti, Dafur and Afar) and can they be traced to the lower mantle? Is the elevation of Eastern and Southern Africa caused by mantle processes? What is the formation mechanism of intracratonic sedimentary basins, such as the Taoudeni Basin on the West African Craton and the Congo Basin (e.g. Hartley & Allen 1994; Giresse 2005)? How do sedimentation and tectonics interact (Burke & Gunnell 2008)? Can we reconstruct this elevation and its impact on climate evolution (e.g. Wichura et al. 2011)?

This special volume contains 18 original contributions about the geology of Africa. It celebrates African geology in two ways. First, it celebrates multidisciplinary Earth Science research, highlighting the formation and evolution of Africa from 18 different angles. Second, this volume celebrates the work of Kevin Burke and Lewis Ashwal. We hope that this 'Burke and Ashwal' volume portrays the wide range of interests and research angles that have characterized these two scientists throughout their careers working in Africa and studying African geology (Ashwal & Burke 1989; Burke 1996; Burke et al. 2003).

Content of the volume

This volume focuses on the formation of Africa as a coherent continent, from the formation of some of the oldest continental crust known today in the Kaapvaal Craton (de Wit et al. 1992) to billions of years of collisions and arc-accretions, amalgamating these old continental fragments (Hoffman 1991; Stern 1994; Zhao et al. 2002; Torsvik 2003) into Gondwana and the supercontinent Pangea. The contributions in this volume cover most of the African continent (Fig. 2), span >2 Ga of its history and approach its complex history from a geophysical, geological, geochemical and physical geographical point of view.

In recent years, the importance of deep mantle processes as a trigger for surface volcanism including world-changing (?) Large Igneous Province emplacement (Burke & Torsvik 2004; Burke et al. 2008), diamond-bearing kimberlite formation (Torsvik et al. 2010) and dynamic topography

From: VAN HINSBERGEN, D. J. J., BUITER, S. J. H., TORSVIK, T. H., GAINA, C. & WEBB, S. J. (eds) *The Formation and Evolution of Africa: A Synopsis of 3.8 Ga of Earth History*. Geological Society, London, Special Publications, **357**, 1–8. DOI: 10.1144/SP357.1 0305-8719/11/$15.00 © The Geological Society of London 2011.

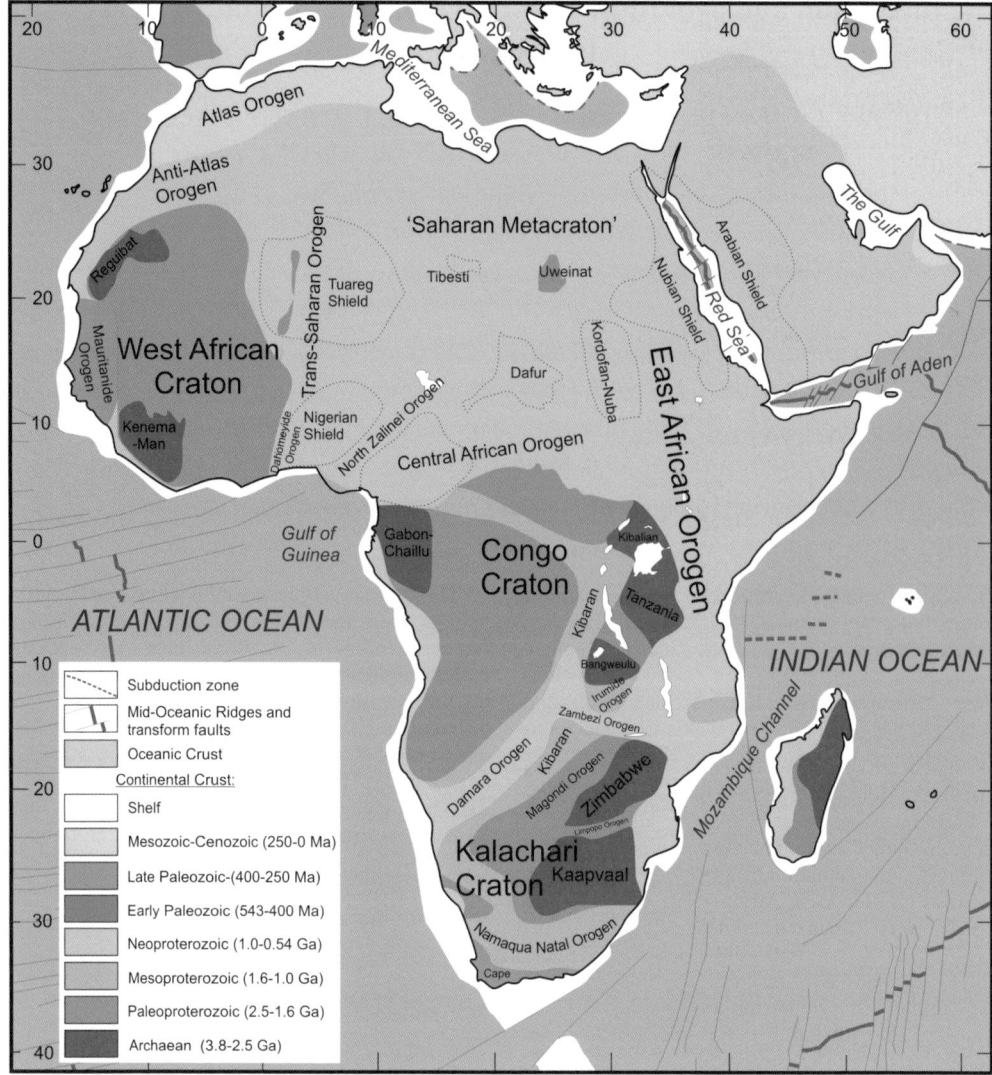

Fig. 1. Age of African crustal basement (after Gubanov & Mooney 2009). The ages are the time of crustal formation or the time of thermal or tectonic crustal reworking.

(e.g. Lithgow-Bertollini & Silver 1998) has become evident. Several contributions in this volume shed new light on these processes and their geological and environmental effects.

Part 1: The making of the African crust: The Archaean–Palaeozoic phases

The first part of this volume covers the formation of Africa from old cratons through assembly in the Pan-African orogeny to the formation of Gondwana.

Letts *et al.* (2011) provide new, high-quality palaeomagnetic poles for *c.* 2 Ga old rocks from the Kaapvaal Craton. By comparing their new results with published information, they demonstrate that the Kaapvaal Craton was not associated with high rates of apparent polar wander during the *c.* 2.1–1.9 Ga interval. Van Schijndel *et al.* (2011) provide new detrital zircon data from sandstones in the Rehoboth province of Namibia that identify three dominant periods of continental crust formation: between 1.3–1.1, 2.0–1.8 and 3–2.7 Ga, the latter of which was previously unknown. Key *et al.* (2011) report an extensive

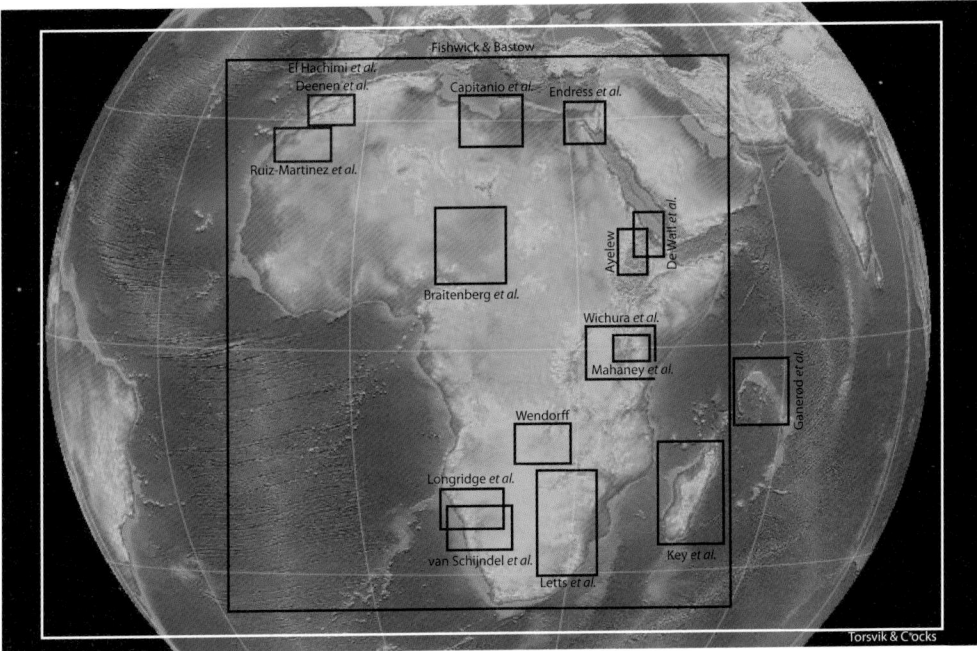

Fig. 2. Study areas covered in this volume.

geological survey of the basement rocks of Madagascar, providing an overview of the five Archaean to Proterozoic basement blocks that are recognized on the island. They review the Neoproterozoic collision and amalgamation history of the Madagascar segment of the East Africa–Antarctica Orogen, which was finalized during the Terminal Pan-African Event 560–490 Ma ago. Wendorff (2011) provides new insights into the history of the Lufilian Arc, which formed as a result of the Pan-African collision between the Kalahari and Congo cratons. The structural and stratigraphic evolution of this belt shows evidence for two previously unidentified rift and foreland basins. De Wall *et al.* (2011) study the metamorphic and tectonic history of the Terminal Pan-African Event in Ethiopia. They provide new metamorphic and magnetic fabric data from the Central Steep Zone, a transpressional belt that can be traced into Eritrea. They argue that the deformation and metamorphism in their study area results from closure of the Mozambique Ocean and the final assembly of Gondwana. Longridge *et al.* (2011) provide a detailed structural geological and geochronological study of the Central Zone of the Pan-African Damara orogen in Namibia. They unravel a history of crustal thickening, heating of the mid-crust, exhumation and orogen-parallel extension between *c.* 540 and 500 Ma. Torsvik & Cocks (2011) provide nine new palaeogeographic maps

of Gondwana between 510 and 250 Ma. They detail the locations of passive and active margins around Gondwana throughout the Palaeozoic, continental shelves, evaporite deposits, volcanism and glaciations, including those affecting Africa and Arabia.

Part 2: Africa since the break-up of Pangea: The Mesozoic–Cenozoic phases

The papers in the second part discuss events from the break-up of Pangea to the late Cenozoic. During this time Africa moved relatively slowly (Burke 1996). El Hachimi *et al.* (2011) focus their attention on the Central Atlantic Magmatic Province (CAMP) that was emplaced around 200 Ma. The mantle plume that led to its emplacement probably triggered Pangea dispersal (Burke & Dewey 1973). El Hachimi *et al.* (2011) study the morphology, internal architecture and emplacement mechanisms of CAMP lavas in the Argana Basin of Morocco. They demonstrate that the emplacement mechanisms are in line with continental flood basalt facies models. Deenen *et al.* (2011) take a stratigraphic approach to the CAMP, and use magnetostratigraphic and cyclostratigraphic techniques to correlate the Moroccan segment of the CAMP to their counterparts on the NW side of the Atlantic Ocean in Canada and the United

States. They provide new age constraints and correlations on the largest Large Igneous Province that
led to the break-up of Pangea and the formation of
Africa as a continent. Ruiz-Martinez *et al.* (2011)
provide a new palaeomagnetic pole for a *c.* 93 Ma
sedimentary section in SW Morocco, which forms
the first Turonian palaeopole for Africa. Given
their large dataset, they can correct their results for
compaction-induced shallowing of the inclination
and provide a reliable pole. They discuss their
results within the context of Africa's apparent
polar wander path for the Cretaceous. Ganerød
et al. (2011) provide new palaeomagnetic, U/Pb
and ^{40}Ar/^{39}Ar data from continental flood basalts
on the Seychelles (Indian Ocean), demonstrating
an age range of *c.* 67–61 Ma. These ages are consistent with an origin related to the Deccan traps
in India. Palaeomagnetic results (after correction
for a vertical axis rotation) confirm that the last
Gondwana fragment (India and the Seychelles)
was split after this event, and the Seychelles microcontinent became part of the African plate again.
Ayelew (2011) provides new Rb/Sr age determinations of bimodal basalt–rhyolite volcanism of
c. 20 Ma from Ethiopia, related to the continental
flood basalt province associated with the Afar
plume. Using Sr and Nd isotopic compositions, he
argues that the rhyolites formed due to fractional
crystallization of mantle-derived basaltic magmas
similar in composition to the exposed flood
basalts. Endress *et al.* (2011) study *c.* 24 Ma old
intraplate magmatism in Egypt and demonstrate
that their lavas are geochemically similar to those
of the Afar plume and to the subcrustal lithosphere.
They speculate that the basalts could be derived
from magmas that come from the edges of the
African Large Low Shear-wave Velocity Province
(LLSVP) at the core–mantle boundary and/or
from small-scale convection at the base of the
upper mantle. The Egyptian magmas found their
way to the surface utilizing incipient rift-related
structures of the Red Sea. Wichura *et al.* (2011)
address the notoriously difficult issue of the determination of palaeo-elevation of continental crust.
They provide an elegant analysis in which they
use the emplacement characteristics of a 350 km
long middle Miocene lava in Kenya to determine a
minimum slope required for its outflow. This
enables them to determine a minimum elevation of
the source region in Kenya in middle Miocene
times of 1400 m. Mahaney *et al.* (2011) provide a
unique, multidisciplinary analysis of 5.5–5.2 Ma
palaeosols preserved between lavas near Mt
Kenya. They use these palaeosols to infer the continental climate history of near-equatorial Africa
during the last 5 Ma, demonstrating generally dry
conditions during the late Miocene followed by
punctuated humid conditions during the Pliocene

and Quaternary. Capitanio *et al.* (2011) study the
Sirte Basin in northern Lybia, a peculiar extensional
domain that has historically been seismically active
and has experienced Pliocene extension. They
correlate this extensional deformation with the
contemporaneous Sicily Channel rift, and argue
that the strong slab-pull gradients of the subducted
African plate in the central Mediterranean region
have tectonic effects up to 1400 km south of the
subduction zone.

Part 3: Current state of the African crust and lithosphere

The last two papers of this volume use gravity and
seismic data to provide an image of the present-day
state of the African lithosphere and upper mantle.
Braitenberg *et al.* (2011) use high-precision global
gravity models to provide images of a sub-Saharan
lithospheric structure in Chad. They argue that this
structure is probably a metamorphic or magmatic
belt within the Saharan Megacraton. Fishwick &
Bastow (2011) provide a review of seismological
observations of the African lithosphere and upper
mantle, providing an observational basis for the
testing of geodynamical modelling studies that
aim to explain Africa's complex topography. They
discuss their overview within the context of the
African LLSVP, small-scale convection and the
role of the sublithospheric mantle.

African geology through the eyes of Kevin Burke and Lewis Ashwal

Kevin Charles Antony Burke was born on 13
November 1929 in London (England). He attended
University College London, where he earned a
BSc in 1951 and a PhD two years later. For two
decades (1961–1981), Kevin held university teaching and research positions in Ghana, Korea,
Jamaica, Nigeria, the United States and Canada.
He met and worked with Tuzo Wilson at the University of Toronto in the early 1970s (Burke & Wilson
1972; Wilson & Burke 1972), an obvious turning
point in Kevin's career. Since 2002 he has been an
Honorary Professor at the School of Geosciences
at the University of the Witwatersrand where he
has been teaching plate tectonics and African
geology during their winter months, and he still
teaches at the University of Houston, Texas.
    Kevin has made fundamental and lasting contri-
butions to our understanding of the origin and evolution of the lithosphere on Earth and other planets.
His influence has been grand and global in its
reach and, as a synthesizer of global geology and
global geological processes, Kevin has few peers.
It is almost impossible to quantify the breadth of

his innovation and knowledge from the oldest remnants in the Archaean to those ongoing today. It was Kevin Burke who coined the term 'Wilson Cycle' for the succession of continental rifting, subsidence and ocean opening, initiation of subduction and ocean closure and eventual continent–continent collision. Kevin was a pioneer in suggesting that Precambrian orogens like the Grenville are the eroded products of Himalayan-style collisions (Dewey & Burke 1973; Burke *et al.* 1976*a*). He was the first to propose that the Archaean auriferous Witwatersrand sedimentary sequence is a foreland basin (Burke *et al.* 1986). He also proposed in the early 1970s that greenstone belts, present in nearly all Archaean regions, are allochthonous volcano-sedimentary packages originally formed as marginal basins, ocean islands and arcs and were later thrust onto older continents (Burke *et al.* 1976*b*). Kevin Burke never stops to amaze the Earth Science community with his innovative and provocative ideas. Over the last eight years he has re-energized his long-lasting interest in mantle plumes. In 2004 Kevin discovered that large igneous provinces from the past 200 million years must have originated as plumes from the edges of the LLSVPs near the core-mantle boundary (Burke & Torsvik 2004; Burke *et al.* 2008). This surprising observation implies that deep-mantle heterogeneities have not changed much for hundreds of millions of years. Recognizing long-term stability of lower-mantle structures and the corresponding parts of the gravity field also fundamentally influences our thinking of how the Earth's moment of inertia and rotation may have changed over geological times (Steinberger & Torsvik 2010).

Lewis David Ashwal was born on 16 November 1949 in New York City (USA) and earned a PhD from Princeton University in 1979. The topic of Lew's PhD thesis was petrogenesis of massif-type anorthosites (Ashwal 1982; Ashwal & Wooden 1983) and he would later become the world leader in the understanding of the origin of anorthosites, a subject of heated theoretical debate for many decades (Ashwal & Burke 1989; Ashwal 1993). Lew's first position from 1978–1980 was postdoctoral research associate at NASA, Johnson Space Center, followed by 9 years as staff scientist at the Lunar and Planetary Institute in Houston. During the Lunar and Planetary Institute period (1980–1989), Lew clearly interacted with his boss (Kevin) but they only wrote one paper together, which was on the topic of African lithospheric structure, volcanism and topography (Ashwal & Burke 1989).

Lew has made fundamental contributions to petrology, mineralogy and geochemistry of anorthosite and related rocks, layered mafic intrusions, origin and evolution of planetary crusts, Precambrian geological history, origin of magmatic ore deposits, the role of fluids in igneous and metamorphic processes, meteorites and their parent bodies, abundance and distribution of crustal radioactivity, thermal and petrologic aspects of granulite metamorphism, geology of Madagascar and other Indian Ocean continental fragments and the Rodinia supercontinent and Gondwana assembly and break-up.

Meteorites and their parent bodies occupied Lew's mind in the late 1970s and early 1980s. In a 1981 groundbreaking paper Chuck Wood and Lew, by the process of elimination, suggested that meteorites of the so-called SNC group (Shergottites–Nakhlites–Chassignites) were derived from a differentiated planetary body, most likely Mars (Wood & Ashwal 1981). The difficulty of blasting material off a planetary surface and into an Earth-crossing orbit made planets such as Venus and Mercury unlikely sources, and chemical comparisons with Lunar samples (collected by the Apollo missions) also eliminated the moon as a potential source; Mars remained the only viable possibility. Lew and colleagues published a follow-up manuscript in 1982 where they demonstrated that petrologic, geochemical and isotopic evidence were inconsistent with an asteroidal origin and concluded that Mars remained the most likely parent body for SNC meteorites; they were later proven correct (Ashwal *et al.* 1982).

In 1990, Lew became Professor of Geology at the Rand Afrikaans University (RAU), Johannesburg, South Africa. He served RAU with distinction for more than 10 years, but in 2001 he moved across the road to the School of Geosciences at the University of the Witwatersrand, where he is an enduring Professor and Director of the African Lithosphere Research Unit. Lew has written several books and hundreds of scientific papers, reports and essays; listing all of his contributions to geology is undoable. We would be negligent if we did not mention Lew's genuine passion for Africa, not only for her geology but also for her inhabitants; as an educator Lew is legendary.

Lew and Kevin have remained friends and colleagues since the Lunar and Planetary Institute days in the 1980s. They still work, converse and passionately argue with each other. Both have enjoyed great scientific success in diverse scientific fields. They have collaborated on projects probing into the world's oldest rocks, the deep continental crust and global characterization of the ancient continents and lithosphere. Joint papers cover diverse subjects such as characterization of terrestrial anorthosites, lithospheric delamination on Earth and Venus, African lithosphere structure and volcanism, identification of old sutures guided by deformed alkaline rocks and carbonatites, Proterozoic mountain building and, most recently, plumes from the deepest mantle (Ashwal & Burke 1989; Burke *et al.* 2003, 2007; Leelanandam *et al.* 2006; Ashwal *et al.* 2007; Torsvik *et al.* 2010). Many more papers are likely to appear in the coming years and we wish them a happy 140th birthday.

This volume was initiated at a conference held in Johannesburg, South Africa in November 2009, honouring the work of Kevin Burke and Lewis Ashwal. We thank all conference participants for their contributions and stimulating discussions. We thank the Geological Society Publishing House and especially Tamzin Anderson, Angharad Hills and Randell Stephenson for their help with the publication of this volume. The editors appreciate financial support from Statoil for 'The African Plate' project.

References

ASHWAL, L. D. 1982. Mineralogy of mafic Fe-Ti oxide-rich differentiates of the Marcy anorthosite massif, Adirondacks, New York. *American Mineralogist*, **67**, 14–27.

ASHWAL, L. D. 1993. *Anorthosites*. Minerals and Rocks 21, Springer-Verlag, Berlin.

ASHWAL, L. D. & WOODEN, J. L. 1983. Isotopic evidence from the eastern Canadian shield for geochemical discontinuity in the proterozoic mantle. *Nature*, **306**, 679–680.

ASHWAL, L. D. & BURKE, K. 1989. African lithospheric structure, volcanism, and topography. *Earth and Planetary Science Letters*, **96**, 8–14.

ASHWAL, L. D., WARNER, J. L. & WOOD, C. A. 1982. SNC meteorites: evidence against an asteroidal origin. *Journal of Geophysical Research*, **87** (Supplement), A393–A400.

ASHWAL, L. D., ARMSTRONG, R. A. *ET AL.* 2007. Geochronology of zircon megacrysts from nepheline-bearing gneisses as constraints on tectonic setting: implications for resetting of the U-Pb and Lu-Hf isotopic systems. *Contributions to Mineralogy and Petrology*, **153**, 389–403.

AYELEW, D. 2011. The relations between felsic and mafic volcanic rocks in continental flood basalts of Ethiopia: Implication for the thermal weakening of the crust. *In*: VAN HINSBERGEN, D. J. J., BUITER, S. J. H., TORSVIK, T. H., GAINA, C. & WEBB, S. J. (eds) *The Formation and Evolution of Africa: A Synopsis of 3.8 Ga of Earth History*. Geological Society, London, Special Publications, **357**, 253–264.

BRAITENBERG, C., MARIANI, P., EBBING, J. & SPRLAK, M. 2011. The enigmatic Chad lineament revisited with global gravity and gravity gradient fields. *In*: VAN HINSBERGEN, D. J. J., BUITER, S. J. H., TORSVIK, T. H., GAINA, C. & WEBB, S. J. (eds) *The Formation and Evolution of Africa: A Synopsis of 3.8 Ga of Earth History*. Geological Society, London, Special Publications, **357**, 329–341.

BURKE, K. 1996. The African plate. *South African Journal of Geology*, **99**, 341–409.

BURKE, K. & WILSON, J. T. 1972. Is the African plate stationary? *Nature*, **239**, 387–390.

BURKE, K. & DEWEY, J. F. 1973. Plume-generated triple junctions – key indicators in applying plate tectonics to old rocks. *Journal of Geology*, **81**, 406–433.

BURKE, K. & TORSVIK, T. H. 2004. Derivation of large igneous provinces of the past 200 million years from long-term hetergeneities in the deep mantle. *Earth and Planetary Science Letters*, **227**, 531–538.

BURKE, K. & GUNNELL, Y. 2008. The African erosion surface: a continental-scale synthesis of geomorphology, tectonics, and environmental change over the past 180 million years. *Geological Society of America Memoir*, **201**, 1–66.

BURKE, K., DEWEY, J. F. & KIDD, W. S. F. 1976a. Precambrian palaeomagnetic results compatible with contemporary operation of the Wilson cycle. *Tectonophysics*, **33**, 287–299.

BURKE, K., DEWEY, J. F. & KIDD, W. S. F. 1976b. Dominance of horizontal movements, arc and microcontinental collisions during the later permobile regime. *In*: WINDLEY, B. F. (ed.) *The Early History of the Earth*. John Wiley & Sons, New York, 113–129.

BURKE, K., KIDD, W. S. F. & KUSKY, T. M. 1986. Archean Foreland basin tectonics in the Witwatersrand, South Africa. *Tectonics*, **5**, 439–456.

BURKE, K., ASHWAL, L. D. & WEBB, S. J. 2003. New way to map old sutures using deformed alkaline rocks and carbonatites. *Geology*, **44**, 391–394.

BURKE, K., ROBERTS, D. & ASHWAL, L. 2007. Alkaline rocks and carbonatites of northwestern Russia and northern Norway: linked Wilson cycle records extending over two billion years. *Tectonics*, **26**, TC4015, doi: 10.1029/2006TC002052.

BURKE, K., STEINBERGER, B., TORSVIK, T. H. & SMETHURST, M. A. 2008. Plume generation zones at the margins of large low shear velocity provinces on the core–mantle boundary. *Earth and Planetary Science Letters*, **265**, 49–60.

CANDE, S. C., PATRIAT, P. & DYMENT, J. 2010. Motion between the Indian, Antarctic and African plates in the early Cenozoic. *Geophysical Journal International*, **183**, 127–149.

CAPITANIO, F. A., FACCENNA, C., FUNICIELLO, F. & SALVINI, F. 2011. Recent tectonics of Tripolitania, Libya: an intraplate record of Mediterranean subduction. *In*: VAN HINSBERGEN, D. J. J., BUITER, S. J. H., TORSVIK, T. H., GAINA, C. & WEBB, S. J. (eds) *The Formation and Evolution of Africa: A Synopsis of 3.8 Ga of Earth History*. Geological Society, London, Special Publications, **357**, 319–328.

DE WALL, H., DIETL, C., JUNGMANN, O., ASHENAFI, T. T. & PANDIT, M. K. 2011. Tectonic evolution of the 'Central Steep Zone', Axum area, northern Ethiopia: inferences from magnetic and geochemical data. *In*: VAN HINSBERGEN, D. J. J., BUITER, S. J. H., TORSVIK, T. H., GAINA, C. & WEBB, S. J. (eds) *The Formation and Evolution of Africa: A Synopsis of 3.8 Ga of Earth History*. Geological Society, London, Special Publications, **357**, 85–106.

DE WIT, M. J., DE RONDE, C. E. J. *ET AL.* 1992. Formation of an archaean continent. *Nature*, **357**, 553–562.

DEENEN, M. H., LANGEREIS, C. G., KRIJGSMAN, W., EL HACHIMI, H. & CHELLAI, E. H. 2011. Palaeomagnetic results from Upper Triassic red beds and CAMP lavas of the Argana basin, Morocco. *In*: VAN HINSBERGEN,

D. J. J., BUITER, S. J. H., TORSVIK, T. H., GAINA, C. & WEBB, S. J. (eds) *The Formation and Evolution of Africa: A Synopsis of 3.8 Ga of Earth History*. Geological Society, London, Special Publications, **357**, 195–209.

DEWEY, J. F. & BURKE, K. 1973. Tibetan, variscan and precambrian basement reactivation: product of continental collision. *Journal of Geology*, **81**, 683–692.

EL HACHIMI, H., YOUBI, N. ET AL. 2011. Morphology, internal architecture and emplacement mechanisms of lava flows from the Central Atlantic Magmatic Province (CAMP) of Argana basin (Morocco). *In*: VAN HINSBERGEN, D. J. J., BUITER, S. J. H., TORSVIK, T. H., GAINA, C. & WEBB, S. J. (eds) *The Formation and Evolution of Africa: A Synopsis of 3.8 Ga of Earth History*. Geological Society, London, Special Publications, **357**, 167–193.

ENDRESS, C., FURMAN, T., ABU EL-RUS, M. A. & HANAN, B. B. 2011. Geochemistry of 24 Ma basalts from northeast Egypt: Source components and fractionation history. *In*: VAN HINSBERGEN, D. J. J., BUITER, S. J. H., TORSVIK, T. H., GAINA, C. & WEBB, S. J. (eds) *The Formation and Evolution of Africa: A Synopsis of 3.8 Ga of Earth History*. Geological Society, London, Special Publications, **357**, 265–283.

FISHWICK, S. & BASTOW, I. R. 2011. Towards a better understanding of African topography: A review of passive-source seismic studies of the African crust and upper mantle. *In*: VAN HINSBERGEN, D. J. J., BUITER, S. J. H., TORSVIK, T. H., GAINA, C. & WEBB, S. J. (eds) *The Formation and Evolution of Africa: A Synopsis of 3.8 Ga of Earth History*. Geological Society, London, Special Publications, **357**, 343–371.

GAINA, C., MULLER, R. D., BROWN, B., ISHIHARA, T. & IVANOV, S. 2007. Breakup and early seafloor spreading between India and Antarctica. *Geophysical Journal International*, **170**, 151–170.

GANERØD, M., TORSVIK, T. H. ET AL. 2011. Palaeoposition of the Seychelles microcontinent in relation to the Decan Traps and the Plume Generation Zone in late Cretaceous–early Palaeogene time. *In*: VAN HINSBERGEN, D. J. J., BUITER, S. J. H., TORSVIK, T. H., GAINA, C. & WEBB, S. J. (eds) *The Formation and Evolution of Africa: A Synopsis of 3.8 Ga of Earth History*. Geological Society, London, Special Publications, **357**, 229–252.

GIRESSE, P. 2005. Mesozoic-Cenozoic history of the congo basin. *Journal of African Earth Sciences*, **43**, 301–315.

GUBANOV, A. P. & MOONEY, W. D. 2009. New global geological maps of crustal basement age. *Eos transactions, AGU*, **90**, Fall Meet. Suppl. Abstract T53B-1583.

HARTLEY, R. W. & ALLEN, P. A. 1994. Interior cratonic basins of Africa: relation to continental break-up and role of mantle convection. *Basin Research*, **6**, 95–113.

HOFFMAN, P. F. 1991. Did the breakout of Laurentia turn Gondwanaland inside-out? *Science*, **252**, 1409–1412.

KEY, R. M., PITFIELD, P. E. J. ET AL. 2011. Polyphase Neoproterozoic orogenesis within the East Africa–Antarctica Orogenic Belt in central and northern Madagascar. *In*: VAN HINSBERGEN, D. J. J., BUITER, S. J. H., TORSVIK, T. H., GAINA, C. & WEBB, S. J. (eds) *The Formation and Evolution of Africa: A Synopsis of 3.8 Ga of Earth History*. Geological Society, London, Special Publications, **357**, 49–68.

KING, S. D. & RITSEMA, J. 2000. African hot spot volcanism: small-scale convection in the upper mantle beneath cratons. *Science*, **290**, 1137–1140.

LABAILS, C., OLIVET, J. L., ASLANIAN, D. & ROEST, W. R. 2010. An alternative early opening scenario for the central Atlantic ocean. *Earth and Planetary Science Letters*, **297**, 355–368.

LEELANANDAM, C., BURKE, K., ASHWAL, L. D. & WEBB, S. J. 2006. Proterozoic mountain building in peninsular India: an analysis based primarily on alkaline rock distribution. *Geological Magazine*, **143**, 195–212.

LETTS, S., TORSVIK, T. H., WEBB, S. J. & ASHWAL, L. D. 2011. New Palaeoproterozoic palaeomagnetic data from the Kaapvaal Craton, South Africa. *In*: VAN HINSBERGEN, D. J. J., BUITER, S. J. H., TORSVIK, T. H., GAINA, C. & WEBB, S. J. (eds) *The Formation and Evolution of Africa: A Synopsis of 3.8 Ga of Earth History*. Geological Society, London, Special Publications, **357**, 9–48.

LITHGOW-BERTOLLINI, C. & SILVER, P. G. 1998. Dynamic topography, plate driving forces and the African superswell. *Nature*, **395**, 269–272.

LONGRIDGE, L., GIBSON, R. L., KINNAIRD, J. A. & ARMSTRONG, R. A. 2011. Constraining the timing of deformation in the south-western Central Zone of the Damara belt, Namibia (520–508 Ma). *In*: VAN HINSBERGEN, D. J. J., BUITER, S. J. H., TORSVIK, T. H., GAINA, C. & WEBB, S. J. (eds) *The Formation and Evolution of Africa: A Synopsis of 3.8 Ga of Earth History*. Geological Society, London, Special Publications, **357**, 107–135.

MAHANEY, W. C., BARENDREGT, R. W., VILLENEUVE, M., DOSTAL, J., HAMILTON, T. S. & MILNER, M. W. 2011. Late Neogene volcanics and interbedded palaeosols near Mount Kenya. *In*: VAN HINSBERGEN, D. J. J., BUITER, S. J. H., TORSVIK, T. H., GAINA, C. & WEBB, S. J. (eds) *The Formation and Evolution of Africa: A Synopsis of 3.8 Ga of Earth History*. Geological Society, London, Special Publications, **357**, 301–318.

MÜLLER, R. D., SDROLIAS, M., GAINA, C. & ROEST, W. R. 2008. Age, spreading rates, and spreading asymmetry of the world's ocean crust. *Geochemistry, Geophysics, Geosystems*, **9**, Q04006, doi:10.1029/2007GC001743.

RUIZ-MARTINEZ, V. C., PALENCIA-ORTAS, A., VILLALAIN, J. J., McINTOSH, G. & MARTIN-HERNANDEZ, F. 2011. Palaeomagnetic and AMS study of the Tarfaya coastal basin, Morocco: an early Turonian palaeopole for the African plate. *In*: VAN HINSBERGEN, D. J. J., BUITER, S. J. H., TORSVIK, T. H., GAINA, C. & WEBB, S. J. (eds) *The Formation and Evolution of Africa: A Synopsis of 3.8 Ga of Earth History*. Geological Society, London, Special Publications, **357**, 211–227.

STEINBERGER, B. & TORSVIK, T. H. 2010. Toward an explanation for the present and past locations of the poles. *Geochemistry, Geophysics, Geosystems*, **11**, Q06W06.

STERN, R. J. 1994. Arc assembly and continental collision in the neoproterozoic east African orogen: implications for the consolidation of Gondwanaland. *Annual Review of Earth and Planetary Sciences*, **22**, 319–351.

TORSVIK, T. H. 2003. The Rodinia jigsaw puzzle. *Science*, **300**, 1379–1381.

TORSVIK, T. H. & COCKS, L. R. M. 2011. The Palaeozoic palaeogeography of central Gondwana. *In*: VAN

8 D. J. J. VAN HINSBERGEN *ET AL.*

HINSBERGEN, D. J. J., BUITER, S. J. H., TORSVIK, T. H., GAINA, C. & WEBB, S. J. (eds) *The Formation and Evolution of Africa: A Synopsis of 3.8 Ga of Earth History.* Geological Society, London, Special Publications, **357**, 137–168.

TORSVIK, T. H., MÜLLER, R. D., VAN DER VOO, R., STEINBERGER, B. & GAINA, C. 2008. Global plate motion frames: toward a unified model. *Reviews of Geophysics*, **41**, RG3004, doi:10.1029/2007RG000227.

TORSVIK, T. H., ROUSSE, S., LABAILS, C. & SMETHURST, M. A. 2009. A new scheme for the opening of the South Atlantic Ocean and the dissection of an aptian salt basin. *Geophysical Journal International*, **177**, 1315–1333.

TORSVIK, T. H., BURKE, K., WEBB, S., ASHWAL, L. & STEINBERGER, B. 2010. Diamonds sampled by plumes from the core-mantle boundary. *Nature*, **466**, 352–355.

VAN SCHIJNDEL, V., CORNELL, D. H., HOFFMANN, K.-H. & FREI, D. 2011. Three episodes of crustal development in the Rehoboth Province, Namibia. *In*: VAN HINSBERGEN, D. J. J., BUITER, S. J. H., TORSVIK, T. H., GAINA, C. & WEBB, S. J. (eds) *The Formation and Evolution of Africa: A Synopsis of 3.8 Ga of Earth History.* Geological Society, London, Special Publications, **357**, 27–48.

WENDORFF, M. 2011. Tectonosedimentary expressions of the evolution of the Fungurume foreland basin in the Lufilian Arc, Neoproterozoic–Lower Palaeozoic, Central Africa. *In*: VAN HINSBERGEN, D. J. J., BUITER, S. J. H., TORSVIK, T. H., GAINA, C. & WEBB, S. J. (eds) *The Formation and Evolution of Africa: A Synopsis of 3.8 Ga of Earth History.* Geological Society, London, Special Publications, **357**, 69–83.

WICHURA, H., BOUSQUET, R., OBERHÄNSLI, R., STRECKER, M. R. & TRAUTH, M. 2011. The middle Miocene East African Plateau: a pre-rift topographic model inferred from the emplacement of the phonolitic Yatta lava flow, Kenya. *In*: VAN HINSBERGEN, D. J. J., BUITER, S. J. H., TORSVIK, T. H., GAINA, C. & WEBB, S. J. (eds) *The Formation and Evolution of Africa: A Synopsis of 3.8 Ga of Earth History.* Geological Society, London, Special Publications, **357**, 285–300.

WILSON, J. T. & BURKE, K. 1972. Two types of mountain building. *Nature*, **239**, 448–449.

WOOD, C. A. & ASHWAL, L. D. 1981. SNC meteorites: Igneous rocks from Mars? *Proceedings of the 12th Lunar and Planetary Science Conference*, 1359–1375.

ZHAO, G., CAWOOD, P. A., WILDE, S. A. & SUN, M. 2002. Review of global 2.1–1.8 Ga orogens: implications for a pre-Rodinia supercontinent. *Earth-Science Reviews*, **59**, 179–204.

New Palaeoproterozoic palaeomagnetic data from the Kaapvaal Craton, South Africa

SHAWN LETTS[1,2], TROND H. TORSVIK[1,3,4,5]*, SUSAN J. WEBB[1] & LEWIS D. ASHWAL[1]

[1]*School of Geosciences, University of Witwatersrand, WITS 2050, South Africa*

[2]*Anglo American, P.O. Box 61587, Marshalltown, Johannesburg 2107, South Africa*

[3]*Physics of Geological Processes, University of Oslo, Norway*

[4]*Centre for Geodynamics, Geological Survey of Norway, N-7491 Trondheim, Norway*

[5]*Centre for Advanced Study, Drammensveien 78, NO-0271 Oslo, Norway*

Corresponding author (e-mail: trond.torsvik@ngu.no)

Abstract: Palaeomagnetic data from the well-dated 2060.6 ± 0.5 Ma Phalaborwa Complex in South Africa (Kaapvaal Craton) are of excellent quality. High unblocking components are carried by magnetite and single polarity remanence directions (mean declination 5.0°, inclination 57.3°, $\alpha_{95} = 5.2°$) yield a palaeomagnetic pole (latitude 27.7°N, longitude 35.8°E, A95 = 6.6°) that overlaps with existing poles from the near coeval 2054.4 ± 1.3 Ma Bushveld Complex. The Phalaborwa and Bushveld complex poles, along with poles from the well-dated Vredefort impact (2023 ± 4 Ma) and Post-Waterberg Dolerites (1874.6 ± 3.9 Ma), define the most reliable poles for the Kaapvaal Craton during this time interval (*c.* 2060–1875 Ma) and witness low rates of Mid-Palaeoproterozoic apparent polar wander. Poorly dated NE–NNE-trending dyke swarms that intrude the Phalaborwa and Bushveld complexes both yield dual-polarity remanence components that share a common mean at the 95% confidence level. Primary palaeomagnetic poles (Phalaborwa dykes pole latitude 7.6°, longitude 12.1°, A95 = 11.8°; Bushveld dykes pole latitude 12.6°, longitude 24.1°, A95 = 10.8°) suggest that they are of the same age as the Post-Waterberg dolerites (*c.* 1875 Ma). They could also be as old as the Phalaborwa and Bushveld Complexes, however; high-precision geochronology is required to resolve this issue and to enlarge the number of Palaeoproterozoic key poles for the Kaapvaal Craton.

Pangea, the youngest and best-documented supercontinent on Earth, formed at the end of the Palaeozoic era. In the deep past, however, Neoproterozoic (Rodinia), Palaeoproterozoic (Columbia) and Archaean (Vaalbara) supercontinents/continents have been postulated (e.g. Hoffman 1991; Dalziel 1992; Torsvik *et al.* 1996; Cheney 1996; Zegers *et al.* 1998; Rogers & Santosh 2002; Zhao *et al.* 2002; Meert & Torsvik 2003; Torsvik 2003; Li *et al.* 2008; de Kock *et al.* 2009; Evans 2009). However, the timing for assembly/dispersal and continental geometry of these older supercontinents/continents differs in almost every paper. The Archaean-aged Vaalbara continent, hypothetically composed of the Kaapvaal (South Africa) and Pilbara (West Australia) cratons, is no exception and we show three different constellations (Fig. 1a) that can be seen in the literature. Vaalbara may have existed between 2.8 and 2.1 Ga (Cheney 1996; Zegers *et al.* 1998; Strik *et al.* 2003; de Kock *et al.* 2009) but Zhao *et al.* (2002) argue that

Vaalbara was also intact at 2.0–1.8 Ga and became the site of two major orogenic events. The first event is that of the Limpopo mobile belt (2.0–1.9 Ga), which formed as a result of the collision between the Kaapvaal and Zimbabwe cratons (Zeh *et al.* 2009). The second orogenic event resulted in the formation of the Capricorn Orogen, in which the Pilbara and Yilgarn cratons were sutured together. These orogens, together with many other 2.1–1.8 Ga orogens (Zhao *et al.* 2002), were argued to be the result of a world-wide amalgamation of cratonic blocks to form the Columbia supercontinent (Fig. 1b).

In order to test the hypothesis of the Columbia supercontinent, high-precision geochronological and palaeomagnetic data from many Palaeoproterozoic cratons are urgently required. In this account we report new palaeomagnetic data from the *c.* 2.06 Ga Phalaborwa Complex and dykes cutting both the Phalaborwa and Bushveld (*c.* 2.05 Ga) complexes in South Africa, and summarize the most

From: VAN HINSBERGEN, D. J. J., BUITER, S. J. H., TORSVIK, T. H., GAINA, C. & WEBB, S. J. (eds) *The Formation and Evolution of Africa: A Synopsis of 3.8 Ga of Earth History*. Geological Society, London, Special Publications, **357**, 9–26. DOI: 10.1144/SP357.2   0305-8719/11/$15.00 © The Geological Society of London 2011.

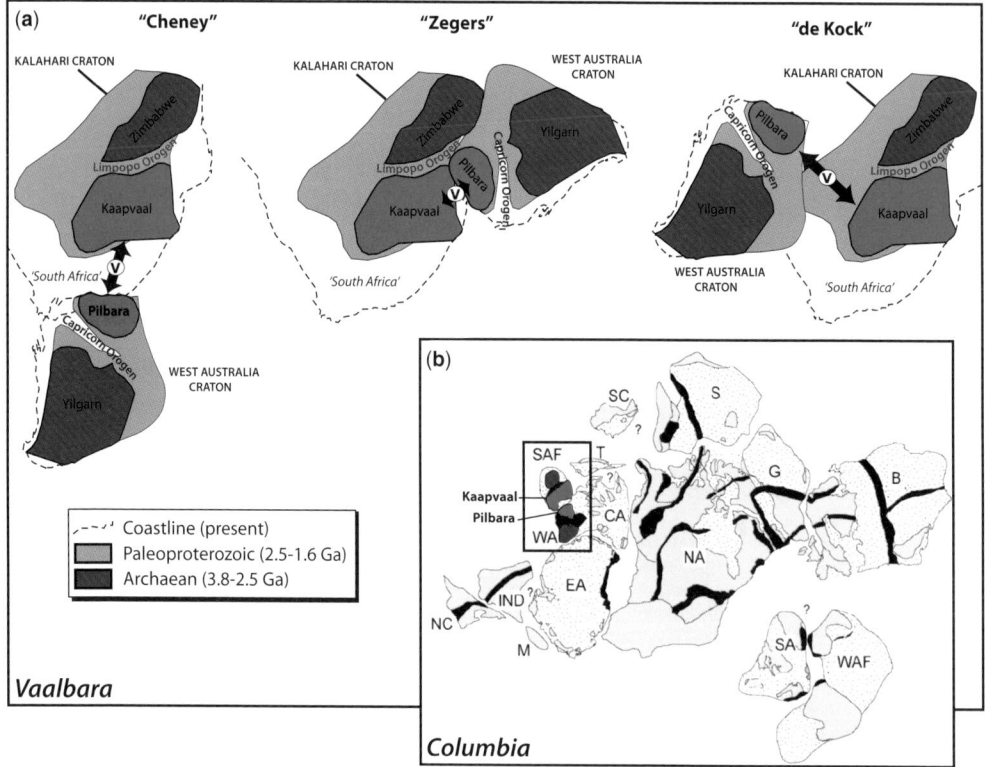

Fig. 1. (**a**) Vaalbara continent showing the proposed Archaean connection between the Kaapvaal (South Africa) and Pilbara (West Australia) cratons according to Cheney (1996), Zegers *et al.* (1998) or de Kock *et al.* (2009). See de Kock *et al.* (2009) for reconstruction parameters for a fixed South Africa (Kaapvaal). Archaean and Palaeoproterozoic cratons that make up the basement of Kalahari and West Australia follow Gubanov & Mooney (2009). (**b**) Reconstruction of the Columbia Supercontinent (Zhao *et al.* 2002), mostly based on correlation of 2.1–1.8 Ga orogens (B: Baltica; CA: Central Australia; EA: East Antarctica; G: Greenland; IND: India; M: Madagascar; NA: North America; NC: North China; S: Siberia; SA: South America; SAF: South Africa (Kalahari); SC: South China; T: Tarim; WA: West Australia; WAF: West Africa). In this reconstruction, the Vaalbara continent is intact at Columbia time and shown in a Cheney (1996) type fit.

reliable palaeomagnetic poles (*c.* 2.06–1.85 Ga) from the Kaapvaal/Kalahari Craton. There are no reliable 2.0 Ga poles from the Pilbara Craton (or Australia at large), and hence proposed connections between the Kaapvaal and Pilbara cratons (Vaalbara continent) remain unproven at this time. A comprehensive global analysis and aspects of the Columbia supercontinent will be dealt with elsewhere.

Phalaborwa and Bushveld

The Phalaborwa Complex (Harmer 2000) outcrops in an elongated kidney shape in the Limpopo Province in northern South Africa (Fig. 2). The Complex was intruded into Archaean basement gneisses at 2060.6 ± 0.5 Ma (U–Pb baddeleyite; Reischmann 1995) in several alkaline cycles, which

emplaced a suite of rocks ranging from ultramafic to peralkaline in character. The first phase, the intrusion of a massive ultramafic pipe-like body, consists mainly of pyroxenites. This was followed by an alkaline phase in which the surrounding syenite plug-like bodies were emplaced. At a later stage, foskorite (olivine + magnetite + apatite + phlogopite rock) and banded carbonatite of the central pegmatoid body were emplaced. The final phase saw the intrusive injection of the transgressive carbonatite (PMC 1976). The Phalaborwa Complex is nearly coeval with the Bushveld Igneous Complex (2054.4 ± 1.3 Ma; Scoates & Friedman 2008), the world's largest layered intrusion.

Regional aeromagnetic data reveal that northern South Africa is riddled with dykes (Fig. 3). Unfortunately, very little is known about the age of those occurring in the Phalaborwa and Bushveld

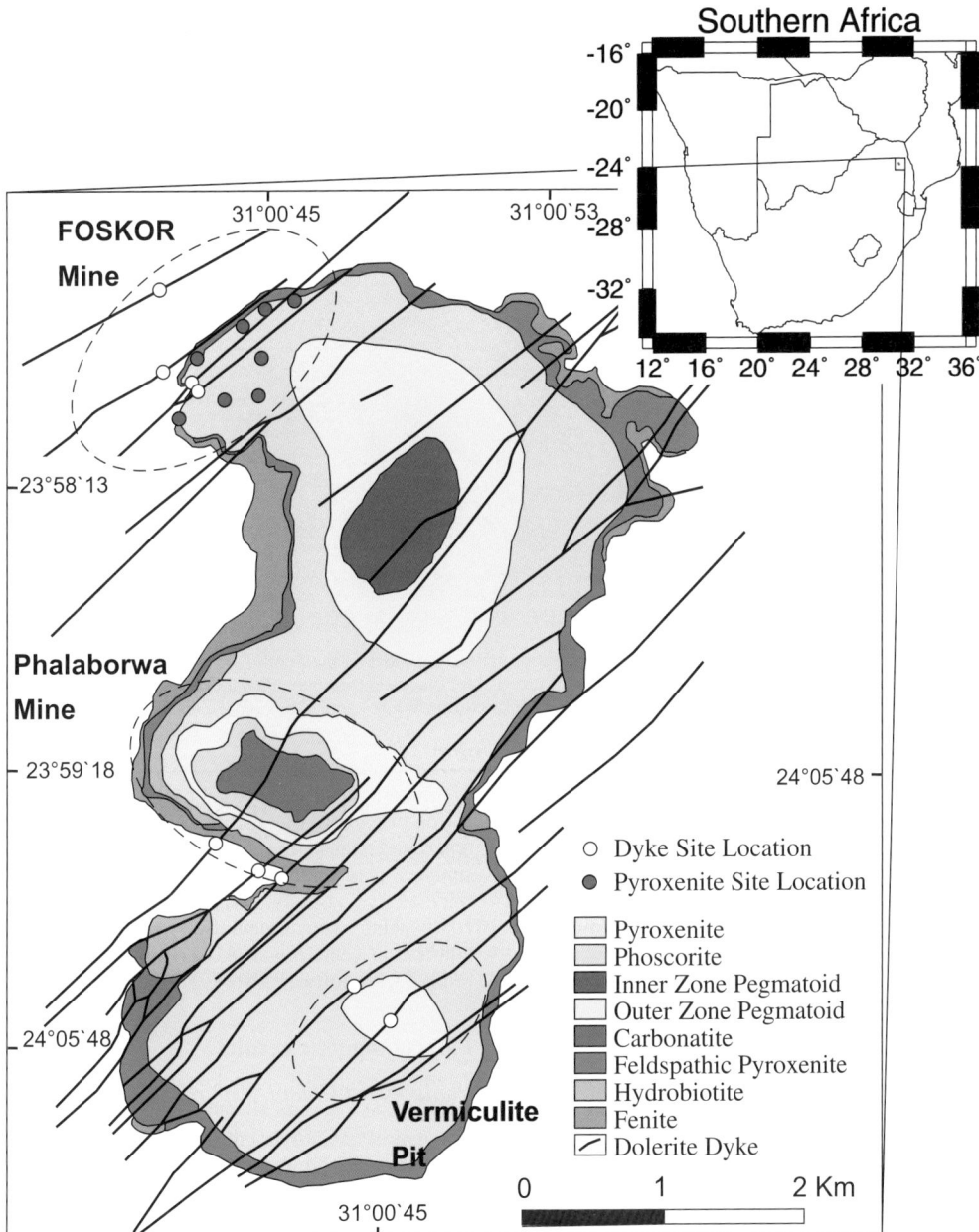

Fig. 2. Simplified geological map of the Phalaborwa Complex with site locations obtained from the pyroxenites (red circles) and dykes (open white circles). The three pits/mines are shown by dashed outline modified from Morgan & Briden (1981).

Complexes. Examination of dyke swarms and their relationship to major mafic magmatic events in the Kaapvaal Craton reveals that dyke swarms have repeatedly exploited certain orientations (Uken & Watkeys 1997). Three main trends have been identified in which dyke swarms were repeatedly

emplaced: NE, NW and east–west. Ages and orientations of major dyke swarms for the entirety of southern Africa are summarized by Hunter & Reid (1987). In this study, only the NE-trending swarm that intrudes the Phalaborwa and Bushveld complexes are considered (Figs 2–4).

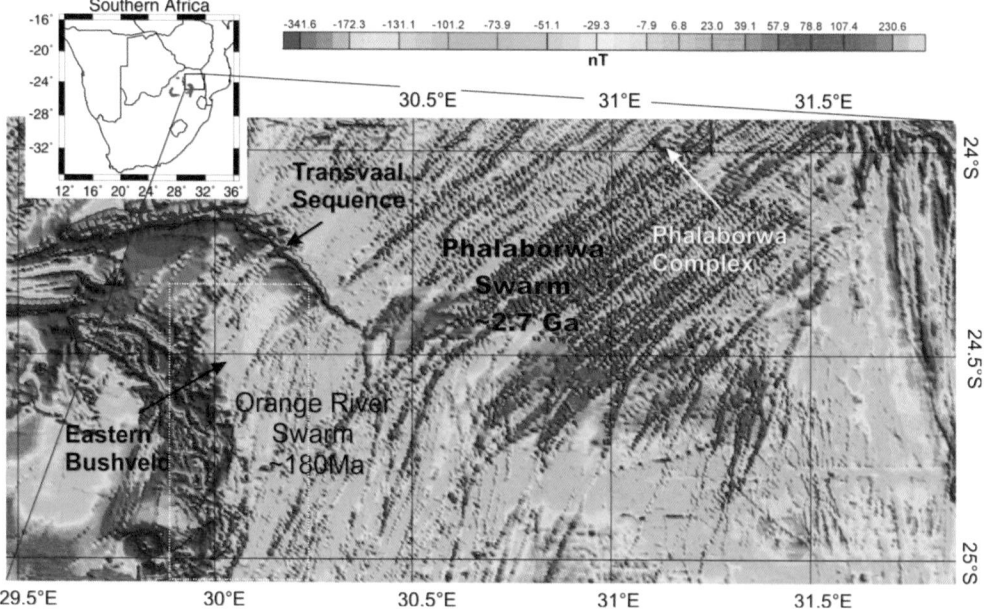

Fig. 3. Regional aeromagnetic data over northern South Africa (provided by the South African Council for Geoscience). The major geological bodies in the area are visible: Bushveld Complex, Phalaborwa Complex, Transvaal Sequence, as well as the Orange River and Phalaborwa dyke swarm. White stippled box represents the area depicted by Figure 4.

A mafic dyke database from the International Institute for Geo-information Science and Earth Observation (Chavez Gomez 2001) suggests that the dykes present in the Bushveld Complex are only from the Orange River Swarm (Fig. 3), that is, Jurassic in age (Karoo, *c.* 180 Ma). However, a palaeomagnetic study conducted on a few of these dykes (Fig. 4) revealed that the palaeomagnetic pole position clearly does not resemble any Karoo pole but instead indicates a Palaeoproterozoic age (Letts *et al.* 2005 and this study).

Sampling and laboratory procedures

All samples were obtained with the use of a portable drill and orientated via sun and magnetic compasses. The natural remanent magnetization (NRM) was measured on a JR6A magnetometer at the Geological Survey of Norway (Trondheim). NRM stability was tested by both thermal (MM-TD-60 furnace) and alternating field (AF; 3-axis *in-house* built tumbler) demagnetization. Bulk susceptibility was measured on a Bartington MS2 system.

Thermomagnetic analysis (TMA) and determination of Curie temperatures were carried out on a horizontal translation bridge, while petrographic analysis was undertaken to identify the opaque mineralogy visually. Characteristic remanence components were calculated with the LineFind algorithm (Kent *et al.* 1983) as implemented in the Super-IAPD software (Torsvik *et al.* 2000).

Palaeomagnetic results

Phalaborwa Complex (pyroxenites)

The main Phalaborwa mine (Fig. 2) was unsafe to enter due to sidewall instability, while in other areas exposure was inadequate for palaeomagnetic sampling. It was therefore only possible to sample eight pyroxenite sites in the northern Foskor pit; a total of 90 core samples were analysed for NRM and bulk susceptibility (Table 1).

Figure 5 shows a representative selection of orthogonal vector plots that demonstrate the demagnetization behaviour. Many samples contain two

Fig. 4. The position of dyke samples collected from the Eastern Bushveld Complex: four sites from this study (yellow-Hackney mine and blue-Modikwa mine) and sites (white dots) from the author's previous study (Letts *et al.* 2005), plotted on high-resolution aeromagnetic data draped on top of regional aeromagnetic data courtesy of Anglo Platinum. Coordinates are in WGS84 (LO 31). See Figure 3 for geographic location.

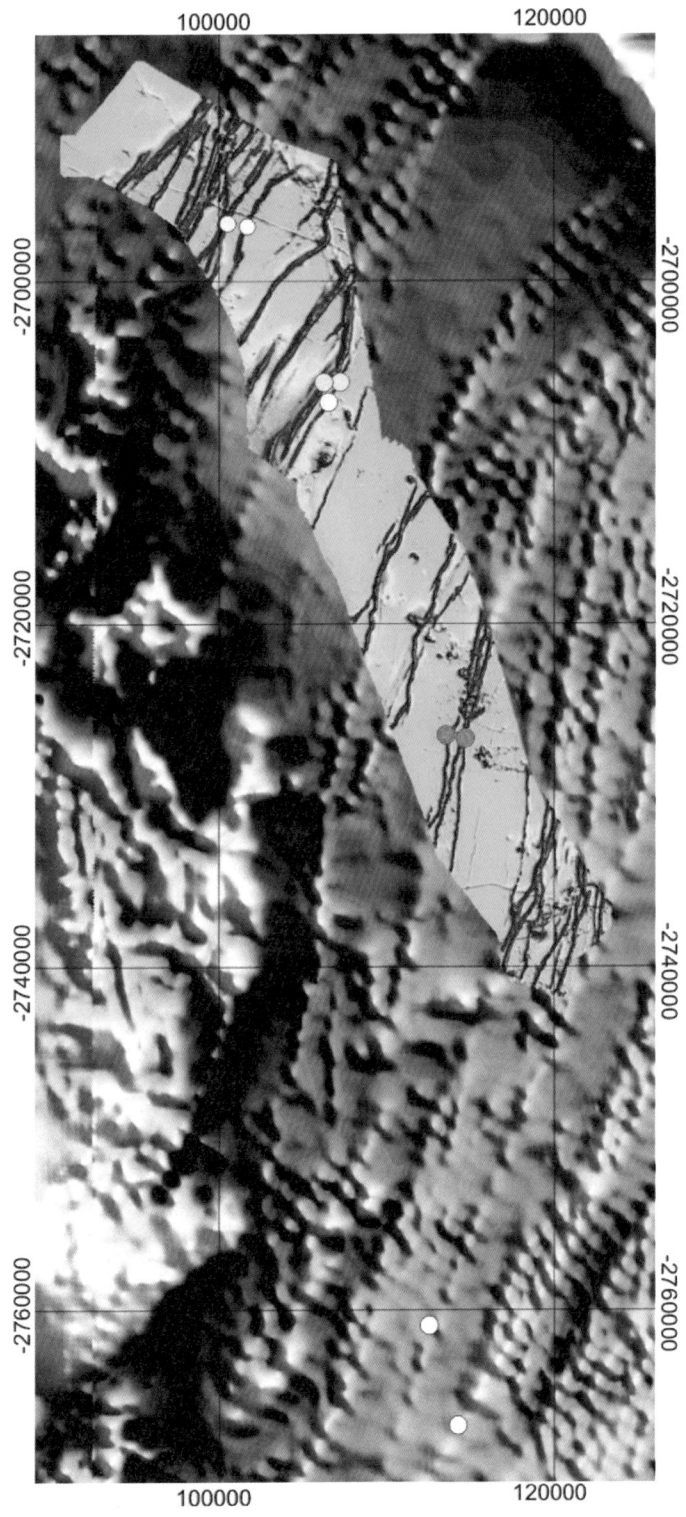

Fig. 4.

Table 1. *Mean NRM and bulk susceptibility for sites from (a) Phalaborwa Complex pyroxenite, (b) Phalaborwa dykes and (c) Bushveld dykes*

Site	NRM Decl.	NRM Incl.	α_{95}	N	Intensity average (A/M)	± (A/m)	Susc. average (×10⁻⁶ SI)	± (×10⁻⁶)	Q ratio
(a) Phalaborwa Complex pyroxenite									
P-1	353.4	48.3	7.9	9	214.5	17.1	226.0	42.0	
P-2	1.4	28.1	30.2	13	64.6	37.6	237.3	56.9	
P-3	356.9	33.1	29.2	9	17.1	4.7	327.3	6.4	
P-4	15.5	41.7	2.9	11	74.7	23.5	248.1	25.1	
P-5	6	34.9	4.7	13	3.5	1.5	228.0	30.4	
P-6	352.8	54.2	50.2	8	12.6	0.8	255.6	89.6	
P-7	13.1	19.2	7.6	15	115.3	37.1	4531.7	1277.6	
P-8	35.4	35.6	14.8	12	14.10	9.9	213.9	18.5	
(b) Phalaborwa dykes (F = FOSKOR pit, P = Phalaborwa pit, V = Vermiculite pit)									
F-1	11.8	50.9	6.3	7	3.15	3.8	28750.2	351.8	4.5
F-2	121.3	−63.9	1.3	10	8.48	3.2	33458.3	628.0	10.6
F-3	129.6	−68.7	3.6	10	8.76	11.7	35131.99	3779.7	10.4
F-4	158.7	−66.8	5.3	10	6.79	19.0	29177.5	837.4	9.7
P-1	167.1	−78.9	3.4	10	15.70	5.7	49172.5	772.5	13.3
P-2	317.3	69.7	20	5	30.83	20.4	15907.4	42325	81.2
P-3	20.6	35.1	73.9	11	1.38	3.3	40557.1	473.2	1.4
V-1	328.3	4.6	29.4	9	2.29	28.5	39453.6	1131.04	2.4
V-2	354.6	54.9	49.6	7	4.00	7.4	21442.0	1169.4	7.8
(c) Bushveld dykes (M = Modikwa Mine, H = Hackney Mine, B = Letts *et al.* (2005).									
M-1	170.4	−62.8	4.6	9	10.96	2130.7	36942	662.6	12.4
M-2	178.3	−63.5	4	10	12.47	13204.1	35449	1896.5	14.7
H-2	186.7	−67.9	6.1	10	3.35	529.4	23159	225.3	6
H-1	211.6	−65.4	12.4	7	3.27	253.7	25422	750	5.3
B-1	212	−73	6.2	8	3.387	1225	17288	9502	9.6
B-2	138	−71	3.8	9	4.282	447	30184	1507	6.4
B-3	6	−79	19.1	5	0.69	31	10473	5988	3.6
B-4	343	84	19.1	3	0.43	18	24154	1523	0.7
B-5	12	65	13.6	6	22.1	1041	27114	6297	36.8

Decl: declination, Incl: inclination, α_{95}: 95% confidence circle around mean direction, N: number of samples, ±: 1-σ error.

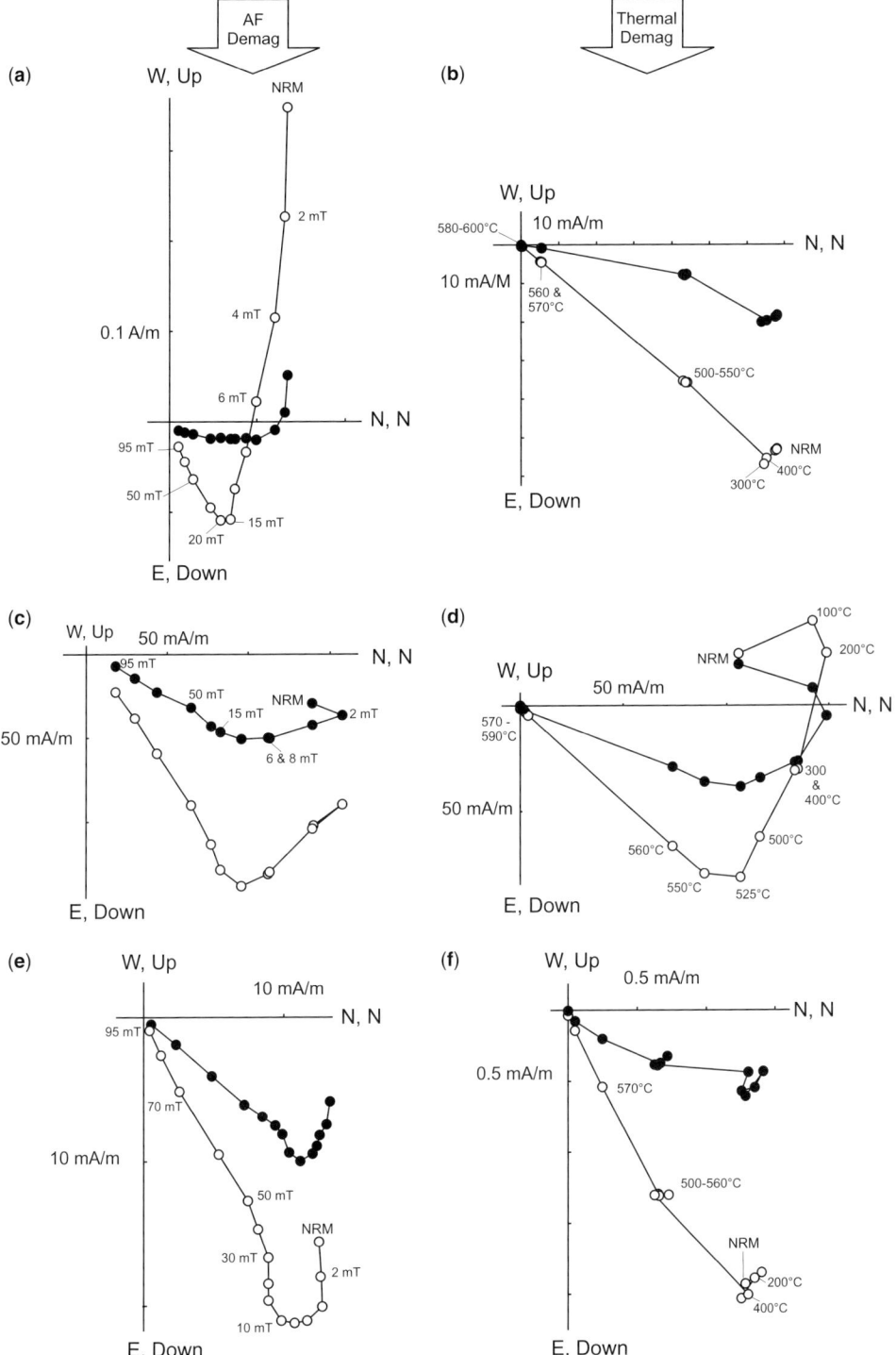

Fig. 5. Characteristic examples of AF (**a, c, d**) and thermal (**b, d, e**) demagnetization of samples (sites P-7 and P-8) from the Phalaborwa Complex (pyroxenite). In orthogonal vector plots, open (solid) symbols represent projections onto the vertical (horizontal) planes.

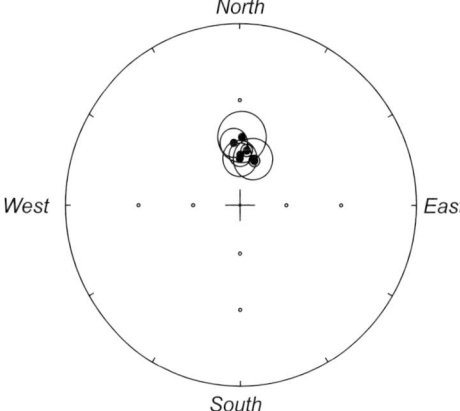

Fig. 6. Site mean directions and α_{95} confidence circles for Phalaborwa pyroxenite samples. Closed symbols denote downwards-pointing (positive) inclinations.

components of magnetization: a low coercivity or low unblocking (LB) component and a well-defined high-stability or high-temperature blocking (HB) component. HB components are typically identified above 500–550 °C or 20–50 mT, and thermal and AF demagnetization yielded similar directional results. HB components showed reasonable within-site grouping and site-mean directions are characterized by northerly declination and steep positive inclination (Fig. 6, Table 2). Samples from site P-6 behaved unstably during demagnetization and have not been included in the findings. LB components are scattered at site level and are not considered further.

Typical response to TMA is shown in Figure 7a, in which samples indicate a Curie temperature of *c.* 580 °C corresponding to magnetite. Petrographic examination discloses abundant sulphides, mostly pyrrhotite, but magnetite occurred as subordinate small discrete grains (Fig. 7b).

Phalaborwa dykes (dolerites)

Nine dolerite dykes were sampled in the Phalaborwa Complex: four in the Foskor pit, three in the main Phalaborwa Pit and two in the Vermiculite Pit (Fig. 2). A total of 79 dolerite dyke cores were analysed and NRM and bulk susceptibility values are listed in Table 1. None of the dykes have been dated.

Samples of the dykes behaved exceptionally well to both thermal and AF demagnetization, showing the presence of two components of magnetization (Fig. 8). LB components are random within sites and typically removed at around 300–500 °C

or 10–30 mT. HB components show reasonable within-site grouping (Table 2). Site-mean directions plot with southerly to SE declination and very steep negative inclination or antipodal to this (Fig. 9). One site plots slightly off the other site-means but is included in the final statistics.

TMA (Fig. 7c, d) indicates Curie temperature at *c.* 580 °C. Analysis of samples from site V-2 revealed that, upon heating, the saturation magnetization first increased at 150–170 °C and then followed a normal decay for a Curie temperature at *c.* 580 °C. The kink at 150–170 °C, a typical low-temperature titanomagnetite (maghemitization) phenomenon (Ade-Hall *et al.* 1971), is probably related to hydrothermal alteration. Most samples are dominated by pure magnetite with a number of grains showing skeletal textures indicative of rapid cooling (Fig. 7e). A minor number of magnetite grains have oxy-exsolution lamellae of ilmenite (Fig. 7f). Samples from the V-2 site showed minor alteration of magnetite that could not be identified due to grain sizes being smaller then the optical resolution. Petrographic observations are in agreement with the findings obtained by thermomagnetic analysis, that is, that magnetite is the dominant carrier of magnetization.

Bushveld dykes

Samples from four underground dyke sites were obtained from the Eastern Bushveld Complex, two from the Hackney mine and two from the Modikwa mine. These sites are combined with five sites previously reported in Letts *et al.* (2005). The locations of all nine sites (Fig. 4) are plotted on high-resolution regional aeromagnetic data over the Bushveld Complex. The dykes are not adequately dated (see below).

Sixty-five core samples were analysed for NRM and bulk susceptibility (Table 1). Samples behaved well to both thermal and AF demagnetization, showing the presence of two components of magnetization (Fig. 10). LB components are typically removed below 500 °C or 30–40 mT. Samples obtained on the surface have stronger LB components relative to those obtained underground. Once again, LB directions are not related to any significant geological event and are therefore not considered. After demagnetization, one of two antipodal HB components is identified (Fig. 11). Most sites show directions with steep negative inclination and south to south-easterly declination. The remaining two sites show opposite polarity. Most sites (except B3, B4 and B5), showed good within-site grouping with α_{95} at around 3° (Fig. 11; Table 2).

TMA (Fig. 7g, h) indicates Curie temperature at *c.* 580 °C, once again suggesting magnetite or

Table 2. *Site mean HB components from (a) Phalaborwa Complex pyroxenite, (b) Phalaborwa dykes and (c) Bushveld dykes*

Site	Dec.	Inc.	α_{95}	N	κ	Lat. (S)	Lon. (E)
(a) Phalaborwa Complex pyroxenite							
P-1	359.8	61.2	10.7	9	74.96	23.962°	31.129°
P-2	1.8	48.9	14.1	13	43.34	23.962°	31.127°
P-3	354	52	8.2	9	127.82	23.969°	31.121°
P-4	7.5	56.3	3.6	11	454.69	23.966°	31.122°
P-5	0.3	59	6.9	13	122.73	23.968°	31.127°
P-7	19	61.2	3.4	15	168.04	23.965°	31.125°
P-8	16.5	60.4	12.2	12	30.61	23.963°	31.125°
Mean	*5.0*	*57.3*	*5.2*	*7*	*137.2*	*23.96*	*31.13*

Pole Latitude 27.9°, Longitude 35.6° (dp/dm = 5.6°/7.6°); based on site directions
      Latitude 27.7°, Longitude 35.8° (A95 = 6.6°); based on site poles

Site	Dec.	Inc.	α_{95}	N	κ	Lat. (S)	Lon. (E)
(b) Phalaborwa dykes							
F-1	14.8	52.8	4.4	8	158.1	23.928°	31.096°
F-2	121	−62.9	1.7	9	959.5	23.965°	31.121°
F-3	132.3	−66.6	3	11	234.7	23.967°	31.122°
F-4	141.4	−66.9	3.2	10	230.2	23.967°	31.122°
P-1	168.1	−77.9	3.2	10	232.0	23.997°	31.125°
P-2	342.4	74.8	6.2	6	401.1	23.998°	31.128°
P-3	321.2	70.5	3.8	10	159.9	23.998°	31.128°
V-1	314.2	61.3	8	7	93.3	24.004°	31.135°
V-2	315.7	72.7	21	7	14.3	24.007°	31.136°
Mean	*328.0*	*69.2*	*7.9*	*9*	*43.6*	*23.99°*	*31.12°*

Pole Latitude 8.3°, Longitude 12.2° (dp/dm = 11.5°/13.5°); based on site directions
      Latitude 7.6°, Longitude 12.1° (A95 = 11.8°); based on site poles

Site	Dec.	Inc.	α_{95}	N	κ	Lat. (S)	Lon. (E)
(c) Bushveld dykes							
H-1	185.4	−70.4	2.8	5	1885.7	24.350°	30.060°
H-2	180.5	−65.2	3.4	10	201.2	24.350°	30.060°
M-1	167.2	−64.3	2.6	9	401.6	24.183°	29.866°
M-2	172.9	−60.2	3.3	10	215.4	24.183°	29.866°
B-1	185	−73	3.2	8	298.54	24.462°	30.051°
B-2	137	−68	3.5	9	220.51	24.949°	30.151°
B-3	110.1	−75.9	20.1	5	21.84	24.883°	30.116°
B-4	359.9	62.2	18	3	48.12	24.413°	29.997°
B-5	10	65.6	12.8	6	28.19	24.370°	30.000°
Mean	*350.9*	*68.6*	*6.4*	*9*	*65.8*	*24.4°*	*30.0°*

Pole Latitude 13.3°, Longitude 24.2° (dp/dm = 9.1°/10.8°); based on site directions
      Latitude 12.6°, Longitude 24.1° (A95 = 10.8°); based on site poles

N, number of samples; α_{95}, 95% confidence circle around the mean direction; κ, precision parameter; Lat. (S) and Lon. (E), mean sampling latitude (south) and longitude (east); A95, 95% confidence around the mean pole.

titanium-poor magnetite as the prime remanence carrier. Most samples showed almost reversible Curie curves upon cooling and heating, except for a small increase in saturation magnetization during cooling. This increase probably indicates secondary production of magnetite. TMA on a sample from site B-3 produced a distinctively different curve to those experienced in other sites. Upon heating, the saturation magnetization first increased at 150–170 °C followed by a distinctive decrease at c. 350 °C, suggesting low-temperature hydrothermal alteration

(maghemitization) as also observed from some Phalaborwa dykes (Fig. 7d).

Petrographic analysis generally revealed quenched magnetite grains (indicating rapid cooling) as the dominant opaque mineralogy (Fig. 7i). A number of magnetite grains contain oxy-exsolution lamellae of ilmenite, possibly due to high-temperature deuteric oxidation during late-stage crystallization of the dyke (Fig. 7j). Samples from site B-3 are dominated by highly altered magnetite; the majority of the alteration mineral could

18   S. LETTS *ET AL.*

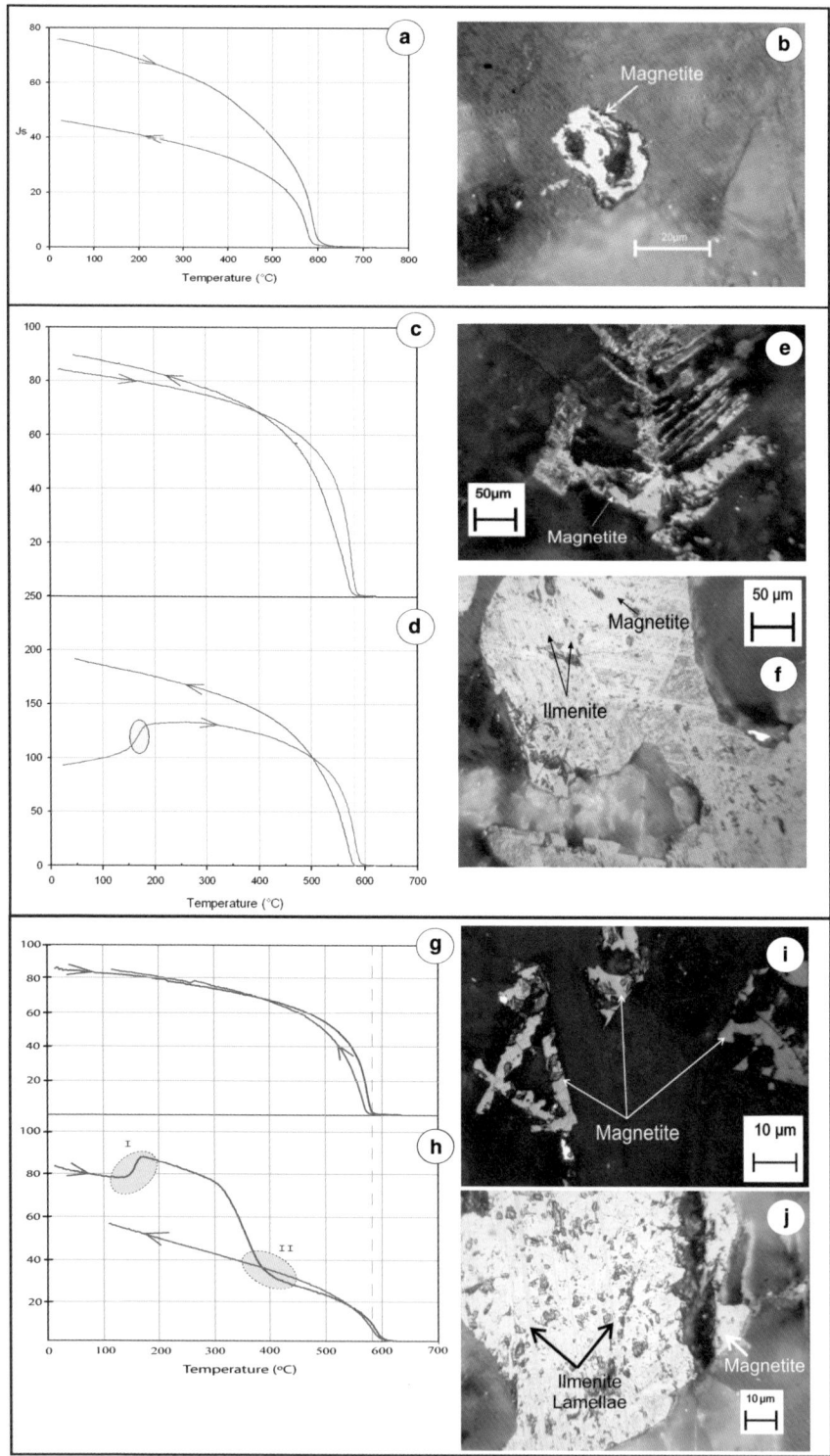

Fig. 7.

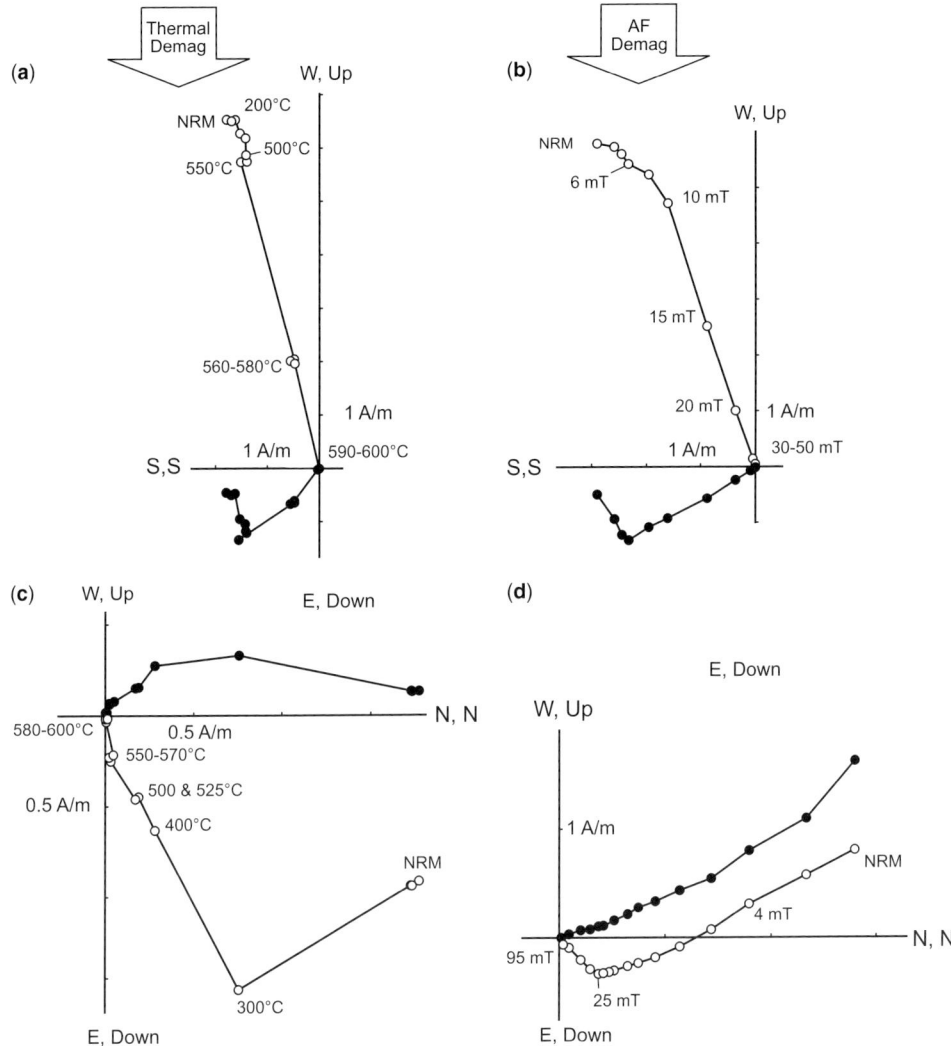

Fig. 8. Phalaborwa dykes: Typical orthogonal vector plots from thermal (**a, c**) and AF demagnetization (**b, d**). (a, b) are from Foskor site 4 (F-4) and (c, d) are from Vermiculite site 1 (V-1). Open (solid) symbols represent projections onto the vertical (horizontal) planes.

Fig. 7. Phalaborwa pyroxenite samples: (**a**) Typical thermomagnetic analysis from a sample from site P-4. Red line and arrow indicate heating phase while blue line and arrow indicates cooling phase. Curie temperature for magnetite at *c.* 580 °C observed (dashed line). (**b**) Discrete magnetite grain from site P-3 (reflected light). Phalaborwa dyke samples: (**c, d**) Thermomagnetic analysis of samples from sites F-3 and V-2. The majority of samples display Curie curves such as in (c), in which a small increase in saturation magnetization is seen during cooling, implying the possible production of more magnetite. In (d) there is irreversible creation of a magnetic phase with higher saturation magnetization at 150–180°, indicated by a grey ellipse. Curie temperature for magnetite is observed at *c.* 580 °C (dashed line). (**e**) Polished thin section showing a magnetite grain with a quenched texture from site F-3 (reflected light). (**f**) Polished thin section showing a magnetite grain with lamellae of ilmenite from site P-1. Bushveld dyke samples: (**g, h**) Thermomagnetic analysis of samples from sites M-1 and B-3. The majority of samples display Curie curves as in (h). In (i) there is irreversible creation of a magnetic phase with higher saturation magnetization at 150–180° (indicated by grey ellipse I) followed by inversion of maghemite to a weaker magnetic phase around 350 °C (indicated by grey ellipse II). (**i**) Polished thin section showing a magnetite grain with a quenched texture from site M-1 (reflected light). (**j**) Polished thin section showing a magnetite grain with lamellae of ilmenite, from site H-2.

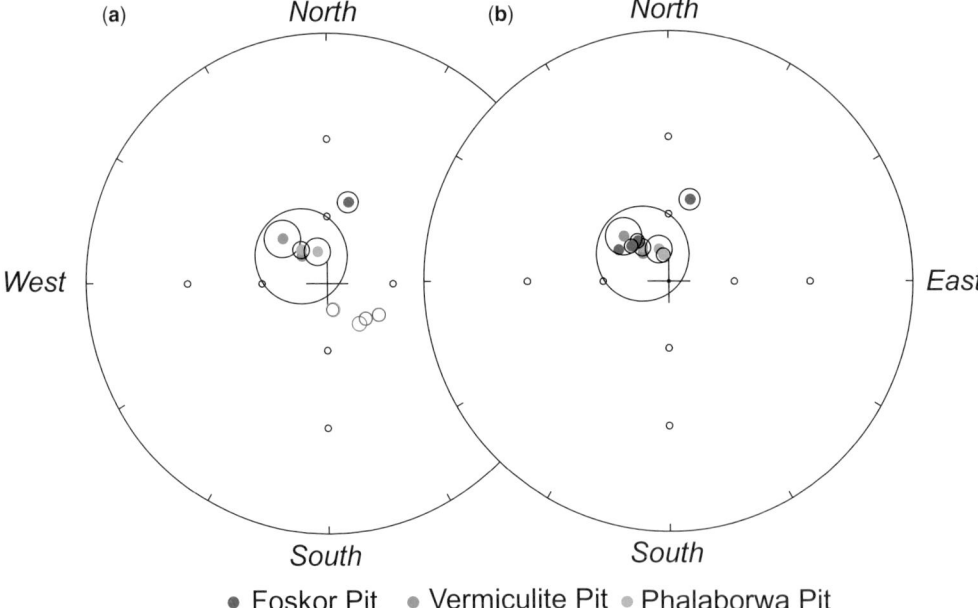

Fig. 9. (**a**) Phalaborwa dyke site-mean directions with α_{95} confidence circles and (**b**) sites inverted to positive inclination. Solid symbols indicate a positive inclination, while open symbols indicate a negative inclination. Note that α_{95} confidence circles for some sites are smaller than the symbol size.

not be determined due to grain sizes below the optical resolution. However, some alteration has been identified as hematite.

Interpretation

Phalaborwa Complex (pyroxenites)

The majority of samples contained minor LB components, successfully removed during demagnetization, and allowed for the identification of well-defined HB components that produce well-clustered results. Thermal demagnetization spectra and TMA suggest that Ti-poor magnetite or nearly pure magnetite (maximum unblocking temperatures of 565–580 °C and Curie temperatures of *c.* 580 °C) is the bulk remanence carrier. Petrographic analysis supports these observations.

Palaeomagnetic data reported in this study (Table 2) are in reasonable agreement with those of Morgan & Briden (1981) for the Phalaborwa Complex. A small observed difference in pole positions is probably due to blanket cleaning at *c.* 400 °C or 50 mT in the previous study. Our new pole for the Phalaborwa Complex pyroxenites (2060.6 ± 0.5 Ma) now plots closer to and statistically overlaps those obtained for the near coeval Bushveld Complex (2054.4 ± 1.8 Ma; Fig. 12)

Phalaborwa and Bushveld dykes

The age of the Phalaborwa and Bushveld dykes is uncertain and the Bushveld dykes have for long been assumed to be of Jurassic (Karoo) age. Palaeomagnetic data from both dyke sets are of excellent quality, and both normal and reversed polarities are recorded. The Phalaborwa dykes share a common mean at 95% confidence according to McFadden & Lowes (1981). The reversal test of McFadden & McElhinny (1990) also indicated a positive reversal test with a 'C' classification (i.e. the angle between the reversals is between 170° and 160°; critical angle 16.6°; observed angle 7.6°).

The Bushveld dykes also share a common mean at 95% confidence according to the method of McFadden & Lowes (1981). The McFadden & McElhinny (1990) reversal test produced a positive reversal test, with a 'C' classification (critical angle 15.1°; observed angle 9.3°). Positive reversal tests suggest that the reversely polarized dykes are of a similar age to the normally polarized dykes. Given the positive reversal tests, HB components isolated from both the Phalaborwa and Bushveld dykes were inverted to the same polarity before calculating mean site-directions and palaeomagnetic poles (Table 2). Given a statistical overlap in uncertainty levels from the two palaeomagnetic poles, the

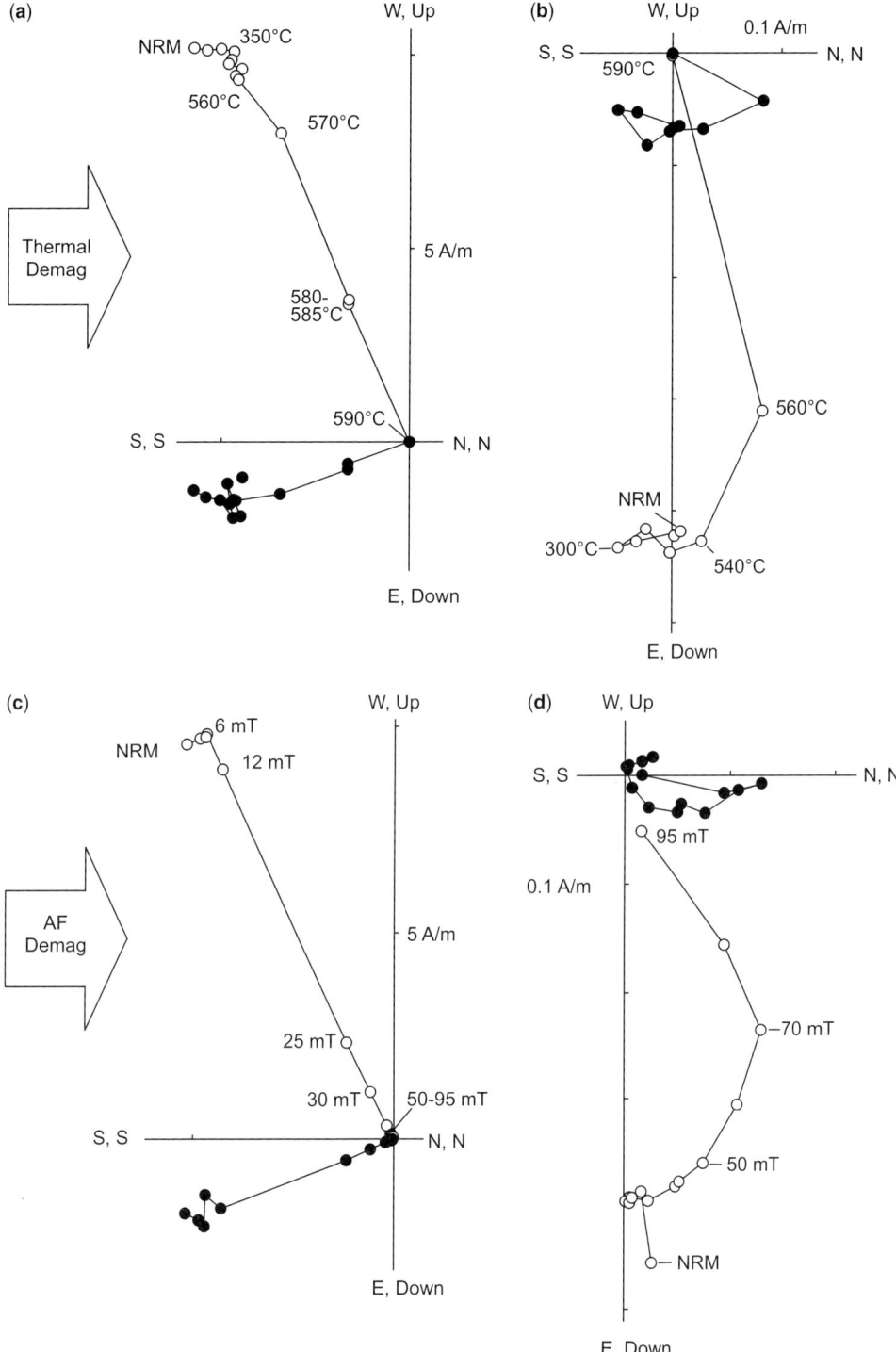

Fig. 10. Bushveld dykes: Typical orthogonal vector plots from thermal (**a, b**) and AF demagnetization (**c, d**). Samples obtained from underground are shown in (a, c). Surface samples obtained from Letts *et al.* (2005) are reflected in (b, d). Open (solid) symbols represent projections onto the vertical (horizontal) planes, respectively.

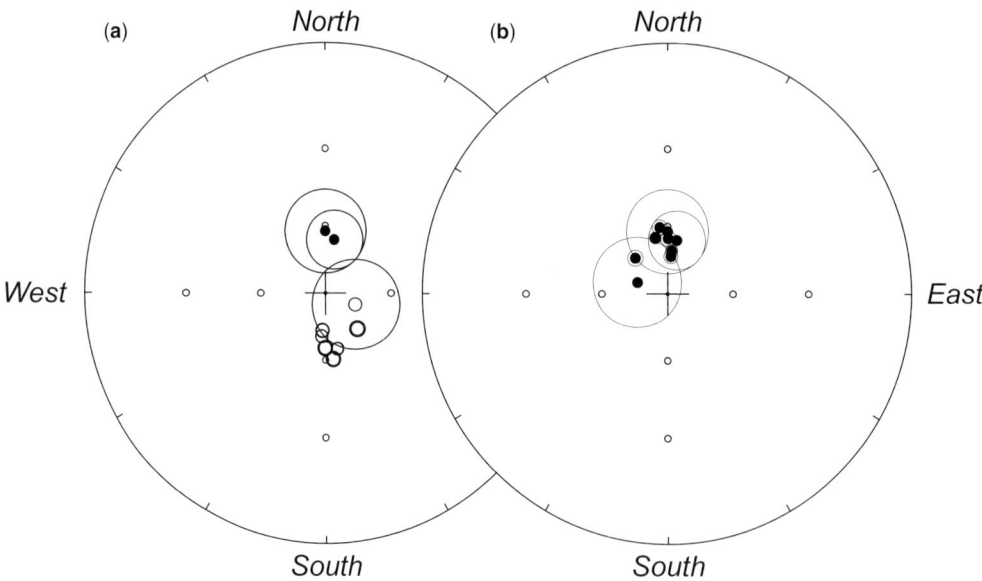

Fig. 11. Bushveld dyke site-mean directions (**a**) and sites inverted to positive inclinations (**b**). Solid symbols indicate a positive inclination, while open symbols indicate a negative inclination.

dykes could indeed be related to the same (but as yet undated) magmatic event.

Conclusions and key palaeomagnetic poles from the Kaapvaal/Kalahari Craton

Palaeomagnetic data from the well-dated Phalaborwa Complex (2060.6 ± 0.5 Ma) are of excellent quality, and HB components yield a palaeomagnetic pole (Fig. 12) that statistically overlaps with the near-contemporaneous Bushveld Complex (2054.4 ± 1.3 Ma). In the latter case normal and reversed directions share a common mean at the 95% confidence level (Letts *et al.* 2009), attesting to a primary magnetic signature. The Bushveld and Phalaborwa Complex poles must be regarded as the most reliable 2.05–2.06 Ga key poles for the Kaapvaal Craton.

   Palaeomagnetic data from the dykes cutting both the Phalaborwa and Bushveld Complexes are also of excellent quality, and normal and reversed polarities from both dyke sets share a common mean at the 95% confidence level. The palaeomagnetic data clearly demonstrate that these dykes are Palaeoproterozoic and not Jurassic (Karoo) in age (Mean Karoo pole is 65.8°S and 83.7°E; A95 = 6.1; $N = 5$ poles; based on Torsvik *et al.* 2008). Compared with palaeomagnetic poles from southern Africa (Fig. 12a), both the Phalaborwa and Bushveld dyke poles statistically overlap with the

well-dated (1874.6 ± 3.9 Ma; Hanson *et al.* 2004) post-Waterberg dolerites and the poorly dated (1878 ± 86 Ma; Barton 1979) Sand River Dykes (Morgan 1985). Based on pole positions, it is therefore possible that the dykes occurring in both the Phalaborwa and Bushveld complexes are the result of the same magmatic event at *c.* 1.9 Ga. However, they also plot close to the Bushveld Complex poles (Fig. 12a); an older age, but obviously ≤2.05–2.06 Ga (e.g. Bushveld age), cannot be precluded. An $^{40}Ar/^{39}Ar$ plagioclase age of 1649 ± 10 Ma was determined from one of the Bushveld dykes (Letts *et al.* 2005), but this age probably represents a low-temperature resetting event and not the emplacement age. Since the true emplacement age for these dykes is not established, the otherwise excellent primary palaeomagnetic poles from these dykes cannot be used as key poles before the dykes are properly dated; we are therefore awaiting geochronologists to elevate our palaeomagnetic poles as key Kaapvaal poles.

   Many different Apparent Polar Wander (APW) paths have been published for the Kaapvaal Craton between *c.* 1.9 and 2.0 Ga. In Figure 12a we show five poles with blue confidence ovals that we consider as the only key poles for this time period. In addition to the Bushveld and Phalaborwa complexes and the Post-Waterberg Dolerites discussed above, we also include poles from the Vredefort Impact (2023 ± 4 Ma) and the Mashonaland Sill (1887.9 ± 1.2 Ma; weighted mean of ages reported

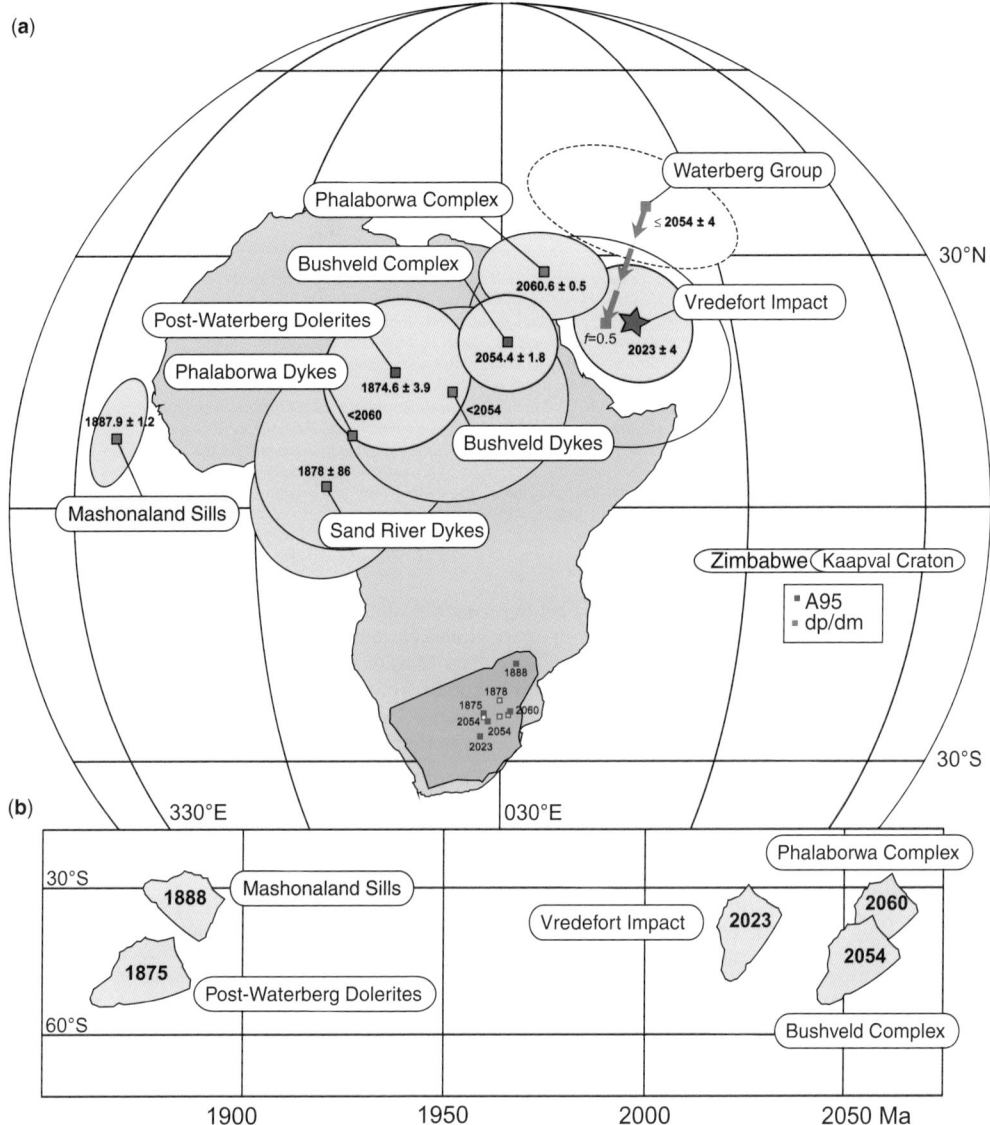

Fig. 12. (**a**) Selected palaeomagnetic poles from the Kaapvaal and Zimbabwe (only Mashonaland Sills) cratons from *c.* 2060 to 1875 Ma. Phalaborwa Complex pyroxenite, Phalaborwa dykes and Bushveld dykes are from this study (see text and Table 3 for other references). The outline of the Kalahari Craton (in green) and mean sampling positions for all poles are also shown. Note that some 95% confidence 'ovals' are reported as A95 circles (blue squares/star) whilst others are d*p*/d*m* ovals (red squares). The Waterberg Group pole has been corrected for sedimentary inclination shallowing using a flattening factor *f* of 0.5. Orthographic projection. (**b**) Reconstruction of the Kalahari Craton (latitude and orientation) according to poles in (a). Galls Projection.

in Söderlund *et al.* 2010). These are plotted along with less reliable poles obtained from the Waterberg Group (≤2054 ± 4 Ma), the Bushveld and Phalaborwa dykes and the Sand River dykes (all three plotted with grey ovals/ellipses). The Sand River pole is not considered as a key pole due to

large age uncertainties; all three dyke poles, however, plot near the well-dated Post-Waterberg Dolerites pole (1874.6 ± 3.9).

Of the nine poles shown in Figure 12a, of which five are reliable in terms of both palaeomagnetism and geochronology, two poles pose some concern.

Table 3. *The most reliable palaeomagnetic poles (2.06–1.87 Ga) from the Kaapval (notes 1–3, 5) and Zimbabwe (note 4) cratons in South Africa (blue ovals in Fig. 12a)*

Group/Lithology	Pole longitude	Pole latitude	A95 (dp/dm)	Age (Ma)	Notes
Phalaborwa Complex pyroxenite	27.7°N	35.8°E	6.6°	2,060.6 ± 0.5	1
Bushveld Complex	19.2°N	30.8°E	5.8°	2,054.4 ± 1.3	2
Vredefort Impact	21.8°N	44.5°E	(11.3/15.4°)	2,023.0 ± 4.0	3
Mashonaland Sills	7.6°N	338.2°E	5.1°	1,887.9 ± 1.2	4
Post-Waterberg Dolerites	15.6°N	17.1°E	8.9°	1,874.6 ± 3.9	5

Notes: (1) This study. (2) Five results obtained for each zone of the Bushveld Complex (Letts *et al.* 2009) have been grouped together to produce a single pole. (3) The 2023 ± 4 Ma (Kamo *et al.* 1996) Vredefort impact structure has been the subject of three palaeomagnetic studies. An initial study conducted by Hargraves (1970) sampled a range of lithologies within the central core and outer overturned collar rocks. A later study by Hart *et al.* (1995) obtained palaeomagnetic data from 16 sites within the central granitic basement of the structure. The most recent study (Carporzen 2006) concentrated on the pseudotachylites and granophyre dykes that are directly related to the impact event and obtained two palaeopoles: 19.9°N, 47.7°E and dp/dm = 6.3°/8.6° for the pseudotachylites and 23.6°N, 41.3°E and dp/dm = 3.6°/4.9° for the granophyre dykes. The pseudotachylites and granophyre results were grouped together to produce a single pole for the impact. (4) The well-dated Mashonaland Sills (Söderlund *et al.* 2010) occur as a suite of early Proterozoic intrusive dolerites that crop out in the north-eastern section of the Archaean Zimbabwe Craton. Here we use a recalculated palaeopole for the Mashonaland sills (Evans *et al.* 2002), in which they combined the dual-polarity results obtained by McElhinny & Opdyke (1964) and Bates & Jones (1996). (5) A number of dolerites occurring as sills, irregular bodies and dykes have intruded the Waterberg group in northern South Africa. Palaeomagnetic analysis conducted on samples located at or near localities from geochronological samples produced well-grouped antipodal directions (Hanson *et al.* 2004).

(i)   The Waterberg Group pole (pole WUBS-I in de Kock *et al.* 2006) is 2.05 Ga or less in age but this pole is statistically different from both the Phalaborwa and Bushveld complexes and the c. 30 Ma younger Vredefort Impact pole (Fig. 12a).

(ii) The Mashonaland Sills pole is now well dated (1887.9 ± 1.2 Ma; Söderlund *et al.* 2010), the palaeomagnetic data (including a positive reversal test) appear robust (e.g. Bates & Jones 1996 and summarized in Evans *et al.* 2002), but the pole plots far away from the key Post-Waterberg Dolerite pole which is only 13.3 Ma (on average) younger.

The Waterberg Group pole is the only pole in our selection that is derived from sedimentary rocks. Inclination shallowing is a well-known phenomena in sedimentary rocks (e.g. Kent & Tauxe 2005), and by using a flattening factor f of 0.5 (a common/realistic value) we calculate a revised pole position for the Waterberg Group (WUBS-I) to 21.4°N and 43.2°E. This revised pole conforms much better with the near-coeval Phalaborwa and Bushveld Complex poles and the somewhat younger Vredefort Impact pole (Fig. 12a).

The Mashonaland Sills pole plots 39° of arc away from the Post-Waterberg Dolerite pole. This amounts to 3°/Ma, which is difficult to explain unless invoking true polar wander. Based on recent estimates of Earth properties, however, a true polar wander speed limit has been estimated as c. 1°/Ma (Steinberger & O'Connell 1997; Tsai & Stevenson 2007) although a more realistic numerical model based on subduction history

suggests that up to 2°/Ma is possible for shorter time periods (Steinberger & Torsvik 2010). The Mashonaland Sills pole is the only pole that comes from the Zimbabwe Craton. An alternative reason for the pole discrepancy with the Post-Waterberg Dolerites pole is that the Zimbabwe and Kaapvaal cratons were not fully sutured at c. 1.9 Ga (as portrayed in Fig. 12b) with the Zimbabwe Craton located at lower latitudes (c. 30°S) than Kaapvaal, which was located to intermediate latitudes.

We thank the South African and Norwegian Research Councils and Anglo American for financial support and logistic assistance. We also thank the reviewers T. Zegers and C. Langereis.

References

ADE-HALL, J. M., PALMER, H. C. & HUBBAR, T. P. 1971. The magnetic and opaque petrological response of basalt to regional hydrothermal alteration. *Geophysical Journal of Royal Astronomical Society*, **24**, 137–174.

BARTON, J. M. 1979. The chemical composition, Rb–Sr isotopic systematics and tectonic setting of certain post-kinematic mafic igneous rocks, Limpopo Mobile belt, Southern Africa. *Precambrian Research*, **9**, 57–80.

BATES, M. P. & JONES, D. L. 1996. A palaeomagnetic investigation of the Mashonaland dolerites, north-east Zimbabwe. *Geophysical Journal of International*, **126**, 513–524.

CARPORZEN, L. 2006. *Meteorite impact crater magnetism in Vredefort (South Africa) and Rochechouart (France)*. Thesis Université Paris VII Denis Diderot, Paris, France.

CHAVEZ GOMEZ, S. 2001. A catalogue of dykes from aero-magnetic surveys in eastern and southern Africa. ITC (The International Institute for Aerospace and Earth Sciences), Enschede.

CHENEY, E. S. 1996. Sequence stratigraphy and plate tectonic significance of Transvaal succession of southern Africa and its equivalent in Western Australia. *Precambrian Research*, **79**, 3–24.

DALZIEL, I. W. D. 1992. On the organization of American plates in the neoproterozoic and the breakout of Lauretia. *GSA Today*, **2**, 237–241.

DE KOCK, M. O., EVANS, D. A. D., DORLAND, H. C., BEUKES, N. J. & GUTZMER, J. 2006. Paleomagnetism of the lower two unconformity-bounded sequences of the Aterberg Group, South Africa: towards a better-defined apparent polar wander path for the Paleoproterozoic Kaapval Craton. *South African Journal of Geology*, **109**, 157–182.

DE KOCK, M. O., EVANS, D. A. D. & BEUKES, N. J. 2009. Validating the existince of the Vaalbara in the Neoarchean. *Precambrian Research*, **174**, 145–154.

EVANS, D. A. D. 2009. The palaeomagnetically viable, long-lived and all-inclusive Rodinia supercontinent reconstruction. *In:* MURPHY, J. B., KEPPIE, J. D. & HYNES, A. (eds) *Ancient Orogens and Modern Analogues.* Geological Society, London, Special Publications, **327**, 371–404.

EVANS, D. A. D., BEUKES, N. J. & KIRSCHVINK, J. L. 2002. Paleomagnetism of a lateritic paleoweathering horizon and overlying Paleoproterozoic red beds from South Africa: implications for the Kaapvaal apparent polar wander path and a confirmation of atmospheric oxygen enrichment. *Journal of Geophysical Research*, **107**, doi:10.1029/2001JB000432.

GUBANOV, A. P. & MOONEY, W. D. 2009. New global maps of crustal basement age. *Eos Trans. AGU*, **90**, Fall Meet. Suppl., Abstract T53B-1583.

HANSON, R. E., GOSE, W. A. *ET AL.* 2004. Paleoproterozoic intraplate magmatism and basin development on the Kaapvaal Craton: age, palaeomagnetism and geochemistry of ~1.93 to ~1.87 Ga post-Waterberg dolerites. *South African Journal of Geology*, **107**, 233–254.

HARGRAVES, R. B. 1970. Palaeomagnetic evidence relevant to the origin of the Vredefort Ring. *Journal of Geology*, **78**, 253–263.

HARMER, R. E. 2000. Mineralisation of the Phalaborwa Complex and the carbonatite connection in iron oxide-Cu-Au-U-REE deposits. *In:* PORTER, T. M. (ed.) *Hydrothermal Iron Oxide Copper–Gold & Related Deposits: A Global Perspective.* Australian Mineral Foundation, Adelaide, 331–340.

HART, R. J., HARGRAVES, R. B., ANDREOLI, A. G., TREDOUX, M. & DOUCOURE, C. M. 1995. Magnetic anomaly near the centre of the Vredefort structure; Implications for impact-related magnetic signatures. *Geology*, **23**, 277–280.

HOFFMAN, P. F. 1991. Did the breakout of Laurentia turn Gondwana inside out? *Science*, **252**, 1409–1411.

HUNTER, D. R. & REID, D. L. 1987. Mafic dyke swarms in Southern Africa. *In:* HALLS, H. C. & FAHRIG, W. F. (eds) *Mafic Dyke Swarms.* Geological Association, Canada, Special Paper, **34**, Toronto, Canada, 445–456.

KAMO, S. L., REIMOLD, W. U., KROGH, T. E. & COLLISTON, W. P. 1996. A 2.023 Ga age for the Vredefort impact event and a first report of shock metamorphosed zircons in pseudotachylitic breccias and granophyre. *Earth and Planetary Science Letters*, **144**, 369–387.

KENT, D. V. & TAUXE, L. 2005. Corrected Late Triassic latitudes for continents adjacent to the North Atlantic. *Science*, **307**, 240–244.

KENT, J. T., BRIDEN, J. C. & MARDIA, K. V. 1983. Linear and planar structure in ordered multivariate data as applied to progressive demagnetisation of palaeomagnetic remanence. *Geophysical Journal of Royal Astronomical Society*, **81**, 75–87.

LETTS, S. A., TORSVIK, T. H., WEBB, S. J., ASHWAL, L. D., EIDE, E. A. & CHUNETT, G. 2005. Palaeomagnetism and $^{40}Ar/^{39}Ar$ geochronology of mafic dykes from the Eastern Bushveld Complex (South Africa). *Geophysical Journal of International*, **162**, 36–48.

LETTS, S. A., TORSVIK, T. H., WEBB, S. J. & ASHWAL, L. D. 2009. Palaeomagnetism of the 2054 Ma Bushveld Complex (South Africa): implications for emplacement and cooling. *Geophysical Journal of International*, **179**, 850–872.

LI, Z. X., BOGDANOVA, S. V. *ET AL.* 2008. Assembly, configuration, and break-up history of Rodinia: a synthesis. *Precambrian Research*, **160**, 179–210.

MCELHINNY, M. W. & OPDYKE, N. D. 1964. The paleomagnetism of the Precambrian dolerites of eastern Southern Rhodesia, an example of geological correlation by rock magnetism. *Journal of Geophysical Research*, **69**, 2465–2475.

MCFADDEN, P. L. & LOWES, F. J. 1981. The discrimination of mean directions drawn from Fisher distributions, *Geophisical Journal of Royal Astronomical Society*, **67**, 19–33.

MCFADDEN, P. L. & MCELHINNY, M. W. 1990. Classification of the reversal test in palaeomagnetism. *Geophysical Journal International*, **103**, 725–729.

MEERT, J. G. & TORSVIK, T. H. 2003. The making and unmaking of a Supercontinent: Rodinia revisited. *Tectonophysics*, **375**, 261–288.

MORGAN, G. E. 1985. The palaeomanetism and cooling history of metamorphic and igneous rocks from the Limpopo Mobile Belt, southern Africa. *Geological Society of America Bulletin*, **96**, 663–672.

MORGAN, G. E. & BRIDEN, J. C. 1981. Aspects of Precambrian palaeomagnetism, with new data from the Limpopo mobile belt and Kaapvaal Craton in Southern Africa. *Physics of the Earth and Planetary Interiors*, **24**, 142–168.

PALABORA MINING COMPANY LIMITED MINE GEOLOGICAL AND MINERALOGICAL STAFF 1976. The geology and the economic deposits of copper, iron, and vermiculite in the Palabora igneous complex; a brief review. *Economic Geology*, **71**, 177–192.

REISCHMANN, T. 1995. Precise U/Pb age determination with baddeleyite (ZrO_2), a case study from the Phalaborwa Igneous Complex, South Africa. *South African Journal of Geology*, **98**, 1–4.

ROGERS, J. J. W. & SANTOSH, M. 2002. Configuration of Columbia, a Mesoproterozoic Supercontinent. *Gondwana Research*, **5**, 5–22.

SCOATES, J. S. & FRIEDMAN, R. M. 2008. Precise age of the platiniferous Merensky Reef, Bushveld Complex, South Africa, by the U-Pb zircon chemical abrasion ID-TIMS technique. *Economic Geology*, **103**, 465–471.

SÖDERLUND, U., HOFMAN, A., KLAUSEN, M. B., OLSSON, J. R., ERNST, R. & PERSSON, P.-O. 2010. Towards a complete magmatic barcode fro the Zimbabwe craton: baddeleyite U-Pb dating of regional dolerite dyke swarms and sill complexes. *Precambrian Research*, **183**, 388–398.

STEINBERGER, B. & O'CONNELL, R. J. 1997. Changes of the Earth's rotation axis owing to advection of mantle density heterogeneities. *Nature*, **387**, 169–173.

STEINBERGER, B. & TORSVIK, T. H. 2010. Toward an explanation for the present and past locations of the poles. *Geochemistry, Geophysics, Geosystems*, **11**, doi: 10.1029/2009GC002889.

STRIK, G., BLAKE, T. S., ZEGERS, T. E., WHITE, S. H. & LANGEREIS, C. G. 2003. Palaeomagnetism of flood basalts in the Pilbara Craton, Western Australia: late Archean continental drift and the oldest known reversal of the geomagnetic field. *Journal of Geophysical Research*, **108**, 2–21.

TORSVIK, T. H. 2003. The Rodinia jigsaw puzzle. *Science*, **300**, 1379–1381.

TORSVIK, T. H., SMETHURST, M. A. *ET AL.* 1996. Continental break-up and collision in the Neoproterozoic and Palaeozoic: a tale of Baltica and Laurentia. *Earth Science Reviews*, **40**, 229–258.

TORSVIK, T. H., BRIDEN, J. C. & SMETHURST, M. A. 2000. Super-IAPD Interactive analysis of palaeomagnetic data. www.geodynamics.no/software.htm.

TORSVIK, T. H., MÜLLER, R. D., VAN DER VOO, R., STEINBERGER, B. & GAINA, C. 2008. Global plate motion frames: toward a unified model. *Reviews of Geophysics*, **46**, RG3004, doi:10.1029/2007RG000227.

TSAI, V. C. & STEVENSON, D. J. 2007. Theoretical constraints on true polar wander. *Journal of Geophysical Research*, **112**, B05415, doi:10.1029/2005JB003923.

UKEN, R. & WATKEYS, M. K. 1997. An interpretation of mafic dyke swarms and their relationship with major mafic magmatic events on the Kaapvaal Craton and Limpopo Belt. *South African Journal of Geology*, **100**, 341–348.

ZEGERS, T. E., DE WIT, M. J., DANN, J. & WHITE, S. H. 1998. Vaalbara, Earth's oldest assembled continent? A combined structural, geochronological, and palaeomagnetic test. *Terra Nova*, **10**, 250–259.

ZEH, A., GERDES, A. & BARTON JR[3], J. M. 2009. Archean accretaion and crustal evolution of the Kalahari Craton – the Zircon age and Hf isotope record of granitic rocks from Barberton/Swaziland to the Francistown Arc. *Journal of Petrology*, doi:10.1093/petrology/egp027.

ZHAO, G., CAWOOD, P. A., WILDE, S. A. & SUN, M. 2002. Review of global 2.1–1.8 Ga orogens: implications for a pre-Rodinia supercontinent. *Earth-Science Reviews*, **59**, 125–162.

Three episodes of crustal development in the Rehoboth Province, Namibia

VALBY VAN SCHIJNDEL[1]*, DAVID H. CORNELL[1], K.-H. HOFFMANN[2] & DIRK FREI[3]

[1]*Department of Earth Sciences, University of Gothenburg, SE-40530, Sweden*

[2]*Geological Survey of Namibia, 1 Aviation Road, Private Bag 13297, Windhoek, Namibia*

[3]*Department of Earth Sciences, Stellenbosch University, Private Bag X1, 7602 Matieland, South Africa*

Corresponding author (e-mail: valby.van.schijndel@gvc.gu.se)

Abstract: The African continental crust was assembled by a series of orogenies over a period of billions of years mainly in Precambrian times. Tracing the build-up history of this stable crust is not always straightforward due to multiphase deformation and regions with poor outcrop. Episodes of metamorphism and magmatism associated with multiple Wilson cycles are recorded in zircons, which found their way into sediments derived from the hinterland. Dating of zircon populations in detrital rocks can hence provide age spectra which reflect the metamorphic and magmatic events of the region. Microbeam dating of detrital zircon is used to characterize the crustal development history of the Rehoboth Province of southern Africa. We investigated a quartzite of the Late Palaeo-Early Mesoproterozoic Billstein Formation, formed in a continental basin, and a quartz-feldspar arenite layer of the late Mesoproterozoic Langberg Formation conglomerates, immature sediments formed within a felsic volcanic system (both close to Rehoboth Town). The combined data indicate three episodes of crustal evolution in the Rehoboth Province. The oldest phase is only documented in the Billstein quartzite by three 2.98–2.7 Ga Archaean zircons. A Palaeoproterozoic phase between 2.2 and 1.9 Ga is older than any known exposures of the Rehoboth Province. The Billstein quartzite shows a main peak at 1.87 Ga, corresponding to the 1863 ± 10 Ma Elim Formation. The Langberg sample reflects magmatism related to the entire Namaqua–Natal Wilson cycle between *c.* 1.32 and 1.05 Ga. The absence of zircons of that age range in the Billstein quartzite indicates a pre-Namaqua age for the Billstein Formation. Our data shows that there were at least three episodes of crustal development at 2.98–2.7 Ga, 2.05–1.75 and 1.32–1.1 Ga. We have documented the existence of a previously unrecognized 2.98–2.7 Ga Archaean crustal component, which was probably exposed in the Rehoboth Province during the Palaeoproterozoic and thus indicates a much longer geological history for the Rehoboth Province than previously known.

The tectonic framework of southern Africa comprises a complex assemblage of Proterozoic structural provinces and older cratons as shown in Figure 1. Most provinces and cratons have well-established ages of crustal development, reflecting the formations of new crust by extraction from the mantle or magmatic and orogenic reworking of the pre-existing protolith. However the evolution of the Rehoboth Province is still poorly known, due to the extensive Kalahari sand cover in the east and Neoproterozoic and Palaeozoic–Mesozoic cover in the west. Available age data from the well-exposed Rehoboth Basement Inlier (Fig. 1) along the northern margin of the Rehoboth Province indicate two major magmatic and deformational episodes during the late Palaeoproterozoic and Mid–Late Mesoproterozoic (Ziegler & Stoessel 1993; Becker *et al.* 2006; Miller 2008). However, Sm–Nd and U–Pb analyses of Palaeoproterozoic

granitoids, amphibolites and basic dykes suggest that the earliest crust within the Rehoboth Basement Inlier was formed between 2.37 and 1.8 Ga (Ziegler & Stoessel 1993). Dating of cobbles in Carboniferous glacial diamictites from Rietfontein and Upington areas in the south-eastern part of the province shows that there could be Archaean (2.7–2.5 Ga) and *c.* 2 Ga Palaeoproterozoic basement components (Cornell *et al.* 2011). This hints at the existence of a late Palaeoproterozoic or even a Neoarchaean crustal influence for the Rehoboth Province. However, there is no direct evidence and we look for confirmation of these data in sedimentary units of the well-exposed Rehoboth Basement Inlier in the north of the province.

In this work we investigate the use of detrital zircons from two stratigraphically different sedimentary units of the Rehoboth Basement Inlier to characterize the crustal development history of the

From: VAN HINSBERGEN, D. J. J., BUITER, S. J. H., TORSVIK, T. H., GAINA, C. & WEBB, S. J. (eds) *The Formation and Evolution of Africa: A Synopsis of 3.8 Ga of Earth History*. Geological Society, London, Special Publications, **357**, 27–47. DOI: 10.1144/SP357.3 0305-8719/11/$15.00 © The Geological Society of London 2011.

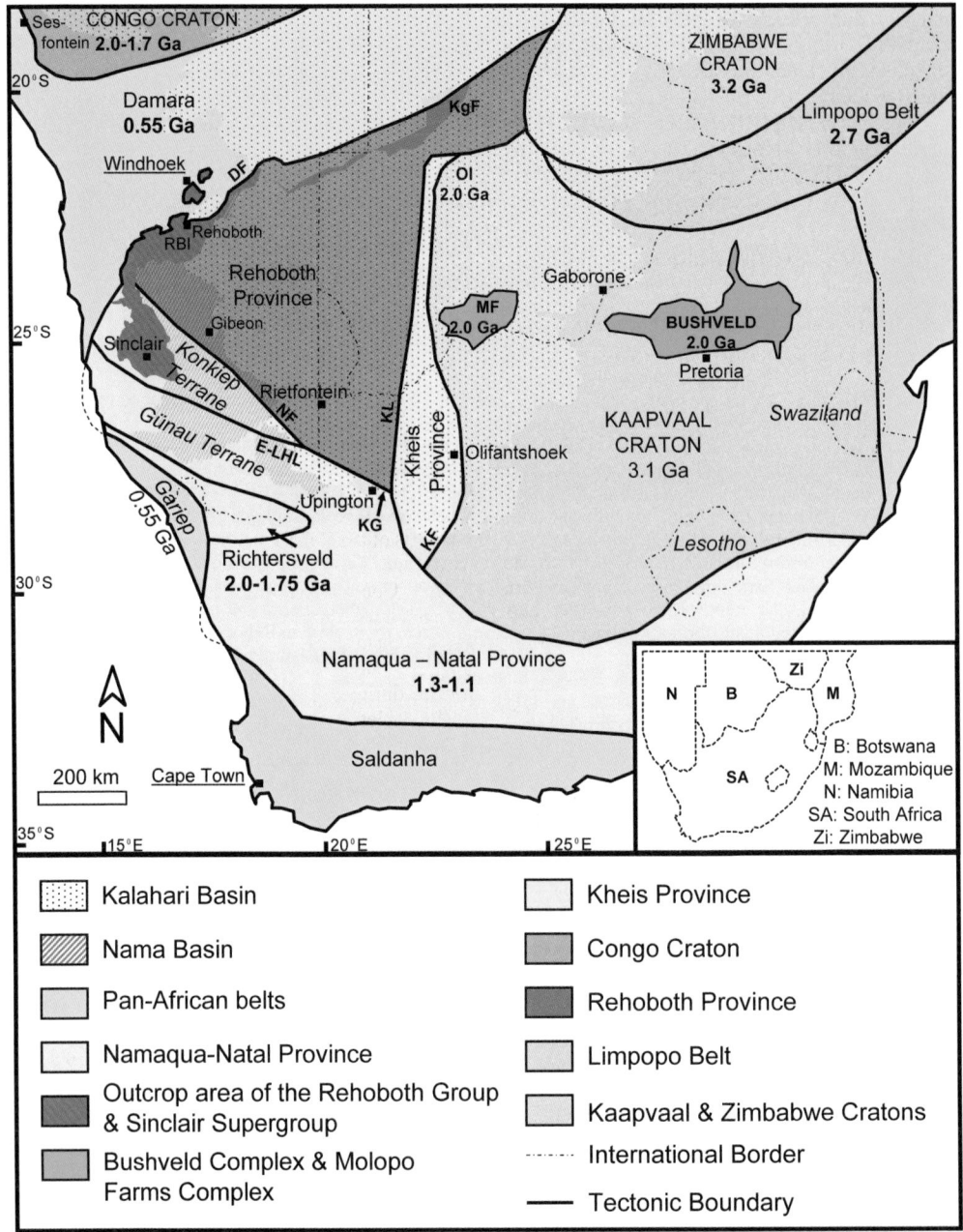

Fig. 1. Tectonic framework of Southern Africa and the distribution of the Palaeoproterozoic Rehoboth Group, Mesoproterozoic Sinclair Supergroup and younger cover sediments (modified after Corner 2003; Cornell *et al.* 2006; Becker *et al.* 2006). Kalahari Basin outline is after Haddon & McCarthy (2005) (DF, Damara Front; E-LHL, Excelsior-Lord Hill Lineament; KF, Kheis Front; KG, Koras Group; KgF, Kgwebe Formation; KL, Kalahari Line; MFC, Molopo Farms Complex; NF, Nama Front; OI, Okwa Inlier; RBI, Rehoboth Basement Inlier).

Rehoboth Province, in particular the presence of suspected Archaean components, with precise microbeam U–Pb methods. Most sedimentary rocks contain a wide range of detrital zircon ages that provide a record of the crustal evolution of their provenance areas, including parts covered by younger sediment as is the case with most of the Rehoboth Province. Our aim is to test the applicability of these methods to assess the crustal evolution of the Rehoboth Province of southern Africa prior to the Pan-African orogeny by documenting the age spectra of detrital zircons of different sedimentary sections, thus gaining insights into successive crustal events of the source regions through time.

The Rehoboth Province and surroundings

The Rehoboth Province is a large domain with c. 700 km long margins defining an equilateral triangle (Fig. 1) and was first defined by Hartnady et al. (1985). The Rehoboth Province of this work has been defined with similar boundaries but referred to by previous authors with names such as the Kheis–Rehoboth Province (Cornell et al. 1998), the Kgalagadi Terrane (van Niekerk 2006), the Nama Province (Corner 2008) and (all or part of) the Kalahari Craton (e.g. Becker et al. 2006).

It is characterized by its smooth magnetic fabric with broad highs and lows and a rough north-eastern trend on total magnetic intensity maps (Corner 2008, figs 2.3 & 2.4). It is bounded to the east by the Kalahari Line and the Kheis Province, which has a marked north–south structural fabric due to its origin as a thin-skinned thrust complex overriding the Kaapvaal Craton (Stowe 1986). It is uncertain when the Rehoboth Province was assembled against the Kaapvaal–Zimbabwe cratons and Limpopo Belt, comprising the Proto-Kalahari Craton (Jacobs et al. 2008) along the Kheis Province. The Rehoboth Province is thought to have been accreted to the Kaapvaal Craton between 1.93 and 1.75 Ga (Tinker et al. 2004) or c. 1.75 Ga as a result of an orogenic event (Cornell et al. 1998; Jacobs et al. 2008). A different interpretation is that the Kheis Province represents a northern branch of the c. 1.2 Ga Namaqua–Natal orogeny (Moen 1999; Moen & Armstrong 2008).

To the SW the Rehoboth Province is bounded by the Mesoproterozoic Namaqua Front, seen as the NW-trending linear magnetic feature beneath the cover sediments of the Nama Basin (e.g. Hoal et al. 1995), the Nama Lineament of Corner (2000). The Konkiep Terrane (preferred nomenclature, or Subprovince of Corner 2008) lies SW of the Nama Lineament and NE of the Excelsior–Lord Hill Lineament and its basement is reworked during the Namaqua–Natal orogeny (Miller 2008, Ch 7, p. 5). Further SW, bounded in the north by the Excelsior–Lord Hill Lineament, lies the accreted Grünau Terrane (Gordonia Subprovince of Miller 2008) of the Namaqua–Natal Province.

To the north the southern margin of the Pan-African Damara Belt forms a major thrust front boundary in which basement slabs have been incorporated. Part of the Rehoboth Province was thrust southwards during this Neoproterozoic orogeny as a large regional antiform. Recent uplift and erosion gave rise to this Rehoboth Basement Inlier, leading to the best exposures of Rehoboth Province rocks. The Congo Craton north of the Damara Belt was not adjacent to the Kalahari Craton (comprising the Kaapvaal and Zimbabwe cratons, Limpopo Belt and Kheis, Rehoboth and Namaqua–Natal Provinces; Jacobs et al. 2008) before Neoproterozoic times, according to Johnson & Oliver (2004).

Age of the Rehoboth Province

The oldest known rocks of the Rehoboth Province are exposed in the Rehoboth Basement Inlier within the southern Damara foreland, and comprise the 1.8–1.7 Ga Rehoboth Group and closely related intrusives (Becker et al. 2006). Samples from various units of the Rehoboth Group have given T_{DM} Nd model ages from 2.37 to 1.66 Ga, where the Weener Igneous Complex has Sm/Nd crustal residence ages combined with zircon $^{207}Pb/^{206}Pb$ ages between 2.3 and 1.7 Ga (Ziegler & Stoessel 1993). This suggests that the major part of the crust near Rehoboth formed during the Palaeoproterozoic.

Re depletion model ages by Hoal et al. (1995) for peridotite xenoliths from the Gibeon kimberlites (Fig. 1) to the south of the Rehoboth Basement Inlier are $2.2-2.0 \pm 0.2$ Ga, indicating the time of partial melt extraction from the mantle (Muller et al. 2009).

Pettersson et al. (2007) found zircon xenocrysts aged 1.74–2.12 Ga in bimodal volcanic and intrusive rocks of the 1.09–1.17 Ga Koras Group near Upington (Fig. 1) along the south-eastern tip of the Rehoboth Province and detrital grains aged 1.82–1.9 Ga in Koras Group sandstone. Together, these data indicate a possible Palaeoproterozoic basement component in the southern Rehoboth Province.

Geology of the Rehoboth Basement Inlier

Palaeoproterozoic

The Rehoboth Basement Inlier (Figs 1 & 2) consists of Palaeoproterozoic and Mesoproterozoic rocks. The oldest known formations in the inlier belong to the Rehoboth Group and include the Gaub Valley, Elim and Marienhof Formations and their related intrusives. Rocks formerly considered to represent possible pre-Rehoboth basement comprise ortho- and paragneisses that are now grouped into the Kangas Metamorphic Complex by Becker & Schalk (2008, Ch 4, p. 24). Their age and possible correlation with the Rehoboth Group are still

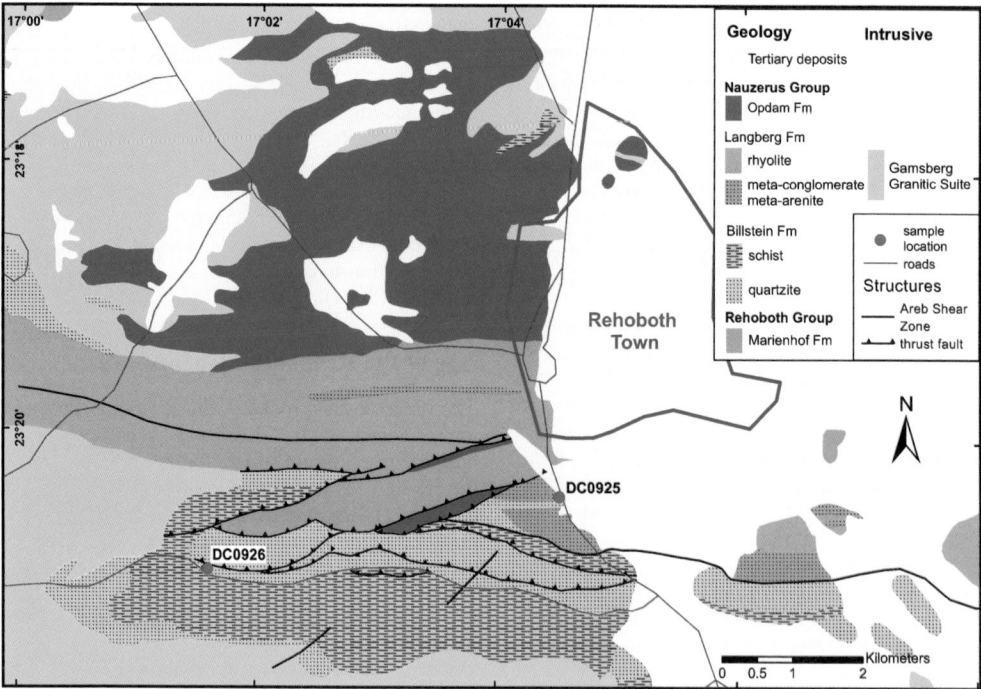

Fig. 2. Geological map of the Rehoboth area and sample locations (modified after Becker *et al.* 1998).

poorly constrained, however. Published zircon age data for the volcanic Gaub Valley Formation and intrusive units such as the Weener Complex, the Piksteel Suite and the Mooirivier Granodiorite range from 1.80 to 1.72 Ga (Table 1). The Weener Igneous Complex, located on the north-western margin of the Rehoboth Basement Inlier, is thought by Becker *et al.* (1994) to represent the intrusive equivalent of the Gaub Valley Formation formed in a subduction-related environment.

In contrast, our unpublished ion probe data for mafic metavolcanic Elim Formation indicate an age of 1863 ± 10 Ma which is significantly older than the zircon dates of the Gaub Valley Formation.

The Palaeoproterozoic rocks are well foliated and the Elim Formation rocks have reached up to amphibolite grade metamorphism (Becker & Schalk 2008, Ch 5, p. 37), but the Gaub Valley Formation rocks have mostly a lower to middle greenschist facies assemblage (Becker & Schalk 2008, Ch 5, p. 29). The Mesoproterozoic rocks show lower grade (low greenschist facies) metamorphism (Becker *et al.* 2006). However, no detailed tectonometamorphic studies of the Rehoboth Basement Inlier have been carried out so relationships between age and metamorphic grade are not well understood.

The Billstein Formation consists entirely of low-grade metasediments that were previously included in the Rehoboth Group (SACS 1980) but are now thought to be post-Rehoboth–pre-Sinclair by Becker & Schalk (2008, Ch 6, p. 1). A large part of their Billstein Formation was previously mapped as the Marienhof Formation but, because of the absence of the Palaeoproterozoic regional gneissic fabric and amphibolite-grade metamorphism, these parts are now included into the Billstein Formation (Becker & Schalk 2008, Ch 5, p. 39).

Mesoproterozoic

The depositional age of the Billstein Formation is uncertain but the formation has now been assigned as a separate stratigraphic unit at the base of the Mesoproterozoic stratigraphy according to Becker & Schalk (2008, Ch 6, p. 1).

The Mesoproterozoic rocks of the Sinclair Supergroup within the Rehoboth Basement Inlier consist of the volcano-sedimentary rocks of the Nauzerus Group and related intrusives of the Gamsberg Granitic Suite (Becker & Schalk 2008). The published ages range from *c.* 1.3–1.0 Ga and are shown in Table 1. The poorly sorted sediments and felsic rhyolites of the Nückopf Formation

Table 1. *Summary of zircon chronological data from the Rehoboth Basement Inlier*

Suite/Group	Formation/Unit	Lithology	Age (Ma)	Technique	Reference
Mesoproterozoic					
Nauzerus Group	Langberg Formation	Rhyolite and quartz-feldspar porphyry	1083 ± 30 1090 ± 15 1100 ± 5	TIMS SHRIMP SHRIMP	Burger & Coertze (1975) Becker *et al.* (2005)
	Nückopf Formation	Rhyolite and quartz-feldspar porphyry	1221 ± 36 1226 ± 10	TIMS TIMS single grain	Hilken (1998) Schneider *et al.* (2004)
			1226 ± 11 1232 ± 30 1770 ± 35	SHRIMP TIMS TIMS	Becker *et al.* (2005) Burger & Coertze (1978)
Gamsberg Granitic Suite	Klein Gamsberg	Granite, gneissic granite, locally mylonitic	1010 ± 30 1078 ± 30 1089 ± 30 1132 ± 26 1079 ± 63 1095 ± 121 1186 ± 54 1336 ± 23 1375 ± 193	TIMS TIMS TIMS TIMS TIMS TIMS TIMS TIMS TIMS	Burger & Coertze (1975) Hilken (1998) Nagel (2000)
	Kobos Granite		1064 ± 20 1150 ± 30 1178 ± 20	TIMS TIMS TIMS	Burger & Coertze (1978)
	Rostock Granite		1084 ± 8 1099 ± 16 1194 ± 26 1207 ± 15 1210 ± 8	TIMS TIMS TIMS TIMS TIMS	Pfurr *et al.* (1991) Ziegler & Stoessel (1993)
Palaeoproterozoic					
Piksteel Granitic Suite	Borodino Swartmodder	Gneissic granite to diorite	1364 ± 11 1476 ± 30 1630 ± 50	TIMS TIMS TIMS	Ziegler & Stoessel (1993) Burger & Coertze (1975)
	NA NA NA NA		1511 ± 51 1627 ± 93 1645 ± 41 1740 ± 144	TIMS TIMS TIMS TIMS	Ziegler & Stoessel (1993) Hilken (1998)
	Opetje Mooirivier granodiorite	Gneissic granodiorite	1782 ± 8 1725 ± 10	TIMS TIMS	Ziegler & Stoessel (1993) Becker *et al.* (1996)
Weener Igneous Complex		Gneissis grano-diorite to tonalite	1743 ± 87 1762 ± 29 1765 ± 21 1767 ± 76	TIMS TIMS TIMS TIMS	Becker *et al.* (1996)
Rehoboth Group	Gaub Valley Formation	Felsic volcani-clastic rocks and sediments	1747 ± 11 1782 ± 18 1793 ± 146	TIMS TIMS TIMS	Nagel *et al.* (1996)
	Eilm Formation	Mafic volcanic rocks and green amphibole schist	1863 ± 10	U–Pb LA–ICPMS	van Schijndel (unpublished data)

Only zircon data that represent the crystallization age are shown. For an extended overview with more dating techniques see Becker *et al.* (2005, 2006), Becker & Schalk (2008) and Miller (2008).

represents the base of the Nauzerus Group with an age of 1226 ± 10 Ma (Schneider *et al.* 2004) and are overlain by the clastic rocks of the Grauwater Formation. The Langberg Formation comprises a later stage of felsic volcanic rocks consisting of rhyolites, porphyries and volcaniclastics with minor interbedded clastic sediments (Becker & Schalk 2008, Ch 8, p. 72).

The Nauzerus Group and the Konkiep Group of the Konkiep Terrane in Namibia are part of the Sinclair Supergroup (Table 1) that occupies the north-western part of the Rehoboth Province, bounded further to the west by the Namaqua Province (Fig. 1). The Sinclair Supergroup also includes the Kgwebe Formation in Botswana and the Koras Group in South Africa according to Becker *et al.* (2005).

Stratigraphy and sampling

Since most of the Rehoboth Province is covered by the recent Kalahari sands, we focused on the Rehoboth Basement Inlier in the north of the province. In this work we present detrital zircon data from two metasedimentary units, represented by the Late Palaeoproterozoic–Early Mesoproterozoic Billstein Formation and the Late Mesoproterozoic Langberg Formation (Fig. 2) which we expected to contain evidence for the major crust formation events of the Rehoboth Province.

The Billstein Formation consists entirely of metasediments, mainly thick orthoquartzites that were deposited unconformably on the Palaeoproterozoic volcanic and intrusive rocks of the Rehoboth Group (Becker & Schalk 2008, Ch 6, p. 2). The Langberg Formation forms a late Mesoproterozoic stage of sedimentation and felsic volcanism, probably started around 1.1 Ma (Becker & Schalk 2008, Ch 8, p. 68). It is therefore expected that zircons from the Billstein Formation quartzites would give a record of the earliest crustal evolution of the Rehoboth Province, whereas zircon ages from the Langberg Formation sediments would include later crustal events.

The following descriptions of the Billstein and Langberg Formations are largely based on Becker *et al.* (2005, 2006), Becker & Schalk (2008) and Miller (2008).

Billstein Formation

The Billstein Formation occurs mainly in the northern part of the Rehoboth Basement Inlier and around the town of Rehoboth (Fig. 2). In the west the formation lies unconformably on the Palaeoproterozoic Rehoboth Group, the Marienhof and Elim formations and the Doornboom Mafic Complex (Becker & Schalk 2008, Ch 6, p. 1). The stratigraphic sequence of the formation based on a coherent succession starts with very pure cross-bedded quartzite (up to 100 m) which grades upwards into (up to 300 m) meta-arenite with interlayered polymict and monomict pebble conglomerates (up to 15 m). The upper part of the succession consists mainly of several hundred metres of reddish Mn-rich garnet–hornblende–quartz–sericite metapelites with thin layers of mafic schist and amphibolites and quartzites near the top. These metasediments represent highly mature sediments formed in a depositional environment that is thought to be fluviatile and lacustrine in an epicontinental setting.

The age of the Billstein Formation is poorly constrained and the formation is only intruded by Mesoproterozoic intrusions of the Gamsberg Granitic Suite; its minimum age is constrained by 1210 ± 8 Ma porphyritic dykes which cut the formation (Ziegler & Stoessel 1993).

Sample DC0926 was taken from the lower part of the Billstein Formation north of the road C24 west of Rehoboth Town (Fig. 3a). The sequence at this location was formerly mapped as Marienhof Formation but is now included in the Billstein Formation on the basis of similar lithology to the Billstein Formation type area (Becker & Schalk 2008, Ch 5, p. 40, fig. 5.27). The sample is a mature, white orthoquartzite with red colour banding (Fig. 3b). The Billstein Formation is tectonically overlain by the felsic volcanics and metasediments of the Langberg Formation, followed by metabasalts and metasediments of the Opdam Formation (Becker & Schalk 2008, Ch 8, p. 75, fig. 2).

Langberg Formation

The Langberg Formation is part of the upper Nauzerus Group of the Mesoproterozoic Sinclair Supergroup within the Rehoboth area. The whole sequence is metamorphosed to lower greenschist facies and very strongly deformed, as is best shown by the highly flattened pebbles of the meta-conglomerate sampled (Fig. 3). The formation lies with a sheared, assumed originally unconformable, contact on the Billstein Formation. The formation occurs only in the eastern part of the Rehoboth Basement Inlier and extends from about 10 km SW of Rehoboth for more than 30 km to the east. The base of the formation consists of poorly sorted matrix supported polymict conglomerate (200 m) with clasts of up to boulder size. This is overlain by several hundred metres of felsic volcaniclastics, quartz-feldspar porphyry and rhyolite which comprise the main part of the Langberg Formation. These are interlayered with thin layers of tuff and greyish pebbly meta-arenite to metaconglomerate

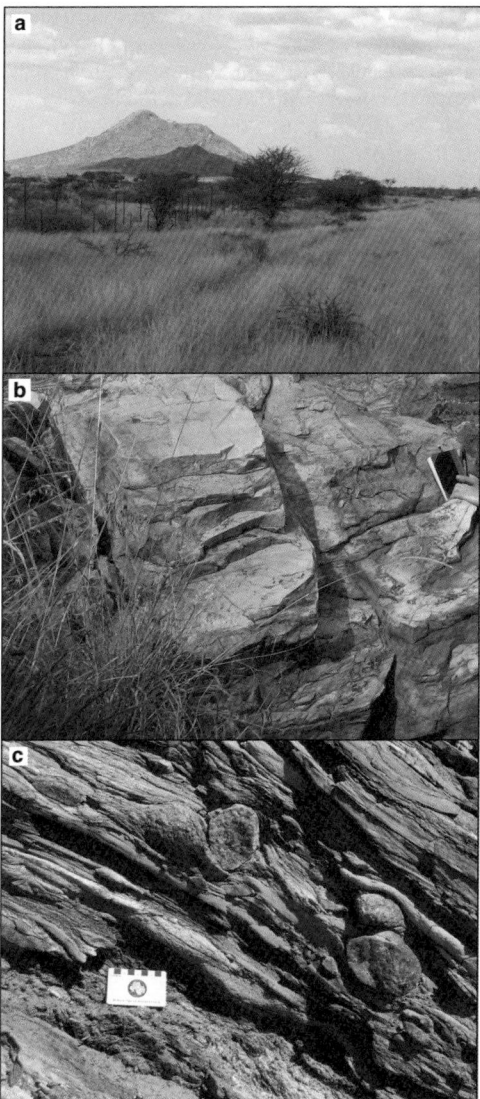

Fig. 3. (**a**) The Billstein Formation quartzite hills seen from the C24, located SW of Rehoboth. (**b**) White orthoquartzite with red colour banding of the Billstein Formation at the sample location. (**c**) Metaconglomerate of the Langberg Formation with flattened granite and quartzite pebbles, road-cut directly south of Rehoboth. The field book is 15 cm wide and the scale card is 9 cm wide.

(Becker *et al.* 2005), indicating contemporaneous volcanism and sedimentation. They point out that the Langberg conglomerates include quartzite pebbles from the underlying Billstein Formation, indicating that the latter was already lithified during the deposition of the Langberg Formation.

The metaconglomerate we sampled is well exposed in a road cutting about 1 km south of Rehoboth (Figs 2 & 3c) and contains pebbles of quartzite, sandstone, granite and foliated mafic rock in a matrix of chlorite phyllite or quartz-feldspar arenite, interlayered with thin greenschist beds. This exposure cannot be directly followed into the more continuous exposures of the Lang-berg Formation to the east, but was correlated with the latter by Becker *et al.* (2006). However, the structure of the area outside Rehoboth is very complex and the relationship of the Langberg metaconglomerate exposed in the road-cut to the Billstein Formation quartzites and the Langberg Formation rhyolites remains uncertain. Sample DC0925 comes from a pebbly, quartz-feldspar arenite layer within the conglomerate-dominated unit.

The Opdam Formation overlies the Langberg with a tectonic contact south of Rehoboth, whereas west of Rehoboth the contact is unconform-able (Becker & Schalk 2008, Ch 8, p. 75, fig. 2).

Analytical methods

Samples were broken into 15 mm blocks, crushed in a Cr-steel swing mill to pass a 400 micron sieve and heavy minerals were concentrated using a gold pan. Zircons were randomly handpicked, mounted in epoxy and imaged using an Hitachi scanning electron microscope as described in detail by Raith *et al.* (2003). The cathodoluminescent (CL) and backscattered electron (BSE) images were used to site microbeam spots in single-age domains, normally in grain cores.

The zircons from the Langberg meta-arenite sample DC0925 were dated by ion microprobe and the zircons of the Billstein quartzite sample DC0926 were analysed by Laser ablation ICPMS, excluding the zircons with large cracks or metamict zones (expected to give discordant results).

Ion probe

Zircons from the Langberg meta-arenite (DC0925) were analysed by secondary ion mass spectrometry on the Cameca IMS1270 ion microprobe at the NordSIM facility, Swedish National Museum of Natural History in Stockholm. The procedure followed is outlined in Whitehouse *et al.* (1999) and Whitehouse & Kamber (2005). A *c.* 5 nA O_2-defocused primary beam is used to illuminate a 100 μm final primary column aperture (Köhler illumination), resulting in an elliptical analysis spot of 20 μm (long-axis). The mass spectrometer operates at a mass resolution ($M/\Delta M$) of *c.* 5000. The U–Th–Pb analyses were performed using automated centring of the beam in the field aperture, optimization of mass calibration using selected high-intensity peaks of the peak-switching routine and optimization of secondary ion energy in the 60 eV energy window. This routine is described in more detail by Whitehouse & Kamber (2005). The Pb/U ratio was calibrated using the Geostandards zircon 91 500 standard with an age of 1065 Ma (Wiedenbeck *et al.* 1995). Common Pb corrections are based on the measured ^{204}Pb signal assuming the present-day model Pb composition of Stacey & Kramers (1975).

LA-ICP-MS

At the Denmark and Greenland Geological Survey (GEUS), zircons of the Billstein quartzite (DC0926) were analysed using a New Wave UP213 UV laser ablation system coupled to an Element 2 magnetic sector field (inductively coupled plasma) mass spectrometer (LA-SF-ICP-MS). All age data presented here were obtained by single spot analyses, with a spot diameter of 20 µm and a crater depth of *c.* 15–20 µm. The methods employed for analysis and data processing are described in detail by Frei & Gerdes (2009) and Gerdes & Zeh (2006). The main difference between ion-probe and LA-SF-ICP-MS measurements is that the laser beam penetrates deeper into the grain whereas the ion probe pits are only a few microns deep.

The accuracy of the laser ablation method and compatibility with ion probe analyses is demonstrated by analyses of 91 500 zircon standard, which has much lower concentrations of U (83 ppm) and Pb (15 ppm) than most samples. It gave a 1069 ± 12 Ma concordia age on 14 points, compared to the TIMS 1065.4 ± 0.3 Ma ^{238}U–^{206}Pb age given by Wiedenbeck *et al.* (1995). Additionally, the Plešovice (Sláma *et al.* 2008) and M127 (Nasdala *et al.* 2008) zircon standards were analysed, and the results were consistently within 1σ of the published ID-TIMS ages.

The analytical results are given in Table 2, all with 1σ errors. Age calculations using ISOPLOT v3.0 by Ludwig (2003) are consistently given with 2σ errors (disregarding decay constant errors) in the figures and text.

Results

DC0926 Billstein Formation

Sample DC0926 contains zircons with sizes between *c.* 50 and 200 µm. The shapes and CL-images are heterogeneous but the majority of zircons were subhedral magmatic zircons with oscillatory zoning, slightly elongated. Very rounded zircons with dark CL-images also occur. Thirty grains were >10% discordant and were excluded from the probability plot with 136 grains shown in Figure 4a. The oldest zircon grains are Archaean, 2742 ± 12, 2749 ± 42 and 2891 ± 28 and one lone grain is 2220 ± 23 Ma. The main population has a smaller peak at 2050 to 1950 Ma, a large peak with a tight cluster of data at 1866 ± 4 Ma and a few younger grains. The youngest five concordant zircon grains have a mean Pb–Pb age of 1770 ± 20 Ma. Together with one extra discordant point, these fall on a discordia plot (not shown) with the upper intercept at 1775 ± 23 Ma (lower intercept 262 ± 220 Ma, MSWD 0.42) which gives a maximum age for the deposition of the Billstein quartzite.

DC0925 Langberg Formation

Sample DC0925 contains euhedral to subhedral zircons ranging in size from 100–400 µm; most of the grains are oscillatory zoned. Forty-eight of 55 grains were concordant within 10%. In the plot of Figure 4b two age groups are seen. A minor group of six grains with ages from 2030 to 1750 Ma reflect the major peaks seen in the Billstein quartzite and might be derived from Billstein Formation sediments. All the other grains plot in a broad 1325 to 1080 Ma composite peak which corresponds in time to the Namaqua–Natal tectogenetic cycle.

This broad peak has a small discrete 5-point group at 1320 Ma and is not smooth in detail. The three oldest grains fall on a discordia plot (not shown) with the upper intercept at 1321 ± 14 (lower intercept at 122 ± 470 and MSWD = 0.74). There is a maximum at about 1220 Ma comprising two poorly defined peaks at 1180 and 1245 Ma. Surprisingly, all seven of the rejected discordant grains conform to a discordia plot (not shown) with intercepts 1226 ± 47 and 81 ± 160 (MSWD 0.12), conforming to the recent lead loss model. The upper intercept also reflects the *c.* 1220 Ma peak in provenance age. The four youngest detrital grains in this sample have a mean Pb–Pb age of 1103 ± 24 Ma.

Stratigraphic implications

Billstein Formation

The depositional age of the Billstein Formation is not accurately constrained. Sample DC0926 is a mature, white quartzite with red colour banding, from the lower part of the formation. It is thought to be part of the first sedimentation pulse since the Billstein Formation has a *c.* 3 km stratigraphic profile from pure quartzite through arkosic sandstone to Mn-rich sericite schists (Becker & Schalk

Table 2. U–Pb isotopic zircon data

Spot in detrital grain	$\frac{207Pb}{206Pb}$	±σ%	$\frac{207Pb}{235U}$	±σ%	$\frac{206Pb}{238U}$	±σ%	Error correl.	Discordance %	f206%[1]	$\frac{207Pb}{206Pb}$	±σ	Age (Ma) $\frac{207Pb}{235U}$	±σ	$\frac{206Pb}{238U}$	±σ	U ppm	Th ppm	Pb ppm	Th/U
DC0925: Langberg Formation 23°20.457 S 17°4.526 E (WGS84)																			
1a	0.119	1.0	5.847	1.3	0.356	0.8	0.65	0.9	0.12	1946	17	1953	11	1961	14.1	42	36	21	0.85
2a	0.085	5.4	2.398	5.5	0.205	1.2	0.22	−8.8	1.65	1309	101	1242	40	1204	13.2	198	154	53	0.78
4a	0.083	1.3	2.602	1.5	0.227	0.8	0.51	4.4	0.22	1270	26	1301	11	1321	9.5	111	32	30	0.29
5a	0.079	0.5	2.281	0.9	0.209	0.8	0.84	4.3	0.03	1177	9.5	1207	6.3	1223	8.4	347	352	101	1.02
6a	0.107	0.7	4.379	1.1	0.296	0.8	0.75	−5.1	0.28	1752	13	1708	8.9	1673	11.8	132	73	50	0.56
7a	0.081	1.5	2.307	1.7	0.207	0.8	0.46	−0.1	2.96	1215	29	1214	12	1214	8.5	206	88	52	0.43
8a	0.085	0.4	2.730	0.9	0.232	0.7	0.86	1.6	0.06	1325	8.6	1337	6.5	1344	9.0	349	170	100	0.49
9a	0.081	5.3	1.871	5.4	0.168	0.7	0.14	−19.6	6.68	1221	101	1071	36	999	6.9	462	450	103	0.97
10a	0.082	0.6	2.405	1.0	0.213	0.8	0.77	0.7	0.07	1239	12	1244	7.1	1247	8.6	261	104	67	0.40
11a	0.077	1.7	2.085	1.9	0.195	0.8	0.41	1.7	1.16	1132	34	1144	13	1150	8.0	103	28	23	0.27
12a	0.125	0.5	6.126	0.9	0.355	0.9	0.83	−4.1	0.40	2031	9.1	1994	8.1	1959	13.0	190	126	89	0.66
13a	0.077	1.3	2.090	1.6	0.196	0.9	0.57	2.6	0.24	1128	26	1145	11	1155	9.7	146	48	33	0.33
13b	0.079	0.9	2.143	1.3	0.198	0.9	0.70	0.1	0.21	1162	18	1163	9.1	1163	9.8	286	82	65	0.29
15a	0.077	1.0	2.137	1.2	0.201	0.8	0.62	5.2	0.15	1126	19	1161	8.6	1180	8.2	117	34	27	0.29
16a	0.080	1.6	2.286	1.9	0.208	0.9	0.50	3.1	0.31	1187	31	1208	13	1220	10.3	95	27	23	0.29
17a	0.085	1.1	2.738	1.4	0.235	0.9	0.64	4.5	0.28	1306	21	1339	11	1359	11.2	233	78	64	0.33
17b	0.082	1.0	2.510	1.3	0.221	0.9	0.70	3.0	0.74	1253	19	1275	9.7	1288	10.8	369	108	93	0.29
18a	0.083	0.6	2.520	1.1	0.221	0.9	0.82	2.2	0.03	1262	13	1278	8.2	1287	10.7	426	290	120	0.68
19a	0.079	0.8	2.305	1.2	0.213	0.9	0.75	7.9	0.29	1160	16	1214	8.7	1244	10.4	362	101	89	0.28
21a	0.080	1.9	2.322	2.2	0.209	0.9	0.43	1.5	0.42	1209	38	1219	15	1225	10.4	81	30	20	0.38
22a	0.081	0.8	2.334	1.3	0.210	0.9	0.74	1.8	0.48	1210	17	1223	9	1230	10.4	474	163	115	0.34
23a	0.124	0.9	6.247	1.3	0.367	0.9	0.70	0.3	0.28	2009	16	2011	12	2013	15.9	123	64	57	0.52
24a	0.080	1.8	2.284	2.0	0.207	0.9	0.46	1.9	0.42	1194	35	1207	14	1215	10.3	125	147	36	1.18
25a	0.077	1.1	2.146	1.4	0.203	0.9	0.65	7.0	0.20	1117	21	1164	9.9	1189	10.0	203	66	48	0.32
26a	0.077	0.9	1.787	1.2	0.169	0.8	0.63	−10.1	2.86	1111	19	1041	7.8	1008	7.0	557	353	119	0.63
27a	0.080	0.5	2.288	1.0	0.207	0.8	0.85	0.9	0.09	1202	10	1209	6.9	1212	9.2	305	348	91	1.14
28a	0.085	0.7	2.474	1.0	0.211	0.8	0.75	−6.6	1.85	1314	13	1264	7.3	1236	8.5	574	33	133	0.06
29a	0.079	0.7	2.252	1.1	0.206	0.8	0.72	2.0	0.15	1184	15	1197	7.5	1205	8.4	155	77	39	0.50
30a	0.083	0.8	2.244	1.1	0.197	0.8	0.72	−9.1	0.69	1263	15	1195	7.9	1158	8.5	445	154	103	0.35
30b	0.080	2.2	1.917	2.3	0.173	0.8	0.34	−16.2	2.05	1184	42	1087	15	1027	7.5	511	181	104	0.35
31a	0.079	1.4	2.082	1.7	0.190	0.9	0.55	−5.7	1.00	1184	27	1143	12	1121	9.4	273	212	67	0.77
34a	0.084	0.4	2.536	0.9	0.219	0.8	0.86	−1.1	0.04	1291	8.7	1282	6.4	1277	8.8	405	354	119	0.87
35a	0.079	0.8	2.153	1.2	0.197	0.9	0.75	−2.1	0.06	1180	16	1166	8.4	1158	9.7	160	44	37	0.28

(Continued)

Table 2. *Continued*

Spot in detrital grain	$\frac{207Pb}{206Pb}$	$\pm\sigma\%$	$\frac{207Pb}{235U}$	$\pm\sigma\%$	$\frac{206Pb}{238U}$	$\pm\sigma\%$	Error correl.	Discordance %	f206%[1]	Age (Ma) $\frac{207Pb}{206Pb}$	$\pm\sigma$	$\frac{207Pb}{235U}$	$\pm\sigma$	$\frac{206Pb}{238U}$	$\pm\sigma$	U ppm	Th ppm	Pb ppm	Th/U
36a	0.081	1.3	2.310	1.6	0.206	0.9	0.56	−2.3	0.54	1232	26	1215	12	1206	10.1	118	65	30	0.55
37a	0.076	1.8	1.897	2.3	0.181	1.4	0.62	−1.5	1.54	1090	36	1080	15	1075	14.1	155	52	32	0.34
38a	0.078	0.6	2.044	1.1	0.190	0.9	0.84	−2.6	0.25	1148	12	1130	7.4	1121	9.4	363	125	80	0.34
40a	0.082	0.5	2.378	1.0	0.211	0.9	0.90	−0.4	0.24	1239	9	1236	7.4	1234	10.4	571	598	167	1.05
42a	0.075	1.8	2.029	2.1	0.195	1.1	0.51	7.1	0.55	1079	36	1125	14	1149	11.3	53	26	13	0.48
43a	0.082	0.7	2.316	1.1	0.204	0.9	0.81	−5.1	0.40	1254	13	1217	8.1	1196	10.0	275	112	66	0.41
45a	0.082	0.7	2.375	1.1	0.211	0.9	0.80	−0.6	0.17	1239	13	1235	8.1	1233	10.1	320	269	90	0.84
46a	0.080	1.0	1.617	1.4	0.146	0.9	0.66	−28.4	5.47	1199	20	977	8.6	881	7.4	930	197	156	0.21
47a	0.077	0.9	2.053	1.3	0.193	0.9	0.71	0.5	0.30	1130	18	1133	8.8	1135	9.4	225	69	51	0.31
48a	0.076	1.1	2.047	1.5	0.194	0.9	0.64	3.7	0.21	1106	22	1131	10	1144	9.8	99	52	24	0.53
51a	0.081	1.1	2.085	1.4	0.186	0.9	0.65	−11.0	2.69	1225	21	1144	9.6	1101	9.2	668	213	147	0.32
52a	0.082	0.7	2.199	1.1	0.194	0.9	0.80	−9.9	0.90	1255	13	1181	8	1141	9.6	413	46	89	0.11
54a	0.080	1.1	2.304	1.4	0.209	1.0	0.67	3.1	0.15	1192	21	1214	10	1226	10.7	102	37	25	0.36
63a	0.080	2.2	1.818	2.4	0.165	0.9	0.40	−18.5	5.70	1191	42	1052	16	986	8.6	437	239	87	0.55
64a	0.080	1.3	1.397	1.6	0.126	0.9	0.56	−38.3	2.09	1201	26	888	9.6	767	6.6	329	78	49	0.24
65a	0.078	0.7	2.100	1.2	0.195	0.9	0.79	0.3	0.55	1147	14	1149	8	1150	9.6	344	160	82	0.47
73a	0.081	1.2	2.316	1.5	0.207	0.9	0.63	−0.9	0.71	1223	23	1217	11	1214	10.3	133	44	32	0.33
78a	0.082	0.8	2.264	1.3	0.201	1.0	0.76	−4.8	0.04	1236	16	1201	8.9	1182	10.4	160	77	40	0.48
80a	0.109	0.6	4.754	1.1	0.316	0.9	0.86	−0.7	0.16	1782	10	1777	9.1	1772	14.4	141	123	60	0.88
98a	0.079	1.8	2.284	2.0	0.209	0.8	0.43	4.4	0.36	1177	35	1207	14	1224	9.4	31	18	8	0.56
107a	0.079	3.2	1.321	3.6	0.121	1.7	0.48	−38.9	2.09	1169	62	855	21	739	12.1	527	60	72	0.11
112a	0.113	0.7	5.240	1.2	0.335	1.0	0.81	0.5	0.13	1855	13	1859	10	1863	15.6	81	57	35	0.70

Spot in detrital grain	$\frac{207Pb}{206Pb}$	$\pm\sigma\text{‰}$	$\frac{207Pb}{235U}$	$\pm\sigma\text{‰}$	$\frac{206Pb}{238U}$	$\pm\sigma\text{‰}$	Error correl.	Discordance %	$\frac{206Pb}{204Pb}$	Age (Ma) $\frac{207Pb}{206Pb}$	$\pm\sigma$	$\frac{207Pb}{235U}$	$\pm\sigma$	$\frac{206Pb}{238U}$	$\pm\sigma$	U ppm	Th ppm	Pb ppm	Th/U
DC0926: Billstein Formation 23°21.033 S 17°1.601 E (WGS84)																			
2	0.114	1.4	5.051	3.8	0.321	3.5	0.92	4.2	3805	1865	26	1828	32	1796	55	153	128	49	0.83
3	0.113	2.0	5.399	4.6	0.346	4.1	0.90	3.8	6323	1853	37	1885	40	1914	68	81	50	28	0.62
4	0.114	1.8	5.311	4.2	0.339	3.7	0.90	1.6	1030	1857	33	1871	36	1883	61	68	46	23	0.68
5	0.114	1.7	4.746	4.1	0.303	3.7	0.91	9.2	29 080	1857	31	1775	35	1707	56	258	243	78	0.94
7a	0.122	1.2	6.056	4.3	0.359	4.1	0.96	1.0	3638	1993	21	1984	37	1976	70	497	94	178	0.19
7b	0.116	2.0	3.778	3.4	0.236	2.8	0.82	31.3	3173	1900	35	1588	27	1364	34	119	296	28	2.48
8	0.115	1.4	5.374	2.9	0.340	2.5	0.87	1.0	5085	1872	26	1881	25	1888	42	237	183	81	0.77
9	0.125	1.5	6.279	3.4	0.364	3.0	0.90	1.9	3002	2032	26	2015	30	1999	52	197	95	72	0.48
10	0.115	1.5	5.477	5.5	0.345	5.2	0.96	1.8	63 778	1882	27	1897	47	1911	87	282	221	97	0.78

11	0.114	2.0	5.294	5.0	0.336	4.6	0.92	0.2	1493	1869	36	1868	43	1866	75	89	134	30	1.51
12	0.114	1.3	5.190	3.2	0.330	2.9	0.91	1.4	3967	1863	24	1851	28	1840	47	228	225	75	0.99
13	0.115	1.5	4.893	4.0	0.309	3.7	0.93	8.4	7297	1876	27	1801	34	1737	57	203	135	63	0.67
14	0.117	2.5	5.571	4.3	0.345	3.5	0.82	0.0	4371	1911	44	1912	37	1912	59	101	94	35	0.94
15	0.119	1.1	4.954	3.9	0.303	3.8	0.96	13.5	8369	1936	19	1811	33	1705	57	344	370	104	1.08
16	0.191	1.3	14.01	3.2	0.532	2.9	0.91	0.1	6154	2749	21	2750	30	2752	65	96	67	51	0.70
18	0.116	1.9	5.224	3.9	0.328	3.5	0.88	3.7	2034	1889	33	1857	34	1828	55	179	149	59	0.83
19	0.125	1.3	6.491	3.8	0.377	3.6	0.94	1.8	1550	2029	23	2045	34	2060	64	256	128	96	0.50
20	0.122	1.4	5.931	3.7	0.353	3.4	0.92	1.8	9227	1982	26	1966	32	1951	57	125	48	44	0.38
21	0.114	1.6	5.332	4.7	0.339	4.5	0.94	1.1	8749	1864	28	1874	41	1883	73	131	123	44	0.94
22	0.121	1.2	5.861	2.8	0.350	2.6	0.91	2.5	5022	1978	21	1955	24	1934	43	129	102	45	0.79
25a	0.122	1.5	4.239	4.4	0.251	4.2	0.94	30.6	1374	1991	27	1682	36	1445	54	196	136	49	0.69
25b	0.113	1.4	4.841	3.6	0.311	3.3	0.92	6.3	28 821	1847	26	1792	31	1745	51	133	136	41	1.02
26	0.114	1.6	3.895	4.1	0.249	3.8	0.92	25.6	8078	1859	29	1613	34	1431	49	220	151	55	0.69
27	0.111	1.5	4.994	3.3	0.326	3.0	0.90	0.3	2016	1816	26	1818	28	1821	47	231	55	75	0.24
28	0.108	1.1	4.216	2.7	0.283	2.5	0.91	10.1	25 536	1766	20	1677	22	1607	35	285	207	81	0.73
29	0.123	1.4	5.191	4.8	0.307	4.6	0.96	15.4	3261	1996	25	1851	41	1725	70	234	290	72	1.24
30	0.109	1.1	4.753	2.3	0.317	2.0	0.87	0.1	2851	1778	20	1777	19	1776	31	173	163	55	0.94
31	0.114	1.0	5.201	5.6	0.330	5.5	0.98	1.6	3738	1866	18	1853	48	1841	89	230	129	76	0.56
32	0.114	1.1	5.205	2.8	0.332	2.6	0.93	0.8	3030	1860	20	1853	24	1848	42	110	93	36	0.85
33	0.115	1.2	5.049	3.3	0.318	3.0	0.93	6.1	4942	1881	22	1828	38	1781	47	202	341	64	1.69
34	0.116	2.3	5.317	3.7	0.331	2.9	0.78	3.4	22 004	1901	42	1872	32	1845	47	60	45	20	0.75
36	0.114	1.8	5.273	2.9	0.335	2.3	0.79	0.3	3459	1867	32	1865	25	1862	38	77	52	26	0.67
37	0.106	1.1	3.399	3.5	0.234	3.3	0.95	23.8	1873	1723	20	1504	27	1354	40	349	317	82	0.91
38	0.112	1.6	4.558	3.8	0.295	3.5	0.91	10.6	1554	1836	28	1742	32	1665	52	250	377	74	1.51
39	0.114	1.1	5.193	2.9	0.330	2.7	0.92	1.7	21 897	1866	20	1851	25	1839	43	144	191	47	1.33
40	0.115	1.6	4.720	4.5	0.298	4.2	0.93	11.8	2052	1877	29	1771	27	1682	62	216	184	64	0.85
41	0.114	1.3	5.218	3.0	0.332	2.7	0.90	1.2	10 954	1865	24	1856	24	1847	44	152	104	50	0.69
42	0.110	1.6	4.910	2.8	0.322	2.3	0.83	0.4	5371	1807	29	1804	27	1801	37	217	122	70	0.56
43	0.123	1.3	5.336	3.2	0.314	2.9	0.92	14.1	3139	2006	22	1875	42	1758	45	478	217	150	0.45
44	0.116	1.4	4.373	5.1	0.273	4.9	0.96	20.3	28 837	1898	25	1707	35	1556	68	694	409	189	0.59
45	0.134	1.0	5.894	4.0	0.319	3.9	0.97	19.6	3416	2153	17	1960	21	1783	61	414	245	132	0.59
47	0.115	1.9	4.947	3.3	0.318	2.7	0.81	3.9	11 180	1844	35	1810	33	1781	42	139	143	44	1.03
49	0.118	1.1	5.447	2.4	0.334	2.2	0.90	4.2	9151	1929	19	1892	21	1859	35	200	113	67	0.57
50	0.114	1.1	5.148	3.8	0.328	3.7	0.96	1.8	99 999	1860	20	1844	33	1830	59	254	191	83	0.75
51	0.208	0.8	15.29	3.9	0.533	3.8	0.98	5.9	14 423	2892	14	2834	37	2753	86	319	116	170	0.36
52	0.114	1.1	4.836	4.0	0.309	3.8	0.96	7.5	5122	1857	20	1791	33	1735	58	93	85	29	0.91
53	0.113	1.4	4.614	3.2	0.297	2.8	0.89	10.3	521	1843	26	1752	26	1676	42	206	208	61	1.01
54	0.116	1.7	5.355	4.0	0.334	3.6	0.90	2.3	9342	1898	30	1878	34	1860	58	91	111	31	1.22
55	0.115	1.3	4.268	4.0	0.270	3.8	0.94	20.2	5236	1876	24	1687	33	1539	52	114	90	31	0.79
56	0.116	1.0	5.359	1.9	0.336	1.7	0.87	1.7	5184	1893	17	1878	16	1865	27	165	195	55	1.18
57	0.118	1.1	5.023	3.1	0.309	2.9	0.93	11.2	17 911	1925	21	1823	26	1735	44	82	99	25	1.20

(Continued)

Table 2. Continued

Spot in detrital grain	$\frac{207Pb}{206Pb}$	$\pm\,\sigma‰$	$\frac{207Pb}{235U}$	$\pm\,\sigma‰$	$\frac{206Pb}{238U}$	$\pm\,\sigma‰$	Error correl.	Discordance %	$\frac{206Pb}{204Pb}$	Age (Ma) $\frac{207Pb}{206Pb}$	$\pm\,\sigma$	$\frac{207Pb}{235U}$	$\pm\,\sigma$	$\frac{206Pb}{238U}$	$\pm\,\sigma$	U ppm	Th ppm	Pb ppm	Th/U
58	0.114	1.0	4.914	3.6	0.312	3.4	0.96	7.3	1526	1869	19	1805	30	1750	53	80	83	25	1.04
60	0.114	1.2	5.227	2.2	0.333	1.8	0.82	0.7	7386	1863	22	1857	19	1852	29	101	101	34	1.00
62	0.125	1.2	6.067	2.1	0.353	1.8	0.82	4.4	4521	2025	22	1986	19	1948	30	261	60	92	0.23
63	0.114	1.2	4.651	2.0	0.297	1.6	0.82	11.3	2686	1860	21	1759	17	1674	24	253	208	75	0.82
64	0.114	1.2	5.242	2.8	0.334	2.5	0.90	0.1	11758	1860	22	1860	24	1859	40	82	100	28	1.22
65	0.115	1.6	5.300	4.8	0.334	4.6	0.95	1.5	9012	1882	28	1869	41	1858	74	187	198	63	1.06
66	0.121	2.0	5.561	3.6	0.333	3.0	0.83	6.9	4531	1972	35	1910	31	1853	48	86	63	29	0.74
67	0.115	0.9	5.072	2.2	0.319	2.0	0.92	6.0	6631	1885	16	1831	19	1785	32	160	199	51	1.24
69	0.114	0.9	4.779	2.3	0.305	2.1	0.91	8.7	3287	1858	17	1781	19	1717	31	199	142	61	0.71
70	0.116	1.7	4.961	2.9	0.311	2.3	0.81	8.7	16202	1890	30	1813	24	1746	36	200	184	62	0.92
71	0.115	0.9	5.249	4.3	0.332	4.2	0.98	1.7	9425	1876	17	1861	37	1847	68	177	160	59	0.90
72	0.126	0.7	6.098	3.1	0.352	3.0	0.97	5.5	5489	2039	13	1990	27	1943	51	321	76	113	0.24
74	0.115	0.7	5.061	1.6	0.320	1.5	0.91	5.1	21878	1874	12	1830	14	1791	23	258	247	83	0.96
75	0.115	0.7	4.999	2.2	0.316	2.1	0.95	6.7	6111	1878	12	1819	19	1768	33	199	298	63	1.49
76	0.113	1.7	5.129	4.0	0.330	3.6	0.91	0.5	1783	1845	30	1841	34	1837	58	100	84	33	0.84
78	0.114	1.1	5.219	1.6	0.331	1.2	0.73	1.7	11025	1870	20	1856	14	1843	19	220	255	73	1.16
79	0.115	1.5	5.340	2.5	0.337	2.1	0.82	0.5	7074	1879	26	1875	22	1872	33	175	189	59	1.08
80	0.113	1.2	4.181	4.5	0.268	4.3	0.96	19.5	24604	1852	21	1670	37	1530	59	160	175	43	1.10
81	0.112	1.1	5.001	2.4	0.324	2.1	0.88	1.3	1893	1831	21	1820	21	1810	34	136	142	44	1.04
82	0.118	0.9	4.941	3.1	0.304	2.9	0.95	12.4	3291	1922	17	1809	26	1713	44	178	252	54	1.41
86	0.114	1.1	4.726	2.9	0.300	2.7	0.92	10.7	1584	1868	20	1772	24	1692	39	106	154	32	1.45
87	0.111	1.5	4.894	2.8	0.320	2.4	0.86	1.4	1632	1813	27	1801	24	1791	38	88	72	28	0.82
90	0.121	1.7	5.674	4.1	0.341	3.7	0.91	4.7	16167	1969	31	1927	36	1889	61	50	39	17	0.77
92	0.116	1.5	5.344	2.7	0.333	2.2	0.83	3.0	33556	1902	26	1876	23	1853	36	185	139	61	0.76
93	0.124	1.1	5.636	2.5	0.331	2.3	0.90	9.4	87431	2008	19	1922	22	1843	37	221	149	73	0.67
95	0.109	0.9	4.766	1.9	0.317	1.7	0.87	0.6	748	1784	17	1779	16	1775	26	149	121	47	0.81
96a	0.114	1.4	5.256	3.0	0.335	2.6	0.88	0.0	15612	1862	26	1862	25	1862	42	171	131	57	0.77
96b	0.113	1.4	5.259	4.0	0.336	3.7	0.93	0.9	5823	1854	26	1862	34	1869	60	86	130	29	1.51
97	0.124	1.0	6.118	3.0	0.357	2.8	0.94	2.6	1065	2016	17	1993	26	1970	48	250	175	89	0.70
98	0.113	0.7	4.240	3.3	0.273	3.2	0.98	17.4	4655	1841	13	1682	27	1557	44	88	93	24	1.06
99	0.114	1.4	4.161	3.8	0.264	3.5	0.93	21.6	4932	1870	26	1666	31	1510	48	97	80	26	0.83
100	0.113	0.9	5.213	1.8	0.334	1.6	0.87	0.2	1330	1853	16	1855	15	1856	25	191	162	64	0.85
101	0.113	0.8	5.151	1.5	0.330	1.2	0.83	1.1	1766	1854	15	1845	12	1836	19	202	138	67	0.68
102	0.113	0.9	5.121	2.2	0.328	2.0	0.91	1.4	42802	1851	17	1840	19	1829	32	218	153	72	0.70
103	0.115	0.5	5.279	1.5	0.334	1.5	0.94	1.1	3075	1875	9	1865	13	1857	23	268	228	89	0.85
104	0.124	0.9	6.171	1.8	0.361	1.5	0.86	1.4	3424	2012	16	2000	15	1989	26	217	151	78	0.69

106	0.113	0.6	4.751	1.4	0.304	1.3	0.90	9.0	3977	1856	11	1776	12	1709	20	211	174	64	0.82
107	0.114	1.2	5.309	2.5	0.338	2.2	0.87	1.0	1628	1862	23	1870	21	1878	35	87	88	29	1.02
108	0.114	1.0	5.047	2.3	0.322	2.1	0.90	3.4	9383	1857	18	1827	19	1801	32	80	84	26	1.05
109	0.114	0.9	5.238	2.7	0.333	2.6	0.95	0.9	5112	1866	16	1859	23	1852	41	70	72	23	1.03
110	0.110	0.9	4.939	3.6	0.325	3.5	0.97	0.7	1026	1803	16	1809	31	1814	55	143	123	46	0.86
111	0.115	0.5	5.246	1.8	0.331	1.7	0.95	1.9	1203	1877	10	1860	16	1845	28	134	118	44	0.89
113	0.114	1.2	5.279	3.5	0.337	3.3	0.94	0.6	520	1860	22	1865	30	1870	53	58	78	19	1.35
114	0.120	1.2	4.902	3.0	0.296	2.7	0.91	16.5	8591	1957	22	1803	25	1672	40	105	98	31	0.93
115	0.114	1.0	5.275	2.2	0.336	2.0	0.90	0.3	6557	1862	18	1865	24	1867	32	113	136	38	1.21
116	0.107	1.5	4.535	2.9	0.309	2.5	0.86	0.5	10 981	1742	27	1737	30	1734	38	141	133	43	0.95
117	0.114	0.9	4.703	3.5	0.299	3.4	0.97	10.6	17 189	1863	16	1768	17	1688	51	177	123	53	0.69
118	0.114	1.1	5.048	2.0	0.321	1.7	0.85	4.3	2030	1865	19	1827	20	1794	27	99	96	32	0.97
120	0.123	0.9	6.316	2.3	0.373	2.1	0.92	2.5	1547	1999	16	2021	28	2042	37	151	54	56	0.36
121	0.114	0.8	5.173	3.2	0.329	3.1	0.97	2.0	19 929	1865	15	1848	15	1833	50	98	91	32	0.93
122	0.113	1.0	5.019	1.7	0.322	1.7	0.83	2.8	9326	1847	16	1822	22	1801	22	158	167	51	1.05
123	0.114	0.9	5.316	2.6	0.338	2.4	0.94	0.6	1641	1866	16	1871	16	1876	39	149	112	50	0.75
124	0.115	0.6	5.372	1.8	0.340	1.7	0.94	0.5	16 586	1876	11	1880	25	1885	28	191	198	65	1.04
125	0.114	1.3	5.306	2.9	0.338	2.6	0.89	1.0	2455	1861	24	1870	18	1878	43	86	102	29	1.18
127	0.114	1.0	5.159	2.1	0.328	1.8	0.88	2.6	8581	1868	18	1846	42	1826	29	205	269	67	1.31
128	0.112	1.2	4.713	2.8	0.304	2.5	0.90	7.6	12 030	1837	22	1770	24	1713	37	77	139	23	1.81
130	0.114	1.2	4.896	5.0	0.310	4.8	0.97	7.9	3896	1871	22	1802	21	1742	74	74	116	23	1.57
131	0.115	1.5	5.088	2.8	0.320	2.3	0.84	5.8	24 780	1885	27	1834	31	1790	36	165	245	53	1.48
132	0.113	0.9	5.036	2.5	0.324	2.3	0.93	2.0	24 876	1843	17	1825	22	1810	37	165	170	53	1.03
133	0.127	1.2	6.230	3.5	0.356	3.3	0.94	5.1	2764	2054	21	2009	29	1964	56	123	94	44	0.76
134	0.139	0.7	7.371	2.4	0.384	2.4	0.96	6.6	4027	2219	12	2157	29	2093	42	112	49	43	0.44
136	0.124	1.3	6.242	3.3	0.366	3.1	0.93	0.2	11 200	2009	29	2010	18	2012	53	323	143	118	0.44
137	0.115	0.8	5.202	2.2	0.329	2.1	0.93	2.2	99 999	1872	14	1853	25	1836	33	207	192	68	0.93
139	0.114	0.9	5.295	1.8	0.336	1.6	0.87	0.1	7736	1869	16	1868	16	1868	26	161	166	54	1.03
140	0.125	0.8	6.073	1.9	0.351	1.8	0.92	5.5	3346	2036	13	1986	17	1939	30	193	63	68	0.32
141	0.113	0.7	3.841	4.0	0.246	3.9	0.98	26.0	3621	1851	13	1601	32	1419	50	216	234	53	1.09
143	0.115	0.8	5.306	1.6	0.335	1.4	0.88	1.2	1599	1880	14	1870	14	1861	23	69	50	23	0.72
147	0.107	1.0	5.341	1.6	0.339	1.3	0.78	0.8	2617	1869	18	1875	25	1882	21	133	162	45	1.21
148	0.114	1.3	4.317	3.0	0.292	2.7	0.90	6.8	6989	1755	23	1697	25	1649	40	65	47	19	0.73
152	0.126	0.9	5.302	2.2	0.337	2.1	0.92	0.1	1077	1868	16	1869	19	1870	34	143	181	48	1.26
153	0.118	0.6	6.490	1.5	0.373	1.3	0.90	0.0	21 336	2045	11	2045	13	2045	23	214	83	80	0.39
154	0.114	1.7	5.321	3.2	0.328	2.7	0.85	5.5	11 904	1920	31	1872	28	1829	43	181	116	59	0.64
155	0.125	0.7	5.329	2.2	0.339	2.1	0.95	1.4	20 296	1862	12	1873	19	1884	34	158	104	54	0.66
156	0.114	0.5	6.378	1.9	0.370	1.8	0.96	0.2	3328	2028	9	2029	17	2031	32	342	231	127	0.68
158	0.124	1.0	5.282	2.0	0.337	1.7	0.86	0.8	9712	1860	18	1866	17	1872	28	98	67	33	0.69
159	0.114	1.3	5.481	4.0	0.320	3.8	0.95	12.8	9183	2016	23	1898	34	1791	59	165	163	53	0.99
161	0.190	0.4	13.81	3.0	0.527	3.0	0.99	0.5	3133	2742	6	2737	29	2730	67	307	202	162	0.66
162	0.121	1.0	5.873	2.1	0.351	1.8	0.87	2.4	12 804	1978	19	1957	18	1937	30	111	56	39	0.51

(Continued)

Table 2. *Continued*

Spot in detrital grain	$\frac{207Pb}{206Pb}$	$\pm\sigma\text{‰}$	$\frac{207Pb}{235U}$	$\pm\sigma\text{‰}$	$\frac{206Pb}{238U}$	$\pm\sigma\text{‰}$	Error correl.	Discordance %	$\frac{206Pb}{204Pb}$	Age (Ma) $\frac{207Pb}{206Pb}$	$\pm\sigma$	$\frac{207Pb}{235U}$	$\pm\sigma$	$\frac{206Pb}{238U}$	$\pm\sigma$	U ppm	Th ppm	Pb ppm	Th/U
163	0.114	1.7	5.238	3.0	0.334	2.4	0.81	0.0	12 093	1859	31	1859	25	1859	39	80	60	27	0.75
164	0.115	0.7	5.045	1.9	0.319	1.7	0.92	5.3	11 769	1873	13	1827	16	1786	27	133	147	43	1.10
165	0.118	1.0	5.585	4.5	0.345	4.4	0.97	0.6	3130	1919	19	1914	39	1909	73	235	129	81	0.55
166	0.114	0.5	5.218	2.0	0.332	2.0	0.97	0.8	10 594	1862	9	1856	17	1850	32	143	177	47	1.24
167	0.115	2.1	5.279	2.8	0.333	1.8	0.65	1.5	7852	1878	38	1865	24	1854	29	110	81	37	0.73
168	0.116	1.0	4.902	2.1	0.308	1.8	0.88	9.6	18 871	1889	17	1803	17	1729	28	192	300	59	1.56
169	0.114	0.6	5.219	2.3	0.332	2.2	0.96	0.7	3132	1862	11	1856	20	1850	36	131	105	44	0.80
170	0.112	0.9	3.831	2.4	0.249	2.2	0.93	23.9	3150	1825	16	1599	19	1433	28	259	217	64	0.84
171	0.121	0.4	6.020	1.8	0.359	1.7	0.97	0.1	7010	1978	7	1979	15	1980	29	410	161	147	0.39
172	0.115	0.6	5.412	1.2	0.340	1.1	0.87	0.3	99 999	1884	11	1887	11	1889	18	219	161	75	0.73
173	0.110	1.2	4.531	2.7	0.298	2.4	0.89	7.5	3036	1802	22	1737	23	1683	36	90	83	27	0.92
174	0.116	2.5	5.413	4.0	0.340	3.1	0.77	0.1	3000	1888	46	1887	34	1886	50	120	91	41	0.76
175	0.114	1.3	2.995	1.6	0.190	0.9	0.56	43.7	9120	1872	24	1406	12	1120	9	280	150	53	0.53
177	0.111	1.2	4.658	2.6	0.305	2.3	0.88	6.1	1324	1813	22	1760	22	1715	35	104	100	32	0.96
178	0.118	0.9	5.381	2.0	0.329	1.7	0.88	5.8	428	1934	16	1882	17	1835	28	88	91	29	1.03
179	0.126	0.9	6.569	2.7	0.379	2.5	0.94	1.6	64 706	2041	17	2055	24	2069	44	162	90	61	0.56
181	0.113	0.8	3.682	2.8	0.236	2.7	0.96	29.2	17 602	1853	14	1568	22	1364	33	316	310	75	0.98
182a	0.114	0.7	5.222	1.6	0.333	1.5	0.90	0.7	3470	1862	13	1856	14	1851	24	125	142	42	1.14
182b	0.114	1.1	5.116	2.4	0.326	2.2	0.90	2.5	1131	1860	19	1839	20	1820	34	78	110	26	1.41
183	0.114	0.7	5.276	1.8	0.335	1.6	0.92	0.2	2182	1867	12	1865	15	1863	26	111	100	37	0.90
184	0.125	0.9	6.376	2.3	0.370	2.1	0.92	0.1	20 051	2030	16	2029	20	2028	37	263	63	97	0.24
185	0.114	1.0	5.106	2.0	0.324	1.8	0.88	3.7	4743	1869	18	1837	17	1809	28	82	64	26	0.79
186	0.113	0.6	5.182	1.5	0.332	1.4	0.93	0.5	21 308	1854	10	1850	13	1846	22	279	188	92	0.68
187	0.115	0.7	5.274	2.3	0.333	2.2	0.96	1.5	8558	1878	12	1865	20	1853	35	149	167	50	1.12
189	0.114	1.6	5.328	2.4	0.339	1.8	0.75	0.9	8017	1866	29	1873	21	1880	30	107	86	36	0.80
190	0.112	0.7	2.527	4.8	0.164	4.8	0.99	49.8	5478	1825	13	1280	35	980	43	809	###	133	1.38
191	0.120	0.8	5.923	1.8	0.357	1.6	0.89	0.2	2313	1963	15	1965	16	1966	28	226	144	81	0.64
192	0.115	0.7	5.378	1.5	0.338	1.3	0.90	0.7	29 902	1888	12	1881	13	1876	22	65	76	22	1.17
193	0.114	1.0	5.163	2.0	0.330	1.7	0.86	1.3	2816	1857	19	1846	17	1837	28	170	97	56	0.57
195	0.114	0.8	5.261	1.6	0.336	1.4	0.85	0.4	9727	1859	15	1863	14	1866	22	174	271	59	1.55
196	0.116	1.1	4.232	2.5	0.264	2.3	0.90	22.7	13 801	1897	20	1680	21	1512	31	136	166	36	1.22
197	0.121	0.7	6.081	2.3	0.364	2.2	0.95	1.3	9922	1976	13	1988	20	1999	37	149	81	54	0.54
198	0.114	0.8	5.305	3.1	0.338	3.0	0.96	0.9	6241	1862	15	1870	26	1877	48	175	165	59	0.94
200	0.115	1.0	5.357	2.2	0.338	1.9	0.88	0.4	77 735	1882	19	1878	19	1875	31	151	240	51	1.59
201	0.126	1.1	6.518	2.1	0.376	1.8	0.86	1.2	6748	2038	19	2048	18	2059	31	209	58	79	0.28
202	0.115	1.2	5.346	2.7	0.337	2.4	0.89	0.4	3666	1880	22	1876	23	1873	38	73	74	24	1.02
203	0.124	1.0	5.703	2.0	0.333	1.7	0.86	9.5	9357	2019	18	1932	17	1852	27	181	121	60	0.67

[1]f206% is the proportion of $^{206}Pb_{common}$ in measured ^{206}Pb.

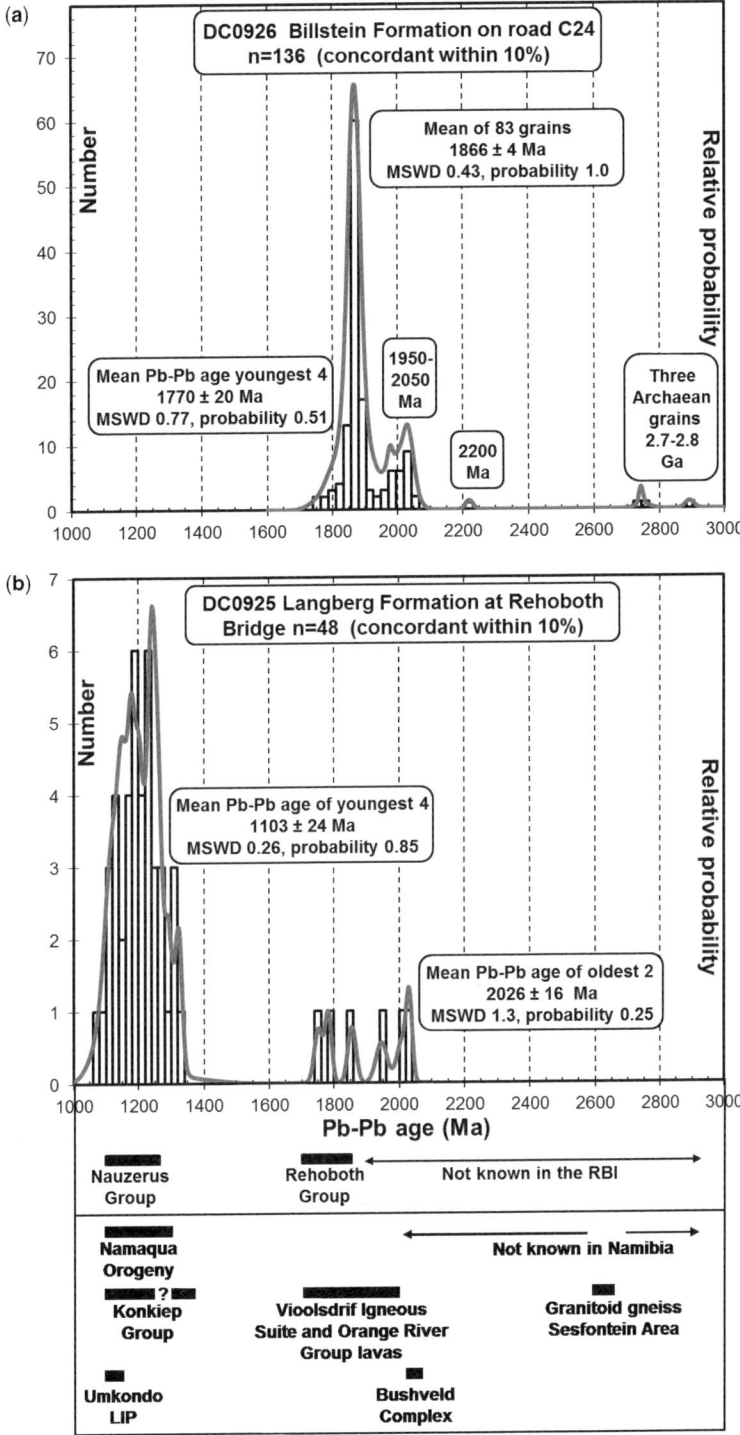

Fig. 4. Detrital zircon probability v. age plot for U–Pb analyses, excluding data with >10% discordance. The combined data point to three episodes of crustal evolution in the Rehoboth Province in Archaean, Palaeoproterozoic and Mesoproterozoic times. (**a**) DC0926 Billstein Formation (**b**) DC0925 Langberg Formation.

2008, Ch 6, p. 2). Our sample has a maximum depositional age of 1770 Ma so its sedimentation therefore began after the c. 1770 Ma Palaeoproterozoic Gaub Valley volcanic sequence of the Rehoboth Group. The fact that zircon ages equivalent to the Gaub Valley Formation and its related intrusives make up such a small proportion of the zircon age population implies that it was not a major source of the Billstein Formation. Another point is that the sample of the Billstein Formation is a mature quartzite, suggesting distal transport distances. Its zircons therefore give an overview of a large source region that likely included parts of the Rehoboth Province now covered by sediments, and not only the Rehoboth Basement Inlier. However, these grains could also have come from other cratons that may have been attached to the province at the time of sedimentation such as the Kaapvaal–Zimbabwe–Limpopo Craton, comprising the Proto-Kahalari Craton (Jacobs *et al.* 2008). It is not well established when the Rehoboth Province and the Kaapvaal Craton became attached; the two main theories are collision around 1.75 Ga (Cornell *et al.* 1998; Jacobs *et al.* 2008) or during the Namaqua orogeny at 1.2 Ga (Moen 1999; Moen & Armstrong 2008). If the accretion occurred at c. 1.75 Ga, the lack of zircons of this age in the Billstein Formation suggests that the provenance did not extend across the suture between the Rehoboth Province and Kaapvaal Craton.

The main provenance of the Billstein Formation was clearly eroded material from a single source, probably time equivalents of the underlying Rehoboth Group which comprises the Gaub Valley, Elim and Marienhof Formations and intrusive granitoids. However, the Gaub Valley Formation has an age of 1780–1750 Ma (known age data in Table 1) so related rocks can only have supplied a few grains to the youngest Billstein component. The main 1866 ± 4 Ma peak of detrital zircons does not correspond to published ages; however, our unpublished 1863 ± 10 Ma age for a chlorite-rich felsic intercalation in the predominantly mafic Elim Formation agrees with the main provenance age. The main part of the zircons of the Billstein quartzite must therefore have been derived from time-equivalent rocks of the Elim Formation located further south and now covered by sediments. The Gaub Valley and Marienhof Formations and their associated intrusives are confined to the northern part of the Rehoboth Basement Inlier north of the Areb Shear Zone (Fig. 2), whereas the Elim Formation occurs mainly to the south. The sediment transport direction was therefore probably from the south.

Constraints on the minimum age of the Billstein Formation are not tight. The gap in detrital zircons ages between 1750 and 1350 Ma seen in the Langberg Formation sample suggests that magmatism was absent in the region during that period. If the Billstein Formation was deposited during that interval, its detrital zircons would not show evidence of that. The only firm minimum age constraint is the 1210 ± 8 Ma porphyritic dykes which cut the Billstein Formation (Ziegler & Stoessel 1993). The absence of any Mesoproterozoic-aged zircons in the Billstein and the presence of 1.32 Ma zircons in the Langberg sample can however be taken as an indication that the Billstein Formation is older than that.

The definition, age and extent of the Rehoboth Group are still problematic, with parts of the formations being transferred to each other based on different lines of reasoning. Due to Neoproterozoic thrusting and later exhumation, the Rehoboth Basement Inlier is well exposed in the Damara Front but that is also the reason why relations between the formations are not straightforward. The original classification of SACS (1980) and Schalk (1988) placed the Billstein Formation beneath the Gaub Valley Formation in the Rehoboth Group. A large part of the Billstein Formation of Becker & Schalk (2008, Ch 5, p. 39) was previously mapped as the Marienhof Formation, but was reclassified because there is no sign of regional gneissic fabric and amphibolite-grade metamorphism which is thought to characterize the Palaeoproterozoic rocks. According to this reasoning, the new Marienhof Formation only includes the higher grade rocks. We think it likely that different parts of the formations may display different degrees of deformation and metamorphism however, depending on their history before exhumation as thrust slices. Similarly, the basal part of the Elim Formation mapped by Schulze-Hulbe (1979) and Schalk (1988) may belong to the upper Gaub Valley Formation on lithological grounds according to Becker & Schalk (2008, Ch 5, p. 31). This stratigraphic succession is complicated by our newfound age difference of c. 100 Ma between the Gaub Valley and Elim formations and it is apparent that the relationship between the two formations should be reconsidered. This is particularly true since there appears to be no direct contact between the Gaub Valley and Elim formations; the latter mainly crops out in the Neoproterozoic Areb Shear Zone while the former largely crops out north of the shear zone indicating a different exhumation history. We propose that a more extensive detrital zircon dating campaign has the potential to resolve this type of problem.

Langberg Formation

The Langberg Formation clearly contains a minor Palaeoproterozoic component derived from the

Billstein Formation which it overlies with a tectonic but probably unconformable contact, and this is also seen in the Billstein Formation quartzite pebbles in the associated conglomerates (Becker *et al.* 2005 and our observations; Fig. 3c).

Mesoproterozoic detrital zircons older than 1.23 Ga were not expected to be present in the sample of the Langberg meta-arenite because rocks with those ages are not exposed in the Rehoboth area at present. The provenance of these zircons could be either the highly metamorphosed terranes in the Namaqua–Natal Province 150 km to the SW, or the lower-grade granites and Sinclair Supergroup which is largely a volcanic cover sequence in the Rehoboth Province and Konkiep Terrane. The oldest Sinclair unit in the Rehoboth area is the 1226 ± 10 Ma Nückopf Formation, however. The related intrusives of the Gamsberg Granite Suite have ages between 1210 and 1010 Ma with only one precise older age of 1336 ± 23 (Table 1). At least the older 1.32–1.23 Ga component was therefore probably derived from the tectonically thickened Namaqua Province which was a mountain belt at that time.

The discordant grains in the Langberg Formation sample conform to a discordia plot with an upper intercept of 1226 ± 47 Ma which is the same age as the dominant histogram age peak at 1.23 Ga. This peak most likely represents the volcanic event which gave rise to the 1220–1230 Ma Nückopf Formation of the Nauzerus Group, Sinclair Supergroup.

There is some doubt as to whether the roadside outcrop we sampled can be followed into the felsic volcanic rocks of the Langberg Formation. The youngest detrital grains in our sample have an age of 1103 ± 24 Ma which is within error of unpublished SHRIMP ages in Becker *et al.* (2006) of 1100 ± 5 and 1090 ± 15 Ma for rhyolites in the upper Langberg Formation. It is therefore likely that our sample is part of the Langberg Formation as shown by Becker (2008, Ch 8, p. 77).

Crustal evolution of the Rehoboth Province

The combined data point to three episodes of crustal evolution in the Rehoboth Province: Archaean, Palaeoproterozoic and Mesoproterozoic times (Fig. 4).

Archaean

The oldest crust-forming event is represented by three 2.98–2.7 Ga Archaean zircons, which comprise about 2% of the Billstein Formation data. These ages do not correspond to known units within the Rehoboth Basement Inlier. Either Archaean units did crop out at the time of Billstein

deposition (and are now covered by younger successions or have been removed by erosion and tectonics) or the older zircons may be present as xenocrysts in Rehoboth Group magmatic rocks, derived from reworking of an Archaean basement. The Archaean grains could have been derived from the Kaapvaal Craton, *c.* 600 km to the east, but it is not yet clear when the Kaapvaal Craton and Rehoboth Province became attached. If the Rehoboth Province was part of the Proto-Kalahari Craton in Palaeoproterozoic times then the 2.98–2.7 Ga zircons indicate a potential source in the Kaapvaal Craton or even the Zimbabwe Craton and Limpopo Belt. The complete absence of 3.2–3.1 Ga zircons (which together make up a much larger source area) and the large distance make this very unlikely, however.

The nearest known Archaean rocks are north of the Damara Belt near Sesfontein (Fig. 1), some 650 km NW of Rehoboth with single zircon ages between 2584 and 2645 Ma and Sm–Nd model ages between 2.6 and 2.8 Ga (Seth *et al.* 1998). It is not known however whether that region, representing the south-western reworked part of the Congo Craton, was adjacent to the Rehoboth Province before the Pan-African tectonic cycle. According to Johnson & Oliver (2004), the Congo Craton north of the Damara Belt was not attached to the Kalahari Craton before Neoproterozoic times and therefore could not have acted as a source for the 2.98–2.7 Ga zircons of the Billstein Formation.

The Archaean (2.7–2.5 Ga) granite cobbles found in Dwyka Formation glacial diamictites around Rietfontein some 450 km SE of Rehoboth were the first evidence that the Rehoboth Province might have an Archaean basement component (Cornell *et al.* 2011). This find instigated an investigation of the metasediments in the Rehoboth Inlier, which led to this second and possibly more direct evidence for an Archaean basement component.

Palaeoproterozoic

Our data confirm published evidence that a major part of the Rehoboth Province formed during the Palaeoproterozoic. This started as early as 2.2 Ga, with a small peak at 2.05 Ga and a major event at 1866 Ma with a minor tail ending at 1770 Ma. The peak ages of 1866 and 2050 Ma suggest two new and previously unknown crustal events. No units in the Rehoboth Basement Inlier have been reliably dated at older than 1863 ± 10 Ma, which is our unpublished age for the Elim Formation. The source of the older Palaeoproterozoic detrital zircon components in the Billstein Formation sample could therefore either be units which previously cropped out or basement rocks brought to the surface as xenoliths and xenocrysts in volcanic rocks, the same alternatives as proposed for the Archaean zircons. However, the large amount of Palaeoproterozoic zircon data in the Billstein Formation sample supports the idea that these grains came from units which cropped out during deposition.

The few Palaeoproterozoic zircons found in the Langberg Formation sample were probably derived from reworked Billstein Formation material, evidenced by the abundant quartzite pebbles in the Langberg conglomerates.

The older Palaeoproterozoic detrital zircons in the Rehoboth Basement Inlier serve to strengthen the link with Rehoboth Province crust to the SE, where 2.1–1.74 Ga zircon xenocrysts were found in the 1.09–1.17 Ga bimodal Koras Group near Upington by Pettersson *et al.* (2007). The 2.0 Ga cobbles in the Dwyka Formation around Rietfontein were found by Cornell *et al.* (2011) and the 2.2–2.0 ± 0.2 Ga Re depletion model ages were found by Hoal *et al.* (1995) for peridotite xenoliths from the Gibeon kimberlites.

Further a-field in South Africa and Botswana to the east, the mafic intrusion and granite plutons of the Bushveld Complex and the Molopo Farms Complex in the Kaapvaal Craton and the granites of the Okwa Basement Complex in the Okwa Inlier formed at 2.05 Ga, reflecting a major thermal event (Mapeo *et al.* 2006). The Rehoboth Province and Kaapvaal Craton might have been attached by that time; the 2.0 Ga event could therefore reflect the addition of new crust to a pre-existing Archaean core. To the south, surrounded by high-grade terranes of the Namaqua–Natal Province, lies the low-grade Richtersveld Terrane (Fig. 1, on the southern border between Namibia and South Africa). This comprises metavolcanic rocks of the 2.0 Ga Orange River Group (Reid 1997) and 1.9–1.7 Ga Vioolsdrif Igneous Suite (Reid 1982), again reflecting a major Palaeoproterozoic crust-forming event. The Richtersveld Terrane could not have been part of the Rehoboth Province however, as it represents an exotic terrane which was probably accreted to the Kaapvaal Craton and Rehoboth Province during the Namaqua–Natal orogeny (Eglington & Armstrong 2004).

Mesoproterozoic

The Mesoproterozoic episode of crustal development is seen only in our Langberg Formation sample as a broad peak with an age range of 1.32–1.05 Ga. This time span corresponds to the complete Namaqua–Natal Wilson cycle, with the main events shown in Figure 5, reviewed for South Africa by Cornell *et al.* (2006) and for Namibia by Miller (2008). Rifting and ocean basin

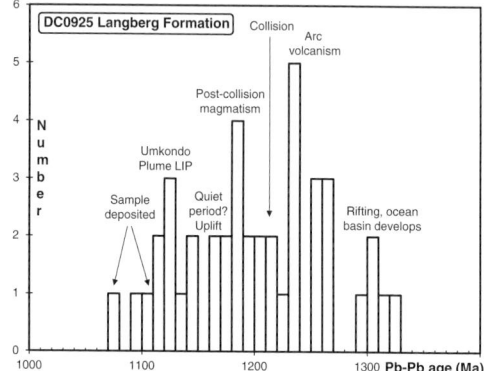

Fig. 5. The Mesoproterozoic episode of crustal development from 1.32 to 1.05 Ga recorded by zircons of the Langberg meta-arenite (DC0925). This time interval corresponds to the complete Namaqua–Natal Wilson cycle, the main events of which are shown (after Cornell *et al.* 2006).

development began at 1350 Ma, and subduction-related island arc and back-arc basin volcanism was in progress between 1300 and 1200 Ma. By 1210 Ma, juvenile arc and older terranes to the west had begun colliding with the Kaapvaal Craton and Rehoboth Province (Pettersson *et al.* 2007). In Namibia the mafic magmatism of the rifting event is reflected in the >1.37 Ga Kairab Formation, emplaced before or during an early Namaqua collision event along a suture now represented by the Namaqua front (Miller 2008, Ch 7, p. 1). Andean-type magmatism reflecting the closure of Namaqua ocean basins by subduction along the Rehoboth Province western margin starts between 1250 and 1220 Ma and is seen in the lower units of the Sinclair Supergroup such as the 1226 ± 10 Ma Nückopf Formation. During the *c.* 1210 Ma collision, parts of the accreted Grünau Terrane were thrust over the Rehoboth Province. This resulted in the reworking of the western margin of the Rehoboth Province and gave rise to the Konkiep Terrane, consisting of pre-tectonic basement overlain by the rocks of the Konkiep Group of the Sinclair Supergroup, but the tectonic fabric belongs to the Namaqua–Natal Province. A cycle of post-collisional granite magmatism followed in the high-grade Namaqua metamorphic terranes from *c.* 1180 until 1150, but in the southeastern Rehoboth Province around Upington the 1170 Ma lower Koras bimodal volcanic rocks were extruded onto stabilized basement and remained unaffected by deformation. A tectonic and magmatic quiet period is seen in most Namaqua–Natal terranes after the post-collision magmatism between 1150 and 1100 Ma, followed by a new pulse of magmatism most likely related to the 1100 Ma Umkondo plume large igneous province (LIP) event which affected the entire Kalahari Craton (Hanson *et al.* 2004). Our Langberg Formation sample was most likely deposited during or after that event.

The provenance of Mesoproterozoic zircons in our sample could be either the highly metamorphosed terranes in the Namaqua–Natal Province 150 km to the SW and the Gamsberg Granitic Suite intrusives or the lower-grade Sinclair Supergroup rocks which are largely a volcanic cover sequence on the Rehoboth Province. The volcanic rocks of the Nauzerus Group, Sinclair Supergroup are considered to represent a Mesoproterozoic Rehoboth volcanic arc which follows the curved north-western boundary of the Rehoboth Province (Watters 1976). The present outcrop pattern is however determined by the tectonic fabric of the Namaqua–Natal Province and the Damara Belt, so the arcuate shape is unlikely to reflect the original geometry. The oldest-known Sinclair unit in the Rehoboth Basement Inlier is the 1226 ± 10 Ma Nückopf Formation and the ages of the Gamsberg Granite Suite lie mostly between 1210 and 1010 Ma with only one precise older age (Table 1). At least the older 1.32–1.23 Ga component was therefore probably derived mainly from the Namaqua Province to the west, which was a mountain belt at that time. Other possible sources in the Rehoboth Basement Inlier are supracrustal and intrusive rocks which have not yet been dated and xenocrysts derived from lower crustal levels.

Conclusions

- Microbeam dating of detrital zircons in two metasedimentary rocks from the Billstein and Langberg Formations in the Rehoboth Basement Inlier provides a powerful method to characterize the crustal development history of the Rehoboth Province of southern Africa.
- The combined data indicate at least three episodes of crustal development in the Rehoboth Province. The oldest is represented by three 2.98–2.7 Ga Archaean zircons which may have been derived from within Rehoboth Province, thus supporting the concept that the Rehoboth Province has an Archaean crustal component.
- A Palaeoproterozoic group of 2.2–1.9 Ga Palaeoproterozoic detrital zircons are also older than any known exposures and confirm published evidence that a significant part of the Rehoboth Province formed during this interval.
- A 1.9–1.77 Ga Palaeoproterozoic phase overlaps the age of the 1.86 Ga Elim and 1.77 Ga Gaub Valley Formations of the Rehoboth Group and associated intrusive granitoids. The youngest Palaeoproterozoic detrital zircons are

1770 ± 20 Ma, which corresponds to the age of the Gaub Valley Formation. However, the data only shows that the Billstein Formation is younger than 1770 ± 20 Ma and it could be considerably younger.
- The upper constraint for the Billstein Formation is provided by 1210 ± 8 Ma porphyritic dykes which intrude the Billstein Formation. There is a hiatus between 1770 and 1325 Ma seen in the Langberg data, indicating that the Billstein formation is older than 1325 Ma.
- A broad peak of Mesoproterozoic detrital zircon ages in the Langberg Formation reflects magmatism related to the entire Namaqua–Natal Wilson cycle, with an age interval of *c.* 1.32–1.05 Ga. The known Sinclair Supergroup exposures within the Rehoboth Basement Inlier probably do not account for the pre-1.22 Ga part of this group, which may have been derived from the Namaqua Province mountain belt situated 150 km to the SW.

This research was supported by the Swedish Research Council through grant no. 2006–2402 to DHC. The ion probe dating was carried out with the help of M. Whitehouse and the staff at the NordSIM facility, supported by the research councils in Denmark, Norway and Sweden, the Geological Survey of Finland and the Swedish Museum of Natural History. This is NordSIM contribution no. 268 and a contribution to IGCP project 509, Paleoproterozoic Supercontinents and Global Evolution. We thank B. Mapani and an anonymous reviewer for their critical and constructive reviews.

References

BECKER, T. & SCHALK, K. E. L. 2008. *In*: MILLER, R. McG. (ed.) *The Geology of Namibia: Archaean to Mesoproterozoic*, Volume **1**. Ministry of Mines and Energy, Geological Survey of Namibia, Windhoek.

BECKER, T., AHRENDT, H. & WEBER, K. 1994. The geological history of the pre-Damara Gaub Valley Formation and the Weener Igneous Complex in the vicinity of Gamsberg. *Communications of the Geological Survey of Namibia*, **9**, 79–91.

BECKER, T., HANSEN, B. T., WEBER, K. & WIEGAND, B. 1996. U–Pb and Rb–Sr isotopic data of the Mooirivier Complex, the Weener Igneous Suite, and the Gaub Valley Formation (Rehoboth Sequence), Nauchas area, and their significance for the Palaeoproterozoic evolution of Namibia. *Communications of the Geological Survey of Namibia*, **11**, 31–46.

BECKER, T., HOFFMANN, K. H. & SCHREIBER, U. M. 1998. *Geological map of Namibia, 1:250000 Geological Series, Sheet 2316 – Rehoboth (provisional)*. Ministry of Mines and Energy, Geological Survey of Namibia, Windhoek.

BECKER, T., GAROEB, H., LEDRU, P. & MILESI, J. P. 2005. The Mesoproterozoic event within the Rehoboth Basement Inlier of Namibia: review and new aspects of stratigraphy, geochemistry, structure and plate

tectonic setting. *South African Journal of Geology,*
108, 465–492.

BECKER, T., SCHREIBER, U., KAMPUNZU, A. B. & ARM-
STRONG, R. 2006. Mesoproterozoic rocks of Namibia
and their plate tectonic setting. *Journal of African
Earth Sciences*, **46**, 112–140.

BURGER, A. J. & COERTZE, F. J. 1975. Age determinations
– April 1972 to March 1974. *Annals of the Geological
Survey of South Africa*, **10**, 135–142.

BURGER, A. J. & COERTZE, F. J. 1978. Summary of age
determinations carried out during the period April
1974 to March 1975. *Annals of the Geological
Survey of South Africa*, **11**, 317–321.

CORNELL, D. H., ARMSTRONG, R. A. & WALRAVEN, F.
1998. Geochronology of the Proterozoic Hartley
Basalt Formation, South Africa: constraints on the
Kheis tectogenesis and the Kaapvaal Craton's earliest
Wilson cycle. *Journal of African Earth Sciences*, **26**,
5–27.

CORNELL, D. H., THOMAS, R. J., GIBSON, R., MOEN,
H. F. G., MOORE, J. M. & REID, D. L. 2006. Namaqua-
Natal Province. *In*: JOHNSON, M. R., ANHAUESSER,
C. R. & THOMAS, R. J. (eds) *Geology of South
Africa*. Geological Society of South Africa and
Council for Geoscience, Pretoria, 325–379.

CORNELL, D. H., VAN SCHIJNDEL, V., INGOLFSSON, O.,
SCHERSTÉN, A., KARLSSON, L., WOJTYLA, J. & KARLS-
SON, K. 2011. Evidence from Dwyka tillite cobbles of
Archaean basement beneath the Kalahari sands of
southern Africa. *Lithos*, **125**, 482–502.

CORNER, B. 2000. Crustal framework of Namibia derived
from magnetic and gravity data. *Communications of
the Geological Survey of Namibia*, **12**, 13–19.

CORNER, B. 2003. Geophysically derived structural framework of sub-equatorial Africa. *In*: 8th Biennial Technical Meeting, South African Geophysical Association
of SAGA, Pilanesberg, South Africa, Poster.

CORNER, B. 2008. The crustal framework of Namibia
derived from an integrated interpretation of geophysical and geological data. *In*: MILLER, R. McG. (ed.)
*The Geology of Namibia: Archaean to Mesoprotero-
zoic, Volume 1*. Ministry of Mines and Energy,
Geological Survey of Namibia, Windhoek, Chapter 2,
1–19.

EGLINGTON, B. M. & ARMSTRONG, R. A. 2004. The Kaapvaal Craton and adjacent orogens, southern Africa: a
geochronology database and overview of the geological development of the craton. *South African Journal
of Geology*, **107**, 13–32.

FREI, D. & GERDES, A. 2009. Precise and accurate in situ
U–Pb dating of zircon with high sample throughput
by automated LA-SF-ICP-MS. *Chemical Geology*,
261, 261–270.

GERDES, A. & ZEH, A. 2006. Combined U–Pb and Hf
isotope LA-(MC)-ICP-MS analyses of detrital
zircons: comparison with SHRIMP and new constraints for the provenance and age of an Armorican
metasediment in Central Germany. *Earth and Planetary Science Letters*, **249**, 47–61.

HADDON, I. G. & MCCARTHY, T. S. 2005. The Mesozoic–
Cenozoic interior sag basins of Central Africa: the
late Cretaceous–Cenozoic Kalahari and Okavango
basins. *Journal of African Earth Sciences*, **43**,
316–333.

HANSON, R. E., CROWLEY, J. L. *ET AL.* 2004. Coeval large-
scale magmatism in the Kalahari and Laurentian
Cratons during Rodinia assembly. *Science*, **304**,
1126–1129.

HARTNADY, C. J., JOUBERT, P. & STOWE, C. 1985. Proterozoic crustal evolution in southwestern Africa. *Episodes*, **8**, 236–244.

HILKEN, U. 1998. *Die Altersstellung von metamorphen
prä-Damara Plutoniten und Vulkaniten am Südrand
des Damara Orogens Namibia (Farm Areb 176), abge-
leitet aus U/Pb-Isotopen-untersuchungen.* Diplomar-
beit University Göttingen, Germany.

HOAL, B. G., HOAL, K. E. O., BOYD, F. R. & PEARSON,
D. G. 1995. Characterisation of the age and nature of
the lithosphere in the Tsumkwe region, Namibia.
*Communications of the Geological Survey of
Namibia*, **12**, 21–28.

JACOBS, J., PISAREVSKY, S., THOMAS, R. J. & BECKER, T.
2008. The Kalahari Craton during the assembly and
dispersal of Rodinia. *Precambrian Research*, **160**,
142–158.

JOHNSON, S. P. & OLIVER, G. J. H. 2004. Tectonothermal
history of the Kaourerea Arc, northern Zimbabwe:
implications for the tectonic evolution of the Irumide
and Zambezi Belts of south central Africa. *Precambrian Research*, **130**, 71–97.

LUDWIG, K. R. 2003. *User's manual for Isoplot/Ex,
Version 3; A geochronological toolkit for Microsoft
Excel*. Berkeley Geochronology Center, Special Pub-
lications, **4**.

MAPEO, R. B. M., RAMOKATE, L. V., CORFU, F., DAVIS, D.
W. & KAMPUNZU, A. B. 2006. The Okwa basement
complex, western Botswana: U–Pb zircon geochronology and implications for Eburnean processes in
southern Africa. *Journal of African Earth Science*,
46, 253–262.

MILLER, R. McG. 2008. Namaqua Metamorphic Complex.
In: MILLER, R. McG. (ed.) *The Geology of Namibia:
Archaean to Mesoproterozoic*, Volume **1**. Ministry of
Mines and Energy, Geological Survey of Namibia,
Windhoek, Chapter 7, 1–55.

MOEN, H. F. G. 1999. The Kheis Tectonic Subprovince,
southern Africa: a lithostratigraphic perspective.
South African Journal of Geology, **102**, 27–42.

MOEN, H. F. G. & ARMSTRONG, R. A. 2008. New age constraints on the tectogenesis of the Kheis Subprovince
and the evolution of the eastern Namaqua Province.
South African Journal of Geology, **111**, 79–88.

MULLER, M. R., JONES, A. G. *ET AL.* & THE SAMTEX
TEAM. 2009. Lithospheric structure, evolution and
diamond prospectivity of the Rehoboth Terrane and
western Kaapvaal Craton, southern Africa: Constraints
from broadband magnetotellurics. *Lithos*, **112**,
93–105.

NAGEL, R. 2000. *Eine Milliarde Jahre geologischer
Entwicklung am NW-Rand des Kalahari Kratons.*
PhD thesis, University of Göttingen (http://webdoc.
sub.gwdg.de/diss/2000/nagel/index.htm).

NAGEL, R., WARKUS, F., BECKER, T. & HANSEN, B. T.
1996. U–Pb–Zirkondatierungen der Gaub Valley
Formation am Südrand des Damara Orogens-Namibia
und ihre Bedeutung für die Entwicklung des Rehoboth
Basement Inlier. *Zeitschrift für Geologische Wis-
senschaft*, **24**, 611–618.

NASDALA, L., HOFMEISTER, W. *ET AL.* 2008. Zircon M257–a homogeneous natural reference material for the ion microprobe U–Pb analysis of zircon. *Geostandards and Geoanalytical Research*, **32**, 247–265.

PETTERSSON, Å., CORNELL, D. H., MOEN, H. F. G., REDDY, S. & EVANS, D. 2007. Ion-probe dating of 1.2 Ga collision and crustal architecture in the Namaqua-Natal Province of southern Africa. *Precambrian Research*, **158**, 79–92.

PFURR, N., AHRENDT, H., HANSEN, B. T. & WEBER, K. 1991. U–Pb and Rb–Sr isotopic study of granitic gneisses and associated metavolcanic rocks from the Rostock massifs, southern margin of the Damara Orogen: implications for lithostratigraphy of this crustal segment. *Communications of the Geological Survey of Namibia*, **7**, 35–48.

RAITH, J. G., CORNELL, D. H., FRIMMEL, H. E. & DE BEER, C. H. 2003. New insights into the geology of the Namaqua Tectonic Province, South Africa, from ion probe dating of detrital and Metamorphic Zircon. *Journal of Geology*, **111**, 347–366.

REID, D. L. 1982. Age relationships within the Vioolsdrif batholith, lower Orange River region: II. A two stage emplacement history and the extent of Kibaran overprinting. *Transactions of the Geological Society of South Africa*, **85**, 105–110.

REID, D. L. 1997. Sm–Nd age and REE geochemistry of Proterozoic arc-related igneous rocks in the Richtersveld Subprovince, Namaqua Mobile Belt, Southern Africa. *Journal of African Earth Sciences*, **24**, 621–633.

SACS (SOUTH AFRICAN COMMITTEE FOR STRATIGRAPHY). 1980. Stratigraphy of South Africa. *In*: KENT, L. E. (ed.) (comp.) *Part 1. Lithostratigraphy of the Republic of South Africa, South West Africa/Namibia, and the Republics of Bophutatswana, Transkei and Venda.* Handbook of the Geological Survey of South Africa, **8**.

SCHALK, K. E. L. 1988. *Pre-Damara basement rocks in the Rehoboth and southern Windhoek districts (areas 2217D, 2316, 2317A-C), a regional description.* Report (Unpublished) Geological Survey of Namibia, **1 & 2**.

SCHNEIDER, T., BECKER, T., BORG, G., HILKEN, U. & WEBER, K. 2004. New U–Pb zircon ages of the Nuckopf Formation and their significance for the Mesoproterozoic event in Namibia. *Communications of the Geological Survey of Namibia*, **13**, 63–74.

SCHULZE-HULBE, A. 1979. *Three reports on pre-Damara and Damara rocks in the western Rehoboth area.* Report (Unpublished) Geological Survey of Namibia.

SETH, B., KRÖNER, A., MEZGER, K., NEMCHIN, A. A., PIDGEON, R. T. & OKRUSCH, M. 1998. Archaean to Neoproterozoic magmatic events in the Kaoko belt of NW Namibia and their geodynamic significance. *Precambrian Research*, **92**, 341–363.

SLÁMA, J., KOŠLER, J. *ET AL.* 2008. Plešovice zircon - a new natural reference material for U–Pb and Hf isotopic microanalysis. *Chemical Geology*, **249**, 1–35.

STACEY, J. C. & KRAMERS, J. 1975. Approximation of terrestrial lead isotope evolution by a two-stage model. *Earth and Planetary Science Letters*, **26**, 207–221.

STOWE, C. W. 1986. Synthesis and interpretation of structures along the north-eastern boundary of the Namaqua Tectonic Province, South Africa. *Transactions of the Geological Society of South Africa*, **89**, 185–198.

TINKER, J. H., DE WIT, M. J. & ROYDEN, L. H. 2004. Old, strong continental lithosphere with weak Archaean margin at ~1.8 Ga, Kaapvaal Craton, South Africa. *South African Journal of Geology*, **107**, 255–260.

VAN NIEKERK, H. S. 2006. *The origin of the Kheis Terrane and its relationship with the Archean Kaapvaal Craton and the Grenvillian Namaqua Province in southern Africa.* PhD thesis, University of Johannesburg.

WATTERS, B. R. 1976. Possible late Precambrian subduction zone in South West Africa. *Nature*, **259**, 471–473.

WHITEHOUSE, M. J. & KAMBER, B. 2005. Assigning dates to thin gneissic veins in high-grade metamorphic terranes: a cautionary tale from Akilia, southwest Greenland. *Journal of Petrology*, **46**, 291–318.

WHITEHOUSE, M. J., KAMBER, B. S. & MOORBATH, S. 1999. Age significance of U–Th–Pb zircon data from early Archaean rocks of West Greenland: a reassessment based on combined ion-microprobe and imaging studies. *Chemical Geology*, **160**, 201–224.

WIEDENBECK, M., ALLE, P. *ET AL.* 1995. Three natural zircon standards for U–Th–Pb, Lu–Hf, trace element and REE analyses. *Geostandards Newsletter*, **19**, 1–23.

ZIEGLER, U. R. F. & STOESSEL, G. F. U. 1993. *Age determinations in the Rehoboth Basement Inlier, Namibia.* Memoirs of the Geological Survey of Namibia, **14**.

Polyphase Neoproterozoic orogenesis within the East Africa–Antarctica Orogenic Belt in central and northern Madagascar

R. M. KEY[1]*, P. E. J. PITFIELD[2], R. J. THOMAS[2], K. M. GOODENOUGH[1], B. DE WAELE[3],
D. I. SCHOFIELD[2], W. BAUER[1], M. S. A. HORSTWOOD[4], M. T. STYLES[2], J. CONRAD[5],
J. ENCARNACION[5], D. J. LIDKE[5], E. A. O'CONNOR[2], C. POTTER[5], R. A. SMITH[1],
G. J. WALSH[5], A. V. RALISON[6], T. RANDRIAMANANJARA[6], J.-M. RAFAHATELO[6] &
M. RABARIMANANA[6]

[1]BGS, West Mains Road, Edinburgh, EH9 3LA, UK

[2]BGS, Keyworth, Notts NG12 5GG, UK

[3]SRK Consulting, 10 Richardson Street, West Perth, WA 6005, Australia

[4]NIGL, Keyworth, Notts NG12 5GG, UK

[5]USGS, Reston, Va, 20192, USA

[6]Projet de Gouvernance des Ressources Minières, Antananarivo, Madagascar

*Corresponding author (e-mail: rmk@bgs.ac.uk)

Abstract: Our recent geological survey of the basement of central and northern Madagascar allowed us to re-evaluate the evolution of this part of the East Africa–Antarctica Orogen (EAAO). Five crustal domains are recognized, characterized by distinctive lithologies and histories of sedimentation, magmatism, deformation and metamorphism, and separated by tectonic and/or unconformable contacts. Four consist largely of Archaean metamorphic rocks (Antongil, Masora and Antananarivo Cratons, Tsaratanana Complex). The fifth (Bemarivo Belt) comprises Proterozoic meta-igneous rocks. The older rocks were intruded by plutonic suites at c. 1000 Ma, 820–760 Ma, 630–595 Ma and 560–520 Ma. The evolution of the four Archaean domains and their boundaries remains contentious, with two end-member interpretations evaluated: (1) all five crustal domains are separate tectonic elements, juxtaposed along Neoproterozoic sutures and (2) the four Archaean domains are segments of an older Archaean craton, which was sutured against the Bemarivo Belt in the Neoproterozoic. Rodinia fragmented during the early Neoproterozoic with intracratonic rifts that sometimes developed into oceanic basins. Subsequent Mid-Neoproterozoic collision of smaller cratonic blocks was followed by renewed extension and magmatism. The global 'Terminal Pan-African' event (560–490 Ma) finally stitched together the Mid-Neoproterozoic cratons to form Gondwana.

The supercontinent of Rodinia was created by c. 1.0 Ga following end-Mesoproterozoic orogenesis along a global network of Grenville–Kibaran–Namaquan orogenic belts (e.g. Torsvik et al. 1996; Dalziel 1997; Kröner 2001; McCourt et al. 2006; Li et al. 2008). Within c. 120 million years of its creation, extensional rift basins started to develop. In some instances these progressed into oceanic basins, partially fragmenting Rodinia into separate crustal plates by the Mid-Neoproterozoic (Unrug 1998; Li et al. 2008; Wendorff & Key 2009). Middle Neoproterozoic plate collision and subsequent extension was followed by multiple collisions of smaller crustal plates between c. 560 and 520 Ma to form the Gondwana supercontinent (Meert 2003; Collins & Pisarevsky 2005; Bingen et al. 2009). The end-Neoproterozoic–Cambrian

collisional tectonics resulted in a network of linear orogenic belts, referred to as 'Pan-African' in Africa (Kennedy 1964; Stern 1994; Gasquet et al. 2008; Michard et al. 2008), including the East Africa–Antarctica Orogenic Belt (EAAO) that can be traced southwards from Arabia through eastern and southern Africa, India, Madagascar and Sri Lanka into Antarctica (Jacobs & Thomas 2004, Fig. 1). Post-collision retrograde metamorphism and shearing continued into Ordovician times (to c. 500 Ma; Emmel et al. 2006).

Recent work has shown that while all parts of the EAAO shared a common end-Proterozoic to earliest Ordovician geological history (Meert 2003; Collins & Pisarevsky 2005; Bingen et al. 2009), the earlier Neoproterozoic tectonothermal events at c. 820–720 Ma and c. 660–610 Ma recognized in East

From: VAN HINSBERGEN, D. J. J., BUITER, S. J. H., TORSVIK, T. H., GAINA, C. & WEBB, S. J. (eds) *The Formation and Evolution of Africa: A Synopsis of 3.8 Ga of Earth History*. Geological Society, London, Special Publications, **357**, 49–68. DOI: 10.1144/SP357.4 0305-8719/11/$15.00 © The Geological Society of London 2011.

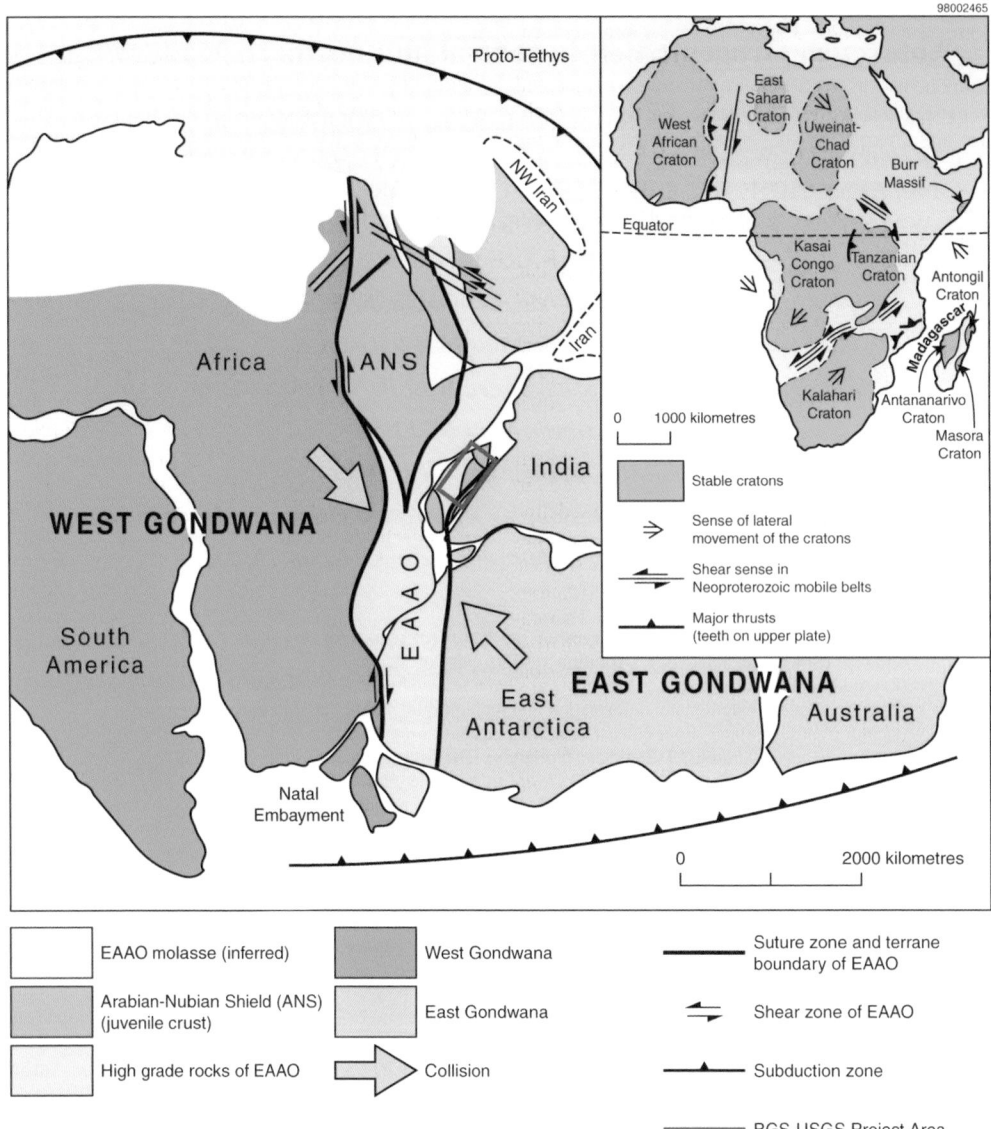

Fig. 1. Location of the BGS-USGS-GLW study area in central and northern Madagascar and its position within the East African–Antarctica Orogen (EAAO); modified from Jacobs & Thomas (2004).

Africa, Arabia, Madagascar and Antarctica are less uniformly within the EAAO (Stern 1994; Meert 2003). We use the phrase 'terminal Pan-African' to describe events that took place in the 560–490 Ma period, following the original definition of Kennedy (1964). It corresponds to the 'Kuunga Orogeny' resulting from India–Australia collision (Meert 2003) and the 'Malagasy Orogeny' caused by India–East Africa collision (Collins & Pisarevsky 2005). The events that make up the wider EAAO have been attributed to progressive closure

of a 'Mozambique Ocean' involving collisions between arc terranes and accretion of micro-continents onto East Africa (Meert 2003; Collins 2006), culminating in the final assembly of all the components of Gondwana during the terminal Pan-African event (e.g. Jacobs *et al.* 1998, 2003; Hanson 2003; Meert 2003; Boger & Miller 2004; Collins & Pisarevsky 2005; Johnson *et al.* 2005).

Early geological work in Madagascar showed that the crystalline basement comprises a number of major crustal segments separated by tectonic contacts and unconformable boundaries (e.g. Besairie 1968–1971; Hottin 1969, 1976; Jourde 1972). Subsequent research confirmed the presence of Archaean crustal components, a range of Proterozoic supracrustal sequences and major periods of Proterozoic reworking, and interpreted the geology in plate tectonic terms (e.g. Windley et al. 1994; Tucker et al. 1997, 1999a, 2010; Cox et al. 1998; Collins et al. 2000; Kröner et al. 2000; Collins & Windley 2002; Collins 2006). The principal tectonic blocks identified in the central-north area include the Archaean Antongil–Masora and Antananarivo cratons and Tsaratanana Complex, the Neoproterozoic Bemarivo Belt and the Itremo Domain comprising Proterozoic metasedimentary rocks (Fig. 2; Collins et al. 2000; Kröner et al. 2000; Collins & Windley 2002; Collins 2006).

The work presented here is a summary of a 4-year World Bank funded project fully reported in BGS-USGS-GLW (2008) that allowed us to redefine the tectonic domains of north-central Madagascar and their boundaries. Following previous studies, we recognize four distinct Archaean domains (Antongil, Masora and Antananarivo cratons and Tsaratanana Complex) and the Neoproterozoic Bemarivo Belt in the north, along with a number of Proterozoic metasedimentary units. The domains are separated by ductile shear zones (Fig. 2) and wide Neoproterozoic belts, including the 'Betsimisaraka Suture' of previous authors (Collins et al. 2000; Kröner et al. 2000) which has now been mapped in greater detail and separated into two belts that are structurally continuous but have rather different characteristics and the Anaboriana and Manampotsy belts.

We define Neoproterozoic periods dominated by extensional tectonics, with resulting magmatism and sedimentation, and intervening periods of compression leading to plate collision and uplift, based on new and existing geochronological data (Tables 1 & 2). These are reported in a number of recently published articles on specific aspects of the geology (e.g. De Waele et al. 2008a; Thomas et al. 2009; Goodenough et al. 2010; Tucker et al. 2010; Schofield et al. 2010) under the umbrella of the project report (BGS-USGS-GLW 2008). The aim of this paper is to draw all the available data together into a consistent new model for the evolution of central-northern Madagascar and its bearing on the evolution of the EAAO as a whole.

Crustal domains of central and northern Madagascar

Antongil and Masora cratons

Two separate, but previously correlated, Archaean crustal blocks are recognized in eastern Madagascar,

known as the Antongil and Masora cratons in the north and south, respectively (Hottin 1976; Collins & Windley 2002). They are separated by high-grade Neoproterozoic rocks belonging to the Manampotsy Belt (see below). Our geological mapping has recognized significant differences in the gross lithological make-up of these two cratons; the Masora Craton is also cut by c. 800 Ma granitoids not present in the Antongil Craton.

Antongil Craton. The northern part of the Antongil Craton is dominated by voluminous, largely unfoliated, Neoarchaean granitoids (Masoala Suite) formed at c. 2550–2520 Ma (Collins et al. 2001; Paquette et al. 2003; Tucker et al. 1999a, b, 2010; Schofield et al. in press). To the south there is a core of older Meso-Palaeoarchaean gneisses (Nosy Boraha Suite) which were formed between c. 3.3 and 3.1 Ga (Vachette & Hottin 1970; Tucker et al. 1999a, b, 2010; Schofield et al. in press). These are interfolded with Meso-Neoarchaean supracrustal rocks (Mananara Group) dated at between c. 3176 and 2597 Ma (Fig. 3). The latter include strongly foliated metabasic rocks and paragneisses, which occur as lenses within the southern granitoids.

The northern part of the Antongil Craton was intruded at c. 2150 Ma by a number of minor metabasic sheets and podiform bodies (Ankavanana Suite; Schofield et al. in press). Previously, Hottin (1969) describes metasedimentary rocks in the Anjiahely Belt in the northern Antongil Craton with many aligned meta-ultramafic lenses of harzburgite, pyroxenite, tremolite–actinolite rocks and soapstone. The northern margin of the craton is overthrust by the Neoproterozoic Bemarivo Belt (Thomas et al. 2009) and locally unconformably overlain by a low-grade metasedimentary sequence of possible Palaeoproterozoic age (Andrarona Group; Hottin 1976). No Neoproterozoic to Cambrian intrusions have been recognized in the Antongil Craton, which is only marginally affected by Pan-African events and structures (Table 2).

Masora Craton. In similar fashion to the Antongil Craton, the Masora Craton in central-east Madagascar comprises a core of Mesoarchaean Nosy Boraha migmatitic orthogneisses with infolds of undated meta-volcanosedimentary rocks (Vohilava Group: Fig. 4). Another metasedimentary sequence (Maha Group) crops out around the western and northern margins of the Masora Craton. A maximum age of c. 1740 Ma is provided for the Maha Group by the youngest age of its detrital zircon population (De Waele et al. 2008a). These rocks were intruded at 840–815 and 780–760 Ma by voluminous Neoproterozoic granitoid sheets (Imorona–Itsindro Suite). Larger intrusions of the latter help define regional

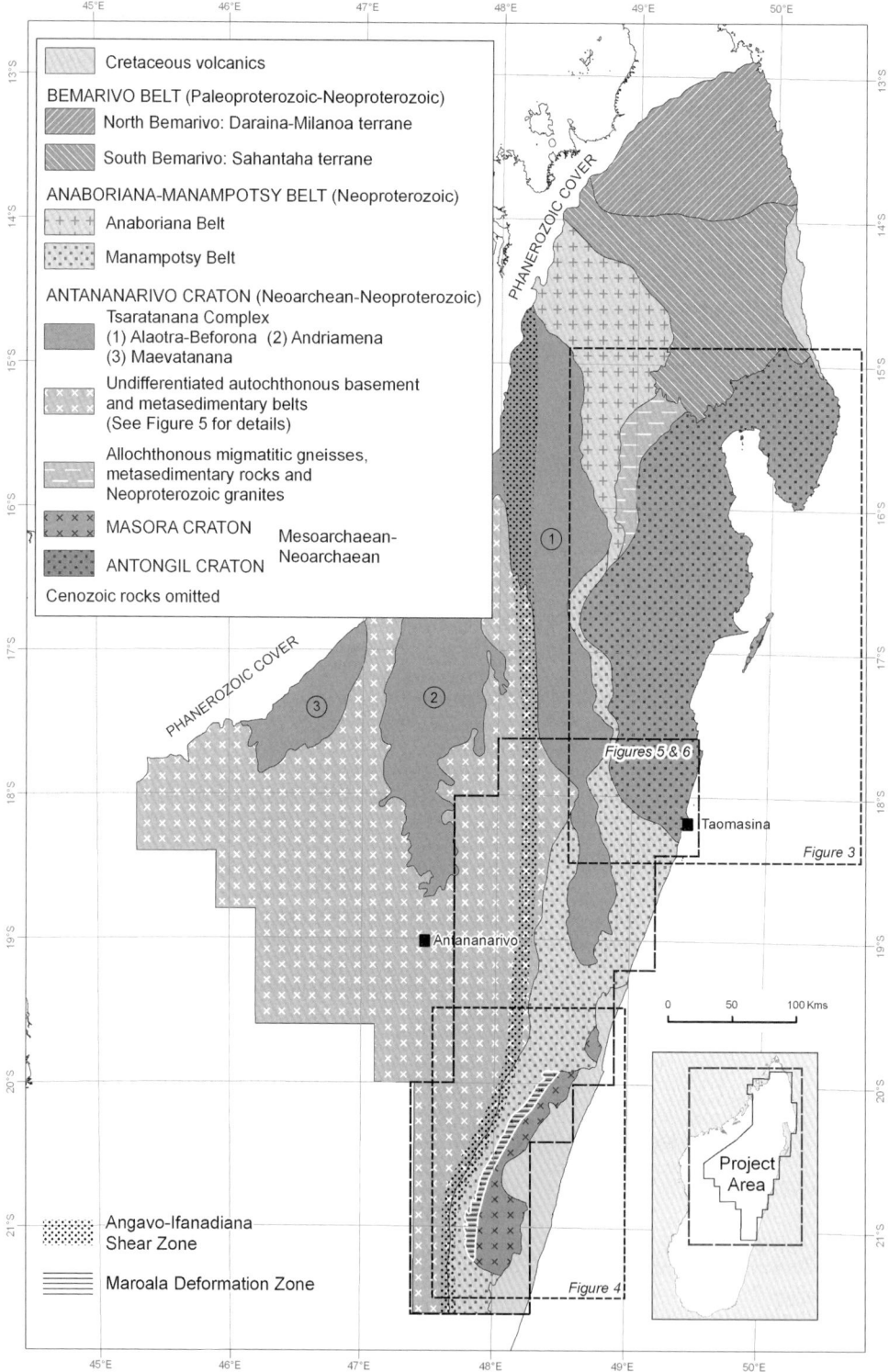

Fig. 2. An outline of the geology of the study area showing the main tectonic domains and main shear zones.

Table 1. *A summary of the Neoproterozoic–Ordovician geotectonic evolution of northern-central Madagascar*

Approximate date (Ma)	Event
500–490	Ductile shearing along major shear zones; final metamorphic overprint
530–510	End stages of terminal Pan-African thermal anomaly including post-collisional magmatism (e.g. Maevarano Suite plutons)
560–530	Terminal Pan-African tectono-thermal events including amalgamation of the Antongil Craton with the already amalgamated Antananarivo and Masora cratons with reactivation of the Manampotsy Belt. Later thrusting of the Bemarivo Belt over the Antongil and Antananarivo cratons and the Anaboriana Belt; the Tsaratanana Complex thrust over both the Antananarivo Craton and the Manampotsy Belt. Initial emplacement of late- to post-tectonic granites including the Maevarano and Ambalavao Suites and Carion Subsuite
630–595	Emplacement of the Kiangara Suite of granites into the Antananarivo and Masora cratons and into the Anaboriana and Manampotsy belts
760–715	Arc volcanicity and associated sedimentation in the Bemarivo Belt
740	Amalgamation of the Antananarivo and Masora cratons and tectonic development of the Manampotsy Belt, possibly during closure of ocean between the Antananarivo and Masora cratons
840–720	Sedimentation on the Archaean foundations of the Antananarivo and Masora cratons and on the surrounding continental shelves and offshore arc terranes (Anaboriana and Manampotsy belts). Emplacement of the Imorona–Itsindro Suite magmatic arc and unroofing of these intrusions

folds that range from open in form in the centre of the craton to tight dome-and-basin interference structures along the western and northern margins (Fig. 4). The observed deformation in these Neoproterozoic intrusions, together with metamorphic ages recorded in zircon grain rims (Table 2), demonstrate that the whole Masora Craton was affected by the terminal Pan-African tectonothermal episode (in contrast to the Antongil Craton). However, earlier recumbent folds and associated thrusts are present to juxtapose low-grade metasedimentary rocks with high-grade migmatitic gneisses. In such zones, normal and inverted metasedimentary sequences are identified from preserved way-up structures to define recumbent folds. Some parts of the metasedimentary sequences appear to be floored by mafic-ultramafic rocks and are thought to represent thrust klippen. The preservation of low metamorphic grade rocks shows that the Pan-African metamorphism was not as intense as found in the adjacent Manampotsy Belt. The age of the early recumbent folds is not known; they could be early Pan-African structures or older (?Mesoproterozoic).

Antananarivo Craton

The Antananarivo Craton, which underlies the central highlands of Madagascar, comprises granulite- to upper amphibolite facies Neoarchaean orthogneisses and paragneisses (Fig. 5). No evidence for Mesoarchaean rocks have been found, in contrast to the Antongil and Masora cratons. Extensive tracts of

Neoarchaean/earliest Palaeoproterozoic tonalitic to granitic gneisses (Betsiboka Suite; Mangoro Complex) have been dated at c. 2490–2590 Ma (Tucker *et al.* 1999a; Kröner *et al.* 2000; BGS-USGS-GLW 2008). These are intimately associated with supracrustal rocks, chiefly migmatitic paragneisses (Vondrozo and Sofia Groups) which form extensive belts (e.g. Sofia Group) or isolated rafts in the orthogneisses. The Vondrozo and Sofia Groups are thus demonstrably Archaean, but there are no constraints on their maximum depositional ages.

In central-west Madagascar, the Antananarivo Craton is overlain by Proterozoic metasedimentary rocks of the Itremo, Molo and Amborompotsy–Ikalamavony groups in the Itremo Domain. These correlated rocks of different metamorphic grades are characterized by significant detrital zircon peaks at c. 2500 and 1800 Ma (Cox *et al.* 1998, 2004; Fitzsimons & Hulscher 2005; Tucker *et al.* 2007; De Waele *et al.* 2008a). The relationship between them and the underlying craton is unclear. Collins *et al.* (2000) recognized a major shear zone forming the contact in some parts, while elsewhere the contact appears unconformable (Collins 2006; Tucker *et al.* 2007).

The Antananarivo Craton and the Itremo Domain are intruded by voluminous Neoproterozoic to Cambrian plutonic rocks. The oldest of these (Dabolava Suite orthogneisses) intruded the Itremo Group at c. 1000 Ma (Tucker *et al.* 2007). Subsequently, plutons of the Imorona–Itsindro Suite, which intrude the whole Antananarivo–Itremo block, have been dated between c. 840 and

Table 2. *Metamorphic ages recorded in rims to zircon grains and associated PT conditions and tectonic styles found in the tectonic domains*

Craton/Belt/Complex	Area/Unit	Approximate metamorphic ages (Ma)	Pan-African PT conditions (based on mineral chemistries; BGS-USGS-GLW 2008)	Terminal Pan-African tectonic style
Antongil	Northern area (Mananara Group)	537–526	600–750 °C; 8–10 kb These conditions may relate to an Archaean metamorphism and most samples did not record Pan-African events	Weakly affected by Pan-African events with relatively open NNW– and north–NE-trending folds
Masora	Southern area	530–515	650–700 °C; 6–7 kb Samples from the central part of this domain did not record a Pan-African metamorphic event	Central area preserves an open SSW-plunging synform that is transposed and cut by ductile thrusts around the margins of the Craton
Antananarivo	Mangoro Ambatolampy Imorona-Itsindro	713, 560–537 560–540 807–801, 767–763, 550–516 535	650–750 °C; 4–6 kb	Dome and basin interference folds aligned c. north–south, and the discordant Antananarivo Virgation Zone
Tsaratanana	Andriamena	819, 779, 750, 535	750–820 °C; 5.5 kb	Strong c. north–south to NNW–SSE structural grain with tight interference fold patterns
Bemarivo	North	526–511	There are very few thermobarometric determinations and mineral assemblages suggest greenschist to upper amphibolite conditions	South-directed horizontal translation and formation of a thrust-imbricated fan. Subsequent shearing and open folding after c. 520 Ma (Buchwaldt et al. 2003)
	South	540–520	500–600 °C; 6–8 kb Higher temperatures up to 850 °C were recorded by the two-feldspar thermometer (previous studies give 800–970 °C, 6.5–9 kb)	
Anaboriana-Manampotsy Belt ('Betsimisaraka Suture Zone')	Anaboriana Ampasary Manampotsy	540–520 551–520 560	West to east decrease in temperature but increase in pressure across the Manampotsy Belt. The south-westernmost part of the belt shows similar PT values to the Antananarivo Craton but slightly higher pressures in the north where 7–8 kb was recorded. Central parts of the Manampotsy Belt record temperatures of 700–750 °C but with higher pressure of 7.5–10 kb, while in the east temperatures of 600–650 °C and higher pressures of 10–12 kb were recorded	South: Strong north–south structural grain with transposed fabrics and dominated by anastomosing ductile shears and the Angavo–Ifanadiana Shear Zone North: NW–SE structural grain within the Anaboriana Belt

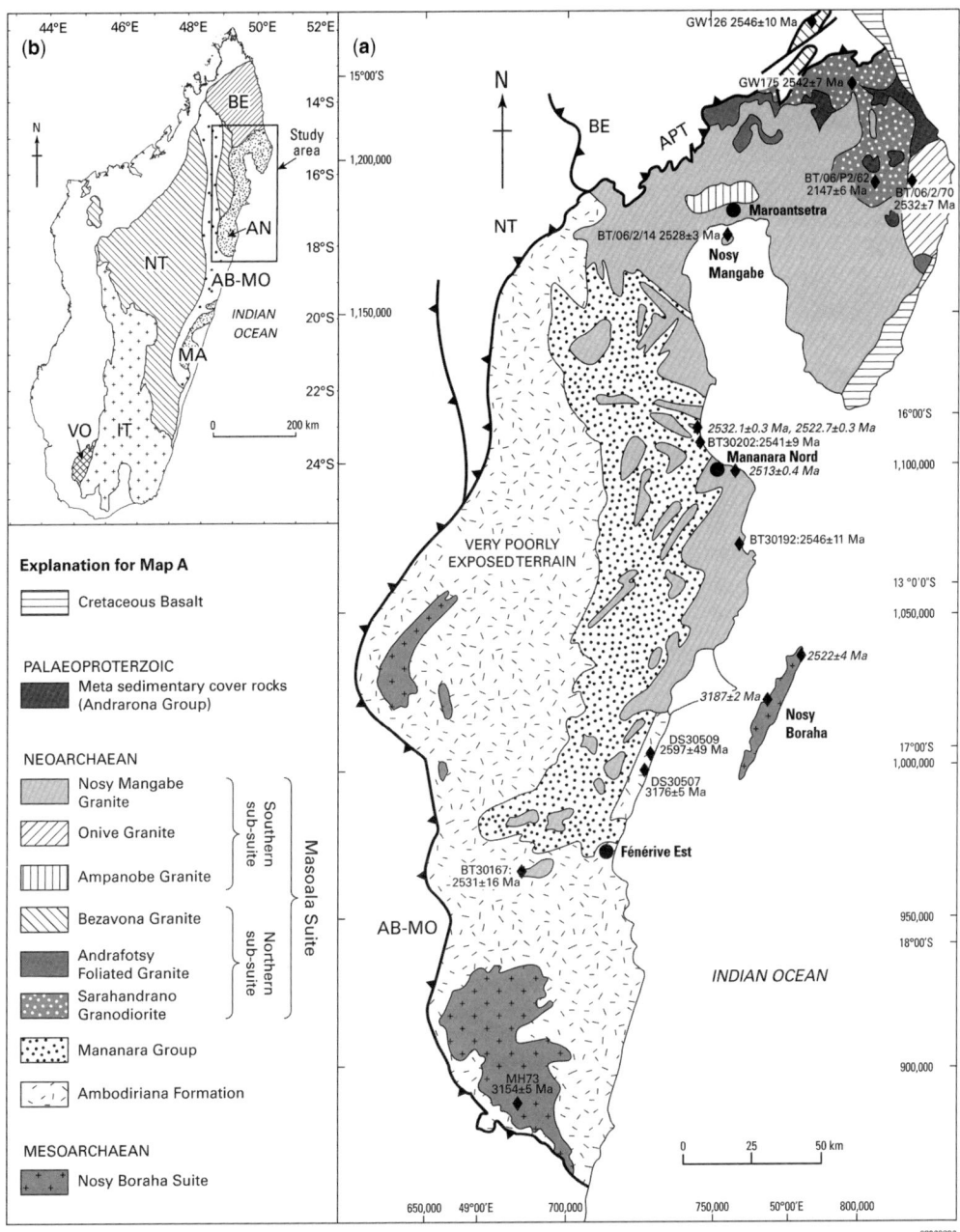

Fig. 3. (**a**) Simplified geological map of the Antongil Craton, showing the main lithodemic units and U–Pb dates. New geochronological ages shown as well as previously published U–Pb dating (shown in italics) of Tucker *et al.* (1999*a, b*) and Paquette *et al.* (2003). Inset map (**b**) shows the major Precambrian crustal terranes of Madagascar, modified after Collins (2006) (APT, Andaparaty Thrust; AB-MO, Anaboriana Belt – Manampotsy Belt; AN, Antongil Craton; BE, Bemarivo Belt; M, Masora Craton; NT, Antananarivo Craton; IT, southern mobile belts including the Itremo Group; VO, Vohibory Unit; from Schofield *et al.* in press).

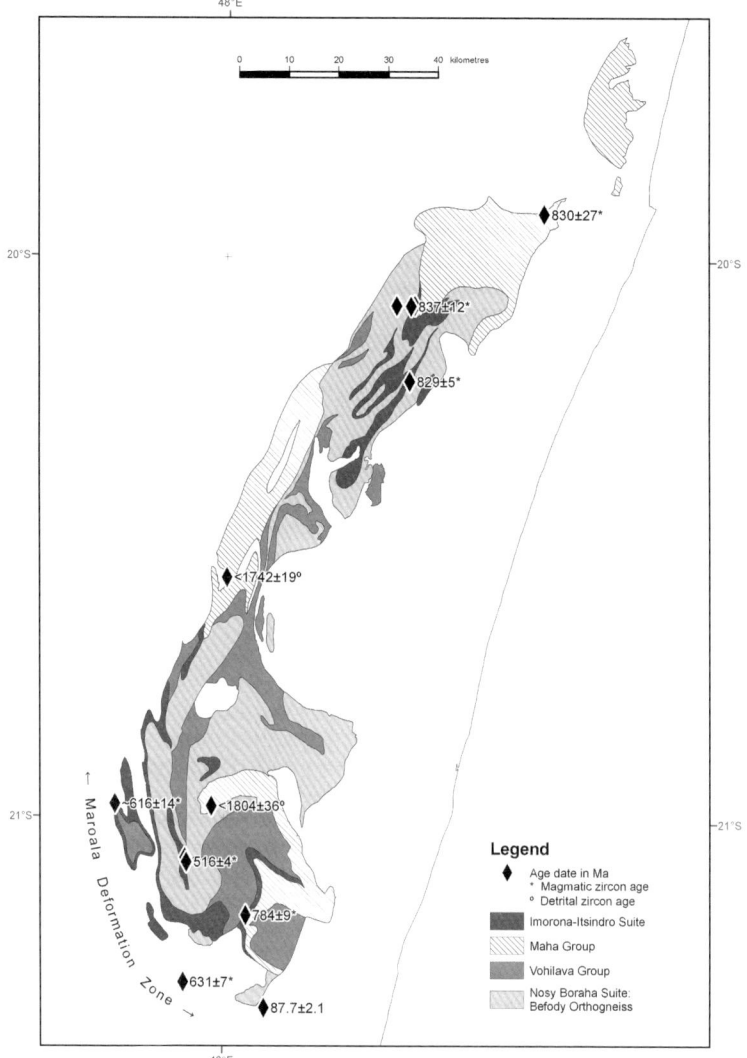

Fig. 4. An outline of the geology of the Masora Craton and surrounding Maroala Deformation Zone with the locations of isotopically dated rock samples. The 87.7 ± 2.1 Ma age in the extreme SE is from the Cretaceous volcanic cover over the Masora Craton.

720 Ma (Handke *et al.* 1999; Tucker *et al.* 1999*a*; Kröner *et al.* 2000; BGS-USGS-GLW 2008). In addition, the northern part of the craton was intruded by sheet-like, alkaline 'stratoid granites' (Kiangara Suite) at *c.* 630 Ma (Nédélec *et al.* 1995; Paquette & Nédélec 1998) and post-collisional granitoids (e.g. Ambalavao and Mahamavy suites and Carion Subsuite) during the period 550–530 Ma (Kröner *et al.* 1999*a*, *b*, 2000; Tucker *et al.* 1999*a*; Meert *et al.* 2001; Goodenough *et al.* 2010).

The Antananarivo Craton is characterized by a penetrative, north to NNW-trending structural grain, marked by parallel ortho- and paragneiss belts. The highest strain is recognized in the east, along the Angavo–Ifanadiana Shear Zone that cuts into the craton from the Manampotsy Belt (Fig. 2; Nédélec *et al.* 2000). Heat transfer associated with rise of magma and fluids along this shear zone continued until *c.* 470 Ma (Grégoire *et al.* 2009).

Tsaratanana Complex

The Tsaratanana Complex, which overlies the northern part of the Antananarivo Craton, comprises three main infolded synformal belts (Fig. 2; Maevatanana, Andriamena and Alaotra–Beforona; Tucker *et al.*

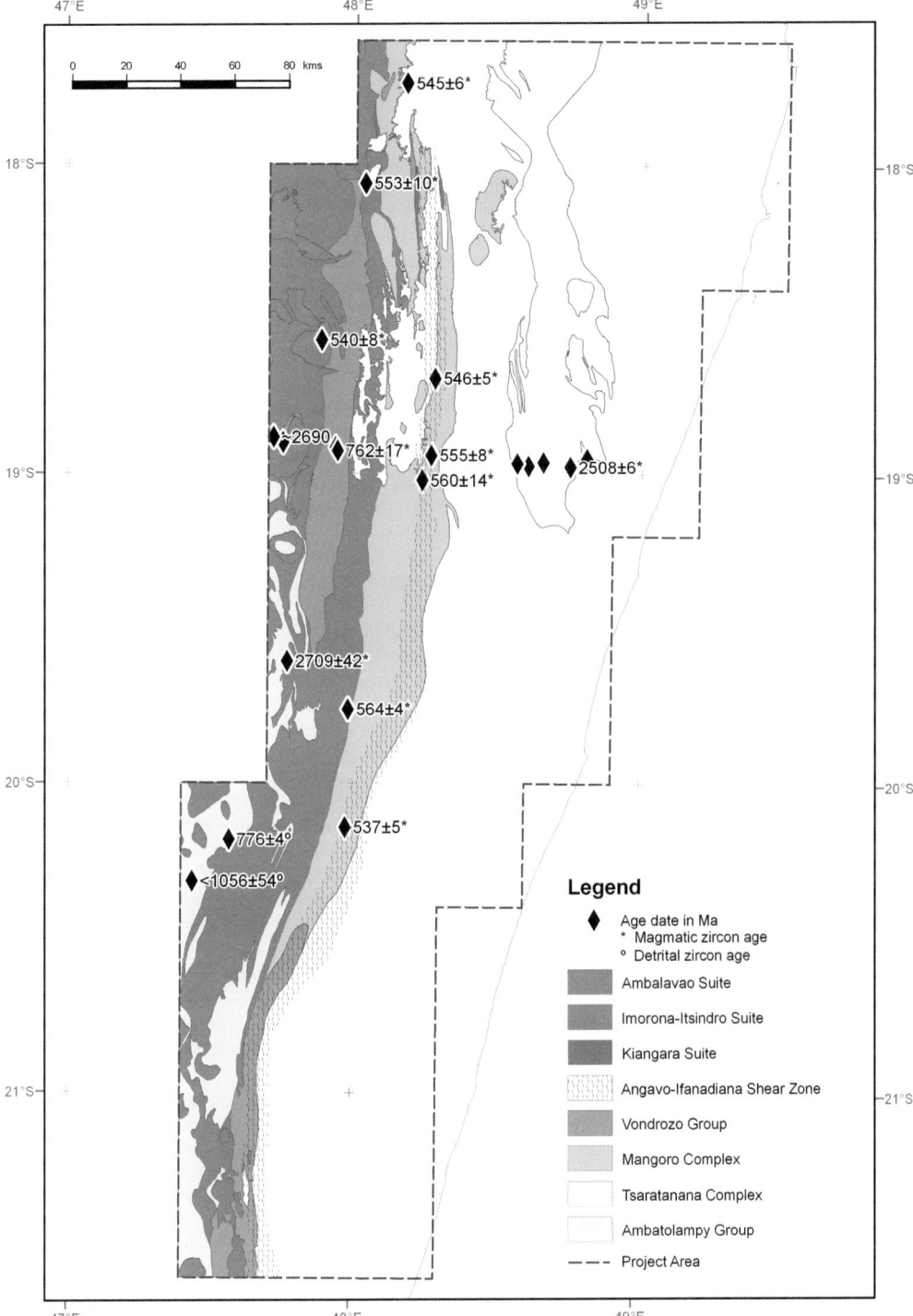

Fig. 5. An outline of the geology of the main outcrop of the Antananarivo Craton with the locations of isotopically dated rock samples.

1999*a*; Collins & Windley 2002; Collins 2006). Each is dominated by Neoarchaean orthogneisses, pelitic schists and paragneisses that have previously been interpreted as greenstone belts (de Wit 2003). The orthogneisses include gabbroic, tonalitic and granodioritic compositions, and have been dated at *c.* 2520–2470 Ma (Tucker *et al.* 1999*a, b*; Kabete *et al.* 2006; BGS-USGS-GLW 2008) with evidence for high-temperature metamorphism of some at *c.* 2500 Ma (Gonçalves *et al.* 2003; Tucker *et al.* 2010). However, a paragneiss from the Andriamena Belt contained detrital zircons in the range 2870–1750 Ma, suggesting the presence of Palaeoproterozoic or younger material (Kabete *et al.* 2006).

In common with the Antananarivo–Itremo block, the Tsaratanana Complex is intruded by gabbroic and granitoid plutons of the Neoproterozoic Imorona–Itsindro Suite (Guérrot *et al.* 1993; Tucker *et al.* 1999*a, b*; Kabete *et al.* 2006; BGS-USGS-GLW 2008). High-grade metamorphism is also recorded in the Andriamena Belt at *c.* 780–730 Ma (Gonçalves *et al.* 2003; Kabete *et al.* 2006; BGS-USGS-GLW 2008).

The Tsaratanana Complex is separated from the underlying rocks by flat-lying to subvertical ductile shears with complex polyphase and polydirectional movement histories (Gonçalves *et al.* 2003; BGS-USGS-GLW 2008). Moreover, the BGS-USGS mapping has shown that parts of the Tsaratanana Complex tectonically overlie not only the Antananarivo Craton, but also the Neoproterozoic Manampotsy Belt (Fig. 5). This observation has critical implications for the timing of emplacement of the Tsaratanana Complex and supports the conclusion of Kröner *et al.* (2000) that (at least some of) the complex is an allochthonous unit and not an integral part (e.g. greenstone belt) of the Antananarivo Craton. However, the precise timing of the final horizontal translation of the Tsaratanana Complex remains uncertain. Although the last period of movement probably occurred in the early Cambrian (Gonçalves *et al.* 2003), at least parts of the Tsaratanana Complex were juxtaposed with the underlying combined Antananarivo–Itremo block before the intrusion of the Imorona–Itsindro Suite (Tucker *et al.* 2010).

Bemarivo Belt

The Neoproterozoic Bemarivo Belt forms the most northerly crustal entity of Madagascar, where it is tectonically juxtaposed against the Antongil and Antananarivo cratons and the Neoproterozoic Anaboriana Belt (Collins & Windley 2002; Collins 2006; Thomas *et al.* 2009). It is divided into northern and southern terranes which, on the basis of lithostratigraphy and geochronology, are considered

to have formed in volcanic arc settings at *c.* 715 and 750 Ma, respectively (Thomas *et al.* 2009).

The northern terrane comprises three volcano-sedimentary successions: the upper amphibolite-facies Milanoa Group, the greenschist-facies to epidote-amphibolite-facies Daraina Group and the Betsiaka Group with measured pressure–temperature (PT) conditions of 12 kb and 650 °C (BGS-USGS-GLW 2008). The first two were derived from Neoproterozoic (probably juvenile) sources, while the Betsiaka Group contains Archaean detrital zircons (Thomas *et al.* 2009). This terrane is intruded by the plutonic Manambato Suite, dated between 718 and 708 Ma (Tucker *et al.* 1999*b*; Thomas *et al.* 2009).

The southern terrane is dominated by the granulite facies metasedimentary Sahantaha Group consisting of quartzites, calc-silicate and metapelitic rocks. Samples from this group have yielded detrital zircon age spectra with ages ranging from *c.* 2800 to 1750 Ma, with the largest peak in the Palaeoproterozoic (Cox *et al.* 2004; De Waele *et al.* 2008*a*). These rocks are intruded by the calc-alkaline plutonic Antsirabe–Nord Suite which was emplaced at *c.* 750–760 Ma (Tucker *et al.* 1999*b*; Thomas *et al.* 2009). Hottin (1969) suggested that the Sahantaha metasedimentary rocks rest unconformably on high-grade 'basement' rocks. However, exposed contacts noted between the Sahantaha Group and other rocks including Antongil Craton rocks were ubiquitously tectonic (BGS-USGS-GLW 2008; Thomas *et al.* 2009).

High-grade metamorphism in the southern terrane of the Bemarivo Belt has been dated at *c.* 560–530 Ma (Jöns *et al.* 2006; Table 2); this is considered to represent the timing of collision of the Bemarivo Belt with the previously amalgamated older 'core' of Madagascar (Jöns *et al.* 2006; Thomas *et al.* 2009). Both terranes of the Bemarivo Belt and the Anaboriana Belt to the south were intruded by voluminous, post-collisional, A-type granitoids (Maevarano Suite) at *c.* 540–520 Ma (Buchwaldt *et al.* 2003; Goodenough *et al.* 2010).

Anaboriana and Manampotsy belts

The north–south trending Anaboriana and Manampotsy belts (Fig. 2) are redefined tectonostratigraphic domains which together broadly correspond to the Betsimisaraka Suture Zone (Kröner *et al.* 2000; Collins & Windley 2002). In the NW, the Anaboriana Belt separates the Bemarivo Belt from the Antananarivo Craton and is bounded by ductile shear zones. It runs southwards into the Antananarivo Craton, close to its eastern margin, and has been linked on geophysical and lithostratigraphical grounds to the Manampotsy Belt which separates the Masora, Antongil and Antananarivo cratons.

The main rock types throughout the Anaboriana and Manampotsy belts are quartzofeldspathic migmatitic paragneisses with varying biotite and hornblende contents, although some graphitic, quartzitic and calc-silicate lithologies are locally present. In the Anaboriana Belt, the paragneisses are termed the Bealanana Group. The Manampotsy Belt paragneisses of the Manampotsy Group are divided into the Sasomanangana Gneiss, Sakanila Paragneiss, Perinet Paragneiss and Ambatondrazaka Paragneiss (Fig. 6) on the basis of their dominant sedimentary protolith. The Ambatolampy Group occurs on the margin of the Antananarivo Craton, while the Ampasary Group crops out on the western margin of the Masora Craton within a tectonic entity known as the 'Maroala Deformation Zone' (Figs 2 & 6). This structural complex comprises a highly deformed imbricate zone, with inter-thrust slices of the Ampasary Group and Masora Craton rocks. Numerous mafic-ultramafic lenses in the Manampotsy Belt range from 1 to 500 m in length and include harzburgite, pyroxenite and actino-tremolitite to hornblendite.

Detrital zircon populations from typical paragneiss samples in both the Manampotsy and Bealanana groups record Neoproterozoic age peaks of 840–780 Ma, with a lead-loss event indicating metamorphism at *c.* 560–510 Ma (BGS-USGS-GLW 2008; Table 2). Older detrital zircons are found in samples from the craton-marginal sequences. Samples from the Ampasary Group therefore contain a range of detrital zircons from Meso- and Neoarchaean sources as well as 2000–1800 Ma and Neoproterozoic (800–780 Ma) grains (Collins *et al.* 2003; BGS-USGS-GLW 2008). Similarly, Ambatolampy Group samples have major detrital peaks at *c.* 2700 and 2500 Ma (BGS-USGS-GLW 2008). These datasets suggest that the original sediments of the Ampasary and Ambatolampy groups were deposited on the underlying (marginal) parts of the Masora and Antananarivo cratons, respectively, while the Manampotsy and Bealanana groups were most likely deposited distal from either of the cratonic domains. A contrary view is presented by Tucker *et al.* (2010).

The Manampotsy Belt also comprises Neoproterozoic magmatic rocks that belong to the Imorona–Itsindro Suite. Rapid unroofing of these intrusions may have provided source material for the spatially associated metasedimentary rocks (indicated by their detrital zircon populations). The Imorona–Itsindro Suite is not present in the Anaboriana Belt, which contains large volumes of post-collisional granitoids of the Maevarano Suite dated between *c.* 540–520 Ma (Goodenough *et al.* 2010). No older, pre-Neoproterozoic magmatic rocks have been found in either the Manampotsy or Anaboriana belts.

The Manampotsy Belt has a strong, curviplanar north–south tectonic grain defined by major polyphase shear zones with an east–west spur separating the Masora and Antongil cratons. It includes the Maroala Deformation Zone around the western perimeter of the Masora Craton (Figs 2, 4 & 6). The western section of the Maroala Deformation Zone is cut by the southern part of the regional Angavo–Ifanadiana Shear Zone. Transposed north–south-oriented planar fabrics, dextral shears and polyphase folds occur in the latter which forms the highest strain tract in the Manampotsy Belt (Raharimahefa 2004).

Geotectonic evolution of central and northern Madagascar

Archaean

The oldest rocks known in Madagascar are the Mesoarchaean orthogneisses of the Antongil and Masora Cratons, known as the Nosy Boraha Suite, and dated at *c.* 3320–3150 Ma. Neoarchaean (about 2500 Ma) granitoid gneisses are a major component of the Antongil Craton. The Maha Group in the Masora Craton contains detrital zircons with ages of *c.* 2500 Ma, providing indirect evidence for rocks of this age in that craton. It has been proposed that the Antongil Craton is a fragment of the Dharwar Craton of western India (Tucker *et al.* 1999a, 2010; Collins & Windley 2002; Collins 2006). The presence of Meso- and Neoarchaean gneisses of similar age in both cratons, common isotopic signatures and their common Palaeoproterozoic igneous activity (Peucat *et al.* 1993; Tucker *et al.* 1999a, 2010; Jayananda *et al.* 2006; Schofield *et al.* in press) provides support for this correlation. The Masora Craton may originally have been part of the Antongil–Dharwar Craton, but has clearly experienced a different Neoproterozoic history.

The Antananarivo Craton and the Tsaratanana Complex are dominated by Neoarchaean gneisses, with no direct evidence for Mesoarchaean magmatism. However, Nd isotope evidence indicates that the Neoarchaean magmas were contaminated by older crustal material, thus raising the possibility that some Mesoarchaean crust may remain to be found in these domains (Tucker *et al.* 1999a, b).

Metasedimentary rocks of probable Neoarchaean age are found in the Antongil Craton (the Mananara Group); the Masora Craton (the Vohilava Group); the Antananarivo Craton (the Vondrozo and Sofia groups); and the Tsaratanana Complex. The sources of the Mananara Group appear to be broadly coeval with the extensive Meso- and Neoarchaean magmatism in the Antongil Craton. Most of the other possible Neoarchaean metasedimentary

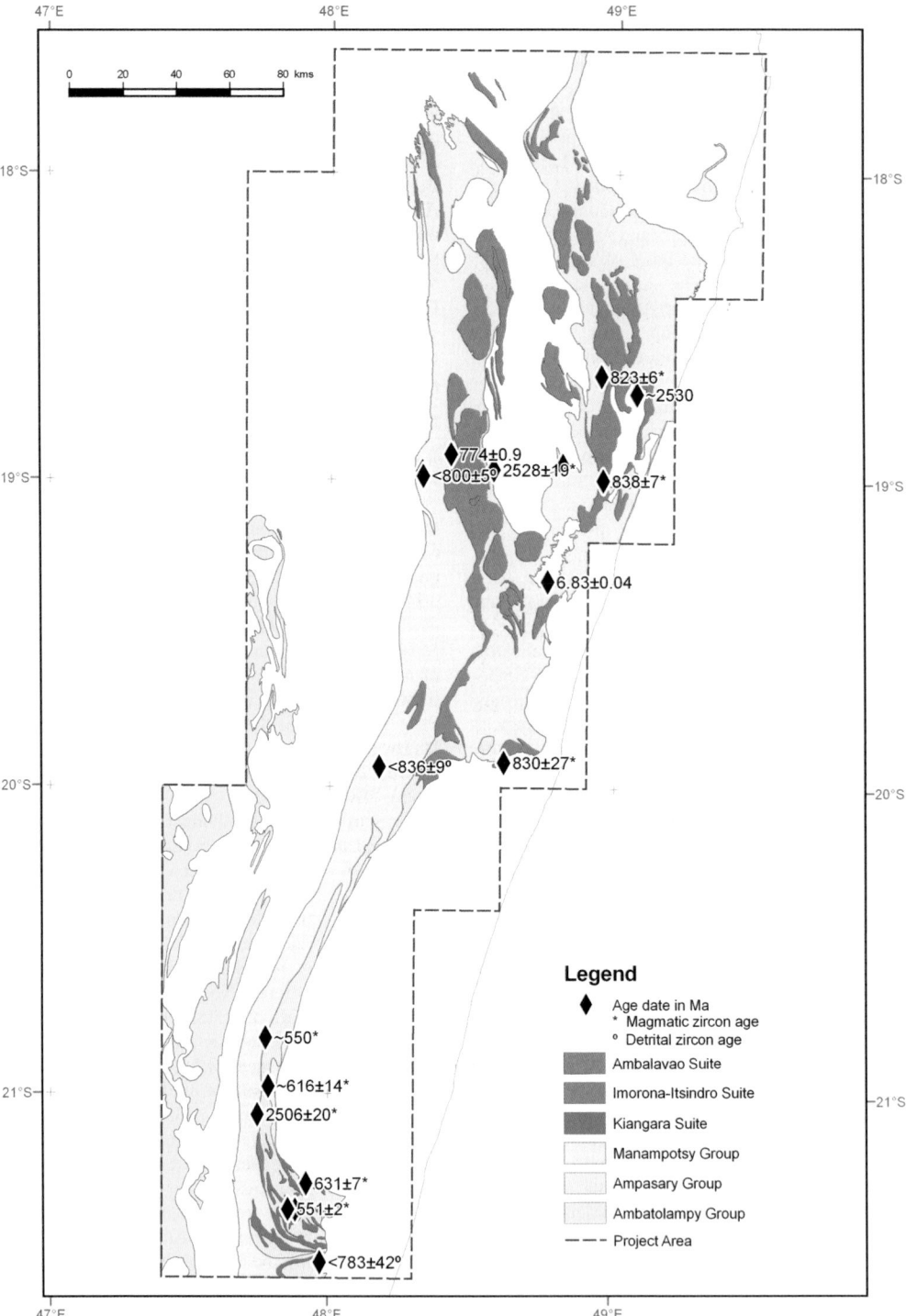

Fig. 6. An outline of the main supracrustal units of the Manampotsy Belt with the locations of isotopically dated rock samples.

sequences have not been investigated for detrital zircons, and so the age of their sources remains uncertain.

Palaeoproterozoic

Palaeoproterozoic magmatic rocks are minor components of the Andriamena Belt (Tsaratanana Complex) and the Antongil Craton (Ankavanana Suite). Mafic dykes of very similar age are also found in the Dharwar Craton (Pandey *et al.* 1997), again supporting previous suggestions that the Antongil and Dharwar Cratons were once linked. However, mafic dyke intrusion of this age is a common feature of many cratons (Ernst & Buchan 2001).

A number of Proterozoic sedimentary sequences are recognized in Madagascar, including the Itremo, Maha and Sahantaha groups, which have evidence for Neoarchaean to Palaeoproterozoic sources with detrital zircon age peaks at *c.* 2500 and 1800 Ma (Cox *et al.* 1998, 2004; Fitzsimons & Hulscher 2005; De Waele *et al.* 2008*a*). The detrital zircon age spectra for the Itremo Group match potential sources in East Africa rather than India, leading to a correlation between the Antananarivo Craton and East Africa (Cox *et al.* 1998; Fitzsimons & Hulscher 2005). Other cratons also contain magmatic rocks dated at *c.* 2500 and 1800 Ma however, so using detrital zircon ages as evidence for provenance is not valid in this case. It could also be argued that the absence of dominant amounts of 2000 Ma detrital zircons from any of the Madagascan metasedimentary sequences would suggest that their source areas were not the African cratons, where igneous rocks of this age are a significant component of the geology (e.g. Hanson 2003; Key *et al.* 2001).

Neoproterozoic–Palaeozoic magmatism and orogenesis

Following a period of apparent quiescence throughout Mesoproterozoic times, renewed tectonothermal activity in Madagascar began *c.* 1000 Ma with the emplacement of granitoids into the Itremo Domain (Tucker *et al.* 2007). The nature of this early magmatism remains uncertain. Subsequently, the voluminous Imorona–Itsindro Suite was emplaced between *c.* 840 and 720 Ma into all crustal domains except the Anaboriana Belt, the Bemarivo Belt and the Antongil Craton. A wide range of rock types from granitic to gabbroic orthogneisses are represented. The suite can be traced for *c.* 450 km along a wide, north–south trending zone in central-northern Madagascar and has been likened to an Andean-type continental magmatic arc on petrological grounds (Handke *et al.* 1999; Kröner *et al.*

2000; Ashwal *et al.* 2002; de Wit 2003). However, more geochemical work is required on this major magmatic suite as it may represent emplacement of plutons in a range of settings along one or more accretionary margins. The Manampotsy Group and associated intrusions may represent a juvenile island arc terrane formed at this time, in a basin between the Antananarivo and Masora cratons (Fig. 7). It has been argued that subsequent collision of the arc system with adjacent continents created the Betsimisaraka Suture Zone (e.g. Cox *et al.* 1998; Collins *et al.* 2000; Kröner *et al.* 2000; Collins & Windley 2002). This scenario was based on the presence of (1) metasedimentary lithologies diagnostic of a marine setting, notably graphitic gneisses; (2) abundant lenses of mafic and ultramafic rocks interpreted as slivers of oceanic crust; (3) Neoproterozoic detrital zircon ages; (4) different Archaean blocks on opposite sides of the belts; (5) meta-igneous rocks of purported supra-subduction origin (Itsindro–Imorona Suite); (6) medial Proterozoic platform sediments on the Antananarivo Craton with detrital zircons of proposed African provenance; and (7) extensive Pan-African overprinting of the Antananarivo Craton but not of the Antongil Craton.

While our work supports many of these arguments it does not support others, and we would question whether the detrital zircon data support an African provenance for the platform sediments on the Antananarivo Craton (see Tucker *et al.* 2010 for a fuller discourse on rebutting the existence of a suture zone). Our work also suggests that the Antananarivo and Masora cratons share a common history prior to terminal Pan-African orogenesis, and that if they collided it took place during an earlier Mid-Neoproterozoic orogenesis (Fig. 7).

The Bemarivo Belt was formed as two juvenile island arc terranes between *c.* 750 and 720 Ma (Tucker *et al.* 1999*b*; Thomas *et al.* 2009). Detrital zircon evidence shows that sediments in these arcs were not sourced from the older rocks of Madagascar, and thus that the arcs are considered to have formed at some distance (Thomas *et al.* 2009).

The Neoproterozoic metasedimentary rocks of the Manampotsy Belt, as well as successions in the Antananarivo Craton and Tsaratanana Complex, record pre-Pan-African structures related to horizontal tectonics and associated high-grade metamorphism. In the Itremo Group, early large-scale recumbent folds may have been formed prior to emplacement of Imorona–Itsindro suite plutons (Collins *et al.* 2003), although this field relationship is disputed by Tucker *et al.* (2007).

Extensional tectonics after *c.* 800 Ma led to the development of new oceanic crust, preserved in the Vohibory volcano-sedimentary succession of southern Madagascar (Collins 2006). This event

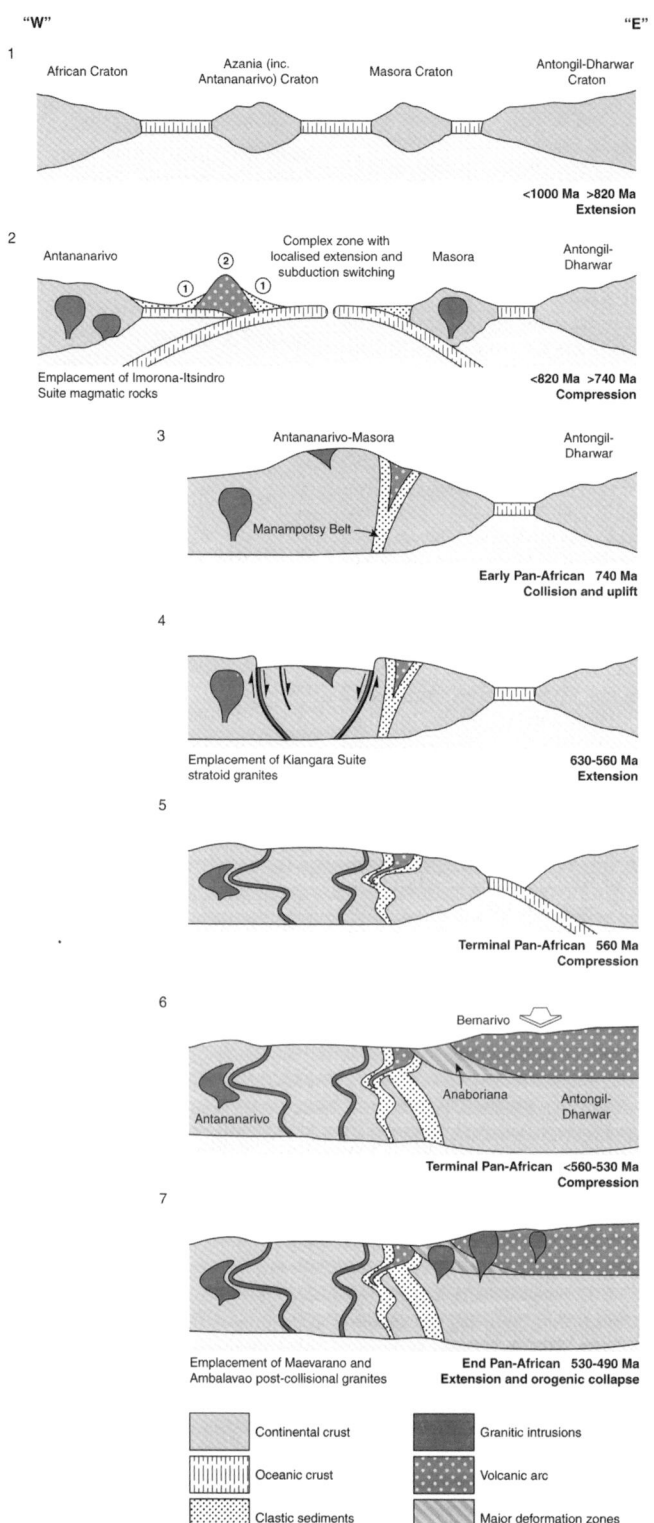

Fig. 7.

has been attributed to the rifting of a microcontinent ('Azania' from East Africa; Collins & Pisarevsky 2005). Subsequent collisional deformation has been recorded in southern Madagascar at c. 650–610 Ma (de Wit et al. 2001). In central Madagascar, the late stages of this event are considered to be manifested by post-collisional alkaline Kiangara Suite granites, which were emplaced into the Antananarivo and Masora Cratons and the Manampotsy Belt between c. 630 and 595 Ma (Nédélec et al. 1995; Paquette & Nédélec 1998). This widespread but volumetrically small suite includes several phases of foliated, typically sheet-like alkaline granites and syenites emplaced within an 800 km long zone north and south of Antananarivo. However, the absence of metamorphic zircon ages in this age bracket (Table 2) in our study implies that there was no collisional event during this period in central and northern Madagascar.

From c. 560 Ma, the major terminal Pan-African collisional event culminated in the amalgamation of the various tectonic blocks of central and northern Madagascar. Table 2 summarizes the inferred ages (and PT conditions) of the Pan-African metamorphism based on U–Pb dates obtained from metamorphic rims to zircon grains (BGS-USGS-GLW 2008). The ages confirm that the terminal Pan-African metamorphism took place over a protracted period between c. 560 and 510 Ma. This long metamorphic period reflects multi-staged collision, subsequent orogenic collapse and high heat flow associated with post-collisional granite emplacement. Estimated PT conditions suggest that amphibolite- to granulite-facies mineral assemblages were prevalent during the Pan-African events (Buchwaldt et al. 2003; Jöns et al. 2005, 2006). An initial high-pressure moderate-temperature metamorphism is localized in the eastern part of the Manampotsy Belt near the tectonic boundaries with the Antongil and Masora cratons and in the Palaeoproterozoic supracrustal rocks (e.g. Maha Group). This metamorphism is considered to be associated with the collision of the Antongil Craton with the previously amalgamated Antananarivo and Masora cratons. No magmatic event is recorded at this time and the emplacement of the Antongil Craton may have been along a major transcurrent ductile shear, rather than by collision across a dipping subduction zone. A later higher temperature metamorphism, probably at lower pressures, caused the commonly seen overprint of kyanite by sillimanite, suggesting a clockwise P–T path (BGS-USGS-GLW 2008). This event is probably associated with the high heat flow caused by the emplacement of the voluminous post-collisional granites and charnockites (Goodenough et al. 2010).

A minimum age for the amalgamation of the Antongil and Antananarivo cratons is given by the subsequent tectonic emplacement of the Bemarivo Belt across the tectonically joined Antananarivo–Antongil cratons and the interleaved Anaboriana Belt. The emplacement of the Bemarivo Belt has been dated by Jöns et al. (2006) from successive monazite rims, which suggest collision at c. 560–530 Ma with (post-collision) peak metamorphic conditions at c. 520–510 Ma. However, Buchwaldt et al. (2003) obtained U–Pb TIMS ages on titanite, monazite and zircon from the southern Bemarivo Belt and suggested that emplacement and metamorphism took place between 520 and 510 Ma, associated with the intrusion of the voluminous post-collisional granitoids of the Maevarano Suite (Goodenough et al. 2010).

The Antongil Craton is least affected by the Pan-African event as fabrics of this age are only present along the western and southern margins (BGS-USGS-GLW 2008; Schofield et al. in press). Metamorphic ages of c. 530 Ma from these margins (Table 2) either date the docking of the Antongil Craton against the older crustal blocks or provide a minimum age for the amalgamation.

Late to post-tectonic, post-collisional granites (c. 550–520 Ma; Tucker et al. 1999b; Kröner et al. 2000; Meert 2003; Collins 2006) are now dated at between c. 537 and 522 Ma (Goodenough et al. 2010). These 'late' granites are potassic in composition and were emplaced into a post-collisional setting, in contrast to all of the older granites that show subduction-related geochemistry.

The final phases of tectono-metamorphic activity associated with the EAAO are shown both by fission track analysis of basement rocks in central Madagascar and by low-temperature shear-controlled hydrothermal mineralization. These data indicate that cooling from the final metamorphic overprint terminated at c. 500 Ma (Emmel et al. 2006) and (emerald) mineralization took place at c. 490 Ma (Moine et al. 2004).

Fig. 7. Schematic representation of the stages recognized in north-eastern Madagascar in the break-up of Rodinia and the eventual formation of Gondwana. Section 1 highlights the early fragmentation of Rodinia into smaller continents. Section 2 shows the development of an arc terrane marginal to the Antananarivo Craton and subduction-related magmatism on the Masora Craton. 1 and 2 in this section refer to sediments and magmatic arc rocks in the Manampotsy Belt, respectively. Sections 3–7 show the various stages in the Pan-African orogeny recorded in central and northern Madagascar with progressive amalgamation of the various continents and arc terranes and successive phases of granitic magmatism.

Discussion and conclusions

Our geological study (BGS-USGS-GLW 2008; De Waele *et al.* 2008*a*; Thomas *et al.* 2009; Goodenough *et al.* 2010; Tucker *et al.* 2010; Schofield *et al.* in press) has confirmed that central and northern Madagascar comprises a number of distinct Archaean domains separated by wide ductile shear zones within the central part of the EAAO. Of the four Archaean domains in north-central Madagascar, the Masora and Antananarivo cratons appear to have different Mesoarchaean histories but to have shared a common history since *c.* 820 Ma (perhaps as early as the Neoarchaean; 2500 Ma). The Antananarivo Craton cannot be matched with an east African craton (e.g. the Tanzania Craton) and we agree with Collins (2006) in interpreting this domain as part of a separate Azania microcontinent within the East African Orogen. The Tsaratanana Complex may originally have formed a continuous sequence that was intruded by Neoarchaean granitic rocks and subsequently tectonically transported *en masse* across the Antananarivo Craton, before segmentation into synformal keels and thrust slices during Pan-African times. Despite this, the similarities in the preserved Archaean histories of the Tsaratanana Complex and tectonically underlying Antananarivo Craton rocks suggests that at least parts of the former were autochthonous parts of the underlying craton.

The Antongil Craton may have a similar Archaean history to the Masora Craton, but it lacks pervasive Neoproterozoic reworking. It therefore probably only docked with the rest of Madagascar during the terminal Pan-African collision. The Bemarivo Belt clearly had a separate Neoproterozoic history as an island arc assemblage which was only amalgamated with the rest of Madagascar in the Late Neoproterozoic to early Cambrian (Thomas *et al.* 2009).

The status of the Anaboriana and Manampotsy Belts as a suture zone (the Betsimisaraka Suture Zone) is not fully resolved (Tucker *et al.* 2010). In support of the suture zone interpretation are (1) the absence of rocks older than Neoproterozoic from the belts; (2) the presence of apparently different Archaean cratons on opposite sides of the Manampotsy Belt with continental margin sediments on its eastern and western margins; (3) the presence of actively eroding Neoproterozoic magmatic arc rocks within the Manampotsy Belt; (4) tectonic slices of mafic/ultramafic rocks interpreted as remnants of oceanic crust; and (5) the 800 Ma detrital zircon population and absence of proximal Archaean crust in the central Manampotsy Group metasediments. Against the interpretation are the similar early Neoproterozoic histories of the Antananarivo and Masora Cratons on either side of the Manampotsy Belt, including the presence of intrusions of the Imorona–Itsindro Suite across the whole area.

We therefore suggest that the Manampotsy Belt first developed as a pre-terminal Pan-African pre-Kiangara Suite collisional/suture zone between the Antananarivo and Masora cratons. Metamorphic fabrics in the Imorona–Itsindro Suite rocks and adjacent metasedimentary rocks that developed during the collisional event are strongly overprinted by terminal Pan-African fabrics associated with the later collision between the Antongil and Masora cratons.

The Antongil Craton only collided with an amalgamated Antananarivo–Masora Craton during the terminal Pan-African. It may have been emplaced along a north–south transcurrent shear that extended into the Manampotsy Belt. Later collision occurred as the Bemarivo Belt was thrust southwards over the older terranes (Thomas *et al.* 2009). The Anaboriana Belt represents the suture between the Bemarivo Belt and the Antananarivo and Antongil cratons. Terminal Pan-African orogenesis in the Manampotsy Belt involved imbrication along the western margin of the Masora Craton in the Maroala Deformation Zone. Post-collisional magmatism (Goodenough *et al.* 2010), retrograde metamorphism and shearing continued until *c.* 490 Ma.

It can be seen that the rocks of northern and central Madagascar record complex, polyphase orogenesis with intervening extensional phases over a period of *c.* 330 million years between *c.* 820 and 490 Ma. This process may have included the creation and subsequent almost complete destruction of oceanic crust that separated the Antananarivo and Masora cratons. Understanding the precise timing of the collisions between these different tectonic blocks partly relies on trying to date specific syn-collision igneous events or tectonic and/or metamorphic fabrics in the suture zones, and less precisely on dating the time when domains on opposite side of a suture start to share a common geological history and were thus 'docked'. In suture zones with polyphase tectono-thermal histories and where the final tectono-thermal event was sufficiently strong to effectively either obliterate or transpose earlier fabrics, it is often difficult to precisely date older structures.

The present study supports and enlarges upon recent studies on the various Neoproterozoic-earliest Palaeozoic orogens of East Africa and Madagascar. These studies describe the partial break-up of Rodinia into smaller continental blocks during the first half of the Neoproterozoic, and the later collision of these blocks to create Gondwana by earliest Phanerozoic times (e.g. Porada & Berhorst 2000; Grantham *et al.* 2003; Johnson *et al.* 2007; De Waele *et al.* 2008*b*; Bingen *et al.* 2009; Thomas *et al.* 2009; Wendorff & Key 2009). The break-up of Rodinia created a number of smaller continental blocks bounded by active and passive margins. A major magmatic arc formed after *c.* 840 Ma to create an uplifted region that was eroded to feed sediments within the Manampotsy Belt and along the margins of the Antananarivo and Masora cratons. The period between *c.* 740 and 630 Ma involved the initial collision between the Antananarivo and Masora cratons with the development of compressional structures in the adjacent Manampotsy Belt. Subsequent extension was accompanied by the emplacement of magmatic rocks (of the Kiangara Suite) that cut across structures formed during the earlier collisions. The global terminal Pan-African orogenic event (560–490 Ma) was common to all the Neoproterozoic orogens and culminated with the amalgamation of all the disparate terrains into Gondwana. The Neoproterozoic Era has the unique distinction of recording the break-up of one supercontinent and the creation of a second; both supercontinents formed by contemporaneous orogenesis in a network of linear orogens. The driving force for the synchronicity of global orogenesis is not known.

The paper is published with the permissions of the Executive Director of the British Geological Survey (NERC) and the Director of the USGS. Some age data in this paper were obtained at the Perth Consortium SHRIMP facilities at the Curtin University of Technology, which are funded by the Australian Research Council. We would also like to thank R. Tucker for his significant input into our project. His different interpretation of the 'Betsimisaraka Suture Zone' as a deformed, Neoproterozoic intracratonic sequence is presented in a separate paper (Tucker *et al.* 2010). The authors would like to thank Professor B. Windley and the other reviewer for their helpful comments on the original manuscript.

References

ASHWAL, L. D., DEMAIFFE, D. & TORSVIK, T. H. 2002. Petrogenesis of Neoproterozoic granitoids and related rocks from the Seychelles: the case for an Andean-type arc origin. *Journal of Petrology*, **43**, 45–83.

BESAIRIE, H. 1968–71. Description géologique du massif ancien de Madagascar. *Document Bureau Géologique Madagascar*, **177**. Bureau Géologique Madagascar, Antananarivo.

BGS-USGS-GLW. 2008. Revision de la cartographie géologique et minière des zones Nord, Centre et Centre-Est de Madagascar. *BGS Report* **CR/08/078**. Keyworth, England.

BINGEN, B., JACOBS, J. ET AL. 2009. Geochronology of the Precambrian crust in the Mozambique belt in NE Mozambique, and implications for Gondwana assembly. *Precambrian Research*, **170**, 231–255.

BOGER, S. D. & MILLER, J. M. 2004. Terminal suturing of Gondwana and the onset of the Ross-Delamerian Orogeny: the cause and effect of an Early Cambrian reconfiguration of plate motions. *Earth and Planetary Science Letters*, **219**, 35–48.

BUCHWALDT, R.,, TUCKER, R. D. & DYMEK, R. F. 2003. Geothermobarometry and U-Pb Geochronology of metapelitic granulites and pelitic migmatites from the Lokoho region, Northern Madagascar. *American Mineralogist*, **88**, 1753–1768.

COLLINS, A. S. 2006. Madagascar and the amalgamation of central Gondwana. *Gondwana Research*, **9**, 3–16.

COLLINS, A. S. & WINDLEY, B. F. 2002. The tectonic evolution of central and northern Madagascar and its place in the final assembly of Gondwana. *Journal of Geology*, **110**, 325–340.

COLLINS, A. S. & PISAREVSKY, S. A. 2005. Amalgamating eastern Gondwana: the evolution of the Circum-Indian Orogens. *Earth Science Reviews*, **71**, 229–270.

COLLINS, A. S., RAZAKAMANANA, T. & WINDLEY, B. F. 2000. Neoproterozoic extensional detachment in central Madagascar: implications for the collapse of the East African Orogen. *Geological Magazine*, **137**, 39–51.

COLLINS, A. S., FITZSIMONS, I. C. W., KINNY, P. D., BREWER, T. S., WINDLEY, B. F., KRÖNER, A. & RAZAKAMANANA, T. 2001. The Archaean rocks of central Madagascar: their place in Gondwana. *In*: CASSIDY, K. F., DUNPHY, J. M. & VANKRANENDONK, M. J. (eds) *Fourth International Archaean Symposium 2001, extended Abstracts*. AGSO-Geoscience Australia, Record 2001/37, 294–296.

COLLINS, A. S., FITZSIMONS, I. C. W., HULSCHER, B. & RAZAKAMANANA, T. 2003. Structure of the eastern margin of the East African Orogen in central Madagascar. *Precambrian Research*, **123**, 111–133.

COX, R., ARMSTRONG, R. A. & ASHWAL, L. D. 1998. Sedimentology, geochronology and provenance of the Proterozoic Itremo Group, central Madagascar, and implications for the pre-Gondwana Palaeogeography. *Journal of the Geological Society, London*, **155**, 1009–1024.

COX, R., COLEMAN, D. S., CHOKEL, C. B., DE OREO, S. B., COLLINS, A. S., KRÖNER, A. & DE WAELE, B. 2004. Proterozoic tectonostratigraphy and paleogeography of central Madagascar derived from detrital zircon U-Pb age populations. *Journal of Geology*, **112**, 379–400.

DALZIEL, I. W. D. 1997. Neoproterozoic-Paleozoic geography and tectonics: review, hypothesis and environmental speculation. *Bulletin of the Geological Society of America*, **109**, 16–42.

DE WAELE, B., HORSTWOOD, M. S. A. ET AL. 2008a. U-Pb detrital zircon geochronological provenance patterns of supracrustal successions in central and northern Madagascar. *In*: *Colloquium of African Geology, Hammamat, Tunisia*, 3.

DE WAELE, B., JOHNSON, S. P. & PISAREVSKY, S. A. 2008b. Palaeoproterozoic to Neoproterozoic growth and evolution of the eastern Congo Craton: its role in the Rodinia puzzle. *Precambrian Research*, **160**, 127–141.

DE WIT, M. J. 2003. Madagascar: heads it's a continent, tails it's an island. *Annual Review of Earth and Planetary Sciences*, **31**, 213–248.

DE WIT, M. J., BOWRING, S. A., ASHWAL, L. D., RANDRIANASOLO, L. G., MOREL, V. P. I. & RAMBELOSON, R. A.

2001. Age and tectonic evolution of Neoproterozoic ductile shear zones in southwestern Madagascar, with implications for Gondwana studies. *Tectonics*, **20**, 1–45.

EMMEL, B., JACOBS, J., KASTOWSKI, G. & GRASER, G. 2006. Phanerozoic upper crustal tectono-thermal development of basement rocks from central Madagascar: an integrated fission-track and structural study. *Tectonophysics*, **412**, 61–86.

ERNST, R. E. & BUCHAN, K. I. (eds) 2001. *Mantle Plumes: their identification through time*. Geological Society of America, Special Papers, **352**.

FITZSIMONS, I. C. W. & HULSCHER, B. 2005. Out of Africa: detrital zircon provenance of central Madagascar and Neoproterozoic terrane transfer across the Mozambique Ocean. *Terra Nova*, **17**, 224–235.

GASQUET, D., ENNIH, N., LIEGEOIS, J.-P., SOULAIMANI, A. & MICHARD, A. 2008. The Pan-African Belt. Continental Evolution. *The Geology of Morocco*. Springer, Berlin. Lecture Notes in Earth Sciences, **116**, 33–64.

GONÇALVES, P., NICOLLET, C. & LARDEAUX, J.-M. 2003. Finite strain patterns in Andriamena unit (north-central Madagascar): evidence for late Neoproterozoic-Cambrian thrusting during continental convergence. *Precambrian Research*, **123**, 135–157.

GOODENOUGH, K. M., THOMAS, R. J. *ET AL.* 2010. Post-collisional magmatism in the central East African Orogen: the Maevarano Suite of north Madagascar. *Lithos*, **116**, 18–34.

GRANTHAM, G. H., MABOKO, M. & EGLINGTON, B. M. 2003. A review of the evolution of the Mozambique Belt and implications for the amalgamation and dispersal of Rodinia and Gondwana. *In*: YOSHIDA, M., WINDLEY, B. F. & DASGUPTA, S. (eds) *Proterozoic East Gondwana: Supercontinent Assembly and Breakup*. Geological Society, London, Special Publications, **206**, 401–425.

GRÉGOIRE, V., NÉDÉLEC, A., MONIÉ, P., MONTEL, J.-M., GANNE, J. & RALISON, B. 2009. Structural reworking and heat transfer related to the late-Panafrican Angavo Shear Zone of Madagascar. *Tectonics*, **477**, 197–216.

GUÉRROT, C., COCHERIE, A. & OHNENSTETTER, M. 1993. Origin and evolution of the west Andriamena Pan African mafic-ultramafic complexes in Madagascar as shown by U-Pb, Nd isotopes and trace elements constraints. *Terra Abstracts*, **5**, 387.

HANDKE, M., TUCKER, R. D. & ASHWAL, L. D. 1999. Neoproterozoic continental arc magmatism in west-central Madagascar. *Geology*, **27**, 351–354.

HANSON, R. E. 2003. Proterozoic geochronology and tectonic evolution of southern Africa. *In*: YOSHIDA, M., WINDLEY, B. F. & DASGUPTA, S. (eds) *Proterozoic East Gondwana: Supercontinent Assembly and Breakup*. Geological Society, London, Special Publications, **206**, 427–463.

HOTTIN, G. 1969. Les terrains cristallins du centre-nord et du nord-est de Madagascar. *Document du Bureau Géologique de Madagascar*, **178**.

HOTTIN, G. 1976. Présentation et essai d'interprétation du Précambrien de Madagascar. *Bulletin du Bureau de Recherches Géologiques et Minières*, **IV** (2nd Series), 117–153.

JACOBS, J. & THOMAS, R. J. 2004. Himalayan-type indenter-escape tectonics model for the southern part of the late Neoproterozoic-early Palaeozoic East African-Antarctic orogen. *Geology*, **32**, 721–724.

JACOBS, J., FANNING, C. M., HENJES-KUNST, F., OLESCH, M. & PAECH, H. J. 1998. Continuation of the Mozambique Belt into East Antarctica: Grenville-age metamorphism and polyphase Pan-African high-grade events in Central Dronning Maud Land. *Journal of Geology*, **106**, 385–406.

JACOBS, J., BAUER, W. & FANNING, C. M. 2003. Late Neoproterozoic/Early Palaeozoic events in central Dronning Maud Land and significance for the southern extension of East African Orogen into East Antarctica. *Precambrian Research*, **126**, 27–53.

JAYANANDA, M., CHARDON, D., PEUCAT, J. J. & CAPDEVILA, R. 2006. 2.61 Ga potassic granites and crustal reworking in the western Dharwar Craton, southern India: tectonic, geochronologic and geochemical constraints. *Precambrian Research*, **150**, 1–26.

JOHNSON, S. P., RIVERS, T. & DE WAELE, B. 2005. A Review of the Mesoproterozoic to early Palaeozoic magmatic and tectonothermal history of south-central Africa: implications for Rodinia and Gondwana. *Journal of the Geological Society, London*, **162**, 433–450.

JOHNSON, S. P., DE WAELE, B., EVANS, D., BANDA, W., TEMBO, F., MILTON, J. A. & TANI, K. 2007. Geochronology of the Zambezi supracrustal sequence, southern Zambia; a record of Neoproterozoic divergent processes along the southern margin of the Congo Craton. *The Journal of Geology*, **115**, 355–374.

JÖNS, N., SCHENK, V., APPEL, P. & RAZAKAMANANA, T. 2005. Two-stage metamorphic evolution of the Bemarivo Belt (northern Madagascar): constraints from reaction textures and in situ monazite dating. *In*: *European Geosciences Union 2005, Graz, Geophysical Research Abstracts*, **7**, 2.

JÖNS, N., SCHENK, V., APPEL, P. & RAZAKAMANANA, T. 2006. Two-stage metamorphic evolution of the Bemarivo Belt of northern Madagascar: constraints from reaction textures and in situ monazite dating. *Journal of Metamorphic Geology*, **24**, 329–347.

JOURDE, G. 1972. Essai de synthèse structurale et stratigraphique du Précambrien malagache. *In*: *Comptes Rendus de la semaine géologique 1971*. Comité national malagache de géologie, 59–70.

KABETE, J., GROVES, D., MCNAUGHTON, N. & DUNPHY, J. 2006. The geology, SHRIMP U-Pb geochronology and metallogenic significance of the Ankisatra-Besakay District, Andriamena belt, northern Madagascar. *Journal of African Earth Sciences*, **45**, 87–122.

KENNEDY, W. Q. 1964. The structural differentiation of Africa in the Pan-African (± 500 m.y.) Tectonic Episode. *In*: *Annual Report on Scientific Results of the Research Institute of African Geology*, University of Leeds, **8**, 48–49.

KEY, R. M., LIYUNGU, A. F., NJAMU, F. J., SOMWE, V., BANDA, J., MOSLEY, P. N. & ARMSTRONG, R. A. 2001. The western arm of the Lufilian Arc in NW Zambia and its potential for copper mineralization. *Journal of African Earth Sciences*, **33**, 503–528.

KRÖNER, A., WINDLEY, B. F., JAECKEL, P., BREWER, T. S. & RAZAKAMANANA, T. 1999a. Precambrian granites, gneisses and granulites from Madagascar: new zircon ages and regional significance for the evolution of the Pan-African orogen. *Journal of the Geological Society, London*, **156**, 1125–1135.

KRÖNER, A., WINDLEY, B. F., JAECKEL, P., COLLINS, A. S., BREWER, T. S., NEMCHIN, A. & AZAKAMANANA, T. 1999*b*. New zircon ages for Precambrian granites, gneisses and granulites from central and southern Madagascar: significance for correlations in East Gondwana. *Gondwana Research*, **2**, 351–352.

KRÖNER, A., HEGNER, E., COLLINS, A. S., WINDLEY, B. F., BREWER, T. S., RAZAKAMANANA, T. & PIDGEON, R. T. 2000. Age and magmatic history of the Antananarivo Block, Central Madagascar, as derived from zircon geochronology and Nd isotopic systematics. *American Journal of Science*, **300**, 251–288.

LI, Z. X., BOGDANOVA, S. V. *ET AL.* 2008. Assembly, configuration, and beak-up history of Rodinia: a synthesis. *Precambrian Research*, **160**, 179–210.

MCCOURT, S., HANSON, R. & KEY, R. M. 2006. Mesoproterozoic orogenic belts in southern and central Africa. *Journal of African Earth Sciences*, **46**, v–xi.

MEERT, J. G. 2003. A synopsis of events related to the assembly of eastern Gondwana. *Tectonophysics*, **362**, 1–40.

MEERT, J. G., NÉDÉLEC, A., HALL, C., WINGATE, M. T. D. & RAKOTONDRAZAFY, M. 2001. Paleomagnetism, geochronology and tectonic implications of the Cambrian-age Carion Granite, central Madagascar. *Tectonophysics*, **340**, 1–21.

MICHARD, A., FRIZON DE LAMOTTE, D., SADDIQI, O. & CHALOUAN, A. 2008. An outline of the geology of Morocco. *In: Continental Evolution; The Geology of Morocco.* Lecture Notes in Earth Sciences, **116**. Springer, Berlin, 1–31.

MOINE, B., CHAN PENG, C. & MERCIER, A. 2004. Rôle du fluor dans la formation des gisements d'émeraude de Mananjary (Est de Madagascar). *C.R. Geoscience*, **336**, 513–522.

NÉDÉLEC, A., STEPHENS, W. E. & FALLICK, A. E. 1995. The Panafrican stratoid granites of Madagascar: alkaline magmatism in a post-collisional extensional setting. *Journal of Petrology*, **36**, 1367–1391.

NÉDÉLEC, A., RALISON, B., BOUCHEZ, J. L. & GRÉGOIRE, V. 2000. Structure and metamorphism of the granitic basement around Antananarivo: a key to the Pan-African history of central Madagascar and its Gondwana connections. *Tectonics*, **19**, 997–1020.

PANDEY, B. K., GUPTA, J. N., SARMA, K. J. & SASTRY, C. A. 1997. Sm–Nd, Pb–Pb and Rb–Sr geochronology and petrogenesis of the mafic dyke swarm of Mahbubnagar, South India: implications for Paleoproterozoic crustal evolution of the Eastern Dharwar Craton. *Precambrian Research*, **84**, 181–196.

PAQUETTE, J.-L. & NÉDÉLEC, A. 1998. A new insight into Pan-African tectonics in the East–West Gondwana collision zone by U–Pb zircon dating of granites from central Madagascar. *Earth and Planetary Science Letters*, **155**, 45–56.

PAQUETTE, J. L., MOINE, B. & RAKOTONDRAZAFY, M. A. F. 2003. ID-TIMS using the step-wise dissolution technique v. ion microprobe U–Pb dating of metamict Archean zircons from NE Madagascar. *Precambrian Research*, **121**, 73–84.

PEUCAT, J. J., MAHABALESHWAR, B. & JAYANANDA, M. 1993. Age of younger tonalitic magmatism and granulite metamorphism in the South India transition zone (Krishnagiri area): comparison with older peninsular gneisses from the Gorur-Hassan area. *Journal of Metamorphic Geology*, **11**, 879–888.

PORADA, H. & BERHORST, V. 2000. Towards a new understanding of the Neoproterozoic-Early Palaeozoic Lufilian and northern Zambezi Belts in Zambia and Democratic Republic of Congo. *Journal of African Earth Sciences*, **30**, 727–771.

RAHARIMAHEFA, T. 2004. *Structure, U-Pb geochronology and geochemistry of the Neoproterozoic crustal scale Angavo shear zone, central Madagascar.* PhD thesis, Saint Louis University, MO 63103, USA.

SCHOFIELD, D. I., THOMAS, R. J. *ET AL.* In press. Geological evolution of the Antongil Craton, NE Madagascar. *Precambrian Research*, **182**, 187–203.

STERN, R. J. 1994. Arc assembly and continental collision in the Neoproterozoic East African Orogen: implications for the consolidation of Gondwanaland. *Reviews of Earth and Planetary Sciences*, **22**, 319–351.

THOMAS, R. J., DE WAELE, B. *ET AL.* 2009. Geological evolution of the Neoproterozoic Bemarivo Belt, northern Madagascar. *Precambrian Research*, **172**, 279–300.

TORSVIK, T. H., SMETHURST, M. A. *ET AL.* 1996. Supercontinent break-up and collision in the Neoproterozoic – a tale of Baltica and Laurentia. *Earth Science Reviews*, **40**, 229–258.

TUCKER, R. D., ASHWAL, L. D.,, HANDKE, M. J. & HAMILTON, M. A. 1997. A geochronologic overview of the Precambrian rocks of Madagascar: a record from the middle Archean to the Late Neoproterozoic. *In*: COX, R. & ASHWAL, L. D. (eds) *Proceedings of the UNESCO-IUGS-IGCP 348/368 International Field Workshop on Proterozoic Geology of Madagascar, Gondwana Research Group.* Rand Afrikaans University Miscellaneous Publication, **5**.

TUCKER, R. D., ASHWAL, L. D., HANDKE, M. J., HAMILTON, M. A., LE GRANGE, M. & RAMBELOSON, R. A. 1999*a*. U-Pb geochronology and isotope geochemistry of the Archean and Proterozoic rocks of north-central Madagascar. *Journal of Geology*, **107**, 135–153.

TUCKER, R. D., ASHWAL, L. D., HAMILTON, M. A., TORSVIK, T. H. & CARTER, L. M. 1999*b*. Neoproterozoic silicic magmatism of northern Madagascar, Seychelles, and NW India: clues to Rodinia's assembly and dispersal. *Geological Society of America, Abstracts with Programs*, **31**, 317.

TUCKER, R. D., KUSKY, T. M., BUCHWALDT, R. & HANDKE, M. J. 2007. Neoproterozoic nappes and superposed folding of the Itremo Group, west-central Madagascar. *Gondwana Research*, **12**, 356–379.

TUCKER, R. D., ROIG, J.-Y. *ET AL.* 2010. Neoproterozoic extension in the Greater Dharwar Craton: a reevaluation of the 'Betsimisaraka Suture' in Madagascar. *Canadian Journal of Earth Science*, **48**, 389–417.

UNRUG, R. 1998. Rodinia to Gondwana: the geodynamic map of Gondwana supercontinent assembly. *Journal of African Earth Sciences*, **26**, I–IX.

VACHETTE, M. & HOTTIN, G. 1970. Age au strontium des granites d'Antongil et del'Androna (Nord-Est et Centre-Nord de Madagascar). *Comptes Rendus Semaine Géologie de Madagascar*, 73–76.

WENDORFF, M. & KEY, R. M. 2009. The relevance of the sedimentary history of the *grand gonglomerat* formation (Central Africa) to the interpretation of the climate during a major Cryogenian glacial event. *Precambrian Research*, **172**, 127–142.

WINDLEY, B. F., RAZAFINIPARANY, A., RAZAKAMANANA, T. & ACKERMAND, D. 1994. Tectonic framework of the Precambrian of Madagascar and its Gondwana connections – a review and reappraisal. *Geologische Rundschau*, **83**, 642–659.

Tectonosedimentary expressions of the evolution of the Fungurume foreland basin in the Lufilian Arc, Neoproterozoic–Lower Palaeozoic, Central Africa

MAREK WENDORFF

AGH University of Science & Technology, Faculty of Geology, Geophysics and Environmental Protection, al. A. Mickiewicza 30, 30-059 Krakow, Poland (e-mail: wendorff@agh.edu.pl)

Abstract: The Lufilian Arc is a part of the Neoproterozoic–Lower Palaeozoic Pan-African orogenic system within southern and central Africa. The succession of the Lufilian orogen, the Katanga Supergroup, contains large bodies of fragmental rocks recently interpreted as syntectonic conglomerates, which reveal the existence of two previously unrecognized basins and shed new light on tectonic evolution of the belt. Rifting between the Congo Craton in the north and the Kalahari Craton in the south at *c.* 880 Ma resulted in the opening of two rift basins: the Roan rift and the succeeding Nguba rift. During post-735 Ma orogenesis, north-advancing nappes supplied detritus into the Fungurume foreland basin in the northern part of the Lufilian Arc. The coarse-clastic sequence of the Fungurume Group includes olistostromes that contain olistoliths of the pre-existing Katangan rocks, rest upon a syntectonic unconformity and are overridden by the Katangan nappes/thrust sheets. Strong tectonic deformations of strata within the olistoliths reflect their provenance from the orogenic source of the Katangan nappes. By contrast, the olistostrome matrix is essentially intact even when olistostrome occurs as a part of a tight fold. These structural relations suggest that nappe overthrusting and further deformation of the foreland occurred soon after deposition of the olistostrome sediments, and prior to their lithification.

The Neoproterozoic–Lower Palaeozoic Katanga Supergroup is exposed in the Lufilian Arc, a segment of the continental system of Pan-African orogenic belts of Africa, and forms a less-deformed to undeformed plateau sequence in the foreland region on the Congo Craton (Fig. 1). Part of the Katanga Supergroup succession hosts the famous Cu and Cu–Co deposits of the Central African Copperbelt. The evolution of the Lufilian orogenic belt involves several tectonic stages from rifting to orogenesis. The successions of the Katangan rifts were folded during the orogenic closure between the Kalahari and Congo cratons and thrust northwards to form a foreland fold-thrust belt in the northern part of the Lufilian Arc (Fig. 2). Inversion from rifting to the first orogenic contractional movements, which resulted in the first foreland basin, affected the southern part of the Katangan system of the Roan and Nguba rifts (Fig. 3a) and occurred between the *Grand Conglomerat* and the *Petit Conglomerat* glacial periods (Wendorff 2005*a*, *c*; Wendorff & Key 2009). It has been traditionally considered that large bodies of fragmental rocks called 'Katangan megabreccias' prominently present in the Lufilian fold-thrust belt in the Democratic Republic of the Congo (DRC) (Fig. 2) originated as tectonic friction breccias due to fragmentation of the Katangan rocks underneath the advancing nappes (Cailteux &

Kampunzu 1995; Wendorff 2003; see revision in Wendorff 2005*c*). Jackson *et al.* (2003) reinterpreted the 'Katangan megabreccias' in terms of salt tectonics. However, Wendorff (2000, 2003, 2005*b*, *c*) provided several lines of evidence to show that the 'megabreccias' are syntectonic sedimentary bodies sourced from uplifted source regions composed of Katangan rocks. Recognition of the sedimentary origin of the 'megabreccia' complexes enabled a radical reinterpretation of the genetic stratigraphy of the Katanga Supergroup and identification of two previously unknown basins within the Lufilian Arc. These are the Nguba rift basin and the foreland basin, in which the newly identified Fungurume Group sediments were deposited in the north of the region (Fig. 2). The identification of these basins and elements of their successions provides new constraints on the tectonic evolution of the Lufilian belt (Wendorff 2000, 2003, 2005*b*, *c*; Wendorff & Key 2009).

The aim of this paper is to integrate recent observations at selected localities in the Fungurume foreland basin in the DRC sector of the Lufilian Arc with earlier, relatively general, interpretations of synorogenic sedimentation (Wendorff 2003, 2005*a*), and to consider their implications for the tectonic evolution of the region at a much more detailed scale.

From: VAN HINSBERGEN, D. J. J., BUITER, S. J. H., TORSVIK, T. H., GAINA, C. & WEBB, S. J. (eds) *The Formation and Evolution of Africa: A Synopsis of 3.8 Ga of Earth History*. Geological Society, London, Special Publications, **357**, 69–83. DOI: 10.1144/SP357.5 0305-8719/11/$15.00 © The Geological Society of London 2011.

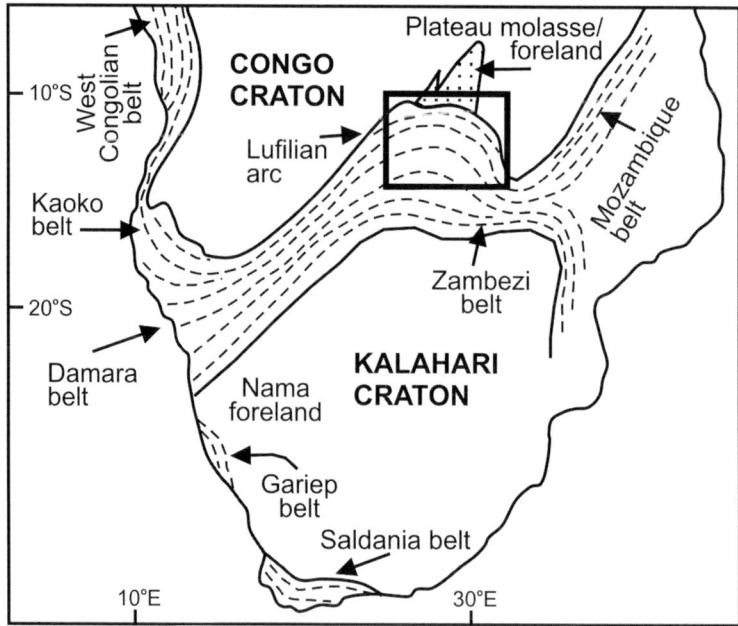

Fig. 1. Position of the Lufilian arc among the Pan-African belts of central and southern Africa. Box outlines the area shown in Figure 2.

Stratigraphy and geodynamic history of the Katanga Supergroup

Following the recently revised stratigraphy (Fig. 3), the Katanga Supergroup is subdivided into the following five groups: Roan, Nguba, Kundelungu, Fungurume and Plateau (Wendorff 2003, 2005a, b). Following a new practice (Master et al. 2005; Batumike et al. 2007; Cailteux et al. 2007), the spelling 'Guba Group' proposed by Wendorff (2003) was replaced by 'Nguba Group' (Wendorff & Key 2009). Two rifting stages resulting from early Neoproterozoic extension of Rodinia are recorded by the Roan Group and the succeeding Nguba Group. The Kundelungu, Fungurume and Plateau Groups were deposited in the succeeding foreland basins related to Pan-African orogenesis, which was diachronous from south to north across the region. The Grand Conglomerat records syn-rift glaciation correlated with the global Sturtian glacial episode (Master & Wendorff 2010). However, it should be noted that this correlation is not certain anymore. The Grand Conglomerat is bracketed in age between 765 and 735 Ma (Key et al. 2002), whereas other sequences correlated with the Sturtian glaciation have ages close to 710 Ma. On the other hand, Halverson et al. (2010) suggested that there was an earlier 'Kaigas' glaciation which precedes the Sturtian glaciation by c. 40–50 Ma. This implies that the Grand Conglomerat should be considered as Kaigas-correlative. The succeeding Petit Conglomerat glaciogenic unit, correlated with the Marinoan glaciation, was deposited after the first orogenic stage, which deformed the southern part of the Lufilian belt in what is now Zambia (Wendorff 2005a; Master & Wendorff 2010).

Strata of the Roan Group (Fig. 2) deposited in the first Katangan rift basin (<880 Ma) nonconformably overlie pre-Katangan basement and form a continuous, transgressive succession grading from terrigenous clastic rocks (the siliciclastic unit or the Mindola Subgroup) at the base into a mixed association of siliciclastic and carbonate deposits (the mixed unit or Kitwe Subgroup rocks). These are succeeded by a carbonate platform sequence (the carbonate unit or Bancroft Subgroup rocks) that prograded from the south (Binda 1994).

Prominent uplift (≥765 Ma) in the southern part of the Roan rift basin (now in Zambia) closed deposition of the Roan Group platform carbonates and resulted in opening of the Nguba rift (Wendorff 2005a). Syn-rift olistostromes of the Mufulira Formation, derived from uplifted Roan rocks, were deposited as products of mass-wasting and sediment gravity-flows at the base of the Mwashya Subgroup (Wendorff 2005a, b). Northwards expansion of the Nguba rift beyond the northern margin of the older Roan rift resulted in progradation of the

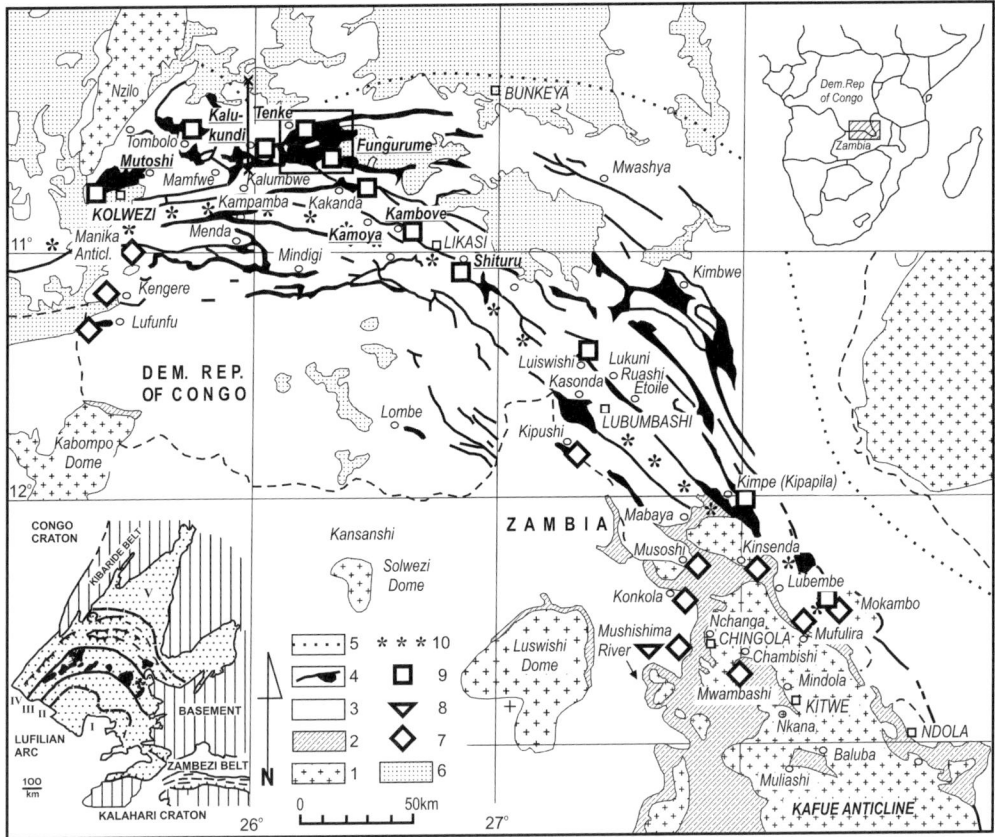

Fig. 2. Northern part of the Lufilian arc in Zambia and the DRC. Map symbols: 1: pre-Katangan basement; 2: Roan Group unconformably overlying basement; 3: undifferentiated post-Roan units of the Katanga Supergroup; 4: undifferentiated complexes of fragmental rocks; 5: boundary between fold-thrust belt and undeformed Plateau Group to north and NE; 6: post-Katangan cover sediments; 7: selected localities within the Mufulira Formation syn-rift olistostromes in Zambia and equivalent complexes in DRC; 8: location of unconformable lower boundary of Petit Conglomerat in Zambia; 9: selected localities within the Fungurume Group foreland succession in the DRC and equivalent complex in Zambia; 10: approximate southern limit of Fungurume Group. Rectangle: position of Tenke-Fungurume map shown in Figure 10. Underlined localities in bold font are described in the text. Lithostratigraphic units are discussed in the text and their stratigraphic position is shown in Figure 3. (After Wendorff 2005c; details 1–3, 5 and 6 modified from Cailteux et al. 1994; inset regional geology of the Lufilian arc modified from Porada & Berhorst 2000.) Structural regions: I: Katanga high; II: synclinorial belt; III: Domes region; IV: external fold-thrust belt; and V: plateau foreland.

olistostromes in the same direction. The overlying deposits of the Mwashya Subgroup are represented by terrigenous siliciclastic rocks, silicified oolitic/pisolitic grainstones, algal dolomites and ironstones of the middle Mwashya. These are succeeded by upper Mwashya shales and siltstones grading upwards to black shales deposited under increasingly anoxic conditions. Basaltic volcaniclastic rocks and lavas form subordinate interbeds in the Mwashya succession but are locally significant. The succeeding glaciogenic sediments of the Grand Conglomerat are overlain by cap carbonates of the Kakontwe Limestone and contain the youngest Katangan occurrences of mafic igneous rocks,

which are interpreted as evidence that rifting continued during the deposition of the Roan and Nguba sediments (Kampunzu et al. 2000; see Hanson 2003 for a review and discussion).

Inversion from an extensional to a compressional tectonic regime was related to the collision of the Congo and Kalahari Cratons during the amalgamation of the Gondwana Supercontinent (Hanson 2003; Master et al. 2005; Wendorff 2005a, b).

The first orogenic episode occurred in the southern Lufilian belt between deposition of the Grand Conglomerat and Petit Conglomerat (Wendorff 2005a), which is suggested by two lines of evidence. The youngest extension-related

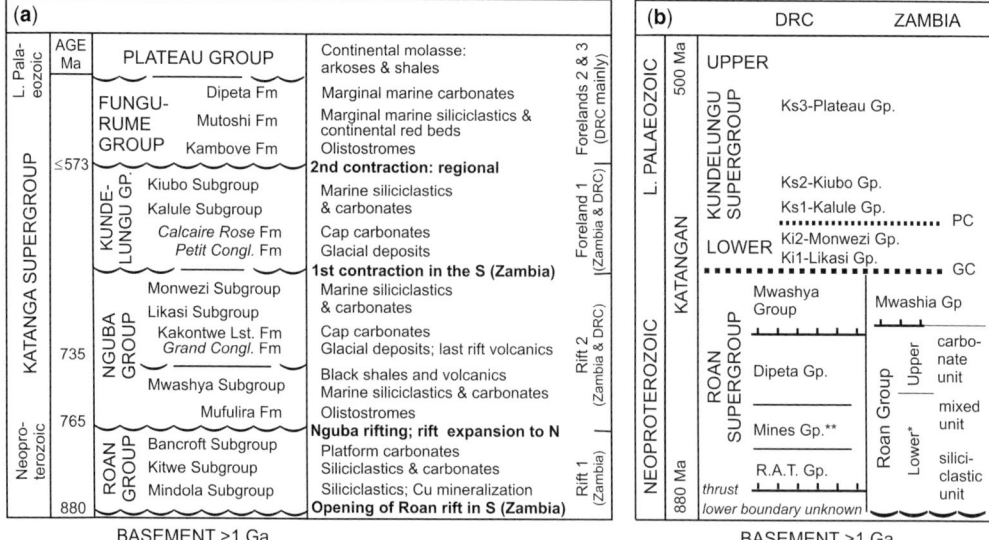

Fig. 3. (**a**) Simplified version of revised stratigraphy of the Katangan succession based upon syntectonic conglomerate complexes and unconformity-bounded megasequences (modified from Wendorff 2003, 2005*a*; Wendorff and Key 2009). Ages from Key *et al.* (2002), Armstrong *et al.* (2005) and Master *et al.* (2005). Wavy lines: major unconformities. (**b**) Traditional/past subdivisions and lithostratigraphic correlation between the DRC and Zambia (simplified from Cailteux *et al.* 1994). Ore horizons: *Cu in Lower Roan in Zambia, **Cu–Co in the Mines Group in the DRC. Glaciogenic horizons GC: Grand Conglomerat; PC: Petit Conglomerat. Note that the Mines Group in the DRC consists of allochthonous megablocks, which resulted from dismemberment of nappes/thrust sheets with the Roan Group strata and gravity-driven emplacement of the Roan megablocks into the foreland basin. The nappes originated south of the orogen (now Zambia) and were thrust towards the foreland region in the north (today's DRC) during the Pan-African orogenesis. Therefore the Mines Group does not exist as an individual stratigraphic unit (Wendorff 2003, 2005*b*), and the megablocks are embedded in the Fungurume Group strata (Fig. 3a).

igneous rocks occur in the Grand Conglomerat (see above). Also, the Petit Conglomerat west of the Kafue Anticline in Zambia rests unconformably on a folded succession of the Roan–Nguba Groups (Wendorff 2005*a*; Wendorff & Key 2009). The Petit Conglomerat defines the base of the Kundelungu Group, which is an infill of the first foreland basin (Wendorff 2005*a*) composed of marine sandy shales, shales and dolomites.

The succeeding Fungurume Group fills the second foreland basin formed in the northern part of the fold-thrust region of the Lufilian belt in the DRC in response to the second orogenic stage (Wendorff 2003, 2005*a*). The lower boundary of the Fungurume Group is unconformable on folded Kundelungu Group strata. Characteristics for this unit are synorogenic conglomerates and megablocks derived from nappes composed of older Katangan rocks uplifted to the south and thrust northwards. The succession comprises sedimentary olistostromes (the Kambove Formation; Fig. 3a), shallow marine and continental red-beds (the Mutoshi Formation) and shallow marine sequences of siliciclastic and carbonate rocks (the Dipeta Formation).

The degree of deformation gradually decreases northwards from the northern sector of the external fold-thrust belt into the overlying Plateau Group. Continental arkoses and shales of the Plateau Group were deposited in the youngest foreland basin, which extended to the north of the Fungurume Group foreland (Figs 2 & 3).

Selected occurrences of the Fungurume Group

Shituru

Background. The strata outcropping in the open pit at Shituru and intersected by borehole cores were described by Lefebvre (1973, 1974) as a Mwashya succession of Cu–Co bearing pure and shaly dolomites containing hematite-enriched beds of pink sandstone, mudstone and ferroan öölitic limestone at the base. Intercalated with the dolomite are layers of tuff, volcano-sedimentary rocks and subordinate interbeds of ironstone and black siliceous öölitic limestone. This succession is underlain by a fragmental rock, which was classified as tectonic breccia containing large blocks of dolomite (Lefebvre 1973, 1974).

Field description. Coarse clastic strata in Shituru occur in the core of a tight, overturned anticline, which forms an inclined fold composed of Nguba strata dipping steeply to the south (Fig. 4). The conglomerate outcrops occur along the anticline axis, both to the SE and NW of the Shituru open pit. The SE occurrence contains a megablock of Cu–Co mineralized Roan dolomite. The rock in the core of the fold exposed in the open pit is massive, polymictic, matrix-supported conglomerate varying from matrix rich to matrix poor. The clasts represent a variety of dolostones ranging from pure and impure crystalline dolomite (similar to the Roan dolostones) to impure pelitic dolomite

and dolomitic shale, together with light grey and pink sandstone and grey, greenish and pinkish siltstone and tuff and mafic rock fragments derived from the Nguba units (including Mwashya strata). Rare fragments of pink dolomite are lithologically similar to the Kakontwe Limestone – the cap carbonate of the Grand Conglomerat glaciogenic complex. The clast size ranges from granules to cobbles (pebble being the prominent class) and the shape varies from subangular to very well rounded (Fig. 5b). Indistinct bedding of the conglomerate is expressed by the presence of crude amalgamated layers differing in maximum clast size, sorting and/or proportion of matrix. Embedded in the matrix are megablocks of Roan dolomite (Fig. 5a).

Except for rare and indistinct shear surfaces that indicate slight NE-oriented movement, the

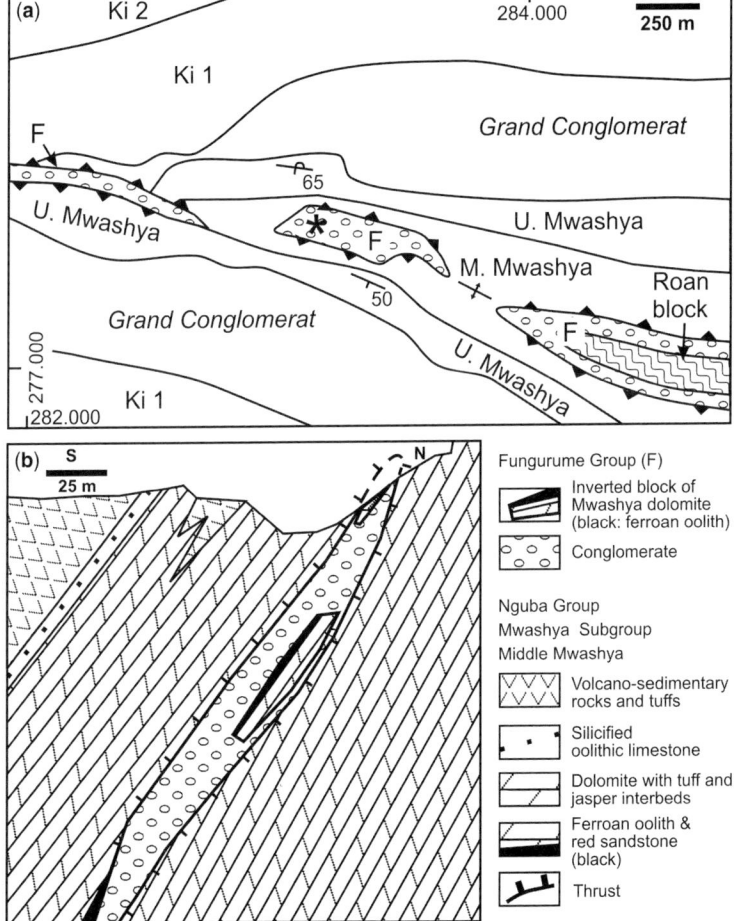

Fig. 4. (a) Geological map of the Shituru area. Note anticlinal structure with Fungurume Group (F) conglomerate in the centre containing megablock of Roan dolomite. (b) South–north cross-section through Shituru anticline through the Shituru open pit (reinterpreted from Lefebvre 1973; no vertical exaggeration; *open pit.).

Fig. 5. Outcrop details at Shituru open pit: (**a**) sheared megablock/olistolith of the Roan dolomite. (**b**) close-up of massive, polymictic, matrix-supported conglomerate in which megablocks/olistoliths are embedded.

conglomerate is tectonically intact. Some pebbles are rotated in the proximity of the shear surfaces, but are not sheared or stretched at all. By contrast, the Roan dolomite megablocks embedded in the conglomerate as well as the dolomite beds of the continuous Mwashya succession near its contact with the conglomerate are all strongly sheared, foliated and locally brecciated (Fig. 5a). The degree of deformation in the Mwashya thrust above the 'breccia' decreases away from the thrust contact; for example, the tuffaceous pebbly sandstone beds *c.* 35 m above are intact.

Interpretation. Crude bedding, disorganized structure, size, shape and provenance of clasts coupled with a lack of any significant tectonic overprint imply that the rock present in the core of the Shituru anticline is not a tectonic breccia but is instead sedimentary in origin. It is a sequence of conglomerate deposited by debris flows of clastic material derived from the uplifted Roan and Nguba strata. The intense deformation of the basal portion of the Mwashya dolomitic succession above is interpreted here as a feature inherited from decollement and thrusting over a competent substratum prior to overriding of the unconsolidated conglomerate. The conglomerate unit was deposited in front of the advancing thrust sheet, which subsequently overrode it. Continuing orogenic movements folded both units jointly, forming the tight anticline. The geometry of this fold is consistent with the regional north vergence of Pan-African folding and thrusting in the Lufilian arc. Some diffuse bed boundaries are still preserved in the conglomerate, which suggests that it was initially unlithified at the time of the thrust sheet emplacement and subsequent folding. The absence of pronounced brittle deformation in the conglomerate suggests it was still prone to plastic deformation.

Folding must also have taken place prior to the final lithification of the conglomerate, therefore soon after thrusting.

The conglomerate unit is interpreted here as an olistostrome, and the derivation of clasts and the genetic reconstruction suggested above imply that it is a part of the Fungurume Group.

Kamoya–Kambove

Background. The coarse clastic complex in the Kamoya–Kambove area (Cailteux 1977) was preliminarily interpreted as an olistostrome by Wendorff (2003). The olistostrome reaches a maximum thickness of *c.* 650 m, contains several olistoliths of the Nguba, Kundelungu and mineralized Roan rocks and is tectonically overridden by an allochthonous sheet of the Nguba strata thrust from the south (Fig. 6a). Stratigraphic relations between Kamoya in the south and Shangulowe in the north, shown in the geological map, point to an unconformity between the jointly folded Nguba and Kundelungu groups and the overlying olistostrome (Fig. 6b). The exploration borehole KYA10 provides an insight into the details of this unconformity and the features of the olistostrome at depth. On the other hand, the outcrops in the Kamoya South open-cast mine west of Kambove show the features of the olistostrome matrix as well as contrasts in the degree of tectonic deformation of olistoliths, the enclosing matrix and the Nguba rocks thrust above. In addition, a cross-section through Kambove East documents the distribution of the olistoliths, the facies trends in the olistostrome and post-depositional deformation of the underlying strata and the overlying olistostrome.

Field description. At depth interval 420–415 m the borehole KYA10 intersects a massive dark

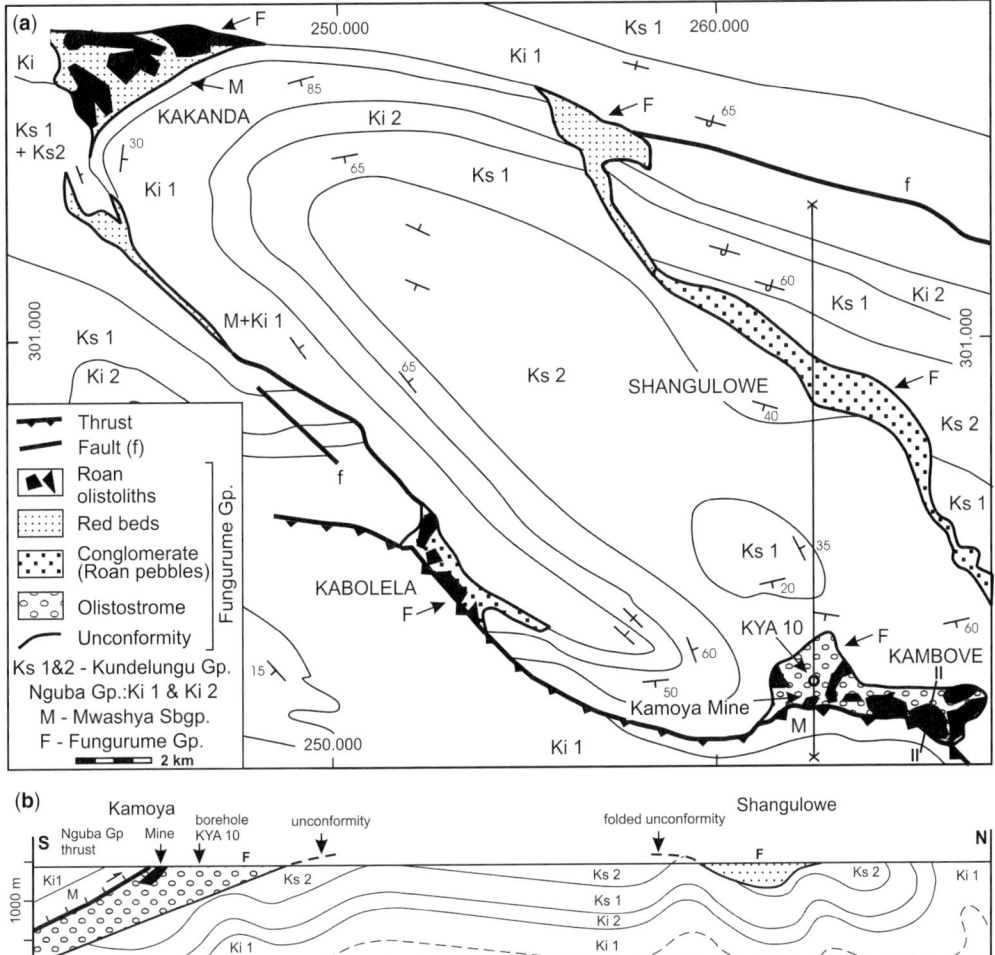

Fig. 6. (**a**) Geological map of Kambove–Kamoya area (reinterpreted from Cailteux 1977). X–X: line of cross-section shown below the map in (b); II–II: line of cross-section shown in Figure 9. (**b**) South–north cross-section showing tectonostratigraphic relations in the area. Note synorogenic unconformity and regional facies gradient of the Fungurume Group strata from coarse clastic olistostrome in the south (Kamoya and Kambove) to sandstone facies in the north (Shangulowe).

mudstone of the Kundelungu Group Ks 2.1 (Fig. 7a). This is overlain with sharp unconformable boundary by thinly bedded basal interval of the Fungurume Group sequence composed of thin beds of light crème fine-grained sandstone and coarse siltstone interlayered with greenish mudstone (Fig. 8a). The sandstone and siltstone beds have sharp lower boundaries and are usually graded into the overlying mudstone, which suggests deposition from low-density turbidity currents. The turbidite strata are overlain along an erosional boundary, sharp and uneven at the borehole scale, by the conglomerate/olistostrome succession. The lowest interval, 1.5 m thick, consists of matrix-supported

monomictic conglomerate, pebbles of which are derived from erosion of the underlying turbidite complex. This is succeeded by a 63.5 m interval of polymictic, massive, matrix-supported conglomerates containing clasts of Katangan carbonate and terrigenous rocks ranging in size from granules to cobbles; some of the clasts are derived from tectonically deformed rocks. The lowest part of this lithology is shown in Figure 8a. The conglomerate is followed with no stratigraphic or tectonic discontinuity by c. 14 m of haematite-rich red sandstone, siltstone and mudstone beds (Fig. 7a), traditionally called 'red RAT'. The red RAT (roches Argilo-Talqueuses), which may also contain interbeds of

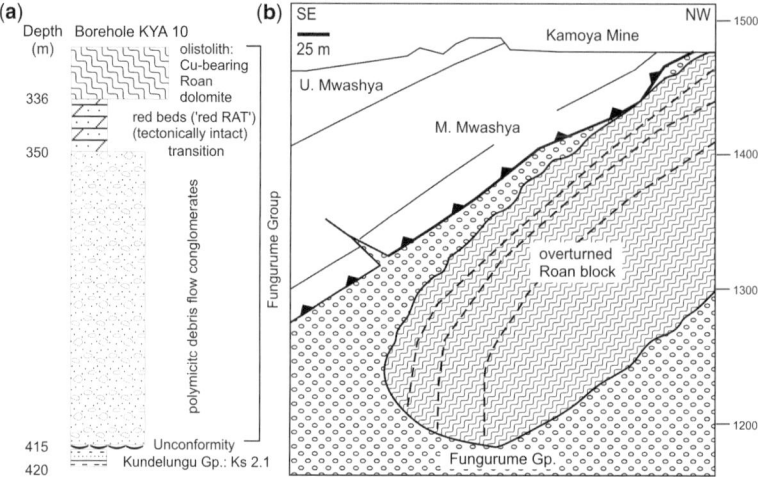

Fig. 7. (**a**) Lithostratigraphic section of the lower part of succession intersected in borehole KYA 10 at Kamoya Mine showing continuous succession of olistostrome (polymictic debris flow conglomerate to dolomitic sandstone) and Roan olistolith (reinterpreted from Cailteux 1983 and based on the present author's own data; compare with details shown in Fig. 8a). (**b**) Cross-section through Kamoya olistostrome with overturned olistolith of the Roan carbonate strata embedded in conglomerate; dashed lines denote bedding (modified from Gecamines 2003).

Fig. 8. (**a**) Oldest part of the succession intersected in borehole KYA 10, interpreted as a coarsening-upwards olistostrome sequence unconformably overlying the Kundelungu strata; this interval is correlative to the basal part of the olistostrome at Kambove, Figure 9. (**b–d**) Kamoya Mine/open pit. Outcrop details: (b) intact dolomite–clast conglomerate, which forms olistostrome matrix; (c) tightly folded Roan dolomite of the olistolith; (d) sheared sole of the Mwashya strata thrust over olistostrome. The strongly sheared rocks (in the lower left) are transitional to the intact strata above, seen in the upper right of the photograph.

carbonate–clast conglomerate, were considered in the past to be stratigraphically followed by the Mines Group (Fig. 3b), a unit considered in the DRC as correlative to the Cu–Co-bearing Roan Group of Zambia (François 1973; Katekesha 1975; Cailteux *et al.* 1994; Wendorff 2005*a*). Above these rocks rests an olistolith of the Roan-derived mineralized strata. No tectonic deformation of the red-bed sediments underlying the olistolith has been observed in the borehole.

In the Kamoya South opencast mine, the conglomerate unit rests upon the Kundelungu Group (unit K2.1), contains a mineralized block of the Roan strata over 200 m long and 100 m thick and is in turn overridden by a sheet of the Mwashya Subgroup thrust from the south (Fig. 7b). An injection of the conglomerate extends upwards into an open joint in the overriding Mwashya slab. The thickness of the conglomerate decreases towards the surface, where a section of this rock (only 2 m thick) crops out between the Roan-derived olistolith and the Mwashya strata thrust above the conglomerate. In outcrop, the conglomerate is polymictic, massive and matrix-supported (Fig. 8b). Similarly to Shituru the conglomerate is an essentially intact rock, affected only by indistinct and rare shear surfaces; the geometry indicates movement towards the north, consistent with the thrusting direction of the Mwashya strata above. The boundary between the conglomerate and the Roan Group olistolith is sharp and uneven, but shows no evidence of deformation. One small injection of conglomerate fills an open joint in the olistolith, extending 50 cm into the megablock. It is apparent in the outcrop that the geometry of the conglomerate–olistolith boundary is defined by an uneven margin of the megablock and locally open fissures in it. The olistolith itself is tightly folded (Fig. 8c), sheared and cut by thrust surfaces. In places, the degree of deformation and brecciation is so high that the rock can be considered a tectonic melange. All these tectonic features terminate abruptly at the boundary of the olistolith with the conglomerate. A strongly foliated dolomitic-talcose tectonic rock at the sole of the thrust slab of Mwahsya strata over the conglomerate is 1.5 m thick and quickly passes upwards into undeformed strata (Fig. 8d).

The north–south section through Kambove East (Fig. 9) shows the lower boundary of the olistostrome unconformable on Ks 2.1 and the tectonic upper boundary marked by a thrust sheet of the Nguba strata, which advanced from the south. The northern termination of the olistostrome (Fig. 6a) is defined by a south-vergent syncline deforming the Kundelungu strata (Ks 2.1), the unconformity and the succeeding olistostrome. This is due to backfolding, which affected the northern regions of the Lufilian arc during the latter stages of

orogenesis (François 1973; Cailteux 1977; François 1995). The olistoliths embedded in the olistostrome are derived from the Nguba and Roan Groups. The largest allochthonous block, referred to as the Kambove syncline, reaches 800 m across and consists of Cu–Co-bearing folded Roan rocks (Cailteux 1977). Large olistoliths cluster in the southern and central sector of the olistostrome but are absent from the northern part, which is composed of carbonate–pebble conglomerate (Wendorff 2003). To the north of Kambove, the pebble conglomerate reappears at Shangulowe and passes into sandstone further to the NW (Fig. 6).

Interpretation. The extremely intense deformation of the Roan megablock in the Kamoya open pit points to a complex tectonic history of the source rock prior to fragmentation and emplacement as an olistolith in the incompetent olistostrome. The deformation at the base of the Mwashya slab thrust above the olistostrome is interpreted as a feature inherited from decollement and thrusting prior to the emplacement of this unit on top of the soft conglomerate. The latter must still have been incompetent at the time of thrusting because it was able to inject upwards into an open fault in the overriding Mwashya block. The turbidites and thin conglomerates together with the conglomeratic lithosome above form a coarsening-upwards succession seen in the KYA 10 borehole, which reflects progradation of the olistostrome. The fine-grained, red-bed-type siliciclastic facies (red RAT lithotype) overlying the conglomerate record a decrease in supply of coarse clastic material from the uplifted orogenic source. The Kambove olistostrome, which contains large olistoliths, grades laterally towards the north and away from the orogenic source into pebble conglomerate and then into sandstone. Such a lateral transition in maximum clast size implies a south–north facies gradient from proximal to distal sediments. This trend is consistent with the regional north-oriented direction of thrusting in the Luflian arc and the related emergence of the orogenic source zone in the south, which supplied coarse clastic material to the foreland basin extending to the north.

Fungurume

Background. In the area between Tenke and Fungurume, megablocks derived from the Roan Group and hosting Cu–Co orebodies are embedded in the lithostratigraphic unit originally classified by Oosterbosch (1950) as the red RAT and the lower part of the succeeding Dipeta (Fig. 10). Oosterbosch (1950) observed that the red RAT strata are hematite-enriched and contain carbonate clasts, which 'were probably derived from a continent

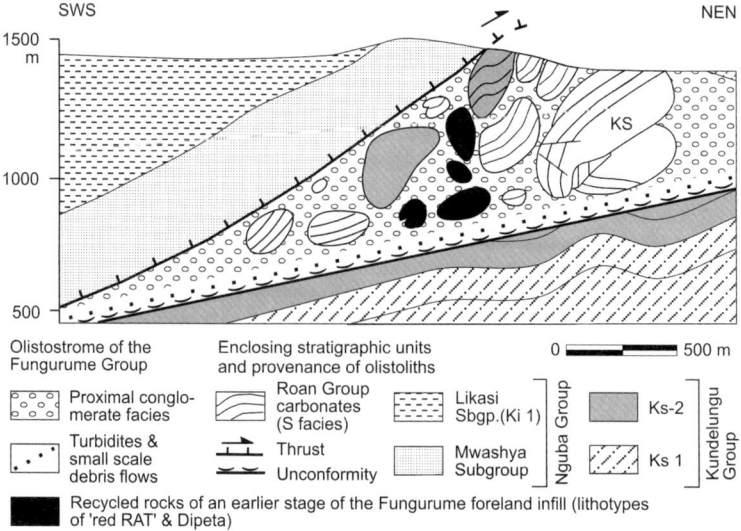

Fig. 9. Cross-section through olistostrome in Kambove (modified from Wendorff 2003). Location II–II is shown in Figure 6a. Note: (i) unconformable lower boundary and Nguba Group strata thrust above the olistostrome; (ii) turbidites and small-scale debris flows underlying the olistostrome (compare with Fig. 8a); (iii) lateral clast-size gradient defined by a cluster of megaolistoliths in the proximal area in the SSW and their absence in the NNE part of the section; this trends continues for several kilometres farther to the north; and (iv) olistoliths derived from RAT and Dipeta and the absence of an unroofing succession, both suggestive of recycling of earlier foreland deposits during northwards propagation of the orogenic front (modified from Wendorff 2003). KS: Kambove Syncline.

composed in large part of carbonate rocks and deposited in a rapidly subsiding basin'. In spite of that, in another paper published a year later (Oosterbosch & Schuiling 1951) he considered these rocks as tectonic breccias. This view has been affirmed and extended even further by Cailteux and Kampunzu (1995), who published a map of the area showing the whole RAT-Dipeta succession as tectonic breccia. Wendorff (2003) reinterpreted this

succession as a continuous synorogenic sedimentary suite, which belongs to the Fungurume Group and rests unconformably upon a jointly folded sequence of the Nguba and Kundelungu groups.

Field description. The Roan-derived mineralized blocks are exposed in the natural slopes of rugged hills and in opencast mines near Fungurume. According to the geological practice of the area,

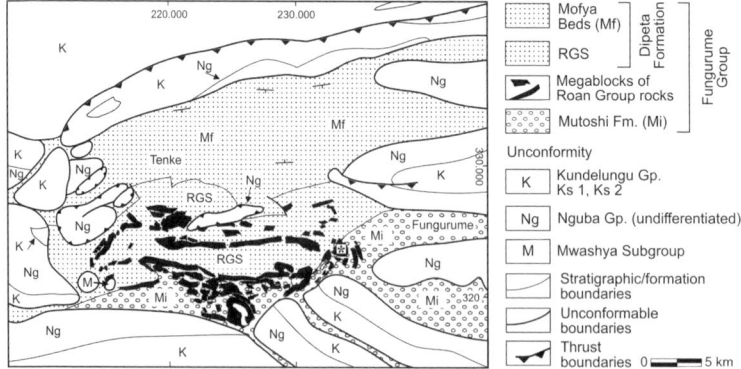

Fig. 10. Geological map of the Tenke–Fungurume area (see Fig. 2). Distribution of units identified by previous authors: RGS: Roches Gréseuse Supérieure (upper sandstone), Mofya Beds and Roan-derived megablocks after Oosterbosch (1950); Mwashya and Kundelungu after Cailteux & Kampunzu (1995). Fungurume Group and revised tectonostratigraphy from Wendorff (2003). Asterisk indicates orebody no. 2 discussed in the text and shown in Figure 11.

the large mined blocks of the Cu–Co mineralized Roan dolomite have been numbered. The conglomerate seen in the outcrop below the megablock labelled as deposit No 2 (Fig. 10) evolves upwards from massive, disorganized, matrix- supported pebbly conglomerate to 4 m of bedded, matrix-supported, mostly massive, granule to cobble conglomerate (Fig. 11a). The predominance of granules is characteristic of some beds of medium thickness. Clasts range from subrounded to well rounded. Bed boundaries range from sharp, uneven and deformed by loading to diffuse erosional/amalgamated, and are defined by changes in clast size, sorting and/or matrix content. Two massive conglomerate beds are quickly graded at the top into 2–3 cm thick parallel laminated beds of medium and fine-grained sandstone. Rare interbeds of silty mudstone less than 5 cm thick are discontinuous and laterally wedge out beneath the erosional bases of the overlying conglomerate beds. Coarse pebbles are sometimes loaded into the topmost part of an immediately underlying bed. Synsedimentary deformation of some beds resulted in load structures, whereas expulsions of liquefaction-prone conglomerate material produced water-sediment escape features erosionally truncated by the overlying bed.

The uppermost 1 m of the conglomerate sequence consists of reddish fine pebble conglomerate and sandstone beds with granules and dolomitic cement (Fig. 11b). Lithologically, these rocks represent the lithotype of the red RAT. They contain subangular and poorly rounded pebbles, cobbles and angular tabular bed fragments of the mineralized Roan rocks that occur as the olistolithic

block immediately above (Fig. 11a, b). The conglomerate sequence is affected by slight tectonic deformation. A few traces of shear surfaces indicate a general orientation of compressional stress consistent with the regional north-vergent folding and thrusting; they offset the bedding at a centimetre–decimetre scale. Several small-scale reverse faults result in vertical displacements not exceeding a decimetre. By contrast, thinly bedded and finely laminated Roan dolomitic rocks in the lowest part of the allochthonous Roan megablock (deposit No. 2) above the conglomerate are very strongly deformed. A 70 cm thick interval of tectonic breccia occurs at the base of the olistolith (thrust sole breccia), succeeded upwards by a large shear fold with associated small parasitic folds (Fig. 11a).

Interpretation. The textural and structural features suggest transport of the conglomerate material by erosive debris flows of varying capacity and competence, and rapid deposition of the transported load. Truncation of the water-escape pipe by a scour above indicates that liquefaction occurred as a synsedimentary process, and was not caused by emplacement of the Roan block above the conglomerate complex. Thin mudstone interbeds record fine suspension settling, possibly from the dilute tails of debris flows. Rare cases of poorly developed normal grading of conglomerate to thin sandstone are interpreted as immature turbidity currents also generated by debris flows. The transition from conglomerate- to dolomite-cemented, hematite-enriched granule sandstone and fine-grained conglomerate beds is inferred to have been related to

Fig. 11. Details of Deposit No. 2 at Fungurume (see asterisk in Fig. 10). (**a**) Heavily tectonized megablock/olistolith of the Cu-mineralized Roan strata rests upon tectonically intact, well-bedded conglomerate and sandstone of the Fungurume Group; sb: sole breccia, sf: shear fold; bedding in conglomerate and sandstone shown by thin white lines; dashed lines indicate minor shear surfaces. (**b**) close-up of boundary between the Fungurume Group sandstone and the overlying Roan olistolith; note angular fragments of the Roan dolomite (outlined) embedded in the Fungurume Group sandstone beds (gs: granule sandstone; ps: pebbly sandstone; sd: sandstone).

a decreasing supply of coarse clastic material into this part of the basin. Clasts of the Roan rocks found in the red-bed strata are interpreted here as fragments shed into the basin by the approaching thrust sheet, from which the overlying Roan olisto-lith was detached.

As in the other localities described in this paper, the conglomerate does not show significant tectonic deformation even near the contact with the overlying Roan block. This again suggests that the underlying sediment was still incompetent at the time of emplacement of the megablock. The strong deformation of the Roan block, contrasting with absence of significant deformation in the underlying conglomerate, is inherited from the Roan nappe.

Discussion and conclusions

Classification of the Katangan coarse fragmentites

During the second orogenic event that affected the northern part of the Lufilian Arc, elevated nappes composed of Katangan strata supplied coarse detri-tus to the Fungurume foreland basin ahead of the orogenic front advancing towards the north.

The units at the examined localities, which were previously considered as tectonic friction breccias, are definitely of sedimentary origin. The pre-dominant facies is represented by disorganized matrix-rich conglomerates emplaced by subaqueous debris flows. Megablocks of various Katangan rocks are embedded in this conglomeratic matrix. Turbi-dites with intercalations of small-scale debris-flow beds occur locally, for example, at the base of the Kambove sequence.

The above features bring to mind the classic work by Flores (1959), who defined olistostrome as '. . . a sedimentary body of lithologically or petro-graphically heterogeneous material. An olistos-trome consists of two parts, intimately admixed: a finer-grained matrix, which supports bodies of more coherent material (clasts or olistoliths). It shows no well defined stratification.' Further, '. . . dispersed bodies of harder rock . . . may range in size from pebble to boulder and further up to several cubic km'. In addition, '. . . the olistostrome emplacement must have occurred in an aqueous medium, as shown by the associated turbidity cur-rents and mudflow depositional phenomena observed'. Crude layering or massive appearance, deposition by debris flows, slides, rockfalls and minor turbidity currents as well as intercalations of 'normal' sediments of the basin are among other features diagnostic for olistostromes (Abbate et al. 1981). The sedimentary package of an olistostrome may range from structurally chaotic (e.g. Hsü 1974; Pini 1999; Cieszkowski et al. 2009) to crudely organized as an association of the above facies that form an olistostrome succession (e.g. Hoedemaeker 1973; Abbate et al. 1981; Bailey et al. 1989; Wendorff 2005c).

In the light of the above features, it is concluded that the discussed occurrences of the coarse frag-mental rocks within the fold-thrust belt of the Lufi-lian Arc are clearly olistostrome bodies.

Tectonosedimentary aspects

Pebbles in the Fungurume Group olistostromes range from subrounded to subangular to angular, which reflects varying distances of transportation and abrasion prior to redeposition into the basin.

The olistoliths are very strongly deformed (e.g. at Shituru), which reflects their origin from nappes or thrust sheets with a long history of tectonic trans-port during the main orogenic phase.

By contrast, the enclosing matrix of debris-flow conglomerate does not show any significant internal tectonic deformation. This applies equally to the conglomerates present in the weakly deformed areas of Kambove and Fungurume as well as to the occurrence in the axial zone of the tight anticline at Shituru. These observations imply that the depo-sition of the olistostromes, emplacement of mega-blocks/olistoliths, overriding by the overlying thrust sheet and, in some cases, tight folding of the overlying allochthonous slab occurred as a conti-nuum within a relatively short time interval, prior to lithification of the olistostrome body.

The lateral gradient from megaolistolith-bearing proximal facies at Kambove to pebble conglomerate at Shangulowe that passes into distal sandstone deposits further to the NW (Fig. 6a, b) implies that an orogenic source elevated to the south of Kambove supplied the local thrust-front trough in coarse clastic material.

Apart from olistoliths derived from the Roan and Nguba strata, the Kambove olistostrome contains blocks of rocks that represent lithotypes character-istic for the Fungurume Group foreland suite, that is, the Mutoshi and Dipeta formations (Figs 3a, b & 9). This illustrates recycling of the foreland basin sediments due to their involvement in thrusting.

The geographical location of the Kambove olis-tostrome to the south of the Fungurume area (Fig. 2) where the entire foreland succession escaped uplift and erosion (Fig. 10) is consistent with advancement of the orogenic front from the south. Furthermore, recycling of the synorogenic deposits within the evolving foreland region characterized by locally formed thrust-front troughs may suggest deposition in a series of piggy-back sub-basins (Ori 1984).

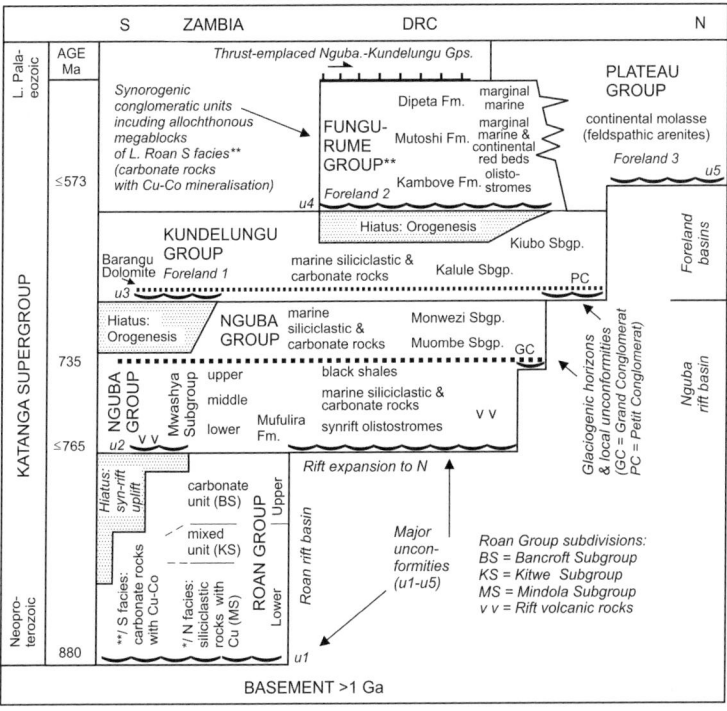

Fig. 12. New stratigraphy of the Katangan basins and successions based upon syntectonic conglomerate complexes, unconformity-bounded sequences and emphasizing the main stages of tectonic evolution (modified from Wendorff, 2005*a*, *c*). Ages from Key *et al.* (2002) and Master *et al.* (2005); other symbols as in Figure 3.

However, this tentative interpretation calls for a more detailed tectonostratigraphic work devoted to temporal relations between tectonics and sedimentation in the Fungurume Group foreland region.

Implications for stratigraphy of the Fungurume Group

The reinterpretation of Katangan megabreccias as sedimentary units revealed previously unrecognized sedimentary basins and tectonic stages of the Lufilian belt evolution, which are reflected in Figure 12. During the second orogenic pulse that affected the northern part of the Lufilian Arc, in response to the load of elevated nappes composed of Katangan strata, the Fungurume Group foreland basin was formed and filled with detritus derived from the north-advancing orogenic front. The foreland sequence evolved from deep marine olistostrome succession (the Kambove Formation) shallowing upwards to detrital marginal marine and continental facies (the Mutoshi Formation) to a mixed association of shallow marine clastic and carbonate deposits (the Dipeta Formation). This

general shallowing-upwards trend is interpreted here as resulting from a gradual decrease of the accommodation space (created by the orogenic load) due to ongoing deposition. The Fungurume Group is characterized by lateral trends of sedimentary facies from proximal in the south to distal in the north. However, this unit also contains solitary interbeds of arkoses that wedge-out towards the south (Intiomale 1982). They were supplied from the Plateau Group continental molasse (Fig. 12) characterized by feldspathic arenites deposited in the basin extending north of the fold-thrust belt and supplied from the adjacent basement. On the other hand, detrital micas in the Plateau Group, which provided an Ar/Ar age of *c.* 575 Ma (Master *et al.* 2005), imply that the Plateau Basin was also fed with detritus derived from erosion of the Katangan thrust sheets. These two relationships indicate that at the late stage of geodynamic evolution of the Lufilian Arc, the Fungurume Basin and the Plateau Basin were connected. Therefore, the Plateau Group deposited in the youngest foreland basin of the Lufilian orogenic belt is late syn- to post-orogenic sequence, partly coeval with the Fungurume Group.

82 M. WENDORFF

Constructive suggestions by the reviewers P. Eriksson, R. Hanson and S. Master facilitated a considerable improvement of the final version of this paper, and are gratefully acknowledged.

References

ABBATE, E., BORTOLOTTI, V. & SAGRI, M. 1981. An approach to olistostrome interpretation. In: RICCI LUCCHI, F. (ed.) Excursion Guidebook with Contributions on Sedimentology of Some Italian Basins. International Association of Sedimentologists, 2nd European Regional Meeting 1981, Bologna, 165–185.

ARMSTRONG, R. A., MASTER, S. & ROBB, L. J. 2005. Geochronology of the Nchanga Granite, and constraints on the maximum age of the Katanga Supergroup, Zambian Copperbelt. In: ROBB, L. J., J., C. & SUTTON, S. J. (eds) Recent Advances in the Geology and Mineralization of the Central African Copperbelt. Journal of African Earth Sciences, 42, 32–40.

BAILEY, R. H., SKEHAN, J. W., DREIER, R. B. & WEBSTER, M. J. 1989. Olistostromes of the Avalonian terrane of Southeastern New England. In: HORTON, J. W. & RAST, N. (eds) Mélanges and Olistostromes of the US Appalachians. Special Paper. The Geological Society of America, Boulder, CO, 228, 93–112.

BATUMIKE, M. J., CAILTEUX, J. L. H. & KAMPUNZU, A. B. 2007. Lithostratigraphy, basin development, base metal deposits, and regional correlations of the Neoproterozoic Nguba and Kundelungu rock successions, central African Copperbelt. Gondwana Research, 11, 432–447.

BINDA, P. L. 1994. Stratigraphy of Zambian Copperbelt orebodies. Journal of African Earth Sciences, 19, 251–264.

CAILTEUX, J. 1977. Particularites stratigraphiques et petrographiques du faisceau inferieur du Groupe des Mines au centre de l'Arc Cuprifère Shabien. Annales de la Société géologique de Belgique, 100, 55–71.

CAILTEUX, J. 1983. Le Roan Shabien dans le région de Kambove (Shaba, Zaire). Etude sédimentologique et métallogénique. PhD thesis, University of Liége, Belgium.

CAILTEUX, J., BINDA, P. L. ET AL. 1994. Lithostratigraphical correlation of the Neoproterozoic Roan Supergroup from Shaba (Zaire) and Zambia, in the central African copper-cobalt metallogenic province. Journal of African Earth Sciences, 19, 265–278.

CAILTEUX, J. & KAMPUNZU, A. B. 1995. The Katangan tectonic breccias in the Shaba Province (Zaire) and their genetic significance. In: WENDORFF, M. & TACK, L. (eds) Late Proterozoic Belts in Central and Southwest Africa. Annales Sciences Géologiques. Musée royal de l'Afrique Centrale, Tervuren, Belgique, 101, 49–62.

CAILTEUX, J. L. H., KAMPUNZU, A. B. & LEROUGE, C. 2007. The Neoproterozoic Mwashya–Kansuki sedimentary rock succession in the central African Copperbelt, its Cu–Co mineralisation, and regional correlations. Gondwana Research, 11, 414–431.

CIESZKOWSKI, M., GOLONKA, J., KROBICKI, M., SLACZKA, A., OSZCZYPKO, N., WASKOWSKA-OLIWA, A. & WENDORFF, M. 2009. The Northern Carpathians plate

tectonic evolutionary stages and origin of olistolites and olistostromes. Geodinamica Acta, 22, 101–126.

FLORES, G. 1959 Evidence of slump phenomena (Olistostromes) in areas of hydrocarbon exploration in Sicily. In: Proceedings 5th World Petroleum Congress, Rome. Sect. 1, paper 13, 259–275.

FRANÇOIS, A. 1973. L'extremité Occidentale de l'Arc Cuprfère Shabien, Etudé Géologique. Gécamines, Lubumbashi.

FRANÇOIS, A. 1995. Problèmes relatifs au Katanguien du Shaba. In: WENDORFF, M. & TACK, L. (eds) Late Proterozoic Belts in Central and Southwest Africa. Annales Sciences Géologiques. Musée Royal de l'Afrique Centrale, Tervuren, Belgique, 101, 1–20.

GECAMINES 2003. Cross-section Kamoya-Sud 2, 1/2.000. Unpublished map, Gécamines, Lubumbashi.

HALVERSON, G. P., WADE, B. P., HURTGEN, M. T. & BAROVICH, K. M. 2010. Neoproterozoic chemostratigraphy. Precambrian Research, 182, 337–350.

HANSON, R. E. 2003. Proterozoic geochronology and tectonic evolution of southern Africa. In: YOSHIDA, M., WINDLEY, B. F. & DASGUPTA, S. (eds) Proterozoic East Gondwana: Supercontinent Assembly and Breakup. Geological Society, London, Special Publications, 206, 427–463.

HOEDEMAEKER, P. J. 1973. Olisthostromes and other delapsional deposits, and their occurrence in the region of Moratalla (Province of Murcia, Spain). Scripta Geologica [Leiden], 19.

HSÜ, K. J. 1974. Melanges and their distinction from olistostromes. In: DOTT, R. H. JR. & SHAVER, R. H. (eds) Modern and Ancient Geosynclinal Sedimentation. Special Publications. Society of Economic Paleontologists and Mineralogists, Tulsa, 19, 321–333.

INTIOMALE, M. M. 1982. Le gisement Zn–Pb–Cu de Kipushi (Shaba, Zaire). Etude Géologique et Métallogénique. PhD thesis, Université Catholique de Louvain.

JACKSON, M. P. A., WARIN, O. N., WOAD, G. M. & HUDEC, M. R. 2003. Neoproterozoic allochthonous salt tectonics during the Lufilian orogeny in the Katangan Copperbelt, central Africa. Geological Society of America Bulletin, 115, 314–330.

KAMPUNZU, A. B., TEMBO, F., MATHEIS, G., KAPENDA, D. & HUNTSMAN-MAPILA, P. 2000. Geochemistry and tectonic setting of mafic igneous units in the Neoproterozoic Katangan basin, Central Africa: implications for Rodinia breakup. Gondwana Research, 3, 125–153.

KATEKESHA, W. M. 1975. Conditions de formation du gisement cupro-cobaltifère de Kamoto Principal (Shaba, Zaire). PhD thesis, University of Liège, Belgium.

KEY, R. M., LIYUNGU, A. K., NJAMU, F. M., SOMWE, V., BANDA, J., MOSLEY, P. M. & ARMSTRONG, R. A. 2002. The western end of the Lufilian arc in NW Zambia and its potential for copper deposits. Journal of African Earth Sciences, 33, 503–528.

LEFEBVRE, J. J. 1973. Identification d'une sédimentation pyroclastique dans le Mwashya inférieur du Shaba méridional (ex-Katanga). Annales de la Société géologique de Belgique, 96, 197–218.

LEFEBVRE, J. J. 1974. Minéralisation cupro-cobaltifère associées aux horizons pyroclastiques situés dans le fasceau supérieur de la Serie de Roan, a Shituru, Shaba, Zaire. In: BARTHOLOMÉ, P. (ed.) Gisements

Stratiformes et Provices Cuprifères. Centenaire de la Société géologique de Belgique, Liege, 103–122.

MASTER, S. & WENDORFF, M. 2010. Neoproterozoic glaciogenic diamictites of the Katanga Supergroup, Central Africa. *In*: ARNAUD, E., HALVERSON, G. P. & SHIELDS-ZHOU, G. (eds) *The Geological Record of Neoproterozoic Glaciations*. The Geological Society, London, Memoirs, **36**, 173–184.

MASTER, S., RAINAUD, C., ARMSTRONG, R. A., PHILLIPS, D. & ROBB, L. J. 2005. Provenance ages of the Neoproterozoic Katanga Supergroup (Central African Copperbelt), with implications for basin evolution. *Journal of African Earth Sciences*, **42**, 41–60.

OOSTERBOSCH, R. 1950. La Série des Mines dans le polygone de Fungurume*Comptes Rendus du Congrès Scientifique, Commemoration du 50ème Anniversaire du Comité Spécial du Katanga. Travaux de la Commission Géographique et Géologique. Comité Spécial du Katanga, Bruxelles*, **2**, 101–118.

OOSTERBOSCH, R. & SCHUILING, H. 1951. Copper mineralization in the Fungurume region, Katanga. *Economic Geology*, **46**, 121–148.

ORI, G. 1984. Sedimentary basins formed and carried piggy back on active thrust sheets. *Geology*, **12**, 475–478.

PINI, G. A. 1999. *Tectonosomes and Olistostromes in the Argille Scagliose of the Northern Apennines, Italy*. Geological Society of America Special Papers, **335**.

PORADA, H. & BERHORST, V. 2000. Towards a new understanding of the Neoproterozoic–Early Palaeozoic Lufilian and northern Zambezi Belts in Zambia and the Democratic Republic of Congo. *Journal of African Earth Sciences*, **30**, 727–771.

WENDORFF, M. 2000. Genetic aspects of the Katangan megabreccias: Neoproterozoic of Central Africa. *Journal of African Earth Sciences*, **30**, 703–715.

WENDORFF, M. 2003. Stratigraphy of the Fungurume Group – evolving foreland basin succession in the Lufilian fold-thrust belt, Neoproterozoic–Lower Palaeozoic, Democratic Republic of Congo. *South African Journal of Geology*, **106**, 17–34.

WENDORFF, M. 2005a. Evolution of Neoproterozoic–Lower Palaeozoic Lufilian arc, Central Africa: a new model based on syntectonic conglomerates. *Journal of the Geological Society, London*, **162**, 5–8.

WENDORFF, M. 2005b. Lithostratigaphy of Neoproterozoic syn-rift sedimentary megabreccia from Mwambashi, Copperbelt of Zambia, and correlation with olistostrome succession from Mufulira. *South African Journal of Geology*, **108**, 505–524.

WENDORFF, M. 2005c. Sedimentary genesis and lithostratigraphy of Neoproterozoic megabreccia from Mufulira, Copperbelt of Zambia. *Journal of African Earth Sciences*, **42**, 61–81.

WENDORFF, M. & KEY, R. M. 2009. The relevance of the sedimentary history of the Grand Conglomerat Formation (Central Africa) to the interpretation of the climate during a major Cryogenian glacial event. *Precambrian Research*, **172**, 127–142.

Tectonic evolution of the Central Steep Zone, Axum area, northern Ethiopia: inferences from magnetic and geochemical data

HELGA DE WALL[1], CARLO DIETL[2], OLGA JUNGMANN[2], ASHENAFI T. TEGENE[1] &
MANOJ K. PANDIT[3]*

[1]*Geozentrum Nordbayern, Universitat Erlangen-Nürnberg, Schlossgarten 5,
D-91054 Erlangen, Germany*

[2]*Institut für Geowissenschaften, Johann Wolfgang Goethe-Universität Frankfurt (Main)*

[3]*Department of Geology, University of Rajasthan, Jaipur – 302004, India*

*Corresponding author (e-mail: manojpandit@gmail.com)

Abstract: Northern Ethiopia is marked by a fanning system of thrust planes with NW-dipping structures in the east and southeast-dipping in the west. The central zone of this large-scale (200 km long) structure is formed by a *c.* 10 km wide zone of localized strain and amphibolites facies metamorphic conditions (680 °C and 3.4 kbar) referred to as the Central Steep Zone (CSZ). The CSZ comprises a mafic rock assemblage of amphibolite, serpentinite showing ocean-floor characteristics and calc-silicate schist. A monzonite intrusion in the central part of the CSZ post-dates the deformation and is related to partial melting of the mafic rocks. Magnetic fabric measurements reveal NE-trending (043°) steep foliations in the CSZ with vertical orientation of lineation, parallel to the axes of micro-folds. This high-strain zone is interpreted as central zone of a positive flower structure on the basis of simultaneous flattening and shear movement, typical for transpressive kinematics. The CSZ has a northern continuation into the Nafka terrane of Eritrea where it can be traced over a distance of 200 km. This high-strain belt forms a major structure in the context of Arabian–Nubian Shield (ANS) collision tectonics during the closure of the Mozambique Ocean and assembly of Gondwana.

High-strain zones are of particular interest in understanding dynamics and kinematics of tectonic events as they provide essential information on P-T conditions, deformation geometry (stress and strain) and timing of the crustal deformation (e.g. Alsop & Holdsworth 2004). The Arabian–Nubian Shield (ANS) offers an appropriate setting to undertake such studies as it archives a complex evolutionary history of long-lived accretion and repetitive magmatism during the Neoproterozoic (900–550 Ma). The ANS is also vital in understanding the evolution of the East African Orogen during the collision between east and west Gondwana blocks.

The ANS comprises a Pan-African basement, which is characterized by zones of localized deformation and higher-grade metamorphism within a milieu of low-grade volcano-sedimentary rock units. This especially the case in the Eastern Desert of Egypt (Miller & Dixon 1992; Abdelsalam & Stern 1996; de Wall *et al.* 2001), the Asmara–Nafka Belt in Eritrea (Ghebreab *et al.* 2009) also referred to as the Central Steep Belt (Drury & de Souza Filho 1998; Teklay 2006) and all-over in the Ethiopian basement. Example include the Central Steep Zone (CSZ) sensu Tadesse *et al.*

(1999, 2000) in the northern terrain; the Baruda–Tulu Dimtu Shear Belt in the western terrain (e.g. Braathen *et al.* 2001; Alemu & Abebe 2007) or in the Adola–Moyale region in the southern terrain (e.g. Tolessa *et al.* 1991; Alene & Barker 1993). The ANS has been envisaged as a collage of arc–arc and arc–continent accreted terranes (Abdelsalam & Stern 1996). Accretion of the ocean floor remnants, thrusting and thrust folding with development of regional schistosity occurred during a long period of formation and consumption of the oceanic crust (Abdelsalam & Stern 1996) and collision with north-central East Africa. The geochronologic data on the East African Orogen (EAO) suggest a final closure of the Mozambique Ocean and the collision of north-central East Africa (Saharan Metacraton, Abdelsalam *et al.* 2002, 2003) and the SLAMIN terrane (Madagascar–India–Somalia–Sri Lanka–the Seychelles) at 650–630 Ma (Meert 2003).

The ANS mainly comprises low-grade volcano-sedimentary and juvenile continental crust decorated by ophiolites (Fig. 1a). The latter mark the boundaries between segments of oceanic crust and island arcs accreted and amalgamated during the closure of the Mozambique Ocean (Schackleton

From: VAN HINSBERGEN, D. J. J., BUITER, S. J. H., TORSVIK, T. H., GAINA, C. & WEBB, S. J. (eds) *The Formation and Evolution of Africa: A Synopsis of 3.8 Ga of Earth History*. Geological Society, London, Special Publications, **357**, 85–106. DOI: 10.1144/SP357.6 0305-8719/11/$15.00 © The Geological Society of London 2011.

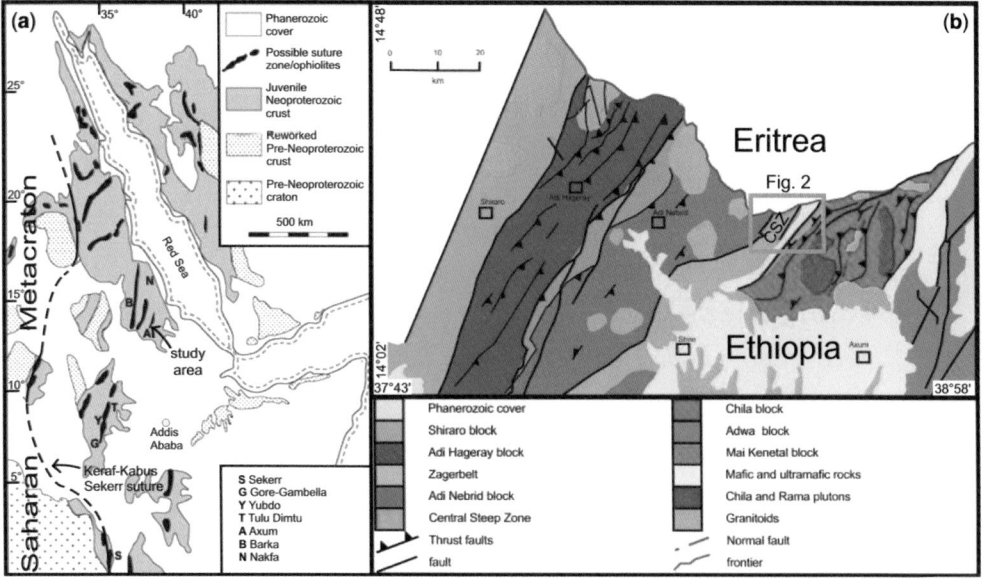

Fig. 1. (**a**) Map showing exposures of Neoproterozoic crust in NE Africa (after Grenne *et al.* 2003) and the eastern boundary of the Saharan Metacraton with the Keraf–Kabus–Sekerr Suture (after Abdelsalam *et al.* 2002). The location of the study area near Axum is indicated. (**b**) Geological map of Northern Ethiopia (after Tadesse 1997) showing location of the Central Steep Zone (CSZ).

1986; Stern & Abdelsalam 1998). The Keraf–Kabus–Sekerr Suture (Stern 1994; Abdelsalam & Stern 1996; Abdelsalam *et al.* 1998, 2003*a*) is regarded as the main suture zone between the pre-Neoproterozoic crust to the east (Saharan Metacraton) and the ANS in the west. An approximately north–south orientation of the suture is realized in its northern part (Keraf Suture) and it trends NW–SE in its southern sector (Kabus–Sekerr Suture). The tectonic models propose a general northwest orientation of crustal convergence during collision and formation of the ANS, resulting in oblique collision with predominant transpressive kinematics (e.g. Worku & Schandelmeier 1996; Abdelsalam *et al.* 1998).

We have studied a high-strain zone, the Central Steep Zone (CSZ), in the vicinity of the city of Axum in the Tigray region in northern Ethiopia (Fig. 1b), and collected field data and samples along a traverse from the volcano-sedimentary units in the east to a monzonite intrusion in the central part (Fig. 2). Structural (including magnetic fabric analysis), petrological, thermo-barometric and geochemical data are presented in this paper to determine the evolution of this high-strain zone, allowing us to discuss its significance in the context of the prevailing tectonic models for the Neoproterozoic evolution of the ANS.

Geological setting

The Precambrian basement in Ethiopia covers the transition zone between two sections of the Neoproterozoic EAO, the ANS in the north and the Mozambique Belt (MB) in the south (e.g. Stern 1994), and is therefore hailed as a key Pan-African terrane in the northeast Africa. ANS and MB are observed to have developed in structural and metamorphic continuity during the Pan-African Orogeny, whereby the difference in metamorphic grade (generally low-grade in the ANS and high-grade in the MB) is related to the increasing intensity of collision from the ANS in the north to the MB in the south (e.g. Stern & Dawoud 1991).

Outcrops of Pan-African basement are exposed in the southern and western parts of Ethiopia (Southwestern Metamorphic Terrain) while its northern prolongation extends further into southern Eritrea (Northern Metamorphic Terrain), as shown in Figure 1a. On a regional scale these exposures are largely masked by East African Rift-related intrusive and extrusive rocks as well as by younger alluvial deposits. The general structural grain of the Pan-African basement in the area of Axum is marked by a fanning system of thrusts with NW-dipping planes in the east and SE-dipping planes in the west (Fig. 1b). The central zone of this large-scale (200 km long) structure comprises

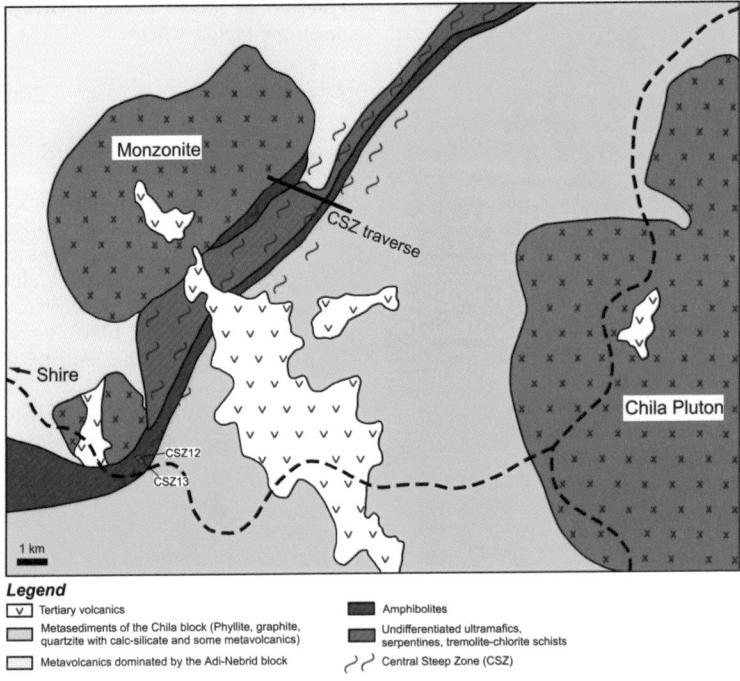

Fig. 2. Geological map of study area showing the lithounits and location of the sampling sites. Samples were collected along a traverse (the CSZ traverse) and at single spots CSZ12 and CSZ13.

c. 10 km wide zone of localized strain, and is referred to as the Central Steep Zone (Tadesse 1997). The CSZ is characterized by a higher metamorphic grade compared to the surrounding areas.

The Neoproterozoic basement in southern Eritrea and northern Ethiopia was initially mapped as a single metavolcanic-sedimentary sequence, the Tsaliet Group (Beyth 1972). However, current models based on more detailed stratigraphic and structural mapping propose a subdivision into units of different structural, metamorphic and magmatic history. Six tectonostratigraphic blocks have been described in northern Ethiopia in the region along the border with Eritrea. From west to east, these are: the Shiharo block, the Adi Hageray block, the Adi-Nebrid block, the Chila block, the Adwa block and the Mai Kental block (Fig. 1b). These blocks are bound by faults which show a general fanning with NW-dipping planes in the east and SE-dipping planes in the west. The central part is defined by a high-strain zone characterized by steeply dipping foliations, the CSZ. Block boundaries, intra-block planar structural features and the CSZ show a NNE–SSW trend. Two belts of mafic and ultramafic rocks, the Zager and the Dahro Tekli belts, respectively, are identified to the west and east of the CSZ (Tadesse 1996;

Tadesse *et al.* 1999). A narrow band of ultramafic rocks is also recognized along the CSZ (Fig. 2). Due to the distinct petrological differences among the blocks, Tadesse *et al.* (1999) inferred a tectonic amalgamation of magmatic sequences formed in intra-oceanic island arc systems.

In this study we focus on the transition from the Chila block in the east towards the Adi Nebrid block in the west (Fig. 1b). The Chila block is comprised predominantly of metapelitic rocks while in the Adi Nebrid block metavolcanic rocks prevail. In these two blocks the metamorphic grade is generally low (greenschist facies).

Age constraints on the geological evolution of the Axum area allow a broad correlation with the 850–650 Ma arc-magmatism, accretion and late-stage intra-crustal melting recorded in other parts of the ANS during the EAO (e.g. Kröner *et al.* 1991; Stern 1994; Abdelsalam & Stern 1996; Cosca *et al.* 1999). Plutonic rocks are abundant in the ANS and occur in all the tectonic blocks of the Northern Metamorphic Terrain of Ethiopia. Available age data for the Azeho and Deset granites (west of the CSZ) and the Chila and Rama granites (east of the CSZ) indicate magmatic activity between 800 and 750 Ma (Tadesse *et al.* 2000). The intrusions are classified as composite island-arc related I-type and range in composition from diorite to granite with locally developed, well-foliated margins and massive internal zones (Tadesse *et al.* 2000). A younger suite of granitoids (*c.* 650–630 Ma, Teklay 1997) exposed in the southern part of Eritrea is referred to as a post-EAO event. The 545–550 Ma Shibta and Shire granites in the Adi Nebrid block mark the latest magmatic event (Tadesse 1997), coeval with the Mareb Granite in the Mai Kental Block. These granites post-date the deformation within the Northern Metamorphic Terrain (Tadesse *et al.* 2000) and might be related to the so-called Kuunga orogeny (Meert 2003); the latter occurred 550–530 Ma ago and defines the final stage of the amalgamation of Gondwana when the Congo and the Kalahari cratons in the western sector and the Indian block and east Antarctica in the eastern sector collided under north–south directed convergence.

Methods and samples

Field observations were documented and samples were collected along the traverse shown in Figure 2. Two additional amphibolites samples (CSZ 12, CSZ 13) from the southern continuation of mafic units in the CSZ were also collected for geochemical analyses (Fig. 2). For anisotropy of magnetic susceptibility (AMS) study 12 samples were drilled in the laboratory for standard-volume 1 inch diameter cylindrical plugs. A total of 62 specimens were measured for AMS using the AGICO kappabridge (KLY-4S). This equipment determines the AMS by measurements in 64 positions for each of three orthogonal sections of the sample. Based on the measurements, the second-order tensor of magnetic susceptibility is calculated, expressed by an ellipsoid with orthogonal principal axes $\kappa_{max} > \kappa_{int} > \kappa_{min}$. Mean susceptibility:

$$\kappa_{mean} = \frac{\kappa_{max} + \kappa_{int} + \kappa_{min}}{3}$$

and shape and anisotropy of AMS-ellipsoids were calculated using the software package 'Anisoft'. The shape is given by the *T*-factor, defined as

$$T = 2 \frac{\ln \kappa_{int} - \ln \kappa_{min}}{\ln \kappa_{max} - \ln \kappa_{int}} - 1$$

which is 0 for neutral, +1 for oblate and −1 for prolate geometries (Jelinek 1981; Tarling & Hrouda 1993). Shape factors in the ranges $-1 < T < 0$ and $0 < T < 1$ correspond to triaxial ellipsoids, and are referred to as dominantly prolate or oblate. Anisotropy of the ellipsoid is

expressed by the P' value, the so-called corrected anisotropy factor, where

$$P' = \{\exp[2(\ln \kappa_{max} - \ln \kappa)^2 + 2(\ln \kappa_{int} - \ln \kappa)^2 + 2(\ln \kappa_{min} - \ln \kappa)^2]\}^{1/2}$$

(Jelinek 1981). More information on method and application can be found in Tarling & Hrouda (1993) and Martín-Hernández *et al.* (2004).

Major oxides and some trace elements (V, Cr, Co, Ni, Zr, Sr and Rb) were measured on fused pellets with a Philips 1480 X-ray fluorescence spectrometer. Rare Earth Elements and some other trace elements (Nb, HF, Y, Ta, Sc, Th, Ga, Ba) analyses were performed with a thermo-elemental Inductively Coupled Plasma Mass Spectrometry (ICP-MS) at the GeoZentrum Nordbayern, Erlangen, following the method described in Brätz & Klemd (2002). Replicate analyses of international geostandards were carried out to maintain the precisions and tolerable accuracies relative to absolute concentrations.

Amphibole and plagioclase mineral pairs in monzonite intrusion and CSZ amphibolite were analyzed with an electron probe micro-analyzer JEOL Superprobe JXA-8900 RL at Frankfurt University. Hornblende measurements were carried out at an accelerating voltage of 20 kV, a beam current of 20 nA and a beam diameter of 1 μm. For plagioclase the conditions were 20 kV, 12 nA and 3 μm, respectively.

Petrofabric characteristics in the CSZ

The CSZ was sampled along a 3 km traverse from its eastern margin to the central part (Fig. 2; Table 1). The lithologies encountered (from east to west) were metapsammitic and metapelitic rocks (CSZ01, 10, 11, 02, where CSZ11 is a graphite-bearing schist) and metabasites comprising amphibolites and ultramafics (CSZ03), quartz-bearing amphibolites (CSZ07) and calc-silicate rocks (CSZ5.2). At site CSZ04 a monzonite is exposed.

Metapelitic rocks (CSZ01) are fine-grained and consist of *c.* 45% volume quartz, 40% micas and 10% feldspar; opaque minerals, such as ilmenite and hematite, are the main accessory phases. Simultaneous growth of biotite and white mica indicates greenschist facies metamorphic conditions. Both these minerals occur as tiny flakes aligned within a prominent foliation. The foliation transposes the material change from quartz-rich to mica-rich layers which is interpreted as sedimentary bedding. Quartz and plagioclase clasts (average size *c.* 300 μm) lie within a fine-grained matrix whereby plagioclase exhibits kinked twin lamellae.

Table 1. *Magnetic susceptibility characteristics of variegated lithologies in the CSZ.* κ_{mean} *is the volume susceptibility in SI units,* P' *is the degree of anisotropy and* T *is the shape factor of AMS ellipsoids. Azimuth and dip are given for the 3 axes of AMS ellipsoids. For location of samples, please refer to Figure 6*

Specimen	κ_{mean}	P'	T	κ_{max} Azimuth	κ_{max} Dip	κ_{int} Azimuth	κ_{int} Dip	κ_{min} Azimuth	κ_{min} Dip	Lithology
CSZ01-1-1	2.66×10^{-4}	1.16	0.46	148	52	317	37	51	5	Metapelite
CSZ01-1-3	2.64×10^{-4}	1.17	0.50	146	56	312	33	46	7	Metapelite
CSZ01-1-2	2.50×10^{-4}	1.17	0.53	155	57	317	31	52	8	Metapelite
CSZ11-1-2	8.09×10^{-5}	1.17	0.68	26	46	176	39	278	15	Graphite-bearing schist
CSZ10-1-1	2.80×10^{-4}	1.08	−0.04	305	60	35	0	125	30	Metapelite
CSZ10-1-2	2.88×10^{-4}	1.07	−0.09	296	62	36	5	129	27	Metapelite
CSZ10-1-3	2.94×10^{-4}	1.07	−0.13	287	66	37	9	131	27	Metapelite
CSZ10-1-4	3.10×10^{-4}	1.07	−0.11	298	63	37	4	129	27	Metapelite
CSZ02-1-1	2.88×10^{-4}	1.06	0.52	263	83	45	5	136	4	Metatuffite
CSZ02-1-2	2.91×10^{-4}	1.05	0.51	230	77	44	13	135	1	Metatuffite
CSZ02-2-1	3.16×10^{-4}	1.15	0.66	243	75	36	14	128	7	Metatuffite
CSZ02-2-2	3.30×10^{-4}	1.15	0.65	229	75	40	15	130	3	Metatuffite
CSZ08-1-1	1.47×10^{-4}	1.10	0.57	92	67	220	15	314	18	Meta-psammopelite
CSZ08-1-2	1.57×10^{-4}	1.10	0.51	99	72	222	10	315	15	Meta-psammopelite
CSZ08-1-3	1.82×10^{-4}	1.10	0.42	114	73	220	5	312	16	Meta-psammopelite
CSZ03-1-1	4.31×10^{-2}	1.31	0.10	213	74	31	16	121	1	Amphibolite
CSZ03-1-2	9.29×10^{-2}	1.38	0.08	230	76	27	13	118	5	Amphibolite
CSZ03-1-5	6.39×10^{-2}	1.30	0.09	236	72	26	16	119	8	Amphibolite
CSZ03-1-4	6.97×10^{-2}	1.32	−0.02	234	78	23	10	114	6	Amphibolite
CSZ03-4-1	1.46×10^{-3}	1.12	0.49	275	84	43	4	133	5	Quartz-bearing amphibolite
CSZ03-5-3A	1.63×10^{-2}	1.31	0.26	30	73	233	15	141	6	Calcsilicate schist
CSZ03-5-3B	2.32×10^{-2}	1.50	0.64	326	79	234	1	144	12	Calcsilicate schist
CSZ03-5-2A	2.12×10^{-2}	1.28	0.30	20	81	230	8	139	5	Calcsilicate schist
CSZ03-5-2B	2.99×10^{-2}	1.52	0.51	34	83	222	7	132	1	Calcsilicate schist
CSZ07-1-1	6.47×10^{-2}	1.31	0.29	320	78	206	5	115	11	Amphibolite
CSZ07-1-2	6.06×10^{-2}	1.30	0.31	311	76	199	5	108	13	Amphibolite
CSZ07-1-3	5.67×10^{-2}	1.38	0.33	320	79	196	6	105	9	Amphibolite
CSZ05-1-1	6.61×10^{-4}	1.04	0.02	175	73	76	3	345	17	Amphibolite
CSZ05-1-2B	6.48×10^{-4}	1.04	0.77	81	30	226	55	341	17	Amphibolite
CSZ05-1-2A	5.14×10^{-4}	1.04	0.64	126	71	252	11	345	15	Amphibolite
CSZ05-1-3	5.88×10^{-4}	1.04	0.51	131	74	246	7	338	15	Amphibolite
CSZ05-2-2	2.94×10^{-4}	1.23	0.35	193	72	289	2	20	18	Calcsilicate schist
CSZ05-2-1	2.24×10^{-2}	1.22	0.31	161	70	282	11	15	17	Calcsilicate schist
CSZ04-3-1B	5.41×10^{-4}	1.06	0.34	190	40	342	46	87	15	Monzonite
CSZ04-3-2A	1.22×10^{-3}	1.17	0.41	241	52	355	17	96	33	Monzonite
CSZ04-3-2B	5.84×10^{-4}	1.06	0.29	209	29	11	60	114	8	Monzonite
CSZ04-3-3A	5.32×10^{-4}	1.06	0.41	215	35	32	55	124	2	Monzonite

Larger quartz clasts, either individual grains or aggregates, exhibit deformation bands and subgrains (Fig. 3a). Small equigranular quartz grains with straight grain boundaries and dihedral angles appear to have formed by dynamic recrystallization and subsequent static recrystallization (Fig. 3b).

A general rise in the grade of metamorphism from CSZ01 to CSZ10 is indicated by an increase in the grain size of the pelitic rocks, particularly the presence of large biotite flakes. In these samples, phyllosilicates (biotite, white mica and chlorite) account for 40%, quartz 50% and feldspar

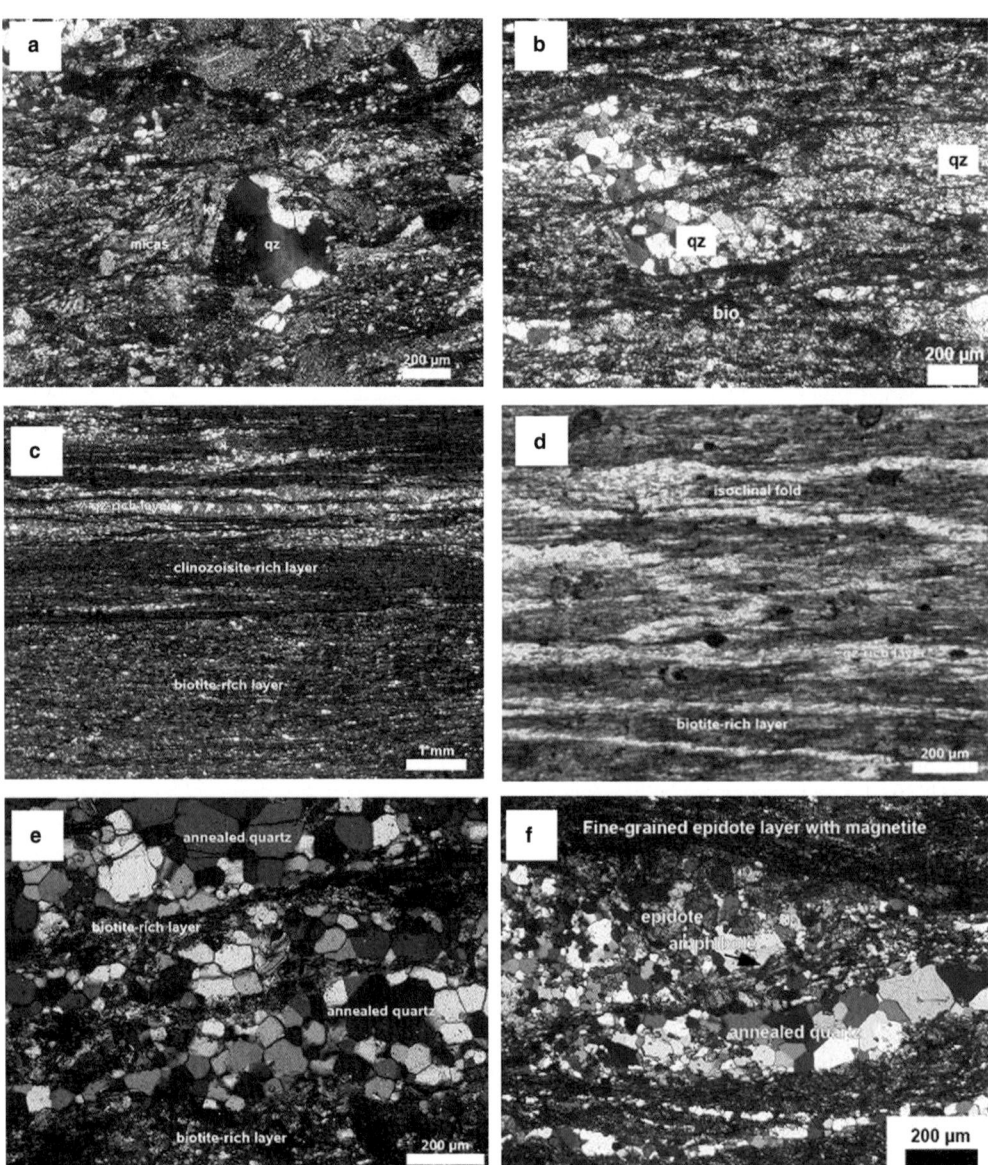

Fig. 3. Deformation fabrics in the CSZ. (**a**) Irregular foliation in metasediments including a quartz clast with undulose extinction (CSZ01). (**b**) Coarse-grained biotite defining a foliation and recrystallized quartz clasts with straight grain boundaries (CSZ10). (**c**) Alternating biotite- and clinozoisite-rich layers (CSZ02). (**d**) Isoclinally folded quartz bands within metasediments (CSZ02). (**e**, **f**) Annealed quartz in psammopelitic rocks (CSZ08) and calc-silicate schist (CSZ03-5).

10%. Ilmenite is the most prominent accessory phase. In sample CSZ02 thin layers of clinozoisite occur parallel to the metamorphic foliation, indicating a tuffaceous protolith (Fig. 3c). White mica is restricted to quartz-rich layers which are either isoclinally folded into intrafolial shear folds (Fig. 3d) or form quartz lenses oblique to the foliation, pointing to a simple shear component accompanying the general flattening character of deformation.

At location CSZ11 metacherts with lamination of quartz-rich and opaque-rich layers are seen intercalated with the metapelites. White mica, hematite and possibly limonite occur as accessory phases (less than 1% each). Some of the quartz layers are intensely folded into isoclinal, intrafolial shear folds with axial planes parallel to the main foliation. The folds can be described as similar folds with clearly thickened hinges. Unfolded quartz layers appear as quartz ribbons, indicating HT conditions and a high finite strain during deformation. Individual quartz grains within the quartz layers show straight grain boundaries indicating annealing; however, deformation bands and subgrain structures are seen in larger grains ($\geq 100 \mu$m).

The psammopelitic rocks (CSZ08) contain millimetre-sized retrograde chlorite blasts, which overgrow the foliated sequence of very-fine-grained, phyllosilicate-rich, pelitic layers and coarser-grained quartz-rich psammitic layers. Individual quartz grains are $100-200 \mu$m in size. Quartz grains do not show any undulose extinction and mark straight grain boundaries (Fig. 3e). The most abundant accessory mineral is hematite (up to 5%) which, together with chlorite, constitutes a possible retrograde mineral assemblage.

Amphibolites and calc-silicate rocks are exposed at sites CSZ03 and CSZ05. Three samples representing lithological variations were collected. Sample CSZ03-1 is mainly composed of blue-green hornblende (65%), intimately intergrown with partly very-fine-grained and probably recrystallized K-feldspar (20%). Moreover, the rock is rich in epidote (10%), which forms up to 500 μm size crystals and occurs in the matrix as well as inclusions within the magnetite crystals. Additionally, some irregularly zoned plagioclase and tempered quartz (together constituting 10%) can also be observed. The rock shows a strong hydrothermal alteration overprint, seen as a high amount of epidote and K-feldspar. This can be attributed to metasomatic substitution of plagioclase (addition of K and removal of Ca and Na). The intergrowth of hornblende and K-feldspar indicates an original magmatic fabric which can be interpreted as a remnant of the ophitic textural relationship between plagioclase (now altered to K-feldspar) and pyroxene (now altered to hornblende). Sample CSZ03-4 is a strongly foliated medium- to fine-grained

quartz-bearing amphibolite. The foliation is defined by blue-green hornblende (80%) which is the major component, while the remaining 20% of the rock comprises quartz, clinozoisite and magnetite. Annealed quartz crystals ($100-300 \mu$m) form ribbons and layers parallel to the foliation. Large and obviously younger amphibole as well as sphene and clinozoisite/epidote crystals are seen growing across the foliation. Sample CSZ03-5 is a fine-grained, relatively quartz- and epidote-rich calc-silicate rock. Individual quartz crystals are strain-free with straight grain boundaries and dihedral angles indicating annealed fabrics (Fig. 3f). Large quartz grains together with epidote crystals are seen aligned along a foliation parallel to the metamorphic layering. Layers rich in very fine-grained epidote contain high amounts (up to 50% of the individual layers) of opaque minerals (magnetite). This succession of alternating amphibolites and calc-silicate rocks continues at locations CSZ07 (quartz-bearing amphibolite), CSZ05-1 (amphibolite; a detailed description follows in the thermobarometry section) and CSZ05-2 (calc-silicate).

The main features observed along the traverse are summarized in the following.

- The first half of the traverse cuts through metapelites and metapsammites while the main lithologies in the second half include amphibolites with intercalated calc-silicate rocks.
- Incipient biotite growth is seen in the easternmost portion of the CSZ (sample CSZ01-1). Biotite and white mica are still tiny there, indicating lower-grade metamorphic conditions and most likely greenschist facies.
- At the site CSZ10, which is located 650 m west of CSZ01, an increase in metamorphic grade towards upper greenschist or even amphibolite facies conditions is indicated by the presence of large biotite laths (50 to 100 μm).
- Occurrence of amphibolite facies rocks in the second half of the traverse (from 1.7 km to the end of the profile) marks a westward increase in the grade of metamorphism.
- Intrafolial folding on meso- and micro-scale can be observed.
- Annealed quartz fabrics (equigranular, straight grain boundaries) indicate static recrystallization at moderate temperatures after folding and shearing (no secondary crystallization was observed).

Magnetic fabric analysis

Bulk magnetic susceptibility data for the analysed samples show a wide range, varying from 10^{-5} to 10^{-2} SI (Table 1) which can be attributed to significant differences in magneto-mineralogy and magnetic behaviour. This fact needs to be considered for interpretation of magnetic fabric data. The metapelites have generally low susceptibility, varying between 1.5 and 3.3×10^{-4} SI, a range typical for pelitic rocks of paramagnetic behaviour (Tarling & Hrouda 1993). Phyllosilicates, predominately bioite, minor white mica and chlorite are the main carriers of magnetic susceptibility and AMS is directly related to the preferred (crystallographic) orientation of phyllosilicates (e.g. Siegesmund *et al.* 1995).

The shape factors (T) of metapelites, except for sample CSZ10, range from 0.4 to 0.7 and the magnetic fabrics are thus characterized as predominantly oblate (Fig. 4a). The degree of anisotropy, however, shows large variations from the lowest values for CSZ02-1 ($P' = 1.05$), intermediate ($P' = 1.1$) for CSZ08 and highest for CSZ01, 11 and 02-2 (P' *c.* 1.15). Such a pattern can be explained by similar structural imprint but variable rock compositions (i.e. higher proportion of clinozoisite in CSZ02-1) and secondary overgrowth of retrograde chlorite (CSZ08) that obliterated the pre-existing foliation. Different shapes of the AMS ellipsoid at site CSZ10 are interpreted to indicate fabric modification, as the mineralogical composition does not differ much from the other pelitic rocks. This observation is also in agreement with the petrographic characteristics. The sample from this location exhibits secondary foliation planes resulting in a phyllosilicate foliation undulating around quartz lenses ('flaser fabrics'). Such a fabric transformation can result in neutral (triaxial) to prolate AMS geometries (Siegesmund *et al.* 1995).

Bulk susceptibilities for calc-silicate and amphibolite samples vary from 10^{-3} to 10^{-2} SI and are within the range of ferrimagnetic mineralogy (magnetite-bearing). Amphibolites CSZ05-1 are an exception with values of 5 to 6×10^{-4}, typical for mafic rocks with paramagnetic constituents (hornblende). The magnitude of anisotropy (Fig. 4b) approaches 1.5 which is definitely higher than the P' values of the metapelites (Fig. 4a). Since this can be due to differences in magneto-mineralogy and in magnetic behaviour, it should therefore not be interpreted in terms of higher strain. In amphibolite and calc-silicate rocks the magnetite grains occur as small layers and form aligned clusters of elongated grains (Fig. 5a) which result in a distinct increase in the degree of anisotropy with increasing magnetite content (Fig. 5b). This increase is caused by the magnetic interaction between individual grains (Hargraves *et al.* 1991; Stephenson 1992). Samples with higher susceptibility show a trend towards a neutral shape of AMS ellipsoids (Fig. 5c).

A consistent orientation of AMS principal axes occurs within the CSZ (Fig. 6). Magnetic lineations are generally steep and in good agreement with field observations. The lineation is parallel to the observed fold axes of centimetre- and decimetre-scale folds in the amphibolites and calc-silicate rocks. The κ_{min} axes (i.e. the poles to the magnetic foliation) plunge preferentially subhorizontally southeast and northwest. Consequently, the magnetic foliation trends NE–SW (043°) and is subvertical. Samples of metabasites near the monzonite intrusion (site 05) show an east–west-trending magnetic foliation. This deflection in foliation can be related to the intrusion of the monzonite (Fig. 6c).

Main features of magnetic fabrics are summarized in the following.

- Paramagnetic fabrics prevail in the metapelites; AMS is related to phyllosilicate preferred orientation and fabrics are predominantly planar.
- Ferrimagnetic fabrics are observed in the metabasic rocks; AMS is related to grain shape anisotropy and distribution anisotropy of magnetite. LS-fabrics (slightly oblate to neutral ellipsoids) are characteristic.

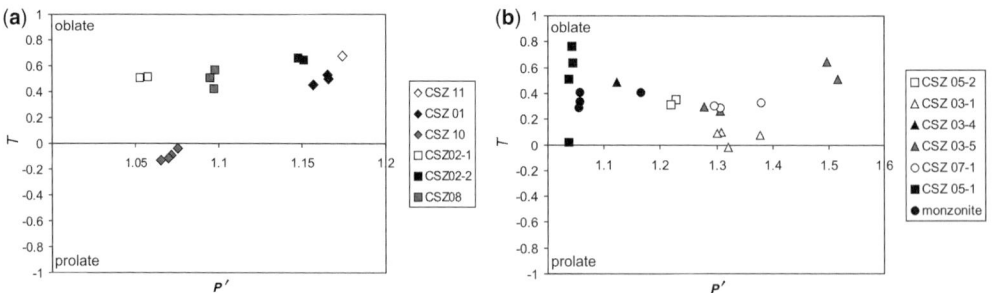

Fig. 4. Shape (*T*) and anisotropy (*P'*) of AMS (Jelinek diagrams) (**a**) for metasediments; and (**b**) for metabasites, calc-silicates and the monzonite.

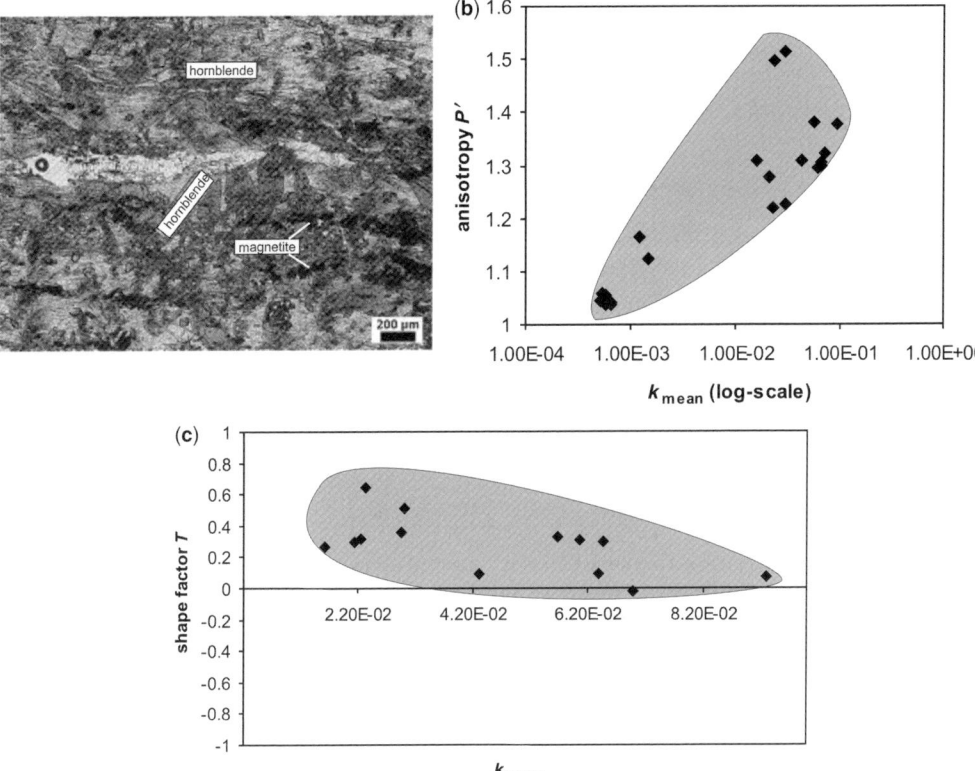

Fig. 5. Magnetomineralogy and related AMS in CSZ. (**a**) Photomicrograph showing bands of fine magnetite grains in amphibolite (CSZ03-4) parallel to the foliation; (**b**) covariance between magnetic anisotropy and volume susceptibility (equivalent to volume proportion of magnetite) and (**c**) a general trend from predominantly oblate towards neutral geometries with increasing magnetic susceptibility.

- Subvertical to steep magnetic foliations and steep magnetic lineations (κ_{max}) are a common feature of all the samples.
- Foliations in the central part of the CSZ trend NE–SW while deflections are seen along the eastern and western boundaries of the CSZ.
- The magnetic fabric of the monzonite is weak and there is no indication for a subsolidus overprint.

Geochemistry of CSZ monzonite and amphibolites

Major and trace element concentrations of ten amphibolite and monzonite samples and REE data on some selected samples are given in Table 2. Amphibolites show silica contents ranging from 44.90 to 52.84% while monzonites have relatively higher silica and a slight overlap with amphibolites. Alkali abundances are highly variable and range from 0.90 to 5.11% in case of amphibolites and

from 4.47 to 8.08% in monzonite. These characteristics indicate the variable degree of alteration which the rocks have undergone. The LOI values range from 1.07 to 3.53%, further substantiating the altered nature of the rocks. MgO shows a large variation (3.6–15%) and lack of any linear relationship with SiO_2. The monzonite samples show a higher average MgO content than the amphibolites. CaO defines a rather restricted range, varying from 5.45 to 10.39% in monzonite and from 8.04 to 10.39% in amphibolites. Such unrelated geochemical variations can be attributed to a variable degree of chemical alteration and serpentinization observed in these rocks. The CSZ amphibolites and monzonite can be clearly discriminated on the basis of TiO_2 concentration. The monzonites have low TiO_2 (0.51–0.56%) while in the amphibolites TiO_2 concentration ranges from 1.39 to 2.25%. Since Ti is known to be geochemically immobile during low-grade metamorphism and alteration, differences in TiO_2 concentration would imply variable source compositions. This contention is also

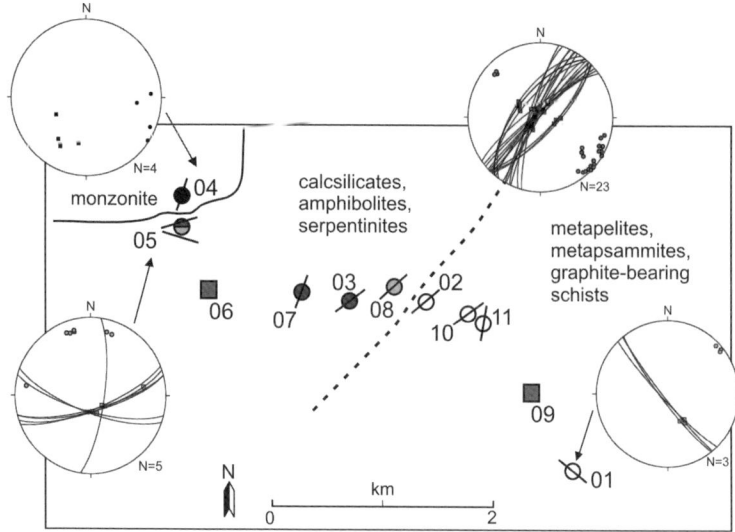

Fig. 6. Sample sites and orientation of magnetic fabrics for structural domains along the CSZ traverse. Additional samples for petrography and geochemistry were collected at sites 06 and 09. The Schmidt net (lower hemisphere) plots show the magnetic lineation (κ_{max}: squares) and magnetic foliation poles (κ_{min}: dots) as well as the great circles of the magnetic foliation (planes normal to κ_{min}). Furtheron the inferred trend of magnetic foliations is indicated at the individual sample sites.

supported by subtle differences in the immobile trace element characteristics (e.g. Zr, Y, Nb, Th, etc.).

In the immobile trace element-based SiO_2–Zr/TiO_2 diagram the CSZ amphibolites plot in the field of subalkaline basalt. The monzonite samples show an array, ranging from subalkaline basalt to trachy-andesite, mainly on account of a variable silica content (Fig. 7a). In the Zr + Y–Ti–Cr diagram (which is considered to mimic the AFM diagram), the amphibolite samples clearly define a tholeiitic trend while a calc-alkaline affinity can be seen in case of the monzonite (Fig. 7b).

The trace elements were plotted in the standard multi-element spiderplots (Fig. 7c). The amphibolites and monzonite samples can be clearly discriminated in terms of behaviour of some critical trace elements. Monzonite samples show peaks for LIL elements (Rb, Ba, K, Sr) and a significant Nb anomaly (crustal input) while the amphibolite patterns are less spiked and characterized by peaks for Nb, Th and Ti. Decoupling between Nb–Ta and Zr–Hf elemental pairs is a characteristic feature of amphibolites. A general enrichment in both LIL and some HFS elements can be interpreted in terms of a mixed ocean floor and arc origin. The $\sum$REE for amphibolites and monzonite are quite comparable (90.32 to 118.93 ppm). In the chondrite normalized REE diagram (Fig. 7d), amphibolite and monzonite samples show a general moderate LREE enrichment and absence of Eu anomaly. However,

the HREE level is slightly higher in the case of amphibolite with a rather flat HREE trend while convex downward patterns are seen in case of monzonite. Notwithstanding the overlap in some major element abundances, distinct differences in critical trace elements and REE clearly underline that the monzonite and the amphibolites were derived from variable sources.

The geochemical signatures of the CSZ metabasites show a wide spectrum that can be interpreted as ocean floor to volcanic arc setting. Similar results have also been reported for the far apart Tulu Dimtu Ophiolite Suite in western Ethiopia (Tadesse & Allen 2005). Tadesse *et al.* (1999) have also inferred MORB to calc-alkaline characteristics for the mafic–ultramafic metavolcanics of Adi Nebrid, Adwa and Adi Hagery blocks and Zager and Daro–Tekli regions in northern Ethiopia. Geochemical characteristics of the CSZ rocks indicate different sources and a complex tectonic evolution, suggesting a possible ophiolite setting, widespread in the ANS. The ophiolites are considered as the remnants of the supra-subduction zones and arc terranes.

Mineral chemistry and thermo-barometry

The monzonite is medium- to coarse-grained (grain size 1–5 mm) and consists of plagioclase (30%),

Table 2. *Whole rock chemical composition of amphibolites (amph) and monzonite (monz) from CSZ. Major oxides are presented as wt% while trace and Rare Earth Elemental abundances are given in ppm*

Sample	03–4.1 amph	03.1–1 amph	03–2 amph	3.5–2a amph	7.1–2 amph	12.1–1 amph	13–4a amph	04–1a monz	04–1b monz	04–4 monz
Major Elements (as wt% oxides)										
SiO_2	44.90	45.50	45.66	49.00	49.30	51.20	52.00	52.84	47.27	56.89
TiO_2	1.39	2.05	1.94	1.56	1.45	2.25	1.16	0.51	0.58	0.56
Al_2O_3	14.01	17.26	16.05	16.81	17.26	15.19	15.26	9.74	8.91	14.22
Fe_2O_3	14.07	13.93	13.01	13.62	12.36	9.48	9.75	8.15	9.66	6.18
MnO	0.31	0.20	0.20	0.19	0.32	0.22	0.14	0.17	0.19	0.11
MgO	11.27	5.25	9.26	3.67	3.60	5.46	5.93	12.57	15.00	6.13
CaO	9.01	10.08	7.83	8.52	8.04	10.08	10.20	8.79	10.39	5.45
Na_2O	0.73	3.57	2.40	3.97	4.74	4.22	3.11	2.13	1.17	3.62
K_2O	0.17	0.28	0.12	0.19	0.37	0.18	0.49	3.17	3.30	4.46
P_2O_5	0.13	0.23	0.21	0.18	0.20	0.32	0.12	0.33	0.35	0.49
LOI	3.53	1.43	3.03	1.72	1.17	1.07	1.09	1.23	1.57	1.08
Sum	99.40	99.70	99.71	99.40	98.70	99.60	99.20	99.63	98.39	99.19
Trace Elements (in ppm)										
V	325	305	272	245	251	262	283	128	158	121
Cr	519	191	174	462	328	123	240	1023	1318	326
Co	79	79	47	80	114	66	79	46	56	26
Ni	243	86	98	113	94	36	67	177	245	90
Zn	91	105	104	98	69	85	74	67	75	60
Ga	16	22	29	16	18	19	14	10	14	24
Rb	bdl[1]	4	bdl	2	3	bdl	10	69	117	87
Sr	158	356	272	188	250	192	211	789	670	1354
Y	24	33	22.6	33	42	45	31	14.5	18.7	17
Zr	79	144	133	83	84	117	72	77	89	177
Nb	6	22	22	7	5	18	4	3.64	3.66	5.88
Ba	40	123	bdl	21	107	61	109	2233	163	2362
Pb	bdl	bdl	bdl	bdl	bdl	bdl	bdl	21	10	19
Th	2	9	177	bdl	bdl	9	1	0.82	0.88	3.02
U	2	2	bdl	4	2	bdl	2	0.67	0.51	1.2
Hf	nd[2]	nd	2.88	nd	nd	nd	nd	2.28	2.43	5.27
Ta	nd	nd	1.56	nd	nd	nd	nd	0.45	0.28	0.73
Rare Earth Elements (in ppm)										
La			16.9					15.1	16.9	20.9
Ce			34.8					34.2	38.7	46.7
Pr			4.4					4.48	5.21	5.95
Nd			19.4					19.6	23.6	25.5
Sm			4.53					4.27	5.49	5.16
Eu			1.56					1.3	1.66	1.59
Gd			4.59					3.81	4.69	4.33
Tb			0.72					0.49	0.65	0.57
Dy			4.64					3.01	3.63	3.5
Ho			0.87					0.54	0.74	0.65
Er			2.44					1.54	1.95	1.77
Tm			0.33					0.23	0.29	0.25
Yb			2.34					1.53	1.99	1.79
Lu			0.38					0.22	0.30	0.26
$\sum REE$			99.42					90.32	105.08	118.93

[1]bdl: below detection limit; [2]nd: not determined.

K-feldspar (20%), quartz (5%), biotite (15%) and hornblende (25%). Sphene, apatite, zircon and magnetite/ilmenite occur as accessory phases. Quartz and the two feldspars show typical hyperso- lidus/magmatic intergrowth fabrics: myrmekitic and micrographic intergrowth of quartz and K-feldspar (generally microcline). Plagioclase is often discontinuously zoned. Both the feldspar types exhibit deformation twinning. The overall fabric does not show any subsolidus deformation and

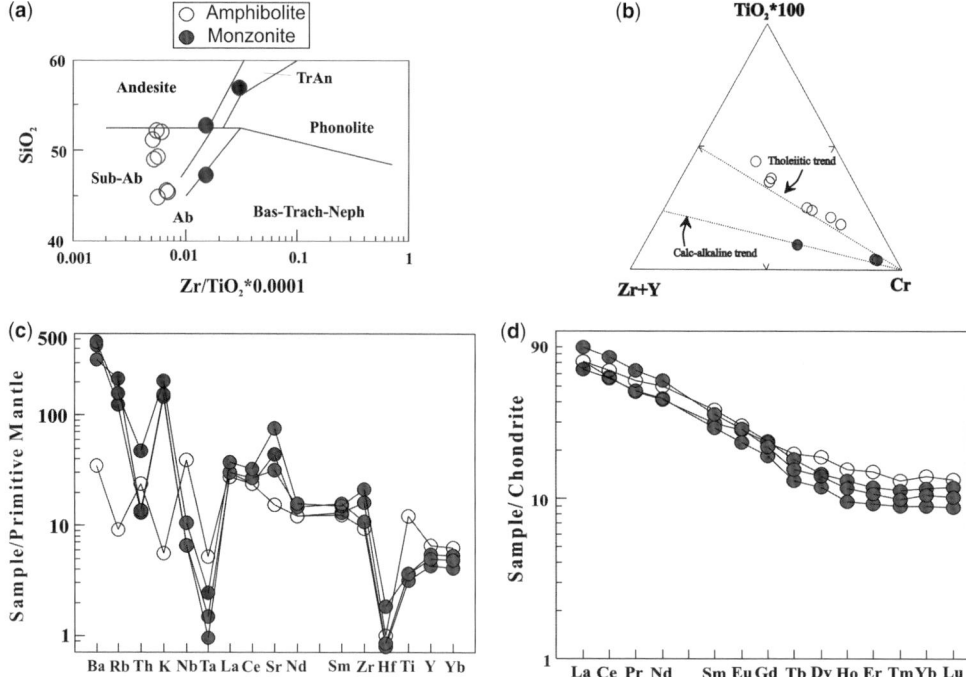

Fig. 7. Geochemical characteristics of amphibolites and monzonite in the CSZ. (**a**) Geochemical classification of CSZ amphibolites and monzonite based on $SiO_2 - Zr/TiO_2$ diagram. (**b**) The $Zr + Y - TiO_2 - Cr$ diagram showing a general tholeiitic trend for CSZ amphibolites. (**c**) Multi-element spiderplots (normalized after Sun & McDonnough 1989) showing behaviour of trace elements in amphibolite and monzonite. (**d**) Chondrite normalized (normalized after Sun & McDonnough 1989) REE patterns for CSZ rocks showing LREE enrichment and almost flat HREE with minor downwards convexity.

fabric formation other than the twinning observed in both the feldspars and subgrain formation in quartz.

Amphibolite CSZ05 is composed of pale hornblende (75%), plagioclase and K-feldspar (together 15%), along with minor biotite (5%) and clinozoisite (5%) with sphene and opaque minerals as main accessory phases. It is medium- to fine-grained with a felty intergrowth of hornblende. Only a very weak foliation, mainly defined by a few biotite flakes, is seen.

In most intermediate plutonic rocks as well as in amphibolites, hornblende, besides ortho- and clinopyroxene, is the only useful mineral for thermobarometric investigations (Anderson 1996). Its Al-content is generally regarded as a direct measure of emplacement depth (Al-in-hornblende barometer). Exchange reactions between plagioclase and hornblende involving Ca, Na, K, Al and Si can be used to determine crystallization temperatures of hornblende-plagioclase pairs (amphibole-plagioclase thermometer). We have analysed amphibole and plagioclase from the monzonite and amphibolites. The results are presented in

Table 3. Hornblende calculation was based on 23 oxygens and standardized on 13 cations (without Ca, Na and K). The average Mg-# (= Mg/[Mg + Fe^{2+}]) is 0.73, Fe$_{tot}$-# (= Fe$_{tot}$/[Mg + Fe$_{tot}$]) is 0.33 and mean Fe^{3+}-# (Fe^{3+}/Fe$_{tot}$ ratio) is 0.26. The average Al$_{tot}$-content of hornblende from the CSZ monzonite is 1.04 a.p.f.u. (atoms per formula unit). Plagioclase in the monzonite sample is consistently of albite to oligoclase composition (ab$_{81-99}$) with anorthitic (ab$_{05}$) cores in some cases. Plagioclase in the amphibolites has ab$_{94}$. Hornblende from both monzonite and amphibolites are classified as magnesio-hornblendes; however, they plot as distinct compositional clusters (Leake *et al.* 1997; Fig. 8a).

Amphibole-plagioclase thermometry

We have employed amphibole-plagioclase thermometry and Al-in-hornblende barometry to estimate depth and temperature of metamorphic conditions of amphibolites and crystallization of monzonite. Holland & Blundy (1994) developed two empirical amphibole plagioclase thermometers: (a) based on

Table 3. *Composition of hornblendes*

Sample CSZ04	Hornblende cores								Hornblende in contact with plag					
Point	CSZ04-1 34	CSZ04-1 38	CSZ04-1 56	CSZ04-1 57	CSZ04-1 58	CSZ04-1 59	CSZ04-1 33	CSZ04-1 40	CSZ04-1 42	CSZ04-1 47	CSZ04-1 48	CSZ04-1 50	CSZ04-1 53	CSZ04-1 113
Oxides (wt%)														
SiO_2	48.47	50.08	48.61	47.94	47.82	48.22	48.91	48.24	47.34	48.72	48.15	47.51	49.30	50.93
TiO_2	0.96	0.82	1.00	1.10	1.02	0.97	0.96	0.90	1.03	0.87	1.01	1.11	0.93	0.44
Al_2O_3	6.35	5.16	6.35	7.01	6.86	6.50	5.88	6.25	7.06	5.99	6.56	7.09	5.93	4.46
Cr_2O_3	0.07	0.18	0.02	0.08	0.10	0.11	0.09	0.22	0.15	0.13	0.08	0.13	0.06	0.14
Fe_2O_3	3.34	3.28	3.90	3.60	3.88	3.85	3.48	3.76	3.92	3.47	3.69	3.50	2.79	3.51
FeO	9.50	9.40	9.34	10.02	9.52	9.67	9.28	9.96	10.14	9.66	9.77	10.17	9.80	8.59
MnO	0.29	0.29	0.28	0.31	0.34	0.32	0.30	0.32	0.29	0.31	0.28	0.29	0.27	0.31
NiO	0.03	0.01	0.02	0.03	0.03	0.01	0.02	0.00	0.01	0.04	0.02	0.03	0.04	0.04
MgO	14.43	14.95	14.47	13.94	14.11	14.24	14.79	14.13	13.57	14.01	14.18	13.72	14.71	15.59
CaO	11.52	11.68	11.33	11.35	11.35	11.37	11.42	11.51	11.33	11.23	11.52	11.41	11.71	11.72
Na_2O	1.50	1.20	1.71	1.79	1.73	1.77	1.68	1.62	1.66	1.37	1.56	1.65	1.37	1.15
K_2O	0.66	0.55	0.66	0.74	0.75	0.68	0.64	0.66	0.77	0.60	0.69	0.76	0.65	0.44
Total	**97.12**	**97.60**	**97.70**	**97.91**	**97.52**	**97.72**	**97.44**	**97.56**	**97.27**	**96.40**	**97.51**	**97.36**	**97.56**	**97.32**
Cations (pfu)														
T spaces														
Si	7.08	7.25	7.07	6.98	6.99	7.03	7.12	7.05	6.96	7.16	7.03	6.97	7.16	7.36
Al	0.92	0.75	0.93	1.02	1.01	0.97	0.88	0.95	1.04	0.84	0.97	1.03	0.84	0.64
Total T	**8.00**	**8.00**	**8.00**	**8.00**	**8.00**	**8.00**	**8.00**	**8.00**	**8.00**	**8.00**	**8.00**	**8.00**	**8.00**	**8.00**
M1–M3 spaces														
Al	0.18	0.13	0.15	0.19	0.17	0.14	0.13	0.13	0.18	0.20	0.16	0.19	0.17	0.12
Ti	0.11	0.09	0.11	0.12	0.11	0.11	0.10	0.10	0.11	0.10	0.11	0.12	0.10	0.05
Cr	0.01	0.02	0.00	0.01	0.01	0.01	0.01	0.02	0.02	0.02	0.01	0.01	0.01	0.02
Fe^{3+}	0.37	0.36	0.43	0.39	0.43	0.42	0.38	0.41	0.43	0.38	0.41	0.39	0.30	0.38
Fe^{2+}	1.16	1.14	1.14	1.22	1.16	1.18	1.13	1.22	1.25	1.19	1.19	1.25	1.19	1.04
Mn	0.04	0.04	0.03	0.04	0.04	0.04	0.04	0.04	0.04	0.04	0.03	0.04	0.03	0.04
Ni	0.00	0.00	0.00	0.00	0.00	0.00	0.00	0.00	0.00	0.01	0.00	0.00	0.00	0.00
Mg	3.14	3.23	3.14	3.03	3.07	3.09	3.21	3.08	2.97	3.07	3.09	3.00	3.18	3.36
Total M1–M3	**5.00**	**5.00**	**5.00**	**5.00**	**5.00**	**5.00**	**5.00**	**5.00**	**5.00**	**5.00**	**5.00**	**5.00**	**5.00**	**5.00**
M4 spaces														
Ca	1.80	1.81	1.76	1.77	1.78	1.78	1.78	1.80	1.78	1.77	1.80	1.79	1.82	1.81
Na	0.20	0.19	0.24	0.23	0.22	0.22	0.22	0.20	0.22	0.23	0.20	0.21	0.18	0.19
Total M4	**2.00**	**2.00**	**2.00**	**2.00**	**2.00**	**2.00**	**2.00**	**2.00**	**2.00**	**2.00**	**2.00**	**2.00**	**2.00**	**2.00**
A spaces														
Na	0.23	0.15	0.25	0.28	0.27	0.28	0.25	0.26	0.26	0.16	0.24	0.26	0.21	0.14
K	0.12	0.10	0.12	0.14	0.14	0.13	0.12	0.12	0.14	0.11	0.13	0.14	0.12	0.08
Total A	**0.35**	**0.25**	**0.37**	**0.42**	**0.41**	**0.40**	**0.37**	**0.38**	**0.40**	**0.27**	**0.37**	**0.40**	**0.33**	**0.22**
Total	**15.35**	**15.25**	**15.37**	**15.42**	**15.41**	**15.40**	**15.37**	**15.38**	**15.40**	**15.27**	**15.37**	**15.40**	**15.33**	**15.22**

(Continued)

Table 3. *Continued*

Sample CSZ04

Point	Hornblende cores							Hornblende in contact with plag						
	CSZ04-1 34	CSZ04-1 38	CSZ04-1 56	CSZ04-1 57	CSZ04-1 58	CSZ04-1 59	CSZ04-1 33	CSZ04-1 40	CSZ04-1 42	CSZ04-1 47	CSZ04-1 48	CSZ04-1 50	CSZ04-1 53	CSZ04-1 113
P (kbar) (Stephenson 1992)														
X_{ab} in plag								0.99	0.99	0.97		0.99	0.99	
T (°C) at 5 kbar (Holland & Blundy 1994; Thermometer (a))								651	651	607		646	618	
T (°C) at 5 kbar (Holland & Blundy 1994; Thermometer (b))														

Sample CSZ04

Point	Hornblende rims in contact with plag			Hornblende rims in contact with plag (continued)				Hornblende in contact with qtz and kf					
	CSZ04-1 115	CSZ04-1 116	CSZ04-1 117	CSZ04-1 119	CSZ04-1 120	CSZ04-1 121	CSZ04-1 124	CSZ04-1 39	CSZ04-1 35	CSZ04-1 112	CSZ04-1 118	CSZ04-1 122	CSZ04-1 123
Oxides (wt%)													
SiO_2	51.03	48.00	47.67	49.50	47.07	47.03	48.24	50.05	48.46	53.49	48.64	47.92	48.09
TiO_2	0.46	0.98	1.06	0.67	1.09	1.14	0.89	0.69	1.00	0.05	0.98	1.01	0.95
Al_2O_3	4.33	6.48	6.55	5.04	7.04	7.07	6.07	5.15	6.35	2.64	6.10	6.60	6.44
Cr_2O_3	0.10	0.09	0.10	0.10	0.11	0.08	0.06	0.13	0.06	0.03	0.06	0.10	0.08
Fe_2O_3	2.95	4.96	4.17	2.94	3.97	3.91	3.33	3.41	3.45	2.79	4.10	4.62	3.92
FeO	9.11	8.60	9.55	9.74	10.01	10.33	9.81	9.26	9.86	7.91	9.42	8.96	9.36
MnO	0.28	0.29	0.30	0.32	0.32	0.31	0.33	0.32	0.30	0.25	0.27	0.33	0.34
NiO	0.05	0.02	0.01	0.03	0.02	0.01	0.03	0.05	0.03	0.00	0.04	0.05	0.03
MgO	15.53	14.43	14.15	14.63	13.71	13.47	14.19	14.94	14.25	17.06	14.44	14.32	14.22
CaO	11.84	11.25	11.37	11.57	11.39	11.40	11.56	11.52	11.50	12.30	11.49	11.29	11.47
Na_2O	1.10	1.65	1.73	1.24	1.74	1.64	1.40	1.39	1.55	0.60	1.52	1.73	1.46
K_2O	0.42	0.72	0.74	0.54	0.79	0.78	0.63	0.51	0.69	0.17	0.66	0.74	0.72
0.00	97.20	97.47	97.39	96.32	97.27	97.18	96.53	97.42	97.50	97.29	97.71	97.66	97.09
Cations (pfu)													
T spaces													
Si	7.39	7.00	6.98	7.27	6.93	6.93	7.10	7.26	7.07	7.64	7.08	6.99	7.04
Al	0.61	1.00	1.02	0.73	1.07	1.07	0.90	0.74	0.93	0.36	0.92	1.01	0.96
Total T	**8.00**	**8.00**	**8.00**	**8.00**	**8.00**	**8.00**	**8.00**	**8.00**	**8.00**	**8.00**	**8.00**	**8.00**	**8.00**

MI–M3 spaces

Al	0.12	0.11	0.11	0.15	0.15	0.16	0.16	0.14	0.16	0.08	0.12	0.12	0.16
Ti	0.05	0.11	0.12	0.07	0.12	0.13	0.10	0.08	0.11	0.01	0.11	0.11	0.10
Cr	0.01	0.01	0.01	0.01	0.01	0.01	0.01	0.02	0.01	0.00	0.01	0.01	0.01
Fe^{3+}	0.32	0.54	0.46	0.32	0.44	0.43	0.37	0.37	0.38	0.30	0.45	0.51	0.43
Fe^{2+}	1.10	1.05	1.17	1.20	1.23	1.27	1.21	1.12	1.20	0.94	1.15	1.09	1.15
Mn	0.03	0.04	0.04	0.04	0.04	0.04	0.04	0.04	0.04	0.03	0.03	0.04	0.04
Ni	0.01	0.00	0.00	0.00	0.00	0.00	0.00	0.01	0.00	0.00	0.00	0.01	0.00
Mg	3.35	3.14	3.09	3.20	3.01	2.96	3.12	3.23	3.10	3.63	3.13	3.11	3.11
Total M1–M3	**5.00**	**5.00**	**5.00**	**5.00**	**5.00**	**5.00**	**5.00**	**5.00**	**5.00**	**5.00**	**5.00**	**5.00**	**5.00**
M4													
Ca	1.84	1.76	1.78	1.82	1.80	1.80	1.82	1.79	1.80	1.88	1.79	1.76	1.80
Na	0.16	0.24	0.22	0.18	0.20	0.20	0.18	0.21	0.20	0.12	0.21	0.24	0.20
Total M4	**2.00**	**2.00**	**2.00**	**2.00**	**2.00**	**2.00**	**2.00**	**2.00**	**2.00**	**2.00**	**2.00**	**2.00**	**2.00**
A spaces													
Na	0.14	0.22	0.28	0.18	0.29	0.27	0.22	0.18	0.24	0.05	0.22	0.25	0.21
K	0.08	0.13	0.14	0.10	0.15	0.15	0.12	0.09	0.13	0.03	0.12	0.14	0.14
Total A	**0.22**	**0.36**	**0.41**	**0.28**	**0.44**	**0.42**	**0.34**	**0.28**	**0.36**	**0.08**	**0.34**	**0.39**	**0.35**
Total	**15.22**	**15.36**	**15.41**	**15.28**	**15.44**	**15.42**	**15.34**	**15.28**	**15.36**	**15.08**	**15.34**	**15.39**	**15.35**
P (kbar) (Stephneson 1992)								2.11	2.82	1.82		2.36	2.29
X_{ab} in plag	0.94			0.85	0.81		0.97						
T (°C) at 5 kbars (Holland & Blundy 1994; Thermometer (a))	663			620	691		635						
T (°C) at 5 kbars (Holland & Blundy 1994; Thermometer (b))				628	697								

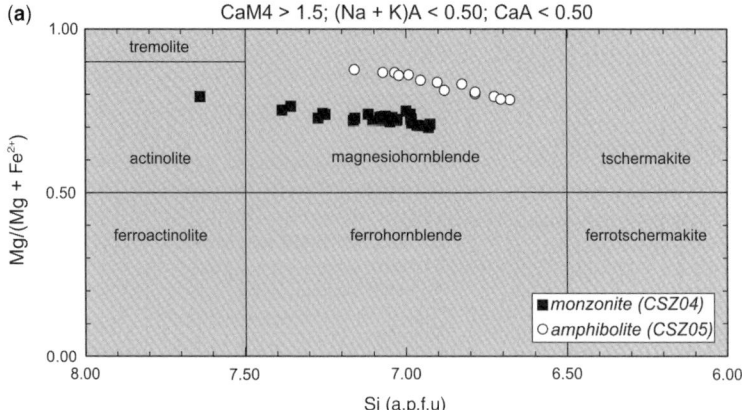

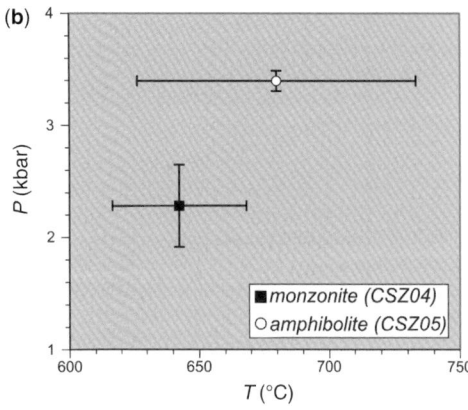

Fig. 8. (a) Classification diagram (after Leake *et al.* 1997) for analysed amphiboles from the CSZ showing magnesio-hornblende characteristics; (b) *P–T* estimates (Schmidt 1992; Holland & Blundy 1994) revealing that the monzonite intruded at an average temperature of *c.* 650 °C (i.e. close to the wet granitic solidus) at a depth of *c.* 6–8 km, corresponding to an average pressure of 2 kbar.

the edenite-tremolite reaction: 4 quartz + edenite = albite + tremolite; and (b) which considers the edenite-richterite reaction: edenite + albite = richterite + anorthite. Both the thermometers are applicable for a temperature range of 400–1000 °C.

The amphibole-plagioclase-thermometers proposed by Holland & Blundy (1994) have been applied to calculate the crystallization temperature. Calculations were carried out with the software 'Hb-Plag' by T. Holland (http://www.esc.cam.ac.uk/research/research-groups/holland/hb-plag). Since both substitution reactions are not only temperature but also pressure dependent, the software calculates temperatures for pressures in steps of 5 kbars (0, 5, 10 and 15 kbars). We chose the temperatures calculated for 5 kbars, because overall geological considerations indicate that the CSZ rocks neither reflect superficial (0 kbars) nor high-pressure conditions (10 to 15 kbars). The software calculates temperatures for individual hornblende plagioclase pairs assuming a lithostatic pressure of *c.* 5 kbar. Average temperatures for both the samples were derived from those *T*-values.

In the monzonite (CSZ04-3), nine hornblende-plagioclase pairs were found suitable for the

edenite-tremolite thermometry and only two for the edenite-richterite thermometry (Table 3). Thermometer (a) (edenite-tremolite) of Holland & Blundy (1994) yields average crystallization temperatures of 642 ± 26 °C (1σ) for the hornblendes from the monzonite (Fig. 8b). According to thermometer (b) (edenite-richterite), an average crystallization temperature of 662 ± 49 °C (1σ) is obtained for hornblende in the monzonite. Consequently, the estimated temperatures lie below the wet monzonitic solidus (*c.* 750 °C), implying subsolidus reequilibration of the monzonite and its hornblendes. This is possibly due to continued magmatism and post-emplacement cooling or is a result of regional metamorphic (amphibolite-facies) overprint postdating the intrusion of the monzonite. The regional geological conditions are equivocal and not distinctive for choosing a suitable interpretation; however, absence of any strong subsolidus deformational overprint in monzonite precludes any significant metamorphic overprint, at least in these rocks.

For the amphibolite sample (CSZ05-1) only thermometer (a) of Holland & Blundy (1994) could be applied. Five hornblende-plagioclase pairs were measured (Tab. 3, Fig. 8b) yielding an average temperature of 680 ± 53 °C (1σ) hornblende crystallization, indicating upper amphibolite facies metamorphic conditions.

Al-in-hornblende barometry

The empirical Al-in-hornblende barometer is based on the observation that the alumina content of hornblende, in rocks with the paragenesis hornblende + biotite + plagioclase (an$_{25-35}$) + K-feldspar + quartz + sphene + magnetite/ilmenite, increases with increasing pressure (Hammerstrom & Zen 1986). Hammerstrom & Zen (1986) and Hollister et al. (1987) formulated the first two calibrations of the Al-in-hornblende barometer based on mineral-chemical investigations of natural rocks. Johnson & Rutherford (1989), Thomas & Ernst (1990) and Schmidt (1992) added three experimental calibrations. The barometer of Schmidt (1992) is particularly useful for granitoids because it has been calibrated at temperatures between 655 and 700 °C (i.e. typical solidus temperatures of granitic to tonalitic systems) and under H_2O-saturated conditions at pressures between 2.5 and 13 kbar. The calibration of the Al-in-hornblende barometer by Schmidt (1992) is:

$$P[\pm 0.6\,\mathrm{kbar}] = -3.01 + 4.76\mathrm{Al_{tot}}.$$

Anderson & Smith (1995) pointed out that the Al content of amphiboles not only depends on the pressure during crystallization and growth but also on the crystallization temperature and the oxygen fugacity. Consequently, they developed a temperature-sensitive calibration for the barometer based on the experimental data of Johnson & Rutherford (1989) and Schmidt (1992). Anderson & Smith (1995) suggest application of the barometer only to hornblendes, which crystallizes at medium to high fO_2 conditions. Consequently, the barometer should be applied only to hornblendes with a Fe_{tot}-# ≤ 0.65 and a Fe^{3+}/Fe_{tot}-ratio ≥ 0.25.

In the present study, the Al-in-hornblende barometer proposed by Schmidt (1992) has been used because it is calibrated for P-T-fH_2O conditions similar to those assumed for the near-solidus history of the investigated monzonite. The calibration by Anderson & Smith (1995) was not considered because it assumes the amphiboles crystallize above the wet granitic solidus of c. 675 °C and do not re-equilibrate at subsolidus temperatures, which is not the case for the monzonite (CSZ04). Nevertheless, the investigated hornblendes in both the samples fulfil the prerequisites for application of the Al-in-hornblende barometer being part of the proper paragenesis and for containing the required amounts of Fe and Mg.

For the monzonite CSZ04, pressure estimates were made for five individual microprobe measurements at the rims of hornblende in contact with quartz or K-feldspar. They range from 1.82 to 2.82 kbar and indicate an average pressure of 2.28 ± 0.37 kbar (1σ) for the crystallization of hornblende (Fig. 8b). For the amphibolite CSZ05, only two hornblende rims in contact with quartz could be measured. However, they provide distinctively higher Al-contents of hornblende than sample CSZ04 (Table 3). Applying the Al-in-hornblende barometer of Schmidt (1992) they yield pressures of 3.33 and 3.46 kbar, respectively (Fig. 8b). Obviously, the hornblendes in the amphibolite CSZ05 grew under a significantly higher pressure than the amphiboles of the monzonite sample CSZ04. The former were also not affected by a thermal imprint of the intruding monzonite.

Discussion and conclusions

The CSZ comprises metasediments of the Chila block in the east and metavolcanics from the Adi Nebrid block in the west and structurally forms the central part of a bivergent fanning thrust system within the northern Pan-African terrain in Ethiopia (Fig. 1b). Furthermore, mafic and ultramafic rocks occur in the central part, which represent remnants of an oceanic crust. Strain concentration in the CSZ has produced steep foliations with vertical lineations and fold axes, and moderately oblate to triaxial fabric geometries. The deformation in the CSZ occurred under higher-grade metamorphic conditions (amphibolite facies) compared to the regional greenschist metamorphism observed in the surrounding area. After metamorphic equilibration, the rocks within the CSZ experienced a considerable uplift as indicated by the thermobarometric estimates of 680 °C and c. 3.4 kbar for syn-kinematic hornblende crystallization (data for the amphibolite sample CSZ05).

Partial melting of mafic rocks (oceanic basalts) generated the monzonitic melt, which has intruded into the high-strain zone. Estimated pressure of around 2 kbar together with the absence of pervasive subsolidus deformation indicate that the monzonite emplacement was a late event in the evolution of the CSZ, which occurred after a considerable uplift of the amphibolites as host rocks. The subsolidus temperatures yielded by the hornblende-plagioclase thermometry are probably the result of slow post-emplacement cooling. There is no evidence of significant post-magmatic regional metamorphism accompanied by regional deformation. It seems likely that the heat transfer by this intrusion was responsible for the ubiquitous annealing of quartz deformation fabrics within the CSZ.

The structural architecture (bivergent thrusting, localized high-strain deformation in the central part) occurs both parallel and transverse to the major tectonic blocks (Fig. 1b) and identifies the structural imprint to post-date the accretion of the blocks. Steep foliations and steep lineations in the ductile high-strain zone of the CSZ are consistent with formation in a transpressive regime at high finite strain (wrench faulting; see also Fossen & Tikoff 1993; Tikoff & Greene 1997) with extrusion towards free boundaries (Sanderson & Marchini 1984; Jones *et al.* 1997). In agreement with the bivergent fanning of the thrust faults, the CSZ is identified as the central zone of a positive flower structure (Fig. 9) where material has been extruding upwards while lateral extrusion causes reverse faulting. The monzonite was emplaced after the positive flower structure had already been formed and the amphibolite facies rocks in the middle part of the CSZ were already uplifted to a green-schist facies level. The low pressure recorded by Al-in-hornblende barometry for the monzonite also supports the inference that the monzonite intruded the central part of the CSZ which had already cooled to the greenschist facies conditions. The late-tectonic character of the monzonite is indicated by

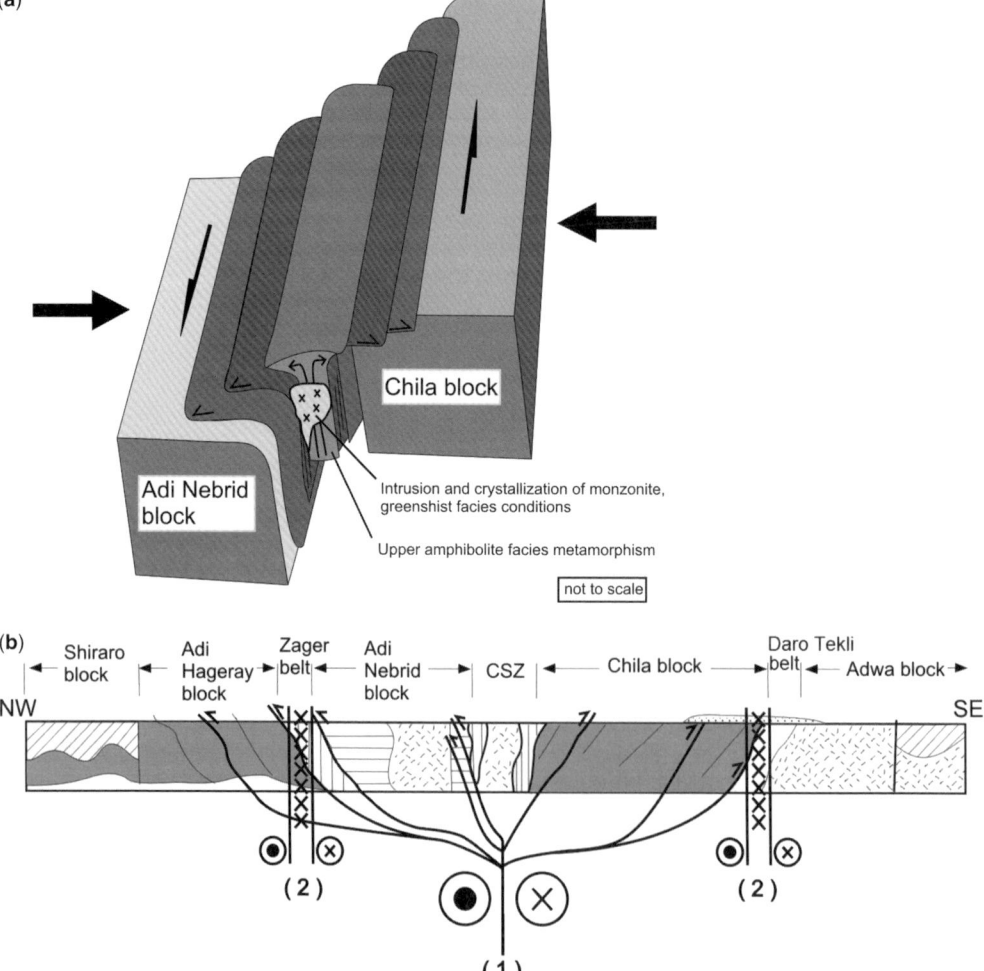

Fig. 9. Cartoon depicting development of CSZ in a tranpressional tectonic setting; (**a**) model showing development of positive flower structure between Chila and Adi Nebrid blocks and shallow level emplacement of monzonite. (**b**) Interpretative cross-section of Axum area (after Tadesse 1997) showing a two-stage development: (1) formation of a positive flower structure with the CSZ as central structure, and (2) development of strike–slip shear zones as transcurrent fault zones.

the magmatic foliation parallel to the general trend of the CSZ and the absence of any subsolidus imprint on its fabric.

The general geological setting of the CSZ in the Axum area of Ethiopia has remarkable similarities to the Central Steep Belt (CSB) in the Nakfa terrane of Eritrea. The Nakfa Terrane, consisting predominately of volcano-sedimentary rocks in which metamorphism generally reached greenschist and lower grades (Drury & De Souza Filho 1998; Teklay 2006; Ghebreab et al. 2009). It also includes some mafic-ultramafic belts. In the CSB migmatites and gneisses occur, possibly derived from high-grade metamorphism and partial anatexis of the volcano-sedimentary rocks. The CSB is decorated with granitoid plutons considered as products of high-grade metamorphism and melting.

Both the belts (CSZ and CSB) show steep planar structures and vertical lineations, a geometry which is in contrast to most high-strain zones in the ANS where strike–slip kinematics with subhorizontal stretching lineations prevail (e.g. Adobha shear belt, Woldehaimanot 2000; Baruda shear belt, Braathen et al. 2001). Vertical lineations are generally realized in wrench-dominated transpression zones with high deformation intensity (Tikoff &

Greene 1997). We therefore interpret the CSZ as a southerly continuation of the CSB and characterize this >300 km traceable structure as a major high-strain zone in the Ethiopian/Eritrean sector of the ANS. The southerly continuation of the CSZ is not known due to phanerozoic cover.

The positive flower structure realized in the Axum area (Fig. 9) is consistent with the geometry of a restraining bend as a consequence of the NE–SW bending of the generally north–south trending zone of sinistral transpression. A positive flower structure has also been described in the Nakfa sector of the high-strain zone in the area of Asmara (Ghebreab et al. 2009) where local rheological heterogeneities are caused by large magmatic intrusions which result in deflection from the north–south geometry of the shear zone.

In addition to the CSZ as a high-strain zone, the regional setting of Axum area (Fig. 1b) also shows localized steep ductile strike–slip shear zones of centimetre scale to more than 100 m in width with a NNE trend. These mylonites with a sinistral sense of movement were mapped and described by Tadesse (1997) to cut and therefore post-date the thrust planes of the flower structure (Fig. 1b) and occur preferentially in the mafic and ultramafic

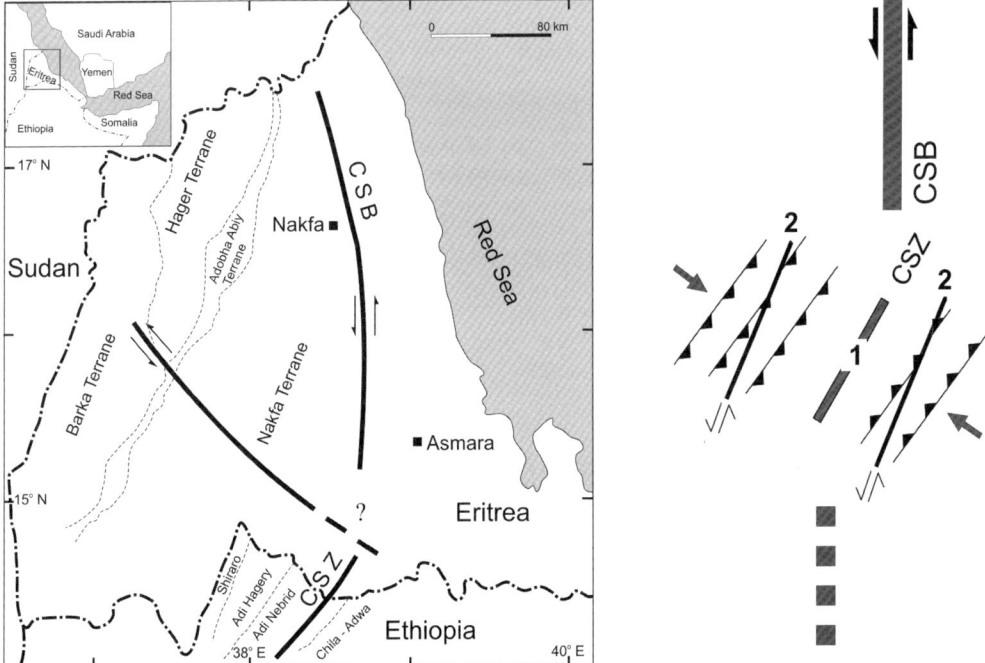

Fig. 10. Interpretative diagram showing the CSZ as southern continuation of a major north–south trending sinistral strike–slip fault zone, the CSB in the Nakfa terrane of Eritrea. Under such kinematics, the NE–SW trend of major structures in the Axum area forms a restraining bend in a transfer zone between north–south trending segments of major transcurrent faults. The numbers (1, 2) refer to the sequence of evolution of structures introduced in Figure 9b.

Zager and Dahro Tekli belts. As the shear zones have not been the objective of our field studies, the interpretation proposed in the following should be considered with some degree of scepticism. It seems likely that during the late stage of deformation the movements in the central uplifted part of the CSZ became locked and further strain was localized along vertical NNE-trending shear zones which were developed preferentially in the units of low strength, such as the low-grade metamorphic mafic sequences. The trend and sense of displacement (sinistral) is in agreement with the kinematics of the transfer zone (Fig. 10).

The north–south trend and the sinistral sense of displacement along this Tigrai–Nakfa shear system (CSZ and CSB) corresponds to the general orientation of transcurrent shear systems in the ANS. The north-eastern part of the Saharan Metacraton formed as a consequence of transpressional suturing in a late stage of the oblique (NW-directed) collision (Abdelsalem *et al.* 2003*b*). A compilation of geochronologic data on the EAO by Meert (2003) suggest a final closure of the Mozambique Ocean and the collision of north-central East Africa (Saharan Metacraton, Abdelsalam *et al.* 2003*a*) and the SLAMIN terrane (Madagascar–India–Somalia–Sri Lanka–the Seychelles) at 650–630 Ma. There are no direct age constraints for the CSZ activity; however, Teklay *et al.* (2001) provided age constraints for the Nakfa area through the 628 Ma syenites which intrude into the fault structures related to the CSB.

The north–south trending structures are cut by later NW–SE oriented faults. Drury & de Souza Filho (1998) mapped large-scale NW–SE strike–slip offsets in Eritrea in the Adobha Abiy and Nakfa Terrane which might also have caused a dextral translation of the CSB/CSZ (Fig. 10). This structure parallels faults described all over the ANS: for example, the Najd fault system in the Eastern Desert of Egypt (Stern 1985) and Saudi Arabia (Johnson & Kattan 2001). Prominent NW–SE-trending regional-scale strike–slip structures in Ethiopia are the Surma and Didesa lineaments (Vail 1983) in the southern part and structures in the Tulu Dimtu area in the Western Ethiopian basement (Alemu & Abebe 2007). The 545–550 Ma Shibta and Shire granites in the Axum area (Tadesse *et al.* 2000) clearly post-date the faulting and are within the time interval of 580–470 Ma inferred for the post-collisional magmatism related to the Late Neoproterozoic Pan-African orogeny in north-eastern Africa (Sudan, Ethiopia, Somalia) and Madagascar (Küster & Harms 1998).

In conclusion, the CSZ is a key area for understanding the Pan-African Orogeny and the interaction of wrench and strike–slip faulting during the formation of Gondwana.

The project was funded by a DFG grant (Wa1010/8-1). A. T. Tegene acknowledges the support from DAAD in form of research fellowship. S. Schöbel is thanked for his help in graphics. We thank M. Abdeen and an anonymous reviewer for constructive comments and suggestions. We are thankful to the guest editors for the handling of the manuscript.

References

ABDELSALAM, M. G. & STERN, R. J. 1996. Sutures and shear zones in the Arabian-Nubian Shield. *Journal of African Earth Science*, **23**, 289–310.

ABDELSALAM, M. G., STERN, R. J., COPELAND, P., ELFAKI, E. M., ELHUR, B. & IBRAHIM, F. M. 1998. The Neoproterozoic Keraf Suture in NE Sudan: sinistral transpression along the eastern margin of West Gondwana. *The Journal of Geology*, **106**, 133–147.

ABDELSALAM, M. G., LIÉGEOIS, J.-P. & STERN, R. J. 2002. The Saharan Metacraton. *Journal of African Earth Science*, **34**, 119–136.

ABDELSALAM, M. G., ABDEL-RAHMAN, E. M., EL-FAKI, E. M., EL-HUR, B. & IBRAHIM, F. M. 2003*a*. Neoproterozoic deformation in the Northeastern part of the Saharan Metacraton. *Precambrian Research*, **123**, 203–221.

ABDELSALAM, M. G., ABDEEN, M. M., DOWAIDAR, H. M., STERN, R. J. & ABDELGHAFFAR, A. A. 2003*b*. Structural evolution of the Neoproterozoic Western Allaqi-Heiani suture, southeastern Egypt. *Precambrian Research*, **124**, 87–104.

ALEMU, T. & ABEBE, T. 2007. Geology and tectonic evolution of the Pan-African Tulu Dimtu Belt, Western Ethiopia. *Online Journal of Earth Sciences*, **1**, 24–42.

ALENE, M. & BARKER, A. J. 1993. Tectonometamorphic evolution of the Moyale region, southern Ethiopia. *Precambrian Research*, **62**, 271–283.

ALSOP, G. I. & HOLDSWORTH, R. E. 2004. Shear Zones – an introduction and overview. *In*: ALSOP, G. I. & HOLDSWORTH, R. E. (eds) *Flow Processes in Faults and Shear Zones*. Geological Society, London, Special Publications, **224**, 1–9.

ANDERSON, J. L. 1996. Status and thermobarometry in granitic batholiths. *Transactions of the Royal Society of Edinburgh: Earth Sciences*, **87**, 125–138.

ANDERSON, J. L. & SMITH, D. R. 1995. The effects of temperature and oxygen fugacity on the Al-in-hornblende barometer. *American Mineralogist*, **80**, 549–559.

BEYTH, M. 1972. *The Geology of Central and Western Tigray*. PhD thesis, Rheinische Friedrich-Wilhams Universitat, Bonn.

BRAATHEN, A., GRENNE, T., SELASSIE, M. G. & WORKU, T. 2001. Juxtaposition of Neoproterozoic units along the Baruda-Tulu Dimtu shear belt in the East African Orogen of western Ethiopia. *Precambrian Research*, **107**, 215–234.

BRÄTZ, H. & KLEMD, R. 2002. *Analysis of Rare Earth Elements in Geological samples by Laser Ablation Inductively Coupled Mass Spectrometry*. Online publication no. 5988-6305EN. http://www.chem.agilent.com.

COSCA, M. A., SCHIMRON, A. & CABY, R. 1999. Late Precambrian metamorphism and cooling in the Arabian-Nubian Shield: petrology and Ar/Ar geochronology of metamorphic rocks of the Elat area (southern Israel). *Precambrian Research*, **98**, 107–127.

DE WALL, H., GREILING, R. O. & SADEK, M. F. 2001. Post-collisional shortening in the late Pan-African Hamisana high strain zone, SE Egypt: field and magnetic fabric evidence. *Precambrian Research*, **107**, 179–194.

DRURY, S. A. & DE SOUZA FILHO, C. R. 1998. Neoproterozoic terrane assemblages in Eritrea: review and prospects. *Journal of African Earth Sciences*, **27**, 331–348.

FOSSEN, H. & TIKOFF, B. 1993. The deformation matrix for simultaneous simple shearing, pure shearing and volume change, and its application to tranpression-translation tectonics. *Journal of Structural Geology*, **15**, 413–422.

GHEBREAB, W., GREILING, R. O. & SOLOMON, S. 2009. Structural setting of Neoproterozoic mineralization, Asmara district, Eritrea. *Journal of African Earth Sciences*, **55**, 219–235.

GRENNE, T., PEDERSEN, R. B., BJERKGARD, T., BRAATHEN, A., SELASSIE, M. G. & WORKU, T. 2003. Neoproterozoic evolution of Western Ethiopia: igneous geochemistry, isotope systematics and U–Pb ages. *Geological Magazine*, **140**, 373–395.

HAMMERSTROM, J. M. & ZEN, E. 1986. Aluminium in hornblende: an empirical igneous geobarometer. *American Mineralogist*, **71**, 1297–1313.

HARGRAVES, R. B., JOHNSON, D. & CHAN, C. Y. 1991. Distribution anisotropy: the cause of AMS in igneous rocks? *Geophysical Research Letters*, **18**, 2193–2196.

HOLLAND, T. & BLUNDY, J. 1994. Non-ideal interactions in calcic amphiboles and their bearing on amphibole-plagioclase thermometry. *Contributions to Mineralogy and Petrology*, **116**, 433–447.

HOLLISTER, L. S., GRISSOM, G. C., PETERS, E. K., STOWELL, H. H. & SISSON, V. B. 1987. Confirmation the empirical correlation of Al in hornblende with pressure of solidification of calc-alkaline plutons. *American Mineralogist*, **72**, 231–239.

JELINEK, V. 1981. Characterization of the magnetic fabrics of rocks. *Tectonophysics*, **79**, 63–67.

JOHNSON, M. C. & RUTHERFORD, M. J. 1989. Experimental calibration of the aluminium-in-hornblende geobarometer with applications to Long Valley caldera (California) volcanic rocks. *Geology*, **17**, 837–841.

JOHNSON, P. R. & KATTAN, F. 2001. Oblique sinistral transpression in the Arabian shield: the timing and kinematics of a Neoproterozoic suture zone. *Precambrian Research*, **107**, 117–138.

JONES, R. R., HOLDSWORTH, R. E. & BAILEY, W. 1997. Lateral extrusion in transpression zones: the importance of boundary conditions. *Journal of Structural Geology*, **19**, 1201–1217.

KRÖNER, A., STERN, R. J., LINNEBACKER, P., MANTON, W., RAISCHMANN, T. & HUSSAIN, I. M. 1991. Evolution of Pan-African island arc assemblages in the south of Red Sea Hills, Sudan, and in SW Arabia as exemplified by geochemistry and geochronology. *Precambrian Research*, **53**, 99–118.

KÜSTER, D. & HARMS, U. 1998. Post-collisional potassic granitoids from the southern and northwestern parts of the Late Neoproterozoic East African Orogen: a review. *Lithos*, **45**, 177–195.

LEAKE, B. E., WOOLLEY, A. R. *ET AL.* 1997. Nomenclature of amphiboles. Report of the Subcommittee on Amphibols of the International Mineralogical Association Commission on New Minerals and Mineral Names. *European Journal of Mineralogy*, **9**, 623–651.

MARTÍN-HERNÁNDEZ, F., LÜNEBURG, C., AUBORG, C. & JACKSON, M. (eds) 2004. *Magnetic Fabric: Methods and Applications*. Geological Society, London, Special Publications, **238**.

MEERT, J. G. 2003. A synopsis of events related to the assembly of eastern Gondwana. *Tectonophysics*, **362**, 1–40.

MILLER, M. M. & DIXON, T. H. 1992. Late Proterozoic evolution of the northern Hamisana zone, northeast Sudan: constraints on Pan-African accretionary tectonics. *Journal of the Geological Society, London*, **149**, 743–750.

SANDERSON, D. J. & MARCHINI, W. R. D. 1984. Transpression. *Journal of Structural Geology*, **6**, 449–458.

SCHACKLETON, R. M. 1986. Precambrian collision tectonics in Africa. *In*: COWARD, M. P. & RIES, A. C. (eds) *Collision Tectonics*. Geological Society, London, Special Publications, **19**, 324–349.

SCHMIDT, M. W. 1992. Amhibole composition in tonalite as a function of pressure: an experimental calibration of the Al-in-hornblende barometer. *Contributions to Mineralogy and Petrology*, **110**, 304–310.

SIEGESMUND, S., ULLEMEYER, K. & DAHMS, M. 1995. Control on magnetic rock fabrics by mica preferred orientation: a quantitative approach. *Journal of Structural Geology*, **17**, 1601–1613.

STEPHENSON, A. 1992. Distribution anisotropy: two simple models for magnetic lineation and foliation. *Physics of the Earth and Planetary Interior*, **82**, 49–53.

STERN, R. J. 1985. The Najd fault system, Saudi Arabia and Egypt: a late Precambrian rift-related transform system? *Tectonics*, **4**, 497–511.

STERN, R. J. 1994. Arc assembly and continental collision in the Neoproterozoic East African orogen: implication of the consolidation of Gondwanaland. *Annual Review of Earth Planetary Sciences*, **22**, 319–351.

STERN, R. J. & ABDELSALAM, M. G. 1998. Formation of juvenile continental crust in the Arabian–Nubian shield: evidence from granitic rocks of the Nakasib suture, NE Sudan. *Geologische Rundschau*, **87**, 150–160.

STERN, R. J. & DAWOUD, A. S. 1991. Late Precambrian (740 Ma) charnockite, enerbite and granite from Jebel Moya, Sudan: a link between the Mozambique Belt and the Arabian-Nubian Shield? *Journal of Geology*, **99**, 648–659.

SUN, S.-S. & MCDONNOUGH, W. F. 1989. Chemical and isotopic systematics of ocean basalts: implications for mantle composition and processes. *In*: SAUNDERS, A. D. & NORRY, M. J. (eds) *Magmatism in the Ocean Basins*. Geological Society, London, Special Publications, **42**, 313–345.

TADESSE, G. & ALLEN, A. 2005. Geology and geochemistry of the Neoproterozoic Tuludimtu Ophiolite suite, western Ethiopia. *Journal of African Earth Sciences*, **41**, 192–211.

TADESSE, T. 1996. Structure across a possible intra-oceanic suture zone in the low-grade Pan-African rocks of northern Ethiopia. *Journal of African Earth Sciences*, **23**, 375–381.

TADESSE, T. 1997. *The Geology of Axum Area (ND 37-6)*. Memoir No. 9. Ethiopian Institute of Geological Surveys, Addis Ababa, Ethiopia.

TADESSE, T., HOSHINO, M. & SAWADA, Y. 1999. Geochemistry of low-grade metavolcanic rocks from the Pan-African of the Axum Area, northern Ethiopia. *Precambrian Research*, **99**, 101–124.

TADESSE, T., HOSHINO, M., SUZUKI, K. & IIZUMI, S. 2000. Sm–Nd, Rb–Sr and Th–U–Pb zircon ages of syn- and post-tectonic granitoids from the Axum area of northern Ethiopia. *Journal of African Earth Sciences*, **30**, 313–327.

TARLING, D. H. & HROUDA, F. 1993. *The Magnetic Anisotropy of Rocks*. Chapman & Hall, London.

TEKLAY, M. 1997. *Petrology, Geochemistry and Geochronology of Neoproterozoic Magmatic Arc Rocks from Eritrea: Implications for Crustal Evolution in the Southern Nubian Shield*. Department of Mines, Asmara, Eritrea, Memoir 1.

TEKLAY, M. 2006. Neoproterozoic arc–back–arc system analog to modern arc–back–arc systems: evidence from tholeiite–boninite association, serpentinite mudflows and across-arc geochemical trends in Eritrea, southern Arabian–Nubian shield. *Precambrian Research*, **145**, 81–92.

TEKLAY, M., KRÖNER, A. & MEZGER, K. 2001. Geochemistry, geochronology and isotope geology of Nakfa intrusive rocks, northern Eritrea: products of a tectonically thickened magmatic arc. *Journal of African Earth Sciences*, **33**, 283–305.

THOMAS, W. M. & ERNST, W. G. 1990. The aluminum content of hornblende in calc-alkaline granitic rocks: a mineralogic barometer calibrated experimentally to 12 kbar. *In*: SPENCE, R. J. & CHOU, I-Ming (eds) *Fluid-mineral Interactions: A tribute to H.P. Eugster*. Geochemical Society Special Publications, **2**, 59–63.

TIKOFF, B. & GREENE, D. 1997. Stretching lineations in transpressional shear zones: an example from the Sierra Nevada Batholith, California. *Journal of Structural Geology*, **19**, 29–39.

TOLESSA, S., BONAVIA, F. F., SOLOMAN, M., HAILE-MESKEL, A. & TEFERRA, E. 1991. Structural pattern of Pan-African rocks around Moyale, southern Ethiopia. *Precambrian Research*, **52**, 179–186.

VAIL, J. R. 1983. Pan-African crustal accretion in northwest Africa. *Journal of African Earth Sciences*, **1**, 285–294.

WOLDEHAIMANOT, B. 2000. Tectonic setting and geochemical characterisation of Neoproterozoic volcanics and granitoids from the Adobha Belt, northern Eritrea. *Journal of African Earth Sciences*, **30**, 817–831.

WORKU, H. & SCHANDELMEIER, H. 1996. Tectonic evolution of the Neoproterozoic Adola Belt of southern Ethiopia: evidence for a Wilson Cycle process and implications for oblique plate collision. *Precambrian Research*, **77**, 179–210.

Constraining the timing of deformation in the southwestern Central Zone of the Damara Belt, Namibia

L. LONGRIDGE[1]*, R. L. GIBSON[1], J. A. KINNAIRD[1] & R. A. ARMSTRONG[2]

[1]*School of Geosciences, University of the Witwatersrand, PVT Bag 3, Wits, 2050, South Africa*

[2]*PRISE, Australian National University, Research School of Earth Sciences, Building 61, Mills Road, Acton, 0200, Australia*

Corresponding author (e-mail: lukelongridge@gmail.com)

Abstract: Structural investigations and U–Pb sensitive high-resolution ion microprobe (SHRIMP) dating of rocks from the southwestern Central Zone of the Damara Belt, Namibia, reveal that a major SE-verging deformation event (D2) occurred at between 520 and 508 Ma. During D2, SE-verging simple shear and NE–SW pure shear extension in a constrictional stress field produced recumbent, south- to SE-verging, kilometre-scale folds and ductile shear zones, a NE–SW extensional lineation and conjugate shear bands, and was coeval with granitoid emplacement and high-grade metamorphism. The timing of this event is constrained by anatectic leucosomes in D2 shear zones (511 ± 18 Ma) and extensional shear bands (508.4 ± 8.7 Ma) as well as by syntectonic grey granites (520.4 ± 4.2 Ma), and is similar to ages for high-grade metamorphism in the Central Zone. An upright folding event (D3) occurred at *c.* 508 Ma, resulting in the formation of basement-cored fold interference domes. The timing of deformation and metamorphism at 520–508 Ma in the mid-crustal SW Central Zone contrasts with ages of 560–540 Ma for shallow crustal NW-verging folding and thrusting elsewhere in the Central Zone that was concomitant with voluminous magmatism. This magmatism led to metamorphism and anatexis of the basement and the emplacement of anatectic red granites at 539 ± 17 to 535.6 ± 7.2 Ma, which contain 1013 ± 21 Ma inherited zircons. The Central Zone therefore contains a record of crustal thickening, heating of the mid-crust, exhumation and orogen-parallel extension over the life of an orogen.

The Damara belt of Namibia is a well-exposed eroded Neoproterozoic (Gray *et al.* 2008) collisional orogen, with an orogenic core (the Central Zone; Miller 1983) that exposes mid-crustal levels of the orogen. Here, high-grade (upper-amphibolite- to granulite-facies; Masberg *et al.* 1992; Nex *et al.* 2001) polydeformed (Smith 1965; Jacob *et al.* 1983; Nex 1997) metasediments have been intruded by voluminous syntectonic (Brandt 1985; Nex 1997; Kisters *et al.* 2004) granitoid magmas during collision, and the Central Zone provides an ideal site to study the nature and timing of mid-crustal processes during continent–continent collision. Suggestions that progressive deformation resulting from SW-vergent, orogen-parallel tectonic escape at granulite-grade mid-crustal levels (Oliver 1994; Poli & Oliver 2001) was coeval with NW-verging thrusting at amphibolite-grade shallow crustal levels (Kisters *et al.* 2004) are re-evaluated using structural mapping and zircon U–Pb sensitive high-resolution ion microprobe (SHRIMP) ages for syntectonic granitoids. It is demonstrated that intense, non-coaxial deformation and extension at high-grade mid-crustal levels shows a temporal progression, but is younger than thrusting and crustal thickening at shallow crustal levels.

The geology of the Damara Orogen

As a Neoproterozoic mobile belt formed during assembly of Gondwana, the Damara Belt is one of many such mobile belts that cross-cut the African continent that have been termed the Pan-African (Kennedy 1964). Resulting from the collision between the Congo Craton and the Kalahari Craton from 542 Ma (Miller 1983, 2008) to between 530 and 500 Ma (Gray *et al.* 2008), the Damara Belt extends NE through Namibia and Botswana into Zambia, where it continues as the Zambezi Belt (Fig. 1a). It was preceded by the collision between the Congo Craton and the Rio de la Plata Craton at 580–550 Ma to form the Kaoko Belt (Prave 1996; Gray *et al.* 2008).

The NE-trending orogenic belt has been divided into a number of tectonometamorphic zones, bounded by major geophysical lineaments (Fig. 1b). These include the foreland basin of the

From: VAN HINSBERGEN, D. J. J., BUITER, S. J. H., TORSVIK, T. H., GAINA, C. & WEBB, S. J. (eds) *The Formation and Evolution of Africa: A Synopsis of 3.8 Ga of Earth History*. Geological Society, London, Special Publications, **357**, 107–135. DOI: 10.1144/SP357.7 0305-8719/11/$15.00 © The Geological Society of London 2011.

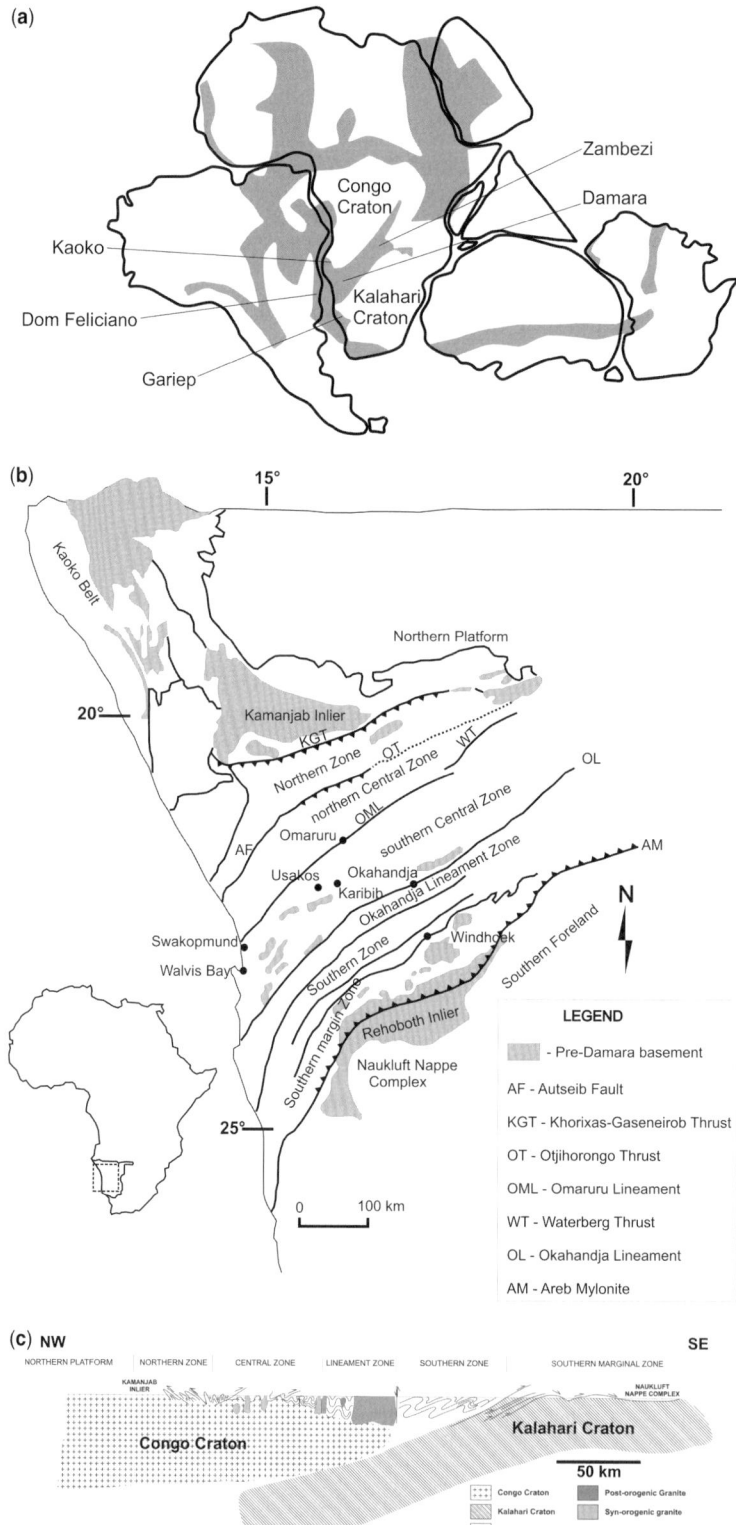

Fig. 1.

Southern Foreland and Platform, the high-pressure low-temperature Southern Zone characterized by SSE-verging structures and metamorphic conditions of c. 0.8 GPa and c. 600 °C (Kasch 1983) and the north-verging thrusts and syntectonic sediments of the Northern Zone and Northern Platform (Miller 2008). The Central Zone forms the core of the orogen, where metasediments have experienced intense deformation (Barnes 1981; Downing & Coward 1981; Poli 1997) and high-temperature low-pressure (HTLP) metamorphism (Masberg et al. 1992; Nex et al. 2001) and have been intruded by voluminous granitoids. Characteristic of the Central Zone are kilometre-scale exposures of pre-Damaran basement in the cores of antiformal structures (Miller 2008). The overall geometry of the orogen indicates that northeastwards-directed subduction of oceanic material beneath the Congo Craton preceded the collision with the Kalahari Craton, closing the Khomas Sea (Miller 2008) which had opened and begun accumulating sediments by 750–760 Ma (although opening could have been as early as 900 Ma; Miller 2008) following the break-up of Rodinia at 970–850 Ma (Yoshida et al. 2003). This closure and subduction resulted in the thrusting of sea-floor basalts and sea-floor sediments (the Matchless Amphibolite and Khomas Schists, respectively; Miller 2008) onto the Kalahari Craton, forming the HPLT Southern Zone (Barnes & Sawyer 1980; Fig. 1c).

The geology of the Central Zone

The Damara Supergroup in the Central Zone comprises a distinctive sequence of quartzites, calc-silicates, marbles, schists and metapelites. This stratigraphic package was deposited on the Palaeoproterozoic–Mesoproterozoic (Kröner et al. 1991; Jacob et al. 2000) quartzofeldspathic basement gneisses of the Abbabis Complex (Fig. 2). The Nosib Group of the Damara Supergroup, which immediately overlies the Abbabis Complex, consists of quartzite and conglomerate of the lower Etusis Formation and diopside-feldspar gneiss, biotite schist and cordierite gneiss of the upper Khan Formation. This is unconformably overlain by the Swakop Group, which begins with the marble, pyritic quartzite, metapelite and calc-silicate

of the lower Rössing Formation (Ugab Subgroup), overlain by the distinctive diamictite and meta-ironstone of the Chuos Formation and marble of the Arandis Formation (Usakos Subgroup). The uppermost Navachab Subgroup comprises minor diamictite at the base (Ghaub Formation), a thick marble and dolomite package of the Karibib Formation. The uppermost unit is a thick package of pelitic schists, the Kuiseb Formation (Smith 1965; Jacob 1974; Hoffmann et al. 2004; Miller 2008).

In the Central Zone, three main phases of deformation (D1–D3) have been recorded (Smith 1965; Nash 1971; Blaine 1977; Barnes 1981; Coward 1983; Jacob et al. 1983) affecting both the Damara Supergroup metasediments and the pre-Damaran Abbabis Complex gneissic basement. D1 macroscopic structures are difficult to recognize and D1 is not always recognized as a separate deformation event (e.g. Jacob 1974). D1 deformation is generally recorded as a widespread bedding-parallel foliation with rare intrafolial folds (Blaine 1977; Barnes 1981), but the orientation and vergence of D1 deformation is uncertain. D2 (the second major phase of deformation) is characterized by high-strain zones and tight to isoclinal, recumbent folds (Miller 2008) which have been suggested to verge to the south (Blaine 1977), SE (de Kock 1989) and SW (Coward 1983), and which have been interpreted as non-cylindrical sheath folds by Coward (1983) and Miller (2008). An intense S2 cleavage associated with this phase of deformation is widespread through the Central Zone, as is a NE–SW extension lineation (Downing & Coward 1981; Coward 1983; Kisters et al. 2009). D3 (the final major phase of deformation) formed upright to SE-verging folds (Jacob 1974; Barnes 1981; Downing & Coward 1981; Sawyer 1981; Downing 1982; Coward 1983; de Kock 1989) on a large scale with few small-scale structures, and it has been suggested that interference between these upright folds and earlier recumbent folding has resulted in the formation of the domal structures that expose basement gneisses of the Abbabis Complex, which characterize the Central Zone (Jacob et al. 1983). Diapirism (Barnes 1981; Kröner 1984) and the effect of ballooning granites (Barnes & Downing 1979; Barnes 1981) have also been invoked to explain the domes of the Central Zone. More recently, dome formation and NE–SW extension lineations have

Fig. 1. (a) Map of the Neoproterozoic mobile belts formed during the assembly of Gondwana showing the location of the Congo and Kalahari cratons, the Damara belt and adjacent orogenic belts (modified after Wilson et al. 1997). (b) Map of the Damara Orogen, Namibia, showing the zones and major lineaments that comprise the orogen and areas of exposed pre-Damaran basement. Note the numerous areas of pre-Damaran basement in the southern Central Zone. The approximate location of this study is shown by the square east of Swakopmund (modified after Miller 1983). (c) Schematic cross-section through the Damara Orogen, showing the vergence of structures in the zones of the orogen and the underplating of the Kalahari Craton beneath the Congo Craton following NE-directed subduction below the Congo Craton (modified from Barnes & Sawyer 1980).

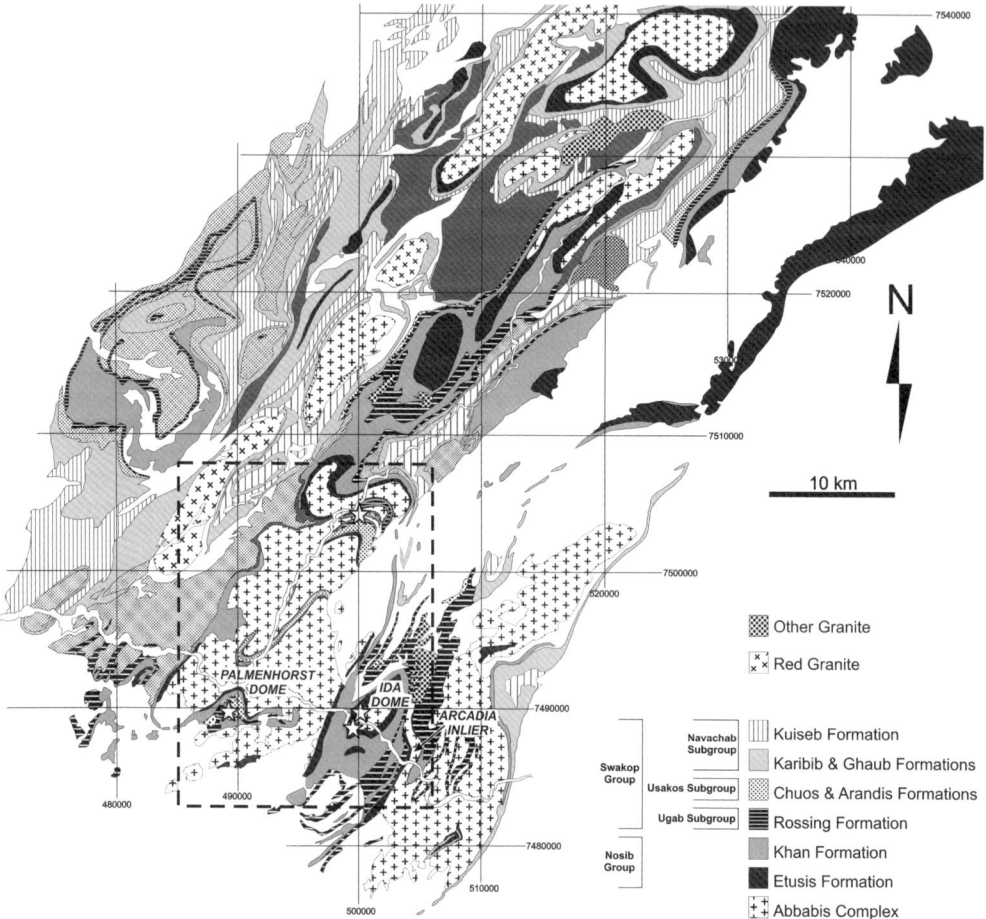

Fig. 2. Map of the Central Zone, showing the distribution of Damara Supergroup rocks and large exposures of pre-Damaran basement (the Palmenhorst Dome, the Ida Dome and the Arcadia Inlier). The area of study is outlined by the dashed rectangle and the locations of samples collected for this study are shown by the white stars (after Lehtonen *et al.* 1995).

been ascribed to the formation of a SW-verging mid-crustal metamorphic core-complex (Oliver 1994, 1995) or to a SW-vergent ductile flow within a laterally constricted mid-crust (Poli 1997; Poli & Oliver 2001) rather than to interference folding. Some recent studies of the southwestern Central Zone have considered deformation as progressive, rather than occurring as discrete (i.e. D1, D2, D3 etc.) deformation events (Poli 1997; Poli & Oliver 2001). NW-verging structures in the Central Zone near Karibib (Kisters *et al.* 2004) have a vergence perpendicular to the suggested south-westwards vergence of structures in the south-western Central Zone. This conflicting vergence of structures in the central v. southwestern parts of the Central Zone has been ascribed to different crustal responses to crustal shortening. Ductile

mid-crust (the highest-grade southwestern Central Zone) experienced SW-vergent tectonic escape while thrusting and crustal thickening occurred at lower-grade shallow crustal levels (the central portions of the Central Zone near Karibib; Kisters *et al.* 2004).

HTLP metamorphism has affected the Central Zone (Jacob 1974; Miller 1983; Puhan 1983; Masberg *et al.* 1992; Jung & Mezger 2003) and metamorphic grade apparently increases along strike to the SW (Hoernes & Hoffer 1979; Goscombe *et al.* 2004), reaching peak upper-amphibolite- to lower-granulite-facies conditions along the Atlantic seaboard. Peak P–T conditions are generally considered to be 650–750 °C and 0.3–0.5 GPa (Nash 1971; Jacob 1974; Sawyer 1981; Puhan 1983; Buhn *et al.* 1995; Poli 1997; Nex *et al.* 2001;

Ward *et al.* 2008). These P–T estimates indicate upper-amphibolite-facies conditions but Masberg *et al.* (1992) made the suggestion (based on micro-fabric analysis) that the rocks reached the granulite facies, despite the lack of metamorphic orthopyroxene in any of the metamorphic assemblages. An issue which remains unresolved is whether metamorphism occurred as a single event (Hartmann *et al.* 1983; Hawkesworth *et al.* 1983; Puhan 1983), was polyphase (Nash 1971; Barnes & Sawyer 1980; Kröner 1982; Kasch 1983; Nex 1997) or should be considered progressive (Poli 1997). Jung & Mezger (2003) obtained Sm–Nd garnet and U–Pb monazite ages from metapelites and migmatites indicating peak regional metamorphism at between 525 ± 2 and 504 ± 3 Ma in addition to some older (540 ± 4 to 530 ± 3 Ma) ages, implying two metamorphic events. Peak metamorphism in the Central Zone is considered to be coeval with intense (D2 and possibly D3) deformation (Nash 1971; Jacob 1974; Oliver 1994; Poli 1997), irrespective of whether this deformation is viewed as a single progression (Oliver 1994; Poli 1997) or as distinct polyphase events (Jacob 1974; Sawyer 1981). However, there is evidence that the thermal effects of the orogen continued post-deformation, leading to annealing of metamorphic textures (Poli 1997; Nex *et al.* 2001).

Timing of deformation and metamorphism in the Central Zone

Although the nature of deformation and metamorphism has been thoroughly investigated by previous workers, the absolute timing of these events is less well established. There is a paucity of U–Pb zircon ages for the Central Zone: ages of 550 to 540 Ma have been suggested for NW-verging folds and thrusts in the Karibib area (Kisters *et al.* 2004), based on ages obtained for the syntectonic Mon Repos quartz diorite (ages of 564 ± 5 and 546 ± 6 Ma; Jacob *et al.* 2000) and Okongava Diorite (age of 558 ± 5 Ma; de Kock *et al.* 2000). These are both part of the syntectonic Goas Intrusive Suite of metagabbros and diorites and the post-tectonic Rotekuppe Granite (539 ± 6 Ma; Jacob *et al.* 2000). A similar U–Pb zircon age of 549 ± 11 Ma for the syntectonic Salem-type (Miller 1983, 2008) Stinkbank granite near Usakos (Johnson *et al.* 2006) confirms these ages for NW-vergent folding and thrusting. This NW-vergent deformation is orthogonal to the SW vergence for deformation noted in the southwestern Central Zone (Downing & Coward 1981; Coward 1983; Oliver 1995; Poli 1997; Poli & Oliver 2001), but both the NW- and SW-verging deformations were considered to be contemporaneous effects of differing

rheological responses of the upper v. mid-crust to crustal shortening (Kisters *et al.* 2004). Peak metamorphism is suggested as being slightly later; Miller (1983, 2008) argues for an age of 535 Ma for post-tectonic metamorphism, although ages of between 534 and 508 Ma are proposed by Nex *et al.* (2001) for M2 and Sm–Nd garnet and U–Pb monazite ages of between 525 and 504 Ma (Jung & Mezger 2003) indicate even younger ages. Observations that metamorphism was coeval with SW-verging deformation in the southwestern Central Zone (Nash 1971; Jacob 1974; Poli 1997) hints at a discrepancy in the ages for deformation, which is suggested to have occurred between 560 and 540 Ma (Kisters *et al.* 2004; Johnson *et al.* 2006) based on syntectonic granitoids. Syntectonic metamorphism may be significantly younger, between 534 to 504 Ma (Nex *et al.* 2001; Jung & Mezger 2003).

A variety of granitoid types exist in the Central Zone which show relationships to deformation; ages for these rocks may be used to constrain the timing of deformation. Red gneissic granites, which are either syn- to late-tectonic (Jacob 1978) or post-tectonic (Miller 2008), are K-feldspar rich, deficient in ferromagnesian minerals and are thought to be the product of anatexis of the pre-Damaran basement and the Etusis Formation (Jacob 1974). They commonly contain xenoliths of migmatitic Abbabis Complex augen paragneisses, are stratigraphically confined to the lower Damara Supergroup and basement-cover contact (Smith 1965; Jacob 1974) and have been dated at 534 Ma (Briqueu *et al.* 1980). The Salem-type granites (Smith 1965; Jacob 1974; Marlow 1981; Miller 1983, 2008; Brandt 1985) are a group of fine- to coarse-grained porphyritic grey granites that are generally confined to stratigraphic levels above the Karibib Formation (Jacob 1974), and which are thought to be syntectonic (Johnson *et al.* 2006). They have been dated at 560–550 Ma (Downing 1982; Kröner 1982; Hawkesworth *et al.* 1983; Johnson *et al.* 2006). Early non-porphyritic granites and leucogranites (Miller 2008), also called 'red and grey homogeneous syntectonic granites' (Brandt 1985) and 'equigranular grey granites' (Nex 1997), are a group of fine- to medium-grained homogeneous granites which occur predominantly in the northern Central Zone (Miller 2008) and are not stratigraphically confined (Brandt 1985; Nex 1997). These granites display a biotite foliation and are thought to be syn-D2 (Haack *et al.* 1980; Miller 2008), with a whole rock Rb–Sr age of 514 ± 22 Ma (Haack *et al.* 1980). Later granites include sheets of leucogranite or alaskite (Jacob 1974; Marlow 1981; Brandt 1985) that are considered to be post-tectonic, host uranium mineralization and are confined to the highest-grade western portions of the Central Zone.

These have been dated at 508 Ma (Briqueu *et al.* 1980) and large stocks, plugs or batholiths of leuco-granite (the Bloedkoppie, Donkerhuk, and Kubas granites; Jacob 1974; Marlow 1981; Brandt 1985) have been dated at 527–505 Ma (Haack & Gohn 1988; Kukla *et al.* 1991).

It is possible, based on the apparent discrepancy in the timing of deformation in the Central Zone, that tectonometamorphic events did not occur sim-ultaneously throughout the Central Zone. Should this be the case, the syntectonic or post-tectonic character of Damaran granitoids, which has been used to constrain the timing of deformation in the Central Zone, should only be applied locally; gran-itoids which are syntectonic in one area may post-date deformation in another.

The geology of the study area

The study area is located in the southwestern Central Zone of the Damara Orogen, *c.* 50 km east of Swakopmund (Fig. 1b). This study is located around three exposures of pre-Damaran basement (Fig. 2). The Palmenhorst Dome is in the west of the study area, the smaller Ida Dome is located in the centre of the study area and the Arcadia Inlier lies to the east of the study area. The Palmenhorst Dome is a large (*c.* 20 km × 10 km) area of pre-Damaran basement, and the large-scale structure is dominated by a kilometre-scale sliver of lower Damara Supergroup rocks in the centre of the dome (first identified by Barnes 1981) and a number of slivers of Damara Supergroup rocks near the northern margin of the dome. Along the southern margin of the Palmenhorst Dome, steeply dipping Damaran metasediments have been complexly folded. The Ida Dome has a core of Abbabis Com-plex gneisses, mantled by Nosib Group rocks. A sliver of Abbabis Complex rocks between folded

Damara Supergroup rocks forms the Arcadia Inlier (Fig. 2).

Deformation in the study area

In the Ida and Palmenhorst Domes and the surround-ing areas along the Khan and Swakop Rivers, excel-lent exposure has enabled a detailed structural investigation and a sequence of deformation can be identified. Pre-Damaran structures are identified in the Abbabis Complex, and early Damaran (D1) structures are rarely preserved. Two main defor-mation events can be identified: an intense, non-coaxial event (D2) which formed kilometre-scale shallow north- or NE-dipping to recumbent folding and south- to SE-verging shear zones on the limbs of kilometre-scale F2 folds, as well as the development of shallow extensional structures and a NE–SW extension lineation. This was followed by NE-trending large-wavelength upright folding (D3) which was largely coaxial and which reoriented earlier formed structures.

Pre-Damaran deformation and D1. Planar to linear pre-Damaran fabrics (S/L- and L/S-tectonites) are commonly found in the quartzofeldspathic gneisses of the Abbabis Complex. Although linear fabrics in the Abbabis Complex have been previously sug-gested to be related to NE–SW extension during Damaran deformation (Oliver 1994; Poli 1997; Poli & Oliver 2001), linear fabrics are seen to be folded by D2 Damaran structures (Fig. 3a), indicat-ing that some of these fabrics predate Damaran deformation. The earliest recognized Damaran deformation in the study area (D1) is preserved as rootless isoclinal intrafolial folds in rocks with a strong fabric (Fig. 3b), and is likely the result of a regional fabric-forming event (Blaine 1977; Barnes 1981; Downing & Coward 1981). This event has been largely overprinted by D2.

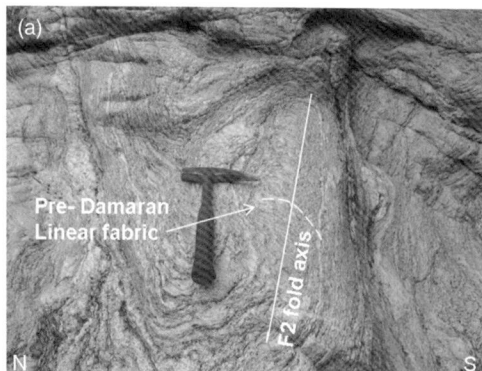

Fig. 3. (**a**) Pre-Damaran linear fabric in quartzofeldspathic Abbabis Complex gneisses, folded by east–west trending north-dipping F2 folds. (**b**) F1 rootless isoclinal intrafolial fold in Chuos Formation biotite-quartz schists.

D2 deformation. D2 is the most intense deformation event in the study area, forming kilometre-scale south- to SE- verging folds and smaller-scale parasitic folds. D2 shear zones are commonly developed on the extending upper limbs of large-scale folds and a regional extensional detachment is seen near the contact between the Abbabis Complex and the Damara Supergroup, which was formed during the late stages of D2. Vergence of D2 shear zones remains to the south or SE, but late-stage extension appears to be largely in a NE–SW direction, forming a conjugate set of shear bands and a shallow NE-plunging mineral stretching lineation. The extension of the upper limbs of large-scale D2 folds and the consistent vergence of F2 folds and later shear zones suggests that D2 was a progressive event, with deformation shifting from largely south- to SE-verging folding to the formation of shear zones and finally NE–SW extensional structures. Furthermore, similar field relationships of both F2 folds and extensional structures to high-grade metamorphism and granitoid magmatism indicate that folding and extension are part of a single deformation event.

South- to SE-verging D2 folding in the southwestern Central Zone. D2 folds are developed in both Abbabis Complex gneisses and Damaran metasediments. Although D2 structures have been reoriented by upright D3 structures, they show a consistent vergence throughout the study area when measured on the hinges of these D3 structures (where the effects of D3 deformation are less pronounced). Along the northern margin of the Palmenhorst Dome (Fig. 4a–c), north-dipping lithological layering is subparallel to axial planes of D2 folds. Mesoscale fold hinge lines show a range of orientations however, suggesting possible sheath folding. Possible sheath folds are locally found (Fig. 4a), but are related to intense shearing in discrete shear zones. Generally mesoscale folds are tight to isoclinal, north-dipping and have subhorizontal hinge lines (Fig. 4b) consistent with a southward vergence for D2. In the centre of the Palmenhorst Dome (Fig. 4d, e), Abbabis Complex gneisses and amphibolites are folded by SE-verging metre-scale folds. These folds have shallow to moderate NE- or SW-plunging fold hinge lines, and moderate NW-dipping axial planes subparallel to the lithological layering. Mesoscale folding elsewhere in the Palmenhorst Dome is parasitic to an isoclinal, north- to NW-dipping, kilometre-scale D2 syncline (the Hook Fold). Limited data show a range of fold hinge line and axial plane orientations (Fig. 4f) which may reflect D2 sheath folding; possible mesoscale sheath folds are locally found (Fig. 4g). However, this major D2 structure has been affected by a kilometre-scale upright D3 synform, and such a

range of axial planar orientations may also reflect the reorientation of D2 structures by D3 folding. In the centre of the Ida Dome (Fig. 4h–j), D2 folds are found on the hinge of a D3 anticline and have not been affected by D3 folding. Here, mesoscale folds are SSE-verging (Fig. 4h) or may be symmetric M-folds with subhorizontal axial planes (Fig. 4i). Axial planes of D2 folds are shallow NNW-dipping while lithological layering dips more steeply to the north, indicating that these mesoscale folds lie on the overturned limb of a major D2 structure. The overall sense of vergence for F2 folding throughout the study area is south- to SE-verging. The general parallelism between F2 fold axial planes and lithological layering indicates the kilometre-scale folds are tight to isoclinal, reflected by the tight to isoclinal nature of mesoscale F2 folds.

D2 shear zones and NE–SW extension in the Central Zone. Continued non-coaxial deformation following D2 folding in the Central Zone led to the development of shear zones on the extending limbs of kilometre-scale D2 folds and at or near the contact between the Abbabis Complex and the Damara Supergroup. One of these shear zones is well exposed near the western margin of the Ida Dome. This major ductile shear zone is 0.5–1 km thick and contains tight to isoclinal F2 folds (Fig. 5a) and D2 shear bands, which typically contain leucosomes (Fig. 5b). These structures are found in both the Abbabis Complex and in the Etusis Formation, although the style of deformation varies according to lithology. In the Etusis Formation, decimetre-scale layering of quartzite and semi-pelite has resulted in a competence contrast, producing tight to isoclinal metre-scale folding of more competent quartzite and less competent semi-pelite layers. These folds have shallow SW-plunging hinge lines and shallow NW-dipping axial planes, which are subparallel to the lithological layering in the area (Fig. 5c). Locally, intersection lineations are noted subparallel to the hinges of F2 folds. In the Abbabis Complex, quartzofeldspathic gneisses form only minor centimetre-scale F2 folds; shear bands are the dominant structures developed, and may be centimetre- to metre-scale. These shear bands are similar to the 'extensional shear zones' noted by Oliver (1994) and develop in conjugate sets: a north-dipping, top-to-the-north-verging set and a west-dipping, SW-verging set (Fig. 5d). Using these two conjugate sets of shear bands, it is possible to calculate principal stress orientations (Fig. 5c): $\sigma 1$ is moderately SE-plunging, $\sigma 2$ is more steeply NW-plunging and $\sigma 3$ is subhorizontal, NE–SW trending.

Evidence for similar late-D2 shearing is found over the entire study area, with similar orientations for shear bands throughout. These shear bands are

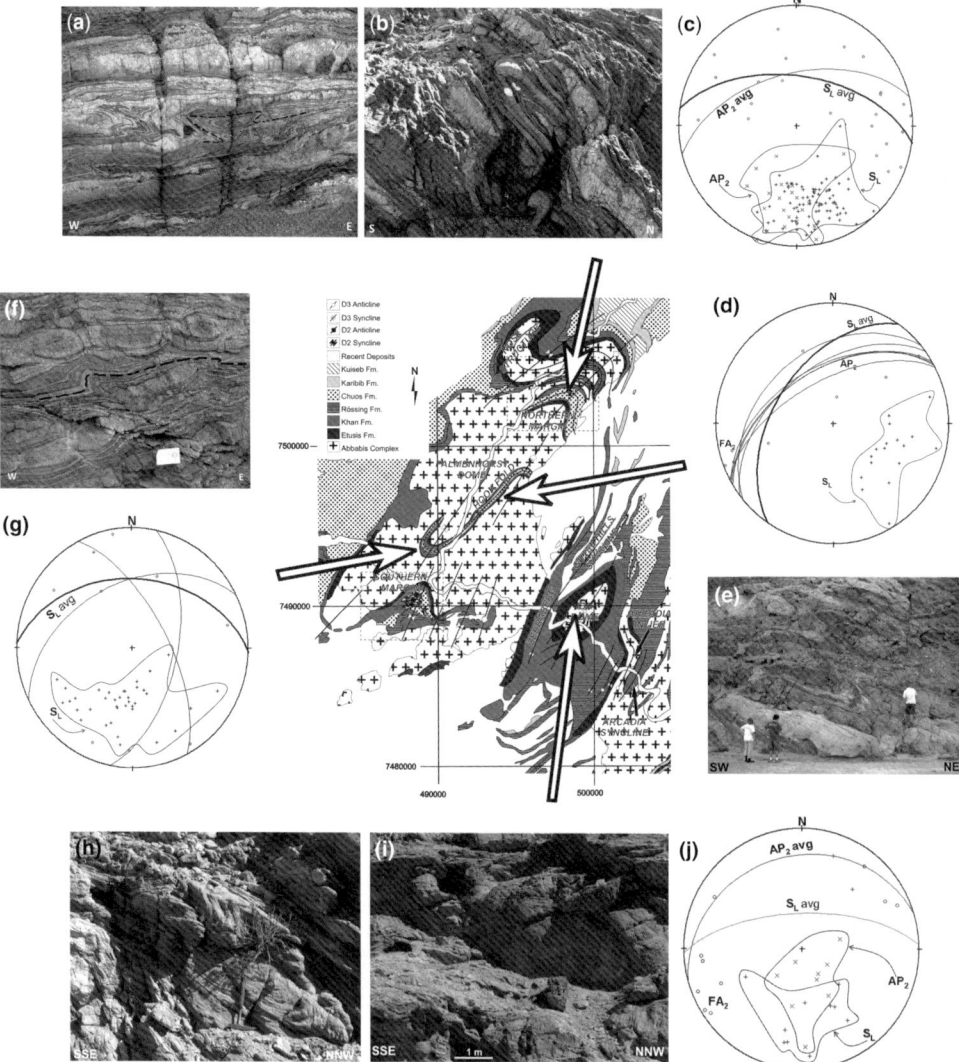

Fig. 4. D2 folding in the study area. (**a**) Possible sheath folds near the northern margin of the Palmenhorst Dome.
(**b**) North-dipping isoclinal fold in Damaran metasediments near the northern margin of the Palmenhorst Dome. (**c**)
Equal area lower hemispheric stereographic projections of structural data from the northern margin of the Palmenhorst
Dome. Open circles are D2 fold hinge lines ($n = 24$); X symbols are poles to axial planes of D2 folds (AP$_2$;
$n = 27$); + symbols are poles to lithological layering (S$_L$; $n = 81$). Average orientations of axial planes (AP$_2$ avg) and
lithological layering (S$_L$ avg) are shown by the great circles. (**d**) Equal area lower hemispheric stereographic projections
of structural data from the centre of the Palmenhorst Dome. Open circles are D2 fold hinge lines (FA$_2$;
$n = 8$); + symbols are poles to lithological layering (S$_L$; $n = 17$); great circles are axial planes of D2 folds (AP$_2$; $n = 6$).
Average orientation lithological layering (S$_L$ avg) is shown by the bold great circle. (**e**) Tight, NE-dipping F2 folds of
amphibolite dykes and Abbabis Complex gneisses in the centre of the Palmenhorst Dome. (**f**) Possible sheath fold in
Damaran metasediments near the southern end of the Hook Fold. (**g**) Equal area lower hemispheric stereographic
projections of structural data from the centre of the Palmenhorst Dome. Open circles are D2 fold hinge lines (FA$_2$;
$n = 8$); + symbols are poles to lithological layering (S$_L$; $n = 37$); great circles are axial planes of D2 folds (AP$_2$; $n = 3$).
Average orientation lithological layering (S$_L$ avg) is shown by the bold great circle. (**h**) Tight NNW-dipping folds in
Abbabis Complex gneisses near the centre of the Ida Dome. (**i**) Open, subhorizontal M-folds in Khan Formation
metasediments, on the hinge of a SSE-verging kilometre-scale D2 anticline near the centre of the Ida Dome. (**j**) Equal
area lower hemispheric stereographic projections of structural data from the northern margin of the Palmenhorst Dome.
Open circles are D2 fold hinge lines (FA$_2$; $n = 10$); X symbols are poles to axial planes of D2 folds (AP$_2$;
$n = 9$); + symbols are poles to lithological layering (S$_L$; $n = 10$). Average orientations of axial planes (AP$_2$ avg) and
lithological layering (S$_L$ avg) are shown by the great circles.

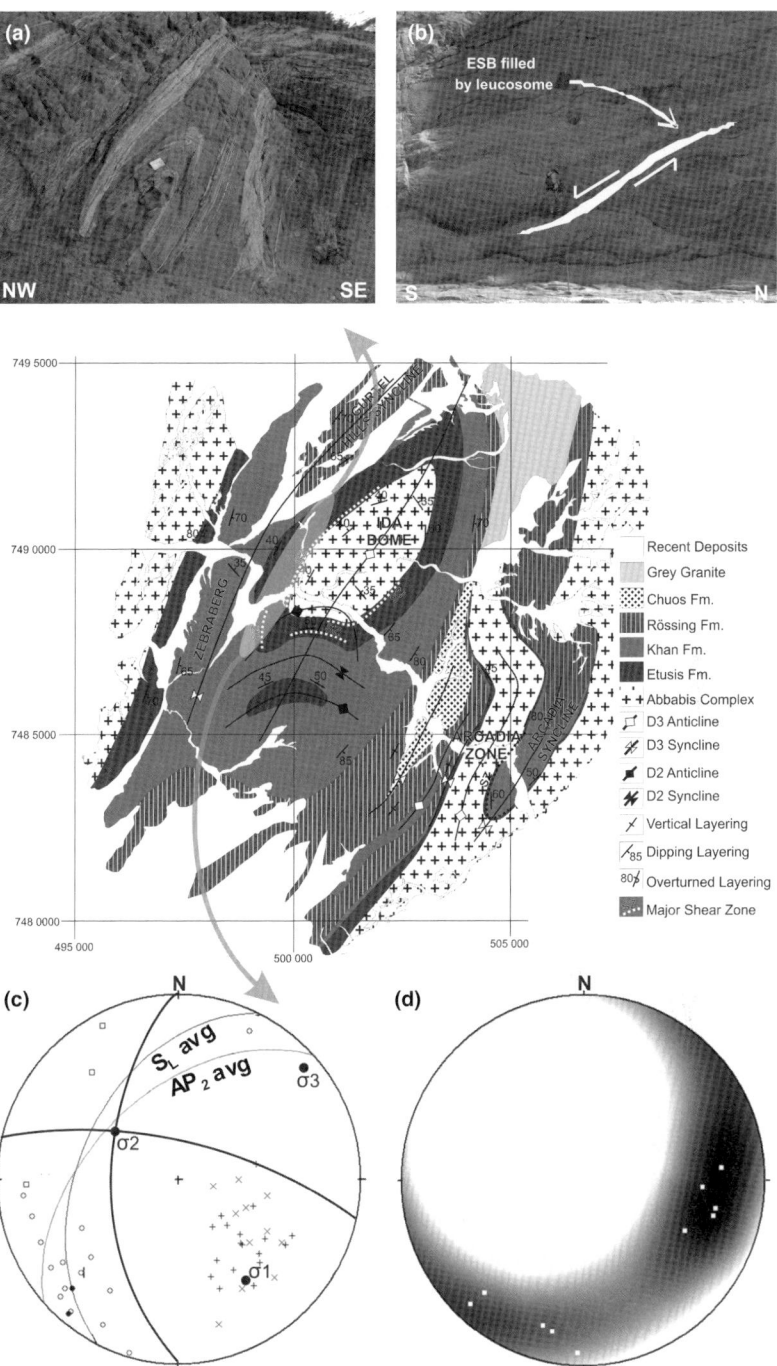

Fig. 5. Structures developed in a major D2 shear zone on the western margin of the Ida Dome. (**a**) NW-dipping isoclinal fold in Damaran metasediments near the contact with the Abbabis Complex. (**b**) Metre-scale extensional shear band in Abbabis Complex gneisses. (**c**) Equal area lower hemispheric stereographic projections of structural data. Open circles are D2 fold hinge lines ($n = 15$); closed circles are intersection lineations ($n = 2$); open squares are boudin axes ($n = 3$); X symbols are poles to axial planes of D2 folds ($n = 11$); + symbols are poles to lithological layering ($n = 17$). Average orientations of axial planes (AP$_2$ avg) and lithological layering (S$_L$ avg) are shown by great circles. Bold great circles are averages for the conjugate set of ESBs shown in (d), and the approximate stress orientations deduced from these ESBs. (**d**) Equal area lower hemispheric stereographic projections of contoured poles to planes for ESBs (white squares), indicating a conjugate set of steep north-dipping and moderate west-dipping shear bands.

common in both Abbabis Complex and Damara Supergroup lithologies, although conjugate sets are better developed in the basement gneisses. A south- to SW-verging set is better developed in north-dipping stratified Damaran metasediments.

In places these shear bands are localized within higher strain zones, but they appear to occur in both Damaran metasediments and Abbabis Complex gneisses. Along the northern margin of the Palmenhorst Dome, two sets are recorded: a steep NNE-dipping north-verging set and a shallower west-dipping SW-verging set (Fig. 6a). These two sets are found in an area with a number of high-strain shear zones, with isoclinal intrafolial folds (Fig. 6b) indicating intense deformation and transposition of bedding in these shear zones.

Further south, near the centre of the Palmenhorst Dome (Fig. 6c–e) and near the southern end of the Hook Fold (Fig. 6f–h), two similar sets of shear bands are also developed: a shallow WSW-dipping SW-verging set (Set 1) and a subvertical to steeply north-dipping north-verging set (Set 2). These form a conjugate set; their conjugate nature is however generally better seen in Abbabis Complex gneisses (Fig. 6c), whereas individual outcrops of Damaran metasediments typically show either the shallow SW-verging set (Fig. 6d) or the steep north-dipping set (Fig. 6f). Principal stress orientations calculated from the average orientations of the shear bands (Fig. 6a, e, g) indicate a shallow north- to NW-plunging $\sigma 3$ and a steeper south- to SE-plunging $\sigma 1$. Also noted locally in the study area are shallow NE-plunging L2 mineral stretching lineations (Fig. 6i). These mineral stretching lineations are usually defined by aligned, elongate porphyroblasts of high-grade metamorphic minerals, but may be pressure shadows adjacent to metamorphic minerals (Fig. 6j). Similar stretching lineations have been noted throughout the southwestern Central Zone by a number of workers (Coward 1983; Oliver 1994; Poli & Oliver 2001; Kisters *et al.* 2009) and suggest a close relationship between high-grade metamorphism and NE–SW extension. The orientations of these two conjugate sets of extensional shear bands, the corresponding stress orientations calculated and the NE-plunging mineral stretching lineations all indicate SW–NE extension during the late stages of D2 deformation (Fig. 6k).

D3 deformation: reorientation of D2 structures. Following the recumbent folding and subhorizontal extension of D2 deformation, D3 formed kilometre-scale upright to steeply NW-dipping shallow NE-plunging folds. These large-scale folds are visible on aerial photographs and satellite images, and led to the formation of the domes characteristic of the Central Zone by interference with the large F2 folds. The effect of these folds is noticeable in the

reorientation of D2 structures by D3 folding, and is well documented in the Ida Dome where a kilometre-scale steep NW-dipping anticline has rotated D2 structures about a shallow NE-plunging axis. The large-scale structure of the Ida Dome is dominated by a kilometre-scale D3 anticline, which refolds earlier south-verging D2 structures. However, small-scale D3 folds are rare. They are distinguishable by the upright form of the folds in contrast to the recumbent or shallowly dipping nature of D2 folds, and are generally more open than D2 folds.

An increase in the intensity of D3 is apparent to the east of the Ida Dome where a number of 100 m-scale upright to steeply WNW-dipping isoclinal folds are developed in Khan and Rössing formation rocks (Fig. 7a). This zone of intense D3 deformation is likely the result of the less competent layered nature of the Khan and Rössing formations; the more competent basement in the Ida Dome resulted in larger wavelength D3 folding. These folds consist of anticlinal cores of Khan Formation diopsidic gneisses, between which Rössing Formation marbles and calc-silicates form tight synclines. However, the effect of D3 deformation in the Ida Dome is best seen in the rotation of D2 structures by the large D3 fold and in the systematic variability of the orientation of D2 structures across the D3 anticline.

The eastern limb of the kilometre-scale D3 anticline which forms the Ida Dome (Fig. 7) has steeply ESE-dipping lithological layering (average 020/79E) (Fig. 7b). A number of open to isoclinal folds are developed in pelitic schist, diopside-plagioclase gneiss, quartz-biotite schist and quartzite of the Rössing, Khan and Etusis formations, and tabular granite sheets have been emplaced axial planar to the folds (Fig. 7c). These folds plunge towards the NE or NNE, and have shallowly east-dipping axial planes (average 009/38E), with similar orientations for granite sheets (350/32E) (Fig. 7b). Despite their NE to NNE plunge, the moderately east-dipping axial planes of these folds is incompatible with their being parasitic to a NE-trending upright to steeply NW-dipping fold (i.e. the kilometre-scale D3 anticline which forms the Ida Dome). They share similar field relationships to granites and high-grade metamorphism elsewhere in the study area (axial planar granites and anatectic leucosomes), and are therefore likely F2 folds which have been reoriented by upright D3 folding.

Plotting all structural data from the Ida Dome reveals that, on a broad scale, fabrics appear to have been folded about a shallow NE- to NNE-plunging axis. Poles to lithological layering (Fig. 7d), poles to granite sheets (Fig. 7e) and axial planes of folds (Fig. 7f) all give almost identical pi-pole girdles. This confirms field observations that recumbent

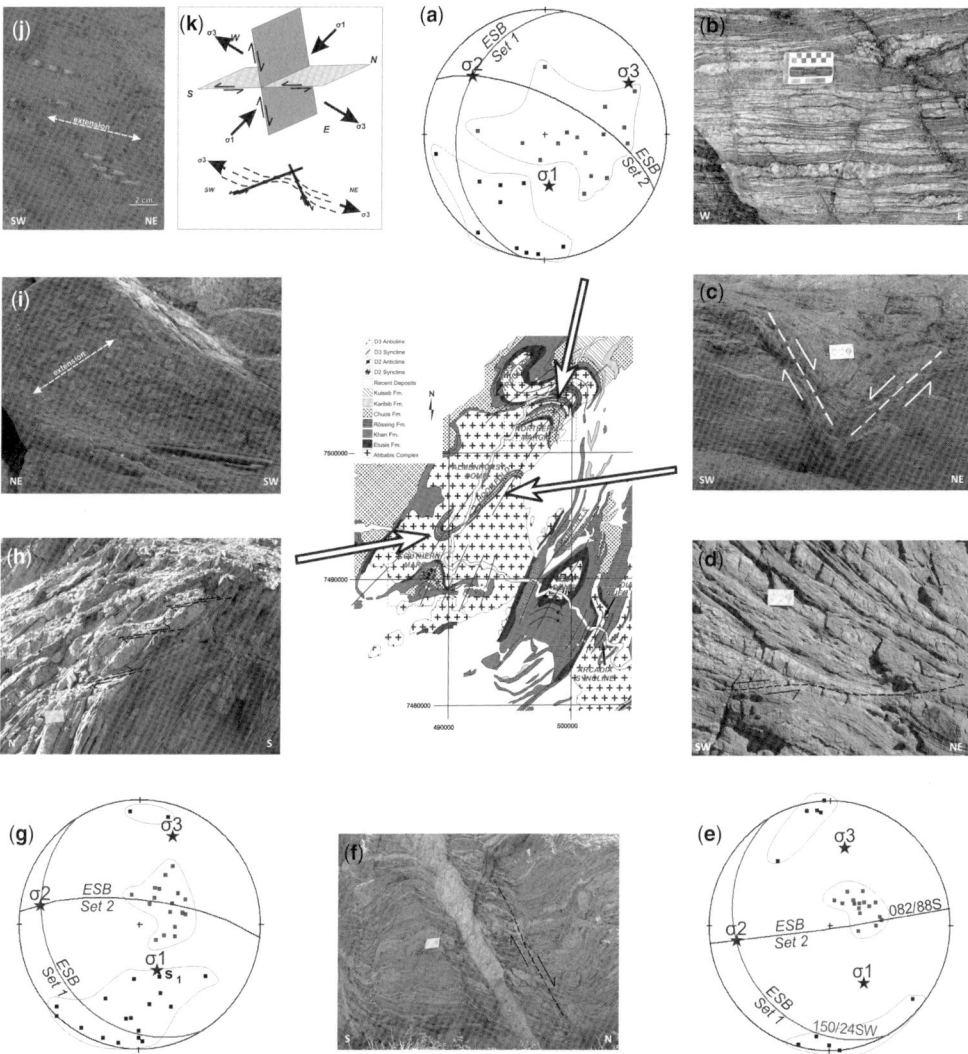

Fig. 6. Evidence for D2b extension in the Central Zone. (**a**) Equal area lower hemispheric stereographic projections of poles to ESBs from the northern margin of the Palmenhorst Dome. Averages for the two sets are shown by great circles, and the orientations of the associated stress directions are indicated. (**b**) Isoclinal intrafolial folds from a shear zone near the northern margin of the Palmenhorst Dome, indicating intense deformation and transposition of lithological layering. (**c**) Conjugate set of ESBs in Abbabis Complex gneisses from the centre of the Palmenhorst Dome. (**d**) Shallow, SW-verging extensional shear band in psammitic beds in Damaran metasediments from the centre of the Palmenhorst Dome. (**e**) Equal area lower hemispheric stereographic projections of poles to ESBs from the centre of the Palmenhorst Dome. Averages for the two sets are shown by great circles, and the orientations of the associated stress directions are indicated. (**f**) Steep north-dipping ESB in Damaran metasediments from the Hook Fold in the Palmenhorst Dome. (**g**) Equal area lower hemispheric stereographic projections of poles to ESBs from the hinge of the Hook Fold in the Palmenhorst Dome. Averages for the two sets are shown by great circles, and the orientations of the associated stress directions are indicated. (**h**) Shallow, south-verging ESBs from the hinge of the Hook Fold in the Palmenhorst Dome. (**i**) NE-plunging stretching lineation defined by elongate porphyroblasts of scapolite in calc-silicate. (**j**) NE-plunging extension lineation defined by elongate leucosomes in pressure shadows between garnet porphyroblasts in pelitic schist. (**k**) Schematic diagram illustrating the orientations of the conjugate set of ESBs relative to the principal stress directions, and showing how the shear bands develop during shallow NE-plunging extension with steep SE-plunging compression.

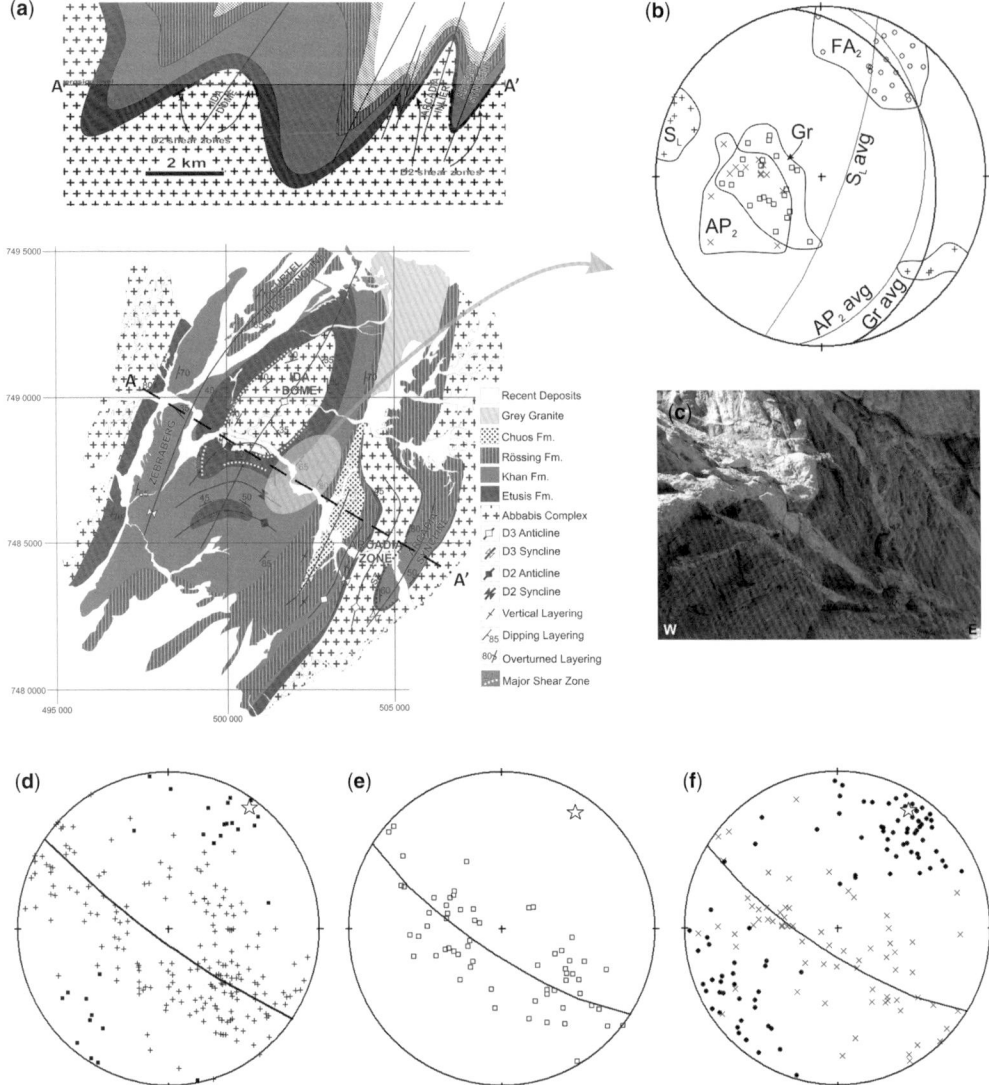

Fig. 7. D3 deformation in the Ida Dome. (**a**) Cross-section A–A′ along the Swakop River, through the Ida Dome. This NW–SE section shows that Damaran metasediments and Abbabis Complex gneisses have been folded by upright to steeply NW-dipping kilometre-scale D3 folds. (**b**) Equal area lower hemispheric stereographic projection from the eastern portion of the Ida Dome, showing D2 structures reoriented on the subvertical eastern limb of a kilometre-scale upright D3 anticline. Open circles are fold hinge lines of D2 folds (FA$_2$; $n = 19$); X symbols are poles to axial planes of D2 folds (AP$_2$; $n = 13$); + symbols are poles to lithological layering (S$_L$; $n = 12$); open squares are poles to orientations of tabular granite sheets (Gr; $n = 23$). The average orientations of fold axial planes (AP$_2$ avg; 009/38E), lithological layering (S$_L$ avg; 020/79E) and tabular granite sheets (Gr avg; 350/32E) are shown by the great circles. (**c**) Open metre-scale D2 fold in quartz-biotite schist, showing leucogranite sheets intruding along the axial plane of the fold. (**d**) Equal area lower hemispheric stereographic projection of poles to lithological layering (+ symbols, $n = 205$) and L2 stretching lineations (black squares, $n = 28$) from across the Ida Dome. (**e**) Equal area lower hemispheric stereographic projection of poles to orientations of granite sheets (open squares; $n = 62$) from across the Ida Dome. (**f**) Equal area lower hemispheric stereographic projection of poles to axial planes of folds (X symbols, $n = 70$) and fold hinge lines (black circles, $n = 90$) from across the Ida Dome. For lithological layering, granite sheets and axial plane data, best-fit pi-pole girdles (great circles) indicate folding plunging shallowly to the NE (7° on 035°, 13° on 033° and 13° on 032°, respectively).

to shallowly north-dipping D2 structures have been refolded by an upright event (D3), and that granite sheets intruded along the axial planes of these D2 folds have also been refolded by D3. L2 extension lineations (Fig. 7d) are also shallowly NE- or SW-plunging, consistent with subhorizontal NE–SW extension, and have not been significantly affected by D3. The interference of these kilometre-scale upright D3 folds with the earlier-formed south- to SE-verging D2 folds has resulted in the formation of the domes characteristic of the Central Zone, via a type-2 (Ramsay 1967) fold interference pattern.

Granites and leucosomes (and their structural relationships)

In the study area, a number of granite types are recognized in addition to anatectic leucosomes. These granites and leucosomes show distinct, consistent relationships with D2 structures. Red granites appear to predate D2 deformation, while grey granites appear to have intruded syntectonically during D2 deformation. Anatectic leucosomes are found in late D2 extensional sites, indicating their syn-extensional timing.

Red granites. Red granites occur in an east–west striking belt along the northern margin of the Palmenhorst Dome, near the contact between the Etusis Formation and the Abbabis Complex, in the vicinity of a number of tight north-dipping D2 folds of Damaran metasediments. The granites have a north-dipping gneissic fabric (with aligned biotite and sillimanite) subparallel to the regional strike and D2 fabric in the adjacent Damaran rocks, and are the oldest granite type based on field relationships. The gneissic fabric suggests that these granites either predate D2 deformation, or may be syn-D2. Similar inhomogeneous K-feldspar-rich red granites with gneissic fabrics have been noted by other workers (Smith 1965; Jacob 1974; Brandt 1985; Nex 1997) and occur near the contact between the Abbabis Complex and Etusis Formation in the highest-grade western portions of the Central Zone, and are generally considered the product of *in situ* anatexis of the Abbabis Complex or Etusis Formation (Smith 1965).

Grey granites. The most voluminous granitoid type found in the study area is a suite of dark- to light-grey fine- to medium-grained equigranular foliated granites termed the grey granites. They display a range of intrusive geometries, from centimetre- to metre-scale cross-cutting dykes to larger bodies (10s to 100s of metres in size). The range in colour is due to variable proportions of biotite and hornblende and a temporal evolution is evident, with darker grey granites (which contain higher proportions of mafic minerals) cross-cut by younger lighter grey granites. The darker grey granites are coarse- to medium-grained; the intermediate and leucogranite phases are uniformly fine- to medium-grained (grain size *c*. 1 mm) and contain large (up to 10 cm) phenocrysts of K-feldspar.

Grey granites are commonly emplaced along the axial planes of D2 folds and, where they are locally perpendicular, they are folded by D2 (Fig. 8a). They contain a fabric which is axial planar to the D2 folds in the country rocks (Fig. 8b). Veins of younger (leucocratic) grey granite within older grey granite are aligned with this D2 fabric, suggesting syndeformational emplacement of the younger granite into the older granite (Fig, 8b). D2 extensional shear bands are seen to partially affect the grey granites. These shear bands locally cross-cut the contact between the grey granites and their country rocks but are better developed in the country rocks than the grey granites, indicating that shear band development took place before the grey granite was entirely crystallized (Fig. 8c).

The emplacement of grey granites into D2 structural sites (as well as their deformation by D2) and the D2 fabric developed in the grey granites are consistent with their emplacement being synchronous with D2 deformation. Furthermore, the axial-planar emplacement of grey granites, which are also cut by D2 extensional shear bands, further constrains the timing of the grey granites to early D2. Measurement of granite sheet orientations across the Ida Dome reveals that, like the axial planes of the D2 folds along which they intrude, they too have been reoriented by a kilometre-scale shallow NE-plunging D3 anticline and are therefore pre-D3 in age. Anatectic leucosomes are also locally developed along the axial planes of D2 folds (in similar structural sites to grey granites), and suggest that emplacement of grey granites was coeval with high-grade metamorphism. Based on the evidence presented in the preceding section, D2 was a progressive, non-coaxial deformation event. It began with recumbent to shallow-dipping SSE-verging folding that was followed by SSE-verging extension during the later stages of deformation. The fact that grey granites were emplaced along early-D2 axial planes but were affected by late-D2 extension further constrains their timing to early D2.

The grey granites appear to be analogous to the early non-porphyritic granites and leucogranites (Miller 2008), which are syn-D2 (Haack *et al.* 1980) and found mostly in the northern Central Zone. Also termed 'red and grey homogeneous syntectonic granites' (Brandt 1985) and 'equigranular grey granites' (Nex 1997), these homogeneous fine- to medium-grained granite types are not stratigraphically confined, occur throughout the Central Zone, typically contain a fabric and display other

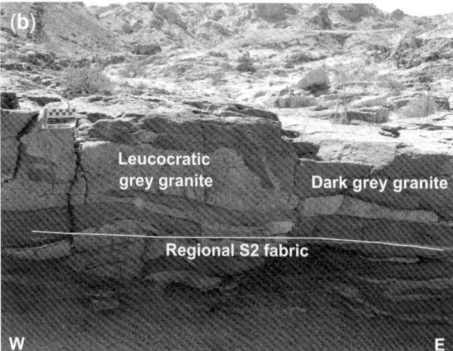

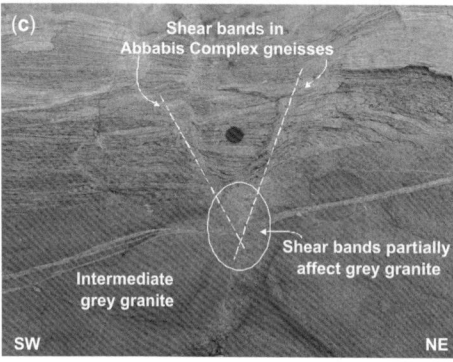

features which has led to description as 'syntectonic' (Brandt 1985). It is clear, however, that the timing of their emplacement can be definitively linked to south- to SE-verging, non-coaxial, recumbent to north- or NW-dipping F2 folding in the study area.

Leucosomes. Throughout the study area, leucosomes occur in ductile shear zones (Fig. 9a). Extensional shear bands containing leucosomes are common in both Abbabis Complex and Damara Supergroup rocks (Fig. 9b). Leucosomes are coarse-grained, comprising plagioclase feldspar and quartz with lesser biotite, K-feldspar and minor amounts of magnetite. They may contain mafic selvedges and

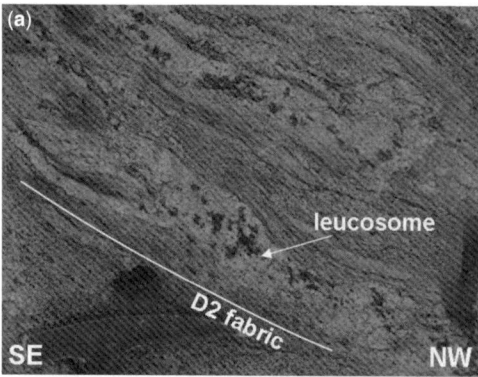

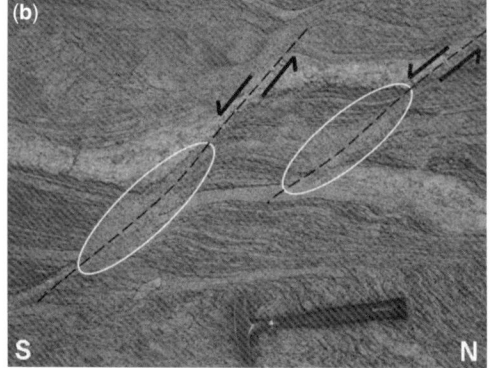

Fig. 8. Characteristics of the grey granites. (**a**) D2 fold in Damara Supergroup schist, with grey granite intruding along the axial plane of the fold and folded grey granite where it has intruded perpendicular to the axial plane of the fold, indicating that the granites were emplaced syn-folding (southern margin of Palmenhorst Dome). (**b**) Outcrop from the centre of the Ida Dome, showing dark grey granite with a strong fabric subparallel to the D2 fabric in the country rocks intruded by veins of light grey granite. Note how the veins are flattened parallel to the fabric, indicating syn-D2 intrusion of the grey granites. (**c**) Outcrop from the western edge of the Ida Dome, showing grey granites intruding gneisses of the Abbabis Complex and being cut by a conjugate set of D2 extensional shear bands. Note that the shear bands are well developed in the gneisses but are poorly developed in the grey granite, suggesting that they developed while the grey granite was partially crystallized.

Fig. 9. Leucosomes developed in D2 structures. (**a**) Anatectic leucosome from a SE-verging D2 shear zone near the basement-cover contact on the western margin of the Ida Dome. Note that the leucosome is aligned with the D2 fabric in the shear zone, suggesting that leucosome formation is the result of syn-D2 high-grade metamorphism leading to anatexis. (**b**) Extensional shear bands in Abbabis Complex gneisses from the western edge of the Ida Dome. Shear bands and vergence are indicated. Note that the leucosomes developed in the extensional shear bands, suggesting that shear bands development was coeval with high-grade metamorphism.

restitic garnet and cordierite, indicating *in situ* partial melting. In the well-exposed shear zone near the western margin of the Ida Dome, anatectic leucosomes intruding into shear bands in Abbabis Complex gneisses form lensoid bodies subparallel to the lithological layering in the Etusis Formation. Shear zones near the northern margin of the Palmenhorst Dome also show numerous anatectic leucosomes subparallel to lithological layering, generated and emplaced into syndeformational structural sites. The migmatitic nature of such D2 shear zones suggests that high-grade (upper-amphibolite- to granulite-facies) metamorphism must have occurred during deformation. This metamorphism would have led to anatexis of Damara Supergroup and Abbabis Complex lithologies to produce the leucosomes seen in D2 shear zones and extensional shear bands. The implication that high-grade metamorphism was coeval with D2 deformation is consistent with work which has demonstrated the synchronicity of high-grade metamorphism and intense non-coaxial deformation in the Central Zone (Poli 1997). However, the predominantly extensional sites of leucosomes and their cross-cutting relationships with the grey granites indicate that peak metamorphism and anatexis was coeval with the late stages of D2 deformation (i.e. late-D2 extension).

Dating of D2 deformation

Red granites, grey granites and leucosomes show relationships to D2 deformation; red granites contain a gneissic fabric subparallel to the regional D2 fabric, suggesting that they predate D2 deformation. Field relationships indicate that emplacement of the grey granites was coeval with D2 folding and occurred during the early stages of D2 deformation, while anatectic leucosomes are found in late-D2 shear zones and extensional shear bands. The ages of these granites and leucosomes may therefore place constraints on the absolute timing of D2 deformation in the study area; that is, D2 should be younger than the red granites and grey granites should constrain the onset of D2 and should be older than leucosomes, which should constrain the late stages of D2.

Samples collected and zircon petrography

In order to date these granites and leucosomes, zircons and monazites from red granite, grey granite and anatectic leucosomes have been dated using the sensitive high-resolution ion microprobe (SHRIMP) at the Australian National University (ANU).

A sample of red granite (LKR016) was collected from the northern margin of the Palmenhorst Dome

(locality 0500360/7504659; units are UTM, WGS84, zone 33S). The granite is made up almost entirely of K-feldspar and quartz, with sillimanite, cordierite and minor amounts of biotite and magnetite. Zircons are clear to pale grey and euhedral to subhedral, with slightly rounded tips. Cathodoluminescence (CL) imaging shows that most zircons contain bright, low-U inherited cores, with narrow, dark, high-U (up to 1800 ppm) magmatic overgrowths (Fig. 10a). Monazites from LKR016 are anhedral and pale yellow. Backscattered electron (BSE) imaging shows that some have an internal structure and a weak zoning (Fig. 10b).

A sample of grey granite (LHA012) was collected from the southern margin of the Palmenhorst Dome. It is representative of the older, dioritic to granodioritic end-member of the grey granite suite, and was taken from within a large outcrop of grey granite where the younger phase is also present. The sample comprises quartz, plagioclase feldspar, hornblende, biotite and minor K-feldspar. Zircons are generally euhedral to subhedral, elongate and prismatic, clear to pale yellow and contain both opaque inclusions, as well as inclusions of elongate, needle-shaped euhedral apatite. CL imaging reveals magmatic oscillatory zoning (Fig. 10c).

Two leucosome samples were also collected from the Ida Dome. Sample CZRL-1 is a sample of anatectic melt from a migmatitic SE-verging shear zone near the western margin of the Ida Dome (locality 0499125/7486844). The sample consists of granitic leucosome which is coarse-grained and made up of plagioclase feldspar and quartz, with lesser biotite and minor amounts of magnetite. Most zircons from CZRL-1 are structured and contain cores with various shapes and sizes, overgrown by high-U poorly zoned rims which have well-developed pyramidal terminations. Sample CZRL-3 is a sample of anatectic melt localized in a D2 extensional shear band developed in Abbabis Complex gneisses near the western edge of the Ida Dome (locality 0500289/7488670). Extensional shear bands are widely developed and represent the final extensional phase of D2 deformation here. The sample comprises coarse-grained quartz, K-feldspar and plagioclase feldspar with minor biotite and magnetite. The zircons are anhedral to euhedral with most grains comprising a low-U core and a high-U overgrowth.

Analytical procedure

In order to extract zircons and monazites, 3–5 kg of each sample was crushed and sieved to 90–250 μm size fraction, processed on a Wilfley table, and the heavy mineral fraction collected. Heavy minerals were further concentrated using heavy liquids at the University of the Witwatersrand, Johannesburg

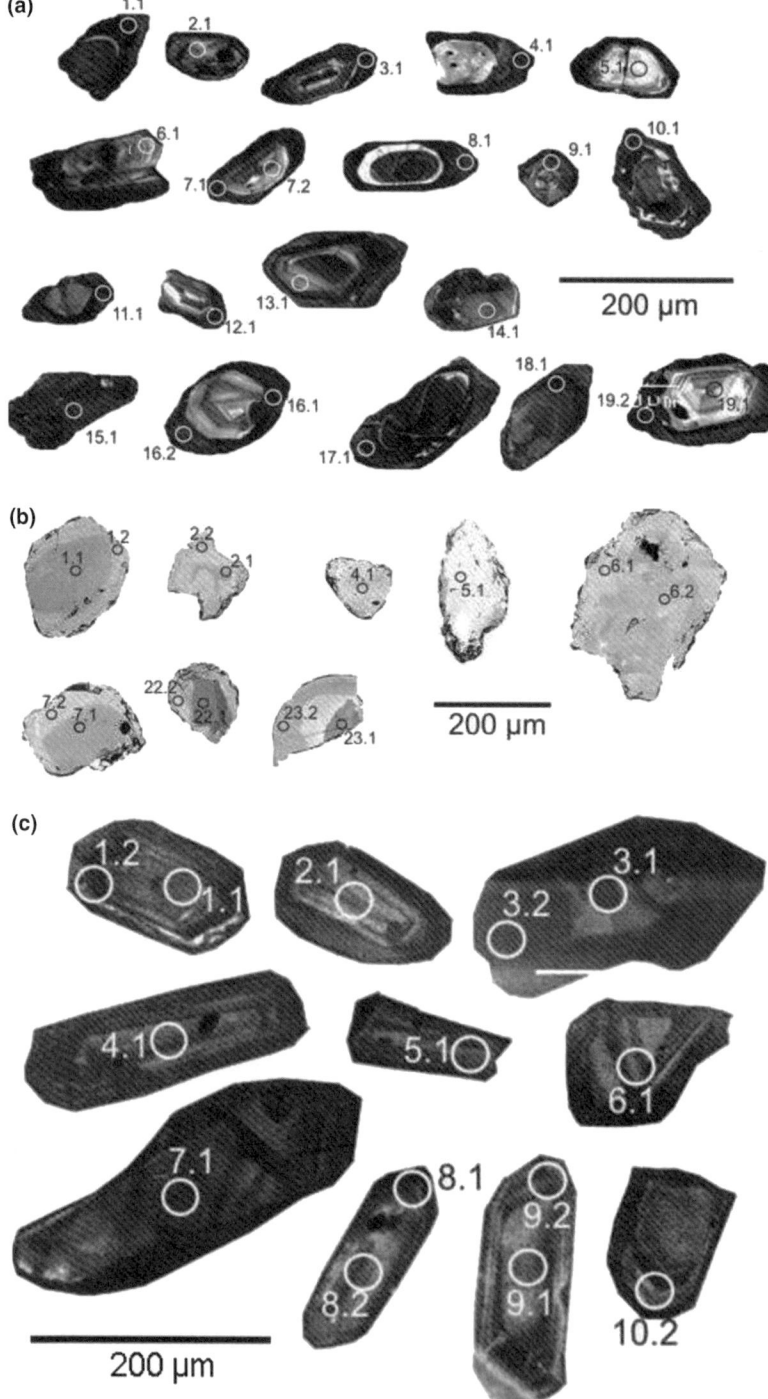

Fig. 10. CL and BSE images of zircons and monazites analysed. (**a**) CL images of zircons from sample LKR016, showing spot locations. Note that almost all zircons analysed have inherited cores. (**b**) BSE images of monazites from LKR016, showing the location of spots analysed. Note the zoning in some grains (1; 7; 22; 23). (**c**) CL images of zircons from sample LHA012, showing locations of spots analysed.

and this concentrate was passed repeatedly through a Franz magnetic separator at various currents to separate the monazite and zircon from the magnetic fraction. Zircons and monazites were then hand-picked under a binocular microscope and the grains mounted in epoxy together with the SHRIMP zircon standards TEMORA and SL13 (RSES, the Australian National University). This grain mount was sectioned, polished, photographed in transmitted and reflected light and imaged under cathodoluminescence on a scanning electron microscope to identify any zircon cores, zonation or metamorphic overgrowths and to select spots for analysis, avoiding cracks and inclusions. The zircons were analysed for U–Pb–Th using the SHRIMP II at the Australian National University, Canberra. The SHRIMP data were reduced following Williams & Claesson (1987), Compston et al. (1992) and Williams (1998) and using the SQUID Excel macro of Ludwig (2001). The U and Th concentrations were determined relative to those measured in the SL13 standard. Ages were calculated using the radiogenic $^{207}Pb/^{206}Pb$ ratios, with the correction for common Pb made using the measured ^{204}Pb and the appropriate common Pb composition assuming the model of Cumming & Richards (1975). Uncertainties in the individual isotopic ratios and ages are reported at the 1σ level, and the final weighted mean ages are reported as 95% confidence limits. The concordia plots and weighted mean age calculations were carried out using Isoplot 3.00 (Ludwig 2003).

Results of U–Pb dating

Red granite. The zircons analysed show a spread of ages between concordant inherited cores and concordant magmatic ages (Table 1). A Model 1 solution through the data gives an upper intercept of 1013 ± 21 Ma, similar to the ages for basement rocks along the Khan River determined by Kröner et al. (1991), and a lower intercept of 539 ± 17 Ma (Fig. 11a). Monazites are concordant, and give a weighted mean age of 535.6 ± 7.2 Ma (Fig. 11b). These ages are within error of previous ages for the Rotekuppe Granite (539 Ma; Jacob et al. 2000) and anatectic red granite from Goanikontes (534 Ma; Briqueu et al. 1980). Pre-D2 red granites from this study are therefore coeval with the post-tectonic Rotekuppe Granite, implying that deformation is not coeval everywhere within the Central Zone and that deformation higher in the Central Zone (in the Karibib area) had ceased while intense deformation was still underway in the lower portions of the orogen.

Grey granite. Sample LHA012 has predominantly concordant zircons, with some of the spots analysed

showing recent Pb-loss (Fig. 12a). However, most analyses are concordant, and sample LHA012 gives a weighted mean age of 520.4 ± 4.2 Ma when discordant zircons are ignored. This is similar to the Rb–Sr age of 514 ± 22 Ma (Haack et al. 1980) for the syn-D2 early non-porphyritic granites and leucogranites, the bulk of which occur in the northern Central Zone (Miller 2008).

Leucosomes. The zircons from sample CZRL1 contain cores which give various ages from Archaean to Neoproterozoic (reflecting the detrital nature of the zircons in the Damara Supergroup, which forms the host for this migmatitic shear zone and the likely source for the leucosome). The high-U rims, developed during migmatite crystallization, plot in a spread from c. 500 Ma showing variable but extensive Pb-loss. A weighted mean $^{207}Pb/^{206}Pb$ age for the four most concordant points (#s 7.1, 8.1, 10.1 and 18.1) is 511 ± 16 Ma (Fig. 12b). Regression of all data for the overgrowths gives a Model 1 solution with a similar upper intercept of 511 ± 18 Ma (Fig. 12b). Analyses from sample CZRL3 shows a cluster of concordant data, with 3 discordant analyses. The concordant zircons give a weighted mean $^{207}Pb/^{206}Pb$ age of 508.4 ± 8.7 Ma (Fig. 12c). Three analyses are slightly discordant and plot away from this cluster, which could indicate an inherited component. Alternatively, they may be the result of subtle mixed core-rim analyses where the beam was burning through into a small older part of the grain, changing the $^{207}Pb/^{206}Pb$ composition during the run. All cores are Palaeoproterozoic in age, reflecting the age of the Abbabis Complex in which the leucosome is found. Data for CZRL1 and CZRL3 are shown in Table 2.

Discussion

The occurrence of an intense, recumbent non-coaxial deformation event (D2) followed by upright NE-plunging folding (D3) has been widely described in the southwestern Central Zone. Tight, recumbent, south-verging folds with folding of the basement-cover interface were described by Jacob et al. (1983). Blaine (1977) described similar recumbent folds, low-angle thrusts and fabric transposition during this event. The upright D3 event, which has reoriented D2 structures, has also been widely described: NE–SW trending upright folds (Nash 1971), NE-plunging, upright folds (Barnes 1981) and large, open, upright concentric folds (Blaine 1977) are similar to the upright NE-trending, kilometre-scale D3 folds found to refold D2 structures in this study.

Although Smith (1965), Nash (1971), Blaine (1977), Barnes (1981), Jacob et al. (1983) and

Table 1. Summary of SHRIMP U–Th–Pb zircon results for zircons and monazites from red granite sample LKR016. Errors are 1-sigma; Pb_c and $Pb*$ indicate the common and radiogenic portions, respectively

Grain spot	% $^{206}Pb_c$	Ppm U	Ppm Th	$^{232}Th/^{238}U$	Ppm $^{206}Pb*$	$^{206}Pb/^{238}U$ age[1]	$^{207}Pb/^{206}Pb$ age[1]	% Discordant	$^{207}Pb*/^{206}Pb*$[1]	±%	$^{207}Pb*/^{235}U$[1]	±%	$^{206}Pb*/^{238}U$[1]	±%	Error corr.
LKR016 – Zircon															
1.1	1.74	1847	247	0.14	136	519.5 ± 5.8	514 ± 71	−1	0.0576	3.2	0.666	3.4	0.08392	1.2	0.340
2.1	0.70	945	883	0.97	114	841.3 ± 8.4	904 ± 25	7	0.06918	1.2	1.33	1.6	0.1394	1.1	0.664
3.1	0.01	866	329	0.39	100	815.6 ± 8.1	886 ± 13	8	0.06857	0.61	1.275	1.2	0.1349	1.1	0.867
4.1	–	976	29	0.03	71.7	529.9 ± 5.5	534 ± 19	1	0.05811	0.85	0.6864	1.4	0.08567	1.1	0.785
5.1	–	128	144	1.16	41.2	2050 ± 23	2037 ± 13	−1	0.12555	0.74	6.481	1.5	0.3744	1.3	0.869
6.1	0.25	481	236	0.51	131	1769 ± 17	1980.3 ± 8.8	11	0.12163	0.5	5.296	1.2	0.3158	1.1	0.909
7.1	–	264	457	1.79	38.7	1018 ± 11	1043 ± 21	2	0.07407	1.1	1.747	1.6	0.171	1.2	0.746
7.2	0.91	1099	698	0.66	113	723 ± 7.2	802 ± 27	10	0.06586	1.3	1.078	1.7	0.1187	1.1	0.630
8.1	0.01	906	20	0.02	67.1	533.3 ± 5.6	528 ± 18	−1	0.05796	0.81	0.6892	1.4	0.08625	1.1	0.804
9.1	–	876	384	0.45	112	895.9 ± 9	952 ± 11	6	0.07081	0.56	1.456	1.2	0.1491	1.1	0.888
10.1	0.96	1461	70	0.05	118	572.8 ± 4.7	627 ± 31	9	0.06065	1.4	0.777	1.8	0.0929	1.1	0.622
11.1	0.43	1371	5580	4.20	87.4	459.7 ± 4.7	485 ± 28	5	0.05682	1.3	0.5792	1.6	0.07392	1.1	0.646
12.1	–	1093	37	0.04	92.6	606.8 ± 6.2	664 ± 19	9	0.0617	0.89	0.84	1.4	0.0987	1.1	0.770
13.1	–	347	207	0.62	48.2	967 ± 10	1003 ± 17	4	0.0726	0.86	1.621	1.4	0.1619	1.1	0.801
14.1	–	594	132	0.23	79.2	930.2 ± 9.5	967 ± 15	4	0.07132	0.72	1.526	1.3	0.1552	1.1	0.834
15.1	–	993	308	0.32	139	975.6 ± 9.5	994 ± 10	2	0.07228	0.5	1.628	1.2	0.1634	1.1	0.903
16.1	0.04	1711	31	0.02	148	618.4 ± 6.1	675 ± 12	8	0.06202	0.55	0.861	1.2	0.1007	1	0.884
16.2	–	774	18	0.02	58	538.7 ± 6.1	570 ± 19	5	0.05908	0.89	0.71	1.5	0.0872	1.2	0.801
17.1	1.25	1135	343	0.31	85.8	536.9 ± 5.6	520 ± 46	−3	0.0577	2.1	0.691	2.4	0.08686	1.1	0.457
18.1	0.03	692	280	0.42	104	1040 ± 10	1027 ± 12	−1	0.07346	0.59	1.773	1.2	0.175	1.1	0.875
19.1	–	206	134	0.67	69.1	2129 ± 22	2049 ± 10	−4	0.12646	0.59	6.822	1.3	0.3912	1.2	0.898
19.2	–	970	38	0.04	79.3	586.1 ± 6.1	719 ± 15	18	0.06333	0.72	0.831	1.3	0.0952	1.1	0.833
LKR016 – Monazite															
1.1	–	4758	51 216	11.1	353	534.3 ± 7.5	544 ± 31	2	0.05837	1.4	0.695	2	0.0864	1.5	0.717
1.2	–	3044	69 737	23.7	228	538.5 ± 8	529 ± 31	−2	0.05799	1.4	0.697	2.1	0.0871	1.5	0.739
2.1	–	2383	61 268	26.6	177	535.8 ± 8	528 ± 34	−1	0.05794	1.5	0.692	2.2	0.0867	1.6	0.712
2.2	–	3081	87 844	29.5	225	526.4 ± 7.7	545 ± 31	3	0.05841	1.4	0.685	2.1	0.0851	1.5	0.734
4.1	0.03	3108	98 199	32.6	221	513.4 ± 7.4	521 ± 33	1	0.05777	1.5	0.66	2.1	0.0829	1.5	0.706
5.1	–	4089	106 549	26.9	309	543.6 ± 7.7	531 ± 25	−2	0.05803	1.2	0.704	1.9	0.088	1.5	0.786
6.1	–	2588	80 750	32.2	184	512.6 ± 9	539 ± 35	5	0.05822	1.6	0.664	2.4	0.0828	1.8	0.748
6.2	–	5990	67 923	11.7	452	543.7 ± 7.6	538 ± 21	−1	0.05775	0.97	0.706	1.7	0.088	1.5	0.831
7.1	0.02	6144	46 491	7.8	469	549.1 ± 7.5	520 ± 18	−6	0.05775	0.82	0.708	1.6	0.0889	1.4	0.866
7.2	–	2268	76 320	34.8	171	543.8 ± 8	546 ± 31	0	0.05843	1.4	0.709	2.1	0.088	1.5	0.731
22.1	–	1233	39 837	33.4	91.8	536.1 ± 8.2	552 ± 37	3	0.05859	1.7	0.701	2.3	0.0867	1.6	0.689
22.2	–	3576	101 830	29.4	270	543.2 ± 7.7	550 ± 26	1	0.05853	1.2	0.71	1.9	0.0879	1.5	0.778
23.1	–	1672	73 565	45.5	126	541.2 ± 8.2	558 ± 33	3	0.05875	1.5	0.709	2.2	0.0876	1.6	0.719
23.2	–	2568	88 033	35.4	186	521.2 ± 7.6	534 ± 27	2	0.05811	1.2	0.675	2	0.0842	1.5	0.773

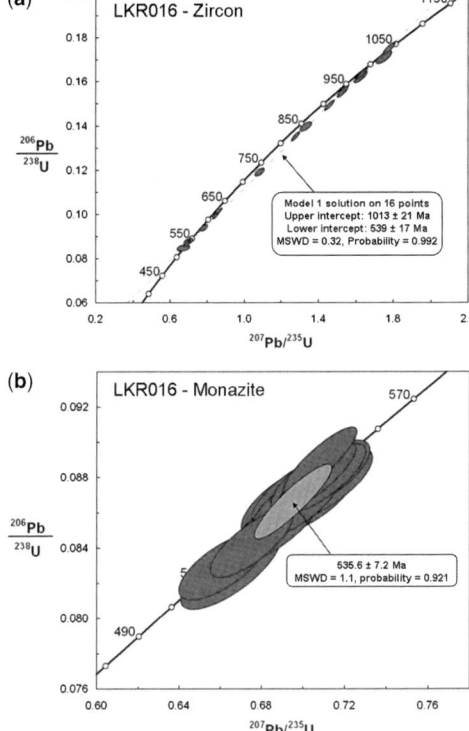

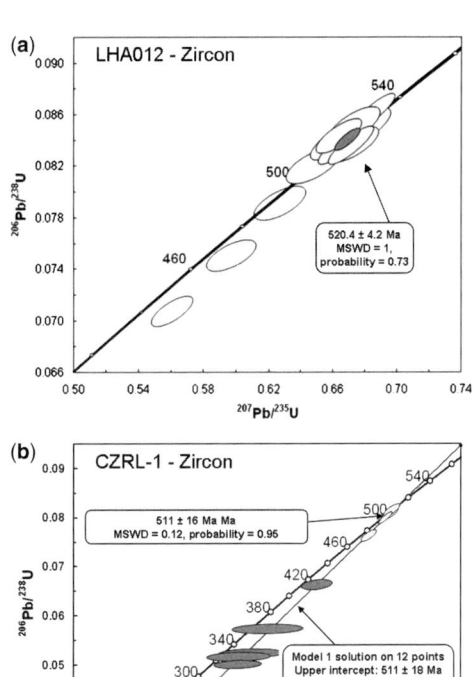

Fig. 11. (**a**) Concordia plot for zircons from LKR016, showing a spread of discordant ages for spots analysed. A Model 1 solution through the data yields an upper intercept 1013 ± 21 Ma and a lower intercept of 539 ± 17 Ma. (**b**) Concordia plot of monazite analyses from LKR016, which give an age of 535.6 ± 7.2 Ma.

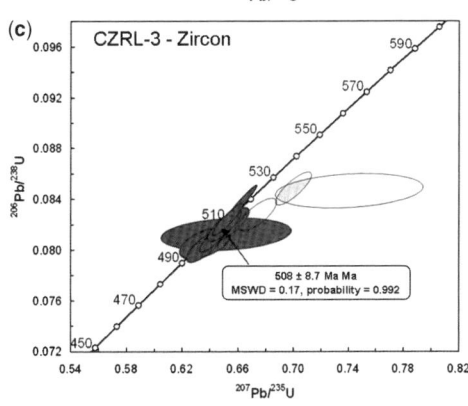

Coward (1983) all describe intense recumbent folding and shearing, the suggested vergence of this deformation varies from the SW to SE. The vergence of D2 folding in this study, based on measurements of fold hinge-lines and fold axial planes, is suggested to be south- to SE-vergent. Mesoscale F2 folds throughout the study area have fold hinge lines that plunge shallowly to the west or SW, and intersection lineations have similar plunges to these hinge lines. F2 folds have shallow to moderate north- to NW-dipping axial planes, subparallel to the north or NW dip of lithological layering. These mesoscale folds are parasitic to kilometre-scale south- to SE-verging, tight to isoclinal, shallow dipping to recumbent folds. D2 shear zones developed on the limbs of these kilometre-scale folds show a similar vergence.

These observations are at odds with widespread evidence for SW-verging structures in the southwestern Central Zone reported by Coward (1983), Oliver (1994, 1995) and Poli & Oliver (2001). All

Fig. 12. (**a**) Concordia plot for grey granite sample LHA012, giving an age of 520.4 ± 3.2 Ma. (**b**) Concordia plot of analyses from CZRL-1, showing the extensive Pb-loss for many of the high-U rims and the results of both the mean ^{207}Pb/^{206}Pb age calculation (511 ± 16 Ma) and a Model 1 solution (upper intercept of 511 ± 18 Ma). (**c**) Concordia plot for CZRL-3, showing the mean ^{207}Pb/^{206}Pb age of 508.4 ± 8.7 Ma. Data-point error ellipses are 68.3% confidence.

interpretations of south-westerly vergence in the southwestern Central Zone are based upon a pronounced NE-plunging extension lineation. Although

Table 2. *Summary of SHRIMP U–Th–Pb zircon results for zircons from grey granite sample LHA012 and leucosome samples CZRL-1 and CZRL-3. Errors are 1-sigma; Pb_c and $Pb*$ indicate the common and radiogenic portions, respectively*

Grain spot	% $^{206}Pb_c$	Ppm U	Ppm Th	$^{232}Th/^{238}U$	Ppm $^{206}Pb*$	$^{206}Pb/^{238}U$ age[1]	$^{207}Pb/^{206}Pb$ age[1]	% Discordant	$^{207}Pb*/^{206}Pb*$[1]	±%	$^{207}Pb*/^{235}U$[1]	±%	$^{206}Pb*/^{238}U$[1]	±%	Error corr.
LHA012 – Zircon															
1.1	0.09	858	353	0.43	62.7	525.8 ± 6.1	529 ± 24	1	0.05798	1.1	0.679	1.6	0.085	1.2	0.743
1.2	0.03	1779	597	0.35	132	533.4 ± 5.5	518 ± 14	−3	0.05769	0.62	0.6861	1.2	0.08626	1.1	0.866
2.1	0.00	1096	595	0.56	79.2	520.7 ± 5.5	526 ± 16	1	0.0579	0.74	0.6716	1.3	0.08412	1.1	0.829
3.2	0.03	4263	182	0.04	308	520.1 ± 5.5	518.7 ± 9	0	0.05771	0.41	0.6686	1.2	0.08402	1.1	0.938
3.1	0.29	1093	120	0.11	77.2	508 ± 5.3	503 ± 28	−1	0.0573	1.3	0.648	1.7	0.08199	1.1	0.656
4.1	0.22	1093	567	0.54	78.7	518.1 ± 5.4	543 ± 24	5	0.05834	1.1	0.673	1.5	0.08368	1.1	0.712
5.1	0.05	842	333	0.41	60.7	519.1 ± 5.5	496 ± 21	−5	0.05712	0.94	0.6605	1.5	0.08386	1.1	0.763
6.1	0.04	1579	325	0.21	113	516.7 ± 5.3	518 ± 15	0	0.05769	0.69	0.6638	1.3	0.08346	1.1	0.844
7.1	0.08	662	237	0.37	48.4	526.7 ± 5.7	505 ± 25	−4	0.05736	1.1	0.673	1.6	0.08514	1.1	0.703
8.1	0.22	960	642	0.69	65.4	490.8 ± 5.2	512 ± 28	4	0.05752	1.3	0.627	1.7	0.07911	1.1	0.653
8.2	0.14	1323	471	0.37	96	522 ± 5.4	495 ± 20	−5	0.05709	0.9	0.6639	1.4	0.08434	1.1	0.769
9.2	0.29	1686	759	0.47	109	466.4 ± 4.9	522 ± 28	11	0.05779	1.3	0.598	1.7	0.07502	1.1	0.648
9.1	0.02	1505	548	0.38	108	517.8 ± 5.9	538 ± 15	4	0.05823	0.68	0.6714	1.4	0.08363	1.2	0.867
10.2	0.14	1185	338	0.29	72.1	440.5 ± 4.7	514 ± 23	14	0.05757	1	0.5614	1.5	0.07072	1.1	0.729
CZRL-1 – Zircon															
1.1	0.10	299	221	0.76	77.5	1698 ± 16	1850.6 ± 9.2	8	0.11315	0.51	4.702	1.2	0.3014	1.1	0.906
2.1	8.18	3050	607	0.21	149	329 ± 3.7	560 ± 170	41	0.0588	7.8	0.425	7.9	0.05236	1.2	0.147
3.1	8.28	3367	370	0.11	181	359 ± 4	488 ± 180	26	0.0569	8	0.449	8.1	0.05727	1.2	0.143
4.1	0.14	123	68	0.57	21.7	1200 ± 13	1183 ± 20	−1	0.07942	1	2.241	1.6	0.2046	1.2	0.756
5.1	0.36	81	55	0.70	13.7	1156 ± 16	1185 ± 32	2	0.0795	1.6	2.154	2.2	0.1965	1.5	0.684
5.2	13.63	4129	165	0.04	182	278.8 ± 3.7	402 ± 310	31	0.0548	14	0.334	14	0.0442	1.4	0.097
6.1	0.08	184	85	0.48	57.8	2008 ± 22	2036 ± 10	1	0.12551	0.56	6.325	1.4	0.3655	1.2	0.912
7.1	0.18	2007	105	0.05	138	495.3 ± 4.9	510 ± 12	3	0.05748	0.54	0.633	1.2	0.07986	1	0.884
8.1	0.27	1902	114	0.06	133	504.7 ± 5	504 ± 19	0	0.05732	0.85	0.6435	1.3	0.08143	1	0.772
9.1	0.00	124	279	2.32	37.7	1948 ± 19	1870 ± 12	−4	0.11436	0.67	5.564	1.3	0.3529	1.1	0.862

10.1	0.58	2345	200	0.09	155	473.7 ± 4.7	520 ± 19	9	0.05774	0.86	0.607	1.3	0.07625	1	0.767
11.1	13.98	4480	966	0.22	188	265.7 ± 3.6	490 ± 310	46	0.057	14	0.33	14	0.04208	1.4	0.098
12.1	7.91	2621	184	0.07	126	324.6 ± 3.7	483 ± 170	33	0.0568	7.8	0.404	7.9	0.05165	1.2	0.146
13.1	2.02	90	106	1.21	10.2	782 ± 10	777 ± 110	−1	0.0651	5.2	1.158	5.4	0.129	1.4	0.254
14.1	13.01	3567	695	0.20	160	285.6 ± 3.8	343 ± 310	17	0.0533	14	0.333	14	0.0453	1.3	0.098
15.1	2.68	2426	762	0.32	142	413.5 ± 4.2	514 ± 63	20	0.0576	2.9	0.526	3.1	0.06625	1.1	0.345
16.1	0.01	224	114	0.53	92.6	2530 ± 28	2678 ± 7.1	6	0.18275	0.43	12.11	1.4	0.4807	1.3	0.952
17.1	6.16	3398	528	0.16	156	315 ± 3.4	544 ± 130	42	0.0584	5.9	0.403	6	0.05008	1.1	0.184
18.1	0.48	1921	111	0.06	134	500.7 ± 5	511 ± 21	2	0.0575	0.96	0.6403	1.4	0.08076	1	0.734
CZRL-3 – Zircon															
2.1	0.11	2815	144	0.05	197	503 ± 5.1	515 ± 13	2	0.0576	0.58	0.6445	1.2	0.08115	1.1	0.876
3.1	1.42	259	170	0.68	78.1	1916 ± 19	1986 ± 21	4	0.122	1.2	5.823	1.7	0.3461	1.2	0.697
4.1	0.47	143	54	0.39	43.5	1948 ± 22	2030 ± 18	4	0.1251	1	6.09	1.7	0.3529	1.3	0.797
5.1	0.10	2913	150	0.05	206	508.3 ± 5.2	522 ± 20	3	0.0578	0.92	0.6539	1.4	0.08204	1.1	0.759
8.1	0.02	145	129	0.92	47.4	2084 ± 20	2119.5 ± 6.7	2	0.13161	0.38	6.924	1.2	0.3816	1.1	0.944
9.1	0.19	2444	104	0.04	173	510.6 ± 5	507.7 ± 9.7	−1	0.05742	0.44	0.6525	1.1	0.08242	1.1	0.916
10.1	0.02	2186	98	0.05	159	524.8 ± 5.1	597 ± 15	12	0.05983	0.7	0.6997	1.2	0.08482	1	0.820
11.1	0.47	2374	110	0.05	164	497.7 ± 4.8	509 ± 19	2	0.05745	0.84	0.6357	1.3	0.08026	1	0.768
11.2	0.30	684	275	0.42	152	1477 ± 13	2030 ± 11	27	0.1251	0.59	4.442	1.2	0.2575	1	0.865
12.1	0.65	4244	239	0.06	304	512.8 ± 5	568 ± 19	10	0.05902	0.89	0.6737	1.4	0.08279	1	0.751
13.1	1.25	2848	142	0.05	201	503.5 ± 5	536 ± 100	6	0.0582	4.7	0.651	4.8	0.08123	1	0.217
14.1	0.53	2365	99	0.04	167	507.5 ± 4.9	513 ± 16	1	0.05756	0.71	0.6501	1.2	0.08191	1	0.818
15.1	0.47	2793	122	0.05	194	499.1 ± 4.9	504 ± 35	1	0.05733	1.6	0.636	1.9	0.0805	1	0.535
16.1	3.78	3762	209	0.06	284	523.7 ± 5.3	723 ± 100	28	0.0634	4.7	0.74	4.8	0.08463	1.1	0.218
17.1	0.01	2831	124	0.05	204	518.8 ± 5.1	505 ± 6.4	−3	0.05735	0.29	0.6627	1.1	0.08381	1	0.962
18.1	0.04	3925	225	0.06	335	610.1 ± 5.9	836.5 ± 6.1	27	0.06696	0.29	0.9164	1	0.09926	1	0.960
19.1	0.01	400	318	0.82	128	2034 ± 18	2045.5 ± 4.5	1	0.12619	0.25	6.455	1.1	0.371	1	0.972

[1]Common Pb corrected using measured ^{204}Pb.

Downing & Coward (1981) noted kilometre-scale SE-facing fold nappes and large SE-facing non-cylindrical folds, consistent with the SE-vergence of F2 folds from this study, they invoked sheath folding due to the presence of a NE–SW extension lineation in zones of intense deformation. This implied a rotation of F2 fold hinges towards the NE–SW extension direction (Downing & Coward 1981), which would reconcile SE-verging folds formed during a SW-verging deformation event. A range of fold closures from NW to south and folds with curvilinear hinges are suggestive of sheath folding.

This study has also shown large variations in hinge-line orientations in some areas (Fig. 4g). Coward (1983) notes a mineral lineation on the axial planar fabric to tight to isoclinal folds with hinges which plunge to the NE, and that mineral lineations are parallel to fold hinges. Such parallelism between hinge lines and lineations suggests SE-verging folding with a NE-plunging intersection, rather than a stretching, lineation. This geometry of structures is explained by Coward (1983) to be due to rotation of fold hinges into the south-westerly transport direction during continued shearing (i.e. due to sheath folding, similar to the explanation given by Downing & Coward 1981). It is possible, however, that intersection lineations parallel to the hinges of SE-verging F2 folds have been mistaken for NE-plunging stretching lineations.

The structural data of Poli (1997), also presented by Poli & Oliver (2001) for the Namibfontein–Vergenog Dome and for the Nose Structure, shows that lineations have similar orientations to fold hinge lines, suggesting that these lineations are intersection lineations between the axial-planar cleavage and hinges of the folds. Although hinge-lines of folds may have been rotated into parallelism with the movement direction, there is an absence of any mesoscale SW-verging sheath folds documented by Coward (1983), Poli (1997) or Poli & Oliver (2001); any possible sheath folds noted in this study are rare.

S_1/S_0 data from the Namibfontein–Vergenoeg Dome (Poli 1997, fig. 3.1; Poli & Oliver 2001, fig. 4) shows that this fabric has been refolded about gently ENE- to east-plunging axes (suggesting that folding was largely cylindrical) and south-to SSE-verging (rather than SW-verging), and may be related to D3. Similarly, the structural data presented for the Nose Structure by Poli (1997) and Poli & Oliver (2001) shows that the general regional fabric in both the basement and cover has been folded about gently ENE-plunging fold axes and that lineations (largely in the basement) are approximately parallel to these fold axes, again suggesting that these are intersection lineations rather than stretching lineations. Although the locally large variability of hinge-line orientations for F2 folds

does suggest possible sheath folding, it appears that D2 folding is generally south- to SE-vergent.

Extensional deformation, found in both Abbabis Complex gneisses and Damaran metasediments, is concentrated in a major extensional detachment near the basement-cover contact, in Etusis Formation metasediments and Abbabis Complex gneisses. A conjugate set of D2 extensional shear bands has formed during this late extension; a steep north-dipping north-verging set and a shallow west-NE SW-verging set have allowed calculation of principal stress directions. A steep SE-plunging $\sigma 1$, shallow west-plunging $\sigma 2$ and shallow NE-plunging $\sigma 3$ are consistently calculated for this conjugate set of shear bands across the study area. NE–SW directed extension also formed mineral stretching lineations. However, shear band orientations and definite NE–SW trending mineral stretching lineations (Fig. 6i, j) do indicate that NE–SW subhorizontal extension has taken place in the Central Zone. Although NE–SW extension (Oliver 1994, 1995; Poli 1997; Poli & Oliver 2001) has previously been based largely on SW–NE trending lineations, which may in some cases be intersection lineations parallel to the hinges of SE-verging F2 folds, a NE–SW extension lineation is consistent with the shallow NE-plunging $\sigma 3$ direction calculated from shear bands.

Field relationships suggest that SE-verging folding and NE–SW extension were near synchronous. Both folding and extension appear to have occurred during the peak of metamorphism; leucosomes are developed along the axial planes of F2 folds in D2 shear bands. Grey granites emplaced during F2 folding are partially affected by shear bands, indicating that the granites were only partially crystallized during shear band development. It appears that isoclinal, recumbent SE-verging folding may immediately precede NE–SW extension, but that folding and extension are part of a progressive sequence of deformation (as suggested by Poli 1997).

Although the formation of folds with hinge lines subparallel to the extension direction has been accounted for by fold axes being reoriented towards the displacement direction with progressive deformation, or due to sheath folding (Downing & Coward 1981; Coward 1983), shear bands in the southwestern Central Zone occur as a conjugate set. This suggests that extension is orogen-parallel, and that orogen-parallel deformation occurred as a result of pure shear rather than simple shear. The vergence of folds to the south–SE however indicates that a component of simple shear was involved in producing orogen-perpendicular deformation. The $\sigma 1$ direction calculated from conjugate shear bands is consistently moderately to steeply south- to SE-plunging (Fig. 6), with $\sigma 2$ shallow west- or NW-plunging and $\sigma 3$ shallowly NE-plunging.

Using the widespread occurrence of prolate porphyroblasts and deformed pebbles, Poli (1997) showed that strain in the southwestern Central Zone was constrictional (i.e. $\sigma 1$ and $\sigma 2$ were almost equivalent). If we consider that NE–SW extension occurred as a result of pure shear in a constrictional field, with SE-verging folds the result of orogen-perpendicular simple shear, then the occurrence of SE-verging folds coeval with NE–SW extension is possible (Fig. 13). This model explains the consistent SE-vergence for folding in the southwestern Central Zone, as recorded by the SE-facing folds of Downing & Coward (1981) and the NE-trending hinge lines of Poli (1997), rather than being formed by SW-verging sheath folds (Downing & Coward 1981; Coward 1983).

There is a lack of mesoscale sheath folds, as well as any NW-verging folds (which would be converse to SE-verging folds suggested to form on the limbs of supposed SW-verging sheath folds). A model of SE-verging simple shear deformation coeval with NE–SW extension in a constrictional field does not require sheath folding as a mechanism, and also explains the lack of SW-verging asymmetrical porphyroblasts or shear sense markers (Oliver 1994). Folds in ductile shear zones generally initiate with hinges normal or at a high angle to the displacement direction (Harris *et al.* 2002), which in this case is perpendicular to the extension direction.

Progression of deformation in the southwestern Central Zone is therefore suggested to have resulted in orogen-perpendicular, tight to isoclinal, SE-verging recumbent folding followed by localization of this intense deformation into SE-verging shear zones. This SE-verging deformation occurred in a constrictional stress field, and was accompanied by NE–SW directed orogen-parallel extension. This led to the formation of a strong mineral stretching lineation and a conjugate set of shear bands. D2 folds and extensional structures were then reoriented by kilometre-scale upright to steep NW-dipping shallow NE-plunging D3 folds.

Timing of deformation

Intense non-coaxial ductile deformation in the southwestern Central Zone (D2), expressed as kilometre-scale south- to SE-verging folds and shear zones, extensional shear bands and NE–SW stretching lineations, is temporally constrained by red granites, grey granites and anatectic leucosomes. Red granites near the northern margin of the Palmenhorst Dome have a strong north-dipping D2 gneissic fabric, suggesting that they predate D2 deformation. These red granites have a concordant U–Pb monazite age of 535.6 ± 7.2 Ma and a zircon U–Pb lower intercept of 539 ± 17 Ma. A 1013 ± 21 Ma upper intercept confirms that they are derived from anatexis of the Abbabis Complex, parts of which have been dated at 1100–1040 Ma (Kröner *et al.* 1991). Ages of 535–539 Ma for the red granites are similar to an anatectic red granite

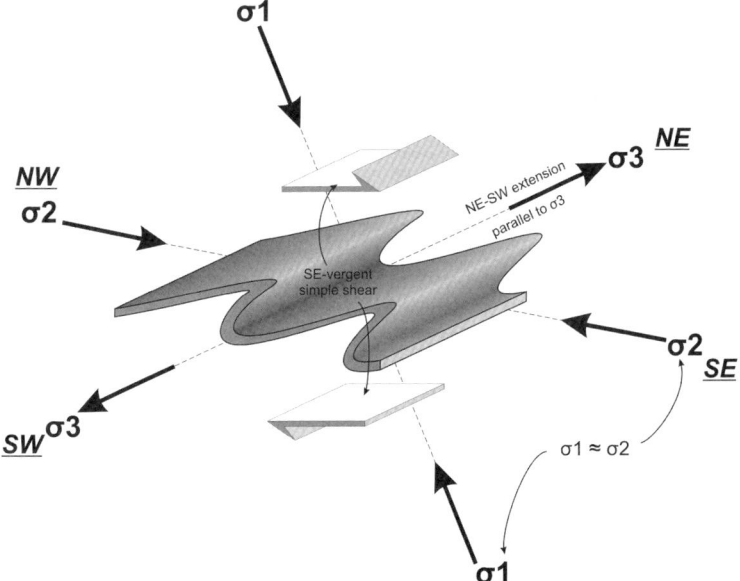

Fig. 13. NE–SW pure shear extension coeval with SE-vergent simple shear in a constrictional field. The resultant structures are tight to isoclinal, recumbent to NW-dipping, SE-verging folds with hinge lines subparallel to NE–SW extensional lineations.

from the Goanikontes area dated at 534 Ma, which also contains a fabric (Briqueu et al. 1980), and to the 539 Ma 'post-tectonic' Rotekuppe granite (Jacob et al. 2000) to the NE near Karibib. The pre- to syn-D2 timing of emplacement of the red granite and the fabric noted in the anatectic red granite from Goanikontes (Briqueu et al. 1980) is in contrast to the post-tectonic nature of similarly aged red granite (the Rotekuppe granite) near Karibib.

Grey granites intruded during early D2; these syntectonic grey granites are emplaced along axial planes of F2 folds and are also folded by D2. The U–Pb zircon age of 520.4 ± 4.2 Ma for such a grey granite dates intense non-coaxial SSE-verging recumbent folding in the southwestern Central Zone at 520 Ma. Anatectic leucosomes are observed in D2 shear zones and in late-D2 shear bands, and anatectic leucosomes are also seen developing at high angles to the NE-plunging stretching lineation (Oliver 1994; Kisters et al. 2009). Thus, ages for leucosomes from such shear zones and shear bands at 511 ± 18 and 508.4 ± 8.7 Ma confirms that SE-verging D2 shearing and NE–SW directed extension are younger than kilometre-scale SE-verging folding and occurred at c. 510 Ma. A continuum of deformation in the southwestern Central Zone, from SE-verging folding to SE-verging shear zones and NE–SW extension occurred over c. 10 Ma between 520 and 510 Ma. The fact that intense deformation occurred simultaneously with high-grade metamorphism was observed by earlier workers (Nash 1971; Jacob 1974; Poli 1997); field relationships show that anatectic leucosomes are localized within late-D2 extensional structures and occur subparallel to the fabric in D2 shear zones.

Previous estimates for the timing of peak metamorphism at 534–508 Ma by Nex et al. (2001), at 525–504 Ma by Jung & Mezger (2003) and at 520–510 Ma by Jung et al. (2000) are similar to the ages for deformation obtained from this study. This confirms that deformation and metamorphism were coeval, and that intense deformation and extension in the southwestern Central Zone occurred at c. 520–510 Ma.

U–Pb SHRIMP dating of structurally constrained granitoids and leucosomes has constrained south- to SE-verging D2 deformation in the study area at c. 520–510 Ma (synchronous with high-grade metamorphism in the southwestern Central Zone).

However, work elsewhere in the Central Zone has indicated much older ages for D2 deformation. Miller (2008) suggests that D2 took place at c. 555 Ma. Kisters et al. (2004) provide evidence that NW-verging fold and thrust tectonics (i.e. crustal thickening) in the lower-grade shallow crustal parts of the Central Zone was coeval with a major mafic to dioritic magmatic event (the Goas Intrusive Suite; Lehtonen et al. 1995) at 560–540 Ma (de Kock et al. 2000; Jacob et al. 2000). A 549 ± 11 Ma age (Johnson et al. 2006) for the syn-deformational Stinkbank granite (one of the Salem-type granites) corroborates this. Additionally, there is geochronological evidence for associated metamorphism immediately postdating this thrusting and magmatism. Jung & Mezger (2003) obtained ages of 540–530 Ma, which they ascribe to intrusion-related migmatization.

A 534 ± 7 Ma age for the anatectic red granite from the Goanikontes area (Briqueu et al. 1980) and similar ages of 536–539 Ma have been obtained from this study for a red granite from the northern margin of the Palmenhorst Dome, shown to be the anatectic product of melting of the Abbabis Complex, further indicates a metamorphic event at c. 535–540 Ma. Metamorphic ages of 540–530 Ma for migmatites and anatectic granites suggest that the thermal effects of a major 560–540 Ma magmatic event in the Central Zone (the Goas Intrusive Suite and Salem-type granites) may have caused metamorphism of Damaran metasediments and Abbabis Complex gneisses, resulting in migmatites and anatectic granites which crystallized at 540–530 Ma.

It appears likely that orthogonal fabrics developed in the upper and lower parts of the orogen (shallow-crustal lower-amphibolite-facies areas in the Central Zone and the mid-crustal upper-amphibolite facies southwestern Central Zone, respectively) represent two separate tectonic episodes in the history of the Damara belt. NW-vergent structures are found in the upper portions of the orogen, and form part of a deeply eroded fold-and-thrust belt with thin-skinned tectonics dominated by thrusting and folding at lower metamorphic grades. It is these NW-verging structures that have been dated at 550–540 Ma (Kisters et al. 2004).

In contrast, the higher-grade (upper-amphibolite- to granulite-facies) lower portions of the orogen are dominated by more ductile SE-verging structures and NW–SE extension, associated with high-grade metamorphism. Kisters et al. (2004) suggested that both NW- and SW-verging deformation in the shallow and mid-crust were contemporaneous, and that the variability in deformation is the result of the different responses of the upper and lower parts of a rheologically layered crust to Pan-African shortening. However, this study has shown that ductile, south- to SE-verging deformation in the lower portions of the orogen occurred at 520–510 Ma, significantly younger than the 550–540 Ma ages suggested for deformation in the upper portions of the orogen. It therefore may be that the evolution of the Central Zone was diachronous in the upper and lower crust, or that older deformation and magmatism is preserved in the upper crust while the higher-grade lower crust preserves only the youngest phase of deformation and magmatism. It may be that rare D1 structures in the study area are preserved remnants of the earlier NW-verging deformation documented by Kisters *et al.* (2004).

The 511–508 Ma ages for late-tectonic leucosomes emplaced into late-D2 extensional sites are similar to the 508 Ma age for post-tectonic alaskites (Briqueu *et al.* 1980), which are considered to post-date upright NE-trending D3 deformation. However, Poli (1997) and Poli & Oliver (2001) did not consider deformation to be polyphase, but envisaged that a continuum of deformation occurred in the southwestern Central Zone. Given the fact that late-D2 structures are only marginally older than post-D3 alaskites, it may be that upright D3 deformation resulted from continued deformation in the same stress field as non-coaxial D2 deformation. In the study area D3 structures clearly post-date D2 (D2 structures are reoriented by upright D3

folds), but the similarity in ages between late-D2 and D3 deformation suggests that such relationships may be less clear elsewhere in the southwestern Central Zone and that D2 and D3 are part of a progressive sequence of deformation. A possible explanation for the switch from non-coaxial recumbent deformation with NE–SW directed extension to upright NE-trending folding could be a change in rheology of the southwestern Central Zone during cooling, combined with a change in principal stress direction during exhumation. The bulk of recumbent south- to SE-verging deformation and NE–SW extension (D2) occurred under upper amphibolite- to granulite-facies conditions, leading to anatexis of the mid-crust in the Central Zone. This deformation was the result of orogen-parallel extension and SSE-vergent tectonic escape in the mid-crust during collision of the Congo and Kalahari cratons.

Later, as escape of material and concomitant extension of the Central Zone continued, cooling of the mid-crust and crystallization of anatectic

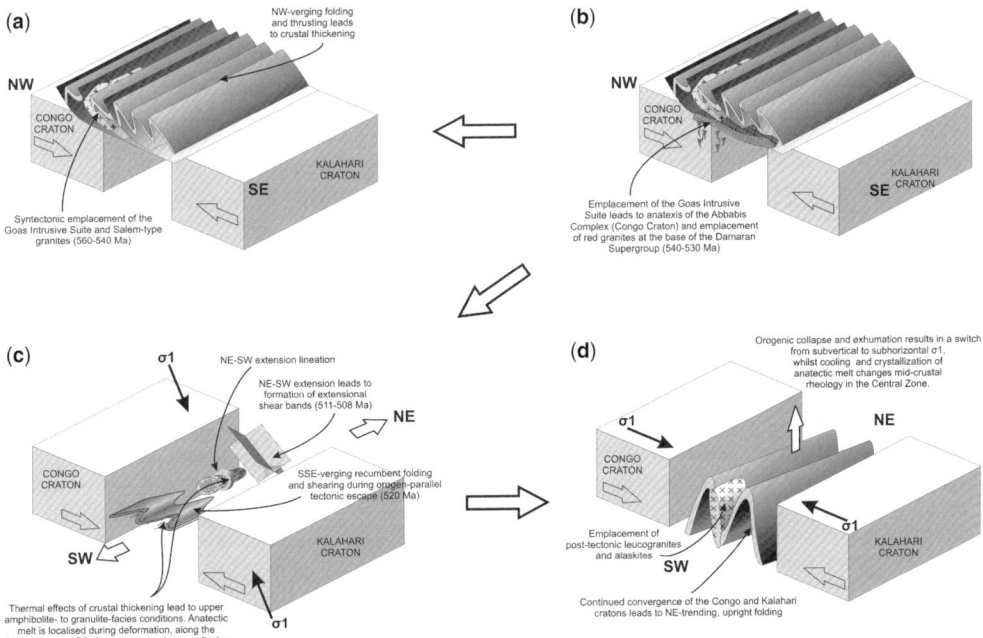

Fig. 14. Proposed tectonic history of the Central Zone. (**a**) At the onset of collision between the Congo and Kalahari cratons, crustal thickening due to NW-vergent folding and thrusting occurs at the same time as emplacement of the mafic to dioritic Goas Intrusive Suite at 560–540 Ma (Kisters *et al.* 2004). (**b**) The thermal effects of the emplacement of the Goas Intrusive Suite results in anatexis of the Abbabis Complex to form the red granites, emplaced at the contact between the Abbabis Complex and Damara Supergroup. (**c**) Heating of the mid-crust following crustal thickening leads to anatexis and a concomitant increase in ductility in the mid-crust. SE-vergent deformation of the ductile mid-crust and NE–SW extension occurs within a constrictional stress field during continued convergence of the Congo and Kalahari cratons. Anatectic leucosomes are emplaced along shear zones and into extensional structural sites. This occurs at 520–508 Ma. (**d**) Cooling of the mid-crust and solidification of melt in the crust changes crustal rheology. A switch in the stress field during the final stages of convergence results in NE-trending upright folding.

leucosomes at *c.* 511–508 Ma caused a change in the rheology from extremely ductile to more brittle. At the same time, orogenic collapse resulted in a switch in σ1 from subvertical to subhorizontal during exhumation of the mid-crust, and in a switch from recumbent SSE-verging deformation and NE–SW extension (subvertical σ1) to upright NE-trending folding (subhorizontal σ3) at 508 Ma. It is also possible that the change in the nature of deformation during D2 from folding to shearing may have been due to rising temperatures. On the prograde path cooler rocks would have retained enough competence contrast to fold (SEE-verging folding), whereas the change in rheology as rocks approached the point of anatexis may have resulted in shear zones and the development of extensional structures.

The Central Zone: a record of crustal thickening and orogenic escape

The conflicting ages for deformation in the northeastern v. southwestern portions of the Central Zone may be explained by considering that a record of both crustal thickening and orogen-parallel escape and extension are preserved in the Central Zone. Crustal thickening occurred at the onset of collision between the Congo and Kalahari cratons, preserved as shallow crustal NW-verging folds and thrusts coeval with the emplacement of large volumes of mafic to granitic magma into the crust (the Goas Intrusive Suite and the Salem-type granites) at 560–540 Ma (Fig. 14a). The thermal effects of this magmatic event led to metamorphism and anatexis of the Abbabis Complex to produce red granites at 540–535 Ma (Fig. 14b). This metamorphic event is also preserved in the 540–530 Ma metamorphic ages of Jung (2000), Jung & Mezger (2003) and Jung *et al.* (2009).

Deformation is however dated at 520–508 Ma in the southwestern Central Zone, coeval with high-grade metamorphism and anatexis of the Abbabis Complex and Damara Supergoup at 525–504 Ma (Jung 2000; Jung & Mezger 2003). Rather than NW-verging thrusting and crustal thickening as observed in the higher portions of the Central Zone to the NE, these ages record the timing of SSE-verging folding and NE–SW orogen-parallel extension during orogenic collapse and tectonic escape (Fig. 14c). It is likely that the upper-amphibolite-to granulite-facies metamorphism of the mid-crust at this time is related to the thermal effects of the crustal thickening which occurred at 560–540 Ma. Finally, a change in stress directions and cooling during exhumation of the Central Zone, with continued convergence of the Congo and Kalahari cratons, resulted in upright D3 folding followed by the emplacement of post-tectonic granites in the southwestern Central Zone (Fig. 14d).

Such a two-stage evolution for the orogen can explain the occurrence of older ages in the southwestern Central Zone, which represents the first tectonic event in the Damara Belt. For example the 542 Ma age for grey granite from the Ida Dome (Tack *et al.* 2002) likely represents an earlier event preserved in the mid-crustal southwestern portions of the Central Zone. It also explains how 540–535 Ma red granites in the upper portions of the orogen (the Rotekuppe granite) appear to post-date deformation, while similar age granites in the southwestern Central Zone are pre- or syntectonic.

Conclusions

Apparently conflicting ages for deformation and metamorphism within the Central Zone of the Damara Belt are resolved by considering that two tectonic events are preserved in the Central Zone. The onset of collision between the Congo and Kalahari cratons led to crustal thickening and NW-verging folding and thrusting at shallow crustal levels. This occurred between 560 and 540 Ma, coeval with emplacement of the voluminous magmas of the Goas Intrusive Suite and Salem-type granites. The thermal effects of this magmatism resulted in anatexis of the Abbabis Complex to form red granites, which crystallized at *c.* 535–540 Ma.

Crustal thickening led to high-grade metamorphism and anatexis of the mid-crust at *c.* 525–504 Ma, coeval with ductile south- to SE-verging folding and NE–SW orogen-parallel extension. SE-verging folding was coeval with the intrusion of grey granites at *c.* 520 Ma. Later shearing and NE–SW directed extension occurred at the same time as widespread anatexis, forming leucosomes at *c.* 510 Ma. Cooling during extension resulted in a change in mid-crustal rheology. Exhumation changed stress directions from a subvertical to a subhorizontal σ1 forming kilometre-scale upright NE-trending folds at *c.* 508 Ma, immediately preceding emplacement of final post-tectonic uraniferous alaskites during the final stages of convergence between the Congo and Kalahari cratons.

The results presented in this paper are the product of work funded by the National Research Foundation (NRF) of South Africa, the Jim and Gladys Taylor Education Trust, and the Research, Education and Investment Fund of the Geological Society of South Africa (GSSA), who are all thanked for their support. Bannerman Resources and Extract Resources provided a great deal of field support, for which we are grateful. Thanks to Joe Aphane for help with zircon extraction. Thanks to Roy Miller and an anonymous reviewer for thorough and thoughtful reviews which greatly improved the manuscript.

References

BARNES, J. F. H. 1981. *Some aspects of the tectonic history of the Khan-Swakop region of the Damara Belt, Namibia.* PhD thesis, University of Leeds.

BARNES, S.-J. & SAWYER, E. W. 1980. An alternative model for the Damara Mobile Belt. Ocean crust subduction and continental convergence. *Precambrian Research*, **13**, 297–336.

BLAINE, J. L. 1977. *Tectonic Evolution of the Waldau Ridge Structure and the Okahandja Lineament in Part of the Central Damara Orogen, West of Okahandja, South West Africa.* Bulletin of the Precambrian Research Unit, University of Cape Town, **21**.

BRANDT, R. 1985. Preliminary report on the geology of the Damara Sequence and the geology and geochemistry of the Damara granites in the area between Walvis Bay and Karibib. *Communications of the Geological Survey of Namibia*, **1**, 31–43.

BRIQUEU, L., LANCELOT, J. P., VALOIS, J. P. & WALGENWITZ, F. 1980. Géochronologie U–Pb et génese d'un type de minéralisation uranifère: les alaskites de Goanikontes (Namibie) et leur encaissant. *Bulletin des Centre de Récherches Exploration Elf-Aquitaine*, **4**, 759–811.

BUHN, B., OKRUSCH, M., WOERMANN, E., LEHNERT, K. & HOERNES, S. 1995. Metamorphic evolution of Neoproterozoic manganese formations and their country rocks at Otjosondu, Namibia. *Journal of Petrology*, **36**, 463–496.

COMPSTON, W., WILLIAMS, I. S., KIRSCHVINK, J. L., ZICHAO, Z. & GUOGAN, M. 1992. Zircon ages for the Early Cambrian time-scale. *Journal of the Geological Society, London*, **149**, 171–184.

COWARD, M. P. 1983. The tectonic history of the Damaran belt. *In*: MILLER, R. McG (ed.) *Evolution of the Damara Orogen of South West Africa/Namibia.* Geological Society of South Africa, Johannesburg, Special Publications, **11**, 409–421.

CUMMING, G. L. & RICHARDS, J. R. 1975. Ore lead isotope ratios in a continuously changing earth. *Earth and Planetary Science Letters*, **28**, 155–171.

DE KOCK, G. S. 1989. *'n Geotektoniese studie van die Damara-orogeen n 'n gebied suidoos van Karibib, Suidwes-Afrika.* PhD thesis, University of the Orange Free State.

DE KOCK, G. S., EGLINGTON, B., ARMSTRONG, R. A., HARMER, R. E. & WALRAVEN, F. 2000. U–Pb and Pb-Pb ages of the Naauwpoort rhyolite, Kawaeup leptite and Okongava Diorite: implications for the onset of rifting and orogenesis in the Damara belt, Namibia. *Communications of the Geological Survey of Namibia*, **12**, 81–88.

DOWNING, K. N. 1982. *The evolution of the Okahandja Lineament and its significance in Damaran tectonics (Namibia).* PhD thesis, Leeds University, Leeds.

DOWNING, K. N. & COWARD, M. P. 1981. The Okahandja lineament and its significance for Damaran tectonics in Namibia. *Geologische Rundschau*, **70**, 972–1003.

GOSCOMBE, B., GRAY, D. & HAND, M. 2004. Variation in metamorphic style along the northern margin of the Damara Orogen, Namibia. *Journal of Petrology*, **45**, 1261–1295.

GRAY, D. R., FOSTER, D. A., MEERT, J. G., GOSCOMBE, B. D., ARMSTRONG, R., TROUW, R. A. J. & PASSCHIER, C. W. 2008. A Damara orogen perspective on the assembly of southwestern Gondwana. *In*: PANKHURST, R. J., TROUW, R. A. J., BRITO NEVES, B. B. & DE WIT, M. J. (eds) *West Gondwana: Pre-Cenozoic Correlations Across the South Atlantic Region.* Geological Society, London, Special Publications, **294**, 257–278.

HAACK, U. & GOHN, E. 1988. Rb–Sr data on some pegmatites in the Damara Orogen, Namibia. *Communications of the Geological Survey of South West Africa/Namibia*, **4**, 13–17.

HAACK, U., GOHN, E. & KLEIN, J. A. 1980. Rb/Sr ages of granitic rocks along the middle reaches of the Omaruru River and the timing of orogenic events in the Damara Belt (Namibia). *Contributions to Mineralogy and Petrology*, **74**, 349–360.

HARRIS, L. B., HEMIN, A. K. & FOSSEN, H. 2002. Mechanisms for folding of high-grade rocks in extensional tectonic settings. *Earth-Science Reviews*, **59**, 163–210.

HARTMANN, O., HOFFER, E. & HAACK, U. 1983. Regional metamorphism in the Damara Orogen: interaction of crustal motion and heat transfer. *In*: MILLER, R. McG (ed.) *Evolution of the Damara Orogen of South West Africa/Namibia.* Geological Society of South Africa, Johannesburg, Special Publications, **11**, 233–241.

HAWKESWORTH, C. J., GLEDHILL, A. R., RODDICK, J. C., MILLER, R. McG. & KRÖNER, A. 1983. Rb–Sr and $^{40}Ar/^{39}Ar$ studies bearing on models for the thermal evolution of the Damara Belt, Namibia. *In*: MILLER, R. McG (ed.) *Evolution of the Damara Orogen of South West Africa/Namibia.* Geological Society of South Africa, Johannesburg, Special Publications, **11**, 323–338.

HOERNES, S. & HOFFER, E. 1979. Equilibrium relations of prograde metamorphic mineral assemblages. A stable isotope study of rocks of the Damara Orogen, from Namibia. *Contributions to Mineralogy and Petrology*, **68**, 377–389.

HOFFMANN, K.-H., CONDON, D. J., BOWRING, S. A. & CROWLEY, J. L. 2004. U–Pb date from the Neoproterozoic Ghaub Formation, Namibia: Constraints on Marinoan glaciation. *Geology*, **29**, 1091–1094.

JACOB, R. E. 1974. *Geology and Metamorphic Petrology of Part of the Damara Orogen along the Lower Swakop River, South West Africa.* Bulletin of the Precambrian Research Unit, University of Cape Town, **17**.

JACOB, R. E. 1978. Granite Genesis and Associated Mineralization in Part of the Central Damara Belt. *In*: VERWOERD, W. J. (ed.) *Mineralization in Metamorphic Terranes.* Geological Society of South Africa, Johannesburg, Special Publications, **4**, 417–432.

JACOB, R. E., SNOWDEN, P. A. & BUNTING, F. J. L. 1983. Geology and Structural Development of the Tumas Basement Dome and its Cover Rocks. *In*: MILLER, R. McG (ed.) *Evolution of the Damara Orogen of South West Africa/Namibia.* Geological Society of South Africa, Johannesburg, Special Publications, **11**, 157–172.

JACOB, R. E., MOORE, J. M. & ARMSTRONG, R. A. 2000. Zircon and Titanite age determination from igneous rocks in the Karibib District, Namibia: implications for Navachab vein-style gold mineralization.

Communication of the Geological Survey of Namibia, **12**, 157–166.

JOHNSON, S. D., POUJOL, M. & KISTERS, A. F. M. 2006. Constraining the timing and migration of collisional tectonics in the Damara Belt, Namibia: U–Pb zircon ages for the syntectonic Salem-type Stinkbank granite. *South African Journal of Geology,* **109**, 611–624.

JUNG, C., JUNG, S., NEBEL, O., HELLEBRAND, E., MASBERG, P. & HOFFER, E. 2009. Fluid-present melting of meta-igneous rocks and the generation of leucogranites – Constraints from garnet major- and trace element data, Lu–Hf whole rock-garnet ages and whole rock Nd-Sr-Hf-O isotope data. *Lithos,* **111**, 220–235.

JUNG, S. 2000. High-temperature, low/medium-pressure clockwise P–T paths and melting in the development of regional migmatites: the role of crustal thickening and repeated plutonism. *Geological Journal,* **35**, 345–359.

JUNG, S. & MEZGER, K. 2003. Petrology of basement-dominated terranes: I. regional metamorphic T-t path and geochronological constraints on Pan-African high-grade metamorphism (central Damara orogen, Namibia). *Chemical Geology,* **198**, 223–247.

JUNG, S., HOERNES, S. & MEZGER, K. 2000. Geochronology and petrology of migmatites from the Proterozoic Damara Belt — importance of episodic fluid-present disequilibrium melting and consequences for granite petrology. *Lithos,* **51**, 153–179.

KASCH, K. W. 1983. Regional P–T variations in the Damara Orogen with particular reference to early high-pressure metamorphism along the Southern Margin. *In*: MILLER, R. McG (ed.) *Evolution of the Damara Orogen of South West Africa/Namibia.* Geological Society of South Africa, Johannesburg, Special Publications, **11**, 243–535.

KENNEDY, W. Q. 1964. *The structural differentiation of Africa in the Pan-African (±500 m.y.) tectonic episode.* Research Institute for African Geology, University of Leeds, 8th Annual Report, 48–49.

KISTERS, A. F. M., JORDAAN, L. S. & NEUMAIER, K. 2004. Thrust-related dome structures in the Karibib district and the origin of orthogonal fabric domains in the south Central Zone of the Pan-African Damara belt, Namibia. *Precambrian Research,* **133**, 283–303.

KISTERS, A. F. M., WARD, R. A., ANTHONISSEN, C. J. & VIETZE, M. E. 2009. Melt segregation and far-field melt transfer in the mid-crust. *Journal of the Geological Society, London,* **166**, 905–918.

KRÖNER, A. 1982. Rb–Sr geochronology and tectonic evolution of the Pan-African Damara belt of Namibia, south-western Africa. *American Journal of Science, London,* **282**, 1471–1507.

KRÖNER, A. 1984. Dome structures and basement reactivation in the Pan-African Damara belt of Namibia. *In*: KRÖNER, A. & GREILING, R. (eds) *Precambrian Tectonics Illustrated.* Schweizerbart, Stuttgart, 191–206.

KRÖNER, A., RETIEF, E. A., COMPSTON, W., JACOB, R. E. & BURGER, A. J. 1991. Single-grain and conventional zircon dating of remobilized basement gneisses in the central Damara belt of Namibia. *South African Journal of Geology,* **94**, 379–387.

KUKLA, C., KRAMM, U., KUKLA, P. A. & OKRUSCH, M. 1991. U–Pb monazite data relating to metamorphism and granite intrusion in the northwestern Khomas Trough, Damara Orogen, central Namibia. *Communication of the Geological Survey of Namibia,* **7**, 49–54.

LEHTONEN, M. I., MANNINEN, T. E. T. & SCHREIBER, U. M. 1995. Geological map sheet 2214 – Walvis Bay, 1:250 000. *Geological Survey of Namibia.*

LUDWIG, K. R. 2001. *SQUID 1.03 - a user's manual.* Berkeley Geochronology Centre, Special Publication, **2**.

LUDWIG, K. R. 2003. *User's manual for Isoplot/Ex version 3.00, A geochronological toolkit for Microsoft Excel.* Berkeley Geochronology Center, Special Publication, **4**.

MARLOW, A. G. 1981. *Remobilisation and primary uranium genesis in the Damaran Orogenic Belt, Namibia.* PhD thesis, University of Leeds.

MASBERG, H. P., HOFFER, E. & HOERNES, S. 1992. Microfabrics indicating granulite-facies metamorphism in the low-pressure central Damara Orogen, Namibia. *Precambrian Research,* **55**, 243–257.

MILLER, R. McG. 1983. The Pan-African Damara orogen of South West Africa/Namibia. *In*: MILLER, R. McG (ed.) *Evolution of the Damara Orogen of South West Africa/Namibia.* Geological Society of South Africa, Johannesburg, Special Publication, **11**, 431–515.

MILLER, R. McG. 2008. *The Geology of Namibia. Volume 2: Neoproterozoic to Lower Palaeozoic.* Geological Survey of Namibia, Windhoek, Namibia.

NASH, C. R. 1971. *Metamorphic Petrology of the SJ Area, Swakopmund District, South West Africa.* Bulletin of the Precambrian Research Unit, University of Cape Town, **9**.

NEX, P. A. M. 1997. *Tectono-metamorphic Setting and Evolution of Granitic Sheets in the Goanikontes Area, Namibia.* PhD thesis, University College Cork, National University of Ireland.

NEX, P. A. M., OLIVER, G. J. H. & KINNAIRD, J. A. 2001. Spinel-bearing assemblages and P–T–t evolution of the Central Zone of the Damara Orogen, Namibia. *Journal of African Earth Sciences,* **32**, 471–489.

OLIVER, G. J. H. 1994. Mid-crustal detachment and domes in the central zone of the Damaran orogen, Namibia. *Journal of African Earth Sciences,* **19**, 331–344.

OLIVER, G. J. H. 1995. The Central Zone of the Damara Orogen, Namibia, as a deep metamorphic core complex. *Communications of the Geological Survey, Namibia,* **10**, 33–41.

POLI, L. C. 1997. *Mid-Crustal Geodynamics of the Southern Central Zone, Damara Orogen, Namibia.* PhD thesis, University of St Andrews.

POLI, L. C. & OLIVER, G. J. H. 2001. Constrictional deformation in the Central Zone of the Damara Orogen, Namibia. *Journal of African Earth Sciences,* **33**, 303–321.

PRAVE, A. R. 1996. Tale of three cratons: Tectonostratigraphic anatomy of the Damara orogen in northwestern Namibia and the assembly of Gondwana. *Geology,* **24**, 1115–1118.

PUHAN, D. 1983. Temperature and pressure of metamorphism in the central Damara Orogen. *In*: MILLER, R. McG (ed.) *Evolution of the Damara Orogen of South West Africa/Namibia.* Geological Society of South Africa, Special Publications, **11**, 219–223.

RAMSAY, J. G. 1967. *Folding and Fracturing of Rocks.* McGraw-Hill, New York.

SAWYER, E. W. 1981. *Damaran Structural and Metamorphic Geology of an Area South-east of Walvis Bay, South West Africa/Namibia.* Memoir of the Geological Survey of South Africa, **7**.

SMITH, D. A. M. 1965. *The Geology of the Area around the Khan and Swakop Rivers in South West Africa.* Memoir of the Geological Survey of South Africa, **3**.

TACK, L., WILLIAMS, I. & BOWDEN, P. (2002) SHRIMP constraints on early post-collisional granitoids of the Ida Dome, central Damara (Pan-African) Belt, western Namibia. *Abstracts of the 11th IAGOD Quadrennial Symposium and Geocongress, Windhoek, Namibia.* Geological Survey of Namibia.

WARD, R., STEVENS, G. & KISTERS, A. 2008. Fluid and deformation induced partial melting and melt volumes in low-temperature granulite-facies metasediments, Damara Belt, Namibia. *Lithos*, **105**, 253–271.

WILLIAMS, I. S. 1998. U–Th–Pb Geochronology by Ion Microprobe. *In*: MCKIBBEN, M. A., SHANKS, W. C., III. & RIDLEY, W. I. (eds) *Applications of Microanalytical Techniques to Understanding Mineralizing Processes.* Reviews in Economic Geology, Society of Economic Geologists, **7**, 1–35.

WILLIAMS, I. S. & CLAESSON, S. 1987. Isotopic evidence for the Precambrian provenance and Caledonian metamorphism of high grade paragneisses from the Seve Nappes, Scandinavian Caledonides. II. Ion microprobe U–Th–Pb. *Contributions to Mineralogy and Petrology*, **97**, 205–217.

WILSON, T. J., GRUNOW, A. M. & HANSON, R. E. 1997. Gondwana assembly: the view from Southern Africa and East Gondwana. *Journal of Geodynamics*, **23**, 263–286.

YOSHIDA, M., KAMPUNZU, A. B., LI, X. & WATANABE, T. 2003. Assembly and Break-up of Rodinia and Gondwana: Introduction. *Gondwana Research*, **6**, 139–142.

The Palaeozoic palaeogeography of central Gondwana

TROND H. TORSVIK[1,2,3]* & L. ROBIN M. COCKS[4]

[1]*PGP and Geosciences, University of Oslo, P.O. Box 1048, 0316 Oslo, Norway*

[2]*Geodynamics Centre, Geological Survey of Norway, Leif Eirikssons vei 39, N-7491 Trondheim, Norway*

[3]*School of Geosciences, University of Witwatersrand, WITS 2050, South Africa*

[4]*Department of Palaeontology, The Natural History Museum, Cromwell Road, London SW7 5BD, UK*

*Corresponding author (e-mail: t.h.torsvik@geo.uio.no)

Abstract: Nine new palaeogeographical maps of central Gondwana are presented at intervals within the Palaeozoic from the Middle Cambrian at 510 Ma to the end of the Permian at 250 Ma. The area covered includes all of Africa, Madagascar, India and Arabia as well as adjacent regions, including parts of southern Europe, much of South America (including the Falkland Isles) and Antarctica. After final assembly in the Late Neoproterozoic the southern margin was largely passive throughout the Palaeozoic, apart from some local orogeny in the Cambrian in the final stages of the largely Neoproterozoic Pan-African Orogeny and during the Late Palaeozoic Gondwanide Orogeny. The northern peri-Gondwana margin was active during the Early Palaeozoic but the NW part became passive by the earliest Ordovician when the Rheic Ocean opened between Gondwana and Avalonia. This was eventually followed by the latest Silurian or Early Devonian opening of the Palaeotethys Ocean between Gondwana and Iberia, Armorica and associated terranes and, much later, the rifting and opening of the Neotethys Ocean near the close of the Permian. In the Late Carboniferous, Gondwana merged with Laurussia to form Pangea. That accretion took place outside the area to the NW, although the consequent orogenic activity extended to Morocco and Algeria. Most of the centre of Gondwana was land throughout the Palaeozoic but with extensive shelf seas over the craton margins, particularly the northern margin from the Cambrian to the Devonian on which the important north African and Arabian hydrocarbon source rocks were deposited in the Lower Silurian (with the chief reservoirs in the adjacent Upper Ordovician) and Upper Devonian. There were also substantial Upper Carboniferous and later non-marine lake basins in central and southern Africa in which the Karroo Supergroup was deposited. The South Pole was located within the area from the Early Palaeozoic to the Mid-Permian and central Gondwana was therefore greatly affected by two ice ages: the short but sharp Hirnantian glaciation at the end of the Ordovician and another lasting sporadically for more than 25 Ma during the later Carboniferous and Early Permian.

Following preliminary global reviews of the Palaeozoic (Cocks & Torsvik 2002; Torsvik & Cocks 2004), we are subsequently describing the palaeogeography of the major continents in separate papers. However, because the supercontinent of Gondwana was so big, our treatment of it is in sectors. The NE sector from Turkey round to New Zealand is published (Torsvik & Cocks 2009), and the present paper deals with the central sector (Fig. 1) which includes Africa, Arabia and India and some of the adjacent South American and Antarctic areas. Parts of Palaeozoic Gondwana which are now in Europe and North America are also included, which we term peri-Gondwana. The Precambrian cratons forming the old basement (>1 Ga) of Gondwana and kimberlite locations (the main carrier of diamonds) are shown in Figure 2.

Figure 3 shows the progressive movements of Africa throughout the Palaeozoic, and there are also Lower Palaeozoic reconstructions of the entire core of Gondwana in Torsvik & Cocks (2009, fig. 2). Gondwana rotated and drifted slightly as Palaeozoic time progressed, but the South Pole lay under this central sector of the continent from the late Precambrian until the middle of the Permian. We have also assessed Africa as having been the continent which has moved the least distance in relation to Earth's underlying mantle during the whole Phanerozoic (Torsvik *et al.* 2008*a*). As can be seen from Figure 3b, the latitudinal displacement rate of Africa remained relatively low throughout the Palaeozoic except for an apparent acceleration just prior to Pangea assembly. The northern Gondwana margin was active from

From: VAN HINSBERGEN, D. J. J., BUITER, S. J. H., TORSVIK, T. H., GAINA, C. & WEBB, S. J. (eds) *The Formation and Evolution of Africa: A Synopsis of 3.8 Ga of Earth History*. Geological Society, London, Special Publications, **357**, 137–166. DOI: 10.1144/SP357.8 0305-8719/11/$15.00 © The Geological Society of London 2011.

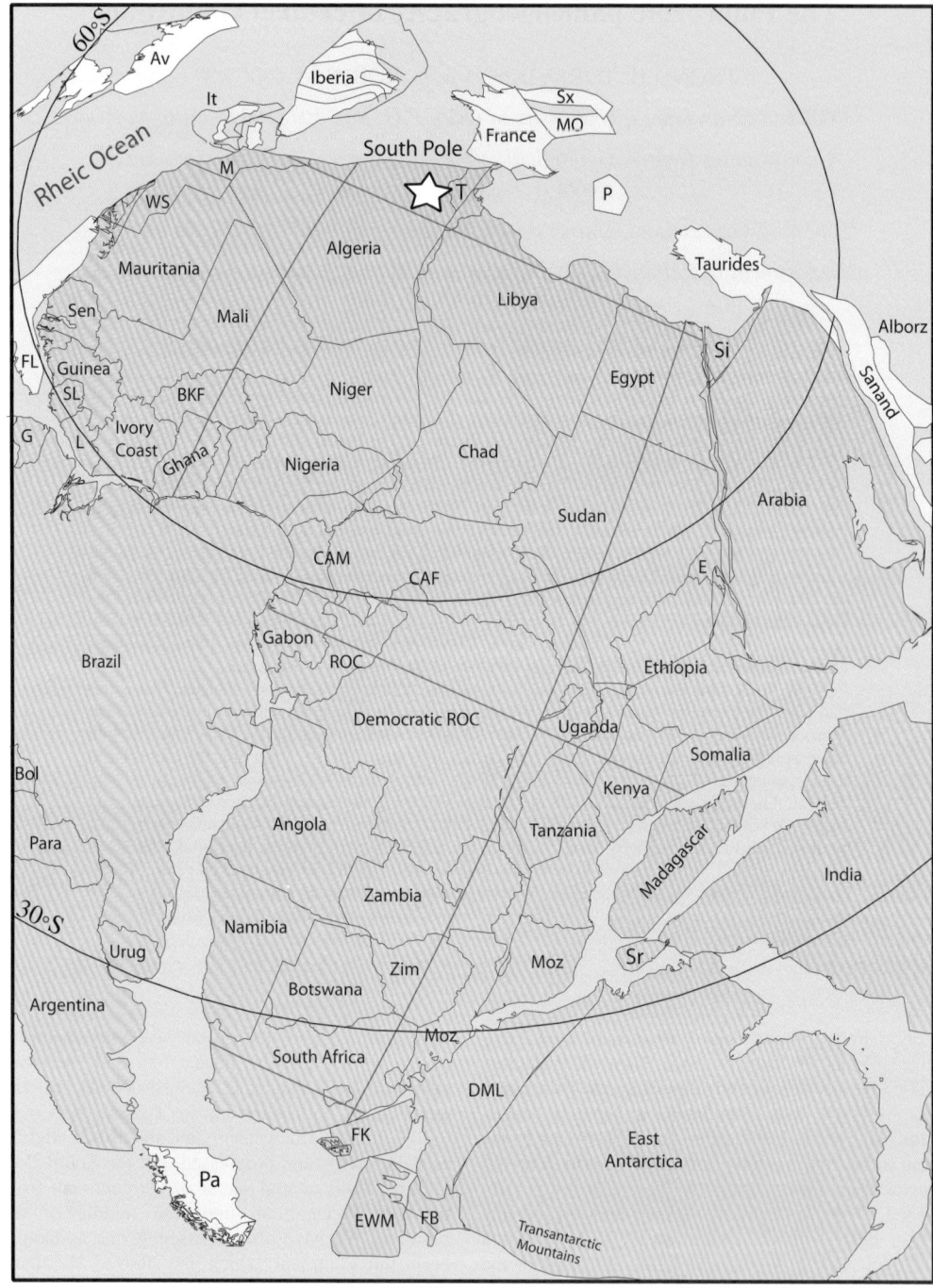

Fig. 1. Map of the central Gondwana area, showing most of the terrane areas discussed here as they were positioned at 480 Ma, and the modern country boundaries within Africa and South America. The boundaries of the separate Precambrian cratons are shown in Africa. Red lines show modern latitudes and longitudes in Africa and black arcs the 480 Ma palaeolatitudes. (Av, Avalonia; BKF, Burkino Fasso; Bol, Bolivia; CAF, Central African Republic; CAM, Cameroon; DML, Dronning Maud Land; E, Eritrea; EWM, Ellsworth–Whitmore Mountains; FB, Filchner Block, Antarctica; FK, Falklands; FL, Florida; G, Guyana; It, Italian terranes; L, Liberia; M, Morocco; MO, Moldanubia;

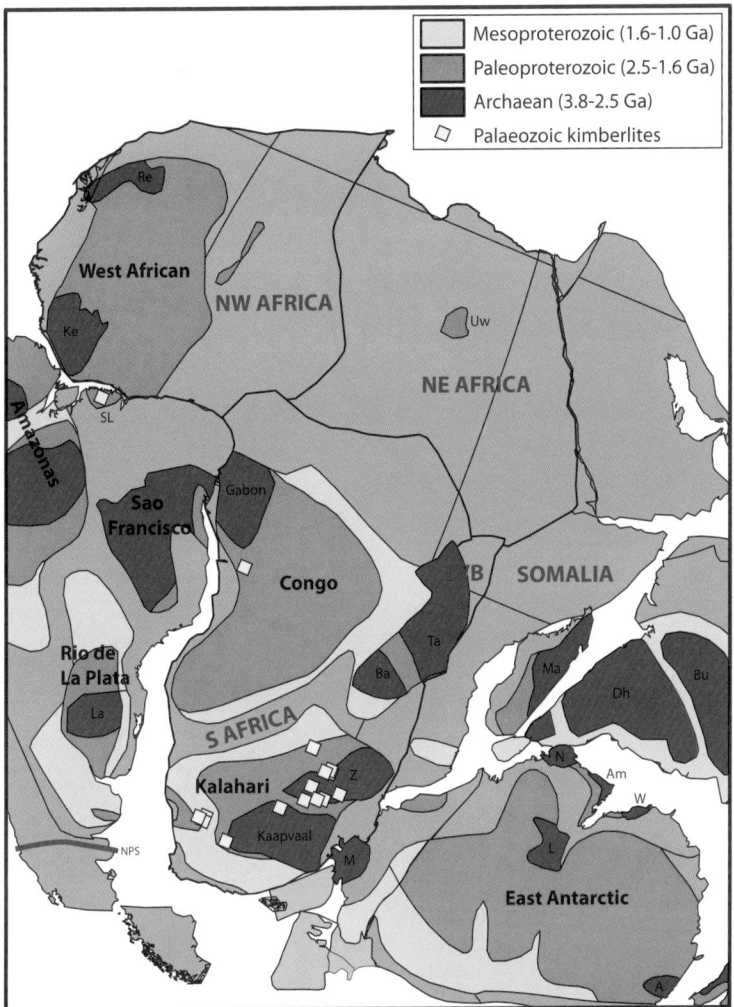

Fig. 2. Older Precambrian (Archaean–Mesoproterozoic) cratons which make up the basement of Central Gondwana (Gubanov & Mooney 2009), plotted on a new 480 Ma reconstruction. Palaeozoic kimberlites are also shown (Torsvik *et al.* 2010). (A, Adélie; Am, Amery; Ba, Bangweulu; Bu, Bundelkhand; Dh, Dharwar; K, Kenema-Man; L, Lambert; Lu, Luis Alves; LVB, Lake Victoria Block; M, Martha; Ma, Madagascar; N, Napier; NPS, North Patagonian shelf; Re, Reguibat; SL, São Luis; Ta, Tanzania; Uw, Uwcinat; W, Westfold; Z, Zimbabwe.)

Mexico to Turkey (e.g. Fig. 6), but passive from Turkey to Australia during most of the Palaeozoic (Torsvik & Cocks 2009). Most of the southern Gondwanan margin was largely passive throughout the Palaeozoic, except at the beginning of the Cambrian and the end of the Palaeozoic. After short summaries of the Palaeozoic geology of the different areas, we present a brief Palaeozoic

history accompanied by palaeogeographical maps for selected times (Figs 5–13).

North and west Africa

This area is underlain by two major intra-cratonic units, NW and NE Africa (Fig. 2). Its Lower Palaeozoic geology has been summarized by Holland

Fig. 1. (*Continued*) Moz, Mozambique; Para, Paraguay; Pa, Patagonia; P, Perunica; ROC, Republic of Congo (there are two different countries, one Democratic ROC); Sen, Senegal; Si, Sinai; SL, Sierra Leone; Sr, Sri Lanka; Sx, Saxothuringia; T, Tunisia; Urug, Uruguay; WS, Western Sahara; Zim, Zimbabwe.)

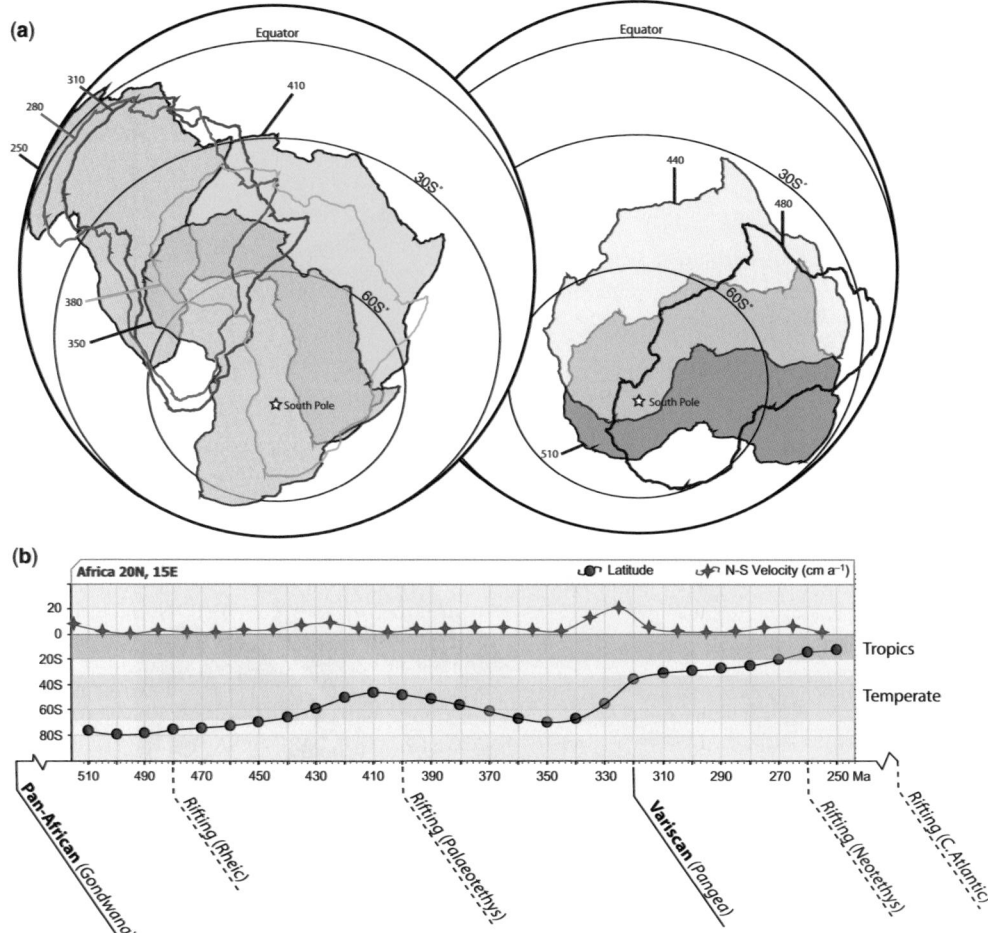

Fig. 3. Progressive palaeomagnetic reconstructions of Africa throughout the Palaeozoic: (**a**) as it migrated over the South Pole and (**b**) its changing latitudinal velocities (above) and the latitudinal position of a location in southern Libya (20°N, 15°E) with time, with the main tectonic episodes (below).

(1981, 1985) and, for the whole Phanerozoic of NW Africa, by Piqué (2001). A set of palaeogeographical maps was published by Guiraud *et al.* (2005), and a useful summary of the geology of all the countries in Africa can be found in Schlüter (2006). The Silurian and Devonian of north Africa contain significant hydrocarbon source rocks (Arthur *et al.* 2003). Summaries of the individual areas within the region are provided in the following.

Morocco and the Meseta

The north-western part of Morocco (about half of the country) is termed the Meseta, which is divided into west and east Meseta by the Middle Atlas Mountains. Although separate from the main African Craton until its accretion in the Late Carboniferous

during the union of Gondwana and Pangea, the Palaeozoic successions in the Meseta are broadly similar to the rest of Morocco; the two were probably not far apart. On the main African Craton there is an important succession of Lower Palaeozoic fossiliferous rocks well exposed in the Anti-Atlas Mountains (Destombes *et al.* 1985), which form the northern margin of the east–west trending Tindouf Basin. The Upper Palaeozoic also includes shallow to deeper-water Devonian and Lower Carboniferous marine rocks with rich faunas, including the varied and beautiful Emsian trilobites which are mined and exported for sale in markets round the world as well as volcanics of Pragian age in the east Anti-Atlas. The Middle and Upper Carboniferous rocks are of terrestrial origin. In the Meseta, the rocks also include Early Cambrian archaeocyathid limestones and Middle Cambrian deeper shelf sequences including turbidites and volcanics, all probably representing the final stages of the Pan-African Orogeny. There are also Ordovician granites, Silurian and Devonian shelf deposits and Carboniferous volcanics.

Algeria and Tunisia

There are extensive Lower Palaeozoic rocks in the several basins of Algeria which extend eastwards, largely in the subsurface, into Tunisia as reviewed by Legrand (1974) and Holland (1985). Although the Cambrian is sparse, the Ordovician is well represented including the spectacular Hirnantian glacial deposits so well illustrated by Beuf *et al.* (1971) which are the clastic reservoir rocks for many of the hydrocarbons in the area. Those are followed by the Llandovery 'hot' shales, which were the prime source rocks in the region. There are also many later Silurian, Devonian and Carboniferous rocks in the various basins, although the Permian is known only from Tunisia.

Libya and Niger

Libya includes several oil-rich basins, including the Murzuq Basin which has varied sediments from the Cambrian to the Carboniferous but with several unconformities (Ramos *et al.* 2006). Mergl & Massa (2000) summarized the rocks and fossils of the Devonian and Carboniferous there, which include varied Middle and Upper Devonian brachiopods and other shelly invertebrates deposited on the shallow shelf. The Murzuq Basin extends southwards from Libya into Niger, and includes Upper Ordovician rocks in which Denis *et al.* (2006) identified two phases within the Hirnantian glaciation. Hallett (2002) has summarized the geology of the whole of Libya and the volumes edited by Salern & Oun (2003) describe the NW part of Libya, which includes the Sirte and Ghademes basins. The latter spans the Algerian–Libyan border (Lüning *et al.* 2000), and has significant Early Cambrian–Late Devonian (Famennian) successions below the unconformity with the Jurassic. In contrast, the Sirt Basin has only Middle Cambrian–Early Silurian (Llandovery) rocks beneath the sub-Mesozoic unconformity.

Northeast Africa

Palaeozoic rocks in Egypt are essentially an eastwards extension of the basins seen in Libya. The substantial Kufra Basin, which largely comprises Lower Palaeozoic rocks, underlies the tripartite junction between Libya, Chad and Sudan and there are also Permo-Triassic continental rocks there.

Eritrea includes some Hirnantian glaciogenic rocks and granites of assumed Lower Palaeozoic age (Schlüter 2006). Ethiopia has Upper Carboniferous glaciogenic sandstones, described by Bussert & Schrank (2007). Schandelmeier & Reynolds (1997) provided palaeogeographical maps for the Upper Palaeozoic and later rocks for all of northeast Africa.

West Africa

The geology of the countries in western Africa is summarized by Schlüter (2006). Deynoux *et al.* (*in* Holland 1985) reviewed the Lower Palaeozoic of western Africa from Mauritania to Niger, including the southern margin of the extensive Tindouf Basin. Willefert (1988) described Mauritania, with particular reference to its Late Ordovician and Early Silurian rocks. Further south, the substantial Taoudeni Basin (whose northern half is depicted in Fig. 6) underlies the boundaries of Mauretania, Mali, southern Algeria and Burkino–Faso and continues into Guinea, Guinea-Bissau and Senegal, where it is termed the Bové Basin. It includes Cambro-Ordovician *Skolithos*-bearing sandstones with inarticulate brachiopods, Hirnantian tillites, Llandovery graptolitic shales and Devonian shales and reef limestones. In Guinea there are Ludlow–Devonian marine sandstones with Emsian, Eifelian and Givetian brachiopods. In Mali the Devonian shales are unconformably overlain by Lower Carboniferous clastics containing brachiopods and which extend upwards into evaporite deposits. In Niger, Cambro-Ordovician clastics are unconformably overlain by Silurian graptolitic shales, Devonian sandstones and shales and Upper Carboniferous–Permian deltaic sandstones, all in basins which are southern extensions of those in Algeria and Libya. In Ghana, Lower and Middle Devonian marine rocks are termed the Accra Group. The geology of Sierra Leone and Guinea was reviewed by Culver & Williams (1979). The Saionia Scarp Group unconformably overlies the Precambrian, and has Early Ordovician (probably Arenig) clastics unconformably overlain by Hirnantian glaciogenic deposits which are in turn unconformably overlain by Jurassic volcanics. In Liberia, there are several small outcrops on the coast near Monrovia of the Paynesville Formation, where a Lower Devonian marine sandstone unconformably overlies the Precambrian.

Southern, central and eastern Africa and Madagascar

The area includes the large south African Precambrian area with two major Mesoproterozoic or older cratons and three much smaller ones, Lake

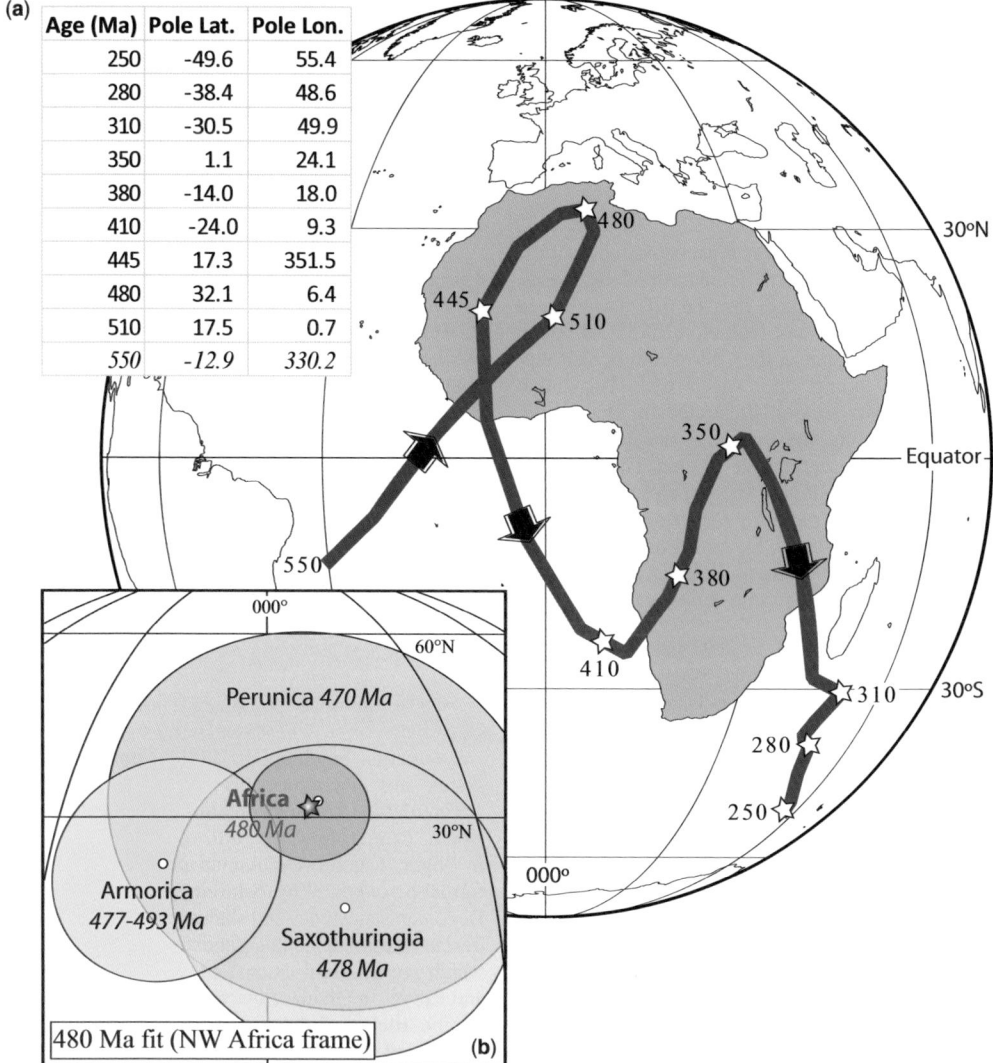

(a)

Age (Ma)	Pole Lat.	Pole Lon.
250	-49.6	55.4
280	-38.4	48.6
310	-30.5	49.9
350	1.1	24.1
380	-14.0	18.0
410	-24.0	9.3
445	17.3	351.5
480	32.1	6.4
510	17.5	0.7
550	*-12.9*	*330.2*

Fig. 4. (a) Apparent Polar Wander (APW) path for Gondwana (550–320 Ma; Torsvik & Van der Voo 2002) and then a global APW path for 320–250 Ma (Torsvik *et al.* 2008c). The APW path is shown and listed in the left-hand table in south African co-ordinates. Open white stars with numbers (in million years) denote reconstruction times illustrated in Figures 1 and 5–13. (b) Early Ordovician poles from Perunica and Saxothuringia (shown with d*p*/d*m* 95% confidence ovals) compared to mean poles from Armorica and Africa (shown with A95 circles). The latter is a mean Gondwana pole and all poles are shown in an Africa reference frame.

Victoria, Somalia and Madagascar, as shown in Figure 2. Summaries of the individual areas within the region are provided in the following.

East Africa and Madagascar

In eastern Africa (Kenya, Uganda and Tanzania), Mozambique and adjacent Madagascar there are no strata known between the Proterozoic and the Late Carboniferous, when deposition started of the non-marine Karroo Supergroup which continued on until the early Jurassic (Schlüter 1997). Madagascar (together with India and the Seychelles) finally left east Africa in the Late Jurassic (*c.* 145 Ma) but soon after became part of Africa again in the Early Cretaceous (*c.* 120 Ma) when seafloor spreading ceased. No Palaeozoic rocks are known there.

Central and south-western Africa

The central African Republic has Palaeozoic rocks in the NW part of this large area, which consist of some Permo-Carboniferous tillites. The adjacent Democratic Republic of Congo (formerly Zaire) has Early Cambrian marine shales with acritarchs unconformably followed by an arkosic, probably terrestrial, sequence of uncertain age, unconformably followed in turn by Late Carboniferous and Permian Karroo rocks (Daly *et al.* 1992). Further to the south, Zambia, Zimbabwe, Malawi, Namibia and Lesotho all have Upper Carboniferous and later continental deposits loosely attributed to the Karroo Supergroup. At the base of this is the Upper Carboniferous Dwyka Group, which has noted fossil amphibians and also ashfall tuffs in Namibia (Bangert *et al.* 1999). Previously in Namibia, the shallow-marine Nama Group had been laid down in the Cambrian.

South Africa

The Neoproterozoic–Lower Cambrian rocks and history of the area are summarized by Gaucher *et al.* (2010). The Palaeozoic of south Africa (including Lesotho and Swaziland) is more extensive and varied than elsewhere in the continent south of the Sahara. It commences with the 2 km thick Klipheuwel (or Klipheuvel) Group which is mostly slightly metamorphosed sediments of Late Neoproterozoic–Early Cambrian age, and there are also Early Cambrian metagranites (Tankard *et al.* 1982, 2009). That group is succeeded by the 4 km thick Ordovician–Devonian Table Mountain Group, which includes the latest Ordovician Hirnantian glaciogenic deposits of the Pakhuis Formation (Rust *in* Holland 1981) followed above by the Cedarberg Formation which carries a *Hirnantia* brachiopod Fauna (although *Hirnantia* itself is not present; Cocks *et al.* 1970). Early Devonian (Pragian) brachiopods occur in the Baviaanskloof Formation near the top of the group. Tankard *et al.* (2009) interpreted most of the Table Mountain Group as progradational fluvial systems with occasional marine incursions, and the Natal Group of SE Africa as contemporaneous with the lower parts of it. The Table Mountain is succeeded by the Devonian Bokkeveld Group which is over 3 km thick and consists largely of deltaic deposits with the occasional marine incursion, one of which has yielded characteristic shallow-water Emsian brachiopod faunas of the Malvinokaffric Province (Boucot *et al.* 1969). This is followed by the Lower Devonian–Lower Carboniferous Witteberg Group, which is a mixture of shallow shelf and deltaic deposits.

After a 25–30 Ma hiatus, the Bokkeveld was followed by deposition of the thick Upper Carboniferous–Jurassic Karroo Supergroup. This consists mainly of non-marine lake sediments although there were some marine incursions, particularly to the east (Tankard *et al.* 2009). Within the Karroo, there is the latest Carboniferous and earliest Permian glaciogenic Dwyka Group (Visser 1997), which was deposited over more than 12 Ma during *c.* 302–290 Ma (Isbell *et al.* 2008a). That is succeeded by the Permian Ecca Group which consists of marine to brackish-water sediments with substantial coals (particularly in the Transvaal) and the first appearance of the characteristic flora of the *Glossopteris* Province. The Ecca contains interlayered ashfall tuffs dated at 275–270 Ma (Bangert *et al.* 1999). It is followed in turn by the 3 km thick Permo-Triassic Beaufort Group, which was deposited in an intracratonic lake basin which covered half of south Africa and is noted for its fossil reptiles (also reviewed by Tankard *et al.* 1982, 2009). The Cape Orogeny was a series of orogenic events in the south of the country and lasted from the Early Permian (290 Ma) to the Trias (220 Ma), and which Tankard *et al.* (2009) interpreted as representing a thick-skinned strike–slip orogen.

Arabia

The Arabian Plate consists of Saudi Arabia, Yemen, Oman, Iraq, Jordan, Israel, Syria, Lebanon, southeast Turkey and southwest Iran, and formed part of the core Gondwana Craton until the Miocene opening of the Red Sea. Stratigraphy within the various parts of the area was reviewed by Sharland *et al.* (2001) and the tectonic history described by Ruban *et al.* (2007). The Sinai–Levant region to the west of the Dead Sea (sometimes termed the Levant Plate) was part of the main Arabian Plate until separated from it by transform faulting in the Neogene. Millson *et al.* (1996) reviewed the extensive Lower Palaeozoic rocks of Oman, which include Hirnantian glaciogenic deposits. Martin *et al.* (2008) reviewed the Upper Palaeozoic there, which also includes glaciogenic deposits. Postglacial Early Silurian (Llanedovery) shales are widespread and form the prime hydrocarbon source rocks. However, strata of Late Silurian–Devonian age are largely absent from the main Arabian plate (Brew *et al.* 2001), but some shales with Devonian spores occur in the foothills of the Zagros Mountains (Bordenave & Hegre 2010). The succeeding Early Permian–Trias shallow marine clastic sediments and carbonates (including Middle Permian reefs) include many reservoirs which host the originally Silurian hydrocarbons.

Adjacent to the main Arabian area and separated from it by the Zagros Thrust is the Sanand Terrane (sometimes termed the Sanandaj–Sirjah Terrane), which stretches from north-eastern Turkey and Armenia through to eastern Iran. The Sanand Terrane has a basement of probably Lower Palaeozoic metamorphic rocks (the Bajgan Complex) which is unconformably overlain by Carboniferous, Permian and later shelf limestones (McCall 1997). Beyond the Sanand Terrane to the NE lies the Alborz Terrane which fringes the Caspian Sea, although some authors (e.g. Ruban et al. 2007; Gaetani et al. 2009) recognized a three-fold division in Iran: the Sanand Terrane and a northerly area divided between a central Iranian Terrane and a smaller Alborz Terrane. Gaetani et al. (2009) described the Carboniferous–Triassic stratigraphy of the Alborz Terrane in Iran. They presented palaeogeographical maps which include the Permian volcanics of different ages associated with the opening of the Neotethys Ocean, which started with rifting between the Sanand Terrane and the main Arabian Plate. Konert et al. (2001) noted the absence of Devonian and Carboniferous rocks from most of Iran.

Konert et al. (2001) summarized the Palaeozoic geology of the whole of Arabia. Schandelmeier & Reynolds (1997) also included the Arabian area in their palaeogeographical maps for the Upper Palaeozoic and later rocks of northeast Africa, which we have used to help create our new figures. The rocks and Palaeozoic history of the parts of Gondwana further east in Asia were reviewed by Torsvik & Cocks (2009). Arabia began its accretion to Eurasia along the Zagros Belt by subduction of the Neotethyan Ocean beneath the Iran block in the Cretaceous, followed by obduction of Neotethyan ophiolites over the northeast Afro-Arabian margin in the Late Cretaceous and finally collision of Afro-Arabia with central Iran in the Miocene (Farzipour-Saein et al. 2009).

The Indian Peninsula

Palaeozoic rocks older than Permian occur in today's northern Himalayan rim of the Indian Peninsula, located at the right-hand edge of the maps presented here. There are numerous small terranes there, many of which had previously left India in the opening of the Neotethys Ocean in the Permian. Torsvik & Cocks (2009) published geological summaries and palaeogeographical maps for the north-eastern Gondwana area. Torsvik et al. (2009a) published palaeomagnetic data from Spiti in the Himalayas, which constrains our positioning of the Gondwana margin there in the Ordovician. There are sporadic outcrops of Early Cambrian marine sediments at several places on the Precambrian craton throughout the Indian

Peninsula (N.C. Hughes, pers. comm. 2009); however, many which have previously been considered Cambrian are actually Neoproterozoic (Gregory et al. 2006). Above those Early Cambrian rocks there is a substantial unconformity between the Neoproterozoic and Cambrian 'basement' and the Gondwana Group of the Permian, from which the *Glossopteris* Flora is recorded at many sites. There are latest Carboniferous to earliest Permian (Gzhelian–Asselian; 304–294 Ma) glaciogenic rocks in the Salt Range of Pakistan. The continental basins of India were reviewed by Veevers (2004).

Adjacent sectors of Gondwana

South America and the Falkland Islands

Only the eastern and southern parts of South America are shown on our maps, together with the Falkland Isles and their associated surrounding submarine areas now to the SE. The South American sector covers most of Brazil, eastern Argentina, Uruguay and Paraguay and a small part of Bolivia. It includes the substantial Palaeozoic Paraná, Parnáiba and Amazon basins, all of which lie unconformably on the Precambrian craton (Caputo 1998). The Neoproterozoic–Early Cambrian palaeogeography is described by Gaucher et al. (2010) and glacial episodes in the Late Devonian (Famennian) and Early Carboniferous (Tournaisian and Visean) by Caputo et al. (2008). Outside the left-hand map margin, the Precordillera (or Cuyania) Terrane of west Argentina appears to have been peri-Laurentian in the Cambrian and Early Ordovician, but drifted across the intervening ocean to be at a high enough palaeolatitude by the end of the Ordovician to bear Hirnantian glacial deposits and a *Hirnantia* brachiopod fauna. Its accretion to Gondwana was certainly before the Mid-Silurian, and may have been as early as the Late Ordovician at 450 Ma (Astini 2003).

The geology of Patagonia (the southern parts of Argentina and Chile) during the Palaeozoic is summarized by Ramos (2008). The northern sector has Ordovician granitoids dated at 475 Ma intruded into lightly metamorphosed Cambrian and Early Ordovician clastics unconformable on a Neoproterozoic basement. Above them lie metamorposed Late Carboniferous amphibolites and undeformed Late Permian granitoids, as well as orthoquartzites deposited on a Silurian–Early Devonian passive margin. The western sector is metamorphosed and includes plutonic rocks with a wide range of Palaeozoic ages from 540 (Early Cambrian) to 280 Ma (Early Permian), which are overlain by terrestrial clastic rocks which have yielded a sequence of Early–Late Permian floras. The southern sector has at its core the Palaeozoic rocks of the Deseado Massif. The tectonic development of Patagonia is uncertain; Ramos (2008) considered that both the northern and southern sectors formed an independent microcontinent that was accreted to southwest Gondwana in the Early Permian. In his model, a northern magmatic arc developed by southwards subduction beneath north Patagonia. In contrast, Pankhurst et al. (2006) concluded that northern Patagonia has always been autochthonous to Gondwana while only southern Patagonia was allochthonous and collided with the North Patagonian Massif in the Late Carboniferous (c. 320–210 Ma) but with deformation lasting until the Early Triassic (see also Vaughan & Pankhurst 2008).

Before the Middle Jurassic, the Falklands Terrane was situated to the SE of South Africa (Marshall 1994; Stone et al. 2008; Torsvik et al. 2008b, 2009b) as shown on our Palaeozoic maps. The geology of the Falkland Isles consists of a Neoproterozoic basement unconformably overlain by possibly Silurian and certainly Devonian and later sediments. From the latter, Emsian brachiopods typical of the Malvinokaffric Province are known from the Fox Bay Formation (Boucot et al. 1969) which conformably underlies Late Devonian and Early Carboniferous rocks. Above this is an unconformity below the latest Carboniferous–Triassic Lafonian Supergroup, which includes early Permian glaciogenic deposits. The old terrane included a substantial area of modern shelf under the south Atlantic.

Antarctica

The varied sectors and geological history of Antarctica are summarized by Torsvik et al. (2008b). Only a part of the continent features on our maps, consisting largely of the substantial East Antarctica Shield which comprises a variety of Archaean and Proterozoic terranes (Harley 2003) including Lützow Holm Bay, Raynor Province and Victoria Land. The adjacent Dronning Maud Land was a separate terrane which now forms a large part of western Antarctica. The smaller but also separate Ellsworth–Whitmore Mountains Terrane, whose geology was reviewed by Duebendorfer & Rees (1998), demonstrates active volcanism in the Cambrian. Marie Byrd Land and the Antarctic Peninsula were also separate from east Antarctica to the east, but we do not show the latter here. The Transantarctic Mountains are an orogenic belt at the margin of east Antarctica which represent much Palaeozoic tectonic activity from the Mid-Cambrian onwards; and they are situated near the lower boundary of our maps (Fig. 1). Isbell et al. (2008b) described the Lower Permian glaciogenic deposits there.

Peri-Gondwana

Avalonia

The geology and history of Avalonia, which included eastern North America north of Cape Cod (Massachusetts), Newfoundland, SE Ireland, England, Wales, Belgium, Holland and NW Germany, was summarized by Cocks & Fortey (2009). The continent left Gondwana with the initial rifting of the Rheic Ocean in the Early Ordovician at c. 485 Ma. It was only an independent terrane during the Ordovician, until its oblique collision with Baltica at about the Ordovician–Silurian boundary time at 443 Ma. Different authors have shown Avalonia outboard of different sectors of core Gondwana, but we think that it lay outside northeast South America, Florida and northwest Africa, all of which remained as parts of the Gondwanan core until Gondwana's Carboniferous amalgamation with Laurussia to form Pangea.

Iberia and Italy

The geology of the Iberian Peninsula of Spain and Portugal is summarized by Gibbons & Moreno (2002). It is an amalgamation of at least six Lower Palaeozoic zones. From north to south, these zones are: the Cantabria, west-Asturian-Leone, Galicia-Tras os Montes, Central Iberian, Ossa Morena and South Portugal zones, all separated by NE–SW-trending Variscan shear zones. Opinions differ as to whether or not the Iberian zones were originally a single part of the Lower Palaeozoic Gondwanan margin, or whether they represent several pre-Variscan terranes. However, since no palinspastic reconstructions are agreed, Iberia is simply shown as a single unit on our reconstructions. Italy was substantially affected by both Variscan and Alpine orogenies and Sardinia and Calabria were probably separate from the Appenines in the Lower Palaeozoic. All formed parts of the Lower Palaeozoic Gondwanan margin and, like Iberia, Sardinia contains the characteristic higher-latitude Gondwanan Mediterranean Province brachiopod faunas in the Lower and Middle Ordovician (Leone et al. 1991). All the Iberian and Italian units left Gondwana when the Palaeotethys Ocean opened near the end of the Silurian.

Armorica

Although the name is often used for the Armorican Peninsula of northwest France, we use the term loosely here to include all of Palaeozoic France which was reviewed by Keppie (1994); some authors refer to it as Cadomia. There are several Palaeozoic terranes there (as many as ten, according

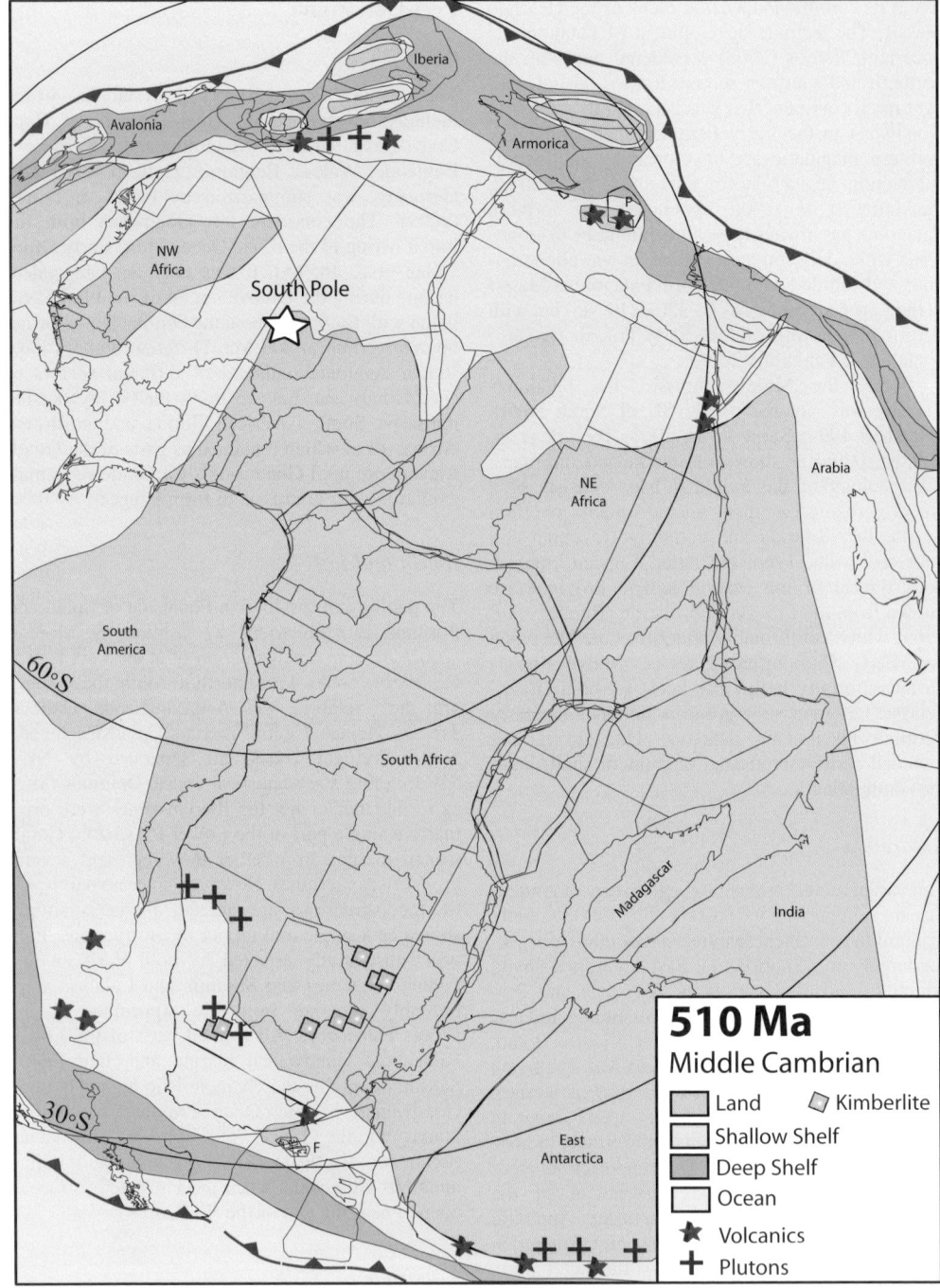

Fig. 5. The palaeogeography of the central Gondwana area in the Middle Cambrian at 510 Ma. Political boundaries are single lines and the boundaries of the Precambrian cratons in Africa (e.g. Northeast Africa) are shown as double lines. Some modern latitude and longitude lines are shown in Africa (F, Falklands; P, Perunica).

to Shelley & Bossière 2000), and we group them into north and south Amorica as shown in Figure 6. Although in many publications Armorica was considered to be part of an assemblage of terranes (including Iberia and others) which left Gondwana very early in the Palaeozoic (including Cocks 2000), we now follow the conclusions of Robardet (2002, 2003) who demonstrated that they did not leave Gondwana until the opening of the Palaeotethys Ocean near the end of the Silurian at the earliest.

Moldanubia and Saxothuringia

The Vosges area of France and the Black Forest of Germany made up Moldanubia. Moldanubia and the Saxothuringia area, largely in Germany, were both relatively small but separate terranes which had complex Palaeozoic histories. Moldanubia may have been an eastwards extension of part of the South Armorican Terrane of France, as shown in Figure 1. The geology of Saxothuringia was reviewed by Linnemann (2003) and, together with all the adjacent areas of central Europe, in the book edited by McCann (2008). Their various parts in the Upper Palaeozoic Variscan Orogeny were described by Franke (2006).

Perunica

This area, often termed Bohemia, occupies the eastern part of the Czech Republic and adjacent areas and includes the classic Barrandian area to the SW of Prague: the Cambrian to Devonian geology was reviewed by Chlupáč *et al.* (1998). Perunica was definitely included within core Gondwana in the Cambrian and Early Ordovician, but its Ordovician progress after 470 Ma is controversial. After analysis of the benthic faunas, some authors (e.g. Havlíček *et al.* 1994) considered that Perunica left Gondwana in the Middle Ordovician as a separate terrane; others (e.g. Robardet 2003) concluded that it remained an integral part of Gondwana until the Palaeotethys Ocean opened in the latest Silurian or even later. Some previous reconstructions have favoured the former option, but it is shown in this paper (e.g. Fig. 7; see also Torsvik & Cocks in press) as remaining peri-Gondwanan until the Early Devonian. Its part in the Variscan Orogeny was described by Franke (2006).

Turkey

The Anatolia Plate today consists of all of Turkey (apart from the Pontides) and its adjacent areas in southern Greece, Syria and Iraq. However, the many Palaeozoic rocks there are divided between a northern Pontides Terrane and a southern Taurides Terrane, with the two separated by a central zone of later rocks. The Taurides, whose Lower Palaeozoic geology and faunas were reviewed by Dean *et al.* (1999) and the Devonian by Wehrmann *et al.* (2010), was an integral part of northern Gondwana until the opening of the Neotethys Ocean in the Permian; Hirnantian glacial deposits are known there (Monod *et al.* 2003). However, the Pontides area is more controversial. Some authors consider the Pontides to have been several terranes, including an Istanbul Terrane, but we follow Dean *et al.* (2000) who described the Lower Palaeozoic stratigraphy there and treated it as a single terrane unit. The Tremadocian is unconformable on basement gneiss, which is undated but possibly Precambrian, and there is no known Cambrian. After description and analysis, Dean *et al.* (2000) noted that the Ordovician faunas in the Pontides were quite different from those in the Taurides but similar to Avalonia, and therefore concluded that the Pontides were much further to the west and at higher latitudes in the Ordovician than the Taurides. However, since their Lower Palaeozoic positions are so poorly known, the Pontides are omitted from our Cambrian and Ordovician maps. The terrane is shown conservatively in its present position attached to the north of the Taurides in the Upper Palaeozoic.

Florida (Suwanee)

The area lying in the SE of the US and to the east of the Mexican terranes (which are outside the maps presented here, located to today's west) was also peri-Gondwanan. This includes most of Florida, and the outcrops consist only of Tertiary and later rocks. However, as known from boreholes, the Mesozoic unconformably overlies Lower Palaeozoic rocks which are termed the Florida or Suwanee Terrane, which is bounded to the north by an east–west trending suture zone in southern Georgia and Alabama. The Upper Proterozoic and Lower Palaeozoic includes volcanic arc rocks, some dated to *c.* 550 Ma and with subsequent 520 Ma granites. Other beds have yielded Ordovician trilobites such as *Plaesiocomia* of undoubted Gondwanan affinity (Whittington 1953). After the Cambrian, Florida was not tectonically active prior to its merger with the rest of North America during the accretion of Gondwana with Laurussia to form Pangea in the Late Carboniferous.

Geological history

We now present a brief Palaeozoic history of central Gondwana, together with some new palaeogeographical maps. The latter were plotted using the SPlates reconstruction system, which is based on digitized modern-day polygons (Labails *et al.* 2009) and relative rotation parameters (mostly listed in Torsvik *et al.* 2008*a*, 2009*b*). Before 320 Ma, the location of Gondwana through time (Figs 3a & 4) is based on the palaeomagnetic data listed in Torsvik & Van der Voo (2002). After that point it is based on a global APW path in southern African co-ordinates taken from Torsvik *et al.* (2008*c*). The movement of the South Pole with respect to southern Africa is shown in Figure 4, in which we show the South Pole location at the times of our reconstructions (Figs 5–13). Note that the reconstructions in Figures 1 and 5–13 and are relative reconstructions against a fixed Africa, but the ancient South Pole and palaeolatitudes are superimposed on the maps.

A short summary of the Precambrian is also included to put the Palaeozoic into perspective. The Lower Palaeozoic time scale and correlation follows Cocks *et al.* (2010).

Precambrian prelude

There are substantial Precambrian shields underlying Africa, South America, Antarctica, Arabia, Madagascar and India (Fig. 2). Africa consists of a cluster of separate intra-cratonic units: north Africa and the Meseta, northeast Africa, south Africa, Somalia and Lake Victoria, all of whose boundaries (derived from Torsvik *et al.* 2009*b*) are shown on the maps here. There is published disagreement on the precise timing within the Neoproterozoic of the assembly of Africa (compare Veevers 2004 with Meert 2003), but it was largely complete before the end of the Precambrian apart from the Meseta of Morocco and the final south Africa–South American welding. However, tectonic activity around several of the Neoproterozoic sutures carried on into the Early Phanerozoic. In particular, the Pan-Gondwanan Orogeny (also termed the Pan-African or Saldanian Orogeny) is now recognized in South America, Africa, Antarctica, India and Australia and lasted from *c.* 600 Ma into the Cambrian at *c.* 530 Ma, as reviewed by Harley (2003). The Late Neoproterozoic (Ediacaran) to Early Cambrian (630–525 Ma) history of southern South America was described by Bossi & Cingolani (2010). They envisaged a tangential collision between elements of the Rio de la Plata block during the Early Cambrian at about 530 Ma, thereby finally closing the ocean which had previously existed between Argentina and south Africa. In the Late Neoproterozoic at *c.* 560 Ma, a change took place from predominantly transpressional tectonics to a transtensional regime on the Arabian Peninsula (Oterdoom *et al.* 1999). This led to the formation of NE–SW trending basins in Oman and adjacent areas which persisted into the Early Cambrian.

Cambrian

We present a map for 510 Ma, in Middle Cambrian times (Fig. 5). The South Pole was located under Mali and India, east Antarctica, the Falkland Isles and Patagonia were all in much warmer latitudes at *c.* 30°S. Although derived primarily from palaeomagnetic data, those latitudes are confirmed by the fossils found (e.g. the rich Cambrian trilobite faunas known from limestones in Antarctica; Shergold 1988).

For the southern Gondwana margin we follow Rapela *et al.* (2003) who described Middle Cambrian orogeny in the Sierra de la Ventana Fold Belt of Argentina, through the Cape Fold Belt of South Africa and continuing eastwards into the Transantarctic Mountains of Antarctica. That orogeny included substantial Middle and Upper Cambrian volcanism in South America and Antarctica, and metamorphism and the intrusion of granites in the Middle Cambrian of South Africa and southern Namibia (Tankard *et al.* 2009; Gaucher *et al.* 2010). How the rifting between that belt and Patagonia, the Antarctic Peninsula and other terranes related to the oblique subduction of the Palaeo-Ocean Plate under Antarctica, in which those terranes were accreted to all of southern Gondwana during the Lower Palaeozoic, is uncertain. Duebendorfer & Rees (1998) identified the Ellsworth–Whitmore Terrane of Antarctica as one of those allochthonous terranes. Bauer *et al.* (2003) described substantial orogenic activity, including the intrusion of a syenite, in a belt on the margin between Antarctica and India. All these Cambrian disturbances were the final phases of the largely Neoproterozoic Pan-African Orogeny.

The northern margin of central Gondwana to the west of Turkey was also active through Cambrian time. In contrast, the sector immediately to its east reviewed by Torsvik & Cocks (2009) was passive once the latest Neoproterozoic–earliest Cambrian tectonics, which included the intrusion of granites (e.g. in the Alborz Terrane of Iran; Zanchi *et al.* 2009), were over. On the Arabian Plate and adjacent Iran there were substantial shallow-marine carbonates, including some Middle Cambrian evaporites (Konert *et al.* 2001). In Saxo-Thuringia there is an unconformity between Neoproterozoic basement rocks and a Lower Ordovician transgressive sequence (Linnemann 2003), so we have portrayed land there which extended westwards into northern Armorica. In Perunica there are continental sediments, shallow- to deep-water clastics and volcanics (Havlíček *in* Chlupáč *et al.* 1998). Álvaro *et al.* (2003) presented nine successive facies maps of the northern margin from Avalonia through Morocco and Iberia to Perunica, which have also helped us to construct Figure 5. In Arabia, El-Araby & Abdel-Motelib (1999) described floodplain deposits with

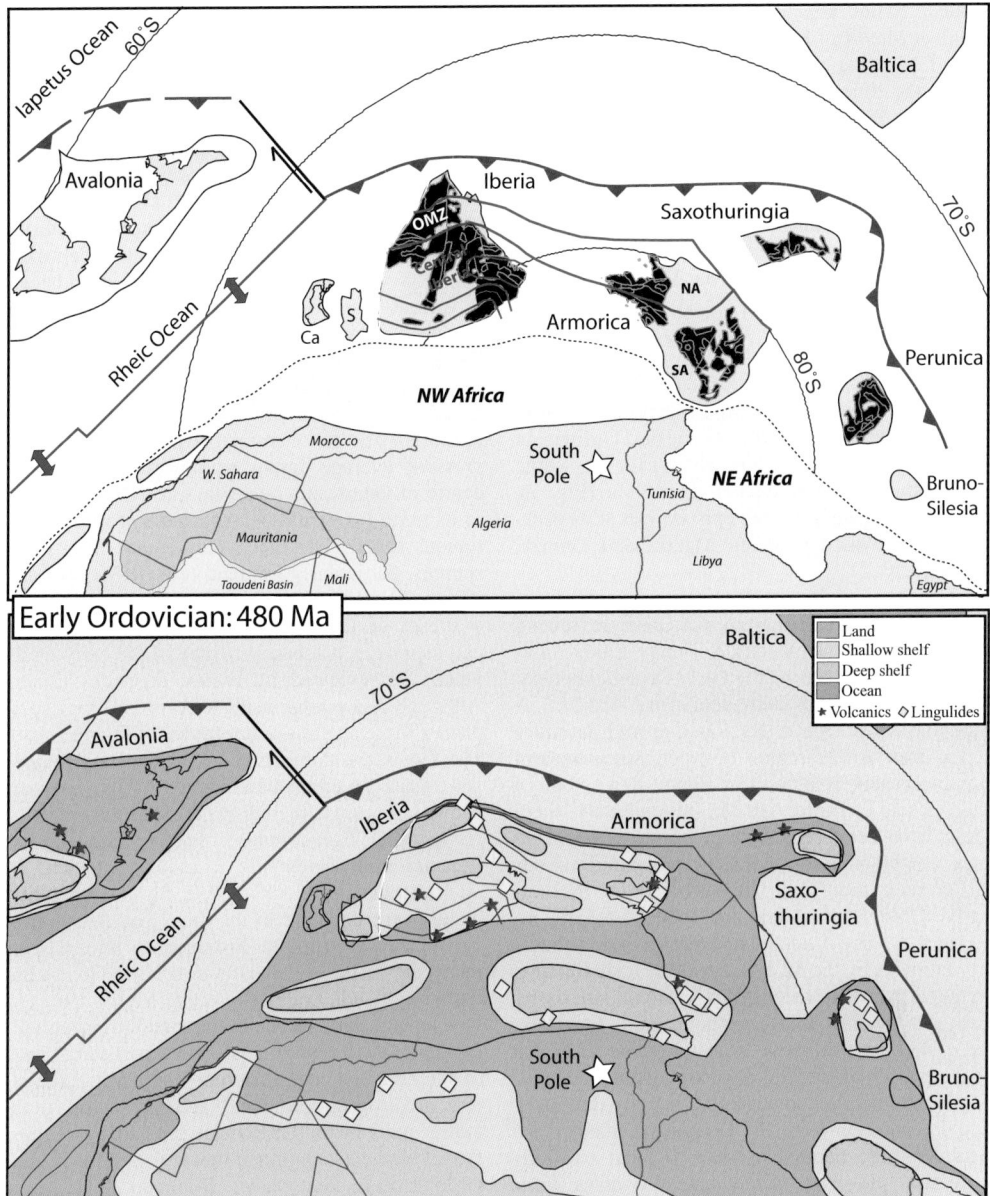

Fig. 6. North-central Gondwana and southern Europe in the Early Ordovician at 480 Ma (near the Tremadocian–Floian boundary). A reconstruction for all of Central Gondwana at the same age is shown in Figure 1. **Above:** map of modern crustal units, with Palaeozoic outcrops shown (in black) in Iberia, Armorica, Saxothuringia and Perunica (Pink: Tindouf Basin; yellow: Taoudeni Basin). **Below:** palaeogeography of the same area, including volcanoes and localities of the large lingulid fauna within the Armorican Quartzite (see text). (Ca, Calabria; NA, North Armorica; OMZ, Ossa Morena Zone; S, Sardinia; SA, South Armorica.)

occasional marine incursions characterized by trilo-bite tracks in the Sinai Peninsula of Egypt. Rift basins developed between Gondwana and Avalonia, heralding the opening of the Rheic Ocean in the earliest Ordovician (von Raumer & Stampfli 2008).

There are 520 Ma granites in Florida which follow Early Cambrian volcanic arc rocks there.

The palaeogeography shown in Figure 5 is highly speculative in many places; for example, in northwest Africa the extensive Tindouf and

Taodeni basins are separated by the substantial Reguibat Massif of Precambrian rocks. There is no way of knowing whether that massif represented land at the time (as we have shown on the map) or whether it was covered by shallow seas, since there are apparently widespread Cambrian limestones on the southern margin of the Tindouf Basin (Deynoux *et al. in* Holland 1985). The palaeogeographical situation in the Early Cambrian is even less certainly known, which is why we have not constructed a map older than for the Middle Cambrian at 510 Ma.

Ordovician

We present maps for the northern part of the region for 480 Ma at the very end of the earliest Ordovician stage (the Tremadocian; Fig. 6) and for the whole area in the latest Ordovician at 445 Ma (Fig. 7), when the Hirnantian glacial episode was at its peak. The South Pole lay under Algeria and Guinea, respectively.

Published opinions differ as to when Avalonia left the Gondwanan margin but, after reviewing the data and faunas, Cocks & Fortey (2009) concluded that it was during the Early Ordovician (probably the Early–Mid-Tremadocian at *c.* 485 Ma). A widening Rheic Ocean developed at high latitudes in the then Arctic region between north-western Gondwana and Avalonia. In Figure 6 we show a wider Rheic Ocean at 480 Ma (alternatively, there could have been considerable pre-drift extension or a combination of the two) compared to our earlier reconstructions. This is in order to limit Early Ordovician Avalonian plate velocity to 20 cm a^{-1}, since Avalonia had moved to mid southerly latitudes by the Mid-Ordovician. It should also be noted that if we reconstruct Avalonia strictly by Early Ordovician palaeomagnetic data (based on the *c.* 485 Ma Trefgarne Volcanics pole in Wales of Trench *et al.* 1992), Avalonia would be geographically inverted and rotating heavily during its initial separation from the Gondwana margin; we attribute those anomalous data to local rotations in Wales. However, palaeomagnetic poles from Armorica (e.g. Nysæther *et al.* 2002), Perunica, Saxothuringia and mean Gondwana (in African co-ordinates and shown in Fig. 4b) match our Early Ordovician reconstruction within error.

The Armorican Quartzite facies, often termed the Grès Armoricain, is of Floian to Darriwilian (Arenig but some Llanvirn) age and stretches across much of north Africa and southern Europe as far northwards as Brittany. This is an amazingly widespread facies whose thickness varies from a few metres to over 600 m (generally 150–300 m thick) and is sporadically distributed over a very large area of shallow shelf (Fig. 6). Much of the

Armorican Quartzite is unconformable (locally termed the Sardic or Toledanian Unconformity) over an irregular variety of Precambrian and early Cambrian rocks, many of which had been deformed in the Late Neoproterozoic Cadomian Orogeny. Although the maturity of the quartzite grains apparently indicates transport some way from a substantial land area, McDougall *et al.* (1987) plausibly concluded that the Armorican Quartzite of the Central Iberia Zone of north Portugal was deposited within a variety of disconnected basins adjacent to various more local land masses. In Figure 6, we therefore show several land areas near the high-latitude edge of Gondwana. That model is supported by the Middle Ordovician diagram shown in Robardet (2002, fig. 4) in which the facies of both north Armorica and the Central Iberia Zone show progressive average fining southwards towards the centre of Gondwana rather than towards the ocean to its north (as would be expected if it were a continuous sheet). The facies data comes from many sources, particularly Legrand (e.g. 1974) for north Africa. The faunas of the Armorican quartzite are a subset of the largely Gondwanan *Neseuretus*-calymenacean trilobite Province of the early Ordovician, as reviewed by Fortey & Cocks (2003). Although *Neseuretus* itself is recorded from some places, those shallow-water faunas within the quartzite facies are dominated by unusual and mainly very large lingulide brachiopods. These sites are shown in Figure 6, including *Lingulobolus brimonti*, *L. hawkei*, *Pseudobolus? salteri*, *Ectenoglossa lesueuri* and *Lingulepis crassipyxis* which are all illustrated by Cocks (2000, fig. 3) and others which were endemic to the area. The *Neseuretus*-calymenacean trilobite Province is known from many sites in the higher-latitude parts of Gondwana, including Saudi Arabia (Fortey & Morris 1982).

The high-latitude Mediterranean Province shelly faunas (Havlíček *et al.* 1994) are found across most of southern and central Europe, including Avalonia (e.g. in Shropshire, England) and north Africa in the earlier parts of the Ordovician, and many of those faunal elements continued upwards into the Middle Ordovician in some places. Dominant in the Mediterranean Province are genera which were mostly originally described from Bohemia, including the brachiopods *Tafilaltia*, *Tissintia* and *Aegiromena* as reviewed by Havlíček *et al.* (1994). The generic diversity of the faunas in that province is (unsurprisingly) lower than the diversity of contemporary faunas from lower latitudes, such as those from Laurentia and the Australasian parts of Gondwana, due largely to the higher palaeolatitudes of north-central Gondwana. An analysis of the faunas across the substantial Cantabrian Zone of Spain (Gutiérrez-Marco *et al.* 1999) shows the Mediterranean Province to be very consistent, right down to the

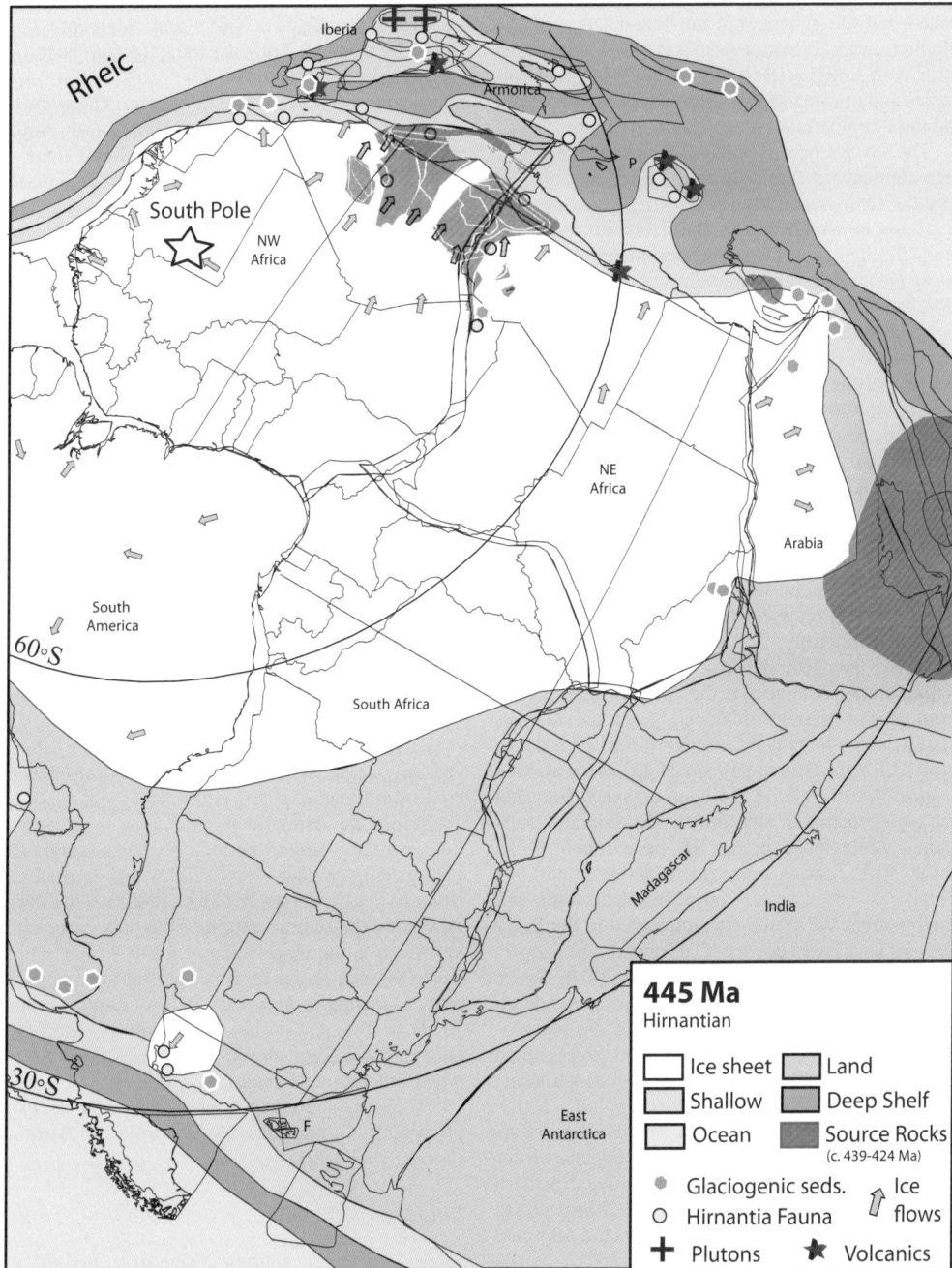

Fig. 7. The palaeogeography of the central Gondwana area near Ordovician–Silurian boundary time at 445 Ma, showing the glacial and peri-glacial features of the Hirnantian glaciation. The north African and Arabian ice sheet is shown at its maximum extent, but its boundary in central and southern Africa is poorly constrained. The ice cap shown in south Africa may have extended westwards to Argentina. All the *Hirnantia* Fauna sites represent shallow marine deposits presumably deposited in one or more interglacial episodes, although some give the false appearance of being within the ice cap on the figure. The Early Silurian (Llandovery) hydrocarbon source rocks in north Africa and Arabia are also shown here (F, Falkland Islands; P, Perunica).

species level. Havlíček & Branisa (1980) established that the genera (but not the species) of the Mediterranean Province also extended westwards into South America (Bolivia). Early Ordovician volcanics and non-marine sediments were deposited in the intra-continental Paraná Basin of Brazil.

The occurrence of many Mediterranean Province elements in Avalonia in the Early but not the Middle Ordovician argues for the proximity of Avalonia to Gondwana in the Tremadocian. This lends further support to the arguments against a latest Neoproterozoic separation of Avalonia from Gondwana, as maintained by a minority of authors as reviewed by Cocks & Fortey (2009). However, the faunal differences between Avalonia and the rest of the peri-Gondwanan terranes of south-central Europe subsequently steadily increased and had become substantial by Mid-Ordovician (Darriwilian–Llanvirn) times. As the Rheic Ocean widened and the distances across the Iapetus Ocean between Avalonia and Laurentia and across the Tornquist Ocean between Avalonia and Baltica steadily dwindled, the Avalonian faunas became progressively more similar to those of Laurentia and Baltica, as reviewed by Fortey & Cocks (2003) and Cocks (2010).

Above the northern sector of the African Craton, sedimentation continued in the many basins, varying from marine through tidal and shoreface to non-marine, all with many local unconformities as documented for the Murzuq Basin of Libya by Ramos et al. (2006). Wells in the west of Syria, Iraq and Turkey penetrated substantial Ordovician sandstones while those in the SE have much added shale, all indicating more open marine conditions to the east (Brew et al. 2001). In the Arabian and Iranian areas the craton was inundated in the Early Ordovician (Tremadocian–Dapingian) and deeper-shelf sediments are found to the north in Syria, Jordan and interior Iran. However, from Sandbian times onwards there were substantial marine prograding clastic sediments deposited over much of the Arabian Plate in inner-neritic to estuarine or deltaic environments (Konert et al. 2001). Paris et al. (2007) analysed the microfossils (chitinozoans and acritarchs) from the Gondwanan Ordovician, and concluded that a distinctive Northern Gondwanan Realm stretched from Morocco to Iran. Some species of these microplankton migrated across the Tornquist Ocean from Baltica to the Taurides of Turkey from the Middle Ordovician (Darriwilian) onwards. Oterdoom et al. (1999) described potassic-mafic volcanics of 460 Ma (Sandbian) age in Oman, which characterize a failed rifting event there.

The close of the Ordovician was heralded by the relatively brief Hirnantian Ice Age, which is best demonstrated in north Africa. The NW edge of the ice sheet on our map (Fig. 7) is largely taken from Le Heron & Craig (2008) and augments data shown by Cocks & Torsvik (2002, fig. 6). The thickest glacial deposits known (over 260 m) are from hydrocarbon wells in southern Algeria. The maps by Ghienne et al. (2007) covering all of north Africa were also used, and Denis et al. (2006) demonstrated that glaciogenic sediments occurred in northeast Niger. Hirnantian glacial deposits are also known from the Taurides of Turkey (Monod et al. 2003), South Africa (Rust in Holland 1981) and the Paraná and other basins in Brazil (Caputo 1998). There were at least two substantial glacial advances over most of the area: for example, in Jordan and Saudi Arabia (Armstrong et al. 2009) and in Niger (Denis et al. 2006). However, the extent of the ice cap across central and southern Africa is not well known and its (current) southern margins on Figure 7 are poorly constrained. It is also uncertain whether or not the ice in south Africa was continuous with the main ice cap in north Africa and South America. During the Hirnantian, the characteristic and relatively shallower-water Hirnantia brachiopod Fauna was widespread. Its sites are also shown on Figure 7, including some data points taken from Rong & Harper (1988) and Sutcliffe et al. (2001). The latter also noted that most of the Hirnantia Fauna localities can be dated as restricted to a single graptolite biozone, the extraordinarius Zone. Paris et al. (1995) plotted the sites of the various distinctive chitinozoan microfloras recovered from the glacial and periglacial areas.

Before the Hirnantian glaciation there was a Late Katian warmer period, however, originally identified by Villas et al. (2002) and termed the Boda Event by Fortey & Cocks (2005). That event facilitated the sedimentation of the only carbonates known from the Ordovician of north Africa, which are to be seen in the form of small bryozoan patch reefs occurring at very high palaeolatitudes (about 70°S) in the Anti-Atlas Mountains of Morocco. During the Katian, the globally distributed Foliomena brachiopod Fauna (reviewed by Rong et al. 1999) occurs sporadically in deeper shelf deposits of central Gondwana, for example in Sardinia (Leone et al. 1991).

Silurian

Since it was the shortest Palaeozoic period, we present no maps of the Silurian here. The Hirnantian 445 Ma map (Fig. 7) shows the area just before its start at 443 Ma, and the Early Devonian 410 Ma map (Fig. 8) soon after its finish at 416 Ma. However, during that 35 Ma period, Gondwana drifted relatively quickly, with the South Pole moving from north Africa (Algeria) to southern Brazil. Following the latest Ordovician Hirnantian glaciation,

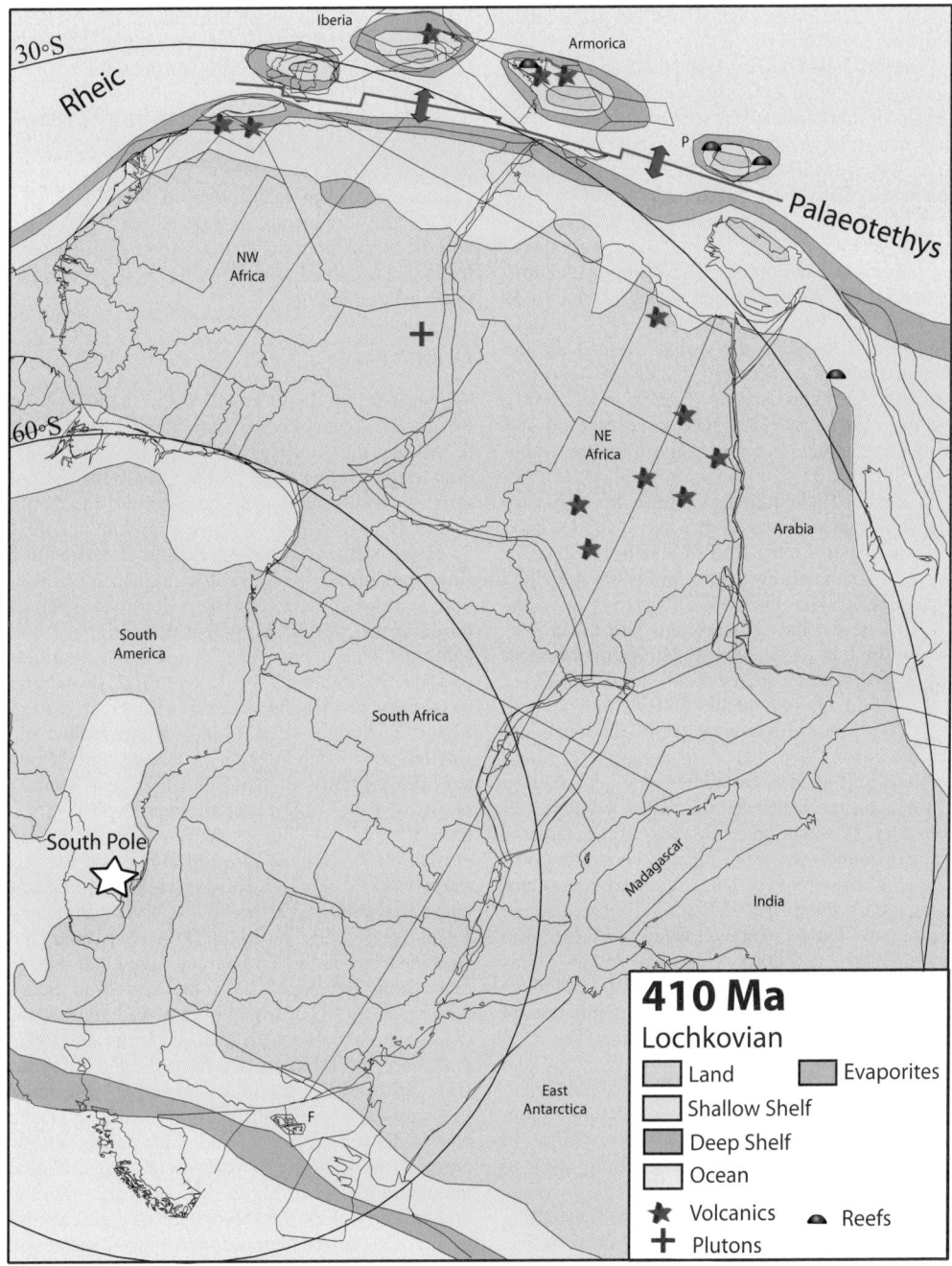

Fig. 8. The palaeogeography of the central Gondwana area in the Early Devonian Lochkovian Stage at 410 Ma (F, Falkland Islands; P, Perunica).

the global climate gradually warmed right from the start of the Silurian. However, glaciation continued in Brazil where there are non-marine diamictites in the Paraná Basin and marine Llandovery and Wenlock diamictites with age-diagnostic chitinozoa in the Amazon Basin (Grahn & Paris 1992; Caputo 1998). Significant amounts of kaolinite in the petroliferous Lower Silurian shales of north Africa indicate colder climates, but no obvious glaciogenic sediments have been found there.

Over much of northern Gondwana and southern Europe there were extensive transgressions, with deposits of thin black shales with graptolites, orthocones and some bivalves. Poor seawater circulation led to widespread anoxia on those sea floors, and there was a relative lack of sediment supply because the north margin of Gondwana was passive until close to the end of the period and there were therefore fewer adjacent uplands. These Upper Llandovery and Lower Wenlock black shales, often termed the Tanezzuft Formation (although it is laterally discontinuous between the various basins) are the estimated source of 80–90% of all the important north African hydrocarbons (Lüning et al. 2000). Legrand (1981) provided successive outcrop and palaeogeographical maps of Algeria for the whole of the Silurian.

In Arabia (including Iran) there are also very extensive shales of Llandovery age – the Qusaiba Shale in Saudi Arabia, the Mudawwara Shale in Jordan, the Sahmah Formation in Oman, the Abba Formation in Syria, the Dadas Formation in southeast Turkey and the Ghakum and Sarchahan Formations in Iran – which together form the most prolific hydrocarbon source rocks of the Palaeozoic in the world (Konert et al. 2001; Bordenave & Hegre 2010). Armstrong et al. (2009) demonstrated that the lower 'hot shales' in Jordan and Saudi Arabia were linked to freshening in a permanently stratified basin caused by the influx of deglacial meltwater. However, in a few areas of the northern sector of Gondwana more aerated conditions prevailed. For example, volcanics occurred in northern Spain, near which one of the few Early Silurian brachiopod faunas known from southern and central Europe and north Africa was found (Villas & Cocks 1996) which, unusually for the Llandovery, contains two endemic genera and several species. Legrand (1994) also documented sea-level and faunal changes during the Late Wenlock and Early Ludlow in Algeria and concluded that, despite evidence of a latest Wenlock regression, there was no significant break in sedimentation in that area (in contrast to the unconformity seen in most of the rest of the Saharan area).

Near the end of the Silurian there was rifting which heralded the opening of the Palaeotethys Ocean between Gondwana and Iberia, Armorica, Moldanubia, Bruno-Silesia, Perunica and other terranes which are now part of central Europe (McCann 2008). We previously (e.g. Cocks & Torsvik 2002) thought that Armorica and the others had left Gondwana at the same Early Ordovician time as Avalonia (as did other authors, e.g. Blakey 2008). However, the arguments by Robardet (2002, 2003) have convinced us that the Palaeotethys did

not open until the latest Silurian at the earliest. The Palaeotethys rifting was accompanied by a fall in sea level which continued into the earliest Devonian, resulting in coastal sand bar, tidal and fluvial deposits in Algeria and Libya. At the southern margin, the low-diversity, higher-latitude Late Silurian Clarkeia brachiopod Fauna is known in clastic rocks from many sites in South America and also less frequently in west Africa. That fauna was the precursor of the main southern high-latitude Devonian Malvinokaffric Province known from the same general region.

Devonian

We present two maps: at 410 Ma at the end of the earliest Devonian Lochkovian Stage (Fig. 8) and at 380 Ma (Fig. 9) during the Upper Devonian Frasnian Stage. During those 30 Ma Gondwana did not drift as rapidly as during the Silurian; the South Pole only moved from southern Brazil to Angola.

In the northern sector, the earliest Devonian low stand was succeeded by a substantial Late Lochkovian to Emsian marine transgression over much of north Africa, which had by that time migrated into subtropical latitudes. Those lower palaeolatitudes of about 30°S enabled the deposition of substantial evaporites in Algeria and Arabia, as shown in Figure 8. Boucot et al. (1983) monographed the shallow-water brachiopods occurring just above the unconformity in the Ghadames and Murzuq basins of Libya and northern Niger, typical representatives of the Old World Province of Boucot et al. (1969). Moreau et al. (1994) dated the Aïr intrusives of Niger, which have yielded significant palaeomagnetic data, as c. 407 Ma (Pragian–Emsian). Konert et al. (2001) described the Arabian Plate in the Emsian and noted that a large delta front developed over the craton in Saudi Arabia, to the NE of which were mixed marine siliciclastics and carbonates. Plusquellec et al. (1997) compared the crinoids, brachiopods and trilobites from north Africa, Iberia and Armorica with those in Laurussia and established that the Rheic Ocean was wide enough for considerable faunal differences to be seen on its opposite margins. Global temperatures had increased slowly in the Silurian, and virtually all of the Devonian was a greenhouse period. This can be demonstrated by the distribution of reefs which Copper (2002) plotted throughout the period. They were present in the Early Pragian in Saudi Arabia, in the Late Pragian–Emsian in Morocco and Mauritania and were numerous in the Givetian and Frasnian in Morocco and Algeria and in the Taurides of Turkey (Wehrmann et al. 2010). However, there were no reefs in the subsequent Famennian of northern Africa (Copper 2002), and some Famennian glaciogenic sediments

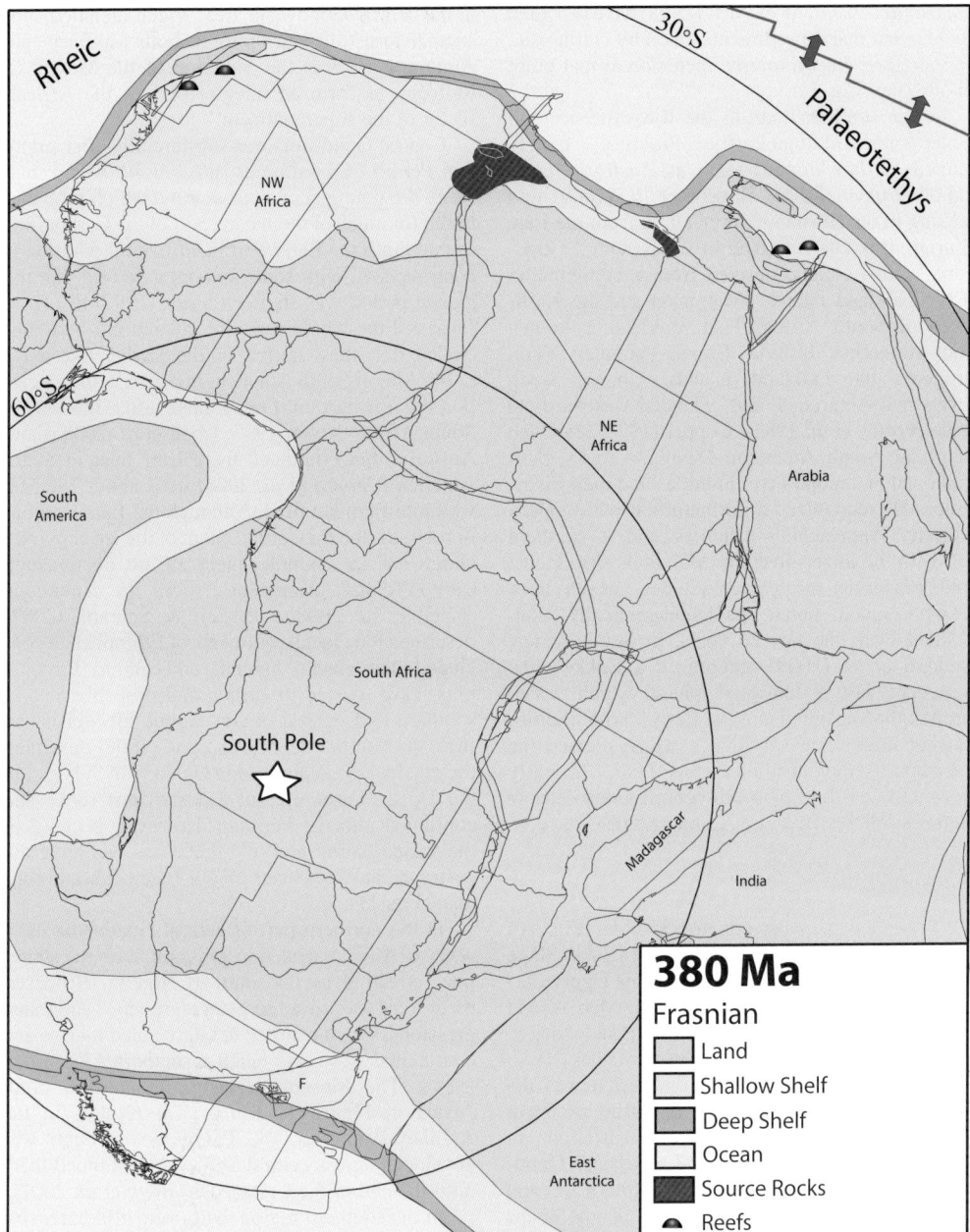

Fig. 9. The palaeogeography of the central Gondwana area in the Late Devonian (Frasnian Stage) at 380 Ma (F, Falkland Islands).

are known from north Africa as well as from the Frasnian and Famennian of South America (Caputo *et al.* 2008). There are extensive Late Devonian (Frasnian and Famennian) black shales in north Africa, particularly in Algeria, forming substantial hydrocarbon source rocks there (Lüning *et al.* 2003). There were several volcanic ring complexes in the uplands of the Sudan area, and the associated tectonic activity caused substantial sandstones to be shed into the local non-marine basins. In the

Amazon Basin, Grahn & Paris (1992) have recorded Lochkovian marine sediments dated by chitinozoa; we therefore show a marine incursion at that point on our map (Fig. 8).

At the southern margin the low-diversity and cooler-water Malvinokaffric Province, largely defined on brachiopods such as *Australospirifer* and *Australocoelia* by Boucot *et al.* (1969) and peaking in the Emsian, continued on from the Late Silurian. It is characteristic of south-central Gondwana, with its name derived from a combination of the Falkland Islands (Malvinas) and the Kaffir tribes of South Africa. That province is known from Argentina, Bolivia, Brazil, Paraguay, Peru, Uruguay, the Falkland Islands, Ghana, South Africa and Antarctica, and extended westwards to Chile (Fortey *et al.* 1982). Copper (1977) reviewed all of the South American Devonian rocks, their contained brachiopod communities and their distribution. He recognized a northern belt with higher-diversity Appalachian affinities and a southern belt with the lower-diversity Malvinokaffric fauna, both mirroring the palaeolatitudes, which have become much better palaeomagnetically constrained over the past 20 years. In South Africa, Tankard *et al.* (2009) recognized a Bokkeveld–Witteberg late extensional phase which lasted through the Devonian into the Early Carboniferous, causing large-scale subsidence which created the substantial sediment-filled basins there. The Early Devonian coastline of south-central Gondwana in Figure 8 follows Hunter & Lomas (2003).

Carboniferous

We present two maps: one for 350 Ma (Fig. 10) during the Lower Carboniferous Tournaisian Stage and the other for 310 Ma (Fig. 11) in the Upper Carboniferous Moscovian Stage. During that 40 Ma period, the position of the South Pole changed from beneath central Africa to below Antarctica.

The main global event of the period was the union of Gondwana with Laurussia to form Pangea, but the initial collision consisted of an oblique soft docking. The prime Laurussian–Gondwanan collision zone was in the southern US and central America, outside the area of the maps in this paper. The southern margin of Laurentia/Laurussia had previously remained passive since the late Neoproterozoic. The Pangean collision was heralded by the Early Carboniferous downwarping of the Ouachita Basin in the US and is reflected directly in the compressional deformations first recorded in the Middle Carboniferous of Oklahoma; these peaked in the Late Carboniferous and the final phase of the accretion was complete in the earliest Permian. It was not until the Late Carboniferous that the Hercynian Orogeny occurred

in the central Gondwana area, which included substantial thrusting and uplift in both Morocco and Algeria as well as the accretion of the Meseta of Morocco to form an integral part of the African sector of the supercontinent.

Central Gondwana was substantially affected by the Permo-Carboniferous glaciation, which had many separate glacial episodes and which probably lasted for much of the second half of the Carboniferous and into the Early Permian (its total duration is controversial, with some authors asserting that the glacial period was shorter). Eyles (1993, fig. 16.1) reviewed the whole series of glaciations, and concluded that they started in the Early Visean (at *c.* 340 Ma) in South America. In contrast, the glaciation did not start until much later (after 320 Ma) in South Africa (Visser 1997; Isbell *et al.* 2008*a*) and Australia, but continued there later than in South America (Caputo *et al.* 2008) until about 280 Ma, well into Permian time. Although the Late Carboniferous saw the maximum extent of the ice caps over Gondwana as a whole, there are no documented Carboniferous glaciogenic rocks in Antarctica (Isbell *et al.* 2008*b*). Bussert & Schrank (2007) described the glacial sediments of Ethiopia and concluded that glacial uplands adjacent to Ethiopia must have existed in Eritrea and probably also in southern and central Saudi Arabia, all well away from the South Pole. Martin *et al.* (2008) described the glaciogenic sediments in Oman, which began in the Upper Carboniferous (Moscovian) rocks and continued into the Permian. However, because of the varied dating and interpretations, we have not portrayed any ice sheets on our Late Carboniferous map (Fig. 11).

In the northern part of central Gondwana there was extensive marine transgression, with the shoreline retreating as far south as Nigeria. However, there were several subsequent regressions and transgressions, which resulted in interbedded marine and continental rocks over much of southern Algeria and Libya. The Moscovian lakes depicted in north Africa in Figure 11 follow Vai (2003). In the Arabian Peninsula, the Palmyrides Trough was developed across central Syria and continued there until the end of the Cretaceous (Brew *et al.* 2001).

In the southern region there were also extensive lakes. The palaeogeography of southern South America, South Africa and adjacent areas follows Augustsson *et al.* (2006) who, from the analysis of detrital zircons within conglomerates, established that Patagonia was autochthonous or very close to the main part of southern Gondwana by the Late Carboniferous. Augustsson *et al.* (2006) also established that subduction outboard of Patagonia must have started before 300 Ma in accordance with Pankhurst *et al.* (2006), who published both Early and Late Carboniferous reconstructions of that

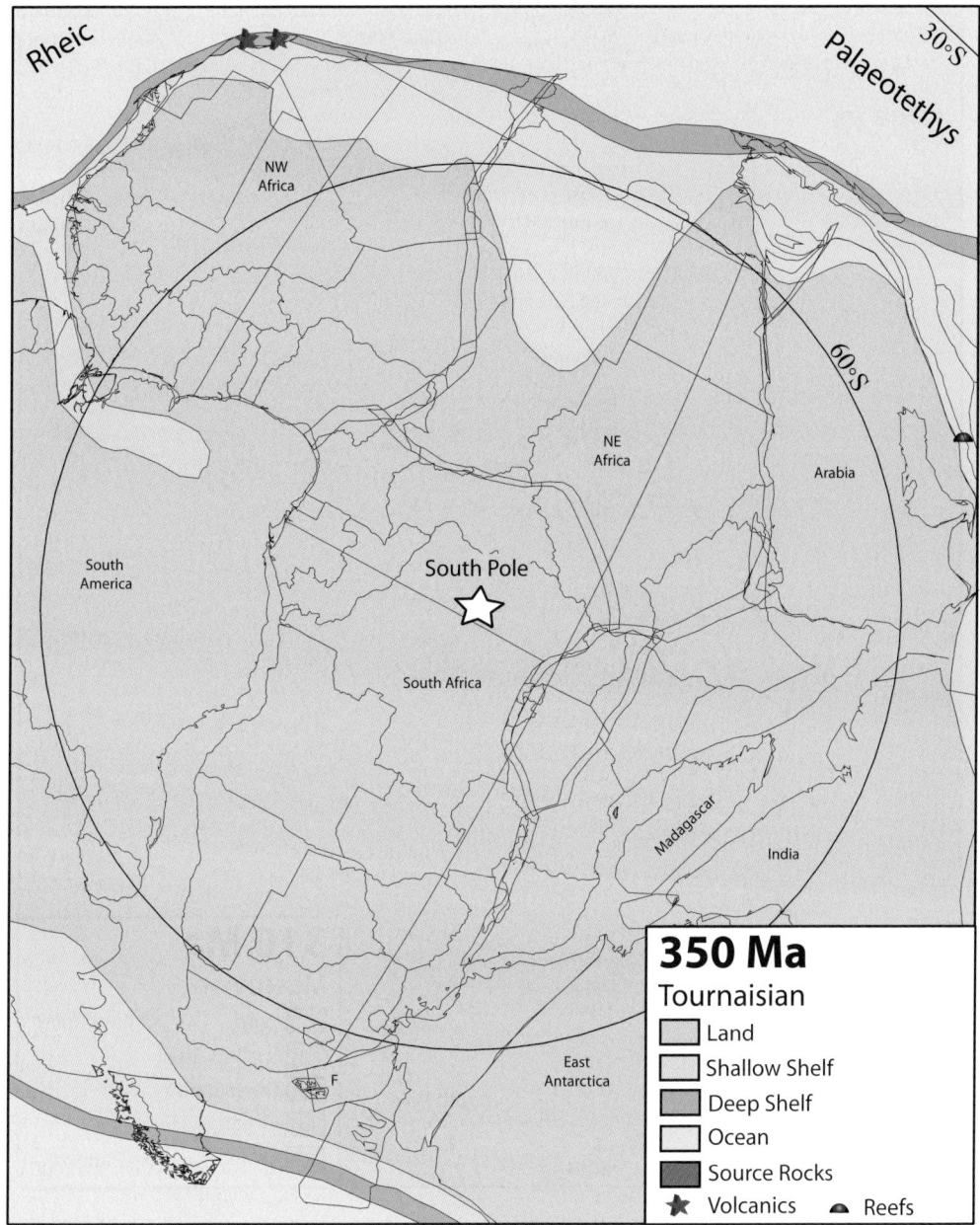

Fig. 10. The palaeogeography of the central Gondwana area in the Tournaisian Stage of the Early Carboniferous at 350 Ma. The southern margin of Gondwana is poorly constrained (see text; F, Falkland Islands).

wider region. In South Africa, Tankard *et al.* (2009) characterized lithospheric subsidence there which created the basins in which the widespread Karroo Supergroup was deposited. Through radiometric ages from the tuffs within the glaciogenic sediments of the Dwyka Group at the base of the Karroo Supergroup, Bangert *et al.* (1999) established that the glaciogenic sediments in Namibia started at 302 Ma and that Dwyka Group sedimentation continued until the end of the Carboniferous. The positions of parts of the old lands that were at the southern margin of the supercontinent are poorly

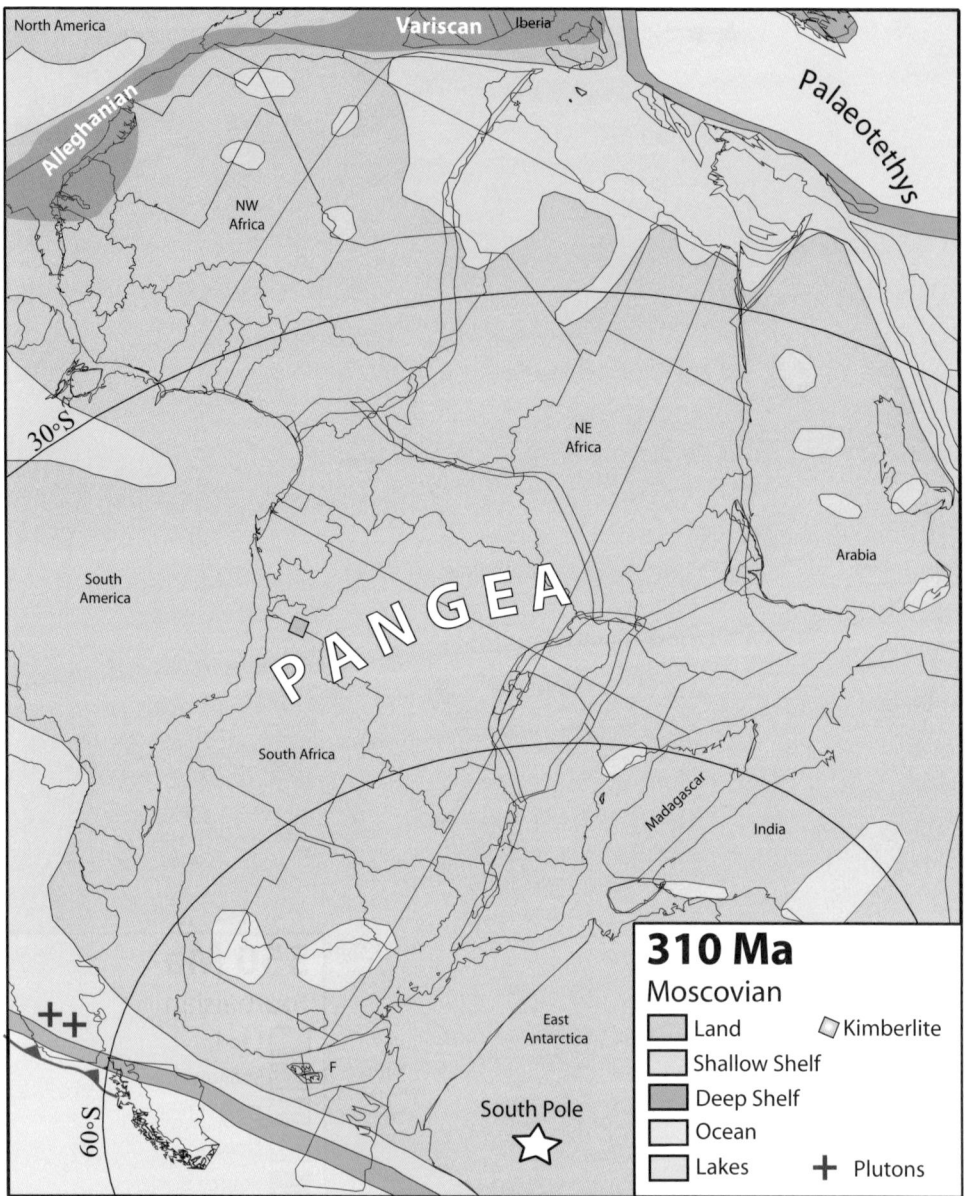

Fig. 11. The palaeogeography of the central Gondwana area in the Late Carboniferous Moscovian Stage at 310 Ma. The southern margin of Gondwana is poorly constrained (see text; F, Falkland Islands). The Variscan and Alleghanian mountain belts are shown in a darker tone.

constrained; there are no Carboniferous marine rocks in south Africa and data from South America are also sparse.

Permian

We present two maps: for the Early Permian Artins-kian Stage at 280 Ma (Fig. 12) and for the end of the

Permian at 250 Ma (Fig. 13). At 280 Ma the South Pole still lay under Antarctica, and the central Gond-wanan part of Pangea spanned all the southern hemi-sphere palaeolatitudes. Before 250 Ma, the South Pole was too far away from Gondwana to appear on our reconstruction. For the first time in the Palaeo-zoic the Equator appears on our maps; much more of central Gondwana was at tropical low latitudes

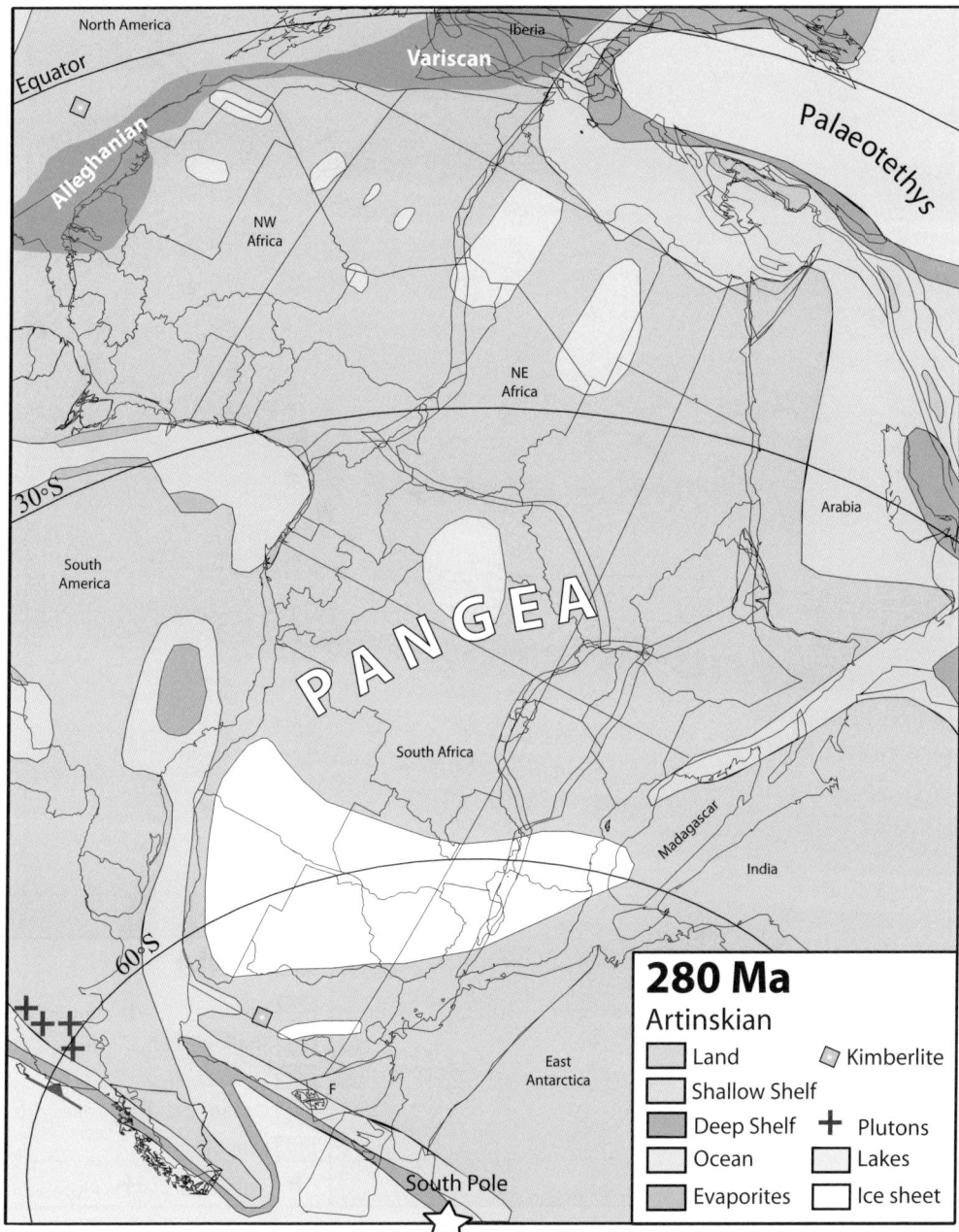

Fig. 12. The palaeogeography of the central Gondwana area in the Early Permian Artinskian Stage at 280 Ma (F, Falkland Islands). The Variscan and Alleghanian mountain belts are shown in a darker tone.

by the end of the Permian than at any previous time in the Palaeozoic.

The widespread glacial event continued on from the Carboniferous and lasted for most of the Early Permian. Veevers (2004, fig. 55) mapped the glacial features for the whole of Gondwana in the Early Permian at *c.* 295 Ma and Isbell *et al.* (2008*a*) for Antarctica; however, there are no Permian glaciogenic rocks known from South America. The subsequent deglaciation was relatively rapid and

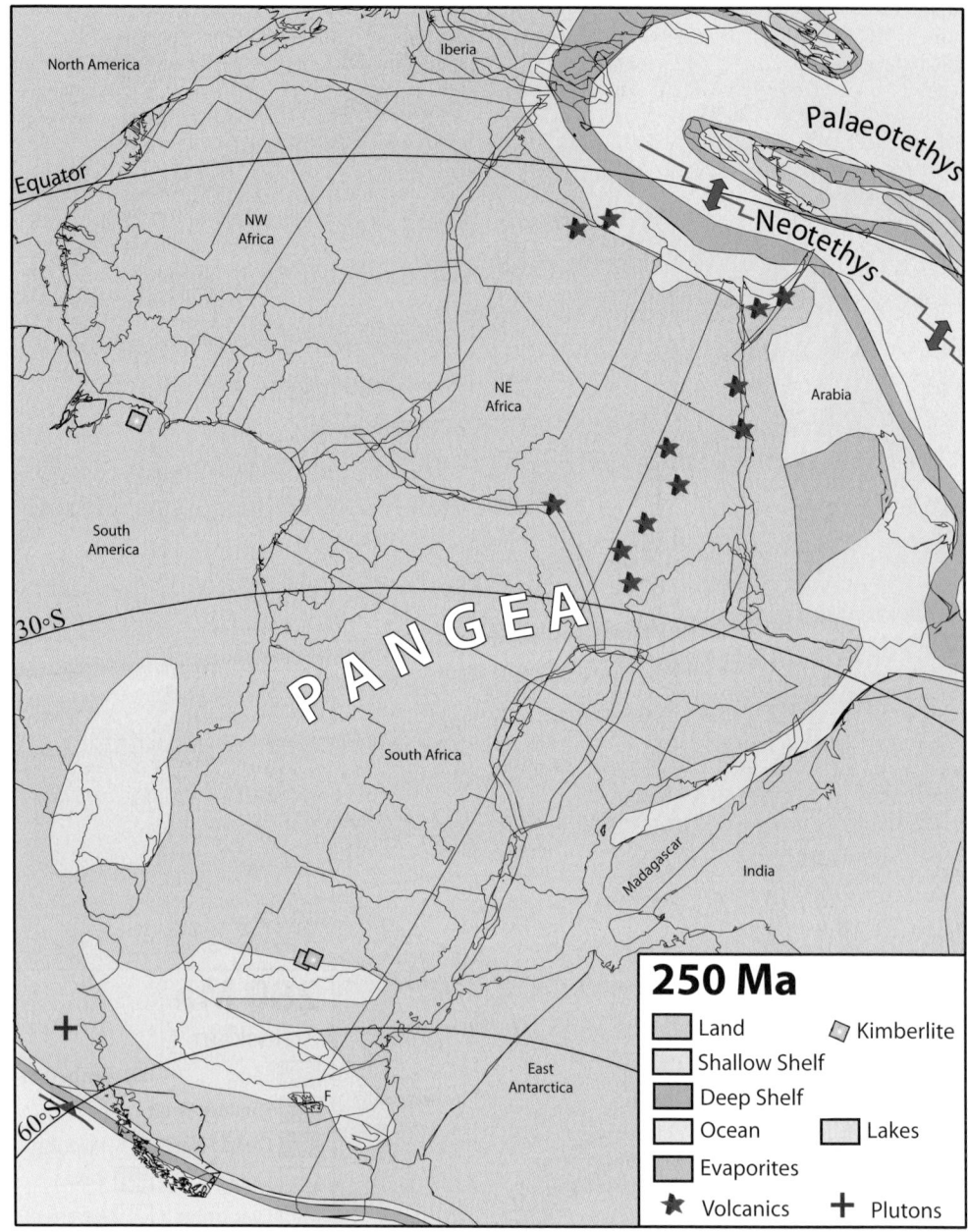

Fig. 13. The palaeogeography of the central Gondwana area at the end of the Permian at 250 Ma (F, Falkland Islands).

occurred during the Sakmarian at *c.* 290 Ma, as described for Oman by Crasquin-Soleau *et al.* (2001). Stephenson *et al.* (2007) reviewed the ages and biota of the post-glacial sediments across the region as the temperature increased. In the southern part of Gondwana a substantial *Glossopteris* forest replaced the ice sheets in much of the area, and the sediments changed rapidly from tillites to coal-swamp deposits (Ziegler *et al.* 1997). However, Crasquin-Soleau *et al.* (2001) also documented the Middle Permian floras in Oman, and discovered that they are a mix of Gondwanan (i.e. *Glossopteris*), Euramerian and Cathaysian elements, indicating that that area was then at the junction between all three major floral provinces. There was an enormous lake in southern Africa in which more Karroo Supergroup sediments were deposited and which contained distinctive non-marine bivalves. We have followed Trewin *et al.* (2002) for the Late Permian (Fig. 13) in the area from Argentina through the Falklands and south Africa to Antarctica, and they show that the Witwatersrand Arch separated the east part of the lake into two basins: the Kalahari–Botswana Basin to the north and the Karroo Basin to the south. Those Waterford Formation sediments extended westwards into the Estrada Nova Formation of Brazil, but to the west of the lake there were mountains from Paraguay to the Falkland Isles.

In the northern sector of Gondwana, which was in equatorial latitudes, the land area of the Pangea continent extended uninterrupted into Europe and North America. Few Permian rocks are known from most of north Africa west of Egypt, although they occur in Tunisia and in the late Permian reefs in the Atlas Mountains of Morocco. However, in the Zagros Fold Belt bordering Arabia, Early Permian fluviatile to shallow-marine clastic sediments were laid down in the transgression there. These are followed by the Mid- to Late Permian (Kungurian to Kazanian) Dalan Formation, which are largely carbonates including reefs and also many substantial evaporite deposits (Bordenave & Hegre 2010).

Rifting from near the end of the Mid-Permian onwards eventually caused the opening of the Neotethys Ocean (Sengor 1990; Ruban *et al.* 2007) and the departure of the Taurides, Sanand, Alborz (and other terranes further to today's east) away from the north African, Turkish and Arabian sectors of Gondwana, which by that time was part of Pangea. The Late Permian geography in northeast Africa and Arabia is taken from Husseini (1992). Substantial volcanics in boreholes in Israel and Jordan were dated at 275 Ma (Kungurian) by Segev & Eshet (2003) at the NW boundary of the Arabian Plate near its junction with the southern end of the Taurides Terrane. The Late Permian development of the Palmyride Trough in Syria and adjacent areas is an aulacogen caused by extension associated with local rifting (Brew *et al.* 2001). The global maps published for the Permian by Vai (2003) and Ziegler *et al.* (1997) were also used to plot the facies and non-marine deposits in north Africa and elsewhere. At the close of the Permian, the greatest biotic extinction event of the whole Phanerozoic occurred.

Conclusions

Gondwana was by far the largest continent (and the only supercontinent) during the Palaeozoic until its merger with Laurussia to form the even larger supercontinent of Pangea in the Carboniferous and

Permian. However, Gondwana's initial assembly, uniting many Precambrian cratons, was only just complete by the start of the Cambrian. The central part of Gondwana described here is today an enormous area which includes the whole continent of Africa and the subcontinent of India, as well as very substantial parts of South America, Asia (Arabia), Antarctica and southern Europe. The position of Africa with respect to the Earth's deep mantle has certainly remained the most stable of all the continents since the Permian, and may also have been so for much longer (perhaps even throughout the Phanerozoic as discussed by Torsvik *et al.* 2008*a*, 2010).

Palaeozoic Gondwana had an active northern margin from South America to as far eastwards as Turkey, leading to the opening of the high-latitude Rheic Ocean in the earliest Ordovician in its (current) western sector and the opening of Palaeotethys at around the beginning of the Devonian in the eastern part. The eastern sector also saw rifting leading to the opening of the Neotethys Ocean during the Permian. The southern margin of central Gondwana was active during the Cambrian, but after that was relatively passive until and beyond the end of the Palaeozoic. Gondwana slowly moved across the South Pole during the Palaeozoic. At the start of the Cambrian the pole lay under north Africa, and thus the central Gondwana area was the most affected by the latest Ordovician (Hirnantian) global glaciation. However, that glacial episode lasted for less than 2 Ma, a complete contrast to the over 25 Ma of the Late Carboniferous and Early Permian glaciation which is also best evidenced in the rocks outcropping in central Gondwana. By then Gondwana had drifted and rotated so that, by the end of the Palaeozoic, the South Pole lay offshore of Antarctica.

As can be seen from the palaeogeographical maps in this paper, varying amounts of Gondwana and (later) Pangea were land but much of the cratonic areas were covered by shallow shelf seas. In addition, there were several very extensive inland lakes during much of the Upper Palaeozoic (especially in central and southern Africa) in which the Karroo Supergroup was deposited and which contain many important faunas and floras. However, since much of Gondwana was at high palaeolatitudes, the biotas were not generally so diverse there as seen elsewhere throughout the Palaeozoic.

We much appreciate discussions and comments from many colleagues, especially N. Hughes (UC Riverside) and R. Van der Voo (Michigan). We are also grateful to Statoil (The African Project) for funding and to The Natural History Museum, London, for the provision of facilities. It is a pleasure to dedicate this paper to Kevin Burke and Lew Ashwal to help celebrate their 80th and 60th birthdays.

References

ÁLVARO, J. J., ELICKI, O., GEYER, G., RUSHTON, A. W. A. & SHERGOLD, J. H. 2003. Palaeogeographical controls on the Cambrian immigration and evolutionary patterns reported in the western Gondwana margin. *Palaeogeography, Palaeoclimatology, Palaeoecology*, **195**, 5–35.

ARMSTRONG, H. A., ABBOTT, G. D. *ET AL.* 2009. Black shale deposition in an Upper Ordovician–Silurian permanently stratified, peri-glacial basin, southern Jordan. *Palaeogeography, Palaeoclimatology, Palaeoecology*, **273**, 368–377.

ARTHUR, T. J., MACGREGOR, D. S. & CAMERON, N. R. (eds) 2003. *Petroleum Geology of Africa: New Themes and Developing Technologies*. Geological Society, London, Special Publications, **207**.

ASTINI, R. A. 2003. The Ordovician Proto-Andean basins. *In*: BENEDETTO, J. L. (ed.) *Ordovician Fossils of Argentina*. Universidad Nacional de Córdoba, Córdoba, Argentina, 1–74.

AUGUSTSSON, C., MÜNKER, C., BAHLBURG, H. & FANNING, C. M. 2006. Provenance of Late Palaeozoic metasediments of the SW South American Gondwana margin: a combined U–Pb and Hf-isotope study of single detrital zircons. *Journal of the Geological Society, London*, **163**, 983–995.

BANGERT, B., STOLLHOFEN, H., LORENZ, V. & ARMSTRONG, R. 1999. The geochronology and significance of ash-fall tuffs in the glaciogenic Carboniferous-Permian Dwyka Group of Namibia and South Africa. *Journal of African Earth Sciences*, **29**, 33–49.

BAUER, W., THOMAS, R. J. & JACOBS, J. 2003. Proterozoic-Cambrian history of Dronning Maud Land in the context of Gondwana assembly. *In*: YOSHIDA, M. & DASGUPTA, S. (eds) *Proterozoic East Gondwana: Supercontinent Assembly and Breakup*. Geological Society, London, Special Publications, **206**, 247–269.

BEUF, S., BIJOU-DUVAL, V., DE CHARPAL, O., ROGNON, P., GARIEL, O. & BENNACEF, A. 1971. Les Grès du Paléozoïque au Sahara. *Publications de l'Institut français de Pétrole*, **18**, 1–464.

BLAKEY, R. C. 2008. Gondwana paleogeography from assembly to breakup – a 500 m.y. odyssey. *Geological Society of America Special Paper*, **441**, 1–28.

BORDENAVE, M. L. & HEGRE, J. A. 2010. Current distribution of oil and gas fields in the Zagros Fold Belt of Iran and contiguous offshore as the result of the petroleum systems. *In*: LETURMY, P. & ROBIN, C. (eds) *Tectonic and Stratigraphic Evolution of Zagros and Makran during the Mesozoic–Cenozoic*. Geological Society, London, Special Publications, **330**, 291–353.

BOSSI, J. & CINGOLANI, C. 2010. Extension and general evolution of the Rio de la Plata craton. *In*: GAUCHER, C., SIAL, A. N., HALVERSON, G. P. & FRIMMEL, H. E. (eds) *Neoproterozoic–Cambrian Tectonics, Global Change and Evolution: Focus on Southwestern Gondwana*. Elsevier, Amsterdam, 73–85.

BOUCOT, A. J., JOHNSON, J. G. & TALENT, J. A. 1969. Early Devonian brachiopod zoogeography. *Geological Society of America Special Papers*, **119**, 1–113.

BOUCOT, A. J., MASSA, D. & PERRY, D. G. 1983. Stratigraphy, biogeography and taxonomy of some Lower and Middle Devonian brachiopod-bearing beds of Libya and northern Niger. *Palaeontographica*, **A180**, 91–125.

BREW, G., BARAZANGI, M., AL-MALEH, A. K. & SAWAF, F. 2001. Tectonic and geologic evolution of Syria. *GeoArabia*, **6**, 573–615.

BUSSERT, R. & SCHRANK, E. 2007. Palynological evidence for a latest Carboniferous–Early Permian glaciation in Northern Ethiopia. *Journal of African Earth Sciences*, **49**, 201–210.

CAPUTO, M. V. 1998. Ordovician–Silurian glaciations and global sea-level changes. *New York State Museum Bulletin*, **491**, 15–25.

CAPUTO, M. V., DE MELO, J. H. G., STREEL, M. & ISBELL, J. L. 2008. Late Devonian and Early Carboniferous glacial records of South America. *Geological Society of America Special Paper*, **441**, 161–173.

CHLUPÁČ, I., HAVLÍČEK, V. & KŘÍŽ, J. 1998. *Palaeozoic of the Barrandian (Cambrian to Devonian)*. Czech Geological Survey, Prague.

COCKS, L. R. M. 2000. The early Palaeozoic geography of Europe. *Journal of the Geological Society, London*, **157**, 1–10.

COCKS, L. R. M. 2010. Caradoc strophomenoid and plectambonitoid brachiopods from Wales and the Welsh Borderland. *Palaeontology*, **53**, 459–500.

COCKS, L. R. M. & TORSVIK, T. H. 2002. Earth geography from 500 to 400 million years ago: a faunal and palaeomagnetic review. *Journal of the Geological Society, London*, **159**, 631–644.

COCKS, L. R. M. & FORTEY, R. A. 2009. Avalonia – a long-lived terrane in the Lower Palaeozoic? *In*: BASSETT, M. G. (ed.) *Early Palaeozoic Peri-Gondwana Terranes: New Insights from Tectonics and Biogeography*. Geological Society, London, Special Publications, **325**, 141–155.

COCKS, L. R. M., BRUNTON, C. H. C., ROWELL, A. J. & RUST, I. C. 1970. The first Lower Palaeozoic fauna proved from South Africa. *Quarterly Journal of the Geological Society, London*, **125**, 583–603.

COCKS, L. R. M., FORTEY, R. A. & RUSHTON, A. W. A. 2010. Correlation for the Lower Palaeozoic. *Geological Magazine*, **147**, 171–180, doi: 1017/S0016756809990562.

COPPER, P. 1977. Paleolatitudes in the Devonian of Brazil, and the Frasnian–Famennian mass extinction. *Palaeogeography, Palaeoclimatology, Palaeoecology*, **21**, 165–207.

COPPER, P. 2002. Silurian and Devonian reefs: 80 million years of global greenhouse between two ice ages. *SEPM Special Publication*, **72**, 181–238.

CRASQUIN-SOLEAU, S., BROUTIN, J., BESSE, J. & BERTHELEIN, M. 2001. Ostracods and palaeobotany from the middle Permian of Oman: implications on Pangaea reconstruction. *Terra Nova*, **13**, 38–43.

CULVER, S. J. & WILLIAMS, H. R. 1979. Late Precambrian and Phanerozoic geology of Sierra Leone. *Journal of the Geological Society, London*, **136**, 605–618.

DALY, M. C., LAWRENCE, S. R., DIEMU-TSHIBAND, K. & MATOUNA, B. 1992. Tectonic evolution of the Cuvette Centra, Zaire. *Journal of the Geological Society, London*, **149**, 539–546.

DEAN, W. T., UYENO, T. T. & RICKARDS, R. B. 1999. Ordovician and Silurian stratigraphy and trilobites, Taurus Mountains near Kemer, southwestern Turkey. *Geological Magazine*, **136**, 373–393.

DEAN, W. T., MONOD, O., RICKARDS, R. B., DEMIR, O. & BULTYNCK, P. 2000. Lower Palaeozoic stratigraphy and palaeontology, Karadere-Zirze area, Pontus Mountains, northern Turkey. *Geological Magazine*, **137**, 555–582.

DENIS, M., BUONCRISTIANI, J. F., KONATE, M., YAHAYA, M. & GUIRAUD, M. 2006. Typology of Hirnantian glacial pavements in SW Djado Basin (NE Niger). *Africa Geoscience Review*, **13**, 145–155.

DESTOMBES, J., HOLLARD, H. & WILLEFERT, S. 1985. Lower Palaeozoic rocks of Morocco. *In*: HOLLAND, C. H. (ed.). *Lower Palaeozoic of North-western and West-central Africa*. Wiley, Chichester, 91–336.

DUEBENDORFER, E. M. & REES, M. N. 1998. Evidence for Cambrian deformation in the Ellsworth-Whitmore Mountains terrane, Antarctica: stratigraphic and tectonic implications. *Geology*, **26**, 55–58.

EL-ARABY, A. & ABDEL-MOTELIB, A. 1999. Depositional facies of the Cambrian Araba Formation in the Taba region, east Sinai, Egypt. *Journal of African Earth Sciences*, **29**, 429–447.

EYLES, N. 1993. Earth's glacial record and its tectonic setting. *Earth-Science Reviews*, **35**, 1–248.

FARZIPOUR-SAEIN, A., YASSAGHI, A., SHERKATI, S. & KOYI, H. 2009. Mechanical stratigraphy and folding style of the Lurestan region in the Zagros Fold-Thrust Belt, Iran. *Journal of the Geological Society, London*, **166**, 1101–1115.

FORTEY, R. A. & MORRIS, S. F. 1982. The Ordovician trilobite *Neseuretus* from Saudi Arabia, and the palaeogeography of the *Neseuretus* fauna related to Gondwanaland in the earlier Ordovician. *Bulletin of the British Museum (Natural History) Geology*, **36**, 63–75.

FORTEY, R. A. & COCKS, L. R. M. 2003. Faunal evidence bearing on global Ordovician–Silurian continental reconstructions. *Earth-Science Reviews*, **61**, 245–307.

FORTEY, R. A. & COCKS, L. R. M. 2005. Late Ordovician global warming – The Boda Event. *Geology*, **33**, 405–408.

FORTEY, R. A., PANKHURST, R. J. & HERVÉ, F. 1982. Devonian trilobites at Buill, Chile (42°S). *Revista Geológica de Chile*, **19**, 133–144.

FRANKE, W. 2006. The Variscan orogen in Central Europe: construction and collapse. *In*: GEE, D. G. & STEPHENSON, R. A. (eds) *European Lithosphere Dynamics*. Geological Society, London, Memoirs, **32**, 333–343.

GAETANI, M., ANGIOLINI, L. *ET AL.* 2009. Pennsylvanian–Early Triassic stratigraphy in the Alborz Mountains (Iran). *In*: BRUNET, M.-F., WILMSEN, M. & GRANATH, J. W. (eds) *South Caspian to Central Iran Basin*. Geological Society, London, Special Publications, **312**, 79–128.

GAUCHER, C., FRIMMEL, H. E. & GERMS, G. J. B. 2010. Tectonic events and palaeogeographic evolution of southwestern Gondwana in the Neoproterozoic and Cambrian. *In*: GAUCHER, C., SIAL, A. N., HALVERSON, G. P. & FRIMMEL, H. E. (eds) *Neoproterozoic-Cambrian Tectonics, Global Change and Evolution:*

Focus on Southwestern Gondwana. Elsevier, Amsterdam, 295–316.

GHIENNE, J.-F., LE HERON, D. P., MOREAU, J., DENIS, M. & DEYNOUX, M. 2007. The Late Ordovician glacial sedimentary system of the North Gondwana platform. *International Association of Sedimentologists Special Publication*, **39**, 297–319.

GIBBONS, W. & MORENO, T. (eds) 2002. *The Geology of Spain*. The Geological Society, London.

GRAHN, Y. & PARIS, F. 1992. Age and correlation of the Trombetas Group, Amazonas Basin, Brazil. *Revue de Micropaléontologie*, **35**, 197–209.

GREGORY, L. C., MEERT, J. G., TAMRAT, E., MALONE, S., PANDIT, M. K. & PRADHAN, V. 2006. A palaeomagnetic and geochronologic study of the Majhgawan kimberlite, India: implications for the age of the Upper Vinghyan Supergroup. *Precambrian Research*, **149**, 69–75.

GUBANOV, A. P. & MOONEY, W. D. 2009. New global maps of crustal basement age. *Eos Transactions, AGU*, **90**, Fall Meet. Suppl., Abstract T53B-1583.

GUIRAUD, R., BOSWORTH, W., THIERRY, J. & DELPLANQUE, A. 2005. Phanerozoic geological evolution of northern and central Africa: an overview. *Journal of African Earth Sciences*, **43**, 83–143.

GUTIÉRREZ-MARCO, J. C., ARAMBURU, C. *ET AL.* 1999. Revision bioestratigrafica de la pizarras del Ordovicico Medio en el noroeste de Espana (zones Cantabrica, Asturoccidental-Leonesa y Centroiberica septentrional). *Acta Geologica Hispanica*, **34**, 3–87.

HALLETT, D. 2002. *Petroleum Geology of Libya*. Elsevier, Amsterdam.

HARLEY, S. L. 2003. Archaean–Cambrian crustal development of East Antarctica: metamorphic characteristics and tectonic implications. *In*: YOSHIDA, M. & DASGUPTA, S. (eds) *Proterozoic East Gondwana: Supercontinent Assembly and Breakup*. Geological Society, London, Special Publications, **206**, 203–230.

HAVLÍČEK, V. & BRANISA, L. 1980. Ordovician brachiopods of Bolivia. *Rozpravy Československé Akademie Věd*, **90**, 1–54.

HAVLÍČEK, V., VANEK, J. & FATKA, O. 1994. Perunica microcontinent in the Ordovician (its position within the Mediterranean Province, series divisions, benthic and pelagic associations). *Sbornik geologickych věd Geologie*, **46**, 25–56.

HOLLAND, C. H. (ed.) 1981. *Lower Palaeozoic of the Middle East, Eastern and Southern Africa, and Antarctica*. Wiley, Chichester.

HOLLAND, C. H. (ed.) 1985. *Lower Palaeozoic of North-western and West-central Africa*. Wiley, Chichester.

HUNTER, M. A. & LOMAS, S. A. 2003. Reconstructing the Siluro-Devonian coastline of Gondwana: insights from the sedimentology of the Port Stephens Formation, Falkland Islands. *Journal of the Geological Society, London*, **160**, 459–476.

HUSSEINI, M. I. 1992. Upper Palaeozoic tectono-sedimentary evolution of the Arabian and adjoining plates. *Journal of the Geological Society, London*, **149**, 419–429.

ISBELL, J. L., COLE, D. I. & CATUNEANU, O. 2008a. Carboniferous-Permian glaciation in the main Karoo Basin, South Africa: stratigraphy, depositional controls, and glacial dynamics. *In*: FIELDING, C. R.,

FRANK, T. D. & ISBELL, J. C. (eds) *Resolving the Late Paleozoic Ice Age: Time and Space*. Geological Society of America Special Papers, **441**, 71–82.

ISBELL, J. L., KOCH, Z. J., SZABLEWESKI, G. M. & LENAKER, P. A. 2008*b*. Permian glaciogenic deposits in the main Transantarctic Mountains, Antarctica. *In*: FIELDING, C. R., FRANK, T. D. & ISBELL, J. C. (eds) *Resolving the Late Paleozoic Ice Age: Time and Space*. Geological Society of America Special Papers, **441**, 59–70.

KEPPIE, J. D. (ed.) 1994. *Pre-Mesozoic Geology in France*. Springer-Verlag, Berlin.

KONERT, G., AFIFI, A. M., AL-HAJRI, S. A. & DROSTE, H. J. 2001. Paleozoic stratigraphy and hydrocarbon habitat of the Arabian Plate. *GeoArabia*, **6**, 407–442.

LABAILS, C., TORSVIK, T. H., GAINA, C. & COCKS, L. R. M. 2009. Global Plate Polygons 2009, SPlates Model (version 2.0). *NGU Report* 2009.047 (confidential).

LEGRAND, P. 1974. Essai sur la paléogéographie de l'Ordovicien au Sahara algérien. *Compagnie français des Pétroles Notes & Mémoires*, **11**, 121–138.

LEGRAND, P. 1981. Essai sur la paléogéographie du Silurien au Sahara algérien. *Compagnie français des Pétroles Notes & Mémoires*, **16**, 9–24.

LEGRAND, P. 1994. Sea level and faunal change during the Late Wenlock and Earliest Ludlow (Silurian): a point of view from the Algerian Sahara. *Historical Biology*, **7**, 271–299.

LE HERON, D. P. & CRAIG, J. 2008. First-order reconstruction of a Late Ordovician Saharan ice sheet. *Journal of the Geological Society, London*, **165**, 19–29.

LEONE, F., HAMMANN, W., LASKE, R., SERPAGLI, E. & VILLAS, E. 1991. Lithostratigraphic units and biostratigraphy of the post-sardic Ordovician sequence in south-west Sardinia. *Bollettino della Società Paleontologica Italiana*, **30**, 201–233.

LINNEMANN, U. (ed.). 2003. Das Saxothuringicum. *Geologica Saxonica*, **48/49**, 1–159.

LÜNING, S., CRAIG, J., LOYDELL, D. K., ŠTORCH, P. & FITCHES, B. 2000. Lower Silurian 'Hot Shales' in North Africa and Arabia: regional distribution and depositional model. *Earth-Science Reviews*, **49**, 121–200.

LÜNING, S., ADAMSON, K. & CRAIG, J. 2003. Frasnian organic-rich shales in North Africa: regional distribution and depositional model. *In*: ARTHUR, T., MACGREGOR, D. S. & CAMERON, N. R. (eds) *Petroleum Geology of Africa: New Themes and Developing Technologies*. Geological Society, London, Special Publications, **207**, 165–184.

MARSHALL, J. E. A. 1994. The Falkland Island: a key element in Gondwana palaeogeography. *Tectonics*, **13**, 499–514.

MARTIN, J. R., REDFERN, J. & AITKEN, J. F. 2008. A regional overview of the late Paleozoic glaciation in Oman. *In*: FIELDING, C. R., FRANK, T. D. & ISBELL, J. C. (eds) *Resolving the Late Paleozoic Ice Age: Time and Space*. Geological Society of America Special Papers, **441**, 175–186.

MCCALL, G. J. G. 1997. The geotectonic history of the Makran and adjacent areas of southern Iran. *Journal of Asian Earth Sciences*, **15**, 517–531.

MCCANN, T. (ed.) 2008. *The Geology of Central Europe. Volume 1: Precambrian and Palaeozoic*. Geological Society, London.

MCDOUGALL, N., BRENCHLEY, P. J., REBELO, J. A. & ROMANO, M. 1987. Fans and fan deltas – precursors to the Armorican Quartzite (Ordovician) in western Iberia. *Geological Magazine*, **134**, 347–359.

MEERT, J. G. 2003. A synopsis of events related to the assembly of eastern Gondwana. *Tectonophysics*, **362**, 1–40.

MERGL, M. & MASSA, D. 2000. A palaeontological review of the Devonian and Carboniferous succession of the Murzuq Basin and the Djado Sub-Basin. *In*: SOLA, M. A. & WORSLEY, D. (eds) *Geological exploration in Murzuq Basin*. Elsevier, Amsterdam, 41–88.

MILLSON, J. A., MERCADIER, C. G. L., LIVERA, S. E. & PETERS, J. M. 1996. The Lower Palaeozoic of Oman and its context in the evolution of a Gondwanan continental margin. *Journal of the Geological Society, London*, **153**, 213–230.

MONOD, O., KOZLU, H., GHIENNE, J.-F., DEAN, W. T., GÜNAY, Y., LE HÉRISSÉ, A. & PARIS, F. 2003. Late Ordovician glaciation in southern Turkey. *Terra Nova*, **15**, 249–257.

MOREAU, C., DEMAIFFE, D., BELLION, Y. & BOULLIER, A. M. 1994. A tectonic model for the location of Palaeozoic ring complexes in Air (Niger). *Tectonophysics*, **234**, 129–146.

NYSÆTHER, E., TORSVIK, T. H., FEIST, R., WALDERHAUG, H. J. & EIDE, E. A. 2002. Ordovician palaeogeography with new palaeomagnetic data from the Montagne Noire (Southern France). *Earth and Planetary Science Letters*, **203**, 329–341.

OTERDOOM, W. H., WORTHING, M. A. & PARTINGTON, M. 1999. Petrological and tectonostratigraphic evidence for a Mid Ordovician rift pulse on the Arabian Peninsula. *GeoArabia*, **4**, 467–500.

PANKHURST, R. J., RAPELA, C. W., FANNING, C. M. & MARQUEZ, M. 2006. Gondwanide continental collision and the origin of Patagonia. *Earth-Science Reviews*, **76**, 235–257.

PARIS, F., ELAOUAD-DEBBAJ, Z., JAGLIN, J. C., MASSA, D. & OULEBSIR, L. 1995. Chitinozoans and Late Ordovician glacial events on Gondwana. *In*: COOPER, C., DROSER, M. L. & FINNEY, S. (eds) *Ordovician Odyssey*. Las Vegas, 171–176.

PARIS, F., LE HÉRISSÉ, A. *ET AL.* 2007. Ordovician chitinozoans and acritarchs from southern and southeastern Turkey. *Revue de Micropaléontologie*, **50**, 81–107.

PIQUÉ, A. 2001. *Geology of Northwest Africa*. Gebrüder Borntraeger, Berlin.

PLUSQUELLEC, Y., BOUMEDJEL, K., MORZADEC, P. & PARIS, F. 1997. Les faunes dévoniennes d'Ougarta dans la paléogéographie des regions Maghrébo-Européennes. *Annales Societé Géologique du Nord*, **5**, 123–128.

RAMOS, V. A. 2008. Patagonia: a Paleozoic continent adrift? *Journal of South American Earth Sciences*, **26**, 235–251.

RAMOS, E., MARZO, M., DE GIBERT, J. M., TAWENGI, K. S., KHOJA, A. A. & BOLATTI, N. D. 2006. Stratigraphy and sedimentology of the Middle Ordovician Hawaz Formation (Murzuq Basin, Libya). *American Association of Petroleum Geologists Bulletin*, **90**, 1309–1336.

RAPELA, C. W., PANKHURST, R. J., FANNING, C. M. & GRECCO, L. E. 2003. Basement evolution of the

Sierra de la Ventana Fold Belt: new evidence for Cambrian continental rifting along the southern margin of Gondwana. *Journal of the Geological Society, London*, **160**, 613–628.

ROBARDET, M. 2002. Alternative approach to the Variscan Belt in southwestern Europe: preorogenic paleobiogeographical constraints. *In*: MARTINEZ CATALÁN, J. R., HATCHER, R. D. JR., ARENAS, R. & DÍAZ GARCIA, F. (eds) *Variscan–Appalachian Dynamics: The Building of the Late Paleozoic Basement.* Geological Society of America Special Papers, **364**, 1–15.

ROBARDET, M. 2003. The Armorican 'microplate': fact or fiction? Critical review of the concept and contradictory palaeobiogeographical data. *Palaeogeography, Palaeoclimatology, Palaeoecology*, **195**, 125–148.

RONG, J.-Y. & HARPER, D. A. T. 1988. A global synthesis of the latest Ordovician Hirnantian brachiopod faunas. *Transactions of the Royal Society of Edinburgh: Earth Sciences*, **79**, 383–401.

RONG, J.-Y., ZHAN, R.-B. & HARPER, D. A. T. 1999. Late Ordovician (Caradoc-Ashgill) brachiopod faunas with Foliomena based on data from China. *Palaios*, **14**, 412–431.

RUBAN, D. A., AL-HUSSEINI, M. I. & IWASAKI, Y. 2007. Review of Middle East Palaeozoic plate tectonics. *GeoArabia*, **12**, 35–56.

SALERN, M. J. & OUN, K. M. (eds) 2003. *The Geology of Northwest Libya.* Earth Science Society of Libya, Tripoli, Libya. 3 vols.

SCHANDELMEIER, H. & REYNOLDS, P. O. (eds) 1997. *Palaeogeographic–Palaeotectonic Atlas of North-Eastern Africa, Arabia, and Adjacent Areas.* A.A. Balkema, Rotterdam.

SCHLÜTER, T. 1997. *Geology of East Africa.* Gebrüder Borntraeger, Berlin.

SCHLÜTER, T. 2006. *Geological Atlas of Africa.* Springer. Berlin.

SEGEV, A. & ESHET, Y. 2003. Significance of Rb/Sr age of Early Permian volcanics, Helez Deep 1A borehole, central Israel. *Africa Geoscience Review*, **10**, 333–345.

SENGOR, A. M. C. 1990. A new model for the late Palaeozoic–Mesozoic tectonic evolution of Iran and implications for Oman. *In*: ROBERTSON, A. H. F., SEARLE, M. P. & RIES, A. C. (eds) *The Geology and Tectonics of the Oman Region.* Geological Society, London, Special Publications, **49**, 797–831.

SHARLAND, P. R., ARCHER, R. *ET AL.* 2001. Arabian plate sequence stratigraphy. *GeoArabia Special Publication*, **2**, 1–371.

SHELLEY, D. & BOSSIÈRE, G. 2000. A new model for the Hercynian Orogen of Gondwanan France and Iberia. *Journal of Structural Geology*, **22**, 757–776.

SHERGOLD, J. H. 1988. Review of trilobite biofacies distributions at the Cambrian–Ordovician boundary. *Geological Magazine*, **125**, 363–380.

STEPHENSON, M. H., ANGIOLINI, L. & LENG, M. J. 2007. The Early Permian fossil record of Gondwana and its relationship to deglaciation: a review. In: WILLIAMS, M., HAYWOOD, A. M., GREGORY, F. J. & SCHMIDT, D. N. (eds) *Deep-Time Perspectives on Climate Change.* The Micropalaeontological Society, Special Publications. The Geological Society, London, 169–189.

STONE, P., RICHARDS, P. C., KIMBELL, G. S., ESSER, R. P. & REEVES, D. 2008. Cretaceous dykes discovered in the Falkland Islands: implications for regional tectonics in the South Atlantic. *Journal of the Geological Society*, **165**, 1–4.

SUTCLIFFE, O. E., HARPER, D. A. T., SALEM, A. A., WHITTINGTON, R. J. & CRAIG, J. 2001. The development of an atypical Hirnantia-brachiopod Fauna and the onset of glaciation in the late Ordovician of Gondwana. *Transactions of the Royal Society of Edinburgh: Earth Sciences*, **92**, 1–14.

TANKARD, A. J., JACKSON, M. P. A., ERIKSSON, K. A., HOBDAY, D. K., HUNTER, D. R. & MINTER, W. E. L. 1982. *Crustal Evolution of Southern Africa: 3.8 Billion Years of Earth History.* Springer-Verlag, Berlin.

TANKARD, A. J., WELSINK, H., AUKES, P., NEWTON, A. R. & STETTLER, E. H. 2009. Tectonic evolution of the Cape and Karoo basins of South Africa. *Marine and Petroleum Geology*, **26**, 1379–1412.

TORSVIK, T. H. & COCKS, L. R. M. 2004. Earth geography from 400 to 250 Ma: a palaeomagnetic, faunal and facies review. *Journal of the Geological Society, London*, **161**, 555–572.

TORSVIK, T. H. & COCKS, L. R. M. 2009. The Lower Palaeozoic palaeogeographical evolution of the northeastern and eastern peri-Gondwanan margin from Turkey to New Zealand. *In*: BASSETT, M. G. (ed.) *Early Palaeozoic Peri-Gondwana Terranes: New Insights from Tectonics and Biogeography.* Geological Society, London, Special Publications, **325**, 3–21.

TORSVIK, T. H. & COCKS, L. R. M. in press. New global palaeogeographical reconstructions for the Lower Palaeozoic and their generation. *In*: HARPER, D. & SERVAIS, T. (eds) *Early Palaeozoic Palaeobiogeography and Palaeogeography.* Geological Society, London, Memoir (in press).

TORSVIK, T. H. & VAN DER VOO, R. 2002. Refining Gondwana and Pangea Palaeogeography: Estimates of Phanerozoic (octupole) non-dipole fields. *Geophysical Journal International*, **151**, 771–794.

TORSVIK, T. H., STEINBERGER, B., COCKS, L. R. M. & BURKE, K. 2008a. Longitude: linking Earth's ancient surface to its deep interior. *Earth and Planetary Science Letters*, **276**, 273–282.

TORSVIK, T. H., GAINA, C. & REDFIELD, T. F. 2008b. Antarctica and global paleogeography: from Rodinia, through Gondwanaland and Pangea, to the birth of the Southern Ocean and the opening of gateways. *In*: COOPER, A. K., BARRETT, P. J., STAGG, H. *ET AL.* (eds) *Antarctica: a Keystone in a Changing World.* National Academies Press, Washington D.C.

TORSVIK, T. H., MÜLLER, R. D., VAN DER VOO, R., STEINBERGER, B. & GAINA, C. 2008c. Global Plate Motion Frames: toward a unified model. *Reviews of Geophysics*, **46**, RG3004, doi: 10.1029/ 2007RG000227.

TORSVIK, T. H., PAULSEN, T. S., HUGHES, N. C., MYROW, P. M. & GANERØD, M. 2009a. The Tethyan Himalaya: palaeogeographical and tectonic constraints from Ordovician palaeomagnetic data. *Journal of the Geological Society, London*, **166**, 679–687.

TORSVIK, T. H., ROUSSE, S., LABAILS, C. & SMETHURST, M. A. 2009b. A new scheme for the opening of the

South Atlantic Ocean and the dissection of an Aptian salt basin. *Geophysical Journal International*, **177**, 1315–1333.

TORSVIK, T. H., BURKE, K., STEINBERGER, B., WEBB, S. J. & ASHWAL, L. D. 2010. Diamonds sampled by plumes from the core-mantle boundary. *Nature*, **466**, 352–355.

TRENCH, A., TORSVIK, T. H., DENTITH, M. C., WALDER-HAUG, H. J. & TRAYNOR, J.-J. 1992. A high southerly palaeolatitude for Southern Britain in early Ordovician times: palaeomagnetic data from the Trefgarne Volcanic Formation SW Wales. *Geophysical Journal International*, **108**, 89–100.

TREWIN, N. H., MACDONALD, D. I. M. & THOMAS, C. G. C. 2002. Stratigraphy and sedimentology of the Permian of the Falkland Isles: lithostratigraphic and palaeoenvironmental links with South Africa. *Journal of the Geological Society, London*, **159**, 5–19.

VAI, G. B. 2003. Development of the palaeogeography of Pangaea from Late Carboniferous to Early Permian. *Palaeogeography, Palaeoclimatology, Palaeoecology*, **196**, 125–155.

VAUGHAN, A. P. M. & PANKHURST, R. J. 2008. Tectonic overview of the West Gondwana margin. *Gondwana Research*, **13**, 150–162.

VEEVERS, J. J. 2004. Gondwanaland from 650–500 Ma assembly through 320 Ma merger in Pangea to 185–100 Ma breakup: supercontinental tectonics via stratigraphy and radiometric dating. *Earth-Science Reviews*, **68**, 1–132.

VILLAS, E. & COCKS, L. R. M. 1996. The first Early Silurian brachiopod fauna from the Iberian Peninsula. *Journal of Paleontology*, **70**, 571–588.

VILLAS, E., VENNIN, E., ÁLVARO, J. J., HAMMANN, W., HERRERA, Z. A. & POVANO, E. L. 2002. The late Ordovician carbonate sedimentation as a major triggering factor of the Hirnantian glaciation. *Bulletin de la Societé géologique de France*, **173**, 569–578.

VISSER, J. N. J. 1997. A review of the Permo-Carboniferous glaciation in Africa. *In*: MARTINI, I. P. (ed.) *Late Glacial and Postglacial Environmental Changes: Quaternary, Carboniferous-Permian, and Proterozoic*. Oxford University Press, Oxford, 169–191.

VON RAUMER, J. F. & STAMPFLI, G. M. 2008. The birth of the Rheic Ocean – Early Palaeozoic subsidence patterns and subsequent tectonic plate scenarios. *Tectonophysics*, **461**, 9–20.

WEHRMANN, A., YILMAZ, I. *ET AL*. 2010. Devonian shallow-water sequences from the North Gondwana coastal margin (Central and Eastern Taurides, Turkey): sedimentology, facies and global events. *Gondwana Research*, **17**, 546–560.

WHITTINGTON, H. B. 1953. A new Ordovician trilobite from Florida. *Breviora*, **17**, 1–6.

WILLEFERT, S. 1988. The Ordovician–Silurian boundary in Mauritania. *Bulletin of the British Museum (Natural History) Geology*, **43**, 177–182.

ZANCHI, A., ZANCHETTA, S., GARZANTI, E., BALINI, M., BERRA, F., MATTEI, M. & MUTTONI, G. 2009. The Cimmerian evolution of the Nakhlak-Anarak area, Central Iran, and its bearing on the reconstruction of the history of the Eurasian margin. *In*: BRUNET, M.-F., WILMSEN, M. & GRANATH, J. W. (eds) *South Caspian to Central Iran Basin*. Geological Society, London, Special Publications, **312**, 261–286.

ZIEGLER, A. M., HULVER, M. L. & ROWLEY, D. B. 1997. Permian world topography and climate. *In*: MARTINI, I. P. (ed.) *Late Glacial and Postglacial Environmental Changes: Quaternary, Carboniferous-Permian, and Proterozoic*. Oxford University Press, Oxford, 111–146.

Morphology, internal architecture and emplacement mechanisms of lava flows from the Central Atlantic Magmatic Province (CAMP) of Argana Basin (Morocco)

HIND EL HACHIMI[1], NASRRDDINE YOUBI[1,2,3]*, JOSÉ MADEIRA[4,5], MOHAMED KHALIL BENSALAH[1,3], LÍNIA MARTINS[3,5], JOÃO MATA[3,5], FIDA MEDINA[6], HERVÉ BERTRAND[7], ANDREA MARZOLI[8], JOSÉ MUNHÁ[3,5], GIULIANO BELLIENI[8], ABDELKADER MAHMOUDI[9], MOHAMED BEN ABBOU[10] & HICHAM ASSAFAR[1]

[1]*Geology Department, Faculty of Sciences-Semlalia, Cadi Ayyad University, Prince Moulay Abdellah Boulevard, P.O. Box 2390, Marrakech, Morocco*

[2]*National Centre for Scientific and Technical Research, Angle avenues des FAR et Allal El Fassi, Madinat Al Irfane, P.O. Box 8027, Nations Unies, 10102 Rabat, Morocco*

[3]*Centro de Geologia da Universidade de Lisboa (CeGUL), Portugal*

[4]*LATTEX, Instituto Dom Luiz – Laboratório Associado (IDL – LA), Lisboa, Portugal*

[5]*Universidade de Lisboa, Faculdade de Ciências, Departamento de Geologia (GeoFCUL), Portugal*

[6]*Laboratory Geotel (URAC 46), Scientific Institute, University Mohammed V-Agdal, Rabat, Morocco*

[7]*Laboratory Sciences de la Terre, ENS de Lyon et UCBL, 46, Allée d'Italie, 69364 Lyon, France*

[8]*Dipt. di Geoscienze, University Padova, I-35137, Italy*

[9]*Geology Department, Faculty Sciences de Meknès, Moulay Ismail University, Meknès, Morocco*

[10]*Geology Department, Faculty Sciences Dhar Al Mahraz, Sidi Mohammed Ben Abdellah University, Fès, Morocco*

*Corresponding author (e-mail: youbi@ucam.ac.ma)

Abstract: The morphology, internal architecture and emplacement mechanisms of the Central Atlantic Magmatic Province (CAMP) lava flows of Argana Basin in Morocco are presented. The volcanic pile was produced by two volcanic pulses. The first, represented by the Tasguint Formation, corresponds to a succession of 3–13 individual flows created by 1–8 eruptions; the second, Alemzi Formation, is composed of 2–7 individual flows formed by 1–4 eruptions. These formations, geochemically distinct, are separated by thin silty or sandy horizons or by palaeosols. They include 'compound pahoehoe flows' and 'simple flows'. The first type is almost exclusive of the lower formation, while the second type dominates the upper formation. The lava flows show clear evidence of endogenous growth or 'inflation'. The characteristics of the volcanic pile suggest slow emplacement during sustained eruptive episodes and are compatible with a continental basaltic succession facies model.

Earth's history has been punctuated over at least the last 3.5 billion years by episodic massive volcanic events on a scale unknown in the recent geological past. Largely unknown mechanical and dynamic processes, with unclear relationships to seafloor spreading and subduction, generated voluminous, predominantly mafic magmas that were emplaced into the Earth's crust. The resultant large igneous provinces (LIPs; Coffin & Eldholm 1994; Ernst & Buchan 2001; Sheth 2007; Bryan & Ernst 2008) were at times accompanied by catastrophic environmental changes (Rampino et al. 1988; Rampino & Stothers 1988; Stothers & Rampino 1990; Stothers 1993; Hallam & Wignall 1997; Self et al. 1997;

From: Van Hinsbergen, D. J. J., Buiter, S. J. H., Torsvik, T. H., Gaina, C. & Webb, S. J. (eds) *The Formation and Evolution of Africa: A Synopsis of 3.8 Ga of Earth History*. Geological Society, London, Special Publications, **357**, 167–193. DOI: 10.1144/SP357.9 0305-8719/11/$15.00 © The Geological Society of London 2011.

Courtillot 1999; Olsen 1999; Wignall 2001; Self *et al.* 2006).

Most previous research on LIPs has focused on chemical stratigraphy of lava piles in order to identify possible mantle sources and crustal or mantle contaminants for such huge volumes of basalt, and investigate how these factors relate to the picture of global plate tectonic and mantle dynamics models. Relatively few recent studies have concentrated on the detailed facies architecture of LIPs and their physical volcanology (e.g. Self *et al.* 1997; Jerram *et al.* 1999*a, b*; Keszthelyi *et al.* 1999; Jerram 2002; Bondre *et al.* 2004*a, b*; Jerram & Widdowson 2005; Waichel *et al.* 2006; Passey & Bell 2007; Bondre & Hart 2008; Duraiswami *et al.* 2008; Jay & Widdowson 2008; White *et al.* 2009).

The Central Atlantic Magmatic Province (CAMP) is one of the largest continental flood basalt (CFB) provinces in the world (Marzoli *et al.* 1999). Considerable attention has been focused on this province in the last years because of the widespread extent and huge volume of magmas produced, and particularly because its age coincides with the Triassic–Jurassic boundary and mass extinction. Voluminous data regarding the CAMP petrology, geochemistry, geochronology and palaeomagnetism have been published in the last decade (e.g. Hames *et al.* 2003; Knight *et al.* 2004; Marzoli *et al.* 2004; Nomade *et al.* 2007; Verati *et al.* 2007; Deenen *et al.* 2010, 2011), but the physical volcanology of the lavas remains only sparsely investigated (Puffer & Student 1992; El Hachimi *et al.* 2005, 2007; Kontak 2008; Martins *et al.* 2008). There is therefore considerable scope for physical volcanology studies focusing on morphology and internal architecture of lava flows. This can help understand how CAMP lava flows and other CFB lavas were emplaced. For example, studies on active lava flows in Hawaii as well as those from Columbia River have shed considerable light on the mechanisms of emplacement of flood basalt lavas. These observations indicate that thick sheets of lava (about 4 m in Hawaii, 10–50 m in the Columbia River) were generated by inflation and coalescence of thin (10–50 cm) pāhoehoe (pahoehoe) lobes (Hon *et al.* 1994; Self *et al.* 1997). Flows formed by this mechanism display diagnostic characters that are pointers to their mode of emplacement. Subsequently, such flows have been reported from several CFB provinces: CAMP (El Hachimi *et al.* 2005, 2007; Kontak 2008; Martins *et al.* 2008); Deccan (Keszthelyi *et al.* 1999; Bondre *et al.* 2004*a, b*; Duraiswami *et al.* 2008; Jay & Widdowson 2008), Paraná–Etendeka (Jerram *et al.* 1999*a, b*; Waichel *et al.* 2006) and the North Atlantic igneous province (Kent *et al.* 1998; Single & Jerram 2004; Passey & Bell 2007).

In Morocco, the Argana Basin is by far the best studied Triassic–Jurassic basin and much research has focused on the petrology, geochronology and palaeomagnetism of the CAMP basalts exposed in this basin (Bertrand & Prioton 1975; Manspeizer *et al.* 1978; Aït Chayeb *et al.* 1998; Marzoli *et al.* 2004; Ruiz-Martínez *et al.* 2007; Deenen *et al.* 2010, 2011). The volcanic succession and facies were first described by Brown (1980), who observed multiple 10–20 m thick basalt flows containing vesicle concentrations both at the top and bottom and presenting oxidized tops, sometimes overlain by thin mudstone beds. Medina *et al.* (1992) described several types of internal structures, including compact (dense) coarse-grained vesicular, platy and columnar jointing on the highly weathered 145 m thick Igounan section. Later, Aït Chayeb *et al.* (1998) carried out a more detailed volcanological study and distinguished a variable number of flows along several sections in the Argana basalts. However, because of the state of the art at that time, no interpretation was given by the previous authors as to the type and emplacement mechanisms of the lava flows.

The main objectives of this paper are to describe the morphology and internal structures of the CAMP basalt lava flows of the Argana Basin, and to define lava flow emplacement processes and the formation of associated structures.

Geodynamic context

Pangean intracontinental rifting, that triggered the opening of the Central Atlantic Ocean, commenced in the Late Permian–Early Triassic (Ruellan 1985; Ruellan *et al.* 1985; Medina 1995, 2000; El Arabi *et al.* 2006) and progressed northwards, following the direction of the late Palaeozoic Alleghenian–Hercynian orogenic chain. The oldest identified magnetic anomaly on conjugate margins from the Central Atlantic have a Mid-Jurassic age (Klitgord & Schouten 1986), but recent reconstructions of the opening of the Central Atlantic Ocean (Sahabi *et al.* 2004) place the age of the earliest oceanic crust at the end of the Sinemurian (196.5–189.6 Ma).

Pangean rifting was accompanied by emplacement of the CAMP (Marzoli *et al.* 1999), a LIP mainly composed of low-Ti tholeiitic basalts. CAMP magmatism is nowadays represented by remnants of intrusive (crustal underplates, layered intrusions, sills, dikes) and, less abundant, extrusive rocks (mainly lava flows) that occur in once-contiguous parts of North and South America, north-western Africa and south-western Europe (Marzoli *et al.* 1999; McHone & Puffer 2003). $^{40}Ar/^{39}Ar$ plateau ages of CAMP lava flows range from 201 to 191 Ma, with a main cluster at *c.* 198–200 Ma (Knight *et al.* 2004; Marzoli *et al.* 2004; Nomade *et al.* 2007; Verati *et al.* 2007; Jourdan *et al.* 2009). This LIP may have extended over 7×10^6 km^2, with a total volume of magma estimated at 2–4×10^6 km^3 (Marzoli *et al.* 1999; Olsen 1999; McHone 2000). The Central Atlantic LIP has variably been considered the result of a mantle superplume (e.g. Hill 1991; Oyarzun *et al.* 1997; Wilson 1997), of lithosphere extension and thinning triggering decompressional melting in the upper asthenosphere (Withjack *et al.* 1998; Medina 2000) or of heat incubation below the Pangea supercontinent (Oyarzun *et al.* 1999; Coltice *et al.* 2007). CAMP magmatism is coeval with the Triassic–Jurassic boundary (e.g. Marzoli *et al.* 2004; Cirilli *et al.* 2009; Deenen *et al.* 2010, 2011; Whiteside *et al.* 2010) and probably triggered the end-Triassic mass extinction (Beerling & Berner 2002; Van de Schootbrugge *et al.* 2009).

Geological setting

The Triassic–Jurassic Argana Basin is located within the western High Atlas, 30 km to the NE of Agadir. It is exposed within a 10–25 km wide corridor, which extends over 85 km from Ameskroud in the south to Imi n'Tanout in the north (Fig. 1). The corridor is bounded to the east by the high

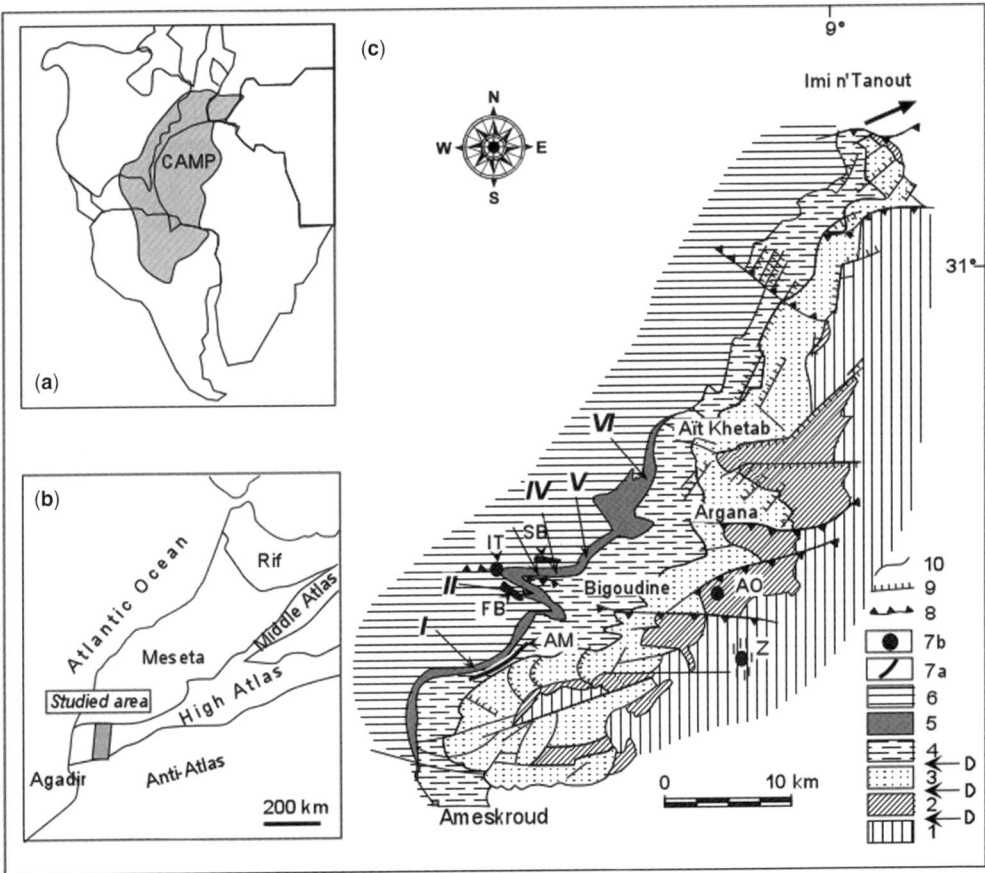

Fig. 1. (**a**) Reconstruction of Africa–South America–North America, Greenland and Europe at time of CAMP emplacement and schematic extent of the CAMP LIP; (**b**) geographical location of the Argana Basin in the northern part of Morocco and (**c**) simplified geological map of the Argana Basin and its surroundings (Tixeront 1973; Brown 1980; Medina 1991, 1994; Aït Chayeb *et al.* 1998). (1, Palaeozoic basement; 2, Ikakern Formation; 3, Timezgadiwine Formation; 4, Bigoudine Formation; 5, Triassic–Jurassic CAMP Basalts; 6, Mesozoic and Tertiary terranes; 7, Jurassic–Cretaceous mafic magmatism; 7a, dykes and sills; 7b, necks (AM, Amelal sill; AO, Aguerd Ouaoudid neck; IT, Imi n'Tiguirt neck; FB, lava flows of the Bigoudine area; SB, sills of the Bigoudine area; Z, necks and dykes of Zerhenrhine); 8, reverse fault; 9, normal fault; 10, stratigraphic contact; I, II, III, IV, V & VI, location of sections through the CAMP volcanic pile; D, angular unconformity.)

reliefs of the Palaeozoic terranes (Bloc Ancien du Haut-Atlas or Old High Atlas Block) and to the west by the Ida-Ou-Bouzia and Ida-Ou-Tanane plateaus of Jurassic and Cretaceous formations. The 2500–5000 m thick Permian–Triassic series consists of alluvial, fluvial, lacustrine, aeolian and playa deposits that accumulated in a large westwards-tilted half-graben during the initial rifting phases of the Central Atlantic Ocean (Manspeizer 1988; Hoffman *et al.* 2000; Youbi *et al.* 2003; Zühlke *et al.* 2004).

Tixeront (1973) proposed a subdivision of the succession in eight lithostratigraphical units (T1–T8) which Brown (1980) and other authors (Medina 1991, 1994; Hoffman *et al.* 2000; Olsen *et al.* 2000) formally assigned to three formations (Fig. 2). From base to top, these are: the Ikakern Formation (T1–T2, 900–1800 m), the Timezgadiouine Formation (T3–T5; 1000–2000 m) and the Bigoudine Formation (T6–T8; 300–1500 m). Each of these is separated from older strata by either an angular or erosional unconformity. The basal Ikakern Formation consists of alluvial fan conglomerates (T1 or Ait Driss Member) grading vertically and laterally into alluvial plain conglomerates, sandstones and mudstones (T2 or Tourbihine Member). The Timezgadiouine Formation is represented by braided river deposits (T3 or Tanamert Member), playa mudstones with intercalations of sheet flood and ephemeral stream sandstones and locally carbonates (T4 or Aglagal Member) and alluvial plain sandstones and mudstones (T5 or Irohalen Member). The upper Bigoudine Formation starts with braided river conglomerates and aeolian sandstones (T6 or Tadrart Ouadou Member) that grade into increasingly fine-grained partially evaporitic playa red-beds (T7 or Sidi Mansour Member; T8 or Ait Hasseine Member). The T6, T7 and T8 members delivered a microflora (pollen and spores) indicating Upper Carnian–Hettangian age (Fowell *et al.* 1996; Lund 1996; Olsen *et al.* 2000; Tourani *et al.* 2000; Marzoli *et al.* 2004; Whiteside *et al.* 2007). The members of Bigoudine Formation are overlain by Jurassic deposits, consisting of limestones and siliciclastic rocks (Bouaouda 2007).

On the western border of the Argana Basin, the upper part of the Bigoudine Formation is capped by a sequence of lava flows that crops out along a narrow, NE–SW trending band. This sequence is part of the CAMP (Fig. 3a, b) and consists of two main units, the Tasguint Formation (63–137 m thick) at the base and the Alemzi Formation (up to 60 m thick) at the top. These two units are separated by a 0.5–3 m thick silty/sandy layer or a clayey palaeosol (Aït Chayeb *et al.* 1998).

The lava flows display a mineralogical composition typical of tholeiitic basalts: plagioclase, clinopyroxene, rare olivine (more common in the lowermost lava flow from the Tasguint Formation) and Fe–Ti oxides. Strong alteration is evidenced by chlorite, zeolite, amorphous silica and calcite often found filling vesicles or substituting for the mesostasis.

The basalts from the two formations yield distinct geochemical compositions (Table 1; Fig. 4). Their major and trace element concentrations and ratios match the composition of the Lower and Intermediate Formations from the Central High Atlas, respectively (Bertrand *et al.* 1982; Aït Chayeb *et al.* 1998; Youbi *et al.* 2003; Marzoli *et al.* 2004; Deenen *et al.* 2010). The Upper and Recurrent Formations, which complete the volcanic sequence in the Central High Atlas, are absent in the Argana Basin.

Argana CAMP basalts are all characterized by normal polarity, but a reversed polarity interval has been described for the infra-basaltic red-beds in this basin (Ruiz-Martinez *et al.* 2007; Deenen *et al.* 2010, 2011) while two very thin zones of reverse polarity occur within the lower half of the Intermediate Formation of the central High Atlas. Palaeomagnetic data suggest that Central High Atlas basalts were erupted as 5 or 6 eruptive pulses, each lasting a few hundred years (Knight *et al.* 2004). Deenen *et al.* (2010, 2011) propose a duration of *c.* 20 ka for the emplacement of the Lower unit at Argana, based on combined cyclostratigraphic and palaeomagnetic correlations with the astronomically dated Newark Basin (Kent & Olsen 1999).

Terminology and methodology used in the present study

Traditionally, basaltic lava flows have been subdivided into pillow, pāhoehoe (pahoehoe), and ʿaʿā (aa) flows (e.g. Macdonald 1953, 1967). This morphological division is important because (i) the mode of lava emplacement for aa and pahoehoe flows is fundamentally different and (ii) despite similar emplacement mechanisms for pahoehoe and pillow lavas, they form in markedly different environments.

Aa flows move like the treads on a bulldozer and are typified by a thermally inefficient mode of emplacement in open channels where they disrupt and mix their upper crusts. Aa flows are characterized by angular, spinose clinkers at both the flow tops and bottoms. Aa clinker primarily forms by the rupture of molten lava at the flow surface. This can happen only when strain rates and/or the viscosity of the fluid lava are high (Peterson & Tilling 1980). This tearing is a result of the non-Newtonian behaviour of lava under these conditions. Large

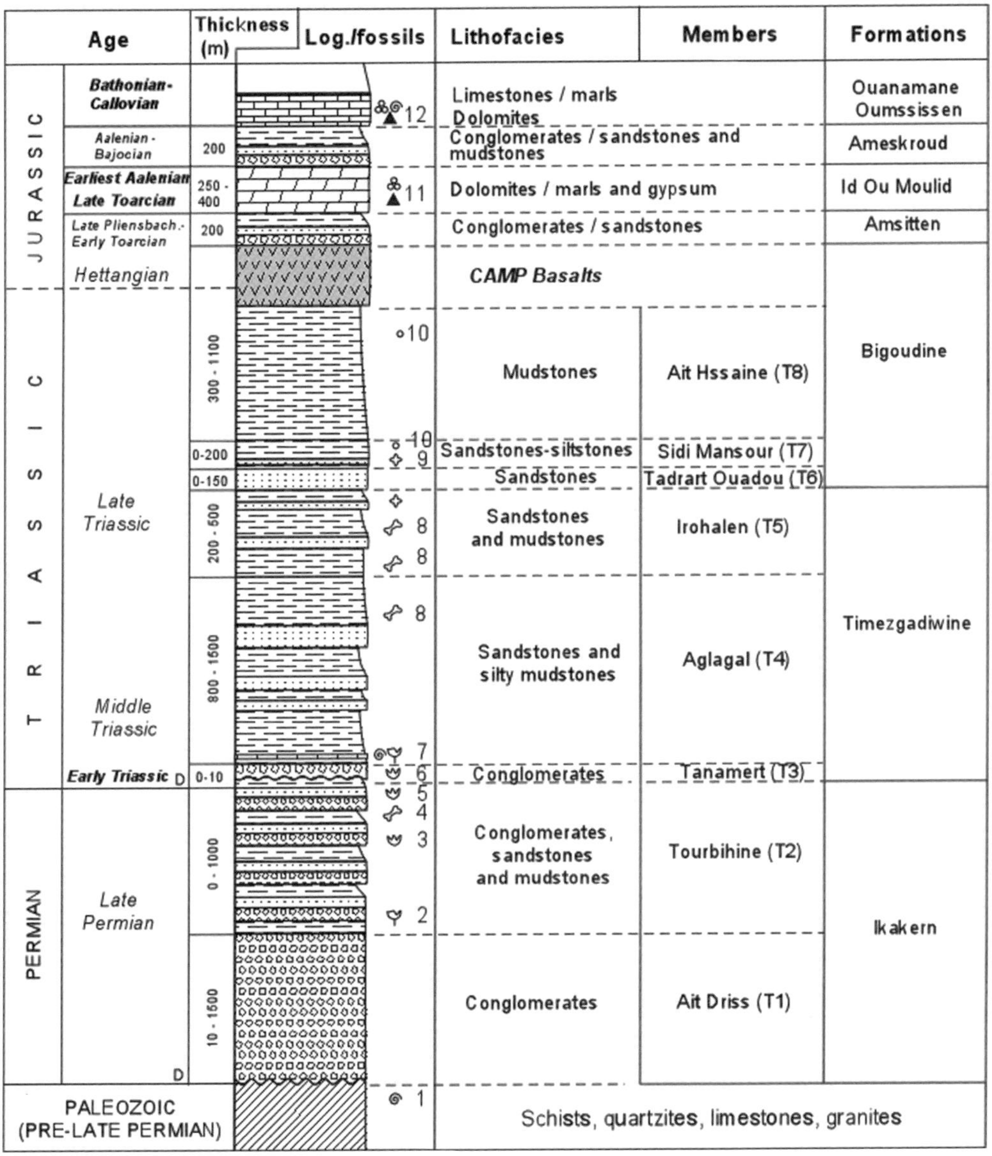

Age		Thickness (m)	Log./fossils	Lithofacies	Members	Formations
JURASSIC	Bathonian-Callovian		12	Limestones / marls Dolomites		Ouanamane Oumssissen
	Aalenian-Bajocian	200		Conglomerates / sandstones and mudstones		Ameskroud
	Earliest Aalenian Late Toarcian	250 - 400	11	Dolomites / marls and gypsum		Id Ou Moulid
	Late Pliensbach.- Early Toarcian	200		Conglomerates / sandstones		Amsitten
	Hettangian			CAMP Basalts		
TRIASSIC	Late Triassic	300 - 1100	○10	Mudstones	Ait Hssaine (T8)	Bigoudine
		0-200	10 / 9	Sandstones-siltstones	Sidi Mansour (T7)	
		0-150		Sandstones	Tadrart Ouadou (T6)	Timezgadiwine
		200 - 500	8 / 8	Sandstones and mudstones	Irohalen (T5)	
	Middle Triassic	800 - 1500	8	Sandstones and silty mudstones	Aglagal (T4)	
	Early Triassic	0-10	7 / 6 / 5 / 4	Conglomerates	Tanamert (T3)	
PERMIAN	Late Permian	0 - 1000	3 / 2	Conglomerates, sandstones and mudstones	Tourbihine (T2)	Ikakern
		10 - 1500		Conglomerates	Ait Driss (T1)	
PALEOZOIC (PRE-LATE PERMIAN)			1	Schists, quartzites, limestones, granites		

Fig. 2. Synoptic stratigraphic column of the formations of the Argana Basin (Tixeront 1973; Brown 1980; Medina 1991; Aït Chayeb *et al.* 1998; Olsen *et al.* 2000; Bouaouda 2007). 1, plants (Jongmans 1950; Broutin *et al.* 1989), pollen (Aassoumi *et al.* 2003, 2006) and insects (Hmich *et al.* 2003); 2, plant specimens of *Voltzia heterophylla* (De Koning 1957) (not confirmed since); 3, reptile footprints of *Rhynchosauroides* sp. (Jones 1975; Brown 1980); 4, vertebrate fauna (Diplocaulidae, Captorhinidae and Pareiasaurs) (Dutuit 1976; Jalil &Dutuit 1996; Jalil & Janvier 2005); 5, reptile footprints of *Hyloidichnus* and *Pachypes* (Voigt *et al.* 2010); 6, reptile footprints of *Protochirotherium, Synaptichnium, Chirotherium, Brachychirotherium, Isochirotherium,* cf. *Rhynchosauroides* (Klein *et al.* 2010; Tourani *et al.* 2010); 7, lacustrine ostracods (*Darwinula*) and Charophyta (Porocharacea) *Stellatochara bulgarica* and *Maslovichara* (Medina *et al.* 2001); 8, vertebrate fauna (*Cyclotosaurus*, Metoposauridae, *Palaeorhinus* and *Metoposaurus*) (Dutuit 1976; Lucas 1998a, b); 9, *Estheria minuta* (Defretin & Fauvelet 1951; Brown 1980); 10, palynomorphs (Fowell *et al.* 1996; Lund 1996; Olsen *et al.* 2000; Tourani *et al.* 2000; Marzoli *et al.* 2004; Whiteside *et al.* 2007); 11, Foraminifera and Brachiopods (Ambroggi 1963; Bouaouda 1987); 12, Foraminifera, Brachiopods and Ammonites (Ambroggi 1963; Bouaouda 1987).

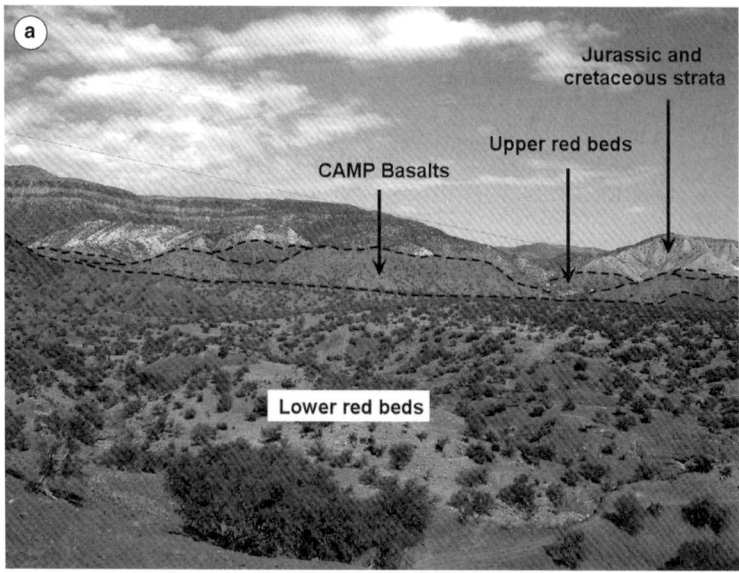

Fig. 3. Some field aspects of the Argana Basin: (**a**) overview of the Argana Basin and (**b**) overview of the CAMP volcanic pile. Photos taken near the town of Bigoudine from the road Argana–Agadir.

(10–200 cm wide) 'fingers' of lava from the massive interior of the flow commonly extend into the breccia, especially along well-defined shear zones (Lockwood & Lipman 1980). Pahoehoe and pillow lava flows are characterized by insulating transport and growth by sequential lobe-by-lobe emplacement. The lava is transported in internal pathways (lava tubes) to the active flow fronts, where they advance by inflating a lobe with a continuous crust (Walker 1991; Hon *et al.* 1994).

A wide range of intermediate flow types occur between these two end-member types. Transitional flows show some of the characteristics of both aa and pahoehoe lava flows. Most subaerial transitional flow types have 'pahoehoe' in their names: for example rubbly pahoehoe, slab pahoehoe and toothpaste pahoehoe (Macdonald 1953, 1967; Rowland & Walker 1987; Keszthelyi & Thordarson 2000; Keszthelyi *et al.* 2006), but some could be considered closer in character to aa than to pahoehoe

Table 1. *Representative major and trace element analyses of CAMP basalts from the Argana Basin*

Formation	Lower formation (Tasguint Formation)							Intermediate formation (Alemzi Formation)					Average (lower formation) (n = 15)	Average (intermediate formation) (n = 9)
Locality	Aguersouane	Alemzi North	Alemzi South	Alemzi South	Alemzi South	Alemzi North	Tasguint	Alemzi North	Alemzi North	Tasguint	Tasguint	Alemzi South		
Sample	AN 132	B3	YAR1.2	YAR3.2	YAR6.2	AN 102	AN 130	AN 123	AN 125	AN 128	AN 129	AN 121		
Major elements (wt%)														
SiO_2 (%)	48.74	52.50	52	51.8	51.2	52.06	54.76	53.08	52.72	52.94	52.95	52.61	51.48	52.45
TiO_2	0.94	1.27	1.36	1.28	1.4	1.31	1.44	1.24	1.19	1.26	1.23	1.09	1.25	1.17
Al_2O_3	9.89	13.25	13.35	13.45	13.65	13.89	13.84	14.21	14.2	14.38	14.27	13.63	13.06	14.02
Fe_2O_{3t}		10.52	10.25	9.95	10.55								11.07	10.82
FeO	14.15					11.53	10.05	11.55	11.72	11.43	11.7	11	11.31	11.51
MnO	0.21	0.15	0.12	0.14	0.16	0.17	0.18	0.16	0.16	0.16	0.16	0.14	0.15	0.17
MgO	16.5	8.00	7.5	7.2	7.15	8.35	6.67	6.6	7.46	7.26	6.59	7.81	9.03	7.35
CaO	7.49	9.15	6.62	7.3	7	9.33	9.51	9.11	7.66	9.94	9.3	10.51	7.65	9.23
Na_2O	1.32	1.95	3.8	3.68	3.65	2.14	2.31	2.51	3.07	1.96	2.48	2.13	2.92	2.35
K_2O	0.64	1.15	2.06	2.08	1.98	1.07	1.07	1.4	1.67	0.51	1.17	0.94	1.48	1.25
P_2O_5	0.12	0.17	0.17	0.16	0.17	0.16	0.17	0.15	0.14	0.15	0.15	0.15	0.15	0.14
LOI	1.74	1.71	2.79	2.62	2.95	1.77	1.53	1.19	2.8	1.84	0.99	1.57	2.40	1.64
Trace elements (ppm)														
Sc (ppm)		32	32	33.5	32.5			36	41		34	38	33.09	35.93
V		270	285	277	288			261	272		265	224	267	257.62
Cr		420	300	282	250			178	196		187	334	339.90	249.62
Co	85.23	47	41	41	39	44.59	36.01	40	43	39.49	40	44	48.90	42.27
Ni		110	85	82	72			67	61		68	110	130.18	85.87
Rb	21.98	28	37	37.5	22.5	14.17	16.48	28	30	10.99	28	28	26.00	27.77
Sr	159.79	225	222	260	354	189.06	162.34	204	213	248.93	213	161	232.87	206.54
Y	18.71	23.5	23.5	22.5	24.5	23.23	20.53	26	23	23.05	25	26	22.61	24.11
Zr	96.70	118	110	110	121	125.87	134.71	109	92	109.28	105	102	111.91	99.03
Nb	8.65	11.5	11.5	10.4	12.4	11.69	11.53	9	9	7.68	9	9	10.94	8.43
Cs	1.82					0.45	0.50			6.68			0.84	6.68
Ba	141.70	220	276	300	800	193.63	189.18	250	270	109.94	254	210	287.40	227.99
La	10.61	16.3	14.5	15.4	17	15.63	13.93	21	11	10.55	9	19	14.16	15.31
Ce	23.75	33.5	32	31.5	37.5	32.47	31.72	31	35	23.63	25	37	30.19	29.45
Pr	3.01					4.07	3.95			3.07			3.67	3.07
Nd	12.60	19	19	18.5	21.5	16.74	16.72	16	18	13.03	13	19	16.61	15.61
Sm	3.13					3.84	4.07			3.25			3.60	3.25
Eu	0.91	1.35	1.33	1.33	1.42	1.27	1.17		18	1.04			1.21	1.11

(*Continued*)

Table 1. *Continued*

Formation	Lower formation (Tasguint Formation)							Intermediate formation (Alemzi Formation)					Average (lower formation) (n = 15)	Average (intermediate formation) (n = 9)
Locality	Aguersouane	Alemzi North	Alemzi South	Alemzi South	Alemzi South	Alemzi North	Tasguint	Alemzi North	Alemzi North	Tasguint	Tasguint	Alemzi South		
Sample	AN 132	B3	YAR1.2	YAR3.2	YAR6.2	AN 102	AN 130	AN 123	AN 125	AN 128	AN 129	AN 121		
Gd	3.28					4.17	4.25			3.71			3.72	3.71
Tb	0.52					0.65	0.68			0.61			0.63	0.61
Dy	3.27	4.35	4.35	4.2	4.65	3.79	3.94			3.69			3.93	3.94
Ho	0.64					0.80	0.85			0.79			0.77	0.79
Er	1.78	2.3	2.4	2.4	2.6	2.19	2.37			2.26			2.20	2.28
Tm						0.24	0.23			0.21			0.24	0.21
Yb	1.63	2.1	2.08	2.06	2.25	1.86	2.02			2.02			1.95	2.08
Lu	0.22					0.27	0.28			0.29			0.26	0.29
Hf	2.39					3.05	3.45	2	1	2.72	8	3	3.85	4.21
Ta	0.54					0.70	0.75			0.50			0.67	0.50
Pb	4.81					1.86	15.88	12	12	3.12	13	17	9.58	12.76
Th	2.50	3.52	4.72	2.16	0.42	3.18	2.84	6	5	2.34	6	6	3.89	4.92
U	0.55	2.36	1.51	1.11	2.85	0.64	0.71	3	2	0.48	4	2	1.50	2.22

Major and trace element data were obtained by Inductively Coupled Plasma-Atomic Emission Spectrometry (ICP-AES). The samples were powdered in an agate grinder. International standards were used for calibrations tests (ACE, BEN, JB-2, PM-S and WS-E). Rb was measured by flame emission spectroscopy. Relative standard deviations are $\pm 1\%$ for SiO_2 and $\pm 2\%$ for other major elements except P_2O_5 and MnO ($\pm 0.01\%$), and c. 5% for trace elements. The analytical techniques are described in detail by Cotten *et al.* (1995) (LOI: loss on ignition determined by weight loss after igniting sample at 1000 °C).

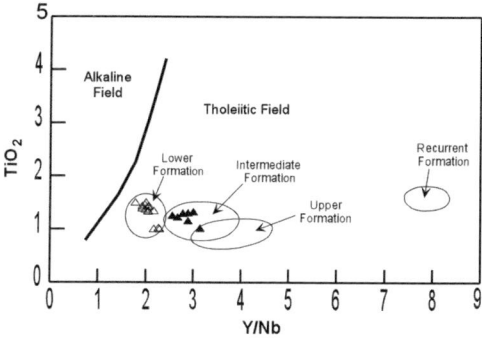

Fig. 4. TiO$_2$ v. Y/Nb plot (Floyd & Winchester 1975) showing the tholeiitic affinity of CAMP basalts from the Argana Basin. Open and filled triangles represent the basalts of the lower and intermediate formations of Argana Basin, respectively. The encircled areas represent the average compositions of the Lower, Intermediate, Upper and Recurrent Formations of the High Atlas, where the CAMP volcanic succession is complete (fields from Marzoli *et al.* 2004).

(such as flows with a breccia top). However, all transitional flows are more akin to typical pahoehoe than to aa.

The terminology for the morphological description of the basalt flows adopted in this study is that of Self *et al.* (1997, 1998) and Thordarson & Self (1998), based on three criteria (Fig. 5): (i) the vesiculation pattern, which is defined by distribution, mode, shape and size of vesicles and other degassing features; (ii) the jointing style, which refers to the arrangement and morphology of cooling joints and columns; and (iii) the petrographic texture which refers to crystallinity and crystal size, properties that are controlled by a number of parameters such as volatile content, cooling rate and crystal nucleation rate.

This terminology has been used by Self and co-workers and subsequently by other authors for the description of the 1783–1784 historical flow of the Laki eruption, Iceland (Guilbaud *et al.* 2005) and several LIPs basalt flows (Kent *et al.* 1998;

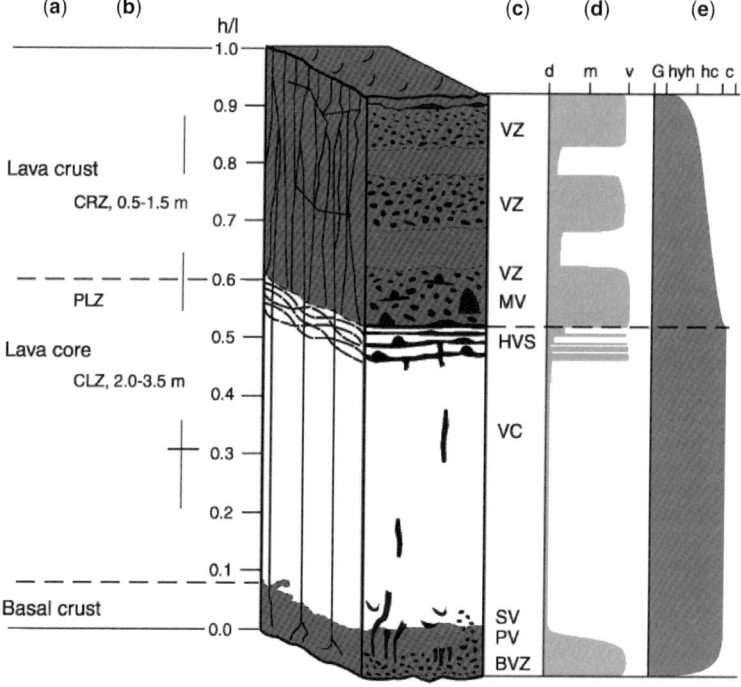

Fig. 5. Composite graphic log illustrating characteristic structures of pahoehoe sheet lobes. The left side of the column shows the characteristic three-part division of (**a**) sheet lobes (CRZ, crustal zone; PLZ, platy zone; CLZ, columnar zone) and (**b**) jointing styles. The right side of the column shows distribution of (**c**) vesiculation structures (VZ, vesicular zone; MV, megavesicle; HVS, horizontal vesicle sheet; VC, vesicle cylinder; SV, segregation vesicle; PV, pipe vesicle; BVZ, basal vesicular zone), (**d**) vesiculation (non- to sparsely vesicular $d = 0$–5 vol%, moderately vesicular $m = 10$–20 vol% and vesicular $v = 30$–40 vol%) and (**e**) degree of crystallinity (G, hyaline; hyh, hypohyaline; hc, hypocrystalline; c, holocrystalline). The scale h/l indicates normalized height above the base of the sheet lobe (h, height in lobe; l, total lobe thickness) (from Thordarson & Self 1998).

Jerram *et al.* 1999*a*, *b*; Keszthelyi *et al.* 1999; Bondre *et al.* 2004*a*, *b*; Single & Jerram 2004; El Hachimi *et al.* 2005, 2007; Waichel *et al.* 2006; Passey & Bell 2007; Bondre & Hart 2008; Duraiswami *et al.* 2008; Kontak 2008; Martins *et al.* 2008).

The products of an effusive eruption are divided into three hierarchic levels: flow lobe, lava flow and lava flow field (Self *et al.* 1997), described in the following:

(i) A *flow lobe* is used to describe an individual package of lava surrounded by a chilled crust. Lobes can be classified (Wilmoth & Walker 1993) as either S-type (spongy) or as P-type (pipe vesicle bearing). S-type lobes lack pipe vesicles and are vesicular throughout their thickness. P-type lobes are characterized by pipe vesicles and display a typical internal structure with a vesicular base and top and a relatively vesicle poor core.

(ii) Individual *lava flows* are the product of a single, more or less continuous outpouring of lava that can be composed of one or more flow lobes. They are typically separated by weathering surfaces and/or clastic lithologies. If a lava flow consists of a single flow lobe it is referred to as a *simple lava flow* (Walker 1971); if the lobe has a sheet-like or tabular geometry it is classified as a *sheet lobe*. Conversely, a *compound lava flow* (Walker 1971) is made up of two or more flow lobes of any geometry or size.

(iii) A *flow field* is the aggregate product of a single eruption or vent, comprises one or more lava flows and is usually identified on the basis of mineralogy and geochemistry of the constituent flows and/or notable weathering of the uppermost lava surface.

Observations

Six detailed sections were investigated from SW to NE across the volcanic pile cropping out in the western flank of the Argana Basin (Figs 1 & 6).

The volcanic succession is subdivided in two formations. The lower Tasguint Formation (63–137 m thick) consists of 3 (section V – Aguersouane) to 13 (section III – Tasguint) individual lava flows. The upper Alemzi Formation (up to 80 m thick) generally consists of one or two flows (sections III – Tasguint and VI – Igounan), but exceptionally (section IV – Alemzi South) 7 lava flows were recognized. These formations are separated by a thin silty or sandy horizon, by slightly–strongly weathered surfaces or by red soil up to 1 m thick (Fig. 7a, b). The soil is structureless or poorly bedded and usually contains rounded, volcanic clasts embedded in a clay- to silt-sized matrix.

According to Walker's (1971) nomenclature, the CAMP flows of the Argana Basin can be grouped into compound pahoehoe flows and simple flows (Fig. 7c, d). However, it is not rare to observe lava flows of intermediate type between pahoehoe and aa such as those with a 'flow top breccia', according to Keszthelyi's (2002) classification (e.g. Alemzi, Aguersouane and Igounan sections).

Compound pahoehoe flows

In the Tasguint Formation, the thickness of lava flows ranges from 4 to 50 m while lateral extent can exceed 100 m for each sheet lobe (Table 2). However, several small lobes have a more limited lateral extent (10 m) (Table 3). The largest lobes (as well as small lobes forming compound pahoehoe flows) are typically characterized by the threefold structure comprising a thin vesicular basal crust, a dense core and a thick vesicular upper crust (Aubele *et al.* 1988; Thordarson & Self 1998). Flow lobe tops often show oxidized rinds which may develop into palaeosols, whereas centimetre-scale silica-filled pipe vesicles are rarely present at the base (Fig. 8a). In the CAMP basaltic pile of Argana Basin, P-type lobes are more common than S-type lobes.

In the Tizi el Hajaj, Alemzi North, Alemzi South and Igounan sections, some flow units are composed of a pile (2–30 m thick) of small coalescent pahoehoe lobes and toes. These flow units are usually found above compound pahoehoe flows formed by large lobes, although locally (Igounan section) they may occur beneath.

Basal crust. The thickness of the basal crust (20–40 cm) is always small with respect to the thickness of the lobe (except in very small lobes). It is relatively less vesicular and exhibits a typical hypocrystalline texture with fine Fe–Ti oxide minerals.

In places, the basal crust shows centimetre-scale cryptocrystalline silica-filled pipe vesicles. The pipe axes are sometimes inclined, thus recording the lava movement and are reliable markers of the flow direction (e.g. Walker 1987). Pipe vesicles are not always restricted to the base but sometimes extend well into the lower part of the lava core. The geometry of the contact with the underlying lobes depends primarily on the topography on which the lobe was emplaced. This is underlined by the fact that in two superimposed lobes the pipe vesicles may be inclined in completely different directions, marking local flow directions. Larger sheet lobes are usually characterized by planar bases; however, the contact can be undulating if they are underlain by hummocky-surfaced flows.

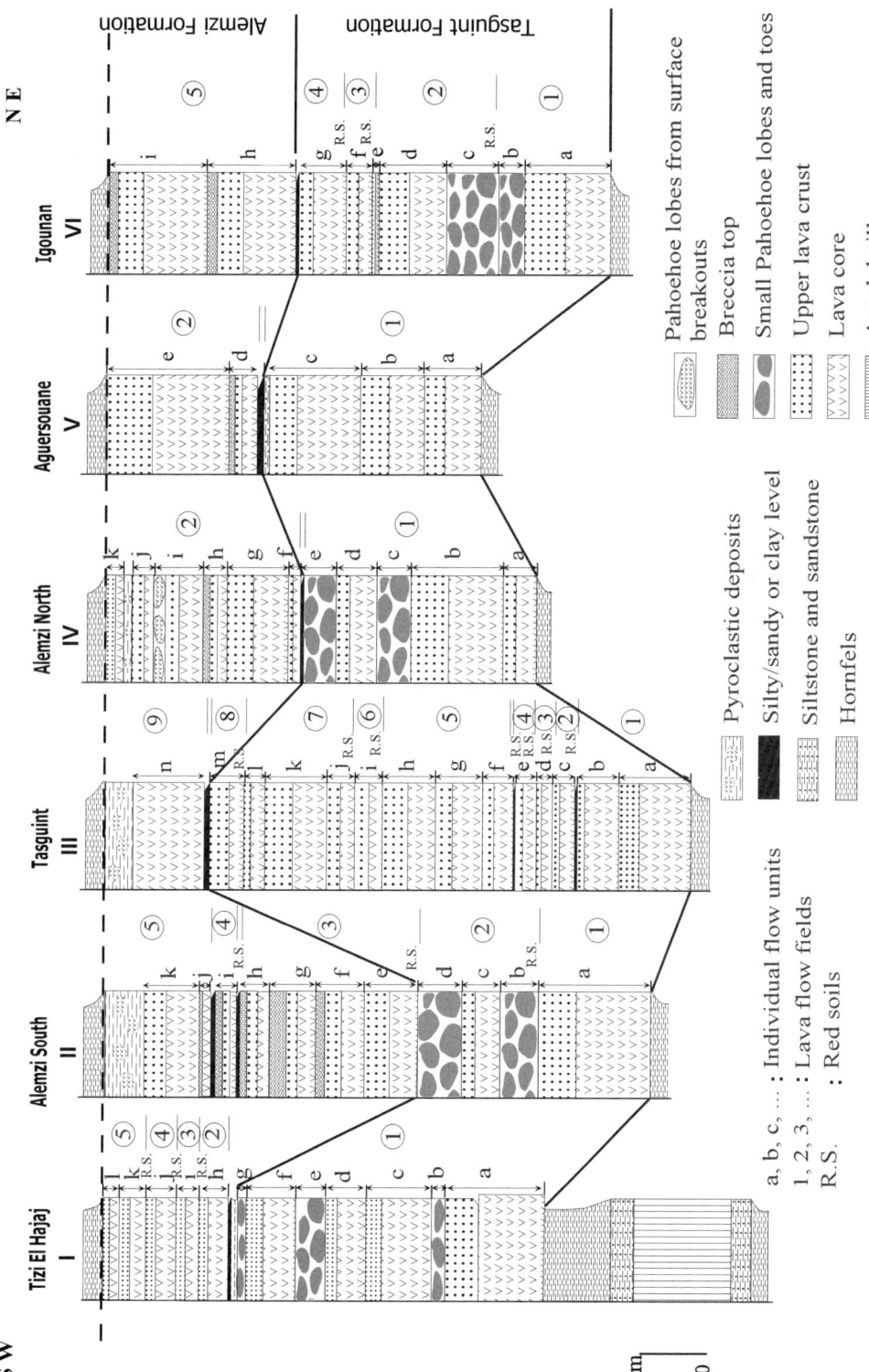

Fig. 6. Lithostratigraphic columns across the CAMP volcanic succession of Argana Basin. See Figure 1 for location of sections.

Fig. 7. (**a**) Sedimentary layer at Tizi el Hajaj section separating the Tasguint and Alemzi formations; (**b**) red soil mantle, up to 1 m thick, developed on flow lobe c of the Tasguint section; (**c**) image illustrating overlapping lobes of a compound pahoehoe flow of the Tasguint Formation (this photo corresponds to a detail of the lower-left corner of photo d); (**d**) simple flows of Alemzi Formation (photos c and d taken near Alemzi North section; LC, lava core; UC, upper crust). See Figures 1 and 4 for location.

Lava core. The lava core is dense and sparsely vesicular and generally represents 40–80% of the lobe thickness. It usually displays ophitic or subophitic textures with plagioclase, clinopyroxene, rare olivine and Fe–Ti oxides.

Characteristic structures are segregation structures, horizontal joints at the boundaries and (rarely) well-developed columnar joints. Spheroidal weathering is common. In thick lobes, the boundary between the upper crust and the core is usually sharp, marked by a change in the style of jointing and the distribution of vesicles.

Two types of segregation structures can be observed in the core: vesicle cylinders (Goff 1996) and vesicle sheets (Thordarson & Self 1998). Vesicle cylinders are observed in the lower and middle parts of the core (Fig. 8b). They are almost always rootless but, unlike the Columbia River Basalts (Thordarson & Self 1998), are not connected with pipe vesicles from the basal crust. Cylinders in thick lobes can be up to 0.80 m long. They are sometimes connected with vesicle sheets. In the core of

some flow lobes, these structures consist of essentially trails of vesicles and are not filled by segregated material. Cylinders are also often bent or at an angle to the flow-top, probably in the flow direction.

Vesicle sheets occur near the interface between the crust and the core. The sheets thickness is 5 cm in average, but can reach up to 40 cm. The vesicle sheets display irregular shapes and branching patterns, which suggests that they may have filled joints that formed at the early stages of cooling. It is also possible that they propagated by hydraulic fracturing (Walker 1993). Their sinuous nature may also indicate that the viscous lava was still mobile during later stages of cooling, which led to the deformation of the vesicle sheets.

Upper crust. The thickness of the upper crust (1–12 m) ranges from about one-quarter to one-half the thickness of the lobe, and commonly consists of a layering of vesicle-poor (dense) and vesicle-rich (vesicular) bands. The vesicles are spherical

Table 2. *Data on representative flow lobes from study areas of Argana Basin (see Fig. 4 for the location of flow lobes in lithostratigraphic columns)*

Area	Lobes	Total thickness (m)	Basal crust (m)	Core (m)	Upper crust (m)	Remarks
Tizi el Hajaj	a	30	—	20	10	Thick, extensive (exposed length >100 m) sheet lobe. Dense lava core with vesicles cylinders. Presence of several injections of baked siltstone, randomly dispersed within the lava core and the upper lava crust ('peperites'?). Multiple small P-type and S-type lobes at the top of the lava flow.
	c	20.2	0.2	15	5	Well-preserved thin basal lava crust. Dense lava core and vesicular upper crust.
	d	12	—	8	4	Dense core. Vesicular upper crust. Multiple small S-type lobes at the top of the lava flow.
Alemzi South	a	30	—	20	10	Thick and extensive sheet lobe (exposed length >150 m). Multiple S-type lobes at the top of the lava flow.
	c	11	0.37	7	3.63	Thin basal lava crust well preserved. Dense core with horizontal jointing. Vesicular crust. Multiple small P-type and S-type lobes at the top of the lava flow.
Tasguint	a	20.5	—	14.5	6	Sheet lobe with horizontal jointing at the upper part of the lava core.
	b	12	—	10	2	Dense core. Vesicular upper crust.
	c	6.5	—	4.5	2	Horizontal jointing in the upper part of the lava core.
	e	6.2	0.2	3	3	Well-preserved thin basal lava crust. The boundary between the lava core and the lava crust is not clear.
	f	9.25	0.25	6	3	Three-tired structure with thin basal lava crust, dense core and vesicular upper crust.
	g	9	—	6	3	Dense core. Vesicular upper crust.
	h	13.75	0.25	8.5	5	Three-tired structure: thin basal lava crust, dense core and vesicular upper crust. Presence of pipe vesicles.
	i	15	—	8	7	Well-preserved thin basal lava crust. Horizontal jointing in lava core.
	m	8	—	4	4	Dense core and vesicular upper crust.
Alemzi North	a	9.5	—	6	3.5	Base not well exposed. Important vesiculation at the upper part of lava core.
	b	27.1	—	16	11.1	Multiple small P-type and S-type lobes at the top of the lava flow. Existence of squeeze-ups, and horizontal squeezes.
	d	12	—	8	4	Multiple small P-type and S-type lobes at the top of the lava flow.
Aguersouane	a	16	—	10	6	Dense core, and vesicular upper crust.
	b	18	—	10	8	Vesicular core and upper crust.
	c	26	—	18	8	Dense core with horizontal jointing.
Igounan	a	25	—	13	12	Thick, extensive sheet lobe. Presence of vesicles cylinders and horizontal vesicles sheets in the lava core. Multiple P-type lobes at the top of the lava flow.
	d	20	—	11	9	Dense core and vesicular upper crust.
	g	14	—	10	4	Dense core and vesicular upper crust.

Table 3. *Dimensions of smaller lobes from Alemzi North section (total thickness of the Alemzi North section is 125 m; the Tasguint Formation in this section is 70 m thick and consists of 5 lobes; the preponderance of relatively small S-type lobes is evident)*

Length of small pahoehoe lobes (m)	Thickness of lobe (m)	Thickness of basal crust (m)[1]	Thickness of upper crust (m)[1]	Description	Lobe type
0.57	0.2	0.02	0.05	The size of vesicles decreases towards the periphery of the lobe	S-type lobe
3.3	0.4	0.04	0.06		S-type lobe
2.2	0.4	0.04	0.08		S-type lobe
0.53	0.2	0.04	0.05		S-type lobe
2	0.32	0.05	0.08		S-type lobe
0.6	0.15	0.02	0.05		S-type lobe
0.65	0.18	0.02	0.03		S-type lobe
1.3	0.37	0.04	0.06	Highly undulating lower surface	S-type lobe
3.1	0.5	0.04	0.09		P-type lobe
0.8	0.12	0.01	0.03		P-type lobe
1.1	0.17	0.01	0.02		P-type lobe
1.8	0.5	0.02	0.08		S-type lobe
0.77	0.16	0.02	0.03	Pipes are present at the distal end of the lobe	S-type lobe
1.2	0.1	0.03	0.03		S-type lobe
0.8	0.27	0.03	0.05	Pipes are present at the distal end of the lobe	S-type lobe
1.6	0.27	0.03	0.04		S-type lobe
1.16	0.37	0.03	0.03		S-type lobe
4.37	0.5	0.08	0.15	Associated with some P-type break-outs	P-type lobe
0.6	0.2	0.03	0.04		S-type lobe
2	0.23	0.02	0.03		S-type lobe

[1]Refers to thickness of glassy rinds in S-type lobes.

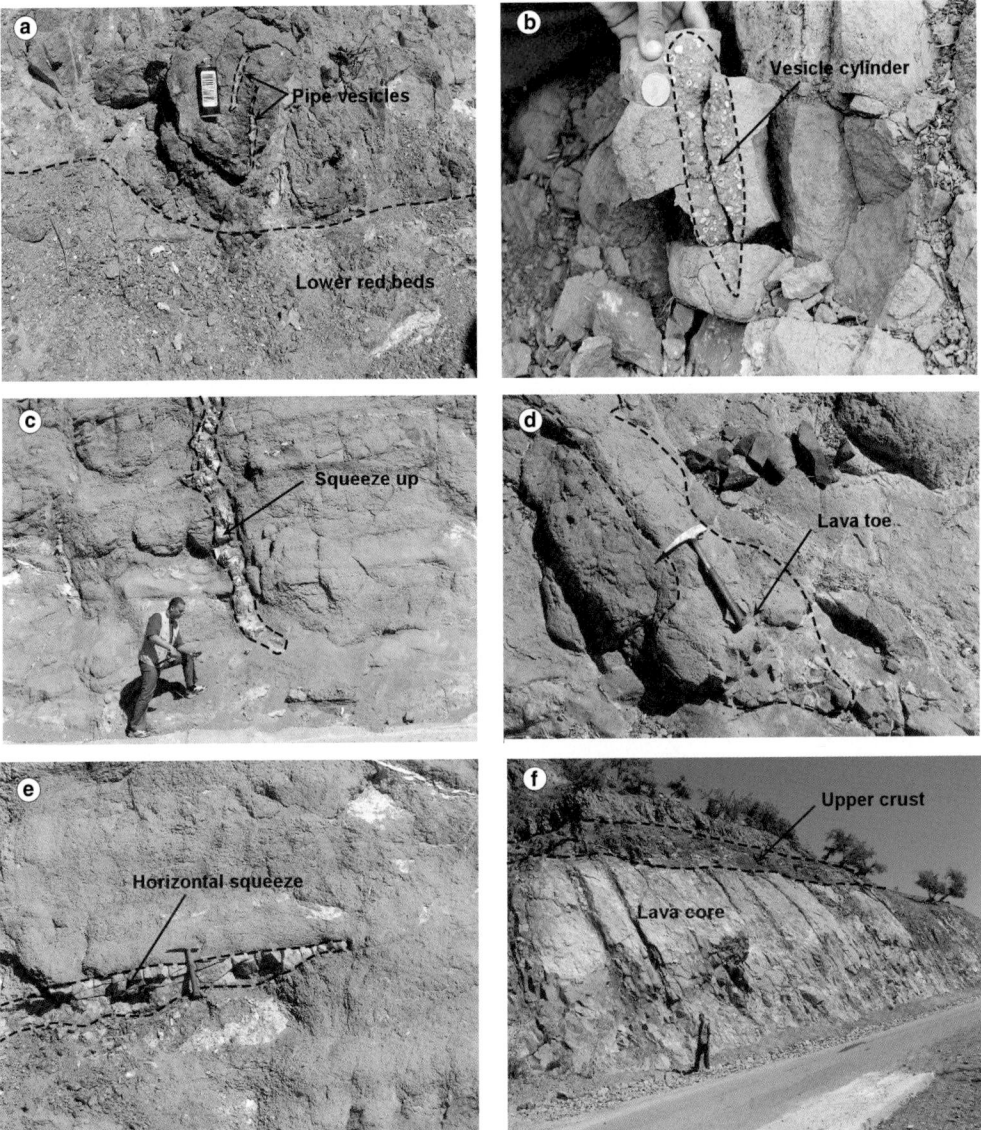

Fig. 8. (**a**) Well-preserved pipe vesicles rising up to 5 cm from the base of flow lobe 'a' from Alemzi North section; (**b**) vesicle cylinder in flow lobe 'c' from Tasguint section; (**c**) squeeze-up in a flow lobe at Alemzi North section; (**d**) small pahoehoe toe formed by extruding squeeze-up at the same locality; (**e**) horizontal squeeze at the same locality; (**f**) lava core and upper crust of a simple flow from Alemzi Formation (Alemzi area). See Figures 1 and 4 for location.

or flat. Thick squeeze-ups often break through the crust (Fig. 8c) and sometimes form what seem to be small pahoehoe toes (Fig. 8d). They may occupy axial clefts of tumuli on the crusts of the lobes. Sometimes, offshoots of squeeze-ups are seen to have intruded horizontally into the crust (Fig. 8e). The upper crust exhibits a typical hypocrystalline texture with fine Fe–Ti oxide minerals.

Along the road from Argana to Imouzzer, the Tasguint Formation shows a tumulus structure *c.* 100 m long and 10 m high (latitude 30°44′34.51″N; longitude 9°15′34.86″W; WGS84). The pahoehoe flow that displays this structure shows a three-tiered structure with a thin (0.40 m) basal crust and thick core and upper crust (40 and 10 m, respectively). The latter shows a squeeze-up structure (up to 0.3 m thick, 10 m horizontal length) oriented north–south (Fig. 8c). The contact with the host lava is sharp and the shape of the squeeze-up is undulated in places. Another north–south-oriented squeeze-up (up to 1.20 m thick and a few metres long) can be seen in a marginal position. Here, the contact with the host lava is not sharp, but is emphasized by centimetre-scale silica-filled vesicle pipes which are perpendicular to the squeeze-up body. The central squeeze-up may have been extruded through the 'axial cleft' of a tumulus. Towards the top of the flow, we observed several horizontal squeeze structures (average 0.3 m thick, 2 m long), which are sometimes offset by N75E-dipping 60° NNW normal faults, contemporaneous to the volcanic activity. Because of the absence of suitable outcrop conditions to observe the three-dimensional (3D) shape of the tumulus, it is not possible to classify it according to the slope-based classification scheme of Walker (1991) and the morphology classification scheme of Duraiswami *et al.* (2001).

In the Tizi el Hajaj section, the core of the lowermost flow unit is injected with red-brown siltstone forming irregular sediments bodies (6 cm to 1 m long) randomly dispersed within the lava (Fig. 9a). These structures can also be found within the upper crust of the same lobe, but their occurrence becomes less notable. This magma/sediment mingling can be interpreted as a peperite (see Skilling *et al.* 2002 for more details). These occurrences indicate that the basal sediments were still soft or slightly consolidated at the time of emplacement of the lava flow, suggesting that the first lava flows are contemporaneous to the deposition of the Ait Hssaine Member (T8) of Bigoudine Formation. In this section, the sediment layer that fed the injections was not observed, but this relation was described in other CAMP sections both in the High Atlas and south Portugal (e.g. Martins *et al.* 2008). Similar structures were observed at the base of a large pahoehoe lobe in the western Deccan Volcanic Province (Duraiswami *et al.* 2003).

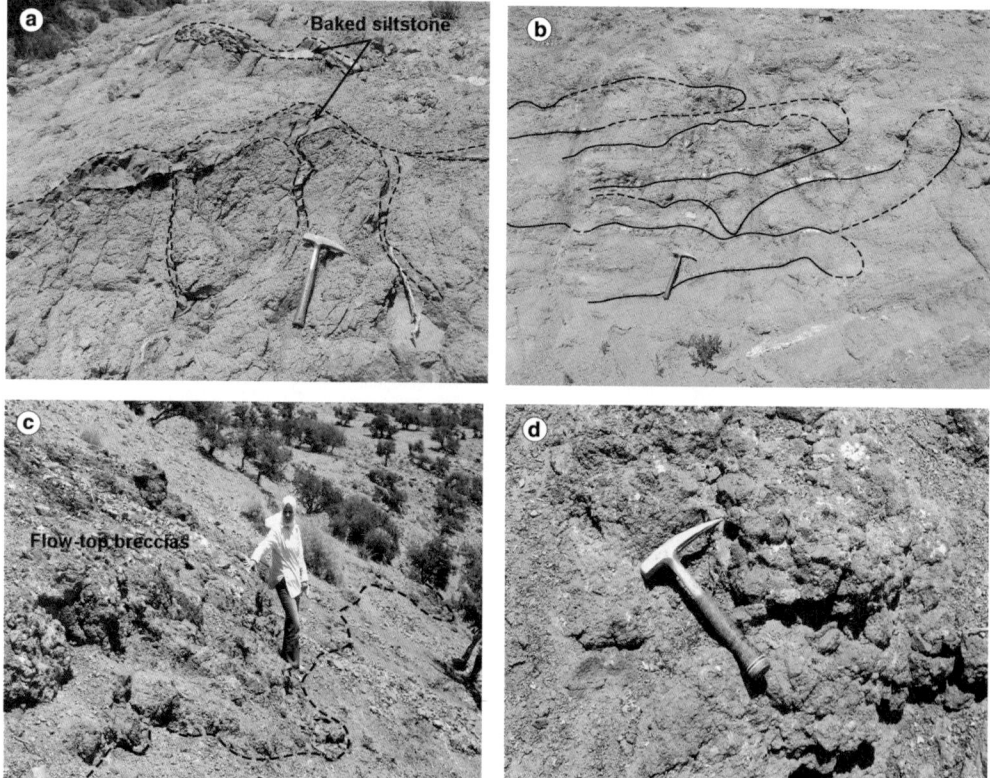

Fig. 9. (**a**) Injections of siltstone, randomly dispersed within flow lobe 'a' of Tizi el Hajaj section; (**b**) accumulation of pahoehoe lobes over lava crust (lower right) in Alemzi North area; each individual lobe is marked by a thin glassy crust; (**c**) horizon of flow-top breccia in the Aguersouane area; and (**d**) detail of flow-top breccia in the Aguersouane area. See Figures 1 and 4 for location.

These structures, called 'enigmatic spiracle-like structures' by the authors, are good indicators of the palaeotopography as they record the injection of sediments within the unconsolidated lava flow. Duraiswami *et al.* (2003) envisage that small, rain-fed puddles may have developed in shallow depressions on larger sheet lobes and between lobes in hummocky flow fields, and became clogged due to the accumulated fine-grained material. Subsequent emplacement of sheet lobes over these puddles produced the slender steam-injection structures, now spectacularly preserved as spiracles within the flow. Waichel *et al.* (2008) also observed sand-deformation features (sand diapirs and peperite-like breccia) on lava flowing over active aeolian dunes in the CFB of Paraná Basin.

Simple flows

At the Argana Basin, simple flows are 3–35 m thick with a more or less constant thickness over large distances (several hundred metres at the outcrop scale and up to a few kilometres at the regional scale). Simple flows appear as single cooling units without multiple flow lobes. Some flows present a top breccia, as can be observed in the Alemzi, Aguersouane and Igounan sections.

As for the compound pahoehoe flows type, the studied simple flows of the Argana Basin show a three-tiered structure (Aubele *et al.* 1988; Thordarson & Self 1998) with a thin basal crust, a dense lava core and an upper crust (Fig. 8f).

Basal crust. The contact of the basal crust of simple flows with underlying sediments or lava flow is sharp or slightly undulated. The lower crust of the flows is 0.26–0.4 m thick, representing 3% of the total thickness of the flow. Although vesicles may be numerous, pipe vesicles were not observed which may be related to the lava rheology and its emplacement style (Walker 1987; Bondre *et al.* 2004*a, b*).

The microscopic study of the basal crust of various flows shows that they usually display hypocrystalline textures with plagioclase, clinopyroxene and Fe–Ti oxides.

Lava core. The lava core is the thickest part of the flow, up to 22 m in the Agersouane area. It usually appears as poorly vesiculated with respect to the upper crust. Segregation structures (spherical vesicles, vesicle sheets and cylinder vesicles), which are frequently observed in the core of compound pahoehoe flows, are rarely observed here. The top of the lava core is often underlined by a dense level with horizontal jointing and silica-filled, elongated or flat vesicles. At the Igounan section, some simple flow cores are vesicle-rich making it

difficult to determine the boundary between the core and the upper crust. This aspect was also observed in one lava flow of the Columbia River CFB (McMillan *et al.* 1989). The lava core usually exhibits ophitic or subophitic textures with plagioclase, clinopyroxene and Fe–Ti oxides.

Upper crust. The upper crusts are 1–13 m thick and strongly vesiculated. Lobes, layering of vesicle-poor ('dense') and vesicle-rich ('vesicular') bands, tumuli and squeeze-up structures, which are characteristic of compound pahoehoe flows, were not seen in the upper crust of simple flows. The contact between two simple flows is clearly observable over long distances (kilometres). Microscopic study of samples collected from the upper crust reveals hypocrystalline texture with plagioclase, clinopyroxene and Fe–Ti oxides. Vesicles have diameters of 1–2 cm and are spherical or oblong. The vesicles are filled with silica or alteration minerals. Locally at Igounan section, spherical or domal vesicles show upwards-decreasing sizes.

Some flows of the Alemzi, Aguersouane and Igounan sections present a top layer formed by accumulations of pahoehoe lobes (20–30 cm thick; Fig. 9b) or breccia (Fig. 9c, d). The contour of individual lobes is marked by glassy crusts. Flow-top breccia consists of rounded, elongated and angular vesicular basalt fragments of centimetre-scale size. Fragment vesiculation is 5–10% in Alemzi area. The presence of a breccia top could suggest an aa type flow; however, the lack of a continuous envelope of breccia (top and base) excludes that interpretation.

At the top of the uppermost simple flows at various sections (Alemzi North, Alemzi South and Tasguint), the red siltstones overlying the volcanic pile include horizons containing vesicular round or fluidly shaped basaltic fragments of lapilli to coarse ash size of probable pyroclastic origin. Similar deposits were described in CAMP sequences from Central Morocco (Chalot-Prat *et al.* 1985; Ouarhache *et al.* 2000; Youbi *et al.* 2003) and the Algarve Basin in southern Portugal (Martins *et al.* 2008).

Discussion

Volcanological evolution and facies model

The CAMP volcanic pile of Argana Basin was formed during two pulses of volcanic activity, represented by the Tasguint and the Alemzi Fms. On the studied sections there is evidence for a variable number of eruptions in each formation. The products of distinct eruptions can be separated by the presence of reddened flow surfaces (slightly weathered surfaces metamorphosed by overlying flows),

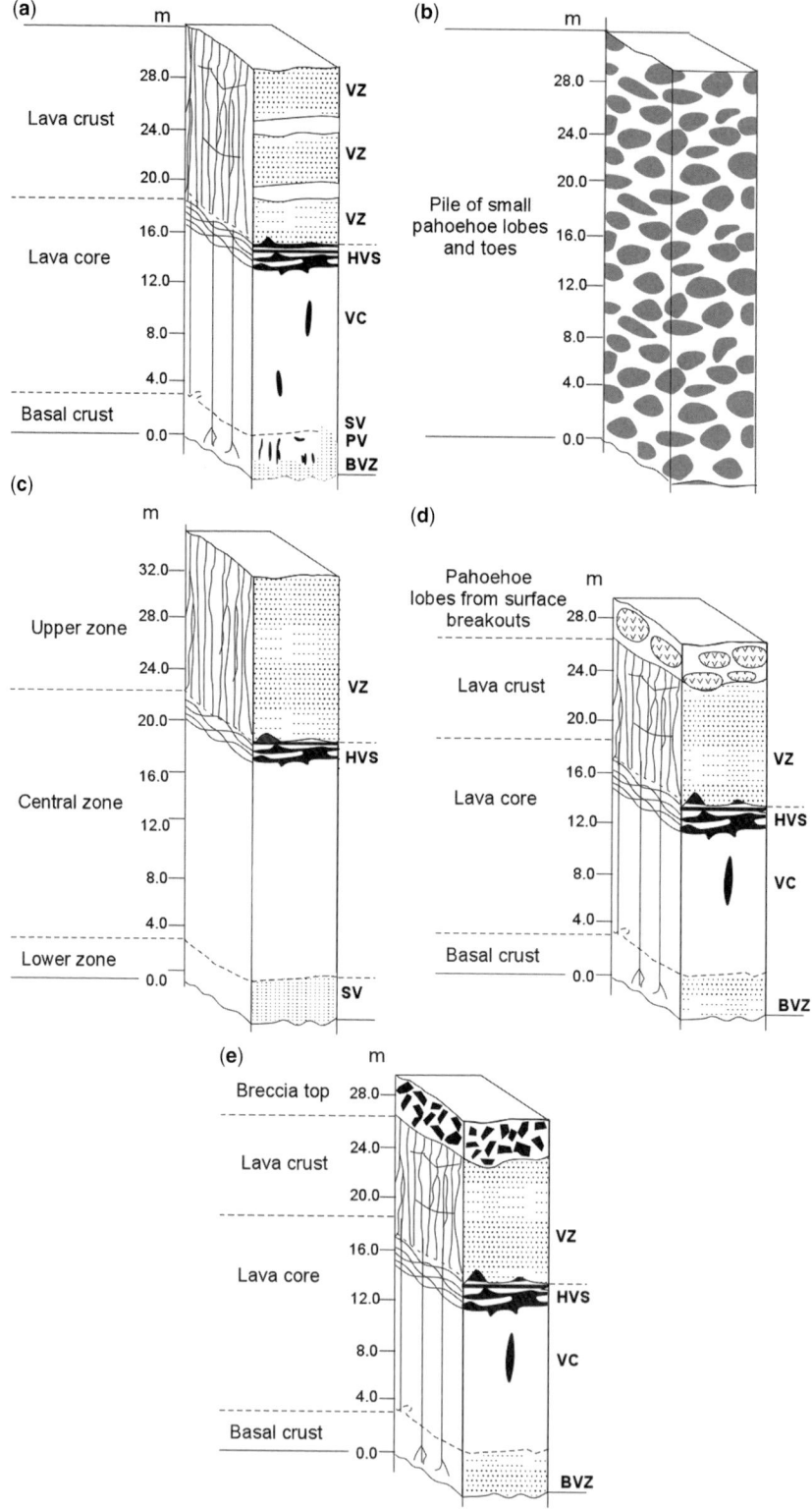

Fig. 10.

development of incipient or more evolved red soils or deposition of fine clastic sediments, indicating significant time intervals separating the emplacement of each package of lava flow-units. On the Tasguint Formation, flow-units composed of piles of pahoehoe lobes and toes may represent a decrease in effusion rate towards the final phase of the eruption; piles of pahoehoe lobes and toes at the base they may represent flow units in a marginal position relative to the lava flow field. The Tasguint Formation was formed by one–eight eruptions, each usually formed by flow fields composed of up to seven flow units. The Alemzi Formation is the result of up to four eruptions mostly composed of simple flows, although locally up to three flow units can be found.

In the most complete CAMP lava sequences of the Central High Atlas, magnetostratigraphic data indicate the occurrence of five short magma pulses (Knight *et al.* 2004).

The CAMP volcanism of the Argana Basin shows the following volcanological characteristics: (i) absence of gradual vertical and lateral facies variations; (ii) the eruptions occurred in subaerial continental environment as inferred from the absence of pillow lavas; (iii) the volcanism is characterized by dominantly basaltic magmas; intermediate and acid lavas are absent (Aït Chayeb *et al.* 1998; Deenen *et al.* 2010, 2011); (iv) eruptions produced compound pahoehoe flows and simple flows (the former is almost exclusively confined to the older formation e.g. Tasguint, while the latter dominates in the younger formation e.g. Alemzi); (v) thin sedimentary beds (claystones, sandstones) and palaeosols occur between the flows, thus attesting for significant time intervals between volcanic episodes, and sedimentary deposits associated with the basalts are detrital and reflect the existence of fluviatile/lacustrine environments; (vi) volcanism occurred on the margin of a continental block (the West African Craton) and was followed by regional uplift, compensated by subsidence (which probably announced the Early Jurassic transgression) after the end of volcanism; (vii) no volcanic cones are preserved; (viii) the basaltic rocks have anorogenic tholeiitic character similar to those described for many CFB (Aït Chayeb *et al.* 1998; Deenen *et al.* 2010, 2011).

The facies models that may be privileged for Argana Basin volcanism are those which favour a continental environment, namely (i) continental basaltic successions; (ii) continental stratovolcanoes; or (iii) continental silicic volcanoes (Cas & Wright 1987; Walker 1987). The continental environment, the architecture of the volcanic succession and the tholeiitic basaltic composition therefore indicate that the most adequate facies model is that of continental basaltic successions.

Emplacement mechanisms

The emplacement of CFB lava flows is currently a topic of considerable interest, following the realization that such flows may have formed by a process analogous to that of small (<1 km^3) pahoehoe sheet flows on Kilauea Volcano, Hawaii (e.g. Hon *et al.* 1994; Self *et al.* 1996, 1997) but on a much larger scale. Observations and measurements of the Kilauea eruption suggest that 1–5 m thick sheets of lava build up by the inflation and coalescence of 10–50 cm thick pahoehoe toes. The lava flows develop over a period of days to years by injection of fresh lava into the molten core of a slow-moving lobe of lava; after months to decades, the lava stagnates and eventually freezes (Hon *et al.* 1994). Evidence for this emplacement history is preserved in the lava structure (e.g. Self *et al.* 1996, fig. 1d). In this section we discuss the emplacement mechanisms of the CAMP lava flows of the Argana Basin. Figure 10 represents different idealized individual lava flow types observed in the CAMP basaltic pile of the Argana Basin.

Compound pahoehoe flows. The CAMP flows of Argana Basin, in particular the compound pahoehoe flows of the Tasguint Formation, show clear evidence of endogenous growth or inflation (e.g. Self *et al.* 1997, 1998). They are very similar to inflated pahoehoe flows from Hawaii (Hon *et al.* 1994), Columbia River Basalt Province (Thordarson & Self 1998), Cenozoic volcanic Province of North Queensland in Australia (Whitehead & Stephenson 1998), Deccan Traps (Keszthelyi *et al.* 1999; Bondre *et al.* 2004*a*, *b*; Jay & Widdowson 2008), Paraná–Etendeka CFB (Jerram *et al.* 1999*a*, *b*; Waichel *et al.* 2006) and CAMP of Fundy (Canada;

Fig. 10. Idealized flow unit types observed in the CAMP basaltic pile of Argana Basin, inspired by Thordarson & Self (1998). (**a**) single pahoehoe flow unit with layered upper crust; (**b**) flow unit composed of a pile of small pahoehoe lobes and toes; (**c**) single pahoehoe flow unit without layered upper crust; (**d**) single pahoehoe flow unit without layered upper crust, covered by coalescing pahoehoe lobes; (**e**) single pahoehoe flow unit with breccia top. Types (a) and (b) are found in the Tasguint Formation while types (c), (d) and (e) occur in the Alemzi Formation. Compound flows from Tasguint Formation are formed by coalescent lobes of types (a) and (b) while simple flows of Alemzi Formation present the structure depicted in (c), (d) and (e). Vesiculation structures (VZ, vesicular zone; HVS, horizontal vesicle sheet; VC, vesicle cylinder; SV, segregation vesicle; PV, pipe vesicle; BVZ, basal vesicular zone).

Kontak 2008) and Algarve basins (Portugal; Martins *et al.* 2008).

The features indicating endogenous growth are: (i) the compound nature of the flows with sheet lobes displaying a three-part structural division of vesicular basal crust, massive lava core and vesicular upper crust (when thick, the latter shows layering of alternating dense and vesicular levels); (ii) the local presence of structure such as break-outs, tumuli and associated structures such as squeeze-ups and horizontal squeezes materializing inflation processes; and (iii) the vertical distribution of vesicles and the presence of segregation structures (spherical vesicles, pipe vesicles, vesicle cylinders and vesicle sheets).

Indeed, the rapid downward decrease in vesicularity, associated with an increase in vesicle size observed in the CAMP basalts, is a characteristic feature of inflated pahoehoe flows (Cashman & Kauahikaua 1997) where the segregation structures are formed when somewhat differentiated liquids flow from host basalts by gas filter-pressing (Anderson *et al.* 1984). According to Goff (1996), basalt containing vesicle cylinders shows positive correlations between increasing lava porosity and increasing ground mass crystalline size; these features may suggest an unusually high water content in the magma.

On the other hand, two alternative mechanisms were proposed for the origin of pipe vesicles. Walker (1987) relates pipe vesicles to bubbles rising through the lower part of cooling lava flow when it has acquired yield strength of the order of 50 N m^2 and a flow rate equivalent to or smaller than the rate of bubble rise, a condition that is favoured by lava flowing on very low angle slopes. Philpotts & Lewis (1987) propose an alternative model which independent of slope and relates pipe vesicles to bubble formation at the cooling front of flows, immediately above glassy rims, in both subaerial and pillow lavas. Observation of horizontal pipe vesicles in squeeze-up structures from Argana sequence favours the second hypothesis.

The predominance of P-type lobes in compound flows can be related to breakouts emerged from larger inflated sheet flows (Hon *et al.* 1994, 2003). According to Wilmoth & Walker (1993), these lobes can be found almost anywhere in the pahoehoe flow field but are common in areas with shallow slopes ($<4°$). The P-type lobes have lower extrusion temperatures, higher density, degassed outer glassy selvage and a well-developed crystalline interior compared to S-type; this characteristic indicates that lavas that form P-type lobes had long enough residence time in a lava distributor system before extrusion in order to exsolve considerable amounts of vapour (Wilmouth & Walker 1993; Oze & Winter 2005). Alternatively, Hon *et al.* (1994)

suggest that P-type lobes may represent pressurized lava containing more dissolved volatiles that S-type; the dense outer crust of P-type would favour a pressure increase leading to solubilization of pre-existing bubbles back into the melt.

The emplacement time of pahoehoe compound flows is still a matter of debate. Anderson *et al.* (1999) suggest a rapid emplacement (1 day to 1 week) while Thordarson & Self (1998) propose a slower emplacement (1–10 years) via endogenous processes. The large thickness of some flows and the abundance of compound flows within the CAMP basalts of the Argana Basin strongly suggest a slow emplacement during sustained eruptive episodes, rather than a fast emplacement.

Simple flows. The simple flows, in particular those of the Alemzi Formation, appear to be single cooling units. In this sense, they are very similar to simple flows described in the CAMP (Kontak 2008; Martins *et al.* 2008), Deccan traps (Keszthelyi *et al.* 1999; Bondre *et al.* 2004a, b), Paraná–Etendeka (Jerram *et al.* 1999a, b; Waichel *et al.* 2006) and Palaeogene Faroe Islands Basalt Group (Passey & Bell 2007). The simple flows also display a characteristic three-tiered structure with a thin basal crust, a dense lava core and a thick, vesicular upper crust. Some are characterized by flow-top breccia, as can be observed in the Alemzi, Aguersouane and Igounan sections. The existence of breccia at the top of some of these flows would suggest that they are aa lava flows. However, their broad (at least a few kilometres) sheet-like morphology and absence of basal breccia and levees or other canalized flow-related features are against the hypothesis of being aa lava flows; their features do not favour an emplacement analogue to aa flows or even a turbulent emplacement at high-discharge rates (Shaw & Swanson 1970; Bondre *et al.* 2004a, b; Waichel *et al.* 2006). Simple flows may initially thicken by inflation and develop the top breccia at a later stage, as a response to an increase in flow rate or viscosity as proposed by Duraiswami *et al.* (2003), Bondre *et al.* (2004a, b) and Keszthelyi *et al.* (2006). On the other hand, Keszthelyi *et al.* (1999) state that most simple flows may be large sheet lobes which, when traced over a considerable distance, will terminate against other lobes. This was observed in CAMP flows from the High Atlas (Tiourjdal region), where flows with large lateral continuity present clefts interpreted as a junction of lobes, but not yet seen in the Argana Basin.

In places, above the upper crust of simple flows, a layer which appears to be constituted by a relatively thin accumulation of pahoehoe lobes is present. This may represent tumuli breakouts piled over the top surface of the flow.

Comparison with other CFB sequences

Some authors (e.g. Thordarson & Self 1998; Guilbaud *et al.* 2005) consider that many pahoehoe flows in CFB sequences may have been formed by inflation and slowly emplaced as those from Hawaii and Iceland. However, other studies (e.g. Anderson *et al.* 1999) propose that CFB flows were emplaced rapidly. Different models were proposed (Anderson *et al.* 2000; Self *et al.* 2000) to explain the process of endogenous growth in basaltic lava flows and lobes, also debating aspects such as the timescales and pulses of inflation.

The morphology of lava flow lobes in CFB provinces can help to understand the mechanisms involved in their emplacement as shown by Self *et al.* (1997). Jerram (2002), using the terminology of Walker (1971), discussed two end-member types of facies architecture for subaerial basalt lava flows. One is the 'compound-braided flow facies architecture', which comprises thin anastomosing pahoehoe flow sheets and lobes up to several metres thick. These compound flows were interpreted as representing lavas emplaced at low effusion rates (Walker 1971; Jerram 2002). This contrasts with the second end-member, the 'tabular-classic facies architecture', which comprises simple flows, typically with sheet geometry, separated by palaeosols and/or various terrestrial clastic lithologies (Walker 1971; Jerram 2002). These were interpreted as indicating rapid emplacement at high effusion rates (Shaw & Swanson 1970; Walker 1971).

In the Argana volcanic pile there are simple and compound flows in both formations, although the former dominate in the upper unit (Alemzi Formation) and the latter in the lower (Tasguint Formation). Similarly, in the Deccan Volcanic Province Bondre *et al.* (2004b) observed that compound pahoehoe flows are almost exclusively confined to older formations (Kalsubai Subgroup), while simple flows dominate the younger formations (Wai Subgroup). It must be emphasized, however, that what is usually observed and described in the field are sections across the lava pile. In ancient volcanic successions it is often impossible to follow laterally all flows for long distances. It is therefore problematic to ensure that a simple flow does not change laterally into a compound flow and vice versa. In modern volcanic sequences this occurs frequently (both in pahoehoe and aa flows). We found that the use of these terms is limiting and, even though we have used it in this work, we believe that it is advisable to use them moderately and in a non-restrictive way.

In the North Mountain CAMP Basalt Formation of the Bay of Fundy, Kontak (2008) describes three laterally continuous lava units: (i) the East Ferry; (ii) Margaretsville; and (iii) Brier Island members. The lower and upper members are the product of single, sustained effusive events rather than numerous shorter-duration eruptive events reflected in the thinner sheet flow lobes of Margaretsville member. Thus, the East Ferry and Brier Island members are simple flows, whereas the Margaretsville member is a compound flow.

According to Martins *et al.* (2008), the volcanic pile of the Algarve CAMP (South Portugal) includes subaerial lava flows, pyroclastic deposits, peperites and intercalated mudstones and conglomerates often containing volcanic fragments. These lithological characteristics and associations are, as in Argana Basin, compatible with a facies model typical of continental basaltic successions. The total thickness of the preserved volcano-sedimentary pile varies between 30 and 50 m. Five–eight lava flows are present in the most complete sections. Lava flows are simple tabular pahoehoe flows 1–5 m thick, but can reach some tens of metres, have massive or platy jointed bases and vesicular upper crusts often with a thin (few centimetres thick) glassy outer shell and occasionally present ill-defined columnar jointing in the lava core. However, unlike the Moroccan CAMP, the Algarve sequences are so variable from section to section that a characteristic type section cannot be established.

In the North Atlantic igneous province, which comprises the British Tertiary and Greenland igneous provinces, Single & Jerram (2004) demonstrated that the Skye lava field (British Tertiary Igneous Province) may be divided into three main architectural sequences each formed by several eruptions: (i) lower compound-braided lavas, interpreted as a sequence formed on the flanks of a low-angle shield volcano; (ii) a transitional lava sequence marking an intermediate eruptive phase between the low-viscosity compound-braided basalts and the architecturally more simple tabular-type basalts; (iii) tabular simple flows in the upper sequence. On the contrary, Passey & Bell (2007) show that the Palaeogene Faroe Islands Basalt Group, located further north in the same LIP, comprises three subaerial eruptive sequences: (i) the first formation has a tabular architecture, and is composed of a sequence of simple flows each comprising a single sheet lobe; (ii) the second formation has a compound-braided facies architecture; and (iii) the third formation consists of a mixture of simple and compound flows.

Compound flows are well described from near-vent settings in currently active basaltic systems and, by analogy, are likely to represent vent proximity when found in a prehistoric succession. In contrast, simple flows where each individual (usually thick) aa or pahoehoe flow represents an eruptive event are commonly found at locations distal to eruption centres (Lesher *et al.* 1999).

In brief, studies on the physical volcanology of CFB provinces indicate that they do not have a simple, 'layer-cake stratigraphy', but contain complex internal and external architectures (Jerram *et al.* 1999*a, b*; Planke *et al.* 2000). Such architectures are governed by the volume of individual eruption events, the location and abundance of volcanic centres and the evolution of these centres through time (Jerram 2002; Jerram & Widdowson 2005; White *et al.* 2009). The architecture of most, if not all, CFB provinces reveals that the production of compound pahoehoe flows was followed by flows with a simpler, sheet-like geometry indicating a fundamental temporal change in the emplacement of flows (see also Jerram 2002; Jerram & Widdowson 2005). Accordingly, it appears that flood basalt volcanism initially starts out at relatively low effusion rate, low-volume eruptions that gradually accelerate to high effusion rate, high-volume eruptions. This worldwide similarity suggests that either the magma genesis and/or magma ascension processes are similar in all CFB provinces.

Concluding remarks

With the exception of Algarve Basin in South Portugal and Fundy Basin in Canada (Kontak 2008; Martins *et al.* 2008), little was known about the physical volcanology of the CAMP successions. In this study we presented data from the extrusive sequence of Argana Basin (Morocco) in order to contribute to the knowledge of the CAMP volcanology.

In Argana Basin (Morocco), the volcanic pile was produced by two eruptive pulses represented by the Tasguint and the Alemzi formations. These are geochemically correlative to the Lower and Intermediate formations of the Moroccan High Atlas, where the sequence is the most complete.

Each formation was the result of several sustained eruptions (1–8 eruptions in the lower formation and up to 4 in the upper formation) that produced inflated pahoehoe flows. Compound flows are dominant in the Tasguint Formation, while simple flows prevail in the Alemzi Formation.

The volcanological characteristics of the Argana volcanic pile are those of continental basaltic successions facies model.

The architecture of the Argana volcanic sequence suggests an increase in effusion rate and volume of eruptions from the first to the second pulse of volcanism, as observed in many other CFB provinces.

Most of this work was carried out at the Department of Geology of the Faculty of Sciences-Semlalia, Cadi Ayyad University of Marrakech. We acknowledge CNRST for funding studentship nr. a 03/034-2005–2007. Financial support for this work was also provided by several research projects: (i) Moroccan PARS (SDU-30) to Fida Medina; (ii) CNRS (France)-CNRST (Morocco) to Hervé Bertrand and Hassan Ibouh; (iii) CNRi (Italy)-CNRST (Morocco) to Giuliano Bellieni, Andrea Marzoli and Nasrrddine Youbi; and (iv) FCT (Portugal)-CNRST (Morocco) to José Munhá, Línia Martins, José Madeira, João Mata and Nasrrddine Youbi. This study is a contribution to research projects PICS, CNRS (France)-CNRST (Morocco) to Hervé Bertrand and Nasrrddine Youbi, and FCT (Portugal)-CNRST (Morocco) to Línia Martins and Nasrrddine Youbi. The pertinent comments and careful corrections of Stephen Self and Ninad Bondre led to major improvements of the manuscript. We thank Douwe van Hinsbergen, Susanne Buiter, Trond Torsvik, Carmen Gaina and Sue Webb for organizing this special publication of the Geological Society of London. Thanks also go to Martijn Deenen for useful comments on an earlier version of this paper.

References

AASSOUMI, H., SABER, H., BROUTIN, J. & EL WARTITI, M. 2003. First spore, pollen and acritarch associations in the Ida Ou Ziki basin (southern slope of the Western High Atlas, Morocco). *In: XVth International Congress on Carboniferous and Permian Stratigraphy,* Utrecht, The Netherlands, Abstract, **195**, 1–2.

AASSOUMI, H., SABER, H., BROUTIN, J. & EL WARTITI, M. 2006. Première étude sporopollinique dans le sous bassin des Ida Ou Zal (Haut Atlas occidental, Maroc). *In: 5ème réunion du Groupe Marocain du Permien et du Trias,* Abstract, 21. El Jadida, Maroc.

AÏT CHAYEB, E. H., YOUBI, N., EL BOUKHARI, A., BOUABDELLI, M. & AMRHAR, M. 1998. Le volcanisme Permien et Mésozoïque inférieur du bassin d'Argana (Haut-Atlas occidental, Maroc): un magmatisme intraplaque associé à l'ouverture de l'Atlantique central. *Journal of African Earth Sciences,* **26**, 499–519.

AMBROGGI, R. 1963. Etude géologique du versant méridional du Haut Atlas occidental et de la plaine du Souss. *Notes et Mémoires du Service Géologique du Maroc,* **74**, 9–11.

ANDERSON, J. A. T., SWIHART, G. H., ARTIOLI, G. & GEIGER, C. H. A. 1984. Segregation vesicles, gas filter-pressing, and igneous differentiation. *The Journal of Geology,* **92**, 55–72.

ANDERSON, S. W., STOFAN, E. R., SMREKAR, S. E., GUEST, J. E. & WOOD, B. 1999. Pulsed inflation of pahoehoe lava flows: implications for flood basalt emplacement. *Earth and Planetary Science Letters,* **168**, 7–18.

ANDERSON, S. W., STOFAN, E. R., SMREKAR, S. E., GUEST, J. E. & WOOD, B. 2000. Reply to: self et al. discussion of 'Pulsed inflation of pahoehoe lava flows: implications for flood basalt emplacement'. *Earth and Planetary Science Letters,* **179**, 425–428.

AUBELE, J. C., CRUMPLER, L. S. & ELSTON, W. E. 1988. Vesicle zonation and vertical structure of basalt flows. *Journal of Volcanology and Geothermal Research,* **35**, 349–374.

BEERLING, D. J. & BERNER, R. A. 2002. Biogeochemical constraints on the Triassic–Jurassic boundary carbon cycle event. *Global Biogeochemical Cycles*, **16**, 10–11.

BERTRAND, H. & PRIOTON, J. M. 1975. *Les dolérites marocaines et l'ouverture de l'Atlantique: étude pétrographique et géochimique*. Thèse de 3ème Cycle, Université de Lyon.

BERTRAND, H., DOSTAL, J. & DUPUY, C. 1982. Geochemistry of Mesozoic tholeiites from Morocco. *Earth and Planetary Sciences Letters*, **58**, 225–239.

BONDRE, N. R. & HART, W. K. 2008. Morphological and textural diversity of the Steens Basalt lava flows, Southeastern Oregon, USA: implications for emplacement style and nature of eruptive episodes. *Bulletin of Volcanology*, **70**, 999–1019.

BONDRE, N. R., DURAISWAMI, R. A. & DOLE, G. 2004a. A brief comparison of lava flows from the Deccan volcanic province and the Columbia–Oregon Plateau flood basalts: Implications for models of flood basalt emplacement. *In*: SHETH, H. C. & PANDE, K. (eds) *Magmatism in India through Time*. Indian Academy of Science, Earth & Planetary Sciences, **113**, 809–817.

BONDRE, N. R., DURAISWAMI, R. A. & DOLE, G. 2004b. Morphology and emplacement of flows from the Deccan volcanic province, India. *Bulletin of Volcanology*, **66**, 29–45.

BOUAOUDA, M. S. 1987. *Biostratigraphie du Jurassique inférieur et moyen des bassins côtiers d'Essaouira et d'Agadir (marge atlantique du Maroc)*. Thèse de 3ème Cycle., Université Paul Sabatier, Toulouse, France, 213.

BOUAOUDA, M. S. 2007. Lithostratigraphie, biostratigraphie et micropaléontologie des formations du Lias au Kimméridgien du bassin atlantique marocain d'El Jadida – Agadir (Maroc). *Travaux de l'Institut Scientifique, Rabat, Série Géologie et Géographie Physique*, **22**, 175.

BROUTIN, J., FERRANDINI, J. & SABER, H. 1989. Implications stratigraphiques et paléogéographiques de la découverte d'une flore permienne euraméricaine dans le Haut Atlas occidental (Maroc). *Comptes rendus de l'Académie des Sciences, Paris, série II*, **308**, 1509–1515.

BROWN, R. H. 1980. Triassic rocks of Argana Valley, Southern Morocco, and their regional structural implication. *American Association Petroleum Geologists Bulletin*, **64**, 988–1003.

BRYAN, S. E. & ERNST, R. E. 2008. Revised definition of Large Igneous Provinces (LIPs). *Earth Science Review*, **86**, 175–202.

CAS, R. A. F. & WRIGHT, J. V. 1987. *Volcanic Successions Modern and Ancient*. Allen and Unwin, London.

CASHMAN, K. V. & KAUAHIKAUA, J. P. 1997. Re-evaluation of vesicle distribution in basaltic lava flows. *Geology*, **25**, 419–422.

CHALOT-PRAT, F., CHARRIERE, A. & OUARHACHE, D. 1985. Découverte d'un volcanisme explosif finitriastique sur la bordure occidentale du Moyen-Atlas (Maroc). *Bulletin de la Faculté des Sciences de Marrakech*, **3**, 127–141.

CIRILLI, S., MARZOLI, A. *ET AL*. 2009. Latest Triassic onset of the Central Atlantic Magmatic Province (CAMP) volcanism in the Fundy Basin (Nova Scotia): new stratigraphic constraints. *Earth and Planetary Science Letters*, **286**, 514–525.

COFFIN, M. F. & ELDHOLM, O. 1994. Large igneous provinces: crustal structure, dimensions, and external consequences. *Reviews of Geophysics*, **32**, 1–36.

COLTICE, N., PHILLIPS, B. R., BERTRAND, H., RICARD, Y. & REY, P. 2007. Global warming of the mantle at the origin of flood basalts over supercontinents. *Geology*, **35**, 391–394.

COTTEN, J., LE DEZ, A. *ET AL*. 1995. Origin of anomalous rare-earth element and yttrium enrichments in subaerially exposed basalts: evidence from French Polynesia. *Chemical Geology*, **119**, 115–138.

COURTILLOT, V. 1999. *Evolutionary Catastrophes: The Science of Mass Extinction*. Cambridge University Press, Cambridge.

DE KONING, G. 1957. Géologie des Ida ou Zal (Maroc). *Leidse Geologische Medelelingen*, **23**, 129–146.

DEENEN, M. H. L., RUHL, M., BONIS, N. R., KRIJGSMAN, W., KÜRSCHNER, W. M., REITSMA, M. & VAN BERGEN, M. J. 2010. A new chronology for the end-Triassic mass extinction. *Earth and Planetary Science Letters*, **291**, 113–125.

DEENEN, M. H. L., LANGEREIS, C., KRIJGSMAN, W., EL HACHIMI, H. & CHELLAI, E. H. 2011. Paleomagnetic research in the Argana basin, Morocco: Trans-Atlantic correlation of CAMP volcanism and implications for the late Triassic geomagnetic polarity time scale. *In*: VAN HINSBERGEN, D. J. J., BUITER, S. J. H., TORSVIK, T. H., GAINA, C. & WEBB, S. J. (eds) *The Formation and Evolution of Africa: A Synopsis of 3.8 Ga of Earth History*. Geological Society, London, Special Publications, **357**, 195–209.

DEFRETIN, S. & FAUVELET, E. 1951. Présence de phyllopodes triasiques dans la région d'Argana-Bigoudine (Haut Atlas occidental). *Notes et Mémoires du Service Géologique du Maroc*, **85**, 129–135.

DURAISWAMI, R. A., BONDRE, N. R., DOLE, G., PHADNIS, V. M. & KALE, V. S. 2001. Tumuli and associated features from the western Deccan volcanic province, India. *Bulletin of Volcanology*, **63**, 435–442.

DURAISWAMI, R. A., DOLE, G. & BONDRE, N. R. 2003. Slabby pahoehoe from the western Deccan Volcanic Province: evidence for incipient pahoehoe-a'a transitions. *Journal of Volcanology and Geothermal Research*, **121**, 195–217.

DURAISWAMI, R. A., BONDRE, N. R. & MANAGAVE, S. 2008. Morphology of rubbly pahoehoe (simple) flows from the Deccan Volcanic Province: implications for style of emplacement. *Journal of Volcanology and Geothermal Research*, **177**, 822–836.

DUTUIT, J. M. 1976. Introduction à l'étude paléontologique du Trias continental marocain: description des premiers Stégocéphales recueillis dans le couloir d'Argana. *Mémoires du Muséum National d'Histoire naturelle de Paris*, **36**, 253.

EL ARABI, E. H., BIENVENID, J. D., BROUTIN, J. & ESSAMOUD, R. 2006. Première caractérisation palynologique du Trias moyen dans le Haut Atlas; implications pour l'initiation du rifting téthysien au Maroc. *Comptes Rendus Geoscience*, **338**, 641–649.

EL HACHIMI, H., YOUBI, N. *ET AL*. 2005. Morphology of the Triassic–Jurassic basaltic lava flows of the Central Atlantic Magmatic Province (CAMP). Example from the Argana Basin volcanic sequence (Western High Atlas, Morocco). *In*: *Meeting of the Geological Society of America Salt Lake City, Abstracts with Programme*, **37**, 529.

EL HACHIMI, H., ASSAFAR, H. *ET AL.* 2007. The CAMP Basalts in the Argana basin (Western High Atlas, Morocco). Physical Volcanology, Petrology and Geochemistry. *In: The First MAPG International Convention, Conference & Exhibition, Marrakech*, 39.

ERNST, R. E. & BUCHAN, K. L. 2001. The use of mafic dike swarms in identifying and locating mantle plumes. *In:* ERNST, R. E. & BUCHAN, K. L. (eds) *Mantle Plumes: Their Identification Through Time*. Geological Society of America, Boulder, Special Paper, **352**, 247–265.

FLOYD, P. A. & WINCHESTER, J. A. 1975. Magma type and tectonic setting discrimination using immobile elements. *Earth and Planetary Science Letters*, **27**, 211–218.

FOWELL, S. J., TRAVERSE, A., OLSEN, P. E. & KENT, D. V. 1996. Carnian and Norian palynofloras from the Newark Supergroup, eastern United States and Canada, and the Argana Basin of Morocco: relationship to Triassic climate zones. *In: 9th International Palynological Congress*, Program and Abstracts: 45–46.

GOFF, F. 1996. Vesicle cylinders in vapor-differentiated basalt flows. *Journal of Volcanology and Geothermal Research*, **71**, 167–185.

GUILBAUD, M. N., SELF, S., THORDARSON, TH. & BLAKE, S. 2005. Flow formation, surface morphology, and emplacement mechanism of the AD 1783–4 Laki lava. *In:* MANGA, M. & VENTURA, G. (eds) *Kinematics and Dynamics of Lava Flows*. Geological Society of America, Boulder, Special Papers, **396**, 81–102.

HALLAM, A. & WIGNALL, P. B. 1997. *Mass Extinctions and their Aftermath*. Oxford University Press, Oxford.

HAMES, W. E., MCHONE, J. G., RUPPEL, C. & RENNE, P. 2003. *The Central Atlantic Magmatic Province.* American Geophysical Union Monograph Series, **136**.

HILL, I. R. 1991. Starting plumes and continental break-up. *Earth and Planetary Science Letters*, **104**, 398–416.

HMICH, D., SCHNEIDER, J. W., SABER, H. & EL WARTITI, M. 2003. First Permocarboniferous insects (blattids) from North Africa (Morocco): implications on palaeobiogeography and palaeoclimatology. *Freiberger Forschungshefte, Hefte C*, **499**, 117–134.

HOFFMANN, A., TOURANI, A. & GAUPP, R. 2000. Cyclicity of Triassic to Lower Jurassic continental red beds of the Argana Valley, Morocco: implications for paleoclimate and basin evolution. *Palaeogeography, Palaeoclimatology, Palaeoecology*, **161**, 229–266.

HON, K., KAUAHIKAUA, J., DENLINGER, R. & MACKAY, K. 1994. Emplacement and inflation of pahoehoe sheet flows: observations and measurements of active lava flows on Kilauea Volcano, Hawaii. *Geological Society of America Bulletin*, **106**, 351–370.

HON, K., GANSECKI, C. & KAUAHIKAUA, J. 2003. The Transition from A'a to Pahoehoe Crust on Flows Emplaced During the Pu'u 'Ō'ō-Kūpaianaha Eruption. *In:* HELIKER, C., SWANSON, D. A. & TAKAHASHI, T. J. (eds) *The Pu'u 'Ō'ō-Kūpaianaha Eruption of Kīlauea Volcano, Hawai'i: The First 20 Years U.S.* Geological Survey Professional Paper, **1676**, 89–103.

JALIL, N. & DUTUIT, J. M. 1996. Permian Capthorinid reptiles from the Argana Formation, Morocco. *Palaeontology*, **39**, 907–918.

JALIL, N. & JANVIER, P. 2005. Les pareiasaures (Amniota, Parareptilia) du Permien supérieur du Bassin d'Argana, Maroc. *Geodiversitas*, **27**, 35–132.

JAY, A. E. & WIDDOWSON, M. 2008. Stratigraphy, structure and volcanology of the SE Deccan continental flood basalt province: implications for eruptive extent and volumes. *Journal of the Geological Society, London*, **165**, 177–188.

JERRAM, D. A. 2002. Volcanology and facies architecture of flood basalts. *In:* MENZIES, M. A., KLEMPERER, S. L., EBINGER, C. J. & BAKER, J. (eds) *Volcanic Rifted Margins*. Geological Society of America, Boulder, Special Paper, **362**, 121–135.

JERRAM, D. A. & WIDDOWSON, M. 2005. The anatomy of Continental Flood Basalt Provinces: geological constraints on the processes and products of flood volcanism. *Lithos*, **79**, 385–405.

JERRAM, D. A., MOUNTNEY, N. & STOLLHOFEN, H. 1999*a*. Facies architecture of the Etjo Sandstone Formation and its interaction with the Basal Etendeka food basalts of NW Namibia: Implications for offshore analogues. *In:* CAMERON, N., BATE, R. & CLURE, V. (eds) *The Oil and Gas Habitats of the South Atlantic.* Geological Society, London, Special Publications, **153**, 367–380.

JERRAM, D. A., MOUNTNEY, N., HOLZFÖRSTER, F. & STOLLHOFEN, H. 1999*b*. Internal stratigraphic relationships in the Etendeka Group in the Huab Basin, NW Namibia: understanding the onset of food volcanism. *Journal of Geodynamics*, **28**, 393–418.

JONES, D. F. 1975. *Stratigraphy, environments of deposition, petrology, age and provenance of the basal red beds of the Argana Valley, western High Atlas Mountains, Morocco.* MSc thesis, New Mexico Institute of Mining and Technology, USA.

JONGMANS, W. J. 1950. Note sur la flore du Carbonifère du versant sud du Haut Atlas. *Notes Mémoires Service Géologique Maroc*, **76**, 155–172.

JOURDAN, F., MARZOLI, A. *ET AL.* 2009. $^{40}Ar/^{39}Ar$ ages of CAMP in North America: implications for the Triassic–Jurassic boundary and the 40 K decay constant bias. *Lithos*, **110**, 167–180.

KENT, D. V. & OLSEN, P. E. 1999. Astronomically tuned geomagnetic polarity time scale for the Late Triassic. *Journal of Geophysical Research*, **104**, 12,831–12,841.

KENT, R. W., THOMSON, B. A., SKELHORN, R. R., KERR, A. C., NORRY, M. J. & WALSH, J. N. 1998. Emplacement of Hebridean Tertiary flood basalts: evidence from an inflated pahoehoe lava flow on Mull, Scotland. *Journal of Geological Society, London*, **155**, 599–607.

KESZTHELYI, L. 2002. Classification of the mafic lava flows from OPD Leg 183. *In:* FREY, F. A., COFFIN, M. F., WALLACE, P. J. & QUALITY, P. G. (eds) *Proceedings of the Ocean Drilling Program. Scientific Results*, **183**, 1–28.

KESZTHELYI, L. & THORDARSON, T. 2000. Rubbly pahoehoe: a previously undescribed but widespread lava type transitional between a'a and pahoehoe. *Meeting of the Geological Society of America Abstracts with Program*, **32**, 7.

KESZTHELYI, L., SELF, S. & THORDARSON, TH. 1999. Application of recent studies on the emplacement of basaltic lava flows to the Deccan Traps. *In:* SUBBARAO,

K. V. (ed.) *Deccan Volcanic Province*. Memoir Geological Society of India, **43**, 485–520.

KESZTHELYI, L., SELF, S. & THORDARSON, TH. 2006. Flood lavas on Earth, Io and Mars. *Journal of the Geological Society, London*, **163**, 253–264.

KLEIN, H., VOIGT, S., HMINNA, A., SABER, H., SCHNEIDER, J. & HMICH, D. 2010. Early Triassic Archosaur-Dominated Footprint Assemblage from the Argana Basin (Western High Atlas, Morocco). *Ichnos*, **17**, 1–13.

KLITGORD, K. D. & SCHOUTEN, H. 1986. Plate kinematics of the Central Atlantic. *In*: TUCHOLKE, B. E. & VOGT, P. R. (eds) *The Geology of North America. Volume M, The Western Atlantic Region. A Decade of North American Geology*, Volume 1. Geological Society of America, Boulder, 351–378.

KNIGHT, K. B., NOMADE, S., RENNE, P. R., MARZOLI, A., BETRAND, H. & YOUBI, N. 2004. The Central Atlantic Magmatic Province at the Triassic–Jurassic boundary: paleomagnetic and ^{40}Ar/^{30}Ar evidence from Morocco for brief, episodic volcanism. *Earth and Planetary Science Letters*, **228**, 143–160.

KONTAK, D. J. 2008. On the edge of CAMP: Geology and volcanology of the Jurassic North Mountain Basalt, Nova Scotia. *In*: DOSTAL, J., GREENOUGH, J. D. & KONTAK, D. J. (eds) *Rift-related Magmatism. Lithos*, **101**, 74–101.

LESHER, C. E., CASHMAN, K. V. & MAYFIELD, J. D. 1999. Kinetic controls on crystallization of Tertiary North Atlantic basalt and implications for the emplacement and cooling history of lava at site 989, southeast Greenland rifted margin. *In*: LARSEN, H. C., DUNCAN, R. A., ALLAN, J. F. & BROOKS, K. (eds) *Proceedings of the Ocean Drilling Program – Scientific Results*, **163**, 135–148.

LOCKWOOD, J. P. & LIPMAN, P. W. 1980. Recovery of datable charcoal from beneath young lava flows-lessons from Hawaii. *Bulletin Volcanologique*, **43**, 609–615.

LUCAS, S. G. 1998*a*. Global Triassic tetrapod biostratigraphy and biochronology. *Paleogeography, Paleoclimatology, Paleoecology*, **143**, 347–384.

LUCAS, S. G. 1998*b*. The aetosaur Longosuchus from the Triassic of Morocco and its biochronological significance. *Comptes Rendus de l'Académie des Sciences de Paris*, **326**, 589–594.

LUND, J. J. 1996. Palynologie der tieferen Bigoudine Fm, Ober-Trias, Marokko (Abstract). *AAP Tagung, AAP Rundbrief*, 14–15.

MACDONALD, G. A. 1953. Pahoehoe, a'a, and block lava. *American Journal of Science*, **251**, 169–191.

MACDONALD, G. A. 1967. Forms and structures of extrusive basaltic rocks. *In*: HESS, H. H. & POLDERVAART, A. (eds) *Basalts: the Poldervaart Treatise on Rocks of Basaltic Composition*, Volume 1. Interscience, New York, 1–61.

MANSPEIZER, W. 1988. Triassic–Jurassic rifting and opening of the Atlantic; an overview. *In*: MANSPEIZER, W. (ed.) *Triassic–Jurassic Rifting, Continental Breakup and the Origin of the Atlantic Ocean and Passive Margins. Developments in Geotectonics, Part A*. Elsevier, Amsterdam, 41–79.

MANSPEIZER, W., PUFFER, J. H. & COUSMIER, H. L. 1978. Separation of Morocco and Eastern North America: A

Triassic–Liassic stratigraphic record. *Geological Society of America Bulletin*, **89**, 901–920.

MARTINS, L. T., MADEIRA, J., YOUBI, N., MUNHÁ, J., MATA, J. & KERRICH, R. 2008. Rift-related magmatism of the Central Atlantic Magmatic Province in Algarve, Southern Portugal. *In*: DOSTAL, J., GREENOUGH, J. D. & KONTAK, D. J. (eds) *Rift-related Magmatism. Lithos*, **101**, 102–124.

MARZOLI, A., RENNE, P. E., PICCIRILLO, E. M., ERNESTO, M., BELLIENI, G. & DE MIN, A. 1999. Extensive 200-million-year-old continental flood basalts of central atlantic magmatic province. *Science*, **284**, 616–618.

MARZOLI, A., BERTRAND, H. ET AL. 2004. Synchrony of the Central Atlantic Magmatic Province and the Triassic–Jurassic boundary climatic and biotic crisis. *Geology*, **32**, 973–976.

MCHONE, J. G. 2000. Non-plume magmatism and tectonics during the opening of the central Atlantic Ocean. *Tectonophysics*, **316**, 287–296.

MCHONE, J. G. & PUFFER, J. H. 2003. Flood basalt province of the Pangean Atlantic rift: regional extent and environmental significance. *In*: LETOURNEAU, P. M. & OLSEN, P. E. (eds) *The Great Rift Valleys of Pangea in Eastern North America, Aspects of Triassic-Jurassic Rift Basin Geoscience*, Volume 1. Columbia University Press, New York, 141–154.

MCMILLAN, K., LONG, P. E. & CROSS, R. W. 1989. Vesiculation in Columbia River basalts. *In*: REIDEL, S. P. & HOOPER, P. R. (eds) *Volcanism and Tectonism in the Columbia River Flood-Basalt Province*. Geological Society of America, Boulder, Special Papers, **239**, 157–167.

MEDINA, F. 1991. Surperimposed extensional tectonics in the Argana Triassic formations (Morocco), related to the Early Rifting of the Central Atlantic. *Geological Magazine*, **128**, 525–536.

MEDINA, F. 1994. *Evolution structurale du Haut Atlas occidental et des régions voisines du Trias à l'actuel, dans le cadre de l'ouverture de l'atlantique Central et de la collision Afrique-Europe*. PhD thesis, Université Mohammed V, Rabat, Maroc.

MEDINA, F. 1995. Syn-and postrift evolution of the El Jadida- Agadir basin (Morocco): contraints for the rifting models of the Central Atlantic. *Canadian Journal Earth Sciences*, **32**, 1273–1291.

MEDINA, F. 2000. Structural styles of the Moroccan Triassic basins. *Epicontinental Triassic International Symposium. Zentralblatt für Geologie und Paläontologie, Teil I*, 1167–1192.

MEDINA, F., EL AMRANI, I. E. & AHMAMOU, M. 1992. Les laves fini-triasiques de la région d'Argana: récisions sur leur gisement et leur pétrologie. *Bulletin de l'Institut Scientifique, Rabat*, **16**, 23–30.

MEDINA, F., VACHARD, D., COLIN, J. P., OUARHACHE, D. & AHMAMOU, M. 2001. Charophytes et ostracodes du niveau carbonaté de Taourirt Imzilen (Membre d'Aglegal, Trias d'Argana); implications stratigraphiques. *Bulletin de l'Institut Scientifique, Rabat*, **23**, 21–26.

NOMADE, S., KNIGHT, K. B. ET AL. 2007. Chronology of the Central Atlantic Magmatic Province: implications for the Central Atlantic rifting processes and the Triassic–Jurassic biotic crisis. *Palaeogeography, Palaeoclimatology, Palaeoecology*, **244**, 326–344.

OLSEN, P. E. 1999. Giant lava flows, mass extinctions and mantle plumes. *Science*, **284**, 604–605.

OLSEN, P. E., KENT, D. V., FOWELL, S. J., SCHLISCHE, R. W., WITHJACK, M. O. & LE TOURNEAU, P. M. 2000. Implications of a comparison of the stratigraphy and depositional environments of the Argana (Morocco) and Fundy (Nova Scotia, Canada) Permian–Jurassic basins. *In*: OUJIDI, M. & ET-TOUHAMI, M. (eds) *Le Permien et le Trias du Maroc, Actes de la Première Réunion su Groupe Marocain du Permien et du Trias*, Hilal Impression, Oujda, 165–183.

OUARHACHE, D., CHARRIERE, A., CHALOT-PRAT, F. & EL WARTITI, M. 2000. Sédimentation détritique continentale synchrone d'un volcanisme explosif dans le Trias Terminal – Infralias du domaine Atlasique (Haute Moulouya – Maroc). *Journal of African Earth Sciences*, **31**, 555–570.

OYARZUN, R., DOBLAS, M., LÓPEZ-RUIZ, J. & CEBRIÁ, J. M. 1997. Opening of the central Atlantic and asymetric mantle upwelling phenomena: implications for long-lived magmatism in western North Africa and Europe. *Geology*, **25**, 727–730.

OYARZUN, R., DOBLAS, M., LOPEZ-RUIZ, J., CEBRIA, J. M. & YOUBI, N. 1999. Tectonically-induced icehouse-greenhouse climate oscillations during the transition from Variscan to Alpine cycle (Carboniferous to Triassic). *Bulletin de la Sociéte Géologique de France*, **170**, 3–11.

OZE, C. & WINTER, D. 2005. The occurrence, vesiculation, and solidification of dense blue glassy pahoehoe. *Journal of Volcanology and Geothermal Research*, **142**, 285–301.

PASSEY, S. R. & BELL, B. R. 2007. Morphologies and emplacement mechanisms of the lava flows of the Faroe Islands Basalt Group, Faroe Islands, NE Atlantic Ocean. *Bulletin of Volcanology*, **70**, 139–156.

PETERSON, D. W. & TILLING, R. I. 1980. Transition of basaltic lava from pahoehoe to a'a, Kilauea Volcano, Hawaii: field observations and key factors. *Journal of Volcanology and Geothermal Research*, **7**, 271–293.

PHILPOTTS, A. R. & LEWIS, C. L. 1987. Pipe vesicles – An alternative model for their origin. *Geology*, **15**, 971–974.

PLANKE, S., SYMONDS, P. A., ALVESTAD, E. & SKOGSEID, J. 2000. Seismic volcanostratigraphy of large-volume basaltic extrusive complexes on rifted margins. *Journal of Geophysical Research*, **105**, 19,335–19,351.

PUFFER, J. H. & STUDENT, J. J. 1992. Volcanic structures, eruptive style, and posteruptive deformation and chemical alteration of the Watchung flood basalts, New Jersey. *In*: PUFFER, J. H. & RAGLAND, P. C. (eds) *Eastern North American Mesozoic Magmatism*. Geological Society of America, Boulder, Special Papers, **268**, 261–277.

RAMPINO, M. R. & STOTHERS, R. B. 1988. Flood basalt volcanism during the past 250 million years. *Science*, **241**, 663–668.

RAMPINO, M. R., SELF, S. & STOTHERS, R. B. 1988. Volcanic Winters. *Annual Review of Earth and Planetary Sciences*, **16**, 73–99.

ROWLAND, S. K. & WALKER, G. P. L. 1987. Toothpaste lava: characteristics and origin of a lava structural type transitional between pahoehoe and a'a. *Bulletin of Volcanology*, **49**, 631–641.

RUELLAN, E. 1985. *Géologie des marges continentales passives: évolution de la marge atlantique du Maroc (Mazagan); étude par submersible seabeam et sismique réflexion. Comparaison avec la marge NW africaine et la marge homologue E américaine.* PhD thesis, Université de Bretagne Occidentale, Brest.

RUELLAN, E., AUZENDE, J. M. & DOSTMANN, H. 1985. Structure and evolution of the Mazagan (El Jadida) plateau and escarpment off central Morocco. *Oceanologica Acta*, **5**, 59–72.

RUIZ-MARTINEZ, V. C., VILLALAN, J. J. & PALENCIA-ORTAS, A. 2007. Paleomagnetic and AMS results of Late Triassic red beds, CAMP-related lava flows and Lower Jurassic limestones from Argana Basin, Morocco: geodynamic implications. 'Earth: Our Changing Planet'. *In*: *Proceedings of IUGG XXIV General Assembly*, Perugia, Italy, 68.

SAHABI, M., ASLANIAN, D. & OLIVET, J. L. 2004. Un nouveau point de départ pour l'histoire de l'Atlantique Central. *Comptes Rendus Géoscience*, **33**, 1041–1052.

SELF, S., THORDARSON, T. *ET AL.* 1996. A new model for the emplacement of Columbia River basalts as large, inflated pahoehoe lava flow fields. *Geophysical Research Letters*, **23**, 2689–2692.

SELF, S., THORDARSON, T. H. & KESZTHELYI, L. 1997. Emplacement of continental flood basalt lava flows. *In*: MAHONEY, J. J. & COFFIN, M. F. (eds) *Large Igneous Provinces: Continental, Oceanic, and Planetary Flood Volcanism*. AGU Geophysical Monograph Series, Washington, **100**, 381–410.

SELF, S., KESZTHELYI, L. & THORDARSON, T. 1998. The importance of pahoehoe. *Annual Review of Earth and Planetary Sciences*, **26**, 81–110.

SELF, S., KESZTHELYI, L. & THORDARSON, T. 2000. Discussion of: 'Pulsed inflation of pahoehoe lava flows: implications for flood basalt emplacement,' by Anderson et al. *Earth and Planetary Science Letters*, **179**, 421–423.

SELF, S., WIDDOWSON, M., THORDARSON, TH. & JAY, A. E. 2006. Volatile fluxes during flood basalt eruptions and potential effects on the global environment: a Deccan perspective. *Earth and Planetary Science Letters*, **248**, 518–532.

SHAW, H. R. & SWANSON, D. A. 1970. Eruption and flow rates of flood basalts. *In*: GILMOUR, E. H. & STRADLING, D. (eds) *Proceedings of the Second Columbia River Basalt Symposium*. Eastern Washington State College Press, Cheney, 271–299.

SHETH, H. C. 2007. "Large Igneous Provinces (LIPs)": Definition, recommended terminology, and a hierarchical classification. *Earth Science Reviews*, **85**, 117–124.

SINGLE, R. T. & JERRAM, D. A. 2004. The 3D facies architecture of flood basalts and their internal heterogeneity: examples from the Skye Lava filed. *Journal of the Geological Society, London*, **161**, 911–926.

SKILLING, I. P., WHITE, J. D. L. & MCPHIE, J. 2002. Peperite: a review of magma-sediment mingling. *Journal of Volcanology and Geothermal Research*, **114**, 1–17.

STOTHERS, R. B. 1993. Flood basalts and extinction events. *Geophysical Research Letters*, **20**, 1399–1402.

STOTHERS, R. B. & RAMPINO, M. R. 1990. Periodicity in flood basalts, mass extinctions, and impacts; a statistical view and a model. *In*: SHARPTON, V. L. & WARD,

P. D. (eds) *Global Catastrophes in Earth History*. Geological Society of America, Boulder, Special Papers, **247**, 9–18.

THORDARSON, T. & SELF, S. 1998. The Roza Member, Columbia River Basalt Group: a gigantic pahoehoe lava flow field formed by endogenous processes? *Journal of Geophysical Research*, **103**, 27411–27445.

TIXERONT, M. 1973. Lithostratigraphie et minéralisation cuprifères et uranifères syngénétiques et familières des formations permo-triasiques du couloir d'Argana (Haut Atlas Occidental, Maroc). *Notes et Mémoires du Service Géologique du Maroc*, **33**, 147–177.

TOURANI, A., LUND, J. J., BENAOISS, N. & GAUP, R. 2000. Stratigraphy of Triassic syn-rift-deposition in Western Morocco. *Zentrablatt fur Geologie und palaontologie*, **Teil I**, 1193–1216.

TOURANI, A., BENAOUISS, N. *ET AL*. 2010. Evidence of an Early Triassic age (Olenekian) in Argana Basin (High Atlas, Morocco) based on new chirotherioid traces. *Comptes Rendus Palevol*, **9**, 201–208.

VAN DE SCHOOTBRUGGE, B., QUAN, T. M. *ET AL*. 2009. Floral changes across the Triassic/Jurassic boundary linked to flood basalt volcanism. *Nature Geoscience*, **2**, 589–594.

VERATI, C., RAPAILLE, C., FÉRAUD, G., MARZOLI, A., BERTRAND, H. & YOUBI, N. 2007. ^{40}Ar/^{39}Ar ages and duration of the Central Atlantic Magmatic Province volcanism in Morocco and Portugal and its relation to the Triassic–Jurassic boundary. *Palaeogeography, Palaeoclimatology, Palaeoecology*, **244**, 308–325.

VOIGT, S., HMINNA, A., SABER, H., SCHNEIDER, J. W. & KLEIN, H. 2010. Tetrapod footprints from the uppermost level of the Permian Ikakern Formation (Argana Basin, Western High Atlas, Morocco). *Journal of African Earth Sciences*, **57**, 470–478.

WAICHEL, P. L., LIMA, E. F., LUBACHESKY, R. & SOMMER, C. A. 2006. Pahoehoe flows from the central Paraná Continental Flood Basalts. *Bulletin of Volcanology*, **68**, 599–610.

WAICHEL, P. L., SCHERER, C. M. S. & FRANK, H. T. 2008. Basaltic lava flows covering active aeolian dunes in the Paraná Basin in southern Brazil: Features and emplacement aspects. *Journal of Volcanology and Geothermal Research*, **171**, 59–72.

WALKER, G. P. L. 1971. Compound and simple lava flows and flood basalts. *Bulletin Volcanologique*, **35**, 579–590.

WALKER, G. P. L. 1987. Pipe vesicles in Hawaiian basaltic lavas: their origin and potential as paleoslope indicators. *Geology*, **15**, 84–87.

WALKER, G. P. L. 1991. Structure, and origin by injection of lava under surface crust, of tumuli, 'lava rises', 'lava-rise pits', and 'lava-inflation clefts' in Hawaii. *Bulletin of Volcanology*, **53**, 546–558.

WALKER, G. P. L 1993. Basaltic-volcano systems. *In*: PRITCHARD, H. M., ALABASTER, T., HARRIS, N. B. W. &

NEARY, C. R. (eds) *Magmatic Processes and Plate Tectonics*. Geological Society, London, Special Publications, **76**, 3–38.

WHITE, J. D. L., BRYAN, S. E., ROSS, P. S., SELF, S. & THORDARSON, T. 2009. Physical volcanology of large igneous provinces: update and review. *In*: THORDARSON, T., SELF, S., LARSEN, G., ROWLAND, S. & HOSKULDSSON, A. (eds) *Studies in Volcanology: The Legacy of George Walker*. Special Publications of IAVCEI, **2**. Geological Society, London, 291–321.

WHITEHEAD, P. W. & STEPHENSON, P. J. 1998. Lava rise ridges of the Toomba basalt flow, north Queensland, Australia. *Journal of Geophysical Research*, **103**, 27,371–27,382.

WHITESIDE, J. H., OLSEN, P. E., KENT, D. V., FOWELL, S. J. & ET-TOUHAMI, M. 2007. Synchrony between the Central Atlantic magmatic province and the Triassic–Jurassic mass-extinction event? *Palaeogeography, Palaeoclimatology, Palaeoecology*, **244**, 345–367.

WHITESIDE, J. H., OLSEN, P. E., EGLINTON, T., BROOKFIELD, M. E. & SAMBROTTO, R. N. 2010. Compound-specific carbon isotopes from Earth's largest flood basalt eruptions directly linked to the end-Triassic mass extinction. *In*: *Proceedings of the National Academy of Sciences of the United States of America*, **107**, 6721–6725.

WIGNALL, P. B. 2001. Large igneous provinces and mass extinctions. *Earth Science Reviews*, **53**, 1–33.

WILMOTH, R. A. & WALKER, G. P. L. 1993. P-type and S-type pahoehoe: a study of vesicle distribution patterns in Hawaiian lava flows. *Journal of Volcanology and Geothermal Research*, **55**, 129–142.

WILSON, M. 1997. Thermal evolution of the Central Atlantic passive margins: continental break-up above a Mesozoic super-plume. *Journal of the Geological Society, London*, **154**, 491–495.

WITHJACK, M. O., SCHLISCHE, R. W. & OLSEN, P. E. 1998. Diachronous rifting, drifting, and inversion on the passive margin of central eastern north america: an analog for other passive margins. *Association of the American Petroleum Geologists Bulletin*, **82**, 817–835.

YOUBI, N., MARTINS, L. T., MUNHÁ, J. M., IBOUH, H., MADEIRA, J., AIT CHAYEB, H. & EL BOUKHARI, A. 2003. The late triassic–early Jurassic volcanism of Morocco and Portugal in the framework of the central Atlantic magmatic province: an Overview. *In*: HAMES, W. E., MACHONE, J. G., RENNE, P. R. & RUPPEL, C. (eds) *The Central Atlantic Magmatic Province: Insights from Fragments of Pangea*. AGU Geophysical Monograph Series, Washington, **136**, 179–207.

ZÜHLKE, R., BOUAOUDA, M.-S., OUAJHAIN, B., BECHSTÄDT, T. & LEINFELDER, R. 2004. Quantitative Meso-Cenozoic development of the eastern Central Atlantic continental shelf, western High Atlas, Morocco. *Marine and Petroleum Geology*, **21**, 225–276.

Palaeomagnetic results from Upper Triassic red-beds and CAMP lavas of the Argana Basin, Morocco

MARTIJN DEENEN[1]*, COR LANGEREIS[1], WOUT KRIJGSMAN[1], HIND EL HACHIMI[2] &
EL HASSANE CHELLAI[2]

[1]*Palaeomagnetic Laboratory Fort Hoofddijk, Utrecht University, The Netherlands*

[2]*Cadi Ayyad University, Faculty of Sciences Semlalia, Marrakech, Morocco*

Corresponding author (e-mail: deenen@geo.uu.nl)

Abstract: The continental Argana Basin of Morocco is the trans-Atlantic counterpart of the extensively studied Fundy, Hartford and Newark basins in north-eastern America, that have provided the astrochronologically tuned geomagnetic polarity timescale (GPTS) for the late Triassic and earliest Jurassic. The Argana red-bed successions also show astronomically driven time control, which allowed trans-Atlantic correlations and revealed that the interval towards volcanism of the Central Atlantic Magmatic Province (CAMP) is without any significant hiatuses. Here, we present palaeomagnetic results from the cyclically bedded upper Triassic red-beds and the intercalated volcanics associated with CAMP. Our composite Argana section comprises an interval of 3.5–4.0 Ma, but its magnetostratigraphic pattern does not allow a straightforward correlation to the Newark GPTS. The continental red-bed deposits of the Bigoudine Formation demonstrate a dominant magnetic overprint that could only be removed at temperatures above 600 °C. We suggest that this overprint could have been caused by a period of (Jurassic, *c.* 170 Ma) magmatism that caused pervasive overprinting of the Triassic palaeomagnetic signal. Correlations between the sections in the Tazantoute region are not straightforward, hampered by the presence of a magmatic sill. The CAMP lava sequences of Tazantoute are all of normal polarity and record secular variation in a manner that agrees with short-lived pulses of CAMP activity in Morocco. Our results indicate that the sedimentary successions of the Argana Basin have the potential to evaluate the Newark GPTS, but that detailed palaeomagnetic analyses of more suitable sections with long(er) cyclostratigraphic records are required.

During the late Triassic, a series of half-graben and strike–slip basins formed along the presently eastern North American, north-western African and south-western European margins, associated with the break-up of Pangea (Fig. 1). The succession of upper Triassic sedimentary facies in the Argana Basin of Morocco is very similar to that found in the Bay of Fundy, Maritime Canada (Smoot & Olsen 1988; Olsen *et al.* 2000; Whiteside *et al.* 2007). Both basins were at that time positioned at 20–25°N latitude where deposition took place under semi-arid to arid conditions (Hay *et al.* 1982; Kent & Tauxe 2005). The Moroccan basins as well as their trans-Atlantic counterparts in North America show thick piles of tholeiitic lavas on top of upper Triassic clastic sediments. These subaerial lavas are part of the largest continental flood basalt province of the Phanerozoic: the Central Atlantic Magmatic Province, CAMP (Marzoli *et al.* 1999). The link between this flood basalt province and the end-Triassic mass extinction, one of the 'big five' of the Phanerozoic (Raup & Sepkoski 1982), was first proposed by Rampino & Stothers (1988) and subsequently by several others (e.g. Courtillot 1994; Courtillot & Renne 2003). It became more widely debated in recent years when several different

intra-CAMP basins correlations were proposed (Olsen *et al.* 2002; Knight *et al.* 2004; Marzoli *et al.* 2004; Whiteside *et al.* 2007; Cirilli *et al.* 2009). Recently, with a multi-disciplinary approach, the onset of CAMP in Morocco has been linked to the major end-Triassic mass extinction documented in the marine realm (Deenen *et al.* 2010). This correlation has been confirmed in continental sections by Whiteside *et al.* (2010) and in marine sections by Ruhl *et al.* (2010).

Integrated stratigraphic studies focusing on cyclostratigraphy, magnetostratigraphy and basalt geochemistry (Deenen *et al.* 2010; Ruhl *et al.* 2010) have suggested that a series of short CAMP volcanic pulses took place 20–120 ka after a characteristic short (*c.* 25 ka) reverse polarity interval (E23r), first recognized in the Newark Basin (Kent & Olsen 1999). The Argana Basin of Morocco forms a key region for trans-Atlantic CAMP correlation and is ideally suited for testing the records of the North American basins. This requires, of course, high-resolution time control on both sides of the Atlantic. Astronomical timing of the magnetostratigraphic record provides such a high-resolution timescale for the volcanic and biologic events straddling the Triassic–Jurassic

From: Van Hinsbergen, D. J. J., Buiter, S. J. H., Torsvik, T. H., Gaina, C. & Webb, S. J. (eds) *The Formation and Evolution of Africa: A Synopsis of 3.8 Ga of Earth History.* Geological Society, London, Special Publications, **357**, 195–209. DOI: 10.1144/SP357.10   0305-8719/11/$15.00 © The Geological Society of London 2011.

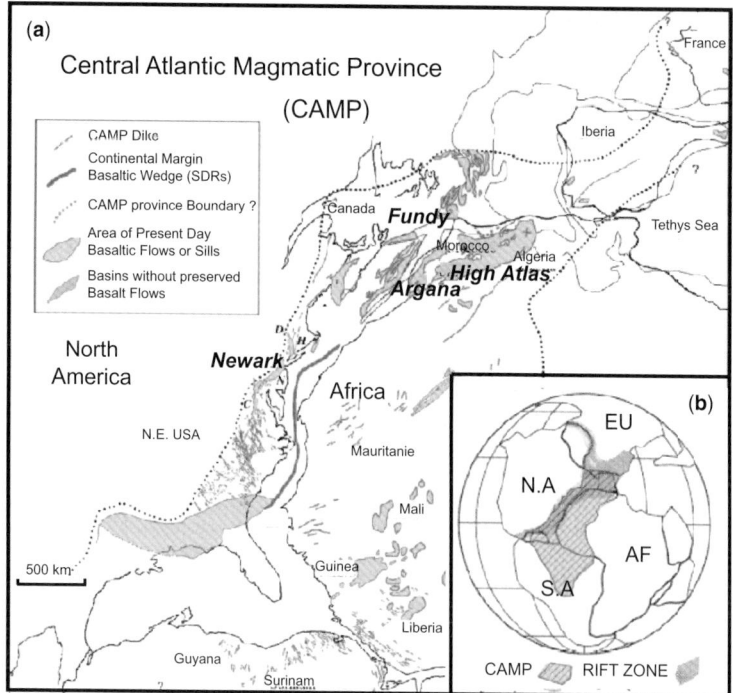

Fig. 1. Palaeoreconstruction (*c.* 200 Ma) of the Central Atlantic Magmatic Province (CAMP). (**a**) The important basins discussed in this study are indicated in large bold font. The extent of this magmatic province on the four continents surrounding the present-day Atlantic Ocean is indicated in (**b**). Maps are modified after McHone (2000).

boundary interval. Here we present palaeomagnetic analyses of upper Triassic red-beds and CAMP basalts on several sections in the Argana Basin, aiming to evaluate the late Triassic geomagnetic polarity timescale (GPTS) derived from the Newark supergroup basins (Kent *et al.* 1995; Kent & Olsen 1999, 2008) and the duration of the volcanic CAMP pulses (Knight *et al.* 2004).

The Argana Basin: geological setting and orbital forcing

Geological setting

Upper Permian, Triassic and lowermost Jurassic continental deposits are well exposed in the Argana Basin, situated between the Moroccan cities of Agadir and Marrakech along the western edge of the High Atlas mountain chain. The Argana Basin is the westwards extension of the Essaouira Basin, now separated from it due to Alpine orogeny (Hofmann *et al.* 2000). Middle- to Upper Triassic rocks in the Argana Valley consist of at least 2500 m of coarse- to fine-grained red-brown fluvial, lacustrine and floodplain clastic deposits (Olsen *et al.* 2003). A succession of eight lithofacies essentially consists of a lower coarse

stream-laid unit derived from nearby uplands, a middle lacustrine and deltaic complex and an upper aggradational mud-plain unit that passed westwards into an extensive salt flat (Hofmann *et al.* 2000). Thickness and lateral continuity of the sedimentary units vary considerably and this is attributed to a rather complex relation between sedimentation and differential movement of basement horsts and grabens during basin development.

Eight lithostratigraphic units prior to CAMP emplacement have been described by Tixeront (1973) and are discussed in more detail by Hofmann *et al.* (2000). In this study we sampled the Bigoudine formation, the CAMP basalts and the red-beds above CAMP. The stratigraphy in the southern part of the Argana Basin is laterally changing on a scale of kilometres (Fig. 2). The thickness of the complete lava sequence is thinning out towards the south of the basin, which indicates that emplacement of CAMP may occur as pulses on a restricted, regional scale. Additionally, the most southern part of the basin is characterized by the presence of intrusives (sills, dykes) of several tens of metres thickness. These intrusives have also been described by Ait Chayeb *et al.* (1998), who argued that they have a middle Jurassic age (K–Ar dating) of between 151 ± 8 and 157 ± 9 Ma (Brown 1980). In our sampled sections the intrusives are intercalated in and on top of the Bigoudine Formation (Fig. 2), and thus stratigraphically below the first CAMP volcanics.

Cyclostratigraphy of the upper Triassic sediments (Bigoudine Formation)

The best-studied and hence best-known basin that formed during the break-up of Pangea is the Newark Basin of New York, New Jersey and Pennsylvania. Scientific coring provided a c. 5000 m thick composite section for the entire Upper Triassic and lowermost Jurassic, therefore comprising one of the longest records of climatic cyclicity. The combination of magnetostratigraphy and cyclostratigraphic control makes it the reference GPTS for the Late Triassic (Kent et al. 1995; Kent & Olsen 1999). The c. 20 ka precession cycles are termed Van Houten cycles and range 3–6 m in thickness in the Newark Basin. The expression of this cycle is modulated by other orbital frequencies, notably by eccentricity variations with periods around 100 ka and – most prominently – of 404 ka, the cycle referred to as the McLaughlin cycle. Since the 404 and 100 ka cycles are constant during geological time, they can be used to obtain time control in cyclic sections, pinpoint events in a relative timescale and correlate to an absolute time frame if the relative timescale is tied to the radio-isotopically dated CAMP volcanics.

The same orbital variations appear to have controlled the depositional environment in the Argana Basin. Hofmann et al. (2000) observed various cycle thicknesses in the field and assumed that the smallest scale cyclicity would correspond to the precessional variation (c. 20 ka). Hence, they derived a sedimentation rate just prior to the onset of CAMP (in unit T8) of c. 9.5 cm ka^{-1} or c. 1.90 m per precession cycle. Deenen et al. (2010) studied a 200 m long uppermost Triassic section (AB-section in Fig. 2) in the Argana Basin for Milankovitch forcing on a high-resolution magnetic susceptibility record. They concluded that their most prominent c. 6 m cycle (Fig. 2, photo AB) was attributed to the c. 100 ka eccentricity cycle, while other peaks in the spectral analyses agreed very well with the orbital periods predicted for the latest Triassic. These results provided a first-order control on sedimentation rates (5.4 cm ka^{-1}) for the south-western part of the Argana Basin (Deenen et al. 2010).

Palaeomagnetic analyses of the Argana Basin

Palaeomagnetic sampling and methods

We densely sampled three red-bed sections in the western Argana Basin (Sections 1–3 in Fig. 2) to obtain a high-resolution composite magnetostratigraphy for the time interval around CAMP emplacement. A short interval of clastic sediments just prior to the first lavas of the CAMP (MO-series in Fig. 2) was earlier analysed to locate the short reverse chron E23r (Deenen et al. 2010). We also sampled the Argana CAMP lava flows near the village of Tazantoute, both along the road near Tazantoute and in the valley along the river including one level of baked sediment. Since the lavas along the road were quite weathered and gave questionable results, we resampled several lavas in a section up-hill (a total of 13 flow units with 4–11 cores per flow; Table 1) to determine polarity as well as palaeosecular variation (PSV) for comparison with the results of lavas in the High Atlas (Knight et al. 2004).

Samples were demagnetized both thermally (TH) and by alternating fields (AF). Thermal demagnetization was performed with temperature increments of 50–100 °C up to c. 500 °C followed by smaller steps (10–20 °C) towards 680 °C, in a magnetically shielded laboratory-built furnace. Alternating field demagnetization has been applied (up to 100 mT with increments of 5–10 mT) on an in-house developed robot, which let the samples pass through a 2G Enterprises SQUID magnetometer (noise level 10^{-12} Am^2). The natural remanent magnetization (NRM) of the thermally demagnetized samples was measured on a horizontal 2G Enterprises DC SQUID cryogenic magnetometer (noise level 3 × 10^{-12} Am^2). Orthogonal projection diagrams (Zijderveld 1967) were interpreted using principal component analysis (Kirschvink 1980). We used Fisher (1953) statistics to derive mean directions, after applying a variable cut-off (Vandamme 1994) to the corresponding virtual geomagnetic pole (VGP) distributions.

Upper Triassic red-beds (Bigoudine Formation)

Demagnetization diagrams in the three sampled sections have similar characteristics as the diagrams of samples from the central Argana Basin (Deenen et al. 2010) and the Fundy Basin in Canada (Kent & Olsen 2000; Deenen et al. 2011). The NRM generally consists of two components (Figs 3 & 4), in addition to an occasional small component removed at low temperature steps (<150 °C) likely reflecting a recent viscous overprint. The first component (A) is the most persistent and is usually observed in linear demagnetization trajectories in the temperature range 150–600/620 °C (or occasionally slightly higher which leads to the conclusion that this component most likely resides in haematite, which is confirmed by thermal decay curves; Fig. 3d). Considering its consistent direction of invariably normal polarity, we conclude that this

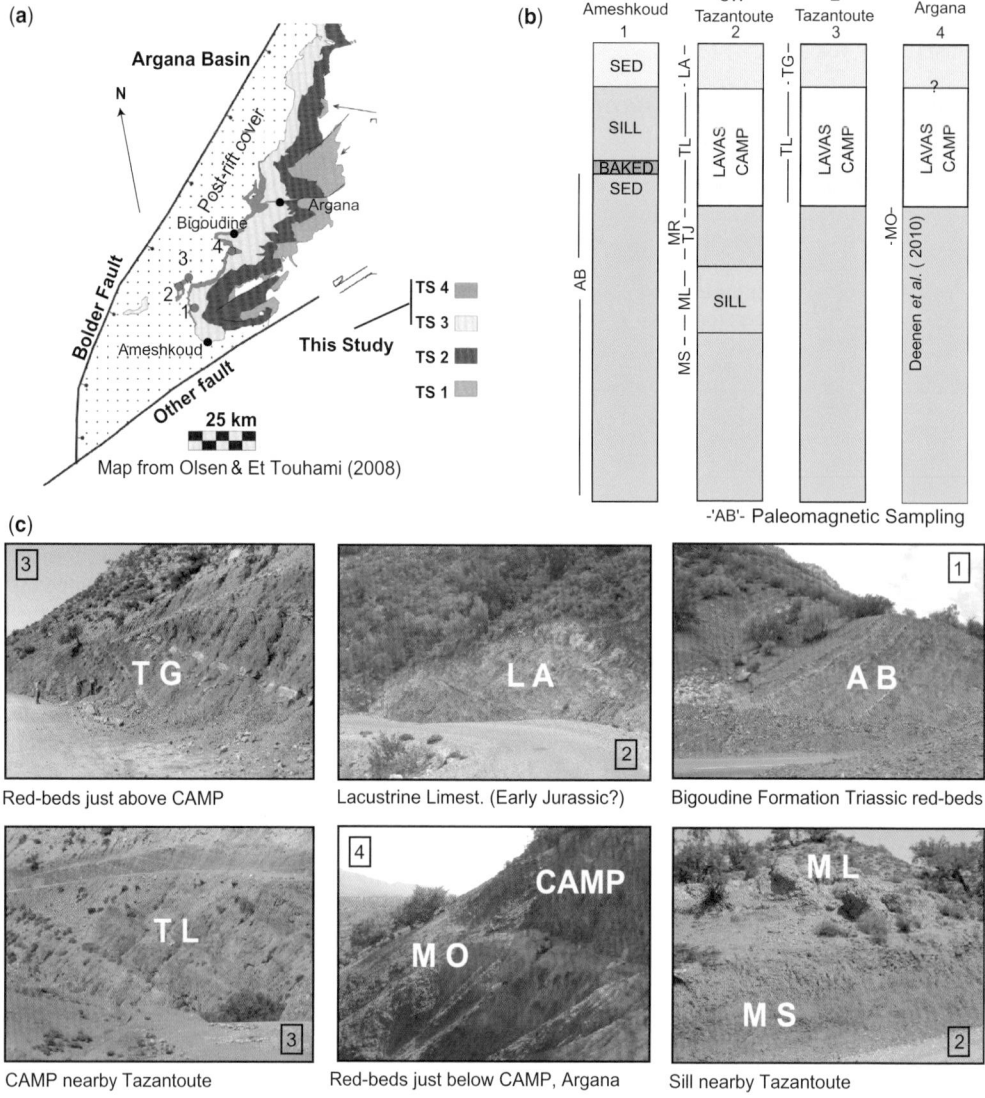

Fig. 2. Overview of Argana sections. (**a**) Geological map of the Argana Basin modified from Olsen & Et Touhami (2008) showing the sampling locations. TS 3 indicates the sediments of late Triassic age deposited prior to CAMP. TS 4 indicates CAMP basalts and the overlying younger sediments (latest Triassic–earliest Jurassic age). TS 1&2 represent Permian to middle–upper Triassic deposits which have not been considered in this study. (**b**) Schematic overview of the studied sections within the Argana Basin. Locations are indicated in (a). Lines and abbreviations along the columns correspond to palaeomagnetic sample tracks and codes used in the sections. (**c**) Photographs of the studied sections, numbers and abbreviations as in (a) and (b).

is a persistent overprint. A present-day or recent origin can be excluded here, because this component shows a substantial anticlockwise rotation and has much lower inclinations than the present-day geocentric axial dipole (GAD) field for the Argana Basin (Fig. 3a). The second component (B) is only removed at the highest temperatures, well above

500–600 °C and up to 680 °C, and was often not reliably determined because intensities became too low or the component was overprinted by the A component too much. This component most likely resides in haematite as well and is considered as the characteristic remanent magnetization (ChRM) for these red-beds (Deenen *et al.* 2010, 2011).

Table 1. *Palaeomagnetic results of the Tazantoute Valley lavas*

Site	Geochemical unit	N_{dem}	N_{sel}	Dec	Inc	k	α95	Strike	Dip	Dec-up	Inc-up	Strike-up	Dip-up	Dec-tc	Inc-tc
05A	IU	7	7	352.5	49.5	119.9	5.5	228	7	352.1	39.5	229.4	12.2	347.9	29.0
07A	IU	9	7	13.5	30.1	171.0	4.6	228	7	11.7	20.9	229.4	12.2	8.8	13.2
08A	IU	9	7	351.6	47.3	24.5	12.4	228	7	351.4	37.3	229.4	12.2	347.6	26.7
09A	LU	11	11	4.6	45.0	264.1	2.8	228	7	2.6	35.3	229.4	12.2	357.9	26.0
13A	IU	8	6	316.1	50.9	136.2	5.8	313	23	321.6	42.3	315.0	14.9	334.2	38.9
20A	IU	9	7	331.0	36.0	110.2	5.8	333	30	332.9	26.5	334.3	20.4	342.9	25.2
20B	IU	6	5	333.0	37.1	312.4	4.3	333	30	334.8	27.5	334.3	20.4	345.0	25.5
TL13	IU	6	3	302.2	51.5	96.3	12.6	313	23	309.9	44.3	315.0	14.9	324.4	43.7
TL14	IU	4	3	319.1	46.6	353.6	6.6	313	23	323.5	37.8	315.0	14.9	334.1	34.2
TL15	IU	7	6	312.9	43.0	564.2	2.8	320	18	317.5	34.8	321.2	9.3	324.0	34.9
TL16	IU	8	7	316.1	38.4	118.7	5.6	320	18	319.7	29.9	321.2	9.3	325.0	29.7
TL17	IU	8	8	333.5	39.8	136.2	4.8	320	18	335.4	30.2	321.2	9.3	340.3	27.5
TL18	IU	8	7	335.1	40.9	163.9	4.7	320	18	336.9	31.2	321.2	9.3	342.0	28.3
TS1+2	sed	6	6	340.9	47.2	119.3	6.2	318	22	342.2	37.3	319.7	13.4	350.4	31.3

Geochemical unit as defined in Deenen *et al.* (2010). N_{dem}, number of demagnetized samples; N_{sel}, number of usable samples; Dec, declinations; Inc, inclinations; *k*, estimate of the precision parameter; α95, cone of confidence derived from the ChRM directions; Strike/Dip, bedding plane used for tectonic correction; Dec-up, Inc-up, Strike-up and Dip-up, declinations, inclinations, inclination and tectonic bedding plane corrected for dipping anticline; Dec-tc and Inc-tc, declination and inclinations corrected for both bedding plane and dipping anticline.

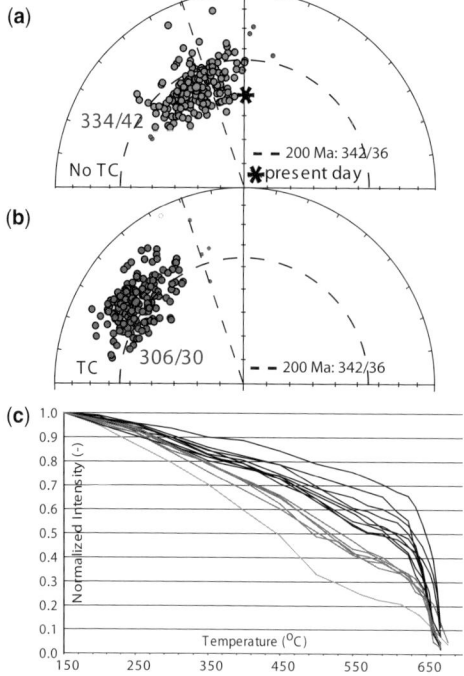

Fig. 3. Palaeomagnetic results of the A-component (overprint) from the Ameshkoud section. (**a, b**) Equal area projections for the uncorrected (No TC) and tectonically corrected (TC) A-component, in general derived from the 100–600 °C interval. Small symbols represent directions rejected by the Vandamme (1994) cut-off. We additionally show (dashed lines) the expected end-Triassic (200 Ma) direction for the Argana Basin calculated from the African palaeopole of Torsvik *et al.* (2008). (**c**) Representative normalised (150 °C) decay curves for thermally demagnetized samples throughout the section.

Several samples of the Ahmeskoud section (Fig. 2, Section 1) show evidence for reversed (R) field behaviour, because they clearly pass the origin of the Zijderveld diagrams (Fig. 5). The B component does not show clear antipodal normal and reverse directions, since it was not possible to reliably determine these high-temperature directions. The polarity, however, can be established with a variable degree of confidence and we interpret this component to be of primary (Triassic) origin since it shows both normal and reverse polarities (Fig. 3).

The palaeomagnetic results for the samples obtained from the red-beds below and above the sill near Tazantoute (MS and MR in Fig. 2; Section 2) show a straightforward behaviour. We find only normal directions for both A and B components for all measured samples (Fig. 4; see Deenen 2010

for details), although some samples (e.g. TJ 14, Fig. 4c) have a tendency to pass the origin at the highest temperatures. The directions of the A and B components are almost indistinguishable, and the persistent A component hampers a reliable determination of ChRM directions.

The palaeomagnetic results for the red-beds overlying the CAMP sequences (Fig. 2; Section 3) are not straightforward. We again observe a pervasive and normal polarity overprint direction in almost all samples (Fig. 6). The high-temperature component is mostly insufficiently resolved, although some samples tend to go to reversed polarity at the highest temperatures (see Deenen 2010 for details). These results are generally of poor quality and we must conclude that a straightforward magnetostratigraphic pattern of the sediments above CAMP cannot be resolved.

The directions of the A component could be reliably established in all three sections and the mean results are presented in Table 2 before (no tc) and after applying tectonic corrections (tc). A fold test applied to these directions is clearly negative, which indicates that the A component overprint is of post-folding origin (Fig. 7).

The CAMP lavas (Tazantoute sections)

Palaeomagnetic results from the lavas have been obtained from two sections near the village of Tazantoute (east- and SW Tazantoute sections, Fig. 2). Both sections form the limbs of a gently north-dipping (*c.* 10°) anticlinal structure. The succession of lavas corresponding to the Lower Unit (LU) and Intermediate Unit (IU) was sampled along the road as well as inside the valley along the river, which surprisingly gave different results. All lava flows of the road section show directions corresponding to the present-day field before tectonic correction suggesting that they have all subrecently been remagnetized. Only two lavas of the later sampled up-hill west section give results (Fig. 8). The Tazantoute lavas in the valley section, however, clearly give different results (Table 1). A small viscous component is removed in the first steps (max. 150 °C or *c.* 20 mT), after which most demagnetization diagrams indicate that the NRM has been completely removed at fields of 100 mT or temperatures of 580 °C. This indicates that (Ti-poor) magnetite is the main carrier of the magnetic signal in the lavas. It provides very consistent ChRM directions, which we interpret as primary components. We only present here the results from the lava flows that recorded primary late Triassic directions (Fig. 8).

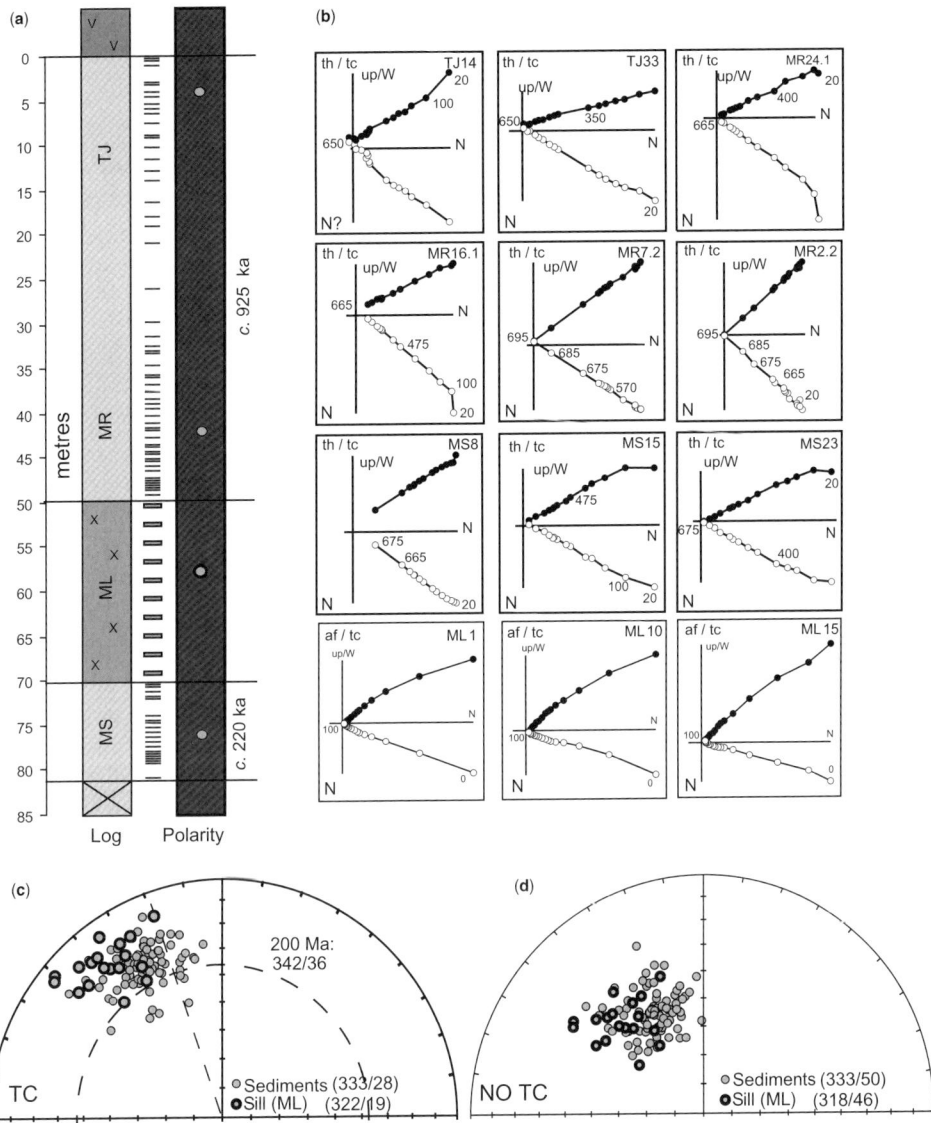

Fig. 4. Palaeomagnetism of the lower SW Tazantoute (sill) section. (**a**) Lithological column with sample levels indicated, all showing normal polarity. Approximate time present in the section is shown on the right and is derived from the average sedimentation rate in the nearby Ameshkoud section (c. 5.4 cm ka^{-1}). (**b**) Representative Zijderveld diagrams (TC) from the red-bed samples (MS, MR, TJ) and for the sill (MS) samples. Abbreviations (MS–ML–MR–TJ) correspond to sample tracks given in Figure 2. (**c**, **d**) Equal area projection of the high-temperature components before (no TC) and after (TC) tilt correction. Different symbols reflect sill and sediment samples. The expected Triassic direction at Argana is indicated by dashed lines.

Discussion

Pervasive overprint in the Argana red-beds: a Jurassic feature?

All sampled red-bed sediments below the first CAMP lavas show a very persistent overprint direction (component A), hampering the reliable determination of the primary B component. The A component must be of post-tilt origin since it is consistent in all three red-bed sections, but becomes randomly directed after tilt correction (Fig. 7). It does not correspond to the GAD direction at the present latitude, since it shows a significant c. 30°

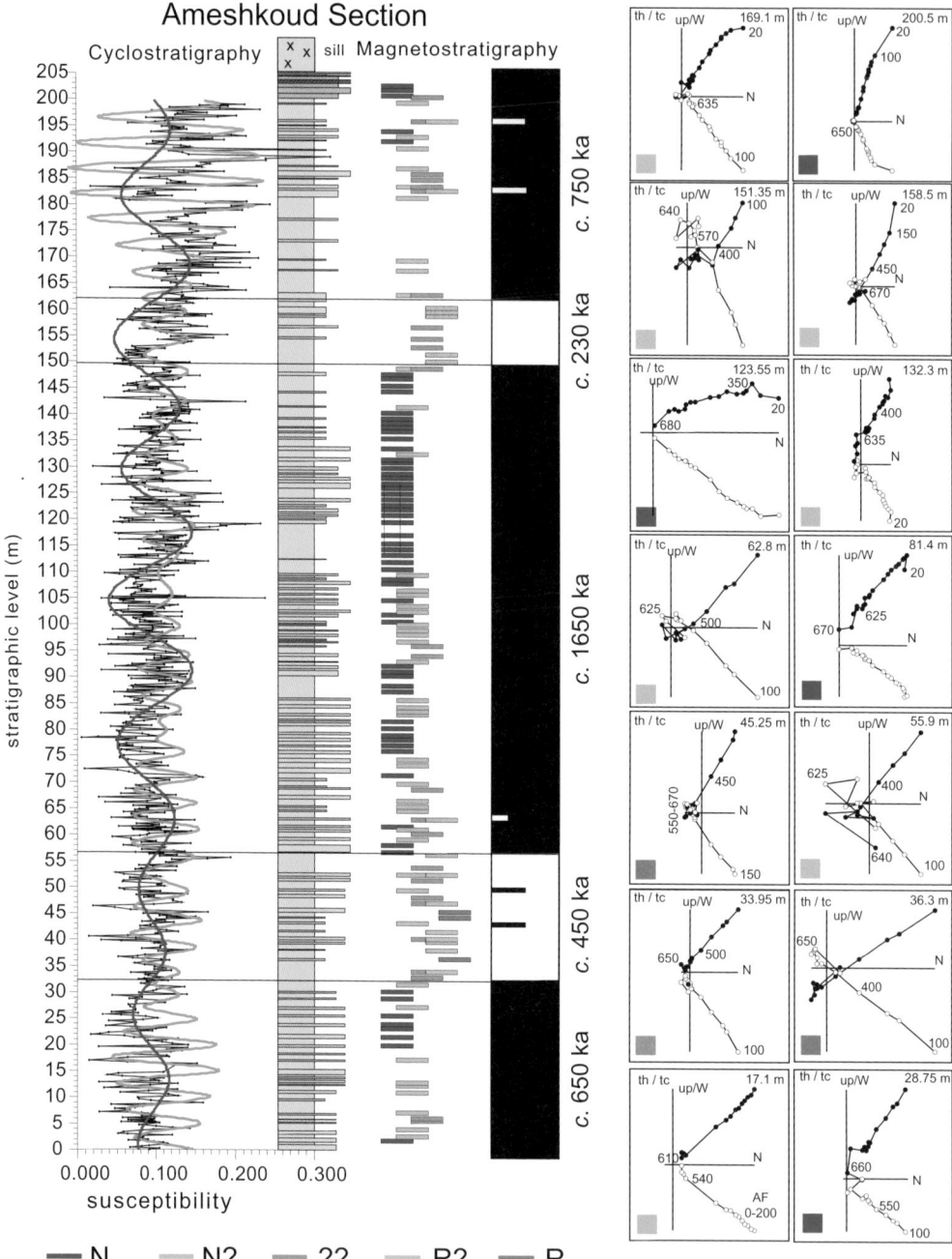

Fig. 5. Magnetostratigraphy for the cyclostratigraphically investigated Ameshkoud section. Magnetic susceptibility data are shown in the left panel together with the filtered 100 and 400 ka eccentricity signal (green and blue lines, respectively) according to Deenen *et al.* (2010). Palaeomagnetic directions cannot accurately be derived because of a pervasive overprint (Fig. 3). Our estimate of the polarity is indicated by N – N? – ?? – R? – R, where N(?) denotes (likely) normal, R(?) denotes (likely) reversed and ?? is uncertain. On the right are representative Zijderveld diagrams (all tectonically corrected) throughout the section; numbers refer to alternating field steps (mT) or temperatures (°C). Durations of the polarity zones are derived from the filtered 100 ka cycles.

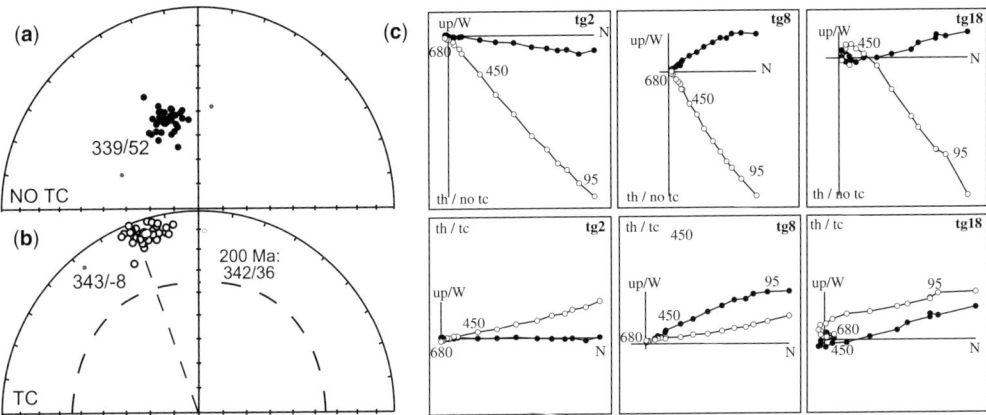

Fig. 6. Palaeomagnetic results for sediments above the CAMP lavas just NE of Tazantoute (TG sampling track in Section 3, Fig. 2). (**a, b**) Equal area projections for the uncorrected (no TC) and tectonically corrected (TC) overprint component, generally derived from the 100–600 °C interval. Small symbols have been rejected by the Vandamme (1994) cut-off. (**c**) Representative Zijderveld diagrams; numbers and symbols as in Figure 5.

anticlockwise declination deviation. Consequently, the overprint must be related to a post-Triassic and post-tilting pervasive remagnetization event affecting the Argana Basin.

To estimate the age of this remagnetization event, we compared the overall mean direction of the A component before tilt correction with the directions derived from the apparent polar wander path (APWP) of the last 200 Ma (Torsvik *et al.* 2008) for the location of the Argana Basin. It shows that two specific time intervals appear as likely periods of remagnetization: the early Jurassic interval of 190–160 Ma and the early Cretaceous interval of 110–100 Ma (Fig. 7). The older period roughly coincides with the assumed age of the sill intrusion (Brown 1980), which has been linked to a major thermal event caused by emplacement of plutonic bodies dated at *c.* 160 Ma in Triassic sedimentary rocks of the High- and Middle Atlas Basins (Rais 2002). We conclude that this event is the most likely candidate to have caused the pervasive overprint A component. This would imply, however, that folding and tilting of the Argana Basin had already occurred in Jurassic times. Alternatively, remagnetization may have been caused by processes related to inversion, folding and thrusting of the Atlas, which has been argued to have occurred between 30 and 20 Ma ago (Beauchamp *et al.* 1999). Indeed, the inclination of component A also coincides with Africa's inclination of the last 30 Ma (Fig. 7). This scenario would then require that our sections underwent a *c.* 25° counter-clockwise Late Oligocene or younger rotation. This is unlikely because our basalt directions – that now fit well with the directions from CAMP basalts in the High Atlas (Knight *et al.* 2004; Fig. 8) – would then have to

be corrected by *c.* 25°. Such corrected directions would then strongly disagree with the High Atlas results (unless the High Atlas has also experienced the same rotation) and both Argana and High Atlas would not have recorded Triassic directions. On the principle of least astonishment, we retain a Jurassic thermal event as the cause for remagnetization.

Trans-Atlantic correlation of upper Triassic red-beds

The very long and supposedly continuous continental record of the US Newark Basin makes it the reference section by choice, to which important events in the late Triassic can be correlated. Several studies, however, have questioned the completeness of the Newark record. In particular, the interval just below the first CAMP lavas has been suggested to contain a major unconformity, mainly because biostratigraphic studies have implied a Norian age for the sediments just below the end-Triassic palynologic turnover event (van Veen 1995; Kozur & Weems 2007; Cirilli *et al.* 2009). Consequently, these authors assumed a major hiatus in the Newark Supergroup Basin approximately one precession cycle below the first CAMP extrusives, for which there is no evidence in the field (Kent & Olsen 1999).

The Triassic–Jurassic palaeomagnetic pattern in the Newark Basin (Fig. 9) is characterized by normal polarity data, with only a very short reverse interval (E23r, *c.* 25 ka) of reverse directions just below the first CAMP lavas. This short reverse polarity chron is therefore of crucial importance

Table 2. *Summary of palaeomagnetic results*

Section	N_{dem}	N_{sel}	Dec	Inc	k	α95	K	A95	ΔD_x	ΔI_x	Dec_tc	Inc_tc	k	α95	K	A95	ΔD_x	ΔI_x
AB	174	169	335.5	42.2	43.3	1.8	38.8	1.8	1.9	2.3	306.1	30.3	49.1	1.6	53.8	1.5	1.6	2.4
TG	34	32	339.3	51.5	170.5	2.0	129.6	2.2	2.6	2.4	343.3	−8.0	170.5	2.0	281.7	1.5	1.5	3.0
ML (sill)	18	18	317.5	46.0	48.1	5.0	42.5	5.4	6.0	6.6	322.2	18.9	48.1	5.0	68.6	4.2	4.3	7.7
MS	22	22	335.6	52.5	176.1	2.3	145.7	2.6	3.1	2.7	332.8	26.7	176.1	2.3	290.2	1.8	2.9	3.1
MR	31	30	330.2	52.9	74.5	3.1	47.8	3.8	4.6	4.0	332.0	26.0	97.6	2.7	121.5	2.4	2.5	4.1
TJ	31	31	330.6	45.8	31.3	4.7	25.9	5.2	5.8	6.4	331.4	30.9	31.3	4.7	35.5	4.4	4.6	7.1
MS + MR + TJ	84	81	332.6	50.3	55.3	2.1	44.5	2.4	2.8	2.7	332.6	27.9	63.4	2.0	79.7	1.8	1.8	3.0
All sediments	3	3	335.8	48.0	216.5	8.4	252.1	7.8	8.9	9.1	328.2	17.5	8.5	45.1	13.8	34.6	35.1	64.4

N_{dem}, number of demagnetized samples; N_{sel}, number of usable samples; Dec, declinations; Inc, inclinations; k, estimate of the precision parameter based on ChRM directions; α95, cone of confidence derived from the ChRM directions; K, estimated of the precision parameter determined from Virtual Geomagnetic Poles (VGPs); A95, cone of confidence determined from VGPs; ΔD_x and ΔI_x, declination and inclination error; _tc, corrected for bedding tilt.

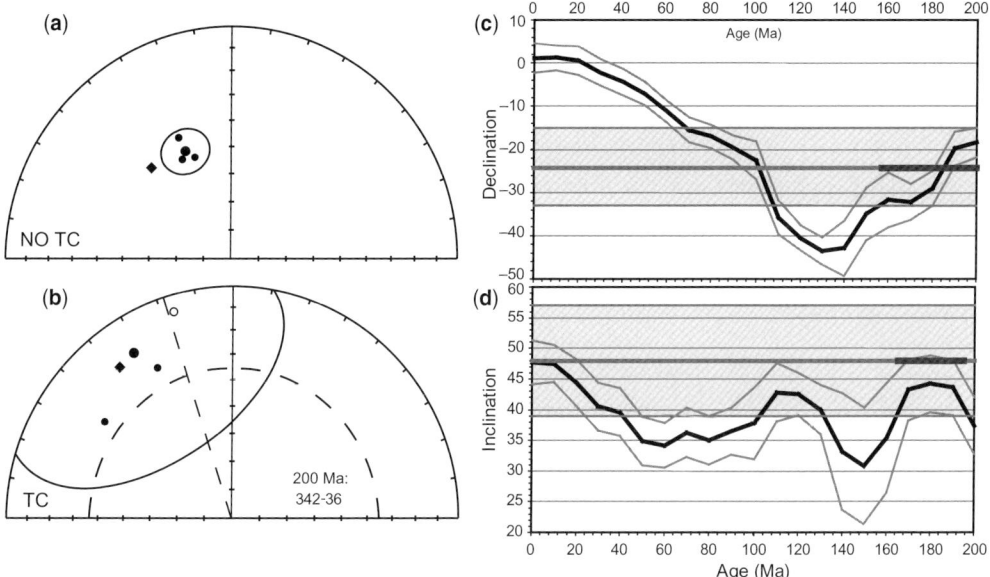

Fig. 7. Comparison of the mean overprint direction (**a**) before and (**b**) after tilt correction from the Ameshkoud section (Fig. 4), the SW Tazantoute (sill) section (Fig. 4) and the SW Tazantoute (above CAMP) section (Fig. 6). The sill mean direction (diamond) is shown separately. The means of the overprint directions are consistent before tilt correction, but highly scattered after tilt correction. (**c, d**) The mean of the (no TC) overprints is compared with the expected directions (declination and inclination) for Argana for the last 200 Ma (from Torsvik *et al.* 2008); shaded areas represent the confidence interval. The most likely fit of the mean overprint direction with the expected directions is indicated by a thick line; shading indicates the errors in declination and inclination. A best fit is found in either Jurassic or early Cretaceous times. The Jurassic magmatic pulse *c.* 160 Ma (Brown 1980; Ait Chayeb *et al.* 1988) seems a good candidate to have caused the overprint.

for intra-CAMP correlation based on palaeomagnetism. The topmost sedimentary interval just prior to the first CAMP in the Argana Basin also comprises an interval with transitional and reversed directions, correlated to E23r (Deenen *et al.* 2010). Cyclostratigraphic estimates further indicate a similar duration for this reversed interval as E23r in Newark. Moreover, chron E23r has also been resolved in the uppermost Triassic sediments of the Fundy Basin, a few centimetres below the transition to the CAMP (Partridge Island) basalt (Deenen *et al.* 2011). These three consistent recoveries of chron E23r indicate that the red-bed sections in Newark, Argana and Fundy are continuous and complete at least until the base of E23r, making further hypotheses on a hiatus in the Newark Supergroup record at the younger faunal turnover obsolete.

Marine–continental correlations for the uppermost Triassic Rhaetian Stage are also in disagreement with palaeomagnetic and cyclostratigraphic data (see Hounslow & Muttoni 2010; Hüsing *et al.* 2011). The hypothesis for a short (*c.* 1 Ma) Rhaetian Stage (Krystyn *et al.* 2002, 2007; Gallet *et al.* 2007; Kozur & Weems 2007), based on palaeontological arguments (conchostracans and equal ammonite duration) is in serious disagreement with the palaeomagnetic correlations from many sections in the Tethys realm that favour a significantly longer Rhaetian of *c.* 9 Ma (Kent *et al.* 1995; Channell *et al.* 2003; Muttoni *et al.* 2004, 2010; Hüsing *et al.* 2011).

The Rhaetian controversy can probably only be solved by detailed cyclostratigraphic correlations that confirm the astronomical duration of the Newark polarity pattern (Hüsing *et al.* 2011). Our composite magnetic polarity stratigraphy for the sediments deposited in the Argana Basin cannot however be correlated straightforwardly with the Newark Basin (Fig. 7). The cyclostratigraphic estimates for the Ameshkoud section suggest a polarity pattern with chron durations of >600 ka (N1), *c.* 450 ka (R1), *c.* 1600 ka (N2), *c.* 230 ka (R2) and an unknown interval (N3) until the base of CAMP. The most logical correlation to E21n, E21r, E22n, E22r and E23n results in unacceptable mismatches of up to *c.* 1 Ma. Potential explanations for these failed correlations are hiatuses in either Newark or Argana, incorrect stratigraphic correlations in the Argana Basin and unresolved

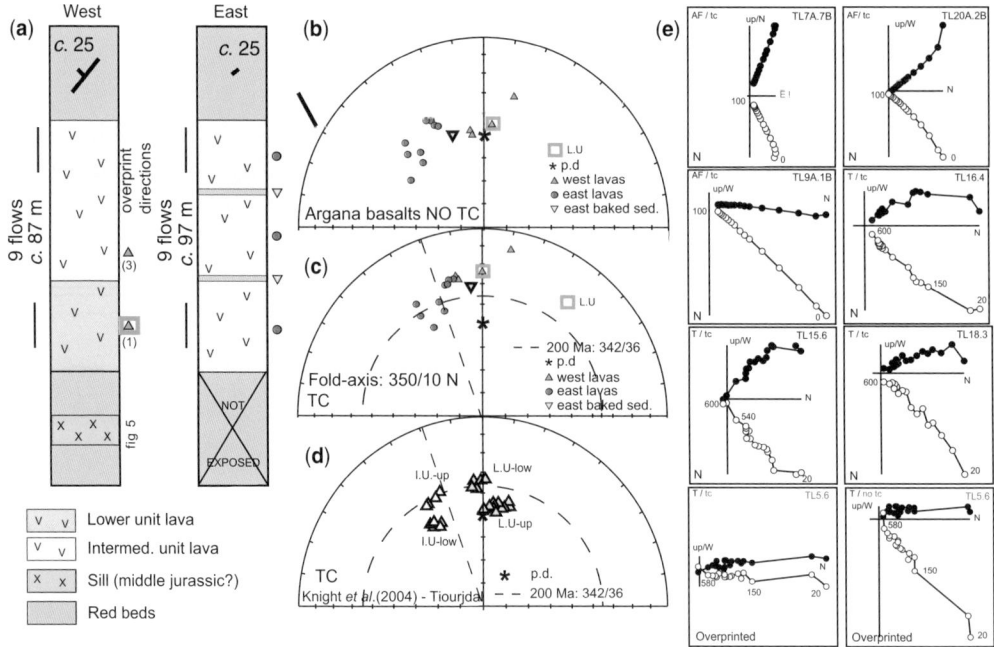

Fig. 8. Palaeomagnetism of the CAMP lavas. (**a**) Schematic overview of the sampled CAMP sections SW and east of Tazantoute. Colours (blue and yellow) correspond to the geochemical signature (lower- and intermediate unit, respectively; Deenen *et al.* 2010). Symbols on the right side of the columns correspond to the symbols used in (b–d). All lavas show normal polarity directions only. Equal area projections (**b**) uncorrected (no TC); (**c**) plunge plus tilt corrected; and (**d**) tilt corrected lavas from the High Atlas of Knight *et al.* (2004) for comparison with the Argana data. Dashed line represents the expected latest Triassic (200 Ma) direction at Argana (Torsvik *et al.* 2008). (**e**) Representative Zijderveld diagrams; numbers and symbols as in Figure 5.

palaeomagnetic signals in the Bigoudine red-beds. A far better match can be made to Newark chrons E15n, E15r, E16n, E16r and E17r, implying a substantial hiatus in our Ameshkoud section of *c.* 8 Ma (most likely at the level of the sill).

We conclude that an unequivocal correlation between the two trans-Atlantic basins is not yet feasible based on the present results. Future attempts should include more careful demagnetization in the 600–680 °C temperature interval and incorporation of more and longer cyclostratigraphic records, at locations where there is a clear continuous transition to the CAMP lavas. Alternatively, high-resolution sampling of the uppermost Triassic red-beds in the Fundy Basin – or in another suitable basin – may further improve the intra-CAMP basins correlation.

Duration of the CAMP pulses

Palaeomagnetic research on the CAMP lavas in the High Atlas, Morocco (Knight *et al.* 2004) together with their geochemistry (Marzoli *et al.* 2004) suggests that CAMP in the High Atlas has been

emplaced in three short consecutive pulses (Lower, Intermediate, Upper Units or LU, IU and UU) followed by one younger pulse (Recurrent Unit or RecU). These pulses each have a characteristic geochemical signature. Marzoli *et al.* (2004) use TiO_2 to discriminate between the different units. Deenen *et al.* (2010) show that these separate pulses are better distinguished/diagnosed with typical Rare Earth Elements (REE) ratios, in particular Y/Nb v. Lu/Hf. Previous geochemical results indicate that the lower- and intermediate units have been emplaced in the Argana Basin (Marzoli *et al.* 2004; Deenen *et al.* 2010), very similar to the emplacement in the High Atlas (Knight *et al.* 2004).

In general, our palaeomagnetic results for the IU-group lavas within the Argana Basin show directions that are very comparable to those predicted by the APWP (Fig. 8; Table 1; Torsvik *et al.* 2008). We therefore conclude that these lavas record the original palaeomagnetic field of the latest Triassic. This has also been shown for the IU-group lavas in the High Atlas (Fig. 8e; Knight *et al.* 2004). Our directional data from the Tazantoute IU-group

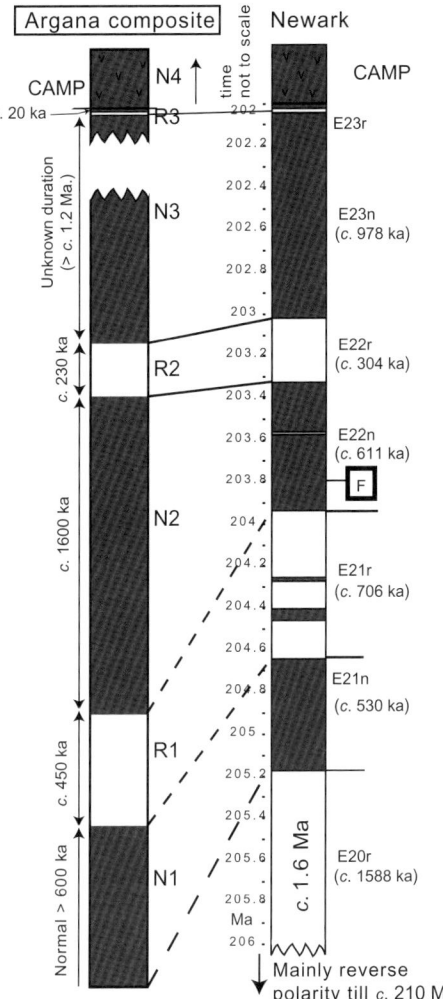

Fig. 9. A composite magnetostratigraphy of the Argana sequences deposited below the first CAMP lavas compared to the Newark GPTS (Kent & Olsen 1999) and Fundy Bay composite (Kent & Olsen 2000; Deenen 2010). Possible faults are indicated with 'F'. Correlations – or the lack thereof – are discussed in the text.

in the Argana Basin are also in very good agreement with those of the same unit from the High Atlas (Knight *et al.* 2004).

We obtained only one reliable direction from a LU-group lava (Fig. 8; Table 1), which is interpreted as a pre-folding direction of most probably latest Triassic origin. This LU lava flow shows a distinctly different direction with no rotation, but very similar to the directions observed in the High Atlas LU lavas (Fig. 8d; Knight *et al.* 2004). In the High Atlas samples, the mean direction of the LU-group

was shown to agree with a Triassic inclination. Secular variation of the field, recorded as distinct directional groups reflecting the short CAMP volcanic pulses, has been put forward to explain the difference between LU and IU directions (Knight *et al.* 2004). Our results from Tazantoute are in good agreement and support the conclusion that the onset of volcanism in Morocco (both in the High Atlas and Argana basins) occurred in many short synchronous pulses.

Conclusions

The palaeomagnetic signal of the upper Triassic red-beds of the Argana Basin in Morocco shows a pervasive overprint, carried by haematite, that is of post-folding origin but significantly different from post-early Cretaceous directions expected at the location of the Argana Basin. It must therefore have been acquired before the late Cretaceous. Plotted against the expected directions of the last 200 Ma it appears to be most likely of Mesozoic origin related to a remagnetization event at 190–160 Ma, in agreement with a Jurassic magmatic pulse around 160 Ma (Brown 1980; Ait Chayeb *et al.* 1988).

A dual-polarity component is revealed at high temperatures of 500–680 °C, also residing in haematite, which is interpreted as the primary magnetic signal. The topmost sedimentary interval just prior to the first CAMP comprises a short interval with transitional and reverse directions. This correlates to chron E23r (Deenen *et al.* 2010), indicating that the red-bed sections in Argana, Newark, and Fundy are continuous and complete at least until the base of E23r. An unequivocal correlation of the upper Triassic Argana magnetostratigraphy of the Ameshkoud and Tazantoute red-bed sections to the Newark GPTS is not yet feasible; future attempts should include more careful demagnetization in the 600–680 °C temperature interval and incorporation of longer cyclostratigraphic records that are demonstrably continuous up to the CAMP lavas.

The palaeomagnetic signal in the CAMP lavas of the Tazantoute sections shows significantly different results for the Lower Unit and Intermediate Unit. These results are in good agreement with the results from CAMP lavas of the High Atlas (Knight *et al.* 2004) and support the conclusion that the onset of volcanism in Morocco (both in the High Atlas and Argana basins) occurred in many short synchronous pulses.

We thank M. Reitsma, M. Haldan and M. Ruhl for their help in the field. The critical and constructive comments of V. C. Ruiz-Martinez and an anonymous reviewer were very welcome and significantly improved the manuscript. This study was funded by the High-Potential project

Earth's and life's history: from core to biosphere (CoBi) of the Utrecht University.

References

AIT CHAYEB, E. H., YOUBI, N., EL-BOUKHARI, A., BOUAB-DELLI, M. & AMRHAR, M. 1998. Permian–Mesozoic volanism of the Argana Basin (western High Atlas, Morocco); intraplate magmatism associated with the opening of the Central Atlantic. *Journal of African Earth Sciences*, **26**, 499–519.

BEAUCHAMP, W., ALLMENDINGER, R. W., BARAZANGI, M., DEMNATI, A., EL ALJI, M. & DAHMANI, M. 1999. Inversion tectonics and the evolution of the High Atlas Mountains, Morocco, based on a geological–geophysical transect. *Tectonics*, **18**, 163–184.

BROWN, R. H. 1980. Triassic rocks of Argana Valley, Southern Morocco, and their regional structural implication. *American Association Petroleum Geologists Bulletin*, **64**, 988–1003.

CHANNELL, J. E. T., KOZUR, H. W., SIEVERS, T., MOCK, R., AUBRECHT, R. & SYKORA, M. 2003. Carnian–Norian biomagnetostratigraphy at Silická Brezová (Slovakia): Correlation to other Tethyan sections and to the Newark Basin. *Palaeogeography, Palaeoclimatology, Palaeoecology*, **191**, 65–109.

CIRILLI, S., MARZOLI, A. *ET AL.* 2009. Latest Triassic onset of the Central Atlantic Magmatic Province (CAMP) volcanism in the Fundy Basin (Nova Scotia): New stratigraphic constraints. *Earth and Planetary Science Letters*, **286**, 514–525.

COURTILLOT, V. 1994. Mass extinctions in the last 300 million years: one impact and seven flood basalts? *Israel Journal of Earth Sciences*, **43**, 255–266.

COURTILLOT, V. E. & RENNE, P. R. 2003. On the ages of flood basalt events. *Comptes Rendus – Geoscience*, **335**, 113–140.

DEENEN, M. H. L. 2010. *A new chronology for the late Triassic to early Jurassic.* PhD thesis, Utrecht, Geologica Ultraiectina No. 323 Universiteit Utrecht, http://igitur-archive.library.uu.nl/dissertations/2010-0414-200155/UUindex.html.

DEENEN, M. H. L., RUHL, M., BONIS, N. R., KRIJGSMAN, W., KUERSCHNER, W. M., REITSMA, M. & VAN BERGEN, M. J. 2010. A new chronology for the end-Triassic mass extinction. *Earth and Planetary Science Letters*, **291**, 113–125.

DEENEN, M. H. L., KRIJGSMAN, W. & RUHL, M. 2011. The quest for chron E23r at Partridge Island, Bay of Fundy, Canada: CAMP emplacement post-dates the end-Triassic extinction event at the North American craton. *Journal of Canadian Earth Sciences*, **48**, 1282–1291.

FISHER, R. A. 1953. Dispersion on a sphere. *Proceedings of the Royal Society London*, **217A**, 295–305.

GALLET, Y., KRYSTYN, L., MARCOUX, J. & BESSE, J. 2007. New constraints on the End-Triassic (Upper Norian–Rhaetian) magnetostratigraphy. *Earth and Planetary Science Letters*, **255**, 458–470.

HAY, W. W., BEHENSKY, J. F., JR., BARRON, E. J. & SLOAN, J. L., II. 1982. Late Triassic–Liassic paleoclimatology of the photo-central North Atlantic rift system. *Palaeogeography, Palaeoclimatology, Palaeoecology*, **40**, 13–30.

HOFMANN, A., TOURANI, A. & GAUPP, R. 2000. Cyclicity of Triassic to Lower Jurassic continental red beds of the Argana Valley, Morocco: implications for palaeoclimate and basin evolution. *Palaeogeography, Palaeoclimatology, Palaeoecology*, **161**, 229–266.

HOUNSLOW, M. W. & MUTTONI, G. 2010. The geomagnetic polarity timescale for the Triassic: linkage to stage boundary definitions. *In*: LUCAS, S. G. (ed.) *The Triassic Timescale*. Geological Society, London, Special Publications, **334**, 61–102.

HÜSING, S. K., DEENEN, M. H. L., KOOPMANS, J. & KRIJGSMAN, W. 2011. Magnetostratigraphic dating of the Rhaetian GSSP at Steinbergkogel (late Triassic, Austria): implications for the Upper Triassic Time Scale. *Earth and Planetary Science Letters*, **302**, 203–216.

KENT, D. V. & OLSEN, P. E. 1999. Astronomically tuned geomagnetic polarity timescale for the Late Triassic. *Journal of Geophysical Research-Solid Earth*, **104**, 12831–12841.

KENT, D. V. & OLSEN, P. E. 2000. Magnetic polarity stratigraphy and paleolatitude of the Triassic–Jurassic Blomidon Formation in the Fundy basin (Canada): implications for early Mesozoic tropical climate gradients. *Earth and Planetary Science Letters*, **179**, 311–324.

KENT, D. V. & TAUXE, L. 2005. Corrected Late Triassic latitudes for continents adjacent to the North Atlantic. *Science*, **307**, 240–244.

KENT, D. V., OLSEN, P. E. & WITTE, W. K. 1995. Late Triassic–earliest Jurassic geomagnetic polarity sequence and paleolatitudes from drill cores in the Newark rift basin, Eastern North-America. *Journal of Geophysical Research*, **100**, 14965–14998.

KIRSCHVINK, J. L. 1980. The least-squares line and plane and the analysis of paleomagnetic data. *Geophysical Journal of the Royal Astronomical Society*, **62**, 699–718.

KNIGHT, K. B., NOMADE, S., RENNE, P. R., MARZOLI, A., BERTRAND, H. & YOUBI, N. 2004. The Central Atlantic Magmatic Province at the Triassic–Jurassic boundary: paleomagnetic and $^{40}Ar/^{39}Ar$ evidence from Morocco for brief, episodic volcanism. *Earth and Planetary Science Letters*, **228**, 143–160.

KOZUR, H. W. & WEEMS, R. 2007. Upper Triassic conchostracan biostratigraphy of the continental rift basins of Eastern North America: it's implication for correlating Newark supergroup events with the Germanic basin and the international geologic time scale. *In*: LUCAS, S. G. & SPIELMANN, J. A. (eds) *The Global Triassic Bulletin, New Mexico Museum of Natural History and Science*, 137–188.

KRYSTYN, L., BOUQUEREL, H., KUERSCHNER, W., RICHOZ, S. & GALLET, Y. 2007. Proposal for a candidate GSSP for the base of the Rhaetian stage. *In*: LUCAS, S. G. & SPIELMANN, J. A. (eds) *The Global Triassic*. New Mexico Museum of Natural History and Science Bulletin, **41**, 189–199.

KRYSTYN, L., GALLET, Y., BESSE, J. & MARCOUX, J. 2009. Integrated Upper Carnian to Lower Norian biochronology and implications for the Upper Triassic polarity timescale. *Earth and Planetary Science Letters*, **203**, 343–351.

MARZOLI, A., BERTRAND, H. *ET AL.* 1999. Extensive 200-million-year-old continental flood basalts of the Central Atlantic Magmatic Province. *Science*, **284**, 616–618.

MARZOLI, A., BERTRAND, H. *ET AL.* 2004. Synchrony of the Central Atlantic magmatic province and the Triassic–Jurassic boundary climatic and biotic crisis. *Geology*, **32**, 973–976.

MCHONE, J. G. 2000. Non-plume magmatism and rifting during the opening of the central Atlantic Ocean. *Tectonophysics*, **316**, 287–296.

MUTTONI, G., KENT, D. V., OLSEN, P. E., DI STEFANO, P., LOWRIE, W., BERNASCONI, S. M. & HERNÁNDEZ, F. M. 2004. Tethyan magnetostratigraphy from Pizzo Mondello (Sicily) and correlation to the Late Triassic Newark astrochronological polarity time scale. *Geological Society of America Bulletin*, **116**, 1043–1058, doi:10.1130/B25326.1.

MUTTONI, G., KENT, D. V., JADOUL, F., OLSEN, P. E., RIGO, M., GALLI, M. T. & NICORA, A. 2010. Rhaetian magneto-biostratigraphy from the Southern Alps (Italy): constraints on Triassic chronology. *Palaeogeography, Palaeoclimatology, Palaeoecology*, **285**, 1–16.

OLSEN, P. E., KENT, D. V. *ET AL.* 2002. Ascent of dinosaurs linked to an iridium anomaly at the Triassic–Jurassic boundary. *Science*, **96**, 1305–1307.

OLSEN, P. E. & ET TOUHAMI, M. 2008. Field trip 1: Tropical to Subtropical Syntectonic Sedimentation in the Permian to Jurassic Fundy Rift Basin, Atlantic Canada. *In: Relation to the Moroccan Conjugate Margin.* Central Atlantic Conjugate Margins Conference, Halifax, Canada.

OLSEN, P. E., KENT, D. V., FOWELL, S. J., SCHLISCHE, R. W., WITHJACK, M. O. & LETOURNEAU, P. M. 2000. Implications of a comparison of the stratigraphy and depositional environments of the Argana (Morocco) and Fundy (Nova Scotia, Canada) Permian–Jurassic basins. *In*: OUJIDI, M. & ET-TOUHAMI, M. (eds) *Le Permien et le Trias du Maroc: Actes de la Premiere Reunion du GroupeMarocain du Permien et du Trias.* Hilal Impression, Oujda, 165–183.

OLSEN, P. E., KENT, D. V., ET-TOUHAMI, M. & PUFFER, J. 2003. Cyclo-, magneto-, and bio-stratigraphic constraints on the duration of the CAMP event and its relationship to the Triassic–Jurassic boundary. *In*: HAMES, W. E., MCHONE, G., RENNE, P. R. & RUPPEL, C. (eds) *The Central Atlantic Magmatic Province: Insights from Fragments of Pangea.* AGU, Geophysical Monographs, **136**, 7–32.

RAIS, N. 2002. *Les roches triasico-liasiques du Maroc septentrional et leur socle hercynien: caractérisation*

pétrologique, minéralogique, cristallochimique et isotopique K–Ar de leur évolution thermique postformationnelle. PhD thesis. University of Saïss-Fès.

RAMPINO, M. R. & STOTHERS, R. B. 1988. Flood basalt volcanism during the past 250 million years. *Science*, **241**, 663–668.

RAUP, D. M. & SEPKOSKI, J. J., JR. 1982. Mass extinctions in the marine fossil record. *Science*, **215**, 1501–1503.

RUHL, M., DEENEN, M. H. L., ABELS, H. A., BONIS, N. R., KRIJGSMAN, W. & KÜRSCHNER, W. M. 2010. Astronomical constraints on the duration of the early Jurassic Hettangian stage and recovery rates following the end-Triassic mass extinction (St. Audrie's Bay/East Quantoxhead, UK). *Earth and Planetary Science Letters*, **295**, 262–276.

SMOOT, J. P. & OLSEN, P. E. 1988. Massive mudstones in basin analysis and paleoclimatic interpretation of the Newark Supergroup. *In*: MANSPEIZER, W. (ed.) *Triassic–Jurassic Rifting, Continental Breakup and the Origin of the Atlantic Ocean and Passive Margins.* Elsevier, New York, 249–274.

TIXERONT, M. 1973. Lithostratigraphie et minéralisation cupriferes et uraniferes stratiformes syngenetiques et familières des formations detriques permo-triasiques du Couloir d'Argana (Haut-Atlas occidental, Maroc). *Notes Serv. Geol. Maroc*, **249**, 147–177.

TORSVIK, T. H., MÜLLER, R. D., VAN DER VOO, R., STEINBERGER, B. & GAINA, C. 2008. Global plate motion frames: toward a unified model. *Reviews of Geophysics*, **46**, RG3004, doi: 10.1029/2007RG000227.

VAN VEEN, P. M. 1995. Time calibration of Triassic/Jurassic microfloral turnover, eastern North America-Comment. *Tectonophysics*, **245**, 93–95.

VANDAMME, D. 1994. A new method to determine paleosecular variation. *Physics of the Earth and Planetary Interiors*, **85**, 131–142.

WHITESIDE, J. H., OLSEN, P. E., KENT, D. V., FOWELL, S. J. & ET-TOUHAMI, M. 2007. Synchrony between the Central Atlantic magmatic province and the Triassic–Jurassic mass-extinction event? *Palaeogeography, Palaeoclimatology, Palaeoecology*, **244**, 345–367.

WHITESIDE, J. H., OLSEN, P. E., EGLINTON, T., BROOKFIELD, M. E. & SAMBROTTO, R. N. 2010. Compound-specific carbon isotopes from Earth's largest flood basalt eruptions directly linked to the end-Triassic mass extinction. *Proceedings of the National Academy of Sciences of the United States of America*, **107**, 6721–6725.

ZIJDERVELD, J. D. A. 1967. A. C. demagnetization of rocks: analysis of results. *In*: RUNCORN, S. K. (ed.) *Methods in Palaeomagnetism.* Elsevier, Amsterdam, New York, 254–286.

Palaeomagnetic and AMS study of the Tarfaya coastal basin, Morocco: an early Turonian palaeopole for the African plate

VICENTE CARLOS RUIZ-MARTÍNEZ[1]*, ALICIA PALENCIA-ORTAS[1],
JUAN JOSÉ VILLALAÍN[2], GREGG MCINTOSH[1] & FÁTIMA MARTÍN-HERNÁNDEZ[1]

[1]*Departamento de Geofísica y Meteorología, Facultad de Física, Universidad Complutense de Madrid, Avda. Complutense s/n, 28040, Madrid*

[2]*Departamento de Física, Escuela Técnica Superior, Universidad de Burgos, Avda. Cantabria s/n, 09006, Burgos*

*Corresponding author (e-mail: vcarlos@fis.ucm.es)

Abstract: An early Turonian (*c.* 93 Ma) anoxic, cyclic marine deposition is registered in the unfolded outcrops from the Tarfaya coastal basin, where very high sedimentation rates enable the investigation of past geomagnetic field record at high temporal resolution. One hundred and fourteen samples have been sampled along a 10.5 m vertical profile (*c.* 200–500 ka) of orbital-scale forced sedimentation. Rock magnetic investigations reveal mineralogy principally controlled by diamagnetic and paramagnetic behaviour, along with very low concentrations of low-coercivity ferromagnetic material which is probably magnetite. A well-defined magnetic fabric can be seen with the minimum susceptibility axis perpendicular to the foliation plane, and magnetic lineation compatible with NW African palaeostress since sedimentation times and/or the palaeocurrent associated with upwelling system deposition. Magnetic signature has the potential for performing reliability checks of reversed tiny wiggles, which were found in four samples not considered for the tectonic analysis. Alternating field demagnetization shows a single, stable, low-coercivity directional component. The new palaeopole ($N = 88$; PLat $= 64.3°$, PLon $= 256.3°$, $A_{95} = 2.5°$; $K = 38.7$), obtained after moderate ($f = 0.8$) inclination flattening correction, is the first early Turonian palaeopole for the NW African Craton. It can contribute to the 90 Ma-centred sliding window of the different proposed synthetic Apparent Polar Wander Paths.

The Tarfaya coastal basin (NW African continental margin, SW Morocco, north–NE oriented between 28°N and 24°N) is limited by the Anti-Atlas mountains to the north, the Mauritanides orogenic domain to the south, the Reguibat shield in the east and the East Canary Ridge to the west; it extends into the Senegal Basin to the south and the Palaeozoic Tindouf Basin to the east. Detailed and comprehensive geological descriptions of this area have been published, including those by Choubert *et al.* (1966, 1972), Martinis & Visintin (1966), Wiedmann *et al.* (1978, 1982), von Rad & Einsele (1980), Ranke *et al.* (1982), Leine (1986) and Heyman (1990). The basin is situated at the tectonically stable western margin of the Saharan Platform, and its basement is a complex of folded Precambrian and Palaeozoic igneous and metamorphic rocks which is discordantly overlain by Mesozoic and Cenozoic sediments (Ranke *et al.* 1982) whose thickness locally exceeds 12 km (Heyman 1990). The sedimentary evolution of the Tarfaya Basin is closely connected to the geological history of the African Craton and the opening of the Atlantic Ocean, resulting in post-Triassic basin subsidence

(Ranke *et al.* 1982; Wiedmann *et al.* 1982). During the Miocene, the Upper Cretaceous succession was partially eroded and the deposition of the relatively thin Moghrebian Formation took place.

Widespread laminated, organic-rich siltstones, limestones and marls were deposited in the basin during Late Cretaceous (Wiedmann *et al.* 1982; Leine 1986) times, registering the Cenomanian/Turonian (C/T) oceanic anoxic event (Bonarelli event, C/T OAE, OAE 2, *c.* 93 Ma). During this transgressive phase, organic-rich strata were deposited in rift shelf basins and slopes across North Africa and in deep-sea basins of the adjacent oceans (e.g. Lüning *et al.* 2004). The Upper Cretaceous pre- to post-OAE 2 deposits in the Tarfaya Basin contain an exceptionally thick (700–800 m) unit of potential petroleum source rock (Leine 1986) and have therefore been intensively investigated. They consist of dark brownish-grey, laminated, kerogenous chalks, alternating with non-laminated lighter-coloured often nodular limestones containing a lower kerogen content and intercalated concretions and lenses of chert and siliceous limestone (e.g. Lüning *et al.* 2004). During late

From: VAN HINSBERGEN, D. J. J., BUITER, S. J. H., TORSVIK, T. H., GAINA, C. & WEBB, S. J. (eds) *The Formation and Evolution of Africa: A Synopsis of 3.8 Ga of Earth History*. Geological Society, London, Special Publications, **357**, 211–227. DOI: 10.1144/SP357.11 0305-8719/11/$15.00 © The Geological Society of London 2011.

Cenomanian and early Turonian times, deposition occurred in an open shelf setting to the south at a depth of 200–300 m (Kuhnt *et al.* 1990), but rapidly shallowed towards the palaeoshoreline to the NE as indicated by increasing terrigenous influx (Gebhardt *et al.* 2004). At several expanded successions (Fig. 1b) from exploration wells (e.g. S13, S75) and individual outcrops along the shoreline or around Oueds and Sebkhas (e.g. Mohammed Plage), Mid-Cenomanian–Early Turonian distal to proximal basin strata have been subdivided into various different microfacies and their palaeoenvironmental evolution has been described (El Albani *et al.* 1999*a*). Different geo-, chemo- and biostratigraphical proxies have been studied for this anoxic period across the C/T boundary, which correlate these Tarfaya Basin sections among themselves and with other successions from the proto-North Atlantic and Tethys Oceans (Kuhnt *et al.* 1990, 1997; Holbourn *et al.* 1999; Luderer 1999; Kolonic *et al.* 2002; Kuypers *et al.* 2002; Tsikos *et al.* 2004; Jenkyns *et al.* 2007; Keller *et al.* 2008; Mort *et al.* 2008; Tantawy 2008). The Tarfaya Basin cyclicity has been interpreted as being forced by the Milankovitch precession (20 ka) and obliquity (39 ka) signals (Kuhnt *et al.* 1997, 2005; Luderer 1999), but short eccentricity (100 ka) origins (Kolonic *et al.* 2005; Kuhnt *et al.* 2009) for some of the observed cycles has also been proposed (each cycle representing a couplet of carbonate-rich

and organic carbon-rich beds). Accordingly, very high sedimentation rates (of the order cm ka^{-1}) at low sea-level proximal settings, increasing up to around 10 cm ka^{-1} at more distal mid- to outer-shelf successions, have been estimated (Luderer 1999; Kolonic *et al.* 2005; Kuhnt *et al.* 2009).

Nevertheless, the magnetic signatures of these key sediments had not been tested. The main goal of this study is to evaluate the magnetic signal of this Upper Cretaceous cyclic deposition close to the palaeocoastline, taking advantage of the available high temporal resolution (independently of the dominant orbital forcing frequency involved) of these expanded sedimentary successions and the absence of any tectonic disruptions affecting them. A Lower Turonian composite vertical profile has been sampled at the Amma Fatma beach cliffs to carry out a detailed palaeomagnetic analysis with the following objectives: (i) to test the response of anisotropy of magnetic susceptibility (AMS) to past tectonic stresses and palaeocurrents (Tarling & Hrouda 1993), (ii) to study ancient geomagnetic field behaviour in a high-resolution Upper Cretaceous temporal bracket, corresponding to the Cretaceous Normal polarity Superchron (CNS, *c.* 83–125 Ma) and (iii) to provide a Lower Turonian palaeomagnetic pole for the NW African Craton, a period that demonstrates a scarcity of good-quality palaeomagnetic poles (e.g. Besse & Courtillot 2002 or Torsvik *et al.* 2008 synthetic Apparent Polar

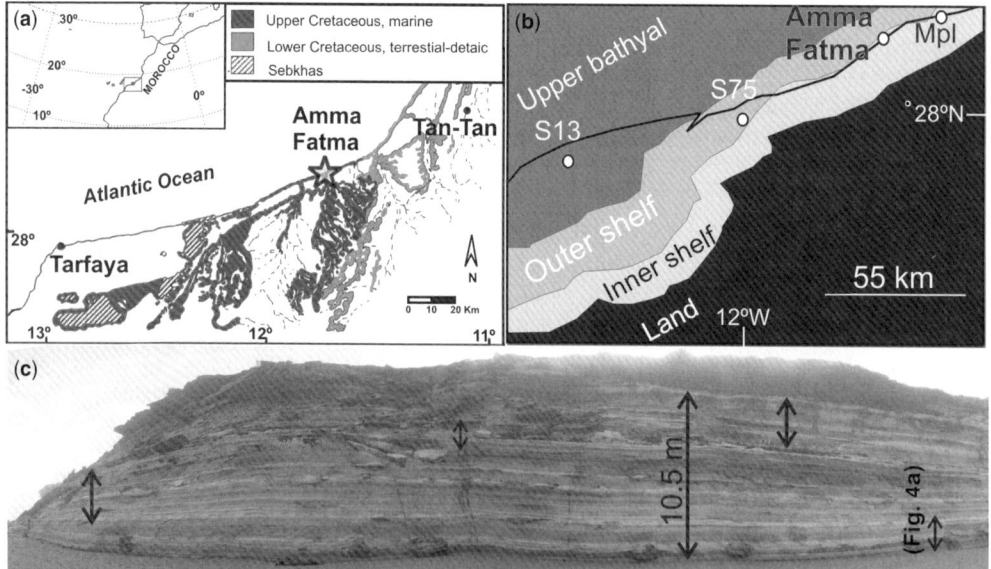

Fig. 1. (**a**) Simplified geological map of Tarfaya Basin (after Choubert *et al.* 1966) showing location of Amma Fatma succession; (**b**) palaeogeographic map of the Tarfaya Basin (modified from Kolonic *et al.* 2002) indicates palaeopositions of other previously studied successions from exploration wells (S13, S75) and individual outcrops (Mpl, Mohammed Plage) and (**c**) palaeomagnetic sampling composite profile (this study) at Amma Fatma beach.

Wander Paths; referred to as BC2002 and T2008 APWPs, respectively). In order to achieve these objectives, rock magnetic experiments have been performed to analyse the magnetic carriers of both the AMS and the observed remanence. The importance of inclination shallowing of the Characteristic Remanent Magnetizations (ChRMs) has been studied and the effects of averaging of geomagnetic palaeosecular variation (PSV) have been examined.

Studied section and sampling

The almost horizontal stratified Upper Cretaceous studied succession ($28.21°N$, $11.78°W$, mouth of Oued Amma Fatma) is situated at the coastal cliff (almost 20 m high) in the northern part of the Tarfaya Basin between the towns of Tan–Tan and Tarfaya (Fig. 1a), where it is overlain by Plio-Pleistocene shallow-marine clastic sediments. At the proximal onshore location of Oued Amma Fatma, the Lower Turonian succession (c. 17 m thick) consists of dark-light alternation of laminated marly shales and marlstones intercalated with coarse bioclastic limestones with silicified nodules (diagenetic concretions aligned parallel to the stratification) in some layers. Based upon the ammonite fauna in the carbonate concretions and the occurrence of planktonic foraminifera (*Helvetoglobotruncana helvetica*), the Amma Fatma section is early Turonian in age (El Albani *et al.* 2001). Bio- and lithofacies distribution suggests maximum depths of 200 m (Kuhnt & Wiedman 1995; El Albani *et al.* 1999a). According to the planktonic microfaunas and the palaeoenvironmental reconstructions of the Tarfaya Basin, the strata were deposited in the offshore area dominated by a wind-induced coastal upwelling system (e.g. Kuhnt *et al.* 1997, 2005; El Albani *et al.* 1999a; Kuypers *et al.* 2002; Mort *et al.* 2008); the water depth during deposition fluctuated frequently and ranged from below to near the storm wave base (El Albani *et al.* 1999b). The laminated biogenic sediments from this shelf basin show a high degree of sulphurization of the organic matter, and were deposited at high average sedimentation rates ($2-5$ cm ka^{-1}; Kolonic *et al.* 2005). This enables the investigation of their magnetic signature at a high temporal resolution with respect to rapid climate change and associated hydrocarbon source–rock formation.

With the aim of studying the magnetic signature of this key sedimentary sequence, 114 samples have been sampled along a composite vertical profile (c. 10.5 m) exposed on the unfolded Amma Fatma beach cliffs (Fig. 1c). Four horizontally separated, parallel, consecutive transects were sampled, each overlapping its predecessor by c. 0.5 m. Both alternated layers of fine laminated, bituminous black

marls and bioclastic and silicic limestones were sampled. They are representative of the orbital-scale forced sedimentation along the palaeocoast of the African Craton platform at c. 93 Ma. With the exception of 34 closely spaced samples (each 3–4 cm apart) at the bottom of the profile, a sample spacing of c. 15 cm was maintained (Fig. 1c). This sampling strategy was adopted bearing in mind previously reported sedimentation rates ($2-5$ cm ka^{-1}) and assuming that geomagnetic secular variation should be averaged over an order of magnitude of c. 10 ka (e.g. Irving 1960). Considering these sedimentation rates, the sampled sections span c. 500–200 ka. In this way the corresponding geomagnetic field record should approximate that of a geocentric axial dipole (GAD). A continuous magnetic signature can therefore be tested along a detailed profile expanding less than (or of the order of) 1 Ma. Samples were cored (2.54 cm in diameter) with a portable gasoline-powered drill and oriented using a magnetic compass and an inclinometer. Cores were cut in the laboratory into standard specimens (2.2 cm length) for palaeomagnetic measurements (2–3 specimens per sample); core end pieces were used for rock magnetic experiments.

Rock magnetic properties

Hysteresis back-field demagnetization of isothermal remanence and thermomagnetic measurements were carried out in the Palaeomagnetic Laboratory at Fort Hoofddijk, Utrecht University. Hysteresis loops, isothermal remanence acquisition and back-field demagnetization curves were measured on sample chips using an alternating gradient magnetometer (Princeton Inc. Micromag 2900 model) with a maximum applied field of 2T. The following parameters were calculated: saturation magnetization (M_s), saturation remanent magnetization (M_{rs}), coercive force (B_c), initial, high-field and ferrimagnetic (high-field–initial) susceptibility (χ_{in}, χ_{hfld} and χ_{ferri} respectively), isothermal remanence (IRM) and remanence coercive force (B_{cr}). Thermomagnetic curves have been calculated in air with a modified horizontal translation Curie balance (Mullender *et al.* 1993). The cycling field varied from $150-300$ mT and the heating and cooling rates were 10 °C min^{-1}. Samples were progressively heated and cooled in $50-100$ °C steps up to 650 °C.

The acquisition and stability of anhysteretic remanence (ARM) and IRM were studied in the Palaeomagnetic Laboratory of the University of Burgos. Representative specimens, previously demagnetized using alternating fields, were subjected to ARM acquisition (peak alternating field of 100 mT in a constant field of 100 mT) followed by alternating field demagnetization of ARM at 40 and 110 mT. They were then subjected to (room temperature) saturation IRM (SIRM) acquisition at 2T and back-field demagnetization at -100 mT. ARM acquisition, demagnetization and measurements were carried out using a cryogenic magnetometer (2G model 755). IRMs were imparted using an impulse magnetizer (Ferronato M2T-1) and measured using the same cryogenic magnetometer.

Of the 40 hysteresis loops measured, 23 were too weak to reliably describe non-reversible behaviour although the reversible parts of the loop were clearly diamagnetic or paramagnetic. The remaining 17 loops have higher magnetizations and well-defined hysteresis which reach reversibility at 270–300 mT (Fig. 2a, b). Very low intensities led to poorly defined IRM curves, so that reliable B_{cr} values have not been determined. After high field correction, both diamagnetic (at high field) and paramagnetic (at high field) samples exhibit similar M_s and M_{rs} values (mean and standard deviation of 0.4 ± 0.3 mAm^2 kg^{-1} and 33 ± 31 μAm^2 kg^{-1}, respectively). The hysteresis parameters have been plotted in a squareness plot (Tauxe *et al.* 2002). The data are relatively tightly grouped in the lower left-hand region of the plot (Fig. 2c), which probably indicates the predominance of a single (ferro)magnetic mineral. The paramagnetic samples had significantly higher B_c and M_{rs}/χ_{ferri} values (12.1 ± 2.7 mT and 11 ± 3 kA m^{-1} respectively, as compared to 9.6 ± 2.0 mT and 7.0 ± 3.3 kA m^{-1} for the diamagnetic samples), which might imply slightly finer grain size assemblages.

In all but one sample, the thermomagnetic curves are noisy and do not show any clear systematic behaviour. In contrast, sample TAN23 (marl) exhibits a thermomagnetic curve that shows a sharp increase in magnetization around 400 °C and a final Curie temperature (T_c) of close to 580 °C (Fig. 2d). This may be interpreted in terms of the transformation of iron sulphides such as pyrite, greigite or pyrrhotite into magnetite. Similar increases in natural remanence (NRM) and magnetic susceptibility (χ) are also seen during thermal demagnetization, starting at c. 320–370 °C (see below). However, the coercivities and magnetization ratios (Fig. 2c) are lower than those typically observed for natural greigites, which tend to be closer to $B_c \approx 50$ mT and $M_{rs}/M_s \approx 0.5$ (e.g. Roberts 1995).

ARM and SIRM intensities have values within the ranges 1.3–5.6 and 11.1–48.2 mA m^{-1}, respectively. The ratio of SIRM/χ varies between 2 and 19 kA m^{-1}, in close agreement with the M_{rs}/χ_{ferri} values calculated from the hysteresis data. Both sets of values are markedly lower than those expected for greigite and pyrrhotite-bearing rocks (Dekkers 1988; Roberts 1995).

The ratio of ARM after demagnetization at 40 mT to its initial value, ARM$_{40mT}$/ARM, shows values

ranging between 0.26 and 0.31. The ratio of IRM acquired in a -100 mT field to that acquired in a 2T field, IRM$_{-100mT}$/SIRM, has values between -0.07 and -0.64. Peters & Thompson (1998) showed that pyrrhotite has relatively high SIRM/χ with respect to ARM$_{40mT}$/ARM, whereas greigite has higher ARM$_{40mT}$/ARM values with respect to SIRM/χ. Combining the results in a pair of bi-plots (Fig. 2e, f) suggests that the data are consistent with (titano)magnetite as the main remanence carrier.

Specimen TAN35 (bioclastic limestone) had a IRM$_{-100 mT}$/SIRM value of -0.07, indicating the presence of a high coercivity phase such as hematite or goethite. The same specimen showed a high coercivity (>100 mT peak alternating field) NRM fraction (as did three others down the profile), suggesting that in rare cases a high coercivity phase may be present and contributing to the NRM.

Together, the relatively low B_c, M_{rs}/M_s, M_{rs}/χ_{ferri} and SIRM/χ values, along with the ARM and IRM stabilities shown in Figure 2e, f, suggest that (titano)magnetite is the most likely carrier of the remanence in the Tarfaya sediments. The magnetic changes seen above 300–400 °C may then be ascribed to the transformation of pyrite. Sulphur and sulphides may be expected to be present in the organic-rich Tarfaya sediments (Kuypers *et al.* 2002), forming during early authigenesis due to anoxic depositional conditions. In this context, the ferromagnetic signal is due to small amounts of (titano)magnetite (and maybe hematite) that have survived dissolution, thereby explaining the weak magnetic signal.

The χ_{hfld} values fall between -0.42×10^{-8} and 0.55×10^{-8} m^3 kg^{-1}, varying in response to the diamagnetic and paramagnetic content of the samples. Diamagnetic samples have χ_{hfld} values of between -0.42 and -0.04×10^{-8} m^3 kg^{-1}, which are slightly higher than those expected for pure calcite (-0.48×10^{-8} m^3 kg^{-1}) and those observed for hydrocarbon reservoir fluids (around -1×10^{-8} m^3 kg^{-1}, Ivakhnenko & Potter 2004). There is no discernable influence of hydrocarbon content on the magnetic properties at the sampled locality. The higher χ_{hfld} may be a consequence of trace amounts of paramagnetic minerals within the predominantly diamagnetic sample matrix.

Anisotropy of magnetic susceptibility

In all the standard (11.15 cm^3) samples along the Amma Fatma section, low-field magnetic susceptibility (χ) and the AMS were measured (Fig. 3a, d) on a KLY-4S instrument manufactured by AGICO (Brno, Czech Republic; Pokorny *et al.* 2004; Hrouda *et al.* 2006). The AMS ellipsoid principal mean directions (Fig. 3a) have been computed

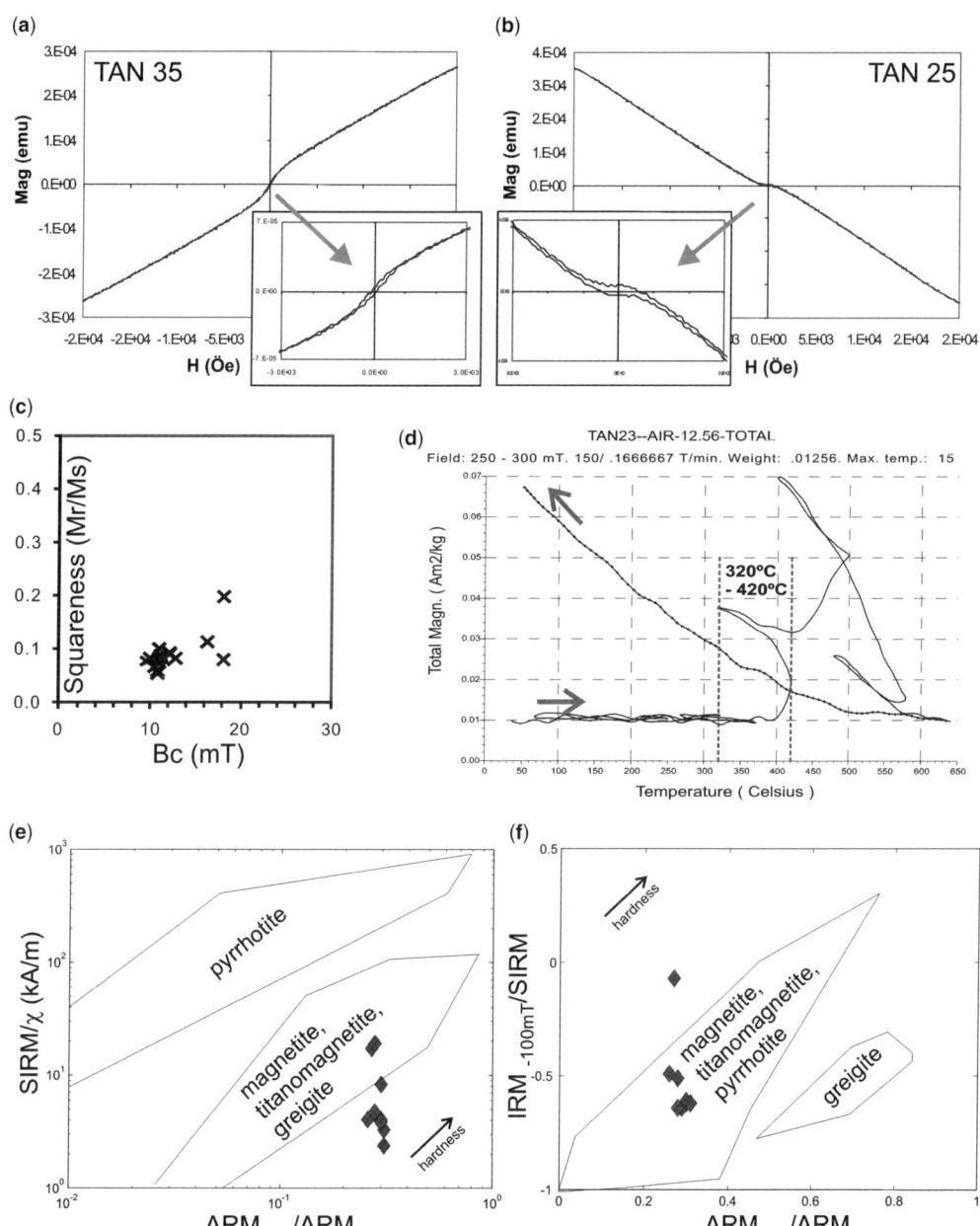

Fig. 2. (**a, b**) Hysteresis loops without high field correction show the importance of the (a) paramagnetic or (b) diamagnetic behaviours. The inset is the same plot at a higher scale to show the hysteretic part of the loop. (**c**) Squareness plot of the high field-corrected, high-quality hysteresis parameters of the analysed samples. (**d**) Thermomagnetic curve measured in air using the incremental heating and cooling segments protocol (Mullender *et al.* 1993). Chemical alteration starts at *c.* 400 °C temperature and shows a sharp increase in magnetization due to magnetite formation (final Curie temperature is close to 580 °C) from pyrite oxidation. Magnetization decay during cooling (dotted line) is interpreted in terms of magnetite oxidation creating hematite after heating up to 650 °C. (**e, f**) Biplots used for qualitative identification of magnetic minerals (modified from Peters & Thompson 1998) in representative samples/lithologies: (e) SIRM/χ v. ARM$_{40mT}$/SARM and (f) IRM$_{-100mT}$/SIRM v. ARM$_{40mT}$/SARM.

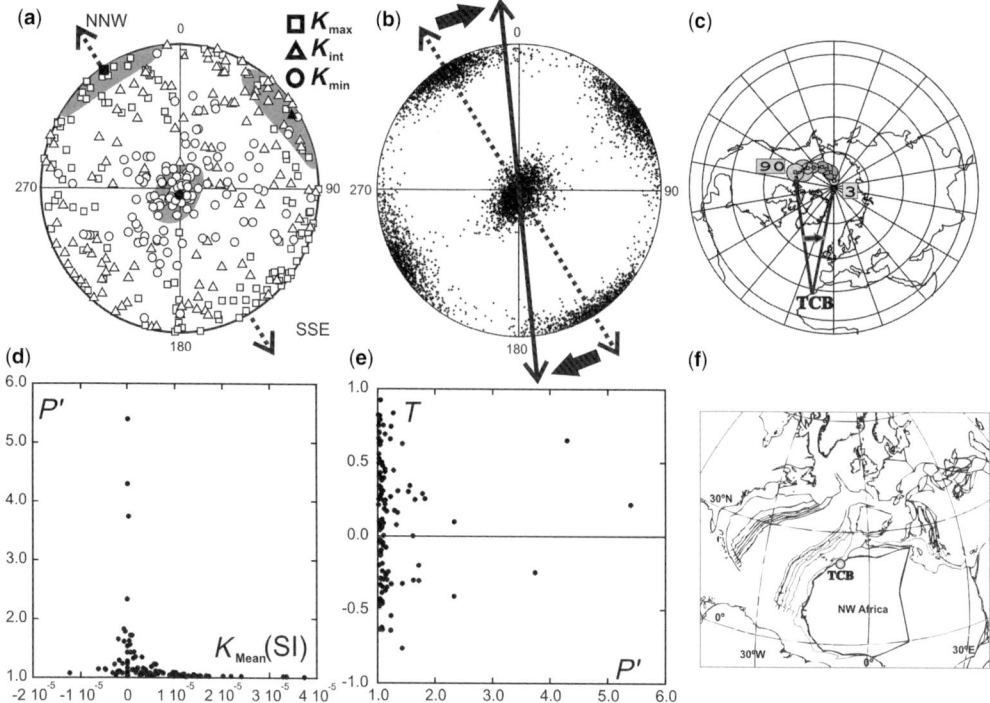

Fig. 3. Equal area projection of the principal directions of AMS: (**a**) open symbols show individual principal maximum, intermediate and minimum (K_{max}, K_{int}, K_{min}) directions of the AMS ellipsoid for each specimen. Full symbols represent mean directions and shaded areas the associated 95% confidence ellipses. (**b**) Bootstrapped eigenvectors for the studied samples. Dotted (solid) lines show magnetic lineation before (after) correction for the early Turonian (*c.* 93 Ma) palaeogeographic position of Africa. (**c**) Tarfaya Coastal Basin (TCB) position and angle used for the magnetic lineation correction (from the geographic pole up to the 90 Ma sliding window pole from the synthetic BC2002 APWP). (**d**) Corrected anisotropy degree P' as a function of the bulk susceptibility (K_{mean}) for all the samples. (**e**) Jelinek plot of the T shape parameter as a function of the degree of anisotropy P'. (**f**) Palaeogegraphic map sketched from GPlates software (http://www.gplates.org) at 93 Ma.

using the statistical methods of Jelinek in spherical coordinates (1978). Along the profile, the boot-strapped eigenvectors (Fig. 3b) have been evaluated following the procedure outlined by Constable & Tauxe (1990).

After correction into geographical coordinates, a well-defined primary fabric can be seen along the profile with the minimum susceptibility axis K_{min} perpendicular to the bedding plane and the magnetic lineation K_{max} oriented 325° NNW–SSE (Fig. 3a, b; dotted line). In order to estimate the direction of this magnetic lineation in its late Cretaceous (*c.* 93 Ma) palaeogeographical position (Fig. 3f), it has been rotated taking into account the BC2002 reference palaeopole of the synthetic APWP for the African plate (Fig. 3c), centred at 90 Ma. At the Tarfaya coastal basin, this pole provides an expected magnetic declination of *c.* 335°. This indicates that since early Turonian times these sediments have undergone a counter-clockwise rotation of *c.* 25°.

As it will be shown, this corrected magnetic lineation value hardly changes if we consider T2008 early Turonian expected direction (*c.* 340°) or even if the ChRM mean declination obtained in this study (*c.* 332°). The magnetic lineation at the late Cretaceous (*c.* 93 Ma) period was therefore oriented 350°–170° (*c.* 325° + 25°) north–south (Fig. 3b, solid line).

The shape has been characterized by the T shape parameter (Fig. 4e), which ranges from 1 (purely oblate) to −1 (purely prolate). The strength of the anisotropy has been calculated using the corrected degree of anisotropy P': isotropic samples giving a value of 1, which increases with increasing anisotropy (Fig. 3d, e). Maximum values of the degree of anisotropy P' correspond to bulk susceptibilities near to zero (Fig. 3d), which is typical of AMS dominated by dia-/paramagnetic minerals (Hrouda *et al.* 1986). When comparing marls and bioclastic limestone layers, the latter display a weaker

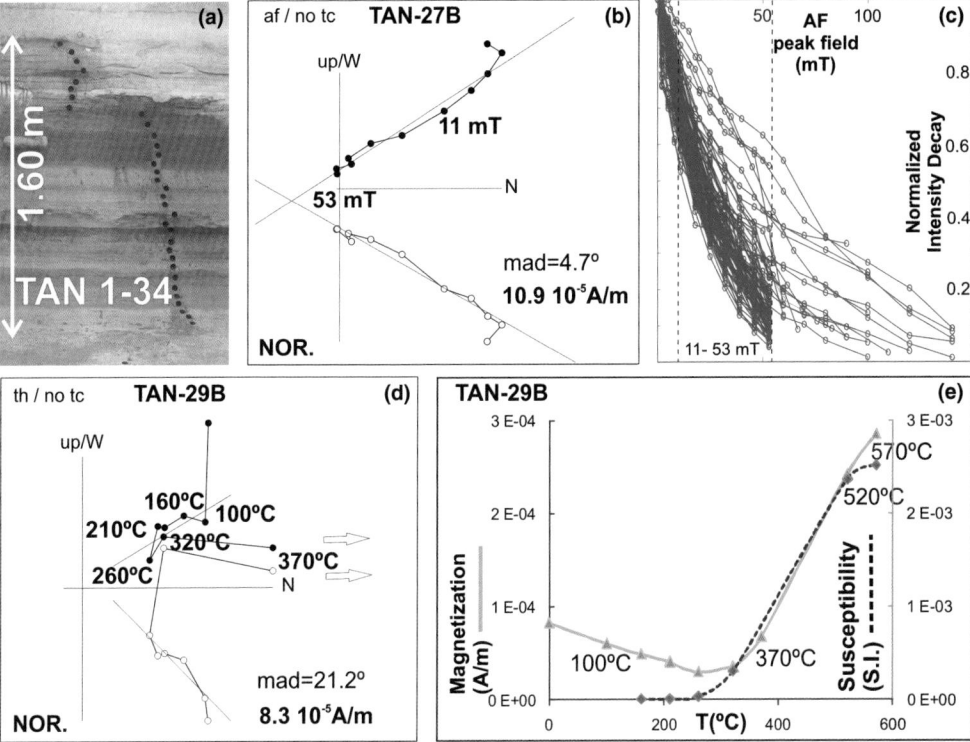

Fig. 4. (**a**) Detailed marls sampling at the bottom of the profile. (**b**) Orthogonal projection plot of a marls-representative AF demagnetization path (closed/open circles denote the projection on the horizontal/vertical plane), showing the initial NRM intensity and the fitted ChRM (mad, ChRM maximum angular deviation). (**c**) Normalized intensity decay during AF demagnetization of samples of the whole (marls-dominated) profile. (**d, e**) Marls alteration during heating. Remanence direction becomes unstable c. 320–370 °C as evidenced by its orthogonal projection plot, where (d) points above 370 °C are not represented for clarity and (e) both NRM intensity and susceptibility increase in values from c. 320–370 °C to 520–570 °C.

anisotropy in terms of both degree of anisotropy and principal directions; a mean magnetic foliation could not be determined for them. The bulk χ is also weaker, being diamagnetic in most samples. Despite the low χ values (mean: 6×10^{-6} SI) due to the dominant dia-/paramagnetic contribution, rapid fluctuations can be observed along the profile. They probably record the variable terrigenous influx which would be related to small-scale eustatic sea-level changes (El Albani *et al.* 1999*b*), suggesting that magnetic signatures could be potentially tested against other geochemical cyclicity proxies (although this is beyond the scope of this study).

Natural remanence demagnetization and magnetostratigraphic column

All remanence measurements were made using 2G cryogenic magnetometers (model 755 DC SQUID,

noise level 5×10^{-12} Am^2). Alternating field (AF) demagnetization was performed (Figs 4b, c & 5) using the automated demagnetization units associated with the cryogenic magnetometers. Thermal (TH) demagnetization was carried out using a TD48-SC (ASC) thermal demagnetizer. χ was monitored after each heating to control possible thermally induced mineralogical changes using an AGICO KLY4 susceptibility metre. For the thermal demagnetization, 11 steps of 50–60 °C from 100 °C up to 575 °C were used (Fig. 4d). Chemical reactions starting c. 320–370 °C were observed during heating, resulting in an increase of both χ and remanence intensity (Fig. 4e) before reaching the original maximum unblocking temperature. For very detailed pilot AF studies, 29–30 steps were used (steps of 2 mT up to 10 mT, then steps of 5 mT up to 50 mT, then steps of 10 mT up to 80–90 mT and then 100, 120, 140 mT; Fig. 4b, c). At least 10 demagnetization steps were applied to the remaining specimens

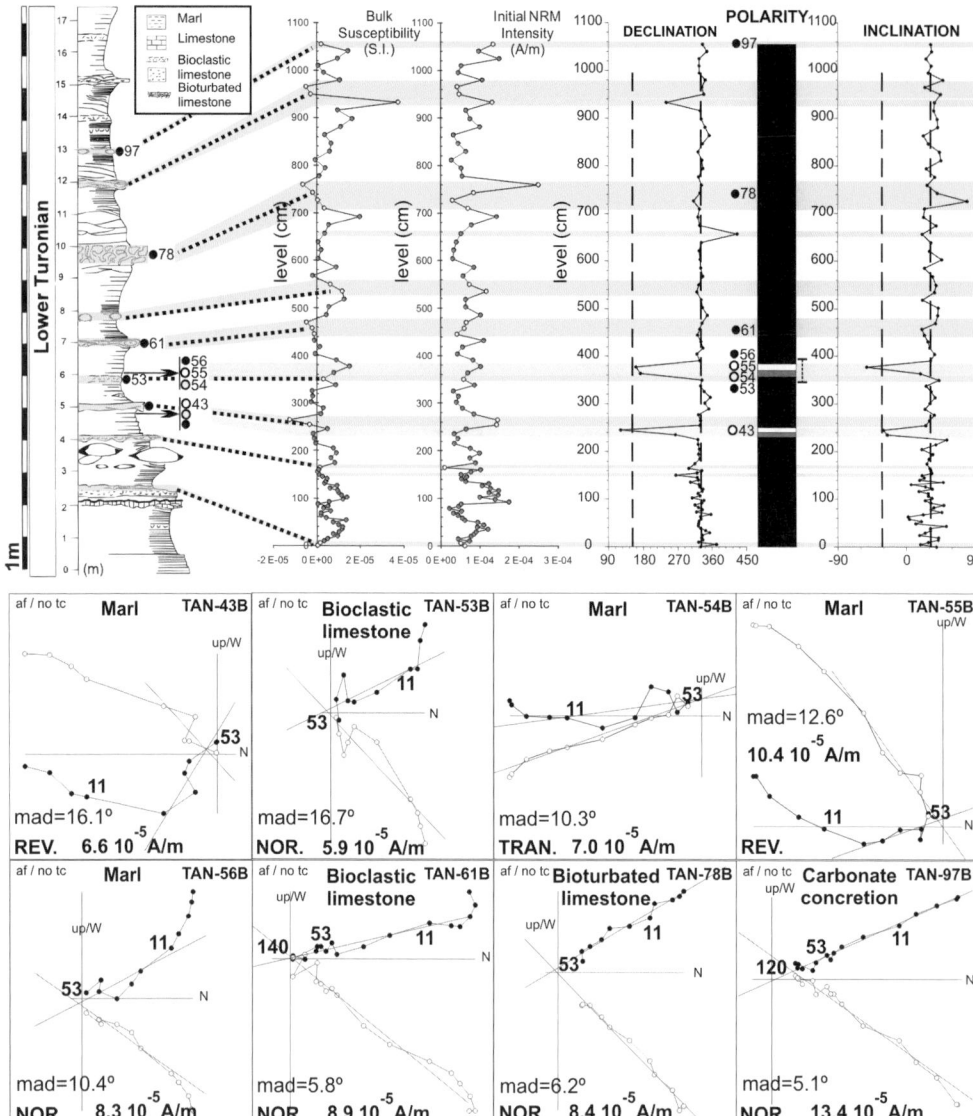

Fig. 5. Upper: Amma Fatma section lithostratigraphic log (from El Albani *et al.* 2001) and corresponding initial magnetic susceptibility, initial NRM intensity and magnetostratigraphic correlated profiles (this study). Along the columns, location and code number of samples representative of the polarities and the lithologies observed throughout the section are marked. Initial magnetic susceptibility and NRM intensity data from bioclastic and bioturbated limestones are distinguished with open circles. Lower: Diagrams of typical AF demagnetization behaviour from these representative samples (closed/open circles denote the projection on the horizontal/vertical plane) indicating assigned polarity, initial NRM intensity value, fitted ChRM component and ChRM maximum angular deviation (mad) for each sample. Numbers close to the horizontal projections of the demagnetization steps refers to AF peak fields in mT. In the upper and lower images, interpreted polarities are represented by black and NOR., transitional by grey and TRAN. and reversed by white and REV., respectively.

over their common, optimum coercivity spectra (2, 5, 8, 11, 18, 25, 32, 39, 46 and 53 mT). The median destructive fields ranged between 11 and 20 mT (Fig. 4c).

The magnetic behaviour of pilot samples indicated that only AF demagnetization was suitable in isolating the ChRM. After removing a soft viscous remanent magnetization a single stable, lineal magnetic component was observed: a low/medium coercitivity direction which is generally isolated starting from 8–11 mT field steps (but sometimes from 2–5 mT or even 18–25 mT; Figs 4b & 5). The ChRM is generally of normal polarity with westerly declinations and relatively low inclinations, ruling out the possibility of a recent overprint. In some specimens, the demagnetization path did not exactly point towards the origin but remained parallel to the general trend. This is probably due to a small, spurious magnetization acquired between demagnetization and measurement in the SQUID magnetometer, which was observed as an offset from the origin of the demagnetization plots due the low NRM intensities involved (Figs 4b & 5). The ChRM direction was calculated by principal component analysis (Kirschvink 1980) with a minimum of 6 or 7 demagnetization steps, without forcing the trends to include the origin (Figs 4b & 5). Maximum angular deviations of the fitted directions had a mean value of 9.9° and a standard deviation of 4.4°. Surprisingly, two specimens exhibited clear reversed magnetizations, and they were each preceded by transitional directions (Fig. 5). There was no apparent lithological or mineralogical variation observed for these specimens, nor did they show unusual magnetic properties (e.g. the two samples carrying reversed polarities are included in Fig. 2e, f). Despite these contrasting directions, we will argue for an early Turonian ChRM age (therefore still belonging to the CNS, at least for the rest of the specimens) related to magnetite.

Late diagenetic processes may lead to the formation of magnetic minerals that carry a younger magnetization. If this is the case for these four specimens, they would record a remagnetization which occurred at least c. 10 Ma after deposition in the reversed polarity chron C33r (83–78 Ma, Cande & Kent 1995) or later. Alternatively, they could represent true, short polarity events or 'tiny wiggles'. Their record may have been favoured by the high sedimentation rate in the Tarfaya Basin, and imply short reversed intervals of c. 2–20 ka. This duration, insufficient to be registered worldwide, would account for their lack of resolution in the marine magnetic anomaly 34 record (Cande & Kent 1992). The existence of Albian (c. 105 Ma) reversed subchrons in the CNS has been discussed (Tarduno et al. 1992) regarding indurated marls and limestones from the Umbrian Apennines, Italy. In addition, a magnetostratigraphic study of the Turonian of SW Morocco reported a reversed polarity in sediments covering the Cenomanian/Turonian boundary (Krumsiek 1982; Stamm & Thein 1982). In any case, the lateral persistence of the observed polarity pattern should be checked at correlated successions along the basin. These four specimens have not been considered for the tectonic analysis. All the ChRM directions are successively plotted (Fig. 5, upper) along with the corresponding Amma Fatma stratigraphic column of El Albani et al. (1999a).

In addition to the four specimens showing contrasting polarity, another three with anomalous ChRM directions/lithologies (Fig. 6a) have been rejected for further discussion. These specimens are interpreted as poorly isolated ChRMs as their demagnetization paths end especially far from the origin. In addition specimen TAN89, which exhibited the highest χ value of the column (Fig. 5,

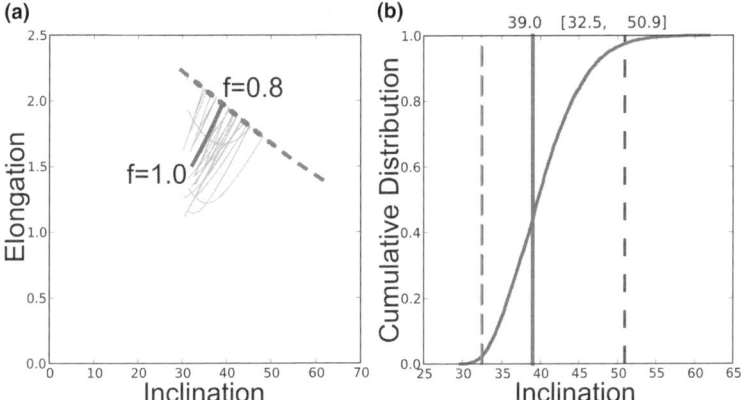

Fig. 6. (**a**) Plot of elongation v. inclination for Amma Fatma data (thick solid line) and for the TK03.GAD model (dashed line). The crossing point represents the inclination/elongation pair most consistent with the TK03.GAD model. As f: 1 → 0.8, the inclination of the unflattened directions increases from 32.4° to 39.0°. Also shown are results from 20 bootstrapped datasets (thin lines). (**b**) Cumulative distribution of crossing points from 5000 bootstrapped datasets (like those shown in the E–I plot) yields an inclination estimate of 39.0° with 95% confidence bounds of 32.5–50.9°.

upper), belongs to a coarse-grained bed at the bottom of a limestone layer which is not representative of the other sampled lithologies. It should be highlighted that the mean directions including or rejecting deviant ChRMs only are 1° apart. A total of 89 samples will be considered for geomagnetic and tectonic analysis.

Testing inclination flattening and secular variation estimates

As many sedimentary palaeomagnetic data (especially but not only from red-beds) have manifested a pronounced bias towards shallow inclinations, the effect has been tested in the ChRMs from the Amma Fatma section. This compaction-induced 'inclination error' is latitudinal dependent and characterized by an empirical 'flattening factor' f which relates observed (I_O) and true or 'unflattened' geomagnetic (I_F) inclinations ($f = \tan I_O / \tan I_F$; King 1955). Flattening factor values ranges from 1 to 0 for unaffected to completely flattened direction estimates.

Assuming that every sample from the Amma Fatma section was affected by an identical flattening factor (principally f would be an estimate of the marl lithology factor), and that the geomagnetic field distribution of directions at early Turonian times can be represented by a statistical PSV field model, we will use the 'elongation–inclination' (E/I) method (Tauxe & Kent 2004) of detecting and correcting inclination shallowing. These authors use a PSV field model (TK03.GAD) which fits the PSV globally distributed lava dataset of the last 5 Ma of McElhinny & McFadden (1997) and produces circularly symmetric Virtual Geomagnetic Pole (VGP) distributions (and therefore north–south elongated directions, with the maximum elongation at the equator). Elongation parameter E, defined as the ratio of the intermediate to least eigenvalues ratio of the 'orientation matrix' of the distribution of directions, is used to quantify their asymmetry in both TK03.GAD and the dataset. In the TK03.GAD model, E varies from c. 3 (rather elongate distribution) at the equator (where quasi-horizontal inclinations I are expected) to unity (approximately symmetric) near the poles (quasi-vertical I values). The E/I method requires a dataset large enough to have sampled geomagnetic field PSV. Following this method, the program '*find_EI*' in the PmagPy-2.40 software distribution available at http://magician.ucsd.edu/Software/PmagPy/ was used to analyse Amma Fatma data (Fig. 6). Using the relation of King (1955), the E/I method inverts the observed directions with decreasing values of the flattening factor f. For each set of corrected data, E and mean I values are

calculated and plotted against each other. The inclination at which the transformed data cross the best-fit polynomial E/I function from TK03.GAD is the unique E/I pair consistent with the field model. The 95% confidence bounds are found using a bootstrap technique.

The E/I method finds an optimal solution at a flattening factor $f = 0.8$ (Fig. 6a). The new corrected ChRM distribution has an unflattened mean inclination of $39.0°_{32.5°}^{50.9°}$ (Fig. 6b) and an elongation value of $1.98_{1.66}^{2.16}$.

The mean observed data distribution has the following mean direction after applying Fisher (1953) statistics: $N = 89$, Dec $= 331.6°$, Inc $= 32.5°$, $\alpha_{95} = 2.7°$; $k = 36.6$ (Fig. 7a). After inverting each sample inclination by the resulting flattening factor ($f = 0.8$) using the relation of King (1955), the inclination value increases by 6° as the unflattened distribution changes to: $N = 89$, Dec $= 331.7°$, Inc $= 38.1°$, $\alpha_{95} = 2.8°$; $k = 34.5$ (Fig. 7b).

As a measure of PSV, geomagnetic field dispersion S of a distribution of N directions, each being an angular distance Δ_i away from the mean, was calculated from: $S^2 = \Sigma \Delta_i^2 / (N - 1)$ (e.g. McElhinny & McFadden 1997). In both the observed and unflattened datasets, the total angular VGP dispersion S and its corresponding 95% bootstrapped bounds have been calculated by applying a fixed cut-off-angle (45°) and after the iterative process which led to the optimum cut-off angle Θ ($\Theta = 1.8 S + 5°$, dependent upon the angular dispersion of each VGP dataset distribution; Vandamme 1994). Dealing with distributions of declinations and inclinations of individual samples, the within-site scatter subtracting correction for S is not taken into account.

Calculated mean directions and palaeopoles are listed in Table 1. The mean observed (flattened, $f = 1$) data distribution mean directions and geomagnetic dispersion S after a fixed 45° cut-off angle coincide with those values obtained applying an optimum Vandamme cut-off angle of 23.5°. In the same way, the mean corrected data distribution (unflattened with $f = 0.8$) yields analogous mean values after the 45° angle cut-off rather than applying the Vandamme cut-off, in this case 25.0°. Figure 7b shows the distributions of the unflattened ($f = 0.8$) directions and their corresponding VGPs (around the mean pole) and discarded samples after applying the Vandamme cut-off.

Independently of the fixed or variable Vandamme cut-off, the effect of the (moderate) correction for inclination error is only observable in the mean inclination values of the distributions; it hardly affects the VGP dispersion estimates around their mean palaeomagnetic poles. Similarly, although application of a 45° cut-off angle results in higher S value with respect to the variable cut-off

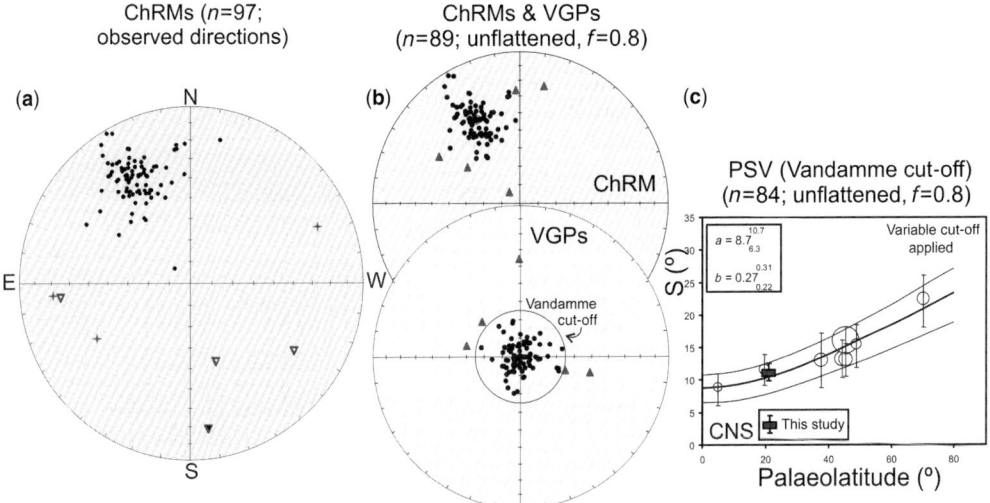

Fig. 7. (a) Equal area projection of observed ChRM directions. Star points/triangles (anomalous/transitional and reversed directions) denote data not considered in the tectonic analyses. (b) Equal area projections of ChRMs corrected for inclination shallowing and corresponding VGPs centred on their mean direction, showing the effect of the Vandamme cut-off angle (triangles denote rejected samples). (c) Amma Fatma VGP dispersion (rectangle), with its 95% bootstrap uncertainty limits, is integrated in the VGP dispersion curve of Biggin *et al.* (2008) for the Cretaceous Normal Superchron (CNS), where open circles refer to other results from lavas at different palaeolatitudes after applying variable cut-offs to high-quality CNS datasets. The shape parameters (*a*, *b*) of the best-fit Model G (McFadden *et al.* 1988) with 95% bootstrap uncertainty limits are given in an inset, and the resulting curves plotted. ('Model G' described VGP dispersion in terms of symmetric and anti-symmetric contributions; the former is constant and the latter dependent on latitude λ: $S = [a^2 + (b\lambda)^2]^{0.5}$).

criteria, Amma Fatma VGP dispersion results do not depend on these cut-off angle selection criteria (Table 1).

Discussion and conclusions

Early recorded magnetizations

Previous geochemical and petrological investigations reported from the Tarfaya Basin suggest that the sediments have been affected by diagenetic processes. Sulphur was incorporated into the

kerogen during early diagenesis, and the presence of molecular fossils from photosynthetic green sulphur bacteria (Kuypers *et al.* 2002) indicates that dissolved sulphide had reached the photic zone at shallow water depths (*c.* 100 m) during times of deposition, favouring sulphurization of organic matter (Kolonic *et al.* 2002). It is suggested that different degrees of sulphurization was controlled by a variable supply of reactive iron (excess iron resulting in greater scavenging by pyrite, whereas iron limitation results in a higher degree of natural sulphurization). In addition, iron

Table 1. *Observed and unflattened mean directions, palaeopoles and VGP dispersion of analysed distributions*

	N	Declination	Inclination	$\alpha 95$	k	PLat	PLon	A95	K	S
Observed (*f* = 1)										
45° cut-off angle	88	331.7	32.0	2.6	43.6	61.9	242.7	2.3	43.6	$11.2°_{9.7°}^{12.8°}$
24° cut-off angle	83	331.2	31.8	2.4	62.7	61.5	242.7	2.0	62.8	$10.3°_{8.9°}^{11.6°}$
Unflattened (*f* = 0.8)										
45° cut-off angle	88	331.7	37.6	2.7	40.5	64.3	256.3	2.5	38.7	$11.8°_{10.3°}^{13.3°}$
25° cut-off angle	84	331.5	37.5	2.5	53.3	64.2	256.1	2.2	49.8	$11.1°_{9.8°}^{12.6°}$

Abbreviations: N, number of sites; PLat, PLon, pole latitude and longitude; k/K and $\alpha 95$/A95, precision parameters and semi-angles of 95% confidence (Fisher 1953) of mean directions/palaeopoles; *S*, VGP dispersion with 95% bootstrap uncertainty limits.

availability estimates (from iron to sulphur ratios at the Amma-Fatma correlated S75 well, Fig. 1b) indicate that not all of the sulphur is fixed in pyrite (Kolonic *et al.* 2002), which could suggest the presence of greigite phases. Carbonate concretions at Amma Fatma section are an early diagenetic product (El Albani *et al.* 2001). Although the timing of this early diagenesis is very difficult to determine, for carbonate concretion levels it is considered a very quick process (*c.* 100–1000 ka) during the period of lower sedimentation rates that characterize sulphate reduction reactions needed for the formation of nodular concretions (which are therefore correlated with relative high-level stands). Higher sedimentation rates from C/T laminated biogenic sediments would favour an earlier halt to the sulphate reduction mechanism in these layers.

The weight of evidence from rock magnetic analyses supports magnetite (or titano-magnetite) as the principal remanence carrier, rather than greigite or pyrrhotite. This suggests that most of the sulphur is fixed in pyrite, and that diagenetic processes that have affected the NRM have been limited to the dissolution of magnetic phases rather than the formation of secondary phases. Rock magnetic properties and ChRM characteristics of the different lithologies are very similar. Samples from bioturbated or bioclastic limestones layers and carbonate concretions show the same ChRM characteristics as marls (which have higher sedimentations rates and are assumed to acquire their diagenetic remanence earlier).

Scattered remanences such as those reported for black shales (Kawasaki *et al.* 2007) are not observed; the ChRM directions do not appear to be overly affected by smoothing effects, such as those observed in the recorded magnetizations carried by early diagenetic greigite (Rowan *et al.* 2009). The dispersion of ChRM directions is similar to those observed in other Cretaceous lavas (Fig. 7c). Finally, the mean palaeopole for the sequence does not correspond to any part of the synthetic APWP for ages younger than 90 Ma (Fig. 8).

The lack of evidence for greigite and important remanence smoothing effects, along with the distribution of ChRM directions and the palaeopole direction, together support a magnetization acquired early in the sedimentary processes. Diagenetic processes therefore lead to the dissolution of some but not all of the magnetic minerals, and the surviving grains preserve the primary remanence of the sequence.

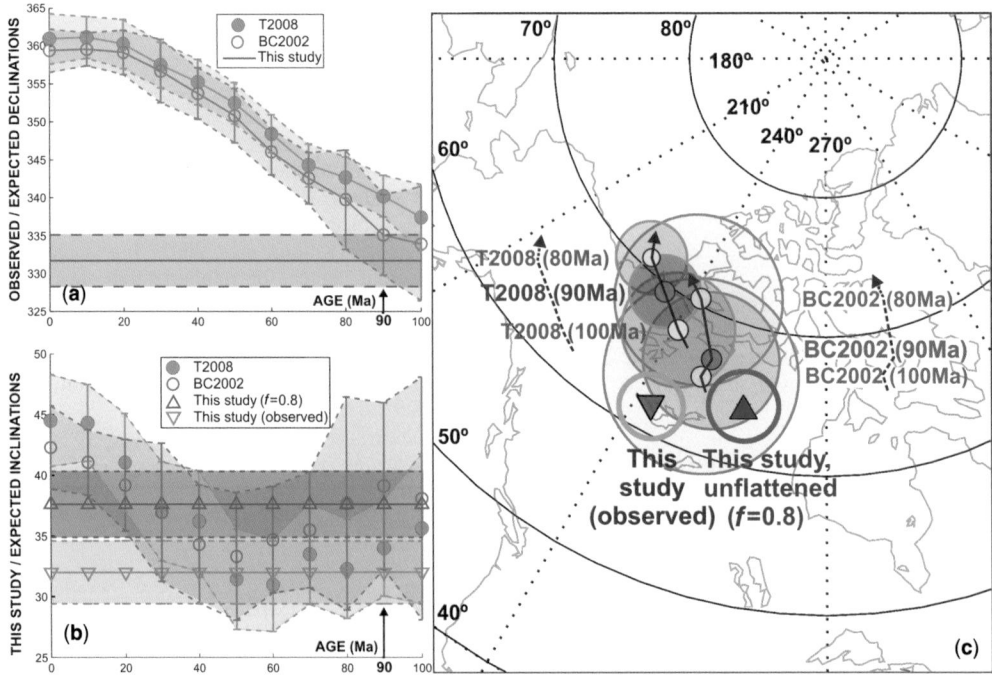

Fig. 8. (**a**, **b**) Amma Fatma mean declination and inclinations compared to the corresponding expected values from BC2002 and T2008 synthetic African APWPs. (**c**) Comparison of this study observed and unflattened (*f* = 0.8) palaeopoles with the running mean paths of BC2002 and T2008 (20 Ma sliding window) in South African coordinates in the range 100–80 Ma.

Magnetic signature potentials

The response of the AMS to past tectonic stresses and palaeocurrents has been tested. From Late Triassic–Early Jurassic times, when the supercontinent Pangaea began to break up, NW Africa was dominated by an extensional regime (e.g. Oyarzun et al. 1997) associated with the opening of the central-north Atlantic (that has been observed in the Tarfaya Basin based on seismics; Heyman 1990), and the associated 'Atlas Gulf' marine seaway 'failed' rift arm. Around C/T times, a general change from extension to compression occurred in North Africa which was related to the closing of the Neotethys and the onset of North Atlantic rifting. This resulted in the inversion of the former rift grabens from the preceding extensional phase, leading to the formation of the Atlas Fold and Thrust Belt as Alpine deformation intensified during the Tertiary.

On the other hand, strong upwelling conditions are thought to have existed along the NW African coast (extending from about Senegal in the south to Morocco in the north) during most of the Cenomanian–Coniacian as documented by palaeoecological analyses (e.g. Kuhnt & Wiedmann 1995) and numerical modelling (Kruijs & Barron 1990). Organic-rich strata were deposited here throughout this interval. Notably, under- and overlying OAE 2 event strata in Tarfaya are characterized by a highly enhanced marine palaeoproductivity fostered by the intensified East Atlantic upwelling (e.g. Lüning et al. 2004).

Anisotropy of magnetic susceptibility is dominated by the dia-/paramagnetic contributions. The magnetic lineation is mainly carried by the marls lithology and, when rotated to early Turonian (c. 93 Ma) African coordinates, seems to be compatible with (i) the resulting direction of the palaeostress record at the Tarfaya Basin since deposition times in the context of the above-mentioned NW African tectonic evolution (bracketed between the present magnetic lineation and that at Early Turonian coordinates) and/or (ii) the direction of the palaeocurrent associated with the coastal wind-driven upwelling system deposition described at the Tarfaya Basin (e.g. Kuhnt et al. 1997, 2005; El Albani et al. 1999a; Kuypers et al. 2002; Mort et al. 2008) and sedimentologically supported in the studied section (El Albani et al. 1999b).

Several cyclostratigraphic studies have been performed throughout Cenomanian–Turonian times in different successions from the Tarfaya coastal basin. Cyclicity has been interpreted as orbitally induced with different, superimposed frequencies and studied proxies include density, total organic carbon, carbon isotope, redox-sensitive/sulphide-forming trace metals and spectrogamma log data

(Kuhnt et al. 1997; Luderer 1999; Kolonic et al. 2005), but the magnetic signature response of these sediments has not been investigated. The organic matter content variations and the lithogenic–biogenic ratio fluctuations related to high-frequency sea-level changes (e.g. Kuhnt et al. 1997; El Albani et al. 1999b) seem to be followed by the initial NRM intensity and magnetic susceptibility records along the studied Amma Fatma succession. It strongly suggests that, together with other environmental-sensible magnetic parameters, they would be suitable to test estimates of forcing frequencies.

Very short polarity intervals or tiny wiggles spanning <30 ka are generally less uniformly well documented than true geomagnetic polarity reversals, and their origin may be due to palaeointensity variations or incomplete reversals of the geomagnetic field. Long wavelength anomaly 34 does not allow any fine detail in the CNS to be modelled which could record short period (2–20 ka) dipole field variations, such as those that may have characterized the geodynamo throughout the Cenozoic (Cande & Kent 1992). The four early Turonian layers from the Tarfaya Basin where transitional to reversed polarities were observed (whose available sedimentation rates convert to durations of less than 20 ka) are key sediments for further investigations dealing with the reliability of these dipole directional fluctuations. Alternatively, if they record a post-CNS reversed field (10 Ma after the deposition or later, which is not supported along the succession), these layers are a potential source of related specific mineralogical investigations. In any case, these four samples have been excluded from the tectonic analysis in this study.

Observed and unflattened directions

Inclination error is maximum at mid-latitudes but negligible at the poles or the equator. In addition, the onshore C/T organic-matter-rich interval of the Tarfaya relatively low compaction is expected at the Tarfaya Basin strata, as maximum depth of burial has probably never been more than 500–600 m (Kolonic et al. 2002). Concordantly, a moderate flattening factor ($f = 0.8$) has been estimated using the E/I method (Tauxe & Kent 2004) of detecting inclination shallowing.

The E/I method corrected ($f = 0.8$) mean inclination of the Amma Fatma dataset ($37.5° \pm 2.5°$) is consistent with those predicted at the sampling site from the 90 Ma centred running paths of BC2002 and T2008 African APWPs ($39.1° \pm 6.9°$ and $34.0° \pm 4.0°$, respectively). Associated 95% uncertainty bounds however overlap with expected APWP inclinations along tens of mega-annum, making any comparison inconclusive. Unflattened (corrected) mean inclination value is closer (even equal) to the BC2002 expected value than the observed (uncorrected) mean inclination, whereas the T2008 expected inclination lies in between the two.

The applicability of the E/I method was tested through time including early Cretaceous (125–133 Ma) results from the Paraná Magmatic Province (Tauxe *et al.* 2008). Elongation distributions (E values ranging from 1.54 to 1.82) of selected Paraná datasets were compatible with the expected E/I trend from the TK03.GAD model, but exhibited large error bars (bootstrapped confidence bounds ranging from 1.22 to 2.75). Unflattened Amma Fatma (c. 38°–39°) mean inclination is comparable with Paraná observed mean inclinations (37.7°–40.5°); the Amma Fatma elongation bracket ($1.98_{1.66}^{2.16}$) can therefore be compared with Paraná Cretaceous results at the same palaeolatitudes. Amma Fatma results also seems to fit the corresponding elongation predicted by TK03.GAD ($E \approx 1.9$), although more and larger datasets would be desirable to enhance the apparent robustness for the E/I method (at least up to Mesozoic times; Tauxe *et al.* 2008).

Palaeosecular variation averaging

Palaeosecular variation has been evaluated in terms of VGP angular dispersion around their mean. VGPs were selected applying an arbitrary fixed cut-off angle (45°) and using the iterative process defined by Vandamme (1994) to obtain optimum latitudinal variable cut-off angles. In both the observed (flattened) and corrected (unflattened, $f = 0.8$) Amma Fatma datasets, the two different cut-offs yield distributions whose mean and geomagnetic dispersions are less than 1 degree distant. PSV analyses of high-quality datasets from the CNS (Biggin *et al.* 2008) indicate that their VGP dispersions were not affected by the choice of the 45° or Vandamme cut-off, and that the dependence of S on latitude differs from that observed during the last 5 Ma. At its corresponding palaeolatitude, Amma Fatma unflattened VGP dispersion ($S = 11.1°_{9.8}^{12.6°}$) is in excellent agreement with the best-fit Model G (McFadden *et al.* 1988) to the quality PSV results from selected CNS datasets of Biggin *et al.* (2008), applying a Vandamme cut-off (Fig. 7c). An additional test to detect potential underrepresentation of PSV was performed by analysing Amma Fatma VGPs in stratigraphic order, following the non-parametric 'Non-Random-Ordering factor' test of Biggin *et al.* (2008) for identifying serial correlation. The serial correlation is not significant at the 95% confidence level along the whole dataset and metre-scaled intervals (excluding samples of reversed polarity).

An early Turonian palaeomagnetic pole

Amma Fatma observed and unflattened ($f = 0.8$) mean inclinations and palaeomagnetic poles are around 6–7° apart, but the mean declination is the same in both distributions (Fig. 8a, b). Mean palaeomagnetic poles do not differ if their VGPs are selected by applying a 45° cut-off angle or the optimum Vandamme cut-off angle. Both observed and unflattened ($f = 0.8$) mean palaeomagnetic poles (or mean directions) obtained following a 45° cut-off angle (the less stringent criterion) are compared to the reference poles (or their expected directions at Tarfaya Basin) from the global BC2002 and T2008 APWPs in South African coordinates (Fig. 8).

The master or global curve referred to South Africa is used because in both APWPs all the (small) angles used for rotations from NW (or NE) Africa onto South Africa are older than the age of deposition (c. 93 Ma) of the studied succession. These APWPs are generated using 20 Ma window length running mean paths with 10 Ma increments. They are affected by the different choices used; it is not only palaeomagnetic data selection (e.g. most of the T2008 Euler rotations used for relative plate motion estimates) which differs from those of BC2002. Figure 8c illustrates the angular discrepancies of BC2002 and T2008 reference poles between 100 and 80 Ma. For instance, at the 90 Ma centred window they have the following number of input poles, means and semi-angle of 95% confidence: $N = 13$, PLat $= 66.7°$, PLon $= 248.7$, $A_{95} = 4.9°$ (BC2002) and $N = 27$, PLat $= 69.5°$, PLon $= 234.7$, $A_{95} = 2.6°$ (T2008). An early Turonian age for the ChRMs observed along the section is assumed (acquired during or soon after deposition) and therefore the APWP sliding windows centred at 90 Ma will be chosen for comparisons. Comparisons between Amma Fatma poles and younger South African APWP poles (e.g. centred at 80 Ma) will be used to check potential signs of a late diagenetic age for the ChRM.

For tectonic purposes, the less stringent cut-off criterion (equal to 45°) has been chosen to compare the observed and unflattened ($f = 0.8$) palaeomagnetic poles with available global APWPs. Due to the overlap of inclinations through time (Fig. 8b), expected declinations are compared to that of Amma Fatma succession (331.7° ± 3.4°) while considering their 95% confidence bounds (Fig. 8a). At the 90 Ma centred window, expected declinations are 335.1° ± 5.3° and 340.2° ± 2.7° (BC2002 and T2008, respectively).

In the 100–80 Ma range, selected palaeomagnetic entries from BC2002 and T2008 compilations are characterized by a considerable scatter between their means (being mostly statistically distinguishable from each other) and by variable, relatively high age uncertainty values (from some Ma up to 19 Ma). This is also applicable to those 3–4 entries from the African plate (none from NW Africa) selected during 100–80 Ma. Amma Fatma palaeopoles does not match with individual poles from the 100–80 Ma interval either. The better fit occurs with the BC2002 90 Ma centred window, but it is out of step with the corresponding T2008 version (note the different number of inputs in both poles) (Fig. 8c). As Amma Fatma palaeomagnetic poles match the T2008 running mean paths which are older (and not younger) than the time of deposition, the imbalance with respect to the T2008 APWP does not support any late diagenetic source for the ChRM acquisition.

The Amma Fatma palaeomagnetic pole ($N = 88$; PLat $= 64.3°$, PLon $= 256.3°$, $A_{95} = 2.5°$, $K = 38.7$) alleviates the need for accurate-aged palaeomagnetic inputs for palaeogeographical reconstructions. It has been obtained from a cyclostratigraphic, marl-dominated succession from a stable, non-deformed craton platform. This palaeopole has been calculated using normal polarity samples, corrected by a moderate inclination error ($f = 0.8$). Adequately averaging its secular variation, it is assumed to represent a record of c. 1 Ma of the geomagnetic field during the early Turonian.

We thank A. El Albani for helpful geological explanations regarding Amma Fatma section and C. Langereis for kind assistance with the 'Pal_vD_s' program for PSV analysis. Financial support was provided by the project CGL2009-10840 of the Dirección General de Enseñanza Superior (DGES), Spanish Ministry of Education. VCRM is grateful to the MEC 'José Castillejo Program, ref. JC2008-00356' (Spain) and to EEA 'Abel extraordinary chairs' for corresponding mobility grants. Sincere thanks to C. Rowan and M. Deenen for their thoughtful reviews, which led to considerable improvement of the manuscript.

References

BESSE, J. & COURTILLOT, V. 2002. Apparent and true polar wander and the geometry of the magnetic field in the last 200 million years. *Journal of Geophysical Research*, **107**, 2300, doi: 10.1029/2000JB000050.

BIGGIN, A. J., VAN HINSBERGEN, D. J. J., LANGEREIS, C. G., STRAATHOF, G. B. & DEENEN, M. H. L. 2008. Geomagnetic secular variation in the Cretaceous Normal Superchron and in the Jurassic. *Physics of the Earth and Planetary Interiors*, **169**, 3–19, doi: 10.1016/j.pepi.2008.07.004.

CANDE, S. C. & KENT, D. V. 1992. A new geomagnetic polarity time scale for the Late Cretaceous and Cenozoic. *Journal of Geophysical Research*, **97**, 13,917–13,951, doi: 10.1029/92JB01202.

CANDE, S. C. & KENT, D. V. 1995. Revised calibration of the geomagnetic polarity timescale for the Late Cretaceous and Cenozoic. *Journal of Geophysical Research*, **100**, 6093–6095, doi: 10.1029/94JB03098.

CHOUBERT, G., FAURE MURET, A. & HOTTINGER, L. 1966. Aperçu géologique du Bassin côtier de Tarfaya (Stratigraphie). *In*: CHOUBERT, G., FAURE MURET, A., HOTTINGER, L., VIOTTI, C. & LECOINTRE, G. (eds) *Le Bassin côtier de Tarfaya (Maroc Meridional)*. Notes et Mémoire Service Géologique du Maroc, **175**, 7–106.

CHOUBERT, G., FAURE MURET, A. & HOTTINGER, L. 1972. La série stratigraphique de Tarfaya Maroc sud occidental et le problème de la naissance de l'Atlantique. *Notes et Mémoire Service Géologique du Maroc*, **31**, 29–40.

CONSTABLE, C. & TAUXE, L. 1990. The bootstrap for magnetic susceptibility tensors. *Journal of Geophysical Research*, **95**, 8383–8395.

DEKKERS, M. J. 1988. Magnetic properties of natural pyrrhotite Part I: behaviour of initial susceptibility and saturation–magnetization-related rock-magnetic parameters in a grain-size related framework. *Physics of the Earth and Planetary Interiors*, **52**, 376–393.

EL ALBANI, A., KUHNT, W., LUDERER, F., HERBIN, J. P. & CARON, M. 1999a. Palaeoenvironmental evolution of the Late Cretaceous sequence in the Tarfaya Basin (southwest of Morocco). *In*: CAMERON, N. R., BATE, R. H. & CLURE, V. S. (eds) *The Oil and Gas Habitats of the South Atlantic*. Geological Society, London, Special Publications, **153**, 223–240.

EL ALBANI, A., VACHARD, D., KUHNT, W. & CHELLAI, H. 1999b. Signature of hydrodynamic activity caused by rapid sea level changes in pelagic organic-rich sediments, Tarfaya Basin (southern Morocco). *Comptes rendus de l'Académie des sciences, Sciences de la terre et des planètes*, **329**, 397–404.

EL ALBANI, A., VACHARD, D., KUHNT, W. & THUROW, J. 2001. The role of diagenetic carbonate concretions in the preservation of the original sedimentary record. *Sedimentology*, **48**, 875–886.

FISHER, R. A. 1953. Dispersion on a sphere. *Proceedings of the Royal Society of London, Series A*, **217**, 295–305.

GEBHARDT, H., KUHNT, W. & HOLBOURN, A. 2004. Foraminiferal response to sea level change, organic flux and oxygen deficiency in the Cenomanian of the Tarfaya Basin, southern Morocco. *Marine Micropaleontology*, **53**, 133–157.

HEYMAN, M. A. W. 1990. Tectonic and depositional history of the Moroccan continental margin. *In*: TANKARD, A. J. & BALKWILL, H. R. (eds) *Extensional Tectonics and Stratigraphy of the North Atlantic Margins*. AAPG Memoir, **46**, 323–340.

HOLBOURN, A., KUHNT, W., EL ALBANI, A., PLETSCH, T., LUDERER, F. & WAGNER, T. 1999. Upper Cretaceous palaeoenvironments and benthonic foraminiferal assemblages of potential source rocks from the western African margin, Central Atlantic. *In*: CAMERON, N. R., BATE, R. H. & CLURE, V. S. (eds) *The Oil and Gas Habitats of the South Atlantic*. Geological Society, London, Special Publications, **153**, 195–222.

HROUDA, F., CHLUPACOVA, M. & POKORNY, J. 2006. Low-field variation of magnetic susceptibility measured by the KLY-4S Kappabridge and KLY-4A magnetic susceptibility meter: accuracy and interpretational programme. *Studia Geophysica et Geodaetica*, **50**, 283–298.

IRVING, E. 1960. Paleomagnetic directions and pole positions, 2. *Geophysical Journal*, **4**, 444–449.

IVAKHNENKO, O. P. & POTTER, D. K. 2004. Magnetic susceptibility of petroleum reservoir fluids. *Physics and Chemistry of the Earth*, **29**, 899–907.

JELINEK, V. 1978. Statistical processing of magnetic susceptibility measured on groups of specimens. *Studia Geophysica et Geodaetica*, **22**, 50–62.

JENKYNS, H. C., MATTHEWS, A., TSIKOS, H. & EREL, Y. 2007. Nitrate reduction, sulfate reduction, and sedimentary iron isotope evolution during the Cenomanian–Turonian oceanic anoxic event. *Paleoceanography*, **22**, PA3208, doi: 10.1029/2006PA001355.

KAWASAKI, K., SYMONS, D. T. A. & COVENEY, R. M., JR. 2007. Current-scattered (?) detrital remanence directions in the Zn-rich Pennsylvanian Stark black shale, USA. *Geophysical Journal International*, **171**, 594–602, doi: 10.1111/j.1365-246X.2007.03552.x.

KELLER, G., ADATTE, T., BERNER, Z., CHELLAI, E. H. & STUEBEN, D. 2008. Oceanic events and biotic effects of the Cenomanian-Turonian anoxic event, Tarfaya Basin, Morocco. *Cretaceous Research*, **29**, 976–994.

KING, R. F. 1955. The remanent magnetism of artificially deposited sediments. *Monthly Notices of the Royal Astronomical Society Geophysical Supplement*, **7**, 115–134.

KIRSCHVINK, J. L. 1980. The least-squares line and plane and the analysis of palaeomagnetic data. *Geophyisical Journal of the Royal Astronomical Society*, **62**, 699–718.

KOLONIC, S., SINNINGHE-DAMSTÉ, J. S. *ET AL.* 2002. Geochemical characterization of Cenomanian/Turonian black shales from the Tarfaya Basin (SW Morocco). *Journal of Petroleum Geology*, **25**, 325–350.

KOLONIC, S., WAGNER, T. *ET AL.* 2005. Mechanisms of black shale deposition at the NW-African shelf during the Cenomanian/Turonian oceanic anoxic event 2: implications for climate coupling and global organic carbon burial. *Paleoceanography*, **20**, 1006, doi: 10.1029/2003PA000950.

KRUIJS, E. & BARRON, E. 1990. Climate model prediction of paleoproductivity and potential source–rock distribution. *In*: HUC, A. Y. (ed.) *Deposition of Organic Facies*. American Association of Petroleum Geologists (AAPG) Studies in Geology, **30**, 195–216.

KRUMSIEK, K. 1982. Cretaceous magnetic stratigraphy of Southwest Morocco. *In*: RAD, U., HINZ, K., SARNTHEIM, M. & SEIBOLD, E. (eds) *Geology of the Northwest African Continental Margin*. Springer, Berlin, 475–497.

KUHNT, W. & WIEDMANN, J. 1995. Cenomanian–Turonian source rocks: paleobiogeographic and paleoenvironmental aspect. *In*: HUC, A. Y. (ed.) *Paleogeography, Paleoclimates and Source Rocks*. AAPG Studies in Geology, **40**, 213–232.

KUHNT, W., HERBIN, J. P., THUROW, J. & WIEDMANN, J. 1990. Distribution of Cenomanian–Turonian organic facies in the western Mediterranean and along the adjacent Atlantic Margin. *In*: HUC, A. Y. (ed.) *Deposition of Organic Facies*. American Association of Petroleum Geologists (AAPG) Studies in Geology, **40**, 133–160.

KUHNT, W., NEDERBRAGT, A. & LEINE, L. 1997. Cyclicity of Cenomanian–Turonian organic-carbon–rich sediments in the Tarfaya Atlantic coastal basin (Morocco). *Cretaceous Research*, **18**, 587–601, doi: 10.1006/cres.1997.0076.

KUHNT, W., LUDERER, F., NEDERBRAGT, S., THUROW, J. & WAGNER, T. 2005. Orbital-scale record of the late Cenomanian–Turonian oceanic anoxic event (OAE-2) in the Tarfaya Basin (Morocco). *International Journal of Earth Sciences*, **94**, 147–159, doi: 10.1007/s00531-004-0440-5.

KUHNT, W., HOLBOURN, A., GALE, A., CHELLAI, E. H. & KENNEDY, W. J. 2009. Cenomanian sequence stratigraphy and sea-level fluctuations in the Tarfaya Basin (SW Morocco). *Geological Society of America Bulletin*, **1697**, doi: 10.1130/B26418.1.

KUYPERS, M. M. M., PANCOST, R. D., NIJENHUIS, I. A. & SINNINGHE DAMSTÉ, J. S. 2002. Enhanced productivity led to increased organic carbon burial in the euxinic North Atlantic basin during the Late Cenomanian oceanic anoxic event. *Paleoceanography*, **17**, 1051, doi: 10.1029/2000PA000569.

LEINE, L. 1986. Geology of the Tarfaya oil shale deposit, Morocco. *Geologie en Mijnbouw*, **65**, 57–74.

LUDERER, F. 1999. *Das Cenoman/Turon Grenzereignis im Tarfaya Becken (SW Marokko)*. PhD thesis, Christian-Albrechts-University, Kiel, Germany.

LÜNING, S., KOLONIC, S., BELHADJ, E. M., BELHADJ, Z., COTA, L., BARIC, G. & WAGNER, T. 2004. Integrated depositional model for the Cenomanian–Turonian organic-rich strata in North Africa. *Earth-Science Reviews*, **64**, 51–117.

MARTINIS, B. & VISINTIN, V. 1966. Données géologiques sur le bassin sédimentaire côtier de Tarfaya (Maroc méridional). *In*: REYRE, D. (ed.) *Bassins Sédimentaires du Littoral Africain*. Assoc. Serv. Géol. Afr., Paris, 13–26.

MCELHINNY, M. W. & MCFADDEN, P. L. 1997. Palaeosecular variation over the past 5 Myr based on a new generalized database. *Geophysical Journal International*, **131**, 240–252.

MCFADDEN, P. L., MERRILL, R. T. & MCELHINNY, M. W. 1988. Dipole quadrupole family modelling of paleosecular variation. *Journal of Geophysical Research*, **93**, 11583–11588, doi: 10.1029/JB093iB10p11583.

MORT, H. P., ADATTE, T. *ET AL.* 2008. Organic carbon deposition and phosphorus accumulation during Oceanic Anoxic Event 2 in Tarfaya, Morocco. *Cretaceous Research* **29**, 1008–1023, doi: 10.1016/j.cretres.2008.05.026.

MULLENDER, T. A. T., VAN VELZEN, A. J. & DEKKERS, M. J. 1993. Continuous drift correction and separate identification of ferrimagnetic and paramagnetic contributions in thermomagnetic runs. *Geophysical Journal International*, **114**, 663–672.

OYARZUN, R., DOBLAS, M., LÓPEZ-RUIZ, J. & CEBRIÁ, J. M. 1997. Opening of the central Atlantic and asymmetric mantle upwelling phenomena: implications for long-lived magmatism in western North Africa and Europe. *Geology*, **25**, 727–730.

PETERS, C. & THOMPSON, R. 1998. Magnetic identification of selected natural iron oxides and sulphides. *Journal of Magnetism and Magnetic Materials*, **183**, 365–374.

POKORNY, J., SUZA, P. & HROUDA, F. 2004. Anisotropy of magnetic susceptibility of rocks measured in variable weak magnetic fields using the KLY-4S Kappabridge. *In*: MARTÍN-HERNÁNDEZ, F., LÜNEBURG, C., AUBOURG, C. & JACKSON, M. (eds) *Magnetic Fabric: Methods and Applications*. Geological Society, London, Special Publications, **238**, 69–76.

RANKE, U., VON RAD, U. & WISSMANN, G. 1982. Stratigraphy, facies and tectonic development of the on- and offshore Aaiun-Tarfaya Basin – A review. *In*: VON RAD, U., HINZ, K., SARNTHEIN, M. & SEIBOLD, E. (eds) *Geology of the Northwest African Continental Margin*. Springer, Berlin, 86–105.

ROBERTS, A. P. 1995. Magnetic properties of sedimentary greigite (Fe$_3$S$_4$). *Earth and Planetary Science Letters*, **134**, 227–236.

ROWAN, C. J., ROBERTS, A. P. & BROADBENT, T. 2009. Reductive diagenesis, magnetite dissolution, greigite growth and paleomagnetic smoothing in marine sediments: a new view. *Earth and Planetary Science Letters*, **277**, 223–235, doi: 10.1016/j.epsl.2008.10.016.

STAMM, R. & THEIN, J. 1982. Sedimentation in the Atlas Gulf III: Turonian carbonates. *In*: RAD, U., HINZ, K., SARNTHEIM, M. & SEIBOLD, E. (eds) *Geology of the Northwest African Continental Margin*. Springer, Berlin, 459–474.

TANTAWY, A. A. 2008. Calcareous nanofossil biostratigraphy and paleoecology of the Cenomanian–Turonian transition in the Tarfaya Basin, southern Morocco. *Cretaceous Research*, **29**, 995–1007, doi: 10.1016/j.cretres.2008.05.021.

TARDUNO, J. A., LOWRIE, W., SLITER, W. V., BRALOWER, T. J. & HELLER, F. 1992. Reversed polarity characteristic magnetizations in the Albian contessa section, Umbrian Apennines, Italy: implications for the existence of a mid-cretaceous mixed polarity interval. *Journal of Geophysical Research*, **97**, 241–271, doi: 10.1029/91JB02257.

TARLING, D. H. & HROUDA, F. 1993. *The Magnetic Anisotropy of Rocks*. Chapman and Hall, London.

TAUXE, L. & KENT, D. V. 2004. A simplified statistical model for the geomagnetic field and the detection of shallowbias in paleomagnetic inclinations: was the ancient magnetic field dipolar? *In*: CHANNELL, J. E. T, KENT, D. V., LOWRIE, W. & MEERT, J. (eds) *Timescales of the Paleomagnetic Field*. American Geophysical Union, Washington, DC, **145**, 101–116.

TAUXE, L., BERTRAM, H. & SEBERINO, C. 2002. Physical interpretation of hysteresis loops: micromagnetic modelling of fine particle magnetite. *Geochemistry, Geophysics, Geosystems*, **3**, doi: 10.129/2001GC000280.

TAUXE, L., KODAMA, K. & KENT, D. V. 2008. Testing corrections for paleomagnetic inclination error in sedimentary rocks: a comparative approach. *Physics of the Earth and Planetary Interiors*, **169**, 152–165.

TORSVIK, T. H., MÜLLER, R. D., VAN DER VOO, R., STEINBERGER, B. & GAINA, C. 2008. Global plate motion frames: Toward a unified model. *Reviews of Geophysics*, **46**, RG3004, doi: 10.1029/2007RG000227.

TSIKOS, H., JENKYNS, H. C. *ET AL*. 2004. Carbon-isotope stratigraphy recorded by the Cenomanian/Turonian Oceanic Anoxic Event: Correlation and implications based on three key-localities. *Journal of the Geological Society, London*, **161**, 711–719.

VANDAMME, D. A. 1994. A new method to determine paleosecular variation. *Physics of the Earth and Planetary Interiors*, **85**, 131–142, doi: 10.1016/0031-9201(94)90012-4.

VON RAD, U. & EINSELE, G. 1980. Mesozoic–Cenozoic subsidence history and palaeobathymetry of the north-west African continental margin (Aaiun Basin to DSDP 397). The Evolution of Passive Continental Margins. *Philosophical Transactions of the Royal Society of London*, **294**, 37–50.

WIEDMANN, J., BUTT, A. & EINSELE, G. 1978. Vergleich von marokkanischen Kreide-Küstenaufschlüssen und Tiefseebohrungen (DSDP): Stratigraphie, Paläoenvironment und Subsidenz an einem passiven Kontinentalrand. *Geology Rundschau*, **67**, 454–508.

WIEDMANN, J., BUTT, A. & EINSELE, G. 1982. Cretaceous stratigraphy, environment, and subsidence history at the Moroccan continental margin. *In*: VON RAD, U., HINZ, K., SARNTHEIN, M. & SEIBOLD, E. (eds) *Geology of the Northwest African Continental Margin*. Springer, Berlin, 366–395.

Palaeoposition of the Seychelles microcontinent in relation to the Deccan Traps and the Plume Generation Zone in Late Cretaceous–Early Palaeogene time

M. GANERØD[1]*, T. H. TORSVIK[1,2,3], D. J. J. VAN HINSBERGEN[2,3], C. GAINA[1,2,3], F. CORFU[4], S. WERNER[2], T. M. OWEN-SMITH[5], L. D. ASHWAL[5], S. J. WEBB[5] & B. W. H. HENDRIKS[1]

[1]*Geodynamikk, Geological Survey of Norway, NO-7491 Trondheim, Norway*

[2]*Physics of Geological Processes (PGP), University of Oslo, P.O. Box 1048, Blindern, NO-0316 Oslo, Norway*

[3]*Center for Advanced Studies, Norwegian Academy of Science and Letters, Drammensveien 78, NO-0271 Oslo, Norway*

[4]*Department of Geosciences, University of Oslo, Pb 1047 Blindern, NO-0316 Oslo, Norway*

[5]*School of Geosciences, University of the Witwatersrand, Private Bag 3, Wits 2050, Johannesburg, South Africa*

Corresponding author (e-mail: morgan.ganerod@ngu.no)

Abstract: The Early Palaeogene magmatic rocks of North and Silhouette Islands in the Seychelles contain clues to the Cenozoic geodynamic puzzle of the Indian Ocean, but have so far lacked precise geochronological data and palaeomagnetic constraints. New $^{40}Ar/^{39}Ar$ and U–Pb dates demonstrate that these rocks were emplaced during magnetochron C28n; however, $^{40}Ar/^{39}Ar$ and palaeomagnetic data from Silhouette indicate that this complex experienced a protracted period of cooling. The Seychelles palaeomagnetic pole (57.55°S and 114.22°E; A9512.3°, $N = 14$) corresponds to poles of similar ages from the Deccan Traps after being corrected for a clockwise rotation of $29.4° \pm 12.9°$. This implies that Seychelles acted as an independent microplate between the Indian and African plates during and possibly after C27r time, confirming recent results based on kinematic studies. Our reconstruction confirms that the eruption of the Deccan Traps, which affected both India and the Seychelles and triggered continental break-up, can be linked to the present active Reunion hotspot, which is being sourced as a deep plume from the Plume Generation Zone.

Supplementary material: Experimental data are available at http://www.geolsoc.org.uk/ SUP18482.

The late Mesozoic and Cenozoic northwards drift of the Indian plate was accommodated by subduction of the Neotethys Ocean below the Eurasian margin and the opening of the Indian Ocean to the south. The Indian Ocean comprises several sub-basins and intervening continental fragments that originated from Gondwana and were stranded by several events of spreading ridge relocation (McKenzie & Sclater 1971; Norton & Sclater 1979; Barron & Harrison 1980; Plummer & Belle 1995). Among these continental fragments is the Seychelles microcontinent, which is an almost entirely submerged and elongated continental fragment (Fig. 1) in the central part of the Indian Ocean (Baker 1963). The timing of separation of the Seychelles continent from India is documented by seafloor spreading

along the Carlsberg Ridge (Fig. 1), where spreading anomalies correspond to chron C28n (64.1 Ma, according to GST2004 by Ogg & Smith 2004) and younger (Chaubey *et al.* 2002; Royer *et al.* 2002; Collier *et al.* 2008).

The Seychelles islands mainly expose coral reefs overlying undeformed Neoproterozoic granites (Velain 1879; Baker 1963), most of which fall within a 755–748 Ma age window (U–Pb; Tucker *et al.* 2001). However, on the islands of Silhouette and the North Island (Fig. 2), Late Cretaceous to Palaeogene alkaline central complexes and mafic volcanic rocks are exposed. These may correspond to the Late Cretaceous–Palaeogene Deccan Traps (Fig. 1), one of the most voluminous of the Large Igneous Provinces (Coffin & Eldholm 1992;

From: VAN HINSBERGEN, D. J. J., BUITER, S. J. H., TORSVIK, T. H., GAINA, C. & WEBB, S. J. (eds) *The Formation and Evolution of Africa: A Synopsis of 3.8 Ga of Earth History*. Geological Society, London, Special Publications, **357**, 229–252. DOI: 10.1144/SP357.12 0305-8719/11/$15.00 © The Geological Society of London 2011.

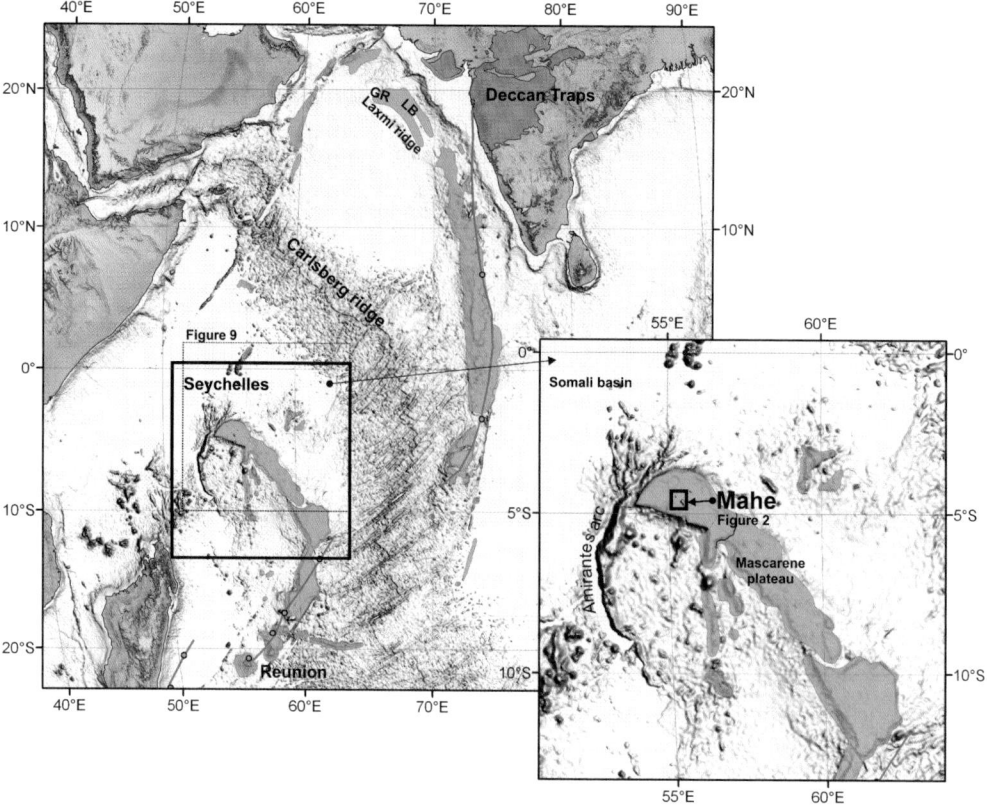

Fig. 1. Location of the Deccan Traps and the Seychelles. The proposed Reunion hotspot track of Duncan (1990) is shown in red. LB and GR denote Laxmi Basin and Gop Rift, respectively. The bathymetry and topography is produced from Smith & Sandwell (1997).

Courtillot *et al.* 1999) where Jay & Widdowson (2008) have provided an estimate of the original Deccan eruptive volume of *c.* 1.3×10^6 km^3. Deccan magmatism has been postulated to be associated with the separation of India from the Seychelles, after which the Seychelles effectively became reunited with the African plate. Establishing high-resolution age control of the mafic volcanics is therefore essential to determine whether volcanism related to the Deccan Traps occurred on both the Indian and Seychelles continents, and whether it was restricted to the Indian Subcontinent.

In order to address this issue we sampled the Late Cretaceous–Palaeogene successions on the Seychelles and carried out ^{40}Ar/^{39}Ar and U–Pb analyses. Additionally, we carried out a palaeomagnetic study to test whether the Seychelles and the Indian Deccan Traps share a common palaeomagnetic pole after correcting for younger sea-floor spreading. We use this information to reconstruct the palaeoposition of the Deccan Traps with respect to the African Large Low Shear-wave Velocity Province (LLSVP) at the core–mantle boundary, which has been referred to a Plume Generation Zone (Burke & Torsvik 2004; Burke *et al.* 2008; Torsvik *et al.* 2008*a*).

Geological setting

Opening of the Indian Ocean

During the Mid-Jurassic, Africa separated from East Antarctica and adjacent India–Seychelles–Madagascar forming the East Somali and Mozambique basins (Coffin & Rabinowitz 1988; König & Jokat 2006). Around the Mid-Cretaceous, Indian–Seychelles–Madagascar separated away from East Antarctica and Australia. Ever since, the Indian–Seychelles–Madagascar trio had a protracted history of rifting and drifting (e.g. Gaina *et al.* 2007). After passing over the assumed 'Marion hotspot' (Mahoney *et al.* 1991; Storey *et al.* 1995, 1997; Torsvik *et al.* 1998), India and the Seychelles drifted away from Madagascar, accommodated by

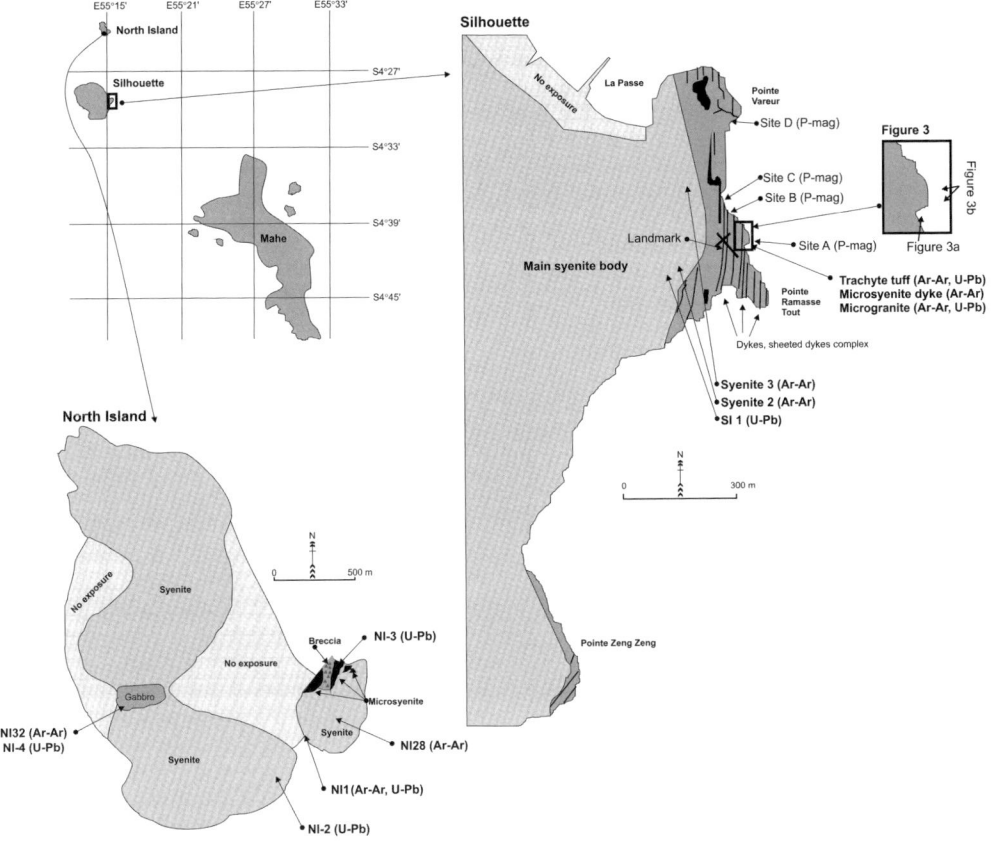

Fig. 2. The location of Silhouette and North Island. Silhouette, the palaeomagnetic sampling sites are divided into sites A–D; most sites measurements were made at site A. Samples for geochronological analysis from Silhouette are the syenite 2 and 3; the trachyte tuff, microsyenite dyke and the microgranite are taken from site A. North Island, samples for geochronologic analysis are the syenite samples NI1, NI28 and the NI32 gabbro. Modified from Stephens & Devey (1992).

the Mascarene spreading ridge. The separation was associated with widespread volcanism between 91.6 and 83.6 Ma, remnants of which are preserved in Madagascar (Storey *et al.* 1995; Torsvik *et al.* 1998). An anticlockwise rotation of the India–Seychelles platform has been proposed for this time window (Plummer 1995; Plummer & Belle 1995). Moving northwards, the impingement of the Réunion mantle plume in the Latest Cretaceous caused massive volcanism in Western India, known as the Deccan Traps. Around the same time, the mid-ocean ridge from the Mascarene Basin gradually relocated northwards between India and Seychelles to become the Carlsberg ridge (e.g. Royer *et al.* 2002), although the plume involvement in this scenario has been questioned (Sheth 2005). This event parted the Indian continent from the Seychelles (Duncan 1990). As a result of this history, microcontinental fragments derived from

Gondwana are now widely distributed across the Indian Ocean. The present-day margin of West India has been described as a conglomerate of highly extended continental crust (Subrahmanyam *et al.* 1995) and small oceanic basins – such as the Gop Rift and possibly the Laxmi Basin (Krishna *et al.* 2006; Collier *et al.* 2008; Yatheesh *et al.* 2009) – that have been affected by volcanism (Deccan and possible pre-Deccan) and underplating (Lane 2006; Minshull *et al.* 2008) and formed several volcanic ridges (Calvès *et al.* 2008, 2011).

Age of the Deccan Traps

The age and the palaeomagnetic signature of the Deccan Traps have been a focus of research for several decades (e.g. Wensink 1973; Wensink *et al.* 1977). The earliest manifestations of Deccan volcanism are alkaline volcanic and intrusive complexes in extensional areas north of the main Deccan province, dated at 68.5 ± 0.16 Ma (Basu *et al.* 1993). The culmination of emplacement of these complexes coincided with the onset of the voluminous tholeiitic Deccan volcanism. The first products of the Deccan flood basalts can be found in the north (Kutch region) and are dated at 67 Ma, whereas the most voluminous phase corresponds broadly in time with the K–T boundary at 65.5 Ma (Allègre *et al.* 1999; Courtillot *et al.* 1999; Courtillot & Renne 2003; Chenet *et al.* 2007), spanning magnetochron 29r and 29n (Jay *et al.* 2009) of the geomagnetic polarity timescale (Ogg & Smith 2004). Based on geochemical and geochronological evidence, a correlation of the Deccan to the Rajahmundry traps (SE India) was proposed (Self *et al.* 2008), dated 64.7 ± 0.57 Ma (Baksi 2005). Due to relative southwards migration of the eruptive centres, the geometry of each pulse has been described as consisting of large-scale clinoforms (e.g. Mitchell & Widdowson 1991). The culmination of the Deccan volcanism can be seen as trachytes in the western Indian rifted continental margin dated *c.* 61.5 ± 0.3 Ma (Sheth *et al.* 2001). This last phase of volcanism has also been documented in the conjugate Seychelles rift margin, as offshore seaward-dipping reflectors (Collier *et al.* 2008).

Geology of the Seychelles

The geology of the Seychelles is dominated by undeformed Neoproterozoic granitic rocks and dolerite dykes (Baker 1963; Ashwal *et al.* 2002) with ages between 755 and 748 Ma (Tucker *et al.* 2001). Paleocene alkaline igneous complexes are restricted to the Silhouette and North Islands (Fig. 2). Most of Silhouette consists of a body of syenite and a small unit of granite (Dickin *et al.* 1986).

On the eastern part of the Silhouette Island (from Pointe Zeng Zeng to Pointe Vareur, Fig. 2) a complex of trachytic tuff units, several pale-coloured microsyenite sheeted dykes, mafic dykes and microgranite dykes are exposed. The lavas contain fragments of syenite and older trachyte tuff, whereas the microsyenite sheeted dykes entrain fragments of trachyte. Two mafic dykes cut the complex close to Pointe Vareur, and are themselves cut by veins of granite (microgranite). Larger granite intrusions are also identified closer to Pointe Ramasse Tout. The trachyte tuffs were therefore emplaced first, followed by mafic dykes and then felsic dykes, microgranite and syenite.

A prominent set of vertical fractures are developed in the volcanic complex (Fig. 3a). Previous structural mapping of the area interpreted the trachytic units to be steeply dipping lavas close to vertical (Stephens & Devey 1992), each lava flow

bound by bedding surfaces. We observed, however, that the alignment of 'bedding planes' follows the structural alignment in the intruding microsyenites. We therefore favour a structural interpretation where the trachytes are bound by fracture planes and not bedding planes (Fig. 3a, b), as no primary structures or other lines of evidence for steep bedding surfaces can be put forward. This implies that the trachytes are structurally bound eruptive units.

Previous dating

The chron 27r age for the oldest oceanic crust NE of Seychelles inferred from magnetic anomaly data was extrapolated to the identified continental–ocean boundary, interpreted from magnetic, gravity and refraction data, and was used to date the age of break-up between Seychelles and Laxmi Ridge/Indian plate as 63.4 Ma (Collier *et al.* 2008). This age correspond to chron C28n (GST2004) and is roughly similar to ages previous obtained from the alkaline igneous complexes of Silhouette and North Island on the continental plateau of Seychelles (Dickin *et al.* 1986). We summarize here some of the results of Dickin *et al.* (1986). A Rb/Sr whole rock isochron was defined by rocks from Silhouette at 63.2 ± 1 Ma (no. 1 in Fig. 4; all ages reported with 2σ errors), while rocks from North Island yielded only an error-chron. A combined regression of Silhouette and North Island samples gave 63 ± 2.2 Ma. The K/Ar analyses from Silhouette indicated a weighted mean age of 63.7 ± 1 Ma (no. 2 in Fig. 4) for the syenite, 62.1 ± 1.3 Ma (no. 3 in Fig. 4) for the trachyte tuffs and a weighted mean age of 60.23 ± 1.1 Ma (no. 4 in Fig. 4) for the granitic suite. The K/Ar ages obtained from North Island ranged from 65 to 62 Ma, but the authors found the feldspar dates less reliable. A weighted mean age for the ferromagnesian minerals gave 63.3 ± 0.9 (no. 5 in Fig. 4). The dates of Dickin *et al.* (1986) have large uncertainties however, too large to tie the formation of these volcanic complexes to a specific magnetochron in the geomagnetic polarity timescale at an acceptable level of probability. An unpublished, combined U–Pb zircon date of 63.3 ± 0.2 Ma is mentioned in Tucker *et al.* (2001) for syenitic units in Silhouette Island.

Sampling

Samples for palaeomagnetic analysis were taken from the island of Silhouette at locations from Pointe Ramasse Tout to Pointe Vareur (Fig. 2) where fresh outcrops of trachytic tuffs and syenitic

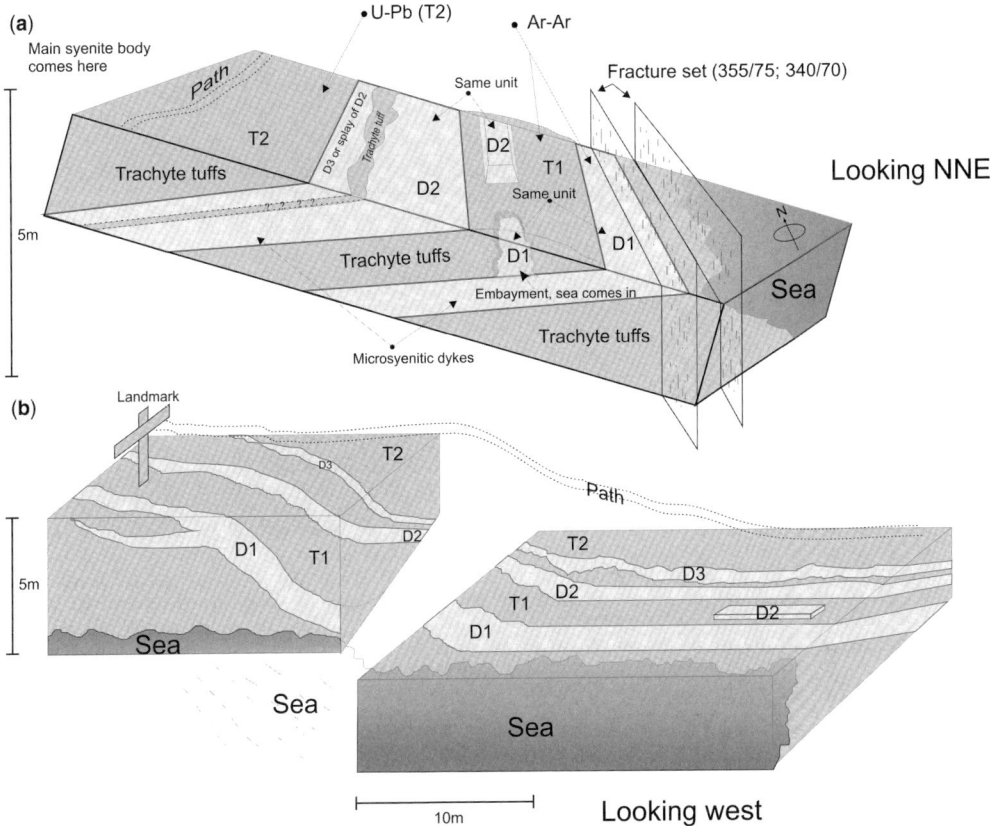

Fig. 3. Structural interpretation of sampling site A looking towards the (**a**) north and (**b**) west. The syenitic dykes (sheeted dykes) becomes more vertical towards the cross (a landmark). Units D1–3 (syenitic dykes) and T1–2 (trachyte tuffs) are the palaeomagnetic sampling sites (see Table 3). The microgranites are not indicated but occur as small patches, one at the cross (Site A). The apparent tilt of the sheeted dykes is less than indicated in the figures.

dykes are exposed. Deccan feeder dykes have also been documented on Prasilin Island (Devey & Stephens 1991), but we had limited time to include those in this study. We have divided the palaeomagnetic sampling locations into sites A to D (Fig. 2). The main syenite body was too coarse-grained to be suitable for palaeomagnetic analysis. A total of 96 25 mm drill cores were extracted from 15 sites with a portable gasoline drill. The orientations of the cores were determined with both magnetic and sun compasses. Due to a very strong local magnetic anomaly (magnetic deviation commonly exceeded 60°), only orientations based on sun azimuth were used. In addition, we sampled the main units (main syenite body, trachyte tuff, microsyenitic sheeted dykes and microgranite) from site A on Silhouette (Figs 2 & 3a, b) for age determination, together with selected sites from North Island (Fig. 2).

^{40}Ar/^{39}Ar age determinations

Samples were crushed and sieved to isolate grains of 180–250 μm. Magnetic separation using a Frantz isodynamic separator, followed by heavy liquid separation with lithium polytungstate, was then employed to concentrate feldspars. Biotite and amphiboles were handpicked after magnetic separation. The mineral separates were washed in acetone several times and finally fresh inclusion-free minerals grains were handpicked under the binocular microscope. Mineral samples were packed in aluminium capsules together with the Taylor Creek rhyolite (TC) flux monitor standard (between each fifth sample, every c. 8 mm) and zero-aged K_2SO_4 and CaF_2 salts. The transformation ^{39}K(n, p)^{39}Ar was performed during irradiation at the McMaster nuclear facility, Hamilton, Canada. The samples were step-heated in the

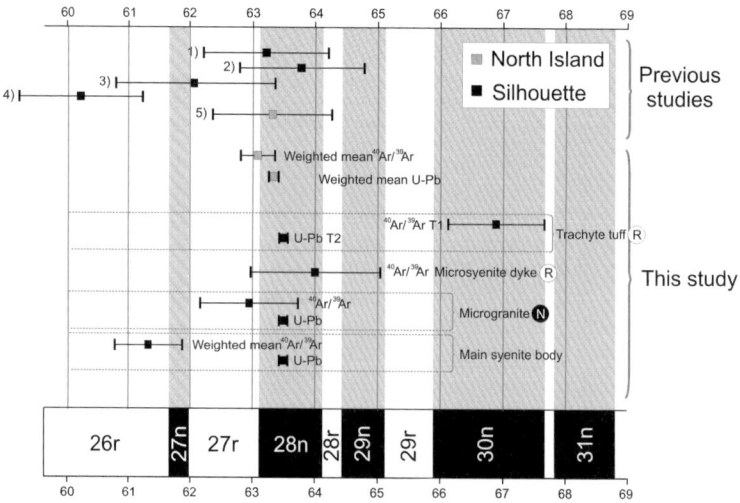

Fig. 4. Dates from North and Silhouette Island with the geomagnetic polarity timescale. Samples labelled 'Previous studies' are from Dickin *et al.* (1986) and are labelled from 1–5 according to the text. Timescale is from Ogg & Smith (2004). The error bars are plotted at the 95% confidence level. The black (N) and white (R) circles are the magnetic polarities found from the palaeomagnetic investigation. The weighted mean ages for North Island are calculated from samples NI1, NI28 and NI32 (^{40}Ar/^{39}Ar) and NI-1, NI-2, NI-3 and NI-4 (^{206}Pb/^{238}U). The weighted mean age for the main syenite body from Silhouette is calculated from samples Syenite 2 and Syenite 3 (^{40}Ar/^{39}Ar). See Table 1 for age summary.

^{40}Ar/^{39}Ar lab at the Geological Survey of Norway using a resistance furnace (Heine type). The extracted gases were swiped over getters (SAES AP-10) for 2 min, and then for 9 min in a separate part of the extraction line. The gas was finally analysed with a MAP 215–50 mass spectrometer. The peaks were determined by peak hopping (at least 8 cycles) on masses ^{41}Ar to ^{35}Ar on a Balzers electron multiplier.

Subtraction of blanks, correction for mass fractionation, correction for ^{37,39}Ar decay and neutron-induced interference reactions produced in the reactor were carried out using in-house software (Age Monster 2010, written by M. Ganerød) which implements the equations in McDougall & Harrison (1999) using the decay constants and the trapped ^{40}Ar/^{36}A ratio of 295.5 ± 0.5 of Steiger & Jäger (1977). The decay constants and the correction factors for the production of isotopes from Ca and K can be found in the Supplementary Material. Uncertainties from the blanks, mass discrimination value, salts, trapped constants and every mass balance calculation are propagated into the final age uncertainty. The blanks were measured at temperatures of 450, 700, 1000, 1130 and 1280 °C. The blanks and associated errors for the respective temperature steps for the unknowns were determined using linear interpolation. We used the age of 28.34 ± 0.16 Ma for the TC monitor (Renne *et al.* 1998) during data reduction. We define a plateau according to the following requirements: at least

three consecutive steps each within 95% confidence level:

$$\mathrm{abs(age_A - age_B)} < 1.96 \times \sqrt{\sigma_A^2 + \sigma_B^2}$$

if ages are quoted at the 1σ level

comprising at least 50% of total ^{39}Ar and mean square of weighted deviates (MSWD) less than the student T critical value. We calculated a weighted mean plateau age (WMPA), weighting by the inverse of the variance.

The step-heating spectra with plateaus are depicted in Figure 5 and the main results from spectrum and inverse isochrons analysis are displayed in Table 1 (the raw experimental data are located in the Supplementary Material). Figure 5 shows that most samples have a slight Ar loss, most likely due to the presence of small amounts of alteration in all age spectra at the lowermost temperature steps, but concordant plateaus are obtained in all analyses. The K/Ca ratios show some variation, which we relate to compositional differences and/or inclusions. However, the variation does not seem to affect the apparent ages. The mean square of weighted deviates (MSWD) indicates a generally good correspondence between expected and estimated errors. Some of the apparent ages from the different units on Silhouette are different at the 95% confidence level so we treat them here as separate cooling ages.

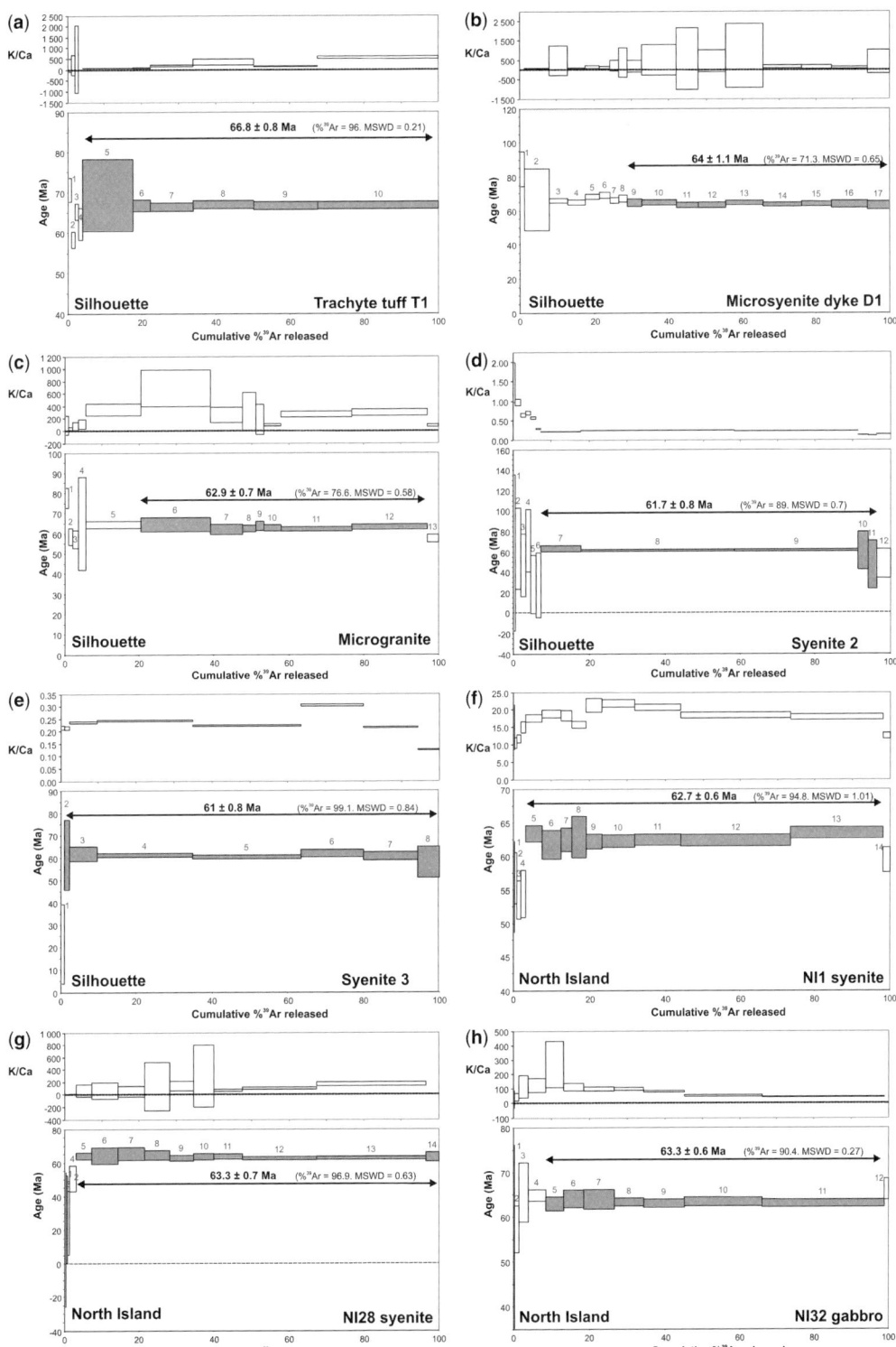

Fig. 5. Step heating release spectra for the samples from Silhouette and North Island. Each age bar is plotted at 95% confidence level. The numbers above the age bars correspond to the row number in the Supplementary Material.

Table 1. *Main results from step heating analysis*

Location	Rock type	Spectrum analysis				MSWD	TFA	K/Ca ±1.96σ	Inverse isochron analysis			
		³⁹Ar %	Steps (N)	Age (Ma)	±1.96σ				Age (Ma) ±1.96σ	MSWD	Intercept ±1.96σ	
North Island	NI1 syenite[K]	94.77	5–13 (9)	62.65	0.36/**0.55**/0.55	1.01	62.37 ± 0.57	18.77 ± 0.7	62.71 ± 0.63	1.15	293.68 ± 12.49	
	NI28 syenite[B]	96.91	5–14 (10)	63.34	0.50/**0.65**/0.65	0.63	62.82 ± 0.78	11.85 ± 2.2	62.79 ± 0.85	0.26	315.48 ± 21.31	
	NI32 gabbro[K]	90.39	5–11 (7)	63.34	0.42/**0.59**/0.59	0.27	63.47 ± 0.63	53.73 ± 3.5	63.27 ± 0.7	0.31	299.12 ± 19.47	
Silhouette	Trachyte tuff[K]	96.02	5–10 (6)	66.84	0.46/**0.75**/0.76	0.21	67.10 ± 1.41	116.14 ± 12.7	66.75 ± 0.82	0.21	299.28 ± 15.37	
	Microsyenite dyke[K]	71.25	9–17 (9)	64.00	0.52/**1.09**/1.10	0.65	65.02 ± 1.62	147.41 ± 78.4	64.08 ± 1.25	0.74	293.02 ± 19.97	
	Microgranite[A]	76.58	6–12 (7)	62.91	0.61/**0.73**/0.74	0.58	63.33 ± 1.02	147.65 ± 31.1	62.81 ± 1.12	0.69	297.24 ± 15.08	
	Syenite 2[B]	89.04	7–11 (5)	61.67	0.64/**0.81**/0.81	0.70	60.03 ± 1.57	0.18 ± 0.003	61.91 ± 1.27	0.87	292.86 ± 11.2	
	Syenite 3[B]	99.08	2–8 (7)	61.01	0.53/**0.80**/0.80	0.84	60.63 ± 0.94	0.20 ± 0.002	60.91 ± 0.96	0.99	296.98 ± 8.05	

The plateau age uncertainties are reported as analytical/internal/external errors where internal includes analytical + experimental error on the J and the fluence age uncertainties; external error includes internal error + the uncertainties on the decay constant. TFA, total fusion age (K/Ar age). The superscript in the Rock type column denotes the mineral used in analysis (K, potassium feldspar; B, biotite; A, amphibole).

The Silhouette samples yield a wide spread of ages (Fig. 5a–e) from 66.8 ± 0.8 Ma (trachyte tuff) to 61 ± 0.8 Ma (main syenite body). The microsyenite dyke (D1) and the microgranite yield 64 ± 1.1 Ma and 62.9 ± 0.7 Ma, respectively. The two samples from the main syenite body in Silhouette (Fig. 5d, e; Table 1) and the three samples from North Island (Fig. 5f–h; Table 1) overlap at the 95% level and we calculate weighted mean ages which give 61.3 ± 0.57 and 63.1 ± 0.34 Ma, respectively. These results are considered in the Discussion (p. 242).

U/Pb age determinations

Zircon was extracted from four samples from North Island and three from Silhouette by using a sequence of jaw crusher, pulverizing mill, Wilfley table, sieving, magnetic separation and heavy liquid methylene iodide (MI) floatation. Suitable grains were selected for analyses by hand-picking under a binocular microscope. The coherent results reported here were obtained by treating the zircon with mechanical abrasion (Krogh 1982) whereas (except for sample microgranite) chemical abrasion (Mattinson 2005) provided only variably discordant data (Corfu 2009). The analyses were carried out by ID-TIMS following Krogh (1973) as described for the Oslo laboratory in Corfu (2004). Decay constants are those of Jaffey et al. (1971) and plotting was carried out using the program of Ludwig (2003).

All the analyses yield overlapping $^{206}Pb/^{238}U$, but all the more precise results plot to the right of the concordia curve (Table 2 & Fig. 6). This discrepancy between $^{206}Pb/^{238}U$ and $^{207}Pb/^{206}Pb$ is confirmed by the precise analyses of the chemical abrasion experiments (Corfu 2009), which will be presented and discussed in detail in a subsequent paper. The deviation towards higher $^{207}Pb/^{206}Pb$ ages cannot be explained by the decay constant bias (Schoene et al. 2006) nor by isotopic disequilibrium, as all the data are corrected for ^{230}Th deficit assuming a Th/U of 4 in the parent magma (the correction increases $^{206}Pb/^{238}U$ by about 0.1 Ma).

The most important observation is the reproducibility of the $^{206}Pb/^{238}U$ ages, which give weighted average values ranging from 63.20 ± 0.12 to 63.31 ± 0.11 Ma for the North Island samples and from 63.46 ± 0.14 to 63.58 ± 0.09 Ma for the Silhouette samples (Fig. 6). Weighted averages for all the samples from North Island and Silhouette gives 63.27 ± 0.05 and 63.54 ± 0.06 Ma, respectively. The consistent results indicate very rapid emplacement of the complexes but with a time gap, with magmatism on Silhouette preceding the events on North Island by 0.27 ± 0.08 Ma.

Palaeomagnetic analysis

Palaeomagnetic laboratory experiments were carried out at the Geological Survey of Norway (Trondheim, Norway). The natural remanent magnetization (NRM) was measured on an AGICO JR6A spinner magnetometer mounted within a Helmholtz coil system. Components of magnetization were identified using stepwise thermal and alternating field demagnetization. All specimens were demagnetized in 15–25 steps. The directions and unblocking temperature spectra of characteristic remanence components (ChRM) were determined using the LineFind algorithm of Kent et al. (1983) as implemented in the Super-IAPD program available at www.geodynamics.no (Torsvik et al. 2000). For those specimens where the ChRM could not be isolated directly where they defined a circle of remagnetization (Khramov 1958), we used the plane fits output from the LineFind algorithm and the great circle principle of McFadden & McElhinny (1988) implemented in the Palaeomag–Tools software (written by Mark W. Hounslow). In the Palaeomag–Tools software we set a fixed point (to force the polarity), but the point is not included in the statistics.

A set of rock magnetic experiments was performed to reveal the magnetic mineral host, using a horizontal Curie Balance for thermomagnetic measurements and a vibrating sample magnetometer (VSM) to produce hysteresis loops. The thermomagnetic investigation (Fig. 7a) on selected samples from all units reveals Curie temperatures close to those of pure magnetite (580 °C). This is supported by the VSM experiments (Fig. 7b), which are indicative of a ferromagnetic mineral phase with low coercivity. Further M_{rs}/M_s against H_{cr}/H_c (Day et al. 1977) shows that the magnetizations are carried by pseudo-single domain and a mixture of pseudo and multidomain grains for all units (Fig. 7c). We therefore conclude that the magnetic mineral host is dominantly magnetite or Ti-poor titanomagnetite.

The site mean directions, estimators, NRM intensities, magnetic susceptibilities and Q-ratios (Königberger) can be found in Table 3. The results from thermal demagnetization were unsuccessful for all sites apart from one because they display a within-site randomness of ChRM directions. Even closely spaced cores showed this behaviour, although most specimens show univectorial demagnetization behaviour towards the origin (Fig. 8a). This behaviour is often seen in rocks that experienced an isothermal remanent magnetization (IRM) caused by lightning strike (Hallimond & Herroun 1933).

Specimens commonly have very high NRM intensities in the range of 5–20 A/m, supporting the inferred influence of lightning strikes.

Table 2. *Zircon U–Pb data*

Characteristics[1]	Weight[2,3] (μg)	U[2,3] (ppm)	Th/U[4]	Pbc[3] (pg)	206Pb/204Pb[5]	207Pb/235U[5]	±2σ (abs)	206Pb/238U[6,7]	±2σ (abs)	rho	207Pb/206Pb[6,7]	±2σ (abs)	206Pb/238U[6,7] (age in Ma)	±2σ	207Pb/235U[6,7]	±2σ
NI-1 syenite, North Island																
fr [1]	46	360	1.04	1.4	7130	0.06433	0.00018	0.00984	0.00002	0.81	0.04740	0.00008	63.20	0.20	63.31	0.20
fr cl [10]	173	746	0.94	1.3	59 764	0.06457	0.00018	0.00987	0.00003	0.97	0.04744	0.00003	63.38	0.20	63.53	0.20
fr cl [3]	97	607	1.04	2.9	12 340	0.06436	0.00017	0.00986	0.00002	0.94	0.04736	0.00004	63.28	0.10	63.34	0.20
NI-2 syenite, North Island																
fr cl [22]	254	866	1.46	4.6	29 315	0.06425	0.00019	0.00983	0.00003	0.97	0.04739	0.00003	63.13	0.17	63.23	0.18
fr y-b [1]	210	517	1.11	7.9	8470	0.06456	0.00018	0.00988	0.00002	0.95	0.04738	0.00004	63.45	0.15	63.53	0.17
NI-3 microsyenite, North Island																
fr cl [37]	45	859	2.04	3.2	7458	0.06429	0.00018	0.00984	0.00002	0.89	0.04739	0.00006	63.17	0.15	63.27	0.17
fr cl-y [12]	49	242	1.36	3.0	2476	0.06430	0.00024	0.00985	0.00003	0.69	0.04735	0.00013	63.22	0.18	63.27	0.23
NI-4 gabbro, North Island																
fr cl [9]	56	1909	2.31	5.0	13 163	0.06448	0.00018	0.00985	0.00003	0.96	0.04747	0.00004	63.24	0.20	63.45	0.20
fr cl [2]	48	1708	1.89	1.3	38 780	0.06450	0.00018	0.00985	0.00003	0.96	0.04748	0.00004	63.25	0.20	63.46	0.20
SI-1 syenite, Silhouette																
fr-eu cl [30]	34	544	1.42	6.6	1757	0.06461	0.00022	0.00990	0.00002	0.72	0.04733	0.00011	63.56	0.20	63.57	0.20
fr-eu cl [33]	63	510	1.56	4.9	4047	0.06480	0.00019	0.00990	0.00003	0.85	0.04750	0.00007	63.53	0.16	63.75	0.18
T2 trachyte, Silhouette																
fr-eu cl [18]	12	616	1.50	1.6	2902	0.06448	0.00030	0.00989	0.00003	0.71	0.04731	0.00015	63.47	0.20	53.45	0.28
fr-eu cl in [17]	8	423	1.43	4.3	507	0.06481	0.00062	0.00988	0.00003	0.48	0.04755	0.00040	63.46	0.20	63.76	0.59
Microgranite, Silhouette																
eu sp cl-tu CA [20]	60	289	0.93	2.3	4649	0.06474	0.00024	0.00990	0.00003	0.73	0.04744	0.00012	63.54	0.19	63.70	0.22
eu sp cl CA [25]	47	363	1.05	2.1	5073	0.06473	0.00019	0.00991	0.00002	0.78	0.04736	0.00009	63.64	0.15	63.69	0.18
eu sp cl [11]	17	580	1.16	12.9	493	0.06439	0.00051	0.00990	0.00003	0.36	0.04718	0.00035	63.55	0.16	63.37	0.49

[1]fr, fragments; eu, euhedral; sp, short prismatic ($l/w = <4$); cl, clear, colourless; r, red; y, yellow; b, brown; tu, turbid spots; incl, in; CA, chemically abraded (Mattinson 2005), all the others mechanically abraded (Krogh 1982); [N], number of grains in fraction.

[2,3]Weight and concentrations are known to better than 10%.

[4]Th/U model ratio inferred from 208/206 ratio and age of sample.

[3]Pbc = total common Pb in sample (initial + blank).

[5]Raw data corrected for fractionation and blank.

[6]Corrected for fractionation, spike, blank and initial common Pb; error calculated by propagating the main sources of uncertainty.

[7]Corrected for initial Th deficit assuming Th/U in the magma = 4 (Schärer 1984).

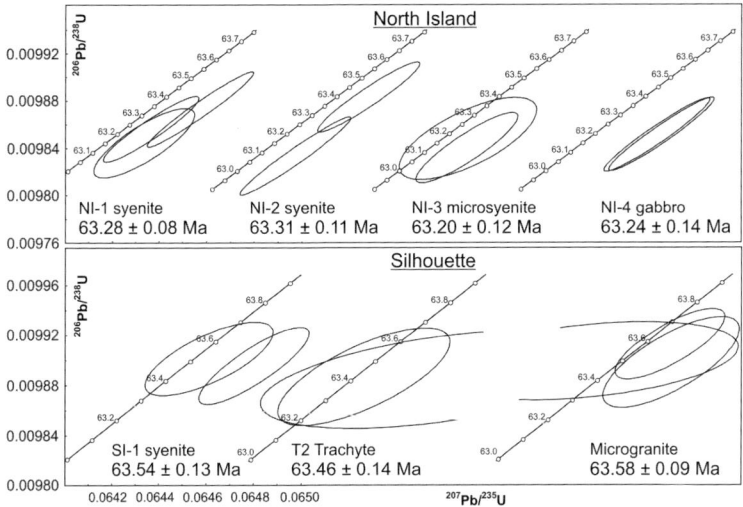

Fig. 6. Composite concordia diagrams with U–Pb data for zircon in samples from the two islands. Ellipses indicate the 2σ uncertainty of the analyses, which are also corrected for initial Th deficit. The ages represent weighted mean $^{206}Pb/^{238}U$ dates for each sample.

Alternating-frequency (AF) demagnetization proved more successful in determining the ChRM from the lightning-induced component. Figure 8a, b shows the difference between thermal and AF demagnetization applied to samples from the same core. The two components partly overlap in most samples, spanning remagnetization great circles.

Because the lightning-induced remagnetization direction varies from sample to sample, the common intersection point of all great circles can be identified and is interpreted as the ChRM (Fig. 8c, d). For most sites the demagnetization curve does not reach a stable endpoint (i.e. Fig. 8c). Figure 8d however shows an example

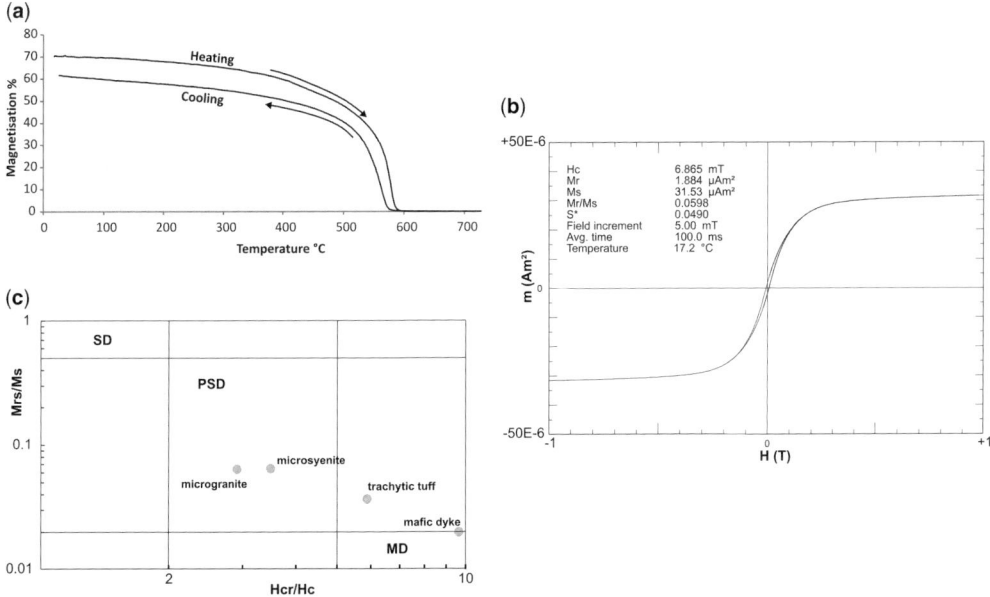

Fig. 7. Rock magnetic experiments: (**a**) Curie-temperature, (**b**) hysteresis and (**c**) Day plot where SD, PSD and MD denote single, pseudo-single and multi-domains, respectively.

Table 3. *Palaeomagnetic sampling sites and results*

Site	Unit	Comment	D	I	N	R	k	A95	NRM$_{int}$	Sus	Q	VGPLat	VGPLong	Dp	Dm
A	MS D1*	Dyke	146.9	6.2	10	9.97	186.7	4	36 718	33 882	27.2	−56.9	146.5	2	4
	T1*	Flow	153.1	47.5	5	4.99	288.68	9.3	60 626	42 058	36.2	−54.9	98.9	7.9	12.1
	T1*	Flow	178.9	45.1	6	5.94	48.13	10.8	42 689	59 761	18	−67.8	57.9	8.7	13.7
	T1*	Flow	142.7	9.7	4	3.98	171.87	7	23 254	32 827	17.8	−52.8	143.1	3.6	7.1
	MS D2*	Dyke	152.6	13.3	5	5.97	80.26	9.8	61 377	30 393	50.8	−62.6	139.3	5.1	10
	MS D3*	Dyke	163	18.5	6	5.99	638.86	2.7	15 284	18 770	20.5	−72.4	127.8	1.5	2.8
	T2*	Flow	140.2	28.4	3	2.99	58.5	8.1	44 539	59 549	18.8	−49.4	127.1	4.9	8.9
	T2*	Flow	124.7	56	4	4	328.97	9.8	13 338	25 400	13.2	−30.2	105.1	10.1	14.1
	Microgranite*	At cross	319.8	−63.8	9	8.97	104.83	6	43 552	25 093	43.6	−36.1	89.4	7.6	9.5
B	Microgranite*		296.3	−28.1	8	7.91	32.24	11.2	38 293	18 821	51.1	−26.5	130.8	6.7	12.3
	Flow*		166.5	11.4	8	7.93	45.66	9.7	159	2071	1.9	−76.5	139.3	5	9.8
C	Mafic dyke*		129.9	27.1	7	6.94	44.28	12.2	6099	57 921	2.6	−39.7	130.3	7.2	13.3
	Dyke*		172.4	16.7	4	3.99	563.67	7.4	789	20 055	1	−81.4	116.7	3.9	7.6
D	Trachyte tuff		131.4	32.1	4	3.98	26.2	26.2	5999	50 786	3	−40.7	126.1	16.6	29.5
	Dyke*		188.6	64.5	5	5	895.2	5.3	1224	22 316	1.4	−47.5	46.5	6.8	8.5

	D	I	ΔD_x	ΔI_x	N										
Mean	150.6	32.4	12.9	19.3	14										

	PLat	Plong	K	A95
	−57.55	114.22	11.4	12.3

D/I, declination/inclination of flow mean remanence directions; N, number of remanence directions; R, length of resultant vector; k and A95 are the Fisher (1953) precision parameter and half angle of the cone of 95% confidence. NRM$_{int}$ denotes moment intensities (mA) and Sus are the susceptibilities ($10-5$ SI) before heating. Q is the Königsberger ratio based on an ambient field strength of 50 000 nT which is 39.79 A m^{-1}. Units marked with an asterisk (*) are used in the calculation of the overall mean direction and the pole.

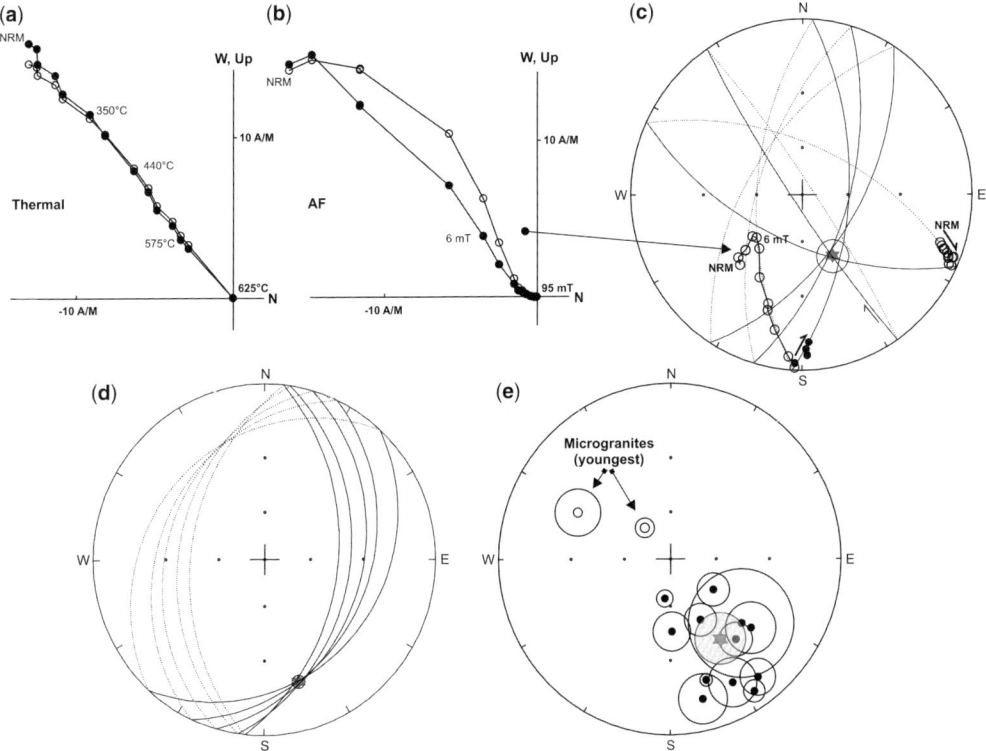

Fig. 8. Stepwise thermal and AF demagnetization data visualized in orthogonal and stereographic plots. For the orthogonal plots solid (open) symbols represent data projected into the horizontal (vertical) plane and for the stereoplots equal area projections are used where black (white) circles represent reverse (normal) polarities (southern hemisphere). (**a**) Thermal demagnetization produces linear decay towards the origin, (**b**) while AF from a specimen from the same core produces a great circle path. The sample in (**b**) is displayed in (**c**) together with several samples from the same site defining a common intersection point but no stable endpoint (b). (**d**) Shows an example where a stable endpoint was reached and also produces the same direction using great circle analysis. All site mean directions are plotted in (**e**) with A95 confidence circles. The overall mean direction is represented by a grey star.

where a stable endpoint was reached, leading to a direction statistically indistinguishable from one derived through great circle analysis.

We interpret all the characteristic component directions obtained from the 15 sites as a reliable record of the Earth's magnetic field at their time of emplacement. The results show that all sites except the microgranites are of reversed polarity (Fig. 8e, Table 3). The microgranites cross-cut the other units, and are therefore the youngest unit in the volcanic complex. Only the syenite is younger, but that was not analysed palaeomagnetically.

Because dispersion within a single lava site should be minimal, a stringent cut-off is normally applied; lava sites with k-values lower than 50 (Biggin *et al.* 2008; Johnson *et al.* 2008; van Hinsbergen *et al.* 2008*a*, *b*) are discarded. Of our 15 sites, 5 have a k-value below 50. Because of the relatively small dataset, we include 4 sites with k-values

between 30 and 50 (Table 3) in our average to increase our statistical power; this does not lead to a significantly different average direction. Averages and cones of confidence were determined using Fisher (1953) statistics applied to Virtual Geomagnetic Poles (VGP), because these are more Fisherian (i.e. a Gaussian dispersion on a sphere) than directions which have a (latitude-dependent) elongated distribution (Tauxe & Kent 2004; Tauxe *et al.* 2008; Deenen *et al.* 2011). Errors in declination and inclination are given separately as ΔD_x and ΔI_x (Butler 1992; Deenen *et al.* in revision) in Table 3. No directions were eliminated by the Vandamme (1994) variable cut-off procedure. We therefore arrive at a $D \pm \Delta D_x = 150.6 \pm 12.9$ and $I \pm \Delta I_x = 32.4 \pm 19.3$ ($n = 14$).

The degree of palaeosecular variation can be expressed through the angular standard deviation (ASD) of the virtual geomagnetic poles (VGP) if the VGPs have a Fisherian distribution. To test whether this is true, we performed a quantile–quantile calculation (Fisher *et al.* 1987). The goodness of fit to the Fisherian distribution for the VGPs is lower than the critical values (Mu = 0.803 < 1.207, Me = 0.509 < 1.094) and the null hypothesis that the distribution of VGPs is Fisherian cannot be rejected at the 95% confidence level (Fisher *et al.* 1987). The ASD of VGPs, corrected for within-flow dispersion Sw = 5.3° (McFadden *et al.* 1991), is 12° with a 95% confidence limit of $9.6 \leq ASD \leq 16$ (Cox 1969). This ASD is within error of the PSV prediction of the G-model of McFadden *et al.* (1991), which indicates an ASD value of c. 11 for this time and palaeolatitude. We therefore conclude that the observed dispersion can be explained by secular variation and the determination of the overall mean direction satisfactorily represents the palaeosecular variation.

Ideally, remanence directions should be Fisherian distributed close to the poles but ellipsoid elsewhere (Tauxe & Kent 2004). To determine a palaeomagnetic pole for this study, we therefore apply Fisher (1953) statistics on the VGPs which leads to an apparent palaeomagnetic pole position at 57.55°S and 114.22°E ($k = 11.4$; A95 = 12.3; Table 3) in Seychelles co-ordinates.

Discussion

Timing of emplacement

The new $^{40}Ar/^{39}Ar$ and U–Pb determinations and 95% confidence bars from North and Silhouette islands are displayed in Figure 4, together with the geomagnetic polarity time scale (Ogg & Smith 2004). The chron boundaries are intercalibrated with an age of 28.34 Ma for the TCR fluence monitor used here to derive the $^{40}Ar/^{39}Ar$ data. The accuracy of the U–Pb ages is dependent largely on the spike calibration; measurements of reference solutions indicate that our spike is within about 1 permil of accepted values, translating into a potential bias of less than 0.07 Ma for this age range.

The $^{40}Ar/^{39}Ar$ dates obtained from K-feldspar in gabbro and syenite N1 and from biotite in syenite NI28 on the North Island are identical within error and yield a weighted mean age of 63.1 ± 0.34 Ma, which overlaps the U–Pb weighted mean of 63.27 ± 0.05 Ma and corresponds to chron C28n (Fig. 4). This also resembles previous age determinations from the North Island (Dickin *et al.* 1986; Tucker *et al.* 2001) and the 63.4 Ma age assigned to the oldest ocean floor anomaly between the Seychelles and the Laxmi ridge by Collier *et al.* (2008).

The $^{40}Ar/^{39}Ar$ dates obtained from the Island of Silhouette range from 66.8 ± 0.8 Ma for K-feldspar in trachyte tuff T1, over 64 ± 1.1 Ma for K-feldspar in microsyenite D1 and 62.9 ± 0.7 Ma for hornblende in microgranite, to 61.7 ± 0.8 and 61 ± 0.8 Ma for biotite in syenites 2 and 3 (Fig. 4). The age of 66.8 ± 0.8 Ma for the trachytic unit contrasts with zircon U–Pb age for a similar tuff at 63.46 ± 0.14 Ma. The latter is slightly younger but within error of the U–Pb ages of 63.58 and 63.54 Ma for syenite and microgranite (Figs 4 & 5). Even though the sample for U–Pb analysis comes from a separate trachyte unit located higher in the sequence (see Fig. 3), all the tuffs share the same macroscopic texture and characteristics and hence the age difference is difficult to explain. Dickin *et al.* (1986) suggested that their 62.1 ± 1.3 Ma age from this unit should be regarded as a minimum age due to the tendency of acid whole-rock samples to lose argon.

Our total fusion age of the K-feldspar in the trachyte (TFA, Table 1) does not resemble the results of Dickin *et al.* (1986) either. We obtained a plateau comprising 96% of the total ^{39}Ar released and the date resembles the inverse isochron and the TFA. In search of any potential systematic errors, we note that the J parameter determination for this sample is also in the range of the other samples. We notice that the trachyte contains xenoliths of a rock type of unknown age with syenitic composition (Owen-Smith *et al.* pers. comm.); one possibility is that the K-feldspar is xenocrystic and was picked up during eruption, thus escaping outgassing. This possibility will have to be tested by additional dating on the trachyte and xenoliths. Until then, we are reluctant to put too much weight on this age which would put volcanism during chron C30n (Fig. 8) at the onset of the main Deccan tholeiitic eruptions. Early acidic and alkaline volcanism is known from the north of the Deccan Province, but trachytic eruptions of this type are more commonly associated with the late-stage volcanics at 60–63 Ma (chron C28n and later).

The $^{40}Ar/^{39}Ar$ ages for K-feldspar in the microsyenite (64 ± 1.1 Ma) and amphibole in the microgranite (62.9 ± 0.7 Ma) overlap the zircon U–Pb ages. By contrast, the weighted mean $^{40}Ar/^{39}Ar$ age of 61.3 ± 0.6 Ma for the biotite in syenites of Silhouette are younger than the U–Pb zircon age 63.54 ± 0.13 Ma and previous age determinations, including the 63.7 ± 1 Ma whole-rock K/Ar age of Dickin *et al.* (1986). Biotite has a relatively low closing temperature for Ar and the age may reflect slow cooling of the area. The alternative is that the older zircon U–Pb age represents pre-emplacement formation in the magma chamber (Crowley *et al.* 2006; Miller *et al.* 2007), although the overlap with the amphibole and K-feldspar $^{40}Ar/^{39}Ar$ ages and the coherence between these systems on the North island would seem to argue against this possibility. Moreover, the generally high solubility of Zr in alkalic magmas tends to lead to late precipitation of zircon, a factor militating against protracted residence times of zircon in such magmatic systems.

The magnetic polarities observed in all samples are reverse, except for the microgranite which is normal (Table 3). These relationships represent a mismatch with the timescale used in Figure 4, where all the U–Pb dates and part of the $^{40}Ar/^{39}Ar$ age determinations plot inside C28n; it would fit the Silhouette biotite ages, however. In the simple case of a short-lived volcanic complex undergoing rapid cooling, the magnetic polarities and the ages should coincide with the timescale. The mismatch between the radiometric age determinations and the remanence directions can be interpreted in terms of closure temperature and blocking temperature for magnetic remanence. The thermomagnetic investigation (Fig. 7a) reveals a magnetic carrier with Curie temperatures close to pure magnetite (580 °C). The Curie temperature of magnetite is between the closure temperature of zircon (excess of 900 °C, e.g. Cherniak & Watson 2001) and most minerals used for the $^{40}Ar/^{39}Ar$ system. The palaeomagnetic investigation also reveals that some secular variation has been recorded (e.g. units T1, microgranite and T2 in Table 3). Altogether, this could be interpreted in terms of a slow cooling for the whole complex or that the later main syenite body has generated enough heat for a considerable amount of time to reset the $^{40}Ar/^{39}Ar$ system and distort the magnetic remanence.

Based on the timescale of Ogg & Smith (2004), Collier et al. (2008) divided the volcanic record of the region into three stages: pre-Deccan (c. 78–67 Ma, 33n–30n), Deccan (c. 67–63 Ma, 30n–27r) and post-Deccan (c. 63–58 Ma, 27n–26n). Our new U–Pb ages fit chron C28n and overlap with the Deccan stage of Collier et al. (2008), which demonstrates that the emplacement of the magmatic rocks of Silhouette was contemporaneous with the Deccan volcanism. An important point to this end is whether these rocks are the product of the same mantle source responsible for the Deccan volcanism. Devey & Stephens (1991) provide a geochemical correlation of dykes from the Seychelles to the Deccan Bushe Formation. A recent study by Owen-Smith et al. (pers. comm.) indicates strong geochemical grounds for a common mantle source for the Seychelles and Deccan magmatism. In terms of Sr- and Nd-isotopic compositions in particular, the Silhouette and North island rocks show strong affinities with the uncontaminated magma groups of the Deccan (Ambenali and Mahabaleshwar

Formations; see Mitchell & Widdowson (1991) for Deccan chemostratigraphy). This raises the question of how much the areal extent of Deccan Traps can be extended. The Rajahmundry Traps in southeast India have recently been linked to the Deccan (Jay & Widdowson 2008; Self et al. 2008) based on geochemical and geochronological evidence (e.g. 64.7 ± 0.57 Ma from Baksi 2005). The new geochronology presented here, and the geochemical link documented in Devey & Stephens (1991) and Owen-Smith et al. (in preparation), certainly extend the Deccan volcanism to the Seychelles.

Regional implications, plate reconstructions and the African Plume Generation Zone

The evolution of the western Indian margin and the formation of oceanic basins between the Madagascar–India/Seychelles, West India/Laxmi Ridge and Seychelles underwent complex tectonic phases that led to several stretched continental crust blocks, volcanic ridges, oceanic basins and possibly incipient subduction zones. Classical studies (Masson 1984) and more recent papers, based on existent and new geological and geophysical data (Bhattacharya et al. 1994; Todal & Eldholm 1998; Bernard & Munschy 2000; Lane 2006; Collier et al. 2008, 2009; Yatheesh et al. 2009), postulate the existence of extended continental crust/ sublithospheric mantle and/or oceanic crust in the proximity of the western Indian margin in the Laxmi Basin and the Gop Rift. Originally, Masson (1984) suggested that a fan-shaped region of extended continental crust (or oceanic crust) should be present along the western Indian margin to explain the age and architecture of the Mascarene Basin (between Madagascar and the Seychelles/ Mascarene Plateau). The 'missing' crust in the NE could have been created NE of the Seychelles, probably contemporaneous or towards the end of the Mascarene Basin opening. Marine magnetic anomalies have been identified in the Laxmi Basin and the Gop Rift by several authors, although the age of these identifications differs from one study to another. Although we would expect extension and spreading in the Gop Rift (between the Laxmi Ridge and the extended continental crust of west India) to have been extinct at the time of the East Arabian Sea opening (around C28n, 64.1–63.1 Ma on GTS2004 timescale; Ogg & Smith 2004; Lane 2006; Collier et al. 2008), a different identification of magnetic anomalies in the Gop Rift (Yatheesh et al. 2009) suggest an ongoing activity in this basin for another few million years (C25r, 58.5 Ma on GTS2004) while oceanic crust formed along the Carlsberg Ridge. In addition, the southern

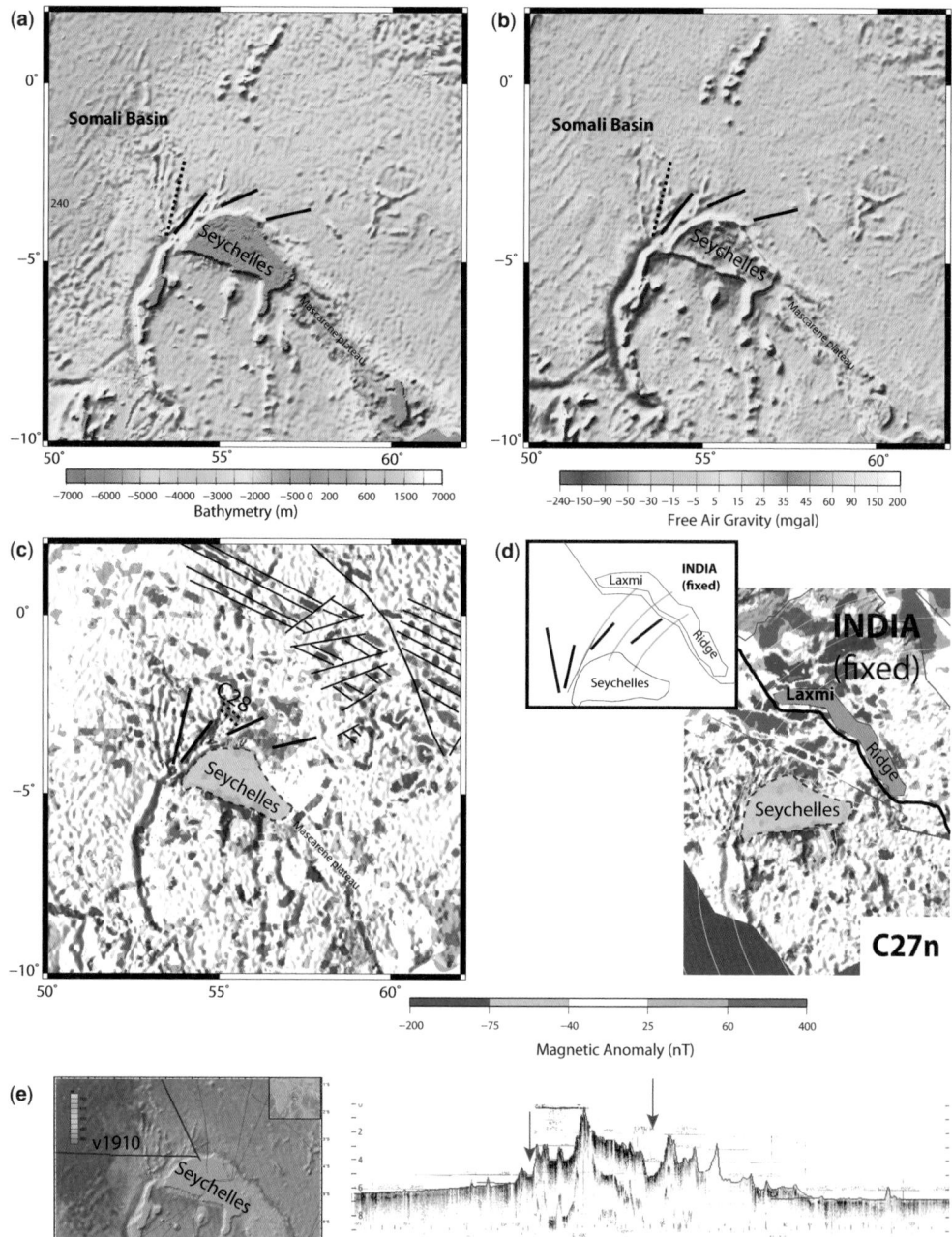

Fig. 9. Seychelles and surrounding areas. The geographical extent of a, b and c are shown in Figure 1. (**a**) Bathymetry (ETOPO1, Amante & Eakins 2009), (**b**) free-air gravity anomaly (Sandwell & Smith 1997) and (**c**) magnetic anomalies from World Digital Magnetic Anomaly Map (Maus *et al.* 2007) superimposed on shaded bathymetry grid (Amante & Eakins 2009). Black, thin lines are isochrons based on magnetic and fracture zone interpretation in the South Arabian Sea (Chaubey *et al.* 2002); dashed, black lines show the approximate location of the oldest chron (C28n) identified by Collier *et al.* (2008); light dashed line represents the outline of the Seychelles microcontinent. Note a fan-shaped set of NNE–SSW to NE–SW features in both bathymetry and free-air gravity (tentatively outlined by thick, black lines) that may indicate an anticlockwise rotation of the Seychelles. (**d**) Reconstructions at chron C27n in a fixed Indian plate reference frame of present-day magnetic anomaly grid (Maus *et al.* 2007) and shaded bathymetry. Reconstructed

extinction of the Mascarene Basin spreading centre occurred *c.* 2 Ma after C27n (Bernard & Munschy 2000), a situation that implies a system of opposed propagating spreading centres in the Mascarene, East Arabian and Gop/Laxmi basins. This scenario has been observed in other regions involving micro-continent formation, for example Jan Mayen in the North Atlantic Ocean (Gaina *et al.* 2009) and the Danakil block in the Red Sea area (e.g. Beyene & Abdelsalam 2005).

Our study shows that the mean palaeomagnetic remanence direction for this part of the Seychelles is translated in an anticlockwise sense of 29.4° ± 12.9°. Plummer (1996) pointed out an asymmetry in the spreading rates of the Mascarene Basin to that of Somali Basin, which led to the triangular shape of the Mascarene and forced the Seychelles to rotate in an anticlockwise sense. Plummer & Belle (1995) assigns a late Cretaceous age for this rotation. The youngest unit sampled for palaeomagnetic analysis in this study is the microgranites where we obtained an $^{40}Ar/^{39}Ar$ age of 62.9 ± 0.7. This implies that the rotation must have occurred during or after the time of remanence acquisition, that is, from C28n during or after C27r. However, this does not rule out the possibility that this rotation could have started earlier. This additional information could fit very well in a tectonic scenario with active rifts/spreading centres at opposing corners of the microcontinent (this seems to be the case with the oceanic crust evolution around the Seychelles between chrons C30–C29 and C27). On a more detailed scale, the crust around the Seychelles shows evidence of deformation both to the north (just south of the oldest magnetic anomaly identified in the east Arabian Sea) and to the south where a rugged oceanic crust which was presumably formed in the Cretaceous Mascarene Basin suffered compression, as described by seismic and bathymetric data (see Masson 1984 and the bathymetric and free-air gravity signature; Fig. 9). More dramatically, the Amirante Ridge (a Cretaceous ridge located along a possible fracture zone SW of the Seychelles) has been described as an extinct trench that was formed as an incipient subduction zone (Miles 1982; Masson 1984; Mart 1988).

Based on earlier studies and our new results that indicate an anticlockwise rotation (29.4° ± 12.9°)

for the Seychelles sometime after magnetochron C28n, we propose that part of this rotation has been accommodated by propagating seafloor spreading in East Somali Basin, NW of Seychelles as documented by magnetic anomalies interpreted as C28n. After magnetochron C27r (when the Seychelles were isolated from the Laxmi Ridge by seafloor spreading) the Seychelles might have continued to rotate as an individual plate, therefore creating limited transpression and transtension along its plate boundaries. We note a series of fan-shaped features visible in bathymetry and free-air gravity NW and north of the Seychelles microcontinent (Fig. 9a, b). When reconstructed at chron 27n time, some of these features correspond to a small circle trend around a stage pole that depicts the anticlockwise rotation of Seychelles between C28n and C27n. This trend aligns well with the low in free-air gravity and trough in bathymetry and vintage seismic data (seismic line collected by Vema 1910 in 1963, published by Lamont Doherty Earth Observatory through GeoMapApp, http://www.geomapapp.org, and seismic line TCO-24, Plummer 1996) situated NW of Seychelles, and as a continuation of the Amirante Trough. A similar interpretation was suggested by Plummer (1996).

We cannot resolve the timing of the Amirante Ridge formation (as a compressional feature), but we emphasize that the rocks that have been described on Silhouette and the North Island on Seychelles have been suggested to be related to incipient subduction (Brown *et al.* 1984). A recent study (Calvès *et al.* 2011) suggests that a Mid–Late Cretaceous compressional event, caused by the relative motion between Madagascar and NW margin of India, led to the formation of the Amirante trench. The trench is suggested to be of age 82 ± 16 Ma from K/Ar analysis of a grab sample (Fisher *et al.* 1968) and formed some volcanic ridges along the western Indian margin.

A similar suggestion comes from a study on anisotropy of the lithosphere and upper mantle beneath Seychelles (Hammond *et al.* 2005). The observed large variation in the magnitude of shear-wave splitting is explained by a lithospheric structure that has been affected by a transpressive regime, possibly during the formation of the Mascarene Basin. If this is correct, then the Amirante Trench might

Fig. 9. (*Continued*) Seychelles microcontinent and fan-shaped lineations are shown in the inset figure, together with small circles around the Seychelles/India–Laxmi Ridge stage pole between 67.7 and 61 Ma (green curved lines). Note that the fan-shaped lineations are parallel or subparallel to the small circle trend that averages the anticlockwise rotation of the Seychelles microcontinent as suggested by the palaeomagnetic data. (**e**) A trough observed in the gravity data and on vintage seismic lines just NW of the Seychelles microcontinent might have been formed or affected by transform or transpressional motion generated by this rotation. Seismic profile (v1910) and location map are courtesy of Lamont Doherty Earth Observatory (LDEO) via the GeoMapApp application.

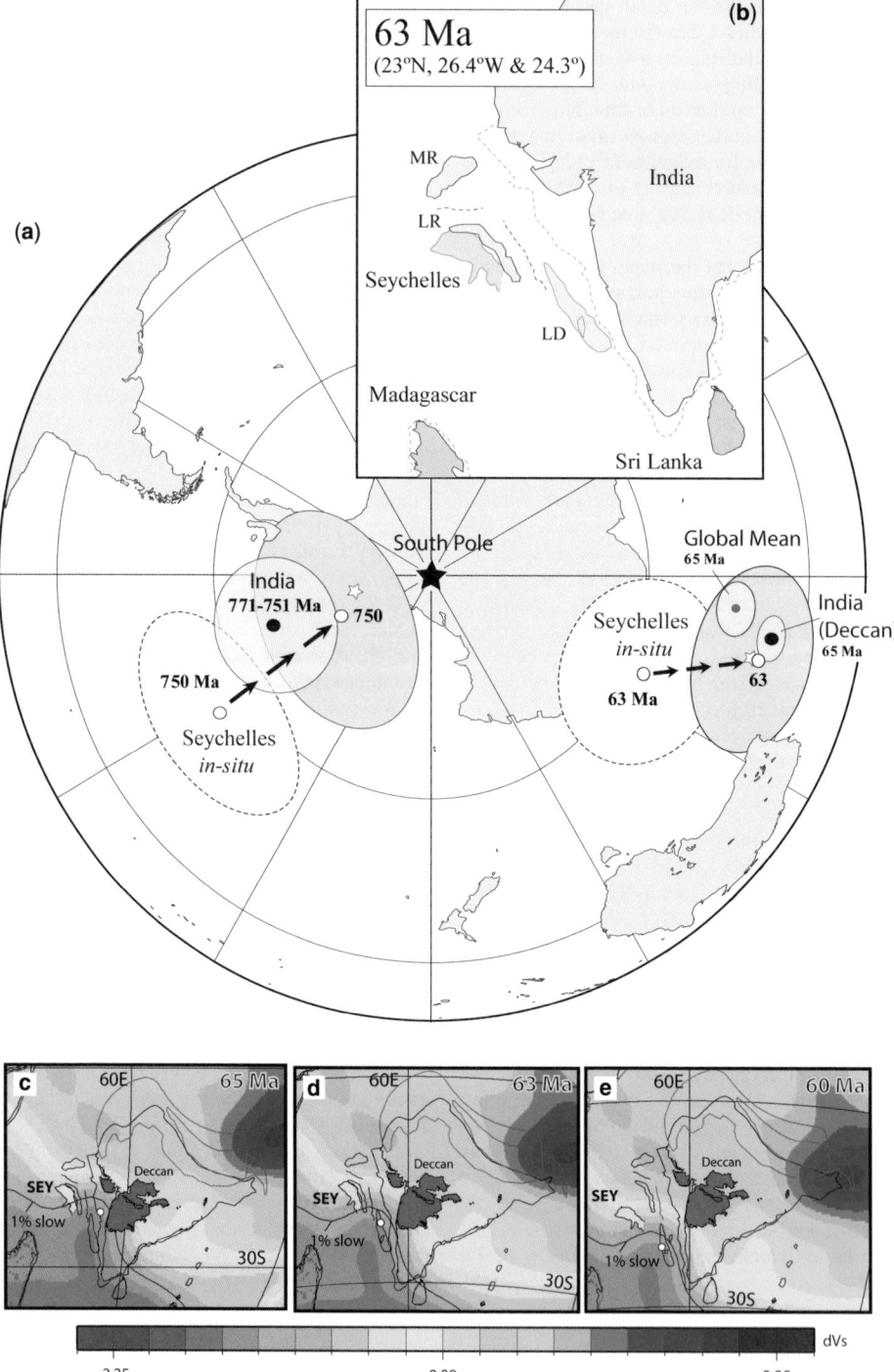

Fig. 10. (**a**) A comparison of the pole position derived from the present study (pole latitude −57.55°, longitude 114.22°, A95 = 12.2°) with a 65 Ma mean pole from India (essentially all from the Deccan Traps; latitutde 37.6°, longitude 100.2°, A95 = 3°, N = 7 poles) and a global 65 Ma mean pole (latitude −45.5°, longitude 96.1°, A95 = 3.5°, N = 29 poles) where all palaeomagnetic data were rotated to Indian co-ordinates and then averaged (Torsvik *et al.* 2008*c*).

have been formed in the Mid-Cretaceous but reactivated in the Paleocene due to the anticlockwise rotation of the Seychelles.

Stephens *et al.* (2009) provide a new insight into the origin of the Amirante ridge-trench. Geochemical and $^{40}Ar/^{39}Ar$ analysis on a gabbro sample, dredged from the southern tip of the arc, indicate a much younger age of 51.4 ± 0.9 Ma and their geochemical data largely rule out a subduction origin for the main phase of this structure. However, their explanation for the genesis of the Amirante trench as an impact crater remanent is refuted by the lack of evidences for strong unconformities that would have been observed in adjacent sedimentary basins (Plummer 1996).

A recent regional study by Cande *et al.* (2010) provides independent evidence for an anticlockwise rotation of the Seychelles microplate. By analysing large datasets of magnetic and fracture zone identifications, they predicted a plate boundary passing through the Amirante Ridge and extending north to the Carlsberg Ridge at least from C26y to C22o. However, they envisage that the independent motion of the Seychelles microplate might have started earlier (S. Cande, pers. comm. 2010).

Torsvik *et al.* (2001) proposed a fit of Seychelles and India at *c.* 750 Ma by matching Seychelles dykes and granites with the Malani pole of Klootwijk (1975). This fit (Euler pole latitude 25.8°N, longitude 30°W and rotation angle 28°) means that both the 750 Ma pole and our new 63 Ma pole (open stars in Fig. 10a) match contemporaneous mean poles from India within error (shown with green shaded A95 confidence oval). This reconstruction is very tight (probably more applicable to the Neoproterozoic); to account for pre-break-up extension along the Western margin of India (Fig. 10b), we therefore present a modified fit for the Early Paleocene (Euler pole latitude 23°N, longitude 26.4°W and rotation angle 28°). This fit also matches both the Neoproterozoic and Paleocene poles from the Seychelles (open white circles with blue shaded A95s). Our Seychelles pole also compares well with the global mean 65 Ma pole (red circle with green A95 oval). We also show the location of the Reunion hotspot (fixed with respect to the mantle) that is commonly linked to the Deccan Large Igneous Province (LIP). The reconstructions differs only by *c.* 4° in latitude compared to a global palaeomagnetic reference frame

(Torsvik *et al.* 2008b), witnessing relatively little drift of the Reunion hotspot.

From cross-cutting relationship of dykes to formation ages in Western India, Hooper *et al.* (2010) argue that significant extension must have started during C29R (Fig. 4). There is also evidence that the intertrappean facies in the Mumbai area (Western India) had a ecosystem resembling a shallow brackish marine gulf environment by *c.* 64 Ma, with associated phreatomagmatic and spilitized flows (Cripps *et al.* 2005). This is in line with our reconstruction (Fig. 10b) where we maintain a tight fit between the Seychelles and the Laxmi Ridge (LR in Fig. 10b). We invoke rifting of both these domains from India between 70 and 65 Ma. After 63 Ma and the main Deccan Trap magmatism, the Seychelles separated from the Laxmi Ridge/India and rotated *c.* 25° anticlockwise during the initial separation. This is illustrated in Figure 10c–e where we reconstruct Seychelles–India from 65 to 60 Ma using an *absolute* Indo-Atlantic moving hotspot frame (O'Neill *et al.* 2005).

These reconstructions (Fig. 10c–e) are draped on shear-wave anomalies (Becker & Boschi 2002) near the core–mantle-boundary (CMB), assuming that these have been stable for several hundred million years (Torsvik *et al.* 2008a). Earlier work has demonstrated that practically all reconstructed LIPs of the past 300 Ma (Burke & Torsvik 2004; Torsvik *et al.* 2006, 2008a; Burke *et al.* 2008) or even >500 Ma (Torsvik *et al.* 2010) and active deep-plume sourced hotspots at the Earth's surface (Montelli *et al.* 2006) project radially down to a narrow stable plume generation zone on the CMB at the edge of the LLSVPs. The 1% slow velocity contour in the SMEAN (Becker & Boschi 2002) model is a fair approximation to the plume generation zones. Our reconstructions (Fig. 10c–e) confirm that India and the Seychelles lay nearly vertically about the Plume Generation Zone during the eruption of the Deccan Traps and that they can be linked to the active Reunion hotspot.

Conclusions

This work demonstrates that the Early Tertiary volcanic rocks of Seychelles were emplaced during magnetochron C28n. A weighted mean of several

---

Fig. 10. (*Continued*) We also compare a Neoproterozoic 750 Ma pole from the Seychelles (Torsvik *et al.* 2001; latitude $-54.8°$, longitude 96.1°, A95 = 3.5°, $N = 29$ poles) with a similar-aged pole from India, that is, the Malani Igneous Suite (Gregory *et al.* 2009); latitude $-67.8°$, longitude 252.5°, A95 = 8.8°). Unrotated (rotated) Seychelles means and 95% confidence ovals are shown as white (blue) colours. (**b**) Seychelles reconstructed. LR, MR and LD denote Laxmi Ridge, Murray Ridge and Lacadives. The Seychelles–Laxmi–Deccan Traps reconstructed. Reconstructions are shown for (**c**) 65 Ma, (**d**) 63 Ma and (**e**) 60 Ma and draped on shear-wave anomalies near the core–mantle-boundary. The 1% slow contour is shown with a black line.

samples from the North Island gave a ^{40}Ar/^{39}Ar age of 63.1 ± 0.34 Ma and a zircon weighted mean U–Pb age of 63.27 ± 0.05 Ma, pointing to a rapid cooling of this complex. The weighted mean U–Pb age obtained from the complex of Silhouette Island is 63.54 ± 0.06 Ma and are 0.27 ± 0.08 Ma older indicating a separate event. The ^{40}Ar/^{39}Ar results record a more prolonged time span with the trachytic tuffs giving an age of 66.8 ± 0.8 Ma, while the youngest unit (syenite) gave a weighted mean age of 61.3 ± 0.57 Ma. Some of these ^{40}Ar/^{39}Ar ages are distinguishable at the 95% confidence level.

The palaeomagnetic mean remanence direction implies that the Seychelles has rotated 29.4° ± 12.9° anticlockwise after their formation, some time after magnetochron C28n. The palaeomagnetic results from the Silhouette indicate that the magnetic remanence carries reverse polarity for all units apart from the microgranites. We observe that some rock units have recorded significant secular variation and that their place in the geomagnetic timescale (GTS2004) does not match the observed magnetic polarities. The difference between the U–Pb and ^{40}Ar/^{39}Ar age determinations and the mismatch with the geomagnetic timescale indicates that the volcanic complex of Silhouette experienced a protracted period of cooling.

The palaeomagnetic pole obtained in this study is 57.55°S and 114.22°E (A95 = 12.3°, N = 14) which, after correcting for the previously mentioned rotation, corresponds very well to poles of similar ages from the Deccan Traps.

Recent detailed studies of early Cenozoic kinematics independently confirmed the presence of a plate boundary between Seychelles and adjacent north and NW oceanic crust (Cande *et al.* 2010). We propose that part of this rotation has been accommodated while the Gop/Laxmi basins were formed north and NE of the Laxmi Ridge/Seychelles composite block, as our results suggest Seychelles motion during the postulated timing for the Gop rifting and seafloor spreading and early seafloor spreading of the East Somali Basin. After magnetochron C27r, when the Seychelles were isolated from the Laxmi Ridge by seafloor spreading, the Seychelles might have continued to rotate as an individual plate therefore creating transpression and transtension along its plate boundaries. This motion may have initiated subduction along the Amirante Trough, or reactivated an already weak zone affected by an earlier tectonic event.

Our palaeomagnetic results and new inferred ages of Seychelles volcanic rocks lead to a reconstructed volcanic region that lay nearly radially above the Plume Generation Zone during the eruption of the Deccan Traps and confirm a hotspot origin (possible Reunion) for this volcanic activity. In the light of recent debates around the timing and extent of the Deccan traps and the role of mantle plumes in relation to rifting of the Seychelles microcontinent (e.g. Sheth 2005; Collier *et al.* 2009; Armitage *et al.* 2010), our study shows that Seychelles volcanism occurred contemporaneously with the Deccan Trap volcanic emplacement and incipient seafloor spreading in the East Somali Basin.

Helpful and constructive reviews were provided by J. Collier and M. Widdowson. MG, DJJvH and THT appreciated funding from Statoil (Splates Project). CG acknowledges Scripps Institute of Oceanography for hosting her while working on the last version of the manuscript. We appreciate the editorial handling by S. Buiter. We thank Linda Vanherck and the staff at the North Island for providing access, guiding and help during sampling.

References

ALLÈGRE, C. J., BIRCK, J. L., CAPMAS, F. & COURTILLOT, V. 1999. Age of the Deccan traps using ^{187}Re–^{187}Os systematics. *Earth and Planetary Science Letters*, **170**, 197–204.

AMANTE, C. & EAKINS, B. 2009. ETOPO1 1 Arc-minute global relief model: procedures, data sources and analysis, Technical Report, NOAA.

ARMITAGE, J. J., COLLIER, J. S. & MINSHULL, T. A. 2010. The importance of rift history for volcanic margin formation. *Nature*, **465**, 913–917.

ASHWAL, L. D., DEMAIFFE, D. & TORSVIK, T. H. 2002. Petrogenesis of Neoproterozoic Granitoids and related rocks from the Seychelles: the case for an Andean-type arc origin. *Journal of Petrology*, **43**, 45–83.

BAKER, B. H. 1963. Geology and mineral resources of the Seychelles Archipelago. *Geological Survey of Kenya Memoir*, **3**, 140.

BAKSI, A. K. 2005. Comment on '^{40}Ar/^{39}Ar dating of the Rajahmundry Traps, eastern India and their relations to the Deccan Traps' by Knight et al. [*Earth and Planetary Science Letters*, **208** (2003) 85–99]; *Earth and Planetary Science Letters*, **239**, 368–373.

BARRON, E. J. & HARRISON, C. G. A. 1980. An analysis of past plate motions the South Atlantic and Inian Oceans. *In*: DAVIES, P. & RUNCORN, S. (eds) *Mechanisms of Plate Tectonics and Continental Drift*. Academic Press, New York, 89–110.

BASU, A. R., RENNE, P. R., DASGUPTA, D. K., TEICHMANN, F. & POREDA, R. J. 1993. Early and Late Alkali Igneous Pulses and a High-^{3}He Plume Origin for the Deccan Flood Basalts. *Science*, **261**, 902–906.

BECKER, W. & BOSCHI, L. 2002. A comparison of tomographic and geodynamic mantle models. *Geochemistry, Geophysics, Geosystems*, **3**, 1003.

BERNARD, A. & MUNSCHY, M. 2000. Le bassin des Mascareignes et le bassin de Laxmi (ocean Indien occidental) se sont-ils formes a l'axe d'un meme centre d'expansion? *Sciences de la Terre et Des Planetes*, **330**, 777–783.

BEYENE, A. & ABDELSALAM, M. G. 2005. Tectonics of the Afar Depression: a review and synthesis. *Journal of African Earth Sciences*, **41**, 41–59.

BHATTACHARYA, G. C., CHAUBEY, A. K., MURTY, G. P. S., SRINIVAS, K., SARMA, K. V. L. N. S., SUBRAHMANYAM, V. & KRISHNA, K. S. 1994. Evidence for seafloor spreading in the Laxmi Basin, northeastern Arabian sea. *Earth and Planetary Science Letters*, **125**, 211–220.

BIGGIN, A., VAN HINSBERGEN, D. J. J., LANGEREIS, C. G., STRAATHOF, G. B. & DEENEN, M. H. 2008. Geomagnetic secular variation in the Cretaceous Normal Superchron and in the Jurassic. *Physics of the Earth and Planetary Interiors*, **169**, 3–19.

BROWN, G. C., THORPE, R. S. & WEBB, P. C. 1984. The geochemical characteristics of granitoids in contrasting arcs and comments on magma sources. *Journal of the Geological Society, London*, **141**, 413–426.

BURKE, K. & TORSVIK, T. H. 2004. Derivation of Large Igneous Provinces of the past 200 million years from long-term heterogeneities in the deep mantle. *Earth and Planetary Science Letters*, **227**, 531–538.

BURKE, K., STEINBERGER, B., TORSVIK, T. H. & SMETHURST, M. A. 2008. Plume Generation Zones at the margins of Large Low Shear Velocity Provinces on the core–mantle boundary. *Earth and Planetary Science Letters*, **265**, 49–60.

BUTLER, R. F. 1992. *Paleomagnetism: Magnetic Domains to Geologic Terranes*. Blackwell Scientific Publications, Boston.

CALVÈS, G., CLIFT, P. D. & INAM, A. 2008. Anomalous subsidence on the rifted volcanic margin of Pakistan: no influence from Deccan plume. *Earth and Planetary Science Letters*, **272**, 231–239.

CALVÈS, G., SCHWAB, A. M. ET AL. 2011. Seismic volcanostratigraphy of the Western Indian rifted margin: the pre-Deccan Igneous Province. *Journal of Geophysical Research*, **116**, B01101, doi: 10.1029/2010JB000862.

CANDE, S. C., PATRIAT, P. & DYMENT, J. 2010. Motion between the Indian, Antarctic and African plates in the early Cenozoic. *Geophysical Journal International*, **183**, 127–149.

CHAUBEY, A. K., GOPALA RAO, D., SRINIVAS, K., RAMPRASAD, T., RAMANA, M. V. & SUBRAHMANYAM, V. 2002. Analyses of multichannel seismic reflection, gravity and magnetic data along a regional profile across the central-western continental margin of India. *Marine Geology*, **182**, 303–323.

CHENET, A., QUIDELLEUR, X., FLUTEAU, F., COURTILLOT, V. & BAJPAI, S. 2007. 40 K–40 Ar dating of the Main Deccan large igneous province: further evidence of KTB age and short duration. *Earth and Planetary Science Letters*, **263**, 1–15.

CHERNIAK, D. J. & WATSON, E. B. 2001. Pb diffusion in zircon. *Chemical Geology*, **177**, 5–24.

COFFIN, M. F. & RABINOWITZ, P. D. 1988. *Evolution of the conjugate East African–Madagascan margins and the western Somali Basin*. Geological Society of America Special Papers, **226**.

COFFIN, M. F. & ELDHOLM, O. 1992. Volcanism and continental break-up: a global compilation of large igneous provinces. *In*: STOREY, B. C., ALABASTER, T. & PANKHURST, R. J. (eds) *Magmatism and the Causes of Continental Break-up*. Geological Society, London, Special Publications, **68**, 17–30.

COLLIER, J. S., SANSOM, V., ISHIZUKA, O., TAYLOR, R. N., MINSHULL, T. A. & WHITMARSH, R. B. 2008. Age of Seychelles–India break-up. *Earth and Planetary Science Letters*, **272**, 264–277.

COLLIER, J. S., MINSHULL, T. A. ET AL. 2009. Factors influencing magmatism during continental breakup: new insights from a wide-angle seismic experiment across the conjugate Seychelles–Indian margins. *Journal of Geophysical Research*, **114**, B03101.

CORFU, F. 2004. U–Pb age, setting, and tectonic significance of the anorthosite–mangerite–charnockite–granite-suite, Lofoten-Vesterålen, Norway. *Journal of Petrology*, **45**, 1799–1819, doi: 10.1093/petrology/egh034.

CORFU, F. 2009. When the CA-TIMS therapy fails: the over-enthusiastic, the mixed-up, and the stubborn zircon. AGU Fall Meeting, Abstract V53B-05.

COURTILLOT, V. E. & RENNE, P. R. 2003. On the ages of flood basalt events. *Comptes Rendus Geosciences*, **335**, 113–140.

COURTILLOT, V., BESSE, J., VANDAMME, D., MONTIGNY, R., JAEGER, J. J. & CAPPETTA, H. 1999. Deccan flood basalts at the Cretaceous/Tertiary boundary? *Memoir – Geological Society of India*, **43**(Part 1), 173.

COX, A. 1969. Confidence limits for the precision parameter k. *Geophysical Journal International*, **17**, 545–549.

CRIPPS, J. A., WIDDOWSON, M., SPICER, R. A. & JOLLEY, D. W. 2005. Costal ecosystem responses to late stage Deccan Trap volcanism: the post K–T boundary (Danian) palynofacies of Mumbai (Bombay), west India. *Palaeogeography, Palaeoclimatology, Palaeoecology*, **216**, 303–332.

CROWLEY, J. L., BOWRING, S. A. & HANCHAR, J. M. 2006. What is a magma crystallization age? Insight from micro-sampling of chemical domains in zircon from the Fish Canyon Tuff. *Geochimica et Cosmochimica Acta*, **70**, A120.

DAY, R., FULLER, M. & SCHMIDT, V. A. 1977. Hysteresis properties of titanomagnetites: grain-size and compositional dependence. *Physics of Earth and Planetary Science Letters*, **13**, 260–267.

DEENEN, M. H., LANGEREIS, C. G. & VAN HINSBERGEN, D. J. J. 2011. Geomagnetic secular variation and the statistics of palaeomagnetic directions. *Geophysical Journal International*, doi: 10.1111/j.1365-246X.2011.05050.x.

DEVEY, C. W. & STEPHENS, W. E. 1991. Tholeiitic dykes in the Seychelles ant eh original spatial extent of the Deccan. *Journal of Geological Society, London*, **148**, 979–983.

DICKIN, A. P., FALLICK, A. E., HALLIDAY, A. N., MACINTYRE, R. M. & STEPHENS, W. E. 1986. An isotopic and geochronological investigation of the younger igneous rocks of the Seychelles microcontinent. *Earth and Planetary Science Letters*, **81**, 46–56.

DUNCAN, R. A. 1990. The volcanic record of the Reunion hotspot. *In*: DUNCAN, R. A., BACKMAN, J., DUNBAR, R.

B. & PETERSON, L. G. (eds) *Proceedings of the Ocean Drilling Project, Scientific Results*, **115**, Ocean Drilling Program, Texas A and M University, 3–10.

FISHER, N. I. 1953. Dispersion on a sphere. *Proceedings of the Royal Society of London*, **A217**, 295–305.

FISHER, N. I., LEWIS, T. & EMBLETON, B. J. J. 1987. *Statistical Analysis of Spherical Data*. Cambridge University Press, Cambridge.

FISHER, R. L., ENGEL, C. G. & HILDE, T. W. 1968. Basalts dredged from the Amirante Ridge, western Indian Ocean. *Deep-Sea Research*, **15**, 521–534.

GAINA, C., GERNIGON, L. & BALL, P. 2009. Paleocene–Recent Plate Boundaries in the NE Atlantic and the formation of Jan Mayen microcontinent. *Journal of Geological Society*, **166**, 601–616, doi: 10.1144/0016-76492008-112.

GAINA, C., MÜLLER, R. D., BROWN, B., ISHIHARA, T. & IVANOV, S. 2007. Breakup and early seafloor spreading between India and Antarctica. *Geophysical Journal International*, **170**, 151–169.

GREGORY, L. C., MEERT, J. G., BINGEN, B., PANDIT, M. K. & TORSVIK, T. H. 2009. Paleomagnetism and geochronology of the Malani Igneous Suite, Northwest India: Implications for the configuration of Rodinia and the assembly of Gondwana. *Precambrian Research*, **170**, 13–26.

HALLIMOND, A. F. & HERROUN, E. F. 1933. Laboratory determinations of the magnetic properties of certain igneous rocks. *Proceedings of the Royal Society of London*, **141**, 302–314.

HAMMOND, J. O. S., KENDALL, J. ET AL. 2005. Upper mantle anisotropy beneath the Seychelles microcontinent. *Journal of Geophysical Research*, **110**, B11401, doi: 10.1029/2005JB003757.

HOOPER, P., WIDDOWSON, M. & KELLY, S. 2010. Tectonic setting and timing of the final Deccan flood basalt eruptions. *Geology*, **38**, 839–842, doi: 10.1130/G31072.1.

JAFFEY, A. H., FLYNN, K. F., GLENDENIN, L. E., BENTLEY, W. C. & ESSLING, A. M. 1971. Precision measurement of half-lives and specific activities of 235U and 238U. *Physical Review, Section C, Nuclear Physics*, **4**, 1889–1906.

JAY, A. E. & WIDDOWSON, M. 2008. Stratigraphy, structure and volcanology of the SE Deccan continental flood basalt province: implications for eruptive extend and volumes. *Journal of Geological Society, London*, **165**, 177–188.

JAY, A. E., MAC NIOCAILL, C., WIDDOWSON, M., SELF, S. & TURNER, W. 2009. Deccan flood basalt province, India: implications for the volcanostratigraphic architecture of continental flood basalt provinces. *Journal of Geological Society, London*, **166**, 13–24.

JOHNSON, C. L., CONSTABLE, C. G. ET AL. 2008. Recent investigations of the 0–5 Ma geomagnetic field recorded by lava flows. *Geochemistry, Geophysics, Geosystems*, **9**, Q04032.

KENT, J. T., BRIDEN, J. C. & MARDIA, K. V. 1983. Linear and planar structure in ordered multivariate data as applied to progressive demagnetisation of palaeomagnetic remanence. *Geophysical Journal of the Royal Astronomical Society*, **81**, 75–87.

KHRAMOV, A. N. 1958. Palaeomagnetism and stratigraphic correlation. *In*: IRVING, E. (ed.) *Paleomagnetism and its Application to Geological and Geophysical Problems*. 1st edn. Wiley, New York.

KLOOTWIJK, C. T. 1975. A note on the Palaeomagnetism of the late precambrian malani rhyolites near Jodhpur – India. *Journal of Geophysics*, **41**, 189–200.

KÖNIG, M. & JOKAT, W. 2006. The Mesozoic breakup of the Weddell Sea. *Journal of Geophysical Research*, **111**, B12102, doi: 10.1029/2005JB004035.

KRISHNA, K. S., GOPALA RAO, D. & SAR, D. 2006. Nature of the crust in the Laxmi Basin (14° –20°N), western continental margin of India. *Tectonics*, **25**, 1–18.

KROGH, T. E. 1973. A low contamination method for hydrothermal decomposition of zircon and extraction of U and Pb for isotopic age determinations. *Geochimica et Cosmochimica Acta*, **37**, 485–494.

KROGH, T. E. 1982. Improved accuracy of U–Pb zircon ages by the creation of more concordant systems using an air abrasion technique. *Geochimica et Cosmochimica Acta*, **46**, 637–649.

LANE, C. I. 2006. *Rifted margin formation in the northwest Indian Ocean: the extensional and magmatic history of the Laxmi Ridge continental margin*. Ph.D. thesis, University of Southampton.

LUDWIG, K. R. 2003. Isoplot 3.0. *A Geochronological Toolkit for Microsoft Excel*. Berkeley Geochronology Center Special Publication No. 4.

MAHONEY, J., NICOLLET, C. & DUPUY, C. 1991. Madagascar basalts: tracking oceanic and continental sources. *Earth and Planetary Science Letters*, **104**, 350–363.

MAUS, S., LUEHR, H., MARTIN, R., HEMANT, K., BALASIS, G., RITTER, P. & CLAUDIA, S. 2007. Fifth-generation lithospheric magnetic field model from CHAMP satellite measurements. *Geochemistry, Geophysics, Geosystems*, **8**, Q05013, doi: 10.1029/2006GC001521.

MART, Y. 1988. The tectonic setting of the Seychelles, Mascarene and Amirante plateaus in the western equatorial Indian Ocean. *Marine Geology*, **79**, 261–274.

MASSON, D. G. 1984. Evolution of the Mascarene Basin, western Indian Ocean, and the significance of the Amirante Arc. *Marine Geophysics Research*, **6**, 365–382.

MATTINSON, J. M. 2005. Zircon U–Pb chemical abrasion ('CA-TIMS') method: combined annealing and multi-step partial dissolution analysis for improved precision and accuracy of zircon ages. *Chemical Geology*, **220**, 47–66.

MCDOUGALL, I. & HARRISON, T. M. 1999. *Geochronology and Thermochronology by the $^{40}Ar/^{39}Ar$ Method*. 2nd edn. Oxford University Press, New York.

MCFADDEN, P. L. & MCELHINNY, M. W. 1988. The combined analysis of remagnetization circles and direct observations in palaeomagnetism. *Earth and Planetary Science Letters*, **87**, 161–172.

MCFADDEN, P. L., MERRILL, R. T., MCELHINNY, M. W. & SUNHEE, L. 1991. Reversals of the earth's magnetic field and temporal variations of the dynamo families. *Journal of Geophysical Research*, **96**, 3923–3933.

MCKENZIE, D. & SCLATER, J. G. 1971. The evolution of the Indian Ocean since the Late Cretaceous. *Geophysical Journal of the Royal Astronomical Society*, **25**, 437–528.

MILES, P. R. 1982. Gravity models of the Amirante Arc, western Indian Ocean. *Earth and Planetary Science Letters*, **61**, 127–135.

MILLER, J. S., MATZEL, J. E. P., MILLER, C. F., BURGESS, S. D. & MILLER, R. B. 2007. Zircon growth and recycling during the assembly of large, composite arc plutons. *Journal of Volcanology Geothermal Research*, **167**, 282–299.

MINSHULL, T. A., LANE, C. I., COLLIER, J. S. & WHITMARSH, R. B. 2008. The relationship between rifting and magmatism in the northeastern Arabian Sea. *Nature Geoscience*, **1**, 463–467.

MITCHELL, C. & WIDDOWSON, M. 1991. A geological map of the southern Deccan Traps, India and its structural implications. *Journal of the Geological Society, London*, **148**, 495–505.

MONTELLI, R., NOLET, G., DAHLEN, F. A. & MASTERS, G. 2006. A catalogue of deep mantle plumes: new results from finite-frequency tomography. *Geochemistry, Geophysics, Geosystems*, **7**, (Q11007).

NORTON, I. O. & SCLATER, J. G. 1979. A model for the evolution of the Indian Ocean and the break-up of Gondwanaland. *Journal of Geophysical Research*, **84**, 6803–6830.

OGG, J. G. & SMITH, A. G. 2004. The geomagnetic polarity time scale. *In*: GRADSTEIN, F. M., OGG, J. G. & SMITH, A. G. (eds) *A Geologic Time Scale 2004*. Cambridge University Press, Cambridge, 63–86.

O'NEILL, C., MÜLLER, R. D. & STEINBERGER, B. 2005. On the uncertainties in hotspot reconstructions, and the significance of moving hotspot reference frames. *Geochemistry, Geophysics, Geosystems*, **6**, Q04003.

PLUMMER, P. S. 1995. Ages and geological significance of the igneous rocks from Seychelles. *Journal of African Earth Sciences*, **20**, 91–101.

PLUMMER, P. S. 1996. The Amirante ridge/trough complex: response to rotational transform rift/drift between Seychelles and Madagascar. *Terra Nova*, **8**, 34–47.

PLUMMER, P. S. & BELLE, E. R. 1995. Mesozoic tectonostratigraphic evolution of the Seychelles microcontinent. *Sedimentary Geology*, **96**, 73–91.

RENNE, P. R., SWISHER, C. C., DEINO, A. L., KARNER, D. B., OWENS, T. L. & DEPAOLO, D. J. 1998. Intercalibration of standards, absolute ages and uncertainties in $^{40}Ar/^{39}Ar$ dating. *Chemical Geology*, **145**, 117–152.

ROYER, J.-Y., CHAUBEY, A. K., DYMENT, J., BHATTACHARYA, G. C., SRINIVAS, K., YATHEESH, V. & RAMPRASAD, T. 2002. Paleogene plate tectonic evolution of the Arabian and Eastern Somali basins. *In*: CLIFT, P. D., KROON, D., GAEDICKE, C. & CRAIG, J. (eds) *The Tectonic and Climatic Evolution of the Arabian Sea Region*. Geological Society, London, Special Publications, **195**, 7–23.

SANDWELL, D. T. & SMITH, W. H. F. 1997. Marine gravity anomaly from Geosat and ERS-1 satellite altimetry. *Journal of Geophysical Research*, **102**, 10 039–10 050.

SCHÄRER, U. 1984. The effect of initial 230Th disequilibrium on young U–Pb ages: the Makalu case, Himalaya. *Earth and Planetary Science Letters*, **67**, 191–204.

SCHOENE, B., CROWLEY, J. L., CONDON, D. J., SCHMITZ, M. D. & BOWRING, S. A. 2006. Reassessing the uranium decay constants for geochronology using ID-TIMS U–Pb data. *Geochimica et Cosmochimica Acta*, **70**, 426–445.

SELF, S., JAY, A. E., WIDDOWSON, M. & KESZTHELYI, L. P. 2008. Correlation of the Deccan and Rajahmundry Trap lavas: are these the longest and largest lava flows on Earth? *Journal of Volcanology and Geothermal Research*, **172**, 3–19.

SHETH, H. C. 2005. From Deccan to Reunion: no trace of a mantle plume. *In*: FOULGER, G. R., NATLAND, J. H., PRESNALL, D. C. & ANDERSON, D. L. (eds) *Plates, Plumes and Paradigms Geological Society of America*. Geological Society of America Special Papers, **388**, Boulder, Colorado, 477–501.

SHETH, H. C., PANDE, K. & BHUTANI, R. 2001. $^{40}Ar-^{39}Ar$ Ages of Bombay Trachytes: evidence for a Palaeocene phase of Deccan Volcanism. *Geophysical Research Letters* **28**, 3513–3516.

SMITH, W. H. F. & SANDWELL, D. T. 1997. Global seafloor topography from satellite altimetry and ship depth soundings. *Science*, **277**, 1957–1962.

STEIGER, R. H. & JÄGER, E. 1977. Subcommission on geochronology: Convention on the use of decay constants in geo- and cosmochronology. *Earth and Planetary Science Letters*, **36**, 359–362.

STEPHENS, W. E. & DEVEY, C. M. 1992. Seychelles and the fragmentation of Gondwana: evidence from the igneous rocks. *In*: PLUMMER, S. (ed.) *Proceedings of the First Indian Ocean Regional Seminar on Petroleum Exploration*, Seychelles. Seychelles National Oil Company (69), Victoria, Seychelles, 211–222.

STEPHENS, W. E., STOREY, M., DONALDSON, C. H., ELLAM, R. M., LELIKOV, E., TARARIN, G. & GARBE-SCHONENBERG, C. 2009. Age and origin of the Amirante ridge-trench structure, western Indian Ocean. *In*: *Proceedings of Fall Meeting of American Geophysical Union*, abstract #T23A-1885.

STOREY, M., MAHONEY, J. J., SAUNDERS, A. D., DUNCAN, R. A., KELLEY, S. P. & COFFIN, M. F. 1995. Timing of hot spot-related volcanism and the break-up of Madagascar and India. *Science*, **267**, 852–855.

STOREY, M., MAHONEY, J. J. & SAUNDERS, A. D. 1997. Cretaceous basalts in Madagascar and the transition between plume and continental lithosphere mantle sources. *In*: MAHONEY, J. J. & COFFIN, M. F. (eds) *Large Igneous Provinces: Continental, Oceanic, and Planetary Flood Volcanism*. American Geophysical Union, Washington, Geophysical Monograph, 95–122.

SUBRAHMANYAM, V., RAO, D. G., RAMANA, M. V., KRISHNA, K. S., MURTY, G. P. S. & GANGADHARARAO, M. 1995. Structure and tectonics of the southwestern continental margin of India. *Tectonophysics*, **249**, 267–282.

TAUXE, L. & KENT, D. V. 2004. A simplified statistical model for the geomagnetic field and the detection of shallow bias in paleomagnetic inclinations: was the ancient magnetic field dipolar? *In*: CHANNEL, J. E. T., KENT, D. V., LOWRIE, W. & MEERT, J. G. (eds) *Timescales of the Paleomagnetic Field*. American Geophysical Union, Washington, Geophysical Monograph, **145**, 101–116.

TAUXE, L., KODAMA, K. P. & &KENT, D. V. 2008. Testing corrections for paleomagnetic inclination error in sedimentary rocks: a comparative approach. *Physics of the Earth and Planetary Interiors*, **169**, 152–165.

TODAL, A. & ELDHOLM, O. 1998. Continental margin off Western India and Deccan Large Igneous Province. *Marine Geophysical Researches*, **20**, 273–291.

TORSVIK, T. H., TUCKER, R. D., ASHWAL, L. D., EIDE, E. A., RAKOTOSOLOFO, N. A. & DE WIT, M. J. 1998. Late Cretaceous magmatism in Madagascar: palaeomagnetic evidence for a stationary Marion hotspot. *Earth and Planetary Science Letters*, **164**, 221–232.

TORSVIK, T. H., BRIDEN, J. C. & SMETHURST, M. A. 2000. Super-IAPD Interactive analysis of palaeomagnetic data. www.Geodynamics.no/software.Htm.2009.

TORSVIK, T. H., ASHWAL, L. D., TUCKER, R. D. & EIDE, E. A. 2001. Geochronology and palaeogeography of the Seychelles microcontinent: the India link. *Precambrian Research*, **100**, 47–59.

TORSVIK, T. H., SMETHURST, M. A., BURKE, K. & STEINBERGER, B. 2006. Large igneous provinces generated from the margins of the large low-velocity provinces in the deep mantle. *Geophysics Journal International*, **167**, 1447–1460.

TORSVIK, T. H., SMETHURST, M. A., BURKE, K. & STEINBERGER, B. 2008a. Long term stability in deep mantle structure: evidence from the ∼300 Ma Skagerrak-Centered Large Igneous Province (the SCLIP). *Earth and Planetary Science Letters*, **267**, 444–452.

TORSVIK, T. H., STEINBERGER, B., COCKS, L. R. M. & BURKE, K. 2008b. Longitude: linking Earth's ancient surface to its deep interior. *Earth and Planetary Science Letters*, **276**, 273–282.

TORSVIK, T. H., MÜLLER, R. D., VAN DER VOO, R., STEINBERGER, B. & GAINA, C. 2008c. Global Plate Motion Frames: Toward a unified model. *Reviews of Geophysics*, **46**, RG3004.

TORSVIK, T. H., BURKE, K., WEBB, S., ASHWAL, L. & STEINBERGER, B. 2010. Diamonds sampled by plumes from the core–mantle boundary. *Nature*, **466**, 352–355, doi:10.1038/nature09216.

TUCKER, R. D., ASHWAL, L. D. & TORSVIK, T. H. 2001. U–Pb geochronology of Seychelles granitoids: a Neoproterozoic continental arc fragment. *Earth and Planetary Science Letters*, **187**, 27–38.

VANDAMME, D. 1994. A new method to determine paleosecular variation. *Physics of the Earth and Planetary Interiors*, **85**, 131–142.

VAN HINSBERGEN, D. J. J., STRAATHOF, G. B., KUIPER, K. F., CUNNINGHAM, W. D. & WIJBRANS, J. R. 2008a. No rotations during transpressional orogeny in the Gobi Altai: coinciding Mongolian and Eurasian apparent polar wander paths. *Geophysical Journal International*, **173**, 105–126.

VAN HINSBERGEN, D. J. J., DUPONT-NIVET, G., NAKOV, R., OUD, K. & PANAIOTU, C. 2008b. No significant post-Eocene rotation of the Moesian Platform and Rhodope (Bulgaria): implications for the kinematic evolution of the Carpathian and Aegean arcs. *Earth and Planetary Science Letters*, **273**, 345–358.

VELAIN, C. 1879. Notes sur la constitution geologique des Iles Seychelles. *Bulletin de la Societe Geologique de France*, **7**, 278.

WENSINK, H. 1973. Newer paleomagnetic results of the Deccan Traps, India. *Tectonophysics*, **17**, 41–59.

WENSINK, H., BOELRIJK, N. A. I. M., HEBEDA, E. H., PRIEM, H. N. A., VERDURMEN, E. A. T. & VERSCHURE, R. H. 1977. Paleomagnetism and radiometric age determinations of the Deccan Traps, India. *IV International Gondwana Symposium*, Calcutta (India). Hindustan Publishing Corporation, Delhi (India). 832–849.

YATHEESH, V., BHATTACHARYA, G. C. & DYMENT, J. 2009. Early oceanic opening off Western India–Pakistan margin: The Gop Basin revisited. *Earth and Planetary Science Letters*, **284**, 399–408.

The relations between felsic and mafic volcanic rocks in continental flood basalts of Ethiopia: implication for the thermal weakening of the crust

DEREJE AYALEW

Department of Earth Sciences, School of Earth and Planetary Sciences,
Addis Ababa University, PO Box 729/1033, Addis Ababa, Ethiopia
(e-mail: dereayal@geol.aau.edu.et)

Abstract: Sr and Nd isotopic compositions are presented for a Miocene bimodal basalt–rhyolite suite from north Shewa, central Ethiopian plateau. Whole-rock Rb–Sr isochron of the rhyolites yields an age of 20.7 ± 2.4 Ma, marking the onset of volcanism in central Ethiopia *c.* 20 Ma, 10 Ma after initial magmatism in the northern Ethiopian plateau. Initial $^{87}Sr/^{86}Sr$ ratios slightly vary in the basalts as well as in the rhyolites, ranging from 0.70440 to 0.70641 and from 0.70563 to 0.70658, respectively. Initial $^{143}Nd/^{144}Nd$ ratios show significant variations in the basalts (0.51248–0.51274), but remain nearly constant in the rhyolites (0.51273–0.51278). The Sr and Nd isotopic ratios of the basalts are interpreted to reflect their derivation from Afar plume contaminated by crustal materials (up to 15% contamination). The rhyolites evolved dominantly by fractional crystallization of mantle-derived basaltic magmas similar in composition to the exposed flood basalts.

It is regarded that the continental flood basalt (CFB) province and associated continental break-up are thought to be related to the upwelling of mantle plume impinged on the base of the lithosphere (e.g. Campbell & Griffiths 1990; Hill 1991; Storey 1995; White & McKenzie 1995; Rooney *et al.* 2007). Nearly all CFB provinces contain significant volumes (up to 20% of the total volcanics) of acidic eruptives, usually in the upper part of the sequence capping the basalts. The volcanism is of a bimodal basalt–rhyolite type, with a lack of intermediate products. Mafic magmas are interpreted as resulting from partial melting of the mantle with variable contributions of plume, asthenosphere and lithosphere (Thompson *et al.* 1982; Arndt *et al.* 1993; Gibson *et al.* 2006). There is, however, considerable controversy on the generation of silicic magmas. Some authors suggest that the rhyolites are crustal anatectic melts (Harris *et al.* 1990; Kirstein *et al.* 2000; Riley *et al.* 2001; Xu *et al.* 2008), partial remelting of underplated basaltic magmas or their derivatives emplaced during extrusion of flood basalts (Lightfoot *et al.* 1987; Garland *et al.* 1995; Miller & Harris 2007; Mahoney *et al.* 2008). Others contend however that rhyolites are closely related to the erupted basalts by fractional crystallization with variable crustal contamination (Baker *et al.* 1996; Barbey *et al.* 2005; Ukstins-Peate *et al.* 2005).

The Ethiopian volcanic plateau (Fig. 1) is one of the youngest, best-exposed CFB provinces related to plume impingement, uplift and extension (Marty *et al.* 1996; Ebinger & Sleep 1998; Pik *et al.* 2008).

The flood volcanism was erupted between 31 and 26 Ma, and the main pulse of magmatism occurred at 30 Ma in the northern part of the northwest Ethiopian plateau (Hofmann *et al.* 1997; Rochette *et al.* 1998; Ukstins *et al.* 2002; Coulié *et al.* 2003). Flood volcanism then continued to occur during the Miocene in central Ethiopia (Fig. 1) with decreasing volumes, forming a thin succession (up to 1 km thick in the Miocene province and over 2 km thick in the Oligocene province; Ayalew *et al.* 2006). The field relations between the Oligocene and Miocene subprovinces are not well known and need further investigation. However, the earliest manifestations of the Afar mantle plume impact are found in southern and south-western Ethiopia where small amounts of tholeiitic (40–45 Ma) and alkali (*c.* 35 Ma) basalts, together with associated rhyolitic magmas, occur (George *et al.* 1998).

Previous petrological and geochemical studies on the northwest Ethiopian volcanic plateau have focused on the Oligocene basalts (Pik *et al.* 1999; Kieffer *et al.* 2004; Meshesha & Shinjo 2007; Beccaluva *et al.* 2009) and rhyolites (Ayalew *et al.* 2002; Ayalew & Yirgu 2003). Little is known about the Miocene volcanism (e.g. Hart *et al.* 1989; Kieffer *et al.* 2004; Furman *et al.* 2006). In this paper, Sr and Nd isotopic data are presented for a Miocene bimodal basalt–rhyolite suite from north Shewa, which were emplaced after the Oligocene Ethiopian CFB province (Fig. 2). The petrogenesis of these rocks, based on major and trace element data, has been studied by Ayalew & Gibson (2009).

From: VAN HINSBERGEN, D. J. J., BUITER, S. J. H., TORSVIK, T. H., GAINA, C. & WEBB, S. J. (eds) *The Formation and Evolution of Africa: A Synopsis of 3.8 Ga of Earth History.* Geological Society, London, Special Publications,
357, 253–264. DOI: 10.1144/SP357.13 0305-8719/11/$15.00 © The Geological Society of London 2011.

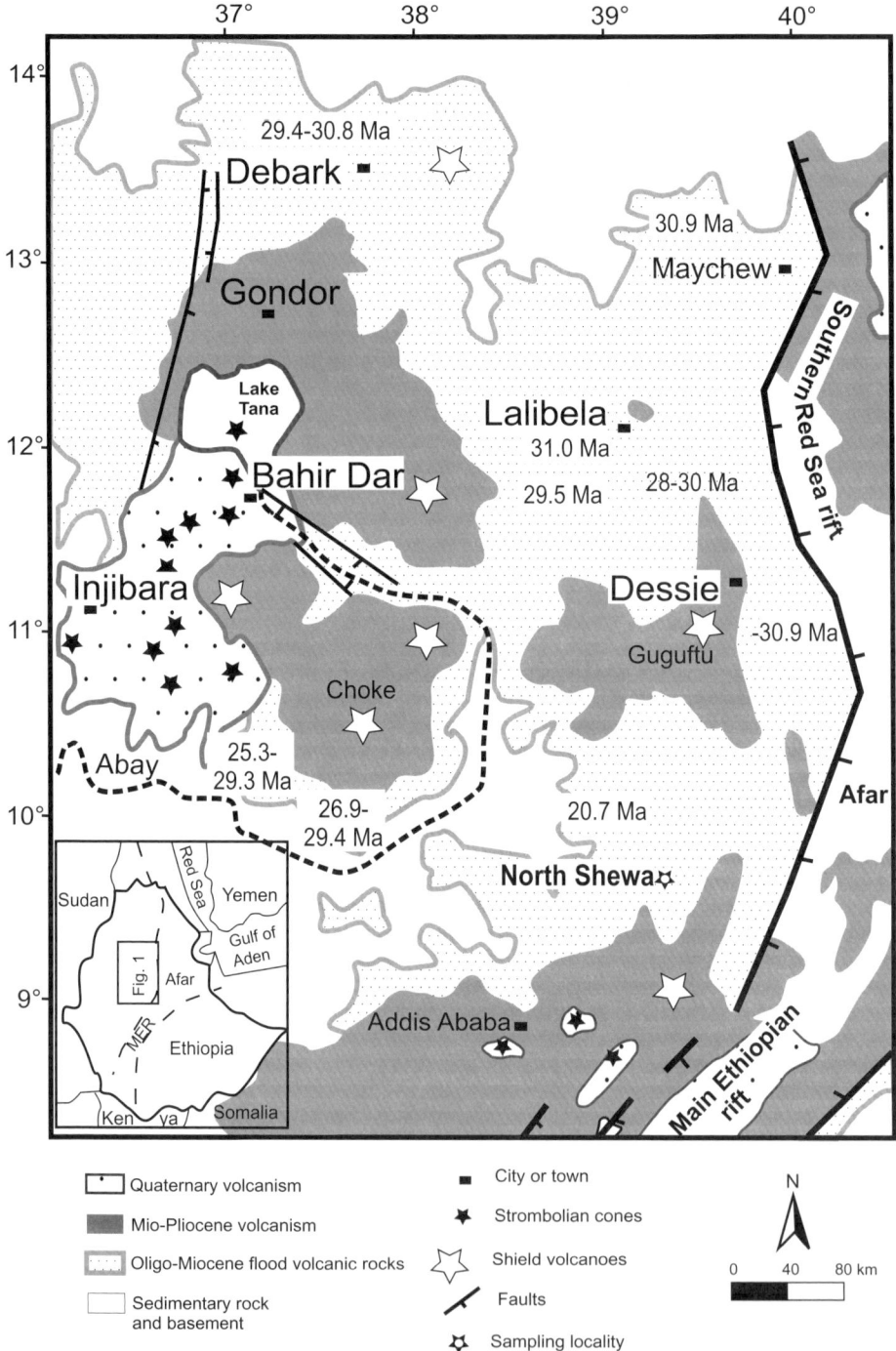

Fig. 1. Geological map of the northern part of the Ethiopian plateau depicting the distribution of the Oligocene–Miocene bimodal basalt–rhyolite flood volcanism and overlying shield volcanoes (after Ayalew & Gibson 2009). Note that the age of flood basalts and rhyolites decreases towards the south in conjunction with the onset of rifting in the main Ethiopian rift (MER) c. 11 Ma. At c. 10°N, the Red Sea rift terminates and northern MER commences. Age data are from Hofmann *et al.* (1997); Rochette *et al.* (1998); Ukstins *et al.* (2002); Ayalew & Yirgu (2003); Coulié *et al.* (2003); Kieffer *et al.* (2004); Meshesha & Shinjo (2007) and this study. Opened star represents the sampling locality. Inset map shows the location of the study area in Africa.

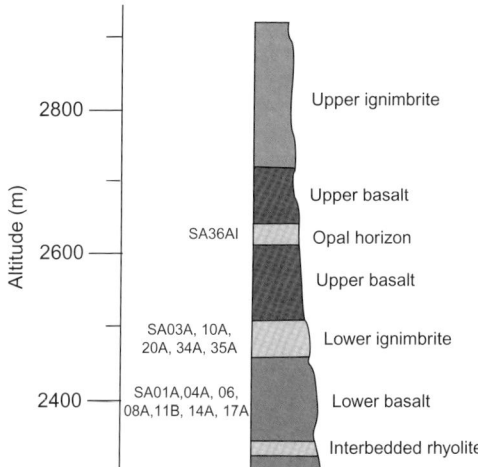

Fig. 2. Composite stratigraphic section of the north Shewa volcanic succession illustrating the intercalation of basalt and rhyolite formations and the location of samples (after Ayalew & Gibson 2009).

Their work demonstrated that the basalts are formed by lower degrees of partial melting (5%) of the Afar mantle plume. They also proposed that the rhyolites are genetically linked to the associated basalts. The aim of this paper is to determine the age of Miocene volcanism and to trace the sources of basalts and rhyolites in north-central Ethiopia's volcanic plateau. This study will allow the temporal evolution of the Afar mantle plume and the thermomechanical properties of the underlying crust to be constrained.

Regional geological setting

Continental flood basalt provinces of Africa

The formation of the oldest CFB province in Africa dates back to the Early Jurassic. The Central Atlantic Magmatic Province (CAMP), which is associated with the break-up of Pangea and the opening of the central Atlantic Ocean, is defined by tholeiitic dikes, sills and lava flows that crop out in once-contiguous parts of eastern North America, southern Europe, west Africa and northernmost South America. Alkaline and silicic rocks are scarce. $^{40}Ar/^{39}Ar$ and palaeomagnetic data indicate that the peak magmatic activity occurred at *c.* 200 Ma (Marzoli *et al.* 1999).

The Karoo is the oldest CFB associated with Gondwana break-up during the Early Jurassic. The lava pile represents a bimodal theleiitic basalt–rhyolites association. Volcanism initiated at *c.* 184 Ma with a peak at 183 Ma (Duncan *et al.* 1997). The rhyolites overlie the main basalt

succession and have been dated at 179 Ma (Cleverly *et al.* 1984).

The Etendeka lava field represents a small fragment of the Paraná–Etendeka province that was associated with the opening of the south Atlantic Ocean during the early Cretaceous. The Etendeka lavas are scattered in north-western Namibia and are strongly bimodal, dominated by tholeiitic basalts with significant quantities of rhyolites (Erlank *et al.* 1984). The lavas were erupted over a longer interval between 137 and 127 Ma, although the main pulse of magmatism lasted from 133 to 131 Ma (Turner *et al.* 1994). The Etendeka rhyolites typically form the uppermost units of the lava sequences and yield ages ranging from 133 to as young as 129 Ma (Turner *et al.* 1994).

The most recent Ethiopian CFB province is dated *c.* 31–26 Ma, with a peak volcanism at 30 Ma (Hofmann *et al.* 1997). Silicic volcanism occurred early during the main basaltic episode and continued to erupt during the lifetime of the flood basalt (31–26 Ma).

Ethiopian continental flood basalt province

The Cenozoic Ethiopian volcanic province can be subdivided into the main Ethiopian rift (MER), Afar and rift-bounding plateau. The MER and Afar represent the northernmost section of the East African Rift System. Extrusion of flood basalts post-dated extension in the Gulf of Aden at *c.* 35 Ma (Audin *et al.* 2004), but predated that in the Red Sea at *c.* 28 Ma (Ukstins *et al.* 2002). Rifting within the southern and central MER is believed to have occurred between 18 and 15 Ma (WoldeGabriel *et al.* 1990) with extension in the northern MER commencing after *c.* 11 Ma with the development of border faults and half grabens (Ebinger & Casey 2001; Wolfenden *et al.* 2004). Sea-floor spreading commenced at *c.* 17.6 Ma in the Gulf of Aden (Leroy *et al.* 2004; d'Acremont *et al.* 2005) and at *c.* 5 Ma in the Red Sea (Girdler & Styles 1974).

The Ethiopian volcanic plateau is dominated by fissure fed basaltic lavas with subordinate rhyolite forming the volcanic plateaus bounding the current Afar and MER. The rhyolites are found throughout the volcanic stratigraphy as ash layers between basaltic lava flows or as individual eruptive units comparable in volume to individual mafic lava units. They become thicker towards the uppermost part of the lava sequences (Fig. 2). The volcanic plateau is overlain by conspicuous low-angle shield volcanoes whose composition and age match that of the underlying flood basalts and rhyolites (Kieffer *et al.* 2004). Quaternary volcanism, related to local rifts and fractures zones (e.g. volcanism south of Lake Tana, Fig. 1), occurs on the plateau (Hofmann *et al.* 1997). The Quaternary basaltic volcanism is rich in mantle xenoliths and characterized by cinder-cone fields, which suggest that the ascent of the magmas may have involved hydraulic fracturing, permitting rapid rise up through the lithosphere without undergoing significant fractional crystallization or crustal contamination.

Analytical procedure

Sr and Nd isotopic compositions for basalts and rhyolites from north Shewa were determined on a Finnigan mass spectrometer 262 at CRPG in Nancy, France. For Sr and Nd isotopic analyses, about 30 mg of sample powder were leached in cold 5.6% acetic acid for 20 min to remove alteration products that normally concentrate on grain boundaries, and then repeatedly rinsed with ultra-clean water. Leached samples were spiked with isotopic tracers and dissolved in a concentrated $HF-HNO_3-HClO_4$ acid mixture heated at 120 °C for 72 h in a closed Teflon beaker. Sr was extracted by conventional cation exchange techniques using AGX 50W resin with 2.5 N HCl acid and then loaded on W filaments. In high-Rb/Sr, low-Sr rhyolites chemical separation of Rb from Sr was improved by separating the Sr three times on the cation columns. Rare earth elements (REE) were separated on cation exchange columns and extracted with 4.4 N HNO_3; Nd and Sm were subsequently isolated from the other REE using HDEP-coated Teflon resin columns with 0.27 N and 0.45 N HCl, respectively. Nd was then loaded on Re filaments. Measured $^{87}Sr/^{86}Sr$ ratios were corrected for isotope fractionation to $^{86}Sr/^{88}Sr = 0.1194$. Nd isotope ratios were normalized to $^{146}Nd/^{144}Nd = 0.7219$. During this work, repeated analyses of standard solution NBS 987 for Sr gave average values of $^{87}Sr/^{86}Sr = 0.710257 \pm 11$ (2σ). Standard solution J.M. and La Jolla for Nd gave average values of $^{143}Nd/^{144}Nd = 0.511106 \pm 12$ (2σ) and 0.511838 ± 12 (2σ), respectively. Total procedural blanks were typically insignificant. Sr, Rb, Nd and Sm concentrations were determined by the isotope dilution technique.

Results

Whole-rock Rb–Sr geochronology

Rb, Sr, Sm and Nd concentrations and $^{87}Sr/^{86}Sr$ and $^{143}Nd/^{144}Nd$ ratios for bimodal basalt–rhyolite suite from north Shewa are reported in Table 1. In a plot of $^{87}Sr/^{86}Sr$ versus $^{87}Rb/^{86}Sr$ (Fig. 3), all analytical points of the north Shewa's rhyolites lie on a straight line defining an isochron (MSWD = 1.3). The good fit of these samples on

the isochron indicates that they had the same age and initial $^{87}Sr/^{86}Sr$ ratio. The slope of the isochron is related to the age of the comagmatic rhyolites and the value of the initial $^{87}Sr/^{86}Sr$ ratio is given by the y-axis intercept. Accordingly, the regression age of these rhyolites is 20.7 ± 2.4 Ma and the initial $^{87}Sr/^{86}Sr$ ratio is 0.7062 ± 10. The mineral assemblage of the rhyolites includes alkali feldspar and quartz with minor ilmenite and alkali pyroxene. The analysed samples have low values of loss on ignition (LOI < 0.11 wt%, Ayalew & Gibson 2009) which confirm that they are fresh, unaltered rocks. The regression is therefore dating a magmatic intrusion, probably representing the time elapsed since their crystallization from a homogeneous magma. The satisfactory fit of the data points to the isochron indicates that the rhyolites have remained closed to Rb and Sr, and the time required for the crystallization of the magma was relatively short. The initial ratio of the isochron is 0.7062, a very high value for a mantle reservoir. The Rb–Sr age is in close agreement with the K–Ar age of 20.7 ± 0.3 Ma obtained by Coulié et al. (2003) for a sample from the base of the basalt pile, c. 70 km north of the study area around Alem Ketema. This result suggests that volcanism commenced in north-central Ethiopia c. 20 Ma, 10 Ma after initial magmatism in the northern Ethiopian plateau.

Initial $^{87}Sr/^{86}Sr$ and $^{143}Nd/^{144}Nd$ ratios

The rhyolites have $^{87}Sr/^{86}Sr$ ratios that overlap only the higher ends of the range of values measured for north Shewa basalts ($0.70563-0.70658$ v. $0.70440-0.70641$ in the basalts). The rhyolites have overall higher $^{143}Nd/^{144}Nd$ ratios than the basalts ($0.51273-0.51278$ v. $0.51248-0.51274$ in basalts), implying that the rhyolites exhibit more mantle-like isotopic signature than the basalts.

In Figure 4, the Sr and Nd isotopic compositions of north Shewa's basalts and rhyolites are compared with the Oligocene basalts (Pik et al. 1999; Kieffer et al. 2004) and rhyolites (Ayalew et al. 2002; Ayalew & Yirgu 2003) from the northern Ethiopian plateau and the local Neoproterozoic crystalline basement rocks (Teklay et al. 1998). It can be seen from Figure 4a that Miocene basalts from north Shewa lie outside the field of Oligocene basalts, exhibiting higher $^{87}Sr/^{86}Sr$ and lower $^{143}Nd/^{144}Nd$ values. They lie on an extension of the high-Ti2 basalt array into the 'enriched mantle' region of the diagram. Miocene rhyolites plot near the lower range of the Oligocene high-Ti rhyolites, forming a horizontal array between the fields of Oligocene high- and low-Ti rhyolites. They lie above the field of the local basement rocks (Fig. 4b), precluding the involvement of crustal materials in their genesis.

Table 1. Rb, Sr, Sm and Nd concentrations (ppm) and Sr and Nd isotopic ratios of basalts and rhyolites from north Shewa ($^{87}Sr/^{86}Sr_i$ and $^{143}Nd/^{144}Nd_i$ represent initial $^{87}Sr/^{86}Sr$ and $^{143}Nd/^{144}Nd$ ratios, respectively, corrected for in situ radioactive decay of ^{87}Rb and ^{147}Sm since crystallization using 20.7 Ma derived from whole-rock Rb–Sr isochron of the rhyolites)

Sample	Rb	Sr	$^{87}Rb/^{86}Sr$	$^{87}Sr/^{86}Sr$	$^{87}Sr/^{86}Sr_i$	Sm	Nd	$^{147}Sm/^{144}Nd$	$^{143}Nd/^{144}Nd$	$^{143}Nd/^{144}Nd_i$
Basalts										
SA 01A	9	332	0.08	0.704851 ± 22	0.704827	5	25	0.133	0.512741 ± 05	0.512723
SA 04A	27	435	0.18	0.706462 ± 21	0.70641	8	36	0.131	0.512509 ± 11	0.512491
SA 06	6	567	0.03	0.704493 ± 16	0.704485	6	31	0.126	0.512695 ± 08	0.512678
SA 08A	28	486	0.17	0.705428 ± 15	0.705378	8	39	0.128	0.512687 ± 11	0.51267
SA 11B	25	308	0.24	0.705958 ± 24	0.705889	9	41	0.127	0.512609 ± 12	0.512592
SA 14A	17	510	0.10	0.704433 ± 22	0.704404	9	40	0.136	0.512756 ± 06	0.512738
SA 17A	20	440	0.13	0.706453 ± 14	0.706414	8	36	0.130	0.512498 ± 12	0.51248
Rhyolites										
SA 03A	145	23	18.30	0.711510 ± 16	0.706130	23	116	0.117	0.512756 ± 13	0.512741
SA 10A	163	16	29.58	0.715089 ± 65	0.706392	6	168	0.021	0.512784 ± 11	0.512781
SA 10A(2)		15		0.715569 ± 13	0.715569					
SA 20A	148	23	18.51	0.711342 ± 53	0.705901	20	103	0.115	0.512758 ± 07	0.512742
SA 20A(2)		23		0.711411 ± 25	0.711411					
SA 34A	167	15	32.94	0.716169 ± 17	0.706485	21	106	0.120	0.512781 ± 10	0.512765
SA 35A	162	11	41.77	0.717820 ± 07	0.705542	17	102	0.098	0.512744 ± 10	0.512731
SA 36AI	160	13	36.89	0.717349 ± 21	0.706504	9	101	0.052	0.512766 ± 15	0.512759

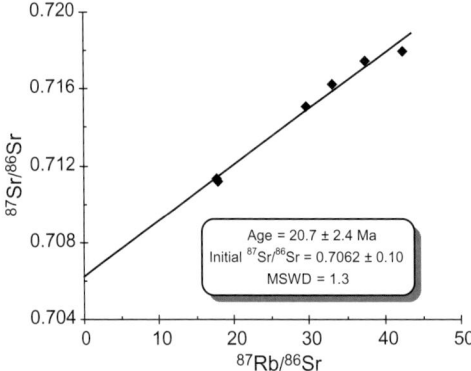

Fig. 3. Whole-rock Rb–Sr model 1 isochron of rhyolites from north Shewa in the central Ethiopian plateau, using the Isoplot program of Ludwing. The regression age indicates that the rhyolites crystallized c. 20.7 ± 2.4 Ma and the whole-rock samples remained closed systems.

Discussion

The focus of this section is to discriminate the effects of mantle versus crustal processes on the north Shewa magma genesis.

The origin of north Shewa basalts

A plot of $^{87}Sr/^{86}Sr$ and $^{143}Nd/^{144}Nd$ ratios of basalts from north Shewa (Fig. 4) forms a negative correlation of $^{87}Sr/^{86}Sr$ and $^{143}Nd/^{144}Nd$, with $^{143}Nd/^{144}Nd$ decreasing and $^{87}Sr/^{86}Sr$ increasing. Such a trend is expected for derivation from enriched mantle source or assimilation/fractional crystallization evolution (e.g. Leeman & Hawkesworth 1986; Hart et al. 1989). The trend toward high Sr and low Nd isotopic values is exhibited by two samples (SA04A and SA17A), which are characterized by high SiO_2 and low MgO contents (Ayalew & Gibson 2009). This trend is clearly compelling evidence for assimilation with concomitant fractional crystallization (AFC) process.

The older terrane in which the Ethiopian plateau developed comprises two principal elements (Cornwell et al. 2010): (1) Neoproterozoic crystalline basement rocks, dominantly low-grade volcano–sedimentary terrane intruded by plutons and ophiolitic remnants (Woldemichael et al. 2009); and (2) carbonate and clastic sedimentary rocks (in excess of 3 km) which range in age from Late Palaeozoic to Cretaceous (Assefa 1991; Russo et al. 1994). In south-eastern Ethiopia, older crust (Pre-Neoprotorozoic high-grade gneissic terrane) that was extensively remobilized during the Pan-African orogenic events (during Neoproterozoic time: 950–450 Ma) is also found (Teklay et al. 1998). The

sediments deposited unconformably on the Proterozoic basement. Due to a maximum thickness of c. 3 km for the sedimentary rocks, the Neoproterozoic rocks are particularly important for assessing the crustal component involved as essential constituents in rhyolite magma genesis or as incidental contaminants in both basaltic and rhyolitic magmas. On a variety of geophysical grounds (Dugda et al. 2005; Maguire et al. 2006; Cornwell et al. 2010), the crust beneath the Ethiopian plateau consists of a mafic lower crust overlain by a more felsic upper crust. Mafic underplate of c. 15 km thickness exists beneath the lower crust (Maguire et al. 2006).

In order to quantify the effect of contamination on the isotopic composition of Nd and Sr, AFC modelling (De Paola 1981) is illustrated in Figure 5. The isotopic compositions and concentrations of Nd and Sr for the end members are listed in Table 2. Two AFC curves are formed of a mixture of the least-evolved contaminated basalt (SA14A) with felsic and mafic rocks, derived from the local Neoproterozoic crystalline basement rocks. They are constructed to account for a very wide range of isotope ratios and concentrations of Nd and Sr in contaminated basalts. AFC modelling yields f values up to 0.5 (where f is the weight fraction of the residual magmas).

The crystallizing assemblage used in the model consisted of olivine, clinopyroxene, plagioclase, Fe–Ti oxides and apatite, which is fully consistent with the phenocryst assemblage observed in the associated flood basalts (Ayalew & Gibson 2009). The resulting curve of mafic crustal rock fits the basaltic data reasonably well and suggests that some of the north Shewa basalts demand up to 15% contamination. It should be noted that this value is probably a maximum estimate for the amount of contamination, because the parental basalt considered appears to have undergone contamination itself. The rhyolites define a flat trend lying outside the AFC curves, indicating that they contain little or no crustal components. The AFC model is in a very good agreement with the major and trace elements geochemistry of the basalts. Ayalew & Gibson (2009) demonstrated that the more evolved basalts show a minor enrichment in Ba and have small troughs at Nb and Ta, which have been interpreted to reflect contamination during fractional crystallization.

The least contaminated sample in the Miocene basalts might provide the clearest information of the mantle signature. As shown in Figure 4a, the Miocene basalts lie outside the fields of the Oligocene basalts and are displaced in an elongated field that trends towards the relatively high $^{87}Sr/^{86}Sr$ and low $^{143}Nd/^{144}Nd$ values. They can be extrapolated back to that of the high-Ti2 basalts (Fig. 4a), which are thought to be representative of

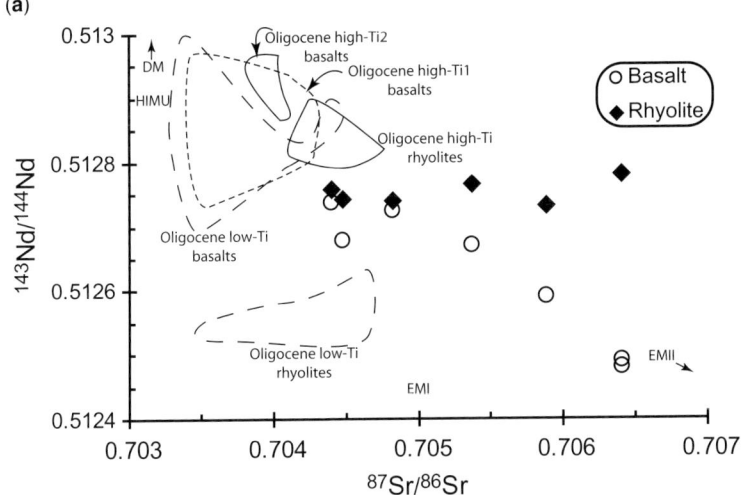

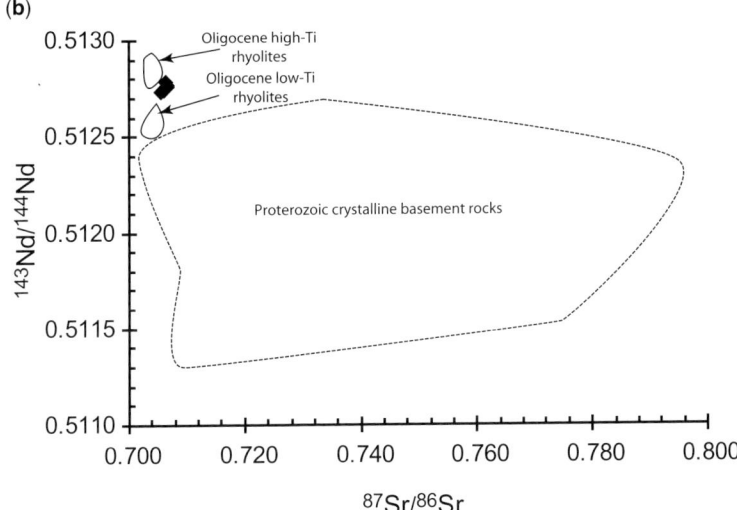

Fig. 4. $^{143}Nd/^{144}Nd$ versus $^{87}Sr/^{86}Sr$ for the north Shewa basalt–rhyolite suite. Also shown for comparisons are fields for (**a**) Oligocene flood basalts (Pik *et al.* 1999; Kieffer *et al.* 2004) and rhyolites (Ayalew *et al.* 2002; Ayalew & Yirgu 2003), and (**b**) Neoproterozoic crystalline basement rocks (Teklay *et al.* 1998). Included for comparison (a) are locations for the accepted values for the main documented mantle reservoirs (DM, HIMU, EMI and EMII) defined by Zindler & Hart (1986).

the primitive mantle intrinsic to the Afar mantle plume (Marty *et al.* 1996; Pik *et al.* 1999). Ayalew & Gibson (2009) interpreted trace element ratios in Miocene basalts in terms of their derivation as a result of a small (5%) amount of melting of the Afar mantle plume. We propose that the Miocene basalts in north Shewa originated from Afar plume source as high-Ti2 basalts and were then contaminated to varying extents during their ascent through the overlying crust.

The origin of north Shewa rhyolites

The $^{87}Sr/^{86}Sr$ and $^{143}Nd/^{144}Nd$ ratios of the rhyolites from north Shewa (Fig. 4) imply that the rhyolites are not likely to be crustal anatectic melts, thus strengthening the fact for entirely mantle contributions to rhyolite genesis. Furthermore, the rhyolites cannot be primary melts derived from basic intrusives or their derivatives associated with the Ethiopian CFB province, because they have very

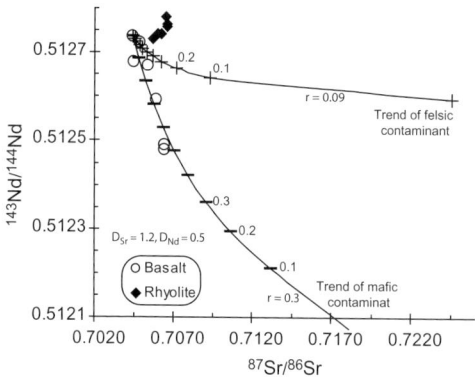

Fig. 5. $^{143}Nd/^{144}Nd$ versus $^{87}Sr/^{86}Sr$ assimilation fractional crystallization (AFC) models for contamination of north Shewa basalts. Solid curves represent AFC models constructed based on the De Paola (1981) equation. Tick marks at interval of 0.1 indicate proportion of the residual magma along each curve. r value is the ratio of the rate of assimilation to the rate of crystallization. D_{Sr} and D_{Nd} represent assumed bulk partition coefficients for Sr and Nd, respectively.

composition from mafic through intermediate to felsic plutons) from Ethiopia have present-day $^{87}Sr/^{86}Sr$ and $^{143}Nd/^{144}Nd$ ratios which overlap significantly those of the fields of north Shewa rhyolites and other Ethiopian rhyolites (e.g. Woldemichael *et al.* 2009). However, all of the intrusions display a marked trough at Nb–Ta in their mantle-normalized trace element variation diagram (Woldemichael *et al.* 2009), reflecting subduction-modified mantle in their source. Such an Nb–Ta trough is not observed in the spider pattern of north Shewa rhyolites, which display a peak at Nb–Ta (Ayalew & Gibson 2009). This contrast clearly demonstrates that the north Shewa rhyolites cannot be derived from partial melting of the Neoproterozoic plutons of any composition.

As observed in Figure 2, the rhyolite layers are found interlayered with basalt layers. This suggests that a rhyolite layer could be genetically related to the underlying basalt layer, the overlying basalt layer or both. Hence, the rhyolitic magmas must have evolved by fractional crystallization from mantle-derived basaltic magmas similar in composition to the exposed flood basalts.

low Sr contents (11–23 ppm) which require extensive feldspar fractionation (e.g. Halliday *et al.* 1991; Ayalew & Ishiwatari 2011). This implies that the rhyolites evolved by fractional crystallization of mantle-derived basaltic magmas.

The lack of intermediate compositions in the volcanic rock suite appears to argue against the fractional crystallization model. Partial melting of mantle-derived basic intrusives or their derivatives may also initiate partial melting of the local crust, generating silicic magmas with isotopic signatures representing mixtures of two components (one crust and the other mantle). Such signatures are not observed in the rhyolites of north Shewa. The radiogenic isotopic characteristics of the north Shewa rhyolites therefore do not support an origin of the rhyolites by partial melting of any source (crustal rocks or mantle-derived igneous rocks).

It is worth mentioning that the isotopic compositions of the Neoproterzoic intrusions (ranging in

Temporal isotopic variations of Ethiopian plateau basalts and rhyolites

The temporal variations of Sr–Nd isotopes of the Ethiopian flood basalts and rhyolites (dated at 30 and 20 Ma) from the northern Ethiopian volcanic plateau are illustrated in Figure 6. For simplicity and for the purpose of our discussion we have concentrated on the Oligocene high-Ti flood basalts and rhyolites. This is because these rocks are found in the same geographic setting as that of the north Shewa basalts and rhyolites. Sr contents (11–23 ppm) in the north Shewa rhyolites are extremely low and hence their Sr isotopic ratio is highly susceptible to even slight modification. In contrast, north Shewa basalts have concentrations of Nd (25–41 ppm) that are lower than that of most potential contaminants. Such low Nd concentrations render the isotopic ratios of Nd more sensitive to crustal contamination. For this purpose, we

Table 2. *Isotopic compositions and concentrations of Nd and Sr in crustal rocks and basalt (Basalt from Table 1, crystalline basement rocks from Teklay et al. 1998)*

	Contaminants		Mantle
	Mafic ET6	Felsic ET9	Basalt SA14A
Nd (ppm)	36.99	57.56	40
$^{143}Nd/^{144}Nd$	0.511414	0.512311	0.512738
Sr (ppm)	271	79	510
$^{87}Sr/^{86}Sr$	0.719783	0.769720	0.704404

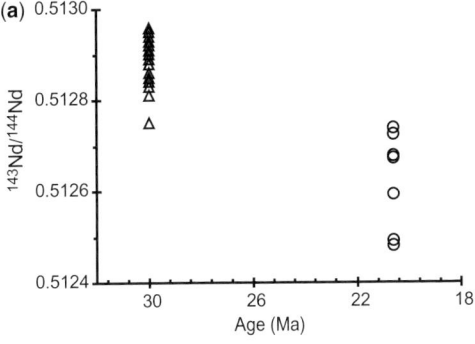

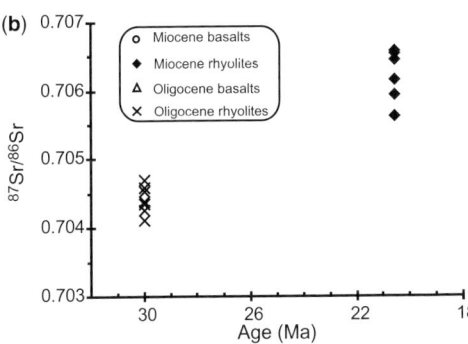

Fig. 6. Temporal variation of (**a**) Nd isotopic ratios of basalts and (**b**) Sr isotopic ratios of rhyolites from northern Ethiopia volcanic plateau. Data sources: Pik *et al.* (1999) for Oligocene basalts and Ayalew & Yirgu (2003) for Oligocene rhyolites.

focused on the variations of $^{143}Nd/^{144}Nd$ ratios in the basalts and $^{87}Sr/^{86}Sr$ ratios in the rhyolites to investigate the involvement of crustal materials in the generation of north Shewa magmas. All of the north Shewa basalts contain Nd isotopic ratios much lower than the high-Ti Oligocene basalts (Fig. 6a). The north Shewa basalts have $^{87}Sr/^{86}Sr$ ratios much greater than those of the Oligocene basalts. The $^{143}Nd/^{144}Nd$ ratios in north Shewa rhyolites are lower than those of the high-Ti Oligocene rhyolites. North Shewa rhyolites have $^{87}Sr/^{86}Sr$ ratios greater than those of the Oligocene rhyolites (Fig. 6b).

The observed Sr and Nd isotopic variations of basalts and rhyolites seemingly represent increasing contributions with time from crustal materials. These features appear to be controlled by the stage of rifting and rate of magma addition. In the earliest stage (*c.* 30 Ma) the crust was unstretched and cold. Consequently, the amount of assimilation of colder crustal rocks was limited. Subsequently, the high heat flows generated as a consequence of emplacement of large volumes of high-temperature basaltic magma into the lithosphere during rifting may

locally elevate crustal temperatures and so promote contamination processes, resulting in the generation of high $^{87}Sr/^{86}Sr$ and low $^{143}Nd/^{144}Nd$ rocks. This would lead to thermal weakening of the crust, resulting in an increase in the rate of contamination with time. Combining the geochemical data with recently acquired geophysical findings (seismic, gravity and magnetotelluric data) allows us to determine how magmatism modified the thermomechanical properties of the Neoproterozoic continental crust in the East African Rift System.

Conclusions

The northwest Ethiopian volcanic plateau which represents the so-called the 'Ethiopian flood basalt province' evolved during at least two major pulses: at 30 Ma in the northern plateau and at 20 Ma in central plateau. Our study suggests that volcanism commenced in north-central Ethiopia c. 20 Ma, 10 Ma after initial magmatism in the northern Ethiopian plateau. The rhyolites evolved by crystal fractionation processes from mantle-derived basaltic magmas similar in composition to the exposed flood basalts. The north Shewa basalts developed from contamination of mantle-derived Ti2-type magma (i.e. Afar plume) with a mafic crustal component. The isotopic compositions of north Shewa basalts and rhyolites indicate the increasing involvement of crustal materials in the evolution of north Shewa magmas with time, probably as a consequence of thermal weakening of the crust triggered by repeated injections of hot basaltic magmas in the lithosphere during rifting.

I am grateful to Addis Ababa University and the Ethiopian Ministry of Science and Technology for financial support. The author is indebted to C. Zimmermann and L. Reisberg for their assistance during the course of the analyses. I thank the two anonymous journal reviewers and the Corresponding Editor of GSL Books D. Van Hinsbergen for their constructive comments that improved the quality of the manuscript. I am also extremely grateful to S. Ali for his assistance during fieldwork.

References

ARNDT, N. T., CZAMANSK, G. K., WOODEN, J. L. & FEDORENKO, V. A. 1993. Mantle and crustal contributions to continental flood volcanism. *Tectonophysics*, **223**, 39–52.

ASSEFA, G. 1991. Lithostratigraphy and environment of deposition of the late Jurassic–early Cretaceous sequence of the central part of northwestern plateau, Ethiopia. *Neues Jahrbuch für Geologie und Paläontologie, Abhandlungen*, **182**, 255–284.

AUDIN, L., QUIDELLEUR, X. ET AL. 2004. Paleomagnetism and K–Ar and $^{40}Ar/^{39}Ar$ ages in the Ali Sabieh area (Republic of Djibouti and Ethiopia): constraints on the mechanism of Aden ridge propagation into south-eastern Afar during the last 10 Myr. *Geophysical Journal International*, **158**, 327–345.

AYALEW, D. & YIRGU, Y. 2003. Crustal contribution to the genesis of Ethiopian plateau rhyolitic ignimbrites: basalt and rhyolite geochemical provinciality. *Journal of the Geological Society, London*, **160**, 47–56.

AYALEW, D. & GIBSON, I. S. 2009. Head-to-tail transition of the Afar mantle plume: geochemical evidence from a Miocene bimodal basalt–rhyolite succession in the Ethiopian large igneous province. *Lithos*, **112**, 461–476.

AYALEW, D. & ISHIWATARI, A. 2011. Comparison of rhyolites from continental rift, continental arc and oceanic island arc: implication for the mechanism of silicic magma generation. *Island Arc*, **20**, 78–93.

AYALEW, D., BARBEY, P., MARTY, B., REISBERG, L., YIRGU, G. & PIK, R. 2002. Source, genesis and timing of giant ignimbrite deposits associated with Ethiopian continental flood basalts. *Geochimica et Cosmochimica Acta*, **66**, 1429–1448.

AYALEW, D., EBINGER, C., BOURDON, E., WOLFENDEN, E., YIRGU, G. & GRASSINEAU, N. 2006. Temporal compositional variation of syn-rift rhyolites along the western margin of the southern Red Sea and northern main Ethiopian rift. *In*: YIRGU, G., EBINGER, C. J. & MAGUIRE, P. K. H. (eds) *The Afar Volcanic Province Within the East African Rift System*. Geological Society, London, Special Publications, **259**, 123–132.

BAKER, J. A., THIRLWALL, M. F. & MENZIES, M. A. 1996. Sr–Nd–Pb isotopic and trace element evidence for crustal contamination of plume-derived flood basalts: Oligocene flood volcanism in western Yemen. *Geochimica et Cosmochimica Acta*, **60**, 2559–2581.

BARBEY, P., AYALEW, D. & YIRGU, G. 2005. Insight into the origin of gabbro-dioritic cumulophyric aggregates from silicic ignimbrites: Sr and Ba zoning profiles of plagioclase phenocrysts from Oligocene Ethiopian plateau rhyolites. *Contributions to Mineralogy and Petrology*, **149**, 233–245.

BECCALUVA, L, BIANCHINI, G., NATALI, C. & SIENA, F. 2009. Continental flood basalts and mantle plumes: a case study of the northern Ethiopian plateau. *Journal of Petrology*, **50**, 1377–1403.

CAMPBELL, I. H. & GRIFFITHS, R. W. 1990. Implications of mantle plume structure for the evolution of flood basalts. *Earth and Planetary Science Letters*, **99**, 79–93.

CLEVERLY, R. W., BETTON, P. J. & BRISTOW, J. W. 1984. Geochemistry and petrogenesis of the volcanic rocks of the Lebombo rhyolites. *Geological Society of South Africa, Special Publications*, **13**, 171–195.

CORNWELL, D. G., MAGUIRE, P. K. H., ENGLAND, R. W. & STUART, G. W. 2010. Imaging detailed crustal structure and magmatic intrusion across the Ethiopian rift using a dense linear broadband array. *Geochemistry Geophysics and Geosystems*, **11**, doi: 10.1029/2009GC002637.

COULIÉ, E., QUIDELLEUR, X., GILLOT, P. Y., COURTILLOT, V., LEFÈVRE, J. C. & CHIESA, S. 2003. Comparative K–Ar and Ar–Ar dating of Ethiopian and Yemen Oligocene volcanism: implications for timing and duration of the Ethiopian traps. *Earth and Planetary Science Letters*, **206**, 477–492.

D'ACREMONT, E., LEROY, S. *ET AL*. 2005. Structure and evolution of the eastern Gulf of Aden conjugate margins from seismic reflection data. *Geophysics Journal International*, **160**, 869–890.

DE PAOLA, D. J. 1981. Trace element and isotopic effects of combined wall rock assimilation and fractional crystallization. *Earth and Planetary Sciences Letters*, **53**, 189–202.

DUGDA, M. T., NYBLADE, A. A., JULIA, J., LANGSTON, C. A., AMMON, C. J. & SIMIYU, S. 2005. Crustal structure in Ethiopia and Kenya from receiver function analysis: implications for rift development in eastern Africa. *Journal of Geophysical Research*, **110**, B01303, doi: 10.1029/2004JB003065.

DUNCAN, R. A., HOOPER, P. R., REHACEK, J., MARSH, J. S. & DUNCAN, A. R. 1997. The timing and duration of the Karoo igneous event, southern Gondwana. *Journal of Geophysical Research*, **102**, 18127–18138.

EBINGER, C. J. & SLEEP, N. H. 1998. Cenozoic magmatism throughout East Africa resulting from impact of a single plume. *Nature*, **395**, 788–791.

EBINGER, C. & CASEY, M. 2001. Continental breakup in magmatic provinces: an Ethiopian example. *Geology*, **29**, 527–530.

ERLANK, A. J., MARSH, J. S., DUNCAN, A. R., MILLER, R. M. & HAWKESWORTH, C. J. 1984. Geochemistry and petrogenesis of the Etendeka volcanic rocks from SWA/Namibia. *Geological Society of South Africa, Special Publications*, **13**, 195–247.

FURMAN, T., BRYCE, J. G., ROONEY, T., HANAN, B. B., YIRGU, G. & AYALEW, D. 2006. Heads and tails: 30 million years of the Afar plume. *In*: YIRGU, G., EBINGER, C. J. & MAGUIRE, P. K. H. (eds) *The Structure and Evolution of the East African Rift System in the Afar Volcanic Province*. Geological Society, London, Special Publications, **259**, 123–132.

GARLAND, F., HAWKESWORTH, C. J. & MANTOVANI, M. S. M. 1995. Description and petrogenesis of the Paraná rhyolites, southern Brazil. *Journal of Petrology*, **36**, 1193–1227.

GEORGE, R., ROGERS, N. & KELLEY, S. 1998. Earliest magmatism in Ethiopia: evidence for two mantle plumes in one flood basalt province. *Geology*, **26**, 923–926.

GIBSON, S. A., THOMPSON, R. N. & DAY, J. A. 2006. Timescales and mechanisms of plume–lithosphere interactions: $^{40}Ar/^{39}Ar$ geochronology and geochemistry of alkaline igneous rocks from the Paraná–Etendeka large igneous province. *Earth and Planetary Science Letters*, **251**, 1–17.

GIRDLER, R. W. & STYLES, P. 1974. Two stage Red Sea floor spreading. *Nature*, **247**, 1–11.

HALLIDAY, A. N., DAVIDSON, J. P., HILDRETH, W. & HOLDEN, P. 1991. Modeling the petrogenesis of high Rb/Sr silicic magmas. *Chemical Geology*, **92**, 107–114.

HARRIS, C., WHITTINGHAM, A. M., MILNER, S. C. & ARMSTRONG, R. A. 1990. Oxygen isotope geochemistry of the silicic volcanic rocks of the Etendeka–Paraná province: source constraints. *Geology*, **18**, 1119–1121.

HART, W. K., WOLDEGABRIEL, G., WALTER, R. C. & MERTZMAN, S. A. 1989. Basaltic volcanism in Ethiopia: constraints on continental rifting and mantle interactions. *Journal of Geophysical Research*, **94**, 7731–7748.

HILL, R. I. 1991. Starting plumes and continental break-up. *Earth and Planetary Science Letters*, **104**, 398–416.

HOFMANN, C., COURTILLOT, V., FÉRAUD, G., ROCHETTE, P., YIRGU, G., KETEFO, E. & PIK, R. 1997. Timing of the Ethiopian flood basalt event and implications for plume birth and global change. *Nature*, **389**, 838–841.

KIEFFER, B., ARNDT, N. ET AL. 2004. Flood and shield basalts from Ethiopia: magmas from the African super-swell. *Journal of Petrology*, **45**, 793–834.

KIRSTEIN, L., PEATE, D. W., HAWKESWORTH, C. J., TURNER, S. P., HARRIS, C. & MANTOVANI, M. S. M. 2000. Early Cretaceous basaltic and rhyolitic magmatism in southern Uruguay associated with the opening of the South Atlantic. *Journal of Petrology*, **41**, 1413–1438.

LEEMAN, W. P. & HAWKESWORTH, C. J. 1986. Open magmatic systems – some isotopic and trace element constraints. *Journal of Geophysical Research*, **91**, 5901–5912.

LEROY, S., GENTE, P. ET AL. 2004. From rifting to spreading in the eastern Gulf of Aden: a geophysical survey of a young oceanic basin from margin to margin. *Terra Nova*, **16**, 185–192.

LIGHTFOOT, P. C., HAWKESWORTH, C. J. & SETHNA, S. F. 1987. Petrogenesis of rhyolites and trachytes from the Deccan trap: Sr, Nd and Pb isotope and trace element evidence. *Contributions to Mineralogy and Petrology*, **95**, 44–54.

MAGUIRE, P. K. H., KELLER, G. R. ET AL. 2006. Crustal structure of the northern main Ethiopian rift from the EAGLE controlled source survey: a snapshot of incipient lithospheric break-up. *In*: YIRGU, G., EBINGER, C. J. & MAGUIRE, P. K. H. (eds) *The Afar Volcanic Province Within the East African Rift System*. Geological Society, London, Special Publications, **259**, 269–291.

MAHONEY, J. J., SAUNDERS, A. D., STOREY, M. & RANDRIAMANANTENASOA, A. 2008. Geochemistry of the volcan de l'Androy basalt–rhyolite complex, Madagascar Cretaceous igneous province. *Journal of Petrology*, **49**, 1069–1096.

MARTY, B., PIK, R. & GEZAHEGN, Y. 1996. Helium isotopic variations in Ethiopian plume lavas: nature of magmatic sources and limit on lower mantle contribution. *Earth Planetary Science Letters*, **144**, 223–237.

MARZOLI, A., RENNE, P. R., PICCIRILLO, E. M., ERNESTO, M., BELLIENI, G. & DE MIN, A. 1999. Extensive 200-million-year-old continental flood basalts of the Central Atlantic Magmatic Province. *Science*, **284**, 616–618.

MESHESHA, D. & SHINJO, R. 2007. Crustal contamination and diversity of magma sources in the northwestern Ethiopian volcanic province. *Journal of Mineralogical and Petrological Sciences*, **102**, 272–290.

MILLER, J. A. & HARRIS, C. 2007. Petrogenesis of the Swaziland and northern Natal rhyolites of the Lebombo rifted volcanic margin, south east Africa. *Journal of Petrology*, **48**, 185–218.

PIK, R., DENIEL, C., COULON, C., YIRGU, G. & MARTY, B. 1999. Isotopic and trace element signatures of Ethiopian flood basalts: evidence for plume–lithosphere interactions. *Geochimica et Cosmochimica Acta*, **63**, 2263–2279.

PIK, R., MARTY, B., CARIGNAN, J., YIRGU, G. & AYALEW, T. 2008. Timing of east African rift development in southern Ethiopia: implication for mantle plume activity and evolution of topography. *Geology*, **36**, 167–170.

RILEY, T. R., LEAT, P. T., PANKHURST, R. J. & HARRIS, C. 2001. Origins of large volume rhyolitic volcanism in the Antarctic Peninsula and Patagonia by crustal melting. *Journal of Petrology*, **42**, 1043–1065.

ROCHETTE, P., TAMRAT, E. ET AL. 1998. Magnetostratigraphy and timing of the Oligocene Ethiopian traps. *Earth and Planetary Science Letters*, **164**, 497–510.

ROONEY, T., FURMAN, T., BASTOW, I., AYALEW, D. & YIRGU, G. 2007. Lithospheric modification during crustal extension in the main Ethiopian rift. *Journal of Geophysical Research*, **112**, B10201, doi: 10.1029/2006JB004916.

RUSSO, A., ASSEFA, G. & ATNAFU, B. 1994. Sedimentary evolution of the Abby River (Blue Nile) Basin, Ethiopia. *Neues Jahrbuch für Geologie und Palaëontologie Monatshefte*, **5**, 291–308.

STOREY, B. C. 1995. The role of mantle plumes in continental breakup: case histories from Gondwanaland. *Nature*, **377**, 301.

TEKLAY, M., KRÖNER, A., MEZGER, K. & OBERHÄNSLI, R. 1998. Geochemistry, Pb–Pb single zircon ages and Nd–Sr isotopic compositions of Precambrian rocks from southern and eastern Ethiopia: implications for crustal evolution in east Africa. *Journal of African Earth Sciences*, **26**, 207–227.

THOMPSON, R. N., DICKIN, A. P., GIBSON, I. L. & MORRISON, M. A. 1982. Elemental fingerprints of isotopic contamination of Hebridean Palaeocene mantle-derived magmas by Archaean sial. *Contributions to Mineralogy and Petrology*, **79**, 159–168.

TURNER, S., REGELOUS, M., KELLEY, S., HAWKESWORTH, C. J. & MONTOVANI, M. S. M. 1994. Magmatism and continental break-up in the south Atlantic: high precision $^{40}Ar/^{39}Ar$ geochronology. *Earth and Planetary Sciences Letters*, **121**, 333–348.

UKSTINS, I. A., RENNE, P. R., WOLFENDEN, E., BAKER, J., AYALEW, D. & MENZIES, M. 2002. Matching conjugate rifted margins: $^{40}Ar/^{39}Ar$ chrono-stratigraphy of pre- and syn-rift bimodal flood volcanism in Ethiopia and Yemen. *Earth and Planetary Science Letters*, **198**, 289–306.

UKSTINS-PEATE, I., BAKER, J. A. ET AL. 2005. Volcanic stratigraphy of large-volume silicic pyroclastic eruptions during Oligocene Afro-Arabian flood volcanism in Yemen. *Bulletin of Volcanology*, **68**, 135–156.

WHITE, R. S. & MCKENZIE, D. 1995. Mantle plume and flood basalts. *Journal of Geophysical Research*, **100**, 17,543–17,585.

WOLDEGABRIEL, G., ARONSON, J. L. & WALTER, R. C. 1990. Geology, geochronology, and rift basin development in the central sector of the main Ethiopia rift. *Geological Society of America Bulletin*, **102**, 439–458.

WOLDEMICHAEL, B. W., KUMURA, J. I., DUNKLEY, D. J., TANI, K. & OHIRA, H. 2009. SHRIMP U–Pb zircon geochronology and Sr–Nd isotopic systematic of the Neoproterozoic Ghimbi-Nedjo mafic to intermediate intrusions of the western Ethiopia: a record of passive margin magmatism at 855 Ma? *International Journal of Earth Sciences*, doi: 10.1007/s00531-009-0481-x.

WOLFENDEN, E., EBINGER, C., YIRGU, G., DEINO, A. & AYALEW, D. 2004. Evolution of the northern main Ethiopian rift: birth of a triple junction. *Earth and Planetary Science Letters*, **224**, 213–228.

XU, Y. G., LUO, Z. Y., HUANG, X. L., HE, B., XIAO, L., XIE, L. W. & SHI, Y. R. 2008. Zircon U–Pb and Hf isotope constraints on crustal melting associated with the Emeishan mantle plume. *Geochimica et Cosmochimica Acta*, **72**, 3084–3104.

ZINDLER, A. & HART, S. R. 1986. Chemical geodynamics. *Annual Review of Earth and Planetary Sciences*, **14**, 493–571.

Geochemistry of 24 Ma basalts from NE Egypt: source components and fractionation history

CHIRA ENDRESS[1,2], TANYA FURMAN[1]*, MOHAMED ALI ABU EL-RUS[3] &
BARRY B. HANAN[4]

[1]*Department of Geosciences, Pennsylvania State University,
University Park, PA 16802, USA*

[2]*Department of Instruction and Learning, University of Pittsburgh,
Pittsburgh PA 15260, USA*

[3]*Geology Department, Assiut University, Assiut 71516, Egypt*

[4]*Department of Geological Sciences, San Diego State University,
San Diego CA 92182, USA*

Corresponding author (e-mail: furman@psu.edu)

Abstract: Subalkaline basalts from NE Egypt represent an episode of magmatism at *c.* 24 Ma, coincident with widespread eruptive activity in northern Africa. New geochemical data provide insight into the mineralogical and isotopic characteristics of the underlying mantle. The basalts show little geochemical variation, with incompatible trace element abundances similar to those of ocean island basalts. They display fairly smooth primitive mantle-normalized incompatible trace element patterns. Trace element abundances and Sr–Nd–Pb–Hf isotopic signatures are consistent with contributions from two distinct source regions, one similar to the Afar plume and the other located within the metasomatized spinel-facies subcontinental lithosphere. Mixing of melts from these two domains was followed by minor crustal contamination during prolonged ascent or emplacement. Integrating the geochemical data with available tomographic information allows us to develop a framework for understanding mid-Tertiary magmatic activity throughout northern Africa. A model for this widespread volcanism involves ascent of upwelling mantle derived from the margins of the South African Superplume rooted at the core–mantle boundary and/or through small-scale convection at the 660 km discontinuity. Ascent of magmas to the surface was facilitated by pre-existing structures within the lithosphere, including those associated with incipient rifting of the Red Sea.

Supplementary material: Mineral chemistry data are available at http://www.geolsoc.org.uk/SUP18483.

Studies of the East African Rift System (Fig. 1) provide insight into magmatic and tectonic processes that occur in a region where extension initiated by a mantle plume has led to the formation of new sea floor. This system includes two ocean basins initiated *c.* 22–25 Ma (Red Sea, Gulf of Aden) and the 3000 km long East African Rift. The NE Egyptian basalts studied here (Fig. 1) were emplaced *c.* 24 Ma (Meneisy 1990) to the west of the northernmost terminus of the Red Sea within a 70 km radius of Cairo. Seismic data and logs of available wells (Williams & Small 1984) indicate the basalt flows continue northwards in the subsurface, reaching thicknesses of 80 m in an area some 60 km west of the Nile delta.

The age of the basalts places them within an important time period of tectonic and magmatic activity in Africa. At *c.* 30 Ma, volcanic activity attributed to the Afar plume produced sequences of flood basalts over 4 km thick in Ethiopia and Yemen (Baker *et al.* 1996; Hofmann *et al.* 1997; Pik *et al.* 1999; Keiffer *et al.* 2004); modern magmatism in Afar and the Main Ethiopian Rift (Barrat *et al.* 1998; Furman *et al.* 2006*a*; Rooney *et al.* 2008) represents the tail of this plume. To the south, volcanic activity began in Turkana, Kenya at *c.* 45 Ma and has continued episodically through the present (George *et al.* 1998; Furman *et al.* 2004, 2006*b*). Perhaps as early as 26 Ma, a major volcanic pulse in Turkana coincided with rift propagation north into Ethiopia and south into Kenya (Furman *et al.* 2006*b*). During this time, magmatism occurred across northern Africa in Hoggar, Algeria (Aït-Hamou *et al.* 2000), Gharyan, Libya (Aboazom *et al.* 2006; Farahat *et al.* 2006; Beccaluva *et al.* 2008), Darfur, Sudan (Franz *et al.* 1999; Lucassen

From: VAN HINSBERGEN, D. J. J., BUITER, S. J. H., TORSVIK, T. H., GAINA, C. & WEBB, S. J. (eds) *The Formation and Evolution of Africa: A Synopsis of 3.8 Ga of Earth History.* Geological Society, London, Special Publications, **357**, 265–283. DOI: 10.1144/SP357.14 0305-8719/11/$15.00 © The Geological Society of London 2011.

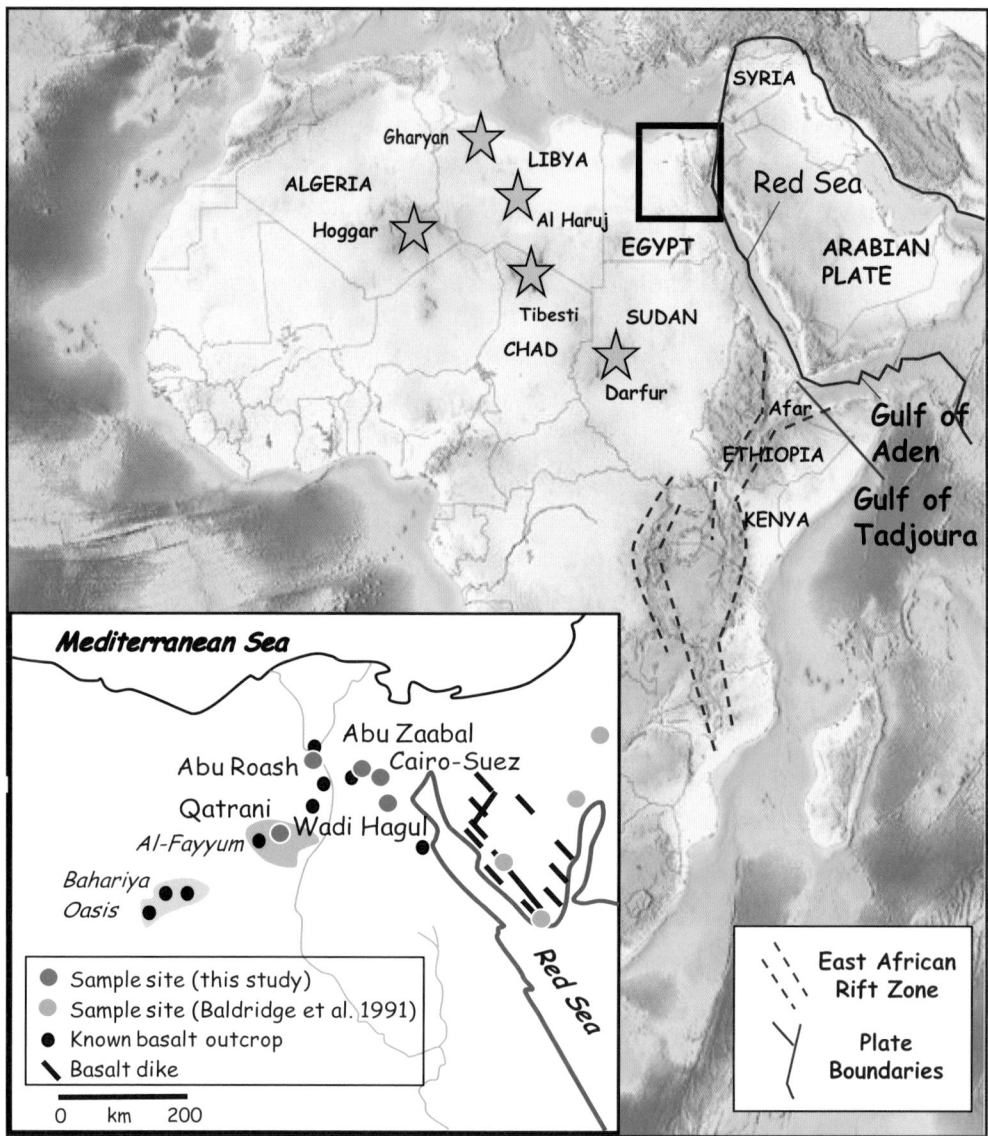

Fig. 1. Map of the East African Rift System showing Miocene and younger volcanic areas; inset shows sample sites for this work as discussed in the text (after Meneisy 1990).

et al. 2008), Tibesti, Chad (Wilson & Guiraud 1992; Pik *et al.* 2006) and in northern Egypt and the Sinai Peninsula (Abdel-Monem & Heikel 1981; Baldridge *et al.* 1991; Moghazi 2003). Rifting of the Red Sea (Coleman & McGuire 1988; Baldridge *et al.* 1991) was initiated and sustained during this widespread volcanic episode.

Our geochemical work fills an important gap in an area of limited prior geochemical and structural analysis. Studies in the Red Sea region have focused on volcanic rocks younger than 13 Ma on the Arabian coast of the Red Sea (Bertrand *et al.* 2003; Shaw *et al.* 2003; Krienitz *et al.* 2006, 2009), and to a lesser degree on basaltic dikes exposed on the Sinai Peninsula (Baldridge *et al.* 1991). We present new geochemical data that provide insight into the thermal, isotopic and mineralogical characteristics of the mantle source region from which the NE Egyptian basalts were derived, and evaluate the depth and degree of melting and fractionation as well as the extent of interaction between the evolving magma and the surrounding crust. These results are evaluated within the context of the regional magmatic environment to develop a broad framework for understanding the widespread tectono-magmatic activity throughout northern Africa since the Late Eocene.

Regional geodynamic setting

The tectonic history of northern Africa is complex. The main crustal units were assembled between 900 and 550 Ma during the Pan African Orogeny as portions of northern and southern Rodinia collided with the Congo Craton during the formation of Gondwana (Stern *et al.* 1994; Coward & Ries 2003; Guiraud *et al.* 2005). In the Mid-Jurassic a set of east–west trending normal faults developed across northern Egypt, likely associated with formation of the Antalya Ocean basin (Schandelmeier & Reynolds 1997; Guiraud & Bosworth 1999). Dextral shear along existing (presumably Phanerozoic) structures created pull-apart basins in NE Libya and northern Egypt that extended into the new Antalya Ocean as transform faults (Schandelmeier & Reynolds 1997; Guiraud *et al.* 2005). Basin formation and subsidence persisted in this region through the Early Eocene (Guiraud & Bosworth 1999); some of the normal faults in northern Egypt were subsequently reactivated as reverse or strike–slip faults (Guiraud & Bosworth 1999). Egyptian basalts, including those studied in this work, are found north of 28°N on both sides of the Nile River; K–Ar dating indicates they were emplaced 16–28 Ma (Fig. 1; Abdel-Monem & Heikel 1981; Meneisy 1990; Schandelmeier & Reynolds 1997). In Sinai, tholeiitic basaltic dikes and flows were emplaced *c.* 18–31 Ma along fissures oriented parallel to the Red Sea (Baldridge *et al.* 1991; Schandelmeier & Reynolds 1997), the dominant modern tectonic feature of the region.

Broadly distributed mafic magmatism began across northern Africa *c.* 35 Ma in Hoggar (Algeria), Gharyan and Al Haruj (Libya), Tibesti (Chad) and Darfur (Sudan) as well as in northern Egypt (Abdel-Monem & Heikel 1981; Ochieng' *et al.* 1988; Baldridge *et al.* 1991; George *et al.* 1998; Franz *et al.* 1999; Aït-Hamou *et al.* 2000; Moghazi 2003; Aboazom *et al.* 2006; Farahat *et al.* 2006; Beccaluva *et al.* 2008; Lucassen *et al.* 2008; Fig. 1). Oligocene through Pliocene basaltic volcanism also occurred across the Arabian Plate in Syria, Jordan, Saudi Arabia and Yemen (Baldridge *et al.* 1991; Bertrand *et al.* 2003; Shaw *et al.* 2003; Krienitz *et al.* 2006, 2009). Contemporaneous with the diffuse Tertiary volcanic activity in northern Africa, voluminous mafic magmatism in Ethiopia and Kenya was associated with initiation of the East African Rift System (Chazot & Bertrand 1993; Baker *et al.* 1996; Hofmann *et al.* 1997; Pik *et al.* 1998, 1999; Keiffer *et al.* 2004). This activity is attributed to the arrival of the Afar mantle plume head beneath the Nubian shield leading to eventual crustal rupture, the onset of sea-floor spreading in two of the three rift arms and the subsequent separation of Nubia from Arabia. Volcanism continued sporadically in the Afar region from 22 Ma to the present day in the Main Ethiopian Rift as well as in Djibouti, Eritrea and Erta 'Ale (Deniel *et al.* 1994; Barrat *et al.* 1998; Furman *et al.* 2006a). Extension in the Red Sea commenced at 22–25 Ma and the Gulf of Suez formed at the northwestern tip of the Red Sea between *c.* 24 and 16 Ma (Coleman & McGuire 1988; Baldridge *et al.* 1991; Schandelmeier & Reynolds 1997). Sea-floor spreading began in the western Gulf of Aden *c.* 10–11 Ma (Schilling *et al.* 1992; Bertrand *et al.* 2003) and *c.* 5 Ma in the central Red Sea (Schandelmeier & Reynolds 1997; Bertrand *et al.* 2003).

Sampling and analytical methods

Basalt lavas were sampled primarily from quarries at five different locations near Cairo, Egypt (Fig. 1). In the Cairo-Suez District of the Eastern Desert, samples were collected from Abu Zaabal, Wadi Hagul and along the Cairo–Suez Road. In the Western Desert, basalts were sampled from Abu Roash and Qatrani in the Fayyum District. Several sites within the broader Cairo region are located in areas of active construction and may not be available for further sampling. Field work revealed widespread massive, blocky and often columnar flows, frequently exhibiting spheroidal weathering in their uppermost portions. There is no evidence for a central volcanic edifice and none of the basalts are exposed at the surface to enable detailed mapping. El-Hinnawi (1965) suggested that basalts visible locally from Abu Roash to Qatrani are likely from the same magmatic phase as those observed in Abu Zaabal. A maximum lava thickness of 60 m has been observed above ground, and well surveys in north Egypt reveal 20–80 m of basalt in the subsurface (Abdel-Monem & Heikel 1981; Williams & Small 1984). Previous studies used the K/Ar method to date basalts from these regions (Meneisy 1990); the Oligo-Miocene age is consistent with palaeontological evidence from underlying and overlying sedimentary units (Morsy & Atia 1983; Williams & Small 1984; Said 1990).

Several of the freshest basalts with low phenocryst contents were chosen for major and trace element and Sr–Nd–Pb–Hf isotopic analysis. Each sample was cut into centimetre-thick slabs to remove weathered surfaces, polished to remove saw marks and crushed into millimetre-sized chips using an alumina ceramic jaw crusher. The chips were powdered in a tungsten carbide disc mill prior to major and trace element analysis. Powdering times were limited to 30 seconds to avoid Ta contamination. Bulk rock major and trace element analyses were conducted at Duke University using direct current plasma (DCP) and inductively coupled plasma mass spectrometry (ICP-MS) following the methods of Klein *et al.* (1991) and Pollock *et al.* (2009). Replicate analyses of samples and natural and synthetic standards indicate reproducibility of *c.* 2–5% for all elements used in interpretation. Samples were oxidized prior to analysis to enable determination of loss on ignition values.

Powders of four basaltic samples from the Sinai Peninsula analyzed previously for major and selected trace elements (Baldridge *et al.* 1991) were also analyzed with ICP-MS to obtain a full complement of incompatible trace element abundances. Bulk rock isotopic analyses were performed at San Diego State University; Nd–Pb–Hf isotopic data were collected using multi-collector- (MC-) ICP-MS on a Nu Plasma 1700 system and Sr isotopic data were analyzed using a VG Sector 54 thermal ionization mass spectrometer (TIMS) using the methods of Hanan *et al.* (2004, 2008). Elemental analyses of phenocrysts, glomerocrysts, inclusions and matrix crystals were determined using a Cameca SX-50 electron microprobe at the Pennsylvania State University with a 15 kV accelerating voltage, 15–20 s counting times, a 12 μA sample current and a 2 μm diameter beam. Mineral chemistry data are available as Supplementary Material.

Results

Petrography and mineral chemistry

The NE Egyptian basalts typically contain phenocrysts and glomerocrysts of plagioclase feldspar and clinopyroxene, although a few samples also have sparse olivine phenocrysts. Basalts from the Abu Roash, Abu Zaabal, Qatrani and Wadi Hagul study areas are greenish-black in colour and exhibit intergranular to ophitic textures. They contain 16–24% phenocrysts (8–16% plagioclase feldspar and 5–10% clinopyroxene) set in a groundmass of plagioclase feldspar, clinopyroxene, olivine, Fe–Ti oxides and devitrified matrix glass. Basalts from the Cairo–Suez Road are dark grey in colour and exhibit intergranular, ophitic and subophitic textures; CS02A contains 35% phenocrysts (21% plagioclase feldspar, 9% clinopyroxene, 4% iddingsitized olivine, 1% Fe–Ti oxides and <1% quartz xenocrysts) set in a fine-grained matrix of plagioclase, clinopyroxene and Fe–Ti oxides.

Plagioclase feldspar occurs as phenocrysts, glomerocrysts, inclusions in clinopyroxene and in the groundmass of all samples. It occurs as generally euhedral phenocrysts up to 7 mm (most *c.* 1 mm) and forms randomly oriented groundmass laths (*c.* 0.3–0.6 mm) and microlites (3–15 μm). Two generations of plagioclase feldspar phenocrysts are commonly evident: large phenocrysts tend to have core compositions of An_{83-77}, while the cores of small phenocrysts and those containing numerous small clinopyroxene inclusions have compositions of An_{65-52}. Plagioclase feldspar glomerocrysts (An_{67-52}) typically exhibit slight normal zoning and show less compositional variation than phenocrysts within a sample. The majority of matrix plagioclase feldspar compositions ranges from An_{68-23} and groundmass phases locally include analyses in the range of $An_{7-3}Ab_{42-60}Or_{34-42}$.

Clinopyroxene is present as anhedral to euhedral prismatic and tabular phenocrysts (1.3 × 1.7 mm) and as groundmass grains in all samples. The crystals are light green to light greenish-brown in plane polarized light, and sometimes display simple twinning. Pyroxenes are predominantly calcic augites with subtle normal zoning: cores have compositions of $Wo_{37-42}En_{37-51}Fs_{13-25}$ (Mg# 73–80; Mg# = $100 \times Mg/[Mg + Fe^{2+}]$) and rims are $Wo_{33-41}En_{37-46}Fs_{14-25}$ (Mg# 60–77). Glomerocrysts are normally zoned; the cores (Wo_{37-40} $En_{41-53}Fs_{13-17}$; Mg# 74–79) have compositions within the range of those in the phenocryst cores, suggesting early formation contemporaneous with the phenocrysts. Glomerocryst rims are slightly more evolved than the cores with compositions of $Wo_{22-40}En_{35-46}Fs_{14-30}$ (Mg# 55–76). Matrix clinopyroxene compositions are $Wo_{17-40}En_{32-48}$ Fs_{14-41} (Mg# 52–77), typically within the range of phenocryst rim values in each sample.

Phenocryst and glomerocryst cores generally have higher Al_2O_3 contents (commonly ≥4.0 wt%) than the rims (1.1–3.7 wt%) suggesting an initial phase of growth at moderate pressure and subsequent ascent to shallower magma chambers. Clinopyroxene from CS02A and WH07 are the most enriched in Al_2O_3 (up to 5.7 wt%), with correspondingly high TiO_2 contents (up to 1.9 wt%). Al_2O_3 contents of phenocryst and glomerocryst clinopyroxene cores are as much as 5 wt% among crystals with high Mg# (Fig. 2). Most of the NE Egyptian clinopyroxene core compositions are close to experimentally determined Fe–Mg equilibrium with the host lava (K_D values of 0.23–0.30; Bryan *et al.* 1981; Tormey *et al.* 1987; Putirka *et al.* 2003). Crystals that exhibit disequilibrium with the whole rock compositions have the highest Al_2O_3 contents. Calculations based on the clinopyroxene barometer of Nimis (1999) indicate that the most aluminous clinopyroxene cores crystallized at 0.4–0.6 GPa (*c.* 12–18 km) while the majority of crystal growth took place at near-surface conditions.

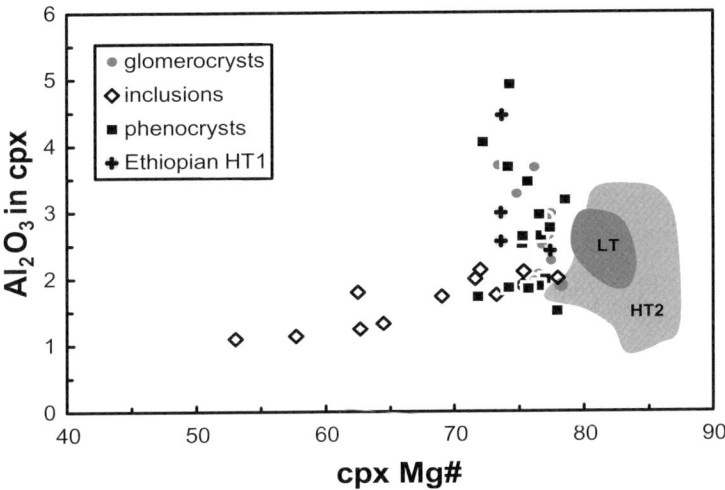

Fig. 2. Concentrations of Al_2O_3 in clinopyroxene cores correlate positively with crystal Mg# and range from values characteristic of near-surface crystallization (clinopyroxene inclusions in plagioclase feldspar) to higher values associated with mid-crustal processes in phenocryst and glomerocryst cores. The Egyptian samples have Al_2O_3 contents that overlap values observed in HT1 basalts from the Oligocene Ethiopian flood basalt province and are much higher than those in LT and HT2 flood basalts (Pik *et al.* 1998; Beccaluva *et al.* 2009).

Plagioclase/clinopyroxene (Plag/Cpx) phenocryst volume ratios of most basalts range from 1.4 to 1.6; CS02A has a Plag/Cpx ratio of *c.* 2:1 that likely reflects the more evolved nature of this sample. These phenocryst proportions are consistent with experimentally determined phase relations for mid-ocean ridge basalts (MORB) crystallizing at 1 atm (Plag/Cpx = 1.4–1.7; Bryan *et al.* 1981; Tormey *et al.* 1987).

Olivine occurs as anhedral to subhedral groundmass crystals, often in clusters of two or more crystals. Their average diameter is between 0.2 and 0.4 mm. Most olivine grains have been altered extensively to reddish-brown iddingsite; fresh olivine crystals have compositions of Fo_{56-39} and do not display zoning. Olivine in CS02A occurs as iddingsitized phenocrysts *c.* 0.2 mm in diameter, many of which exhibit rims of clinopyroxene, Fe–Ti oxides and reaction products.

Quartz xenocrysts *c.* 0.3–0.4 mm in diameter were observed in sample CS02A. Each xenocryst is surrounded by a corona of K–Al-rich glass interfingered with radiating elongate clinopyroxene microlites and bordered by numerous, tiny Fe–Ti oxide crystals and glass.

Bulk rock major and trace element geochemistry

The Egyptian basalts exhibit a high degree of compositional homogeneity (Table 1). All samples contain 5.6–6.3 wt% MgO (Mg# 45–49) and 50.5–51.5 wt% SiO_2 and their bulk compositions are hypersthene to quartz-normative (Endress 2010). The basalts are sub-alkaline tholeiites, broadly similar to contemporaneous dikes from the Sinai Peninsula (Baldridge *et al.* 1991) and post-Miocene tholeiites from NW Syria (Krienitz *et al.* 2006) (Fig. 3). Although the compositional variations are limited, it is apparent that the Egyptian and Syrian lavas lack the high MgO contents and elevated incompatible major and trace element abundances observed in Syrian alkali basalts (Krienitz *et al.* 2009; Fig. 4). Loss on ignition values from 0.3–2.3 wt% indicate that the rocks have experienced only minor alteration to hydrous minerals.

The basalts contain 44–55 ppm Ni, 56–98 ppm Cr and 28–32 ppm Sc (Table 1). These values, with the exception of Sc, are significantly lower than those expected for primary mantle melts (*c.* 300 ppm Ni, *c.* 750 ppm Cr, *c.* 33 ppm Sc), indicating that the samples have undergone extensive olivine loss and relatively minor clinopyroxene fractionation. The samples also exhibit tight compositional ranges of incompatible trace elements (ITE) that correlate weakly at best with lava MgO; abundances of individual ITE are positively correlated over narrow ranges (Fig. 4).

The study samples have ocean island basalt (OIB)-like enrichments in the most highly incompatible trace elements, but distinct and rather flat primitive-mantle normalized ITE patterns from Rb to La (Fig. 5). Most samples have homogeneous ITE patterns with small but consistent positive Pb

Table 1. *Major and trace element geochemistry*

Sample	AR05	AZ03	CS02A	FQ06B	FQ08	FQ10B	FQ11A	WH01	WH07	WH08	BT-352	BT-1049	BT-1142D	BT-1152
Location	Abu Roash	Abu Zaabal	Cairo Suez Rd	Qatrani	Qatrani	Qatrani	Qatrani	Wadi Hagul	Wadi Hagul	Wadi Hagul	Thamad Plug	Ashosh Plug	Sharem-e-Sheikh dike	Santa Catarina dike
	NE Egypt	NE Egypt	NE Egypt	NE Egypt	NE Egypt	NE Egypt	NE Egypt	NE Egypt	NE Egypt	NE Egypt	Sinai peninsula	Sinai peninsula	Sinai peninsula	Sinai peninsula
Latitude (N)	30.029	30.279	30.279	29.733	29.660	29.660	29.660	29.979	29.979	29.979				
Longitude (E)	30.887	31.359	31.626	30.663	30.597	30.597	30.597	31.674	31.674	31.674				
SiO_2	50.57	50.75	51.49	50.54	50.69	50.76	50.73	51.00	50.62	50.82	49.68	48.58	48.76	48.89
TiO_2	2.29	2.35	3.11	2.42	2.44	2.40	2.41	2.32	2.37	2.33	2.89	3.48	3.09	2.89
Al_2O_3	14.69	14.30	13.41	14.53	14.36	14.42	14.47	14.40	14.45	14.52	15.26	13.17	15.21	15.25
Fe_2O_3*	12.71	12.98	13.66	12.90	12.93	12.87	12.88	12.61	12.80	12.71	12.93	15.07	13.46	12.37
MnO	0.17	0.18	0.20	0.17	0.16	0.17	0.16	0.18	0.17	0.17	0.21	0.21	0.21	0.16
MgO	6.06	6.27	5.59	5.89	5.64	6.09	5.91	5.96	5.78	6.00	4.39	5.33	4.72	4.11
CaO	9.54	9.62	9.51	9.49	9.48	9.64	9.81	9.51	9.42	9.57	8.75	9.72	9.55	9.26
Na_2O	2.95	2.84	2.98	2.95	3.09	2.92	2.82	2.84	2.94	3.02	2.98	2.37	2.5	2.49
K_2O	1.03	0.96	0.51	1.24	1.16	1.03	1.17	1.15	1.26	1.08	1.09	0.54	0.87	1.25
P_2O_5	0.30	0.30	0.41	0.32	0.32	0.30	0.30	0.30	0.30	0.29	0.96	0.53	0.53	0.38
Total	100.31	100.55	100.86	100.44	100.28	100.60	100.67	100.27	100.11	100.50	99.14	99.00	98.90	97.05
LOI	2.3	1.9	1.1	2.0	0.3	1.7	1.5	1.2	1.1	1.5	N/A	N/A	N/A	N/A
Cs	0.18	0.16	0.21	0.17	0.19	0.17	0.17	0.19	0.17	0.15	0.42	0.16	0.51	0.41
Rb	20.03	20.65	12.09	20.72	21.77	20.52	20.29	21.77	21.16	20.33	28.06	13.39	23.16	34.35
Ba	226	231	295	228	252	229	225	236	237	228	729	237	263	496
Th	2.84	2.90	2.68	2.89	2.98	2.84	2.81	3.04	2.98	2.84	4.87	2.86	2.95	3.93

U	0.71	0.76	0.65	0.74	0.77	0.74	0.71	0.78	0.76	0.75	1.48	0.63	0.76	0.72
Nb	22.96	23.51	21.81	23.01	22.88	23.03	24.17	25.65	25.38	23.93	39.72	26.77	28.72	26.15
Ta	1.45	1.50	1.43	1.47	1.50	1.46	1.54	1.61	1.60	1.54	2.53	1.57	1.62	1.44
Pb	2.70	3.13	2.25	3.10	3.27	3.14	3.06	3.21	3.25	3.11	5.38	5.28	3.83	5.77
Sr	376	374	448	367	383	372	377	366	370	370	708	374	460	550
Zr	219.41	228.74	193.85	222.19	235.02	224.79	226.16	236.99	231.16	225.53	716.41	287.81	257.01	226.05
Hf	5.39	5.76	4.89	5.49	5.77	5.47	5.56	5.84	5.74	5.67	15.37	6.51	5.81	5.25
Y	35.24	36.04	33.85	35.96	36.42	36.07	36.66	37.07	37.12	35.50	38.60	43.13	37.71	33.01
Co	56.74	55.02	60.75	56.87	58.37	58.47	61.05	60.79	54.63	55.11	44.92	42.86	42.65	34.75
Cr	83.79	85.10	56.34	90.29	80.26	92.41	97.58	95.54	90.27	93.11	56.54	63.58	150.75	120.79
Ni	52.12	55.08	44.45	47.73	43.43	48.55	49.79	48.71	47.37	48.59	27.20	45.04	518.56	569.77
V	290.40	294.91	329.00	306.37	299.84	302.37	310.18	300.36	304.76	289.10	235.87	404.68	295.58	326.70
Sc	30.20	30.62	28.14	31.58	30.91	31.36	31.75	30.95	31.10	29.85	19.68	33.50	27.89	26.64
La	22.49	23.28	22.28	23.14	23.59	22.56	22.98	24.11	23.89	22.36	42.02	24.73	25.84	29.74
Ce	50.13	51.37	50.16	51.84	52.95	50.38	51.10	53.31	52.72	50.06	94.26	56.17	59.05	65.76
Pr	6.89	7.09	7.10	7.17	7.32	7.01	7.04	7.42	7.36	6.90	13.34	8.14	8.41	8.95
Nd	28.28	29.49	30.18	29.57	30.11	28.99	29.33	30.30	30.40	29.05	53.69	33.40	33.98	34.82
Sm	6.73	6.83	6.99	6.78	6.95	6.86	6.92	7.05	7.16	6.75	11.64	8.08	7.83	7.62
Eu	2.08	2.15	2.42	2.14	2.22	2.16	2.15	2.13	2.16	2.10	4.42	2.75	2.71	2.60
Gd	6.74	6.96	7.05	6.98	7.16	7.04	7.00	7.10	7.12	6.81	10.71	8.44	8.04	7.68
Tb	1.10	1.14	1.13	1.13	1.17	1.13	1.15	1.19	1.19	1.11	1.64	1.46	1.33	1.26
Dy	6.42	6.69	6.37	6.58	6.84	6.64	6.63	6.77	6.86	6.56	8.14	8.05	7.34	6.74
Ho	1.29	1.35	1.28	1.34	1.36	1.37	1.37	1.38	1.40	1.35	1.49	1.57	1.35	1.24
Er	3.50	3.65	3.33	3.61	3.73	3.65	3.63	3.80	3.80	3.67	3.65	4.05	3.40	3.04
Yb	3.06	3.18	2.77	3.11	3.16	3.18	3.23	3.35	3.32	3.20	3.04	3.74	3.03	2.63
Lu	0.43	0.45	0.39	0.45	0.45	0.42	0.45	0.46	0.46	0.45	0.44	0.55	0.45	0.38

Note: Major element data for Sinai peninsula basalts from Baldridge *et al.* 1991.

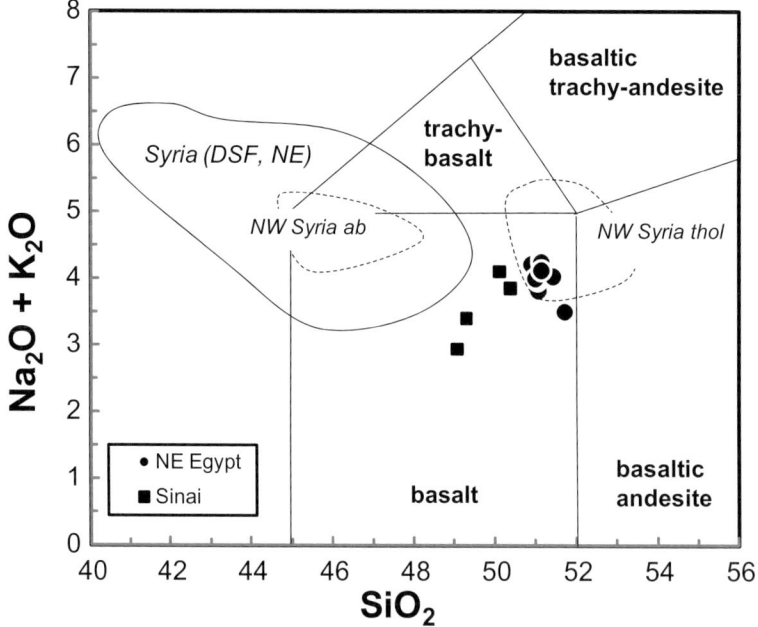

Fig. 3. Total alkalis–silica classification diagram (after LeBas *et al.* 1985). Samples from NE Egypt are sub-alkaline basalts that overlap the field of post-Miocene tholeiites from NW Syria (Krienitz *et al.* 2006). Shown for comparison are post-Miocene basalts from the Dead Sea Fault (DSF) and NE Syria as well as alkali basalts from NW Syria (Krienitz *et al.* 2006, 2009).

and negative P and Ti anomalies that may be associated with crustal contamination. Overall, the ITE patterns are however quite unremarkable as they lack the pronounced enrichments and depletions that characterize crustally contaminated melts or liquids derived from metasomatized lithosphere. The Egyptian lavas have ITE abundances similar to those observed in contemporaneous Sinai basalts as well as post-Miocene Syrian tholeiites (Krienitz *et al.* 2006; Fig. 5). Sample CS02A is unique in exhibiting negative Rb, K and P and positive Ba anomalies. The unusual petrographic nature of this sample suggests that crustal contamination and alteration have contributed to its geochemical features.

Chondrite-normalized Rare Earth Elements (REE) patterns of the NE Egyptian basalts are homogeneous and smooth with shallow negative slopes (Fig. 5). The level of REE enrichment is broadly similar to that of OIB (Sun & McDonough 1989), although La/Yb_n of average OIB (*c.* 12) is considerably greater than what we observe (*c.* 5). Although the basalts are evolved, their Tb/Yb_n values (*c.* 1.6) approximate those of the parental basalt because Tb and Yb have similar compatibilities in the fractionating mineral assemblages (Rollinson 1993). On the basis of their Tb/Yb_n values, we suggest that the NE Egyptian samples

were derived within the spinel stability field (Wang *et al.* 2002).

Several geochemical features suggest a limited role for selective interaction between a parental mafic magma and the lower crust. Ubiquitous but small positive Pb anomalies are indicative of minor assimilation of crustal material. Values of Ce/Pb (16–22) are close to the range of global OIB (Fig. 6), although only CS02A plots within the range of mantle-derived liquids (25 ± 5; Hofmann *et al.* 1986). The presence of quartz xenocrysts in this sample suggests a fortuitous coincidence with mantle-derived liquids. Overall, Ce/Pb values that decrease with decreasing MgO and the clustering of $Nb/U–SiO_2$ and $Nb/U–La/Sm_n$ values that trend towards the composition of average lower continental crust (Rudnick & Fountain 1995; Taylor & McLennan 1995; Wedepohl 1995) indicate some degree of interaction between a parental mafic magma and the surrounding lower crust (Fig. 6). The samples define a tight cluster in La/Nb (0.9–1.0) and Ba/Nb (9–14), well within the range of global OIB (not shown); higher La/Nb–Ba/Nb values in the Sinai Peninsula dikes are typical of interaction between mafic magma and a more enriched material. In each case, the trace element composition of the likely parental basalt

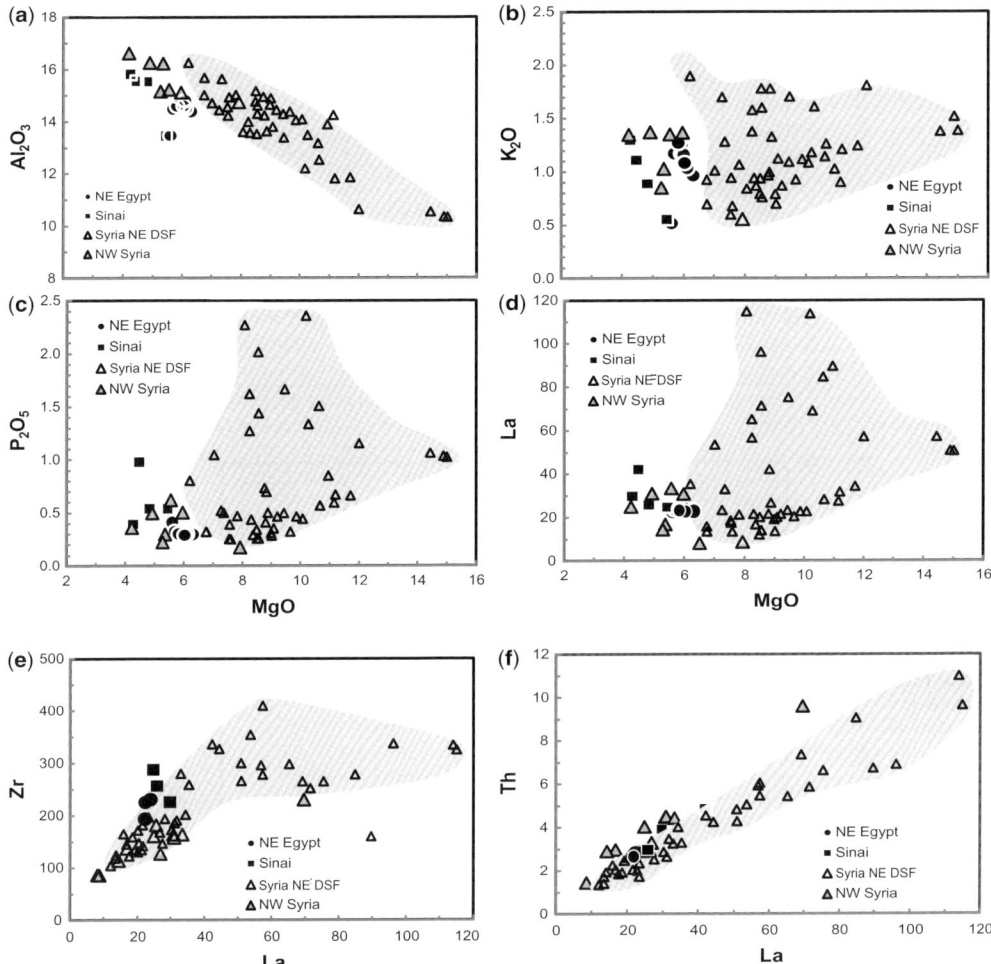

Fig. 4. NE Egyptian basalts show limited variation in major element compositions; dikes and plugs from the Sinai Peninsula (Baldridge *et al.* 1991) extend these ranges to slightly more evolved compositions. Abundances of incompatible trace elements La, Zr and Th show positive correlations suggesting comparable bulk partition coefficients and providing no indication of element mobility. Egyptian basalts have major and trace element characteristics similar to post-Miocene tholeiites from NW Syria (Krienitz *et al.* 2006) that are broadly consistent with removal of the observed phenocryst phases (olivine + plagioclase feldspar + clinopyroxene) at low pressure. The shaded field encompasses alkali mafic lavas from northeastern Syria and the Dead Sea Fault region for comparison (Krienitz *et al.* 2006, 2009).

is similar to values observed in basalts associated with Central Atlantic mantle plumes.

Radiogenic isotopes

The Egyptian samples, including basalts from the Sinai Peninsula (Baldridge *et al.* 1991) have some of the most radiogenic Sr values observed across the African and Arabian Plates (Fig. 7). The basalts define a steep, tight trend of increasing $^{87}Sr/^{86}Sr$ and decreasing $^{143}Nd/^{144}Nd$ that overlaps the most radiogenic Sr isotope values observed in

tholeiitic basalts from NW Syria (Krienitz *et al.* 2006), but at higher $^{143}Nd/^{144}Nd$ for a given $^{87}Sr/^{86}Sr$ value. The new data (Table 2) trend towards the array defined by rejuvenated mantle lithosphere beneath the craton in Uganda (Davies & Lloyd 1989). Similar negative Sr–Nd isotopic correlations are observed among several sample suites from the Middle East, North Africa and the East African Rift and are generally interpreted as indicating interaction between mantle-derived magmas and more radiogenic crustal materials (e.g. Deniel *et al.* 1994; Pik *et al.* 1999; Furman

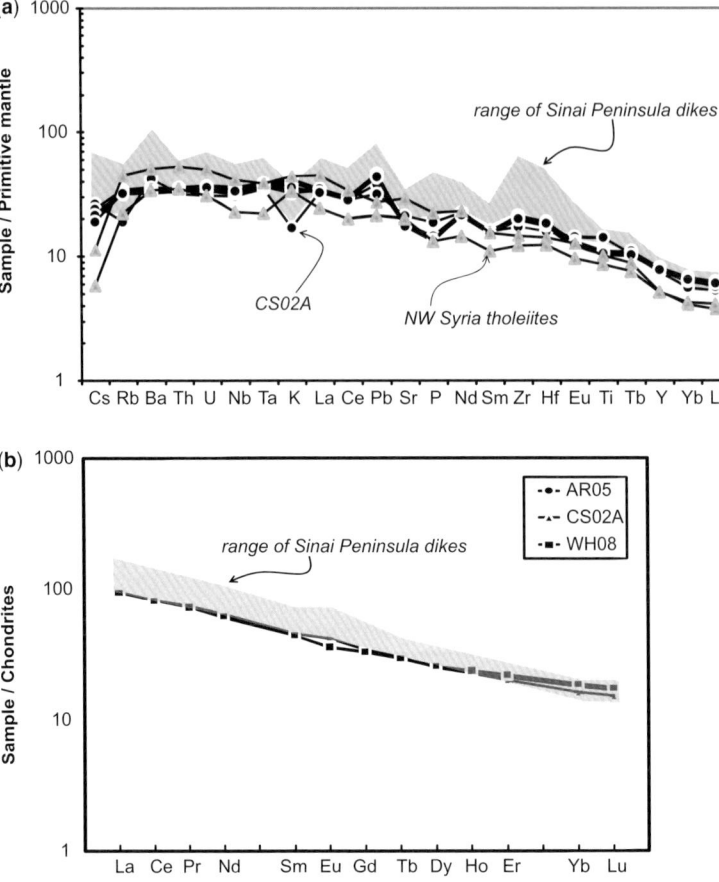

Fig. 5. Trace element abundances measured in Egyptian and Sinai Peninsula mafic rocks are similar to those observed in ocean island basalts and thus provide very limited evidence for contamination by crustal materials. (**a**) Primitive-mantle normalized incompatible trace element abundance patterns of NE Egyptian basalts are tightly parallel with the exception of sample CS02A (normalizing values of Sun & McDonough 1989). Hypabyssal intrusive from the Sinai Peninsula (shaded field) and post-Miocene tholeiites from NW Syria (Krienitz *et al.* 2006) have similar patterns. (**b**) Chondrite-normalized rare earth element abundances display shallow slopes consistent with formation in the spinel stability field, and lack crossing patterns that might indicate a heterogeneous source area.

et al. 2006*a*; Krienitz *et al.* 2006, 2009; Shaw *et al.* 2003). The NW Syrian tholeiites and basalts from the Asal Rift and the Gulf of Tadjoura form a broad field that encompasses the region defined by Red Sea MORB south of 21°N, Afar plume-like compositions ($^{87}Sr/^{86}Sr$ c. 0.7035, $^{143}Nd/^{144}Nd$ c. 0.5129; Barrat *et al.* 1998) and several mafic lower crustal materials (Saudi Arabian gabbros from Hegner & Pallister 1989, Yemeni spinel peridotites from Baker *et al.* 1998). It is worth noting that the majority of spinel lherzolite xenoliths sampled in the region have Sr–Nd isotopic values that are in or close to the field of the Afar plume, making it difficult to distinguish contributions from these two possible source domains.

The Pb isotopic ratios of the Egyptian basalts are generally similar to those found in samples from the Asal Rift and the Gulf of Tadjoura (Schilling *et al.* 1992; Barrat *et al.* 1998) and NW Syria (Krienitz *et al.* 2006; Fig. 7). Basalts from these areas has more highly radiogenic Pb isotopic values and plots above the mantle array defined by MORB from the Red Sea (Eissen *et al.* 1989; Volker & McCulloch 1993), the Afar plume (Barrat *et al.* 1998) and the range of mantle xenoliths from the Red Sea region (Baker *et al.* 1998; Shaw *et al.* 2003). In each case the samples with the lowest $^{87}Sr/^{86}Sr$ (and correspondingly highest $^{143}Nd/^{144}Nd$) isotopic values also have $^{206}Pb/^{204}Pb-^{208}Pb/^{204}Pb$ values that approach

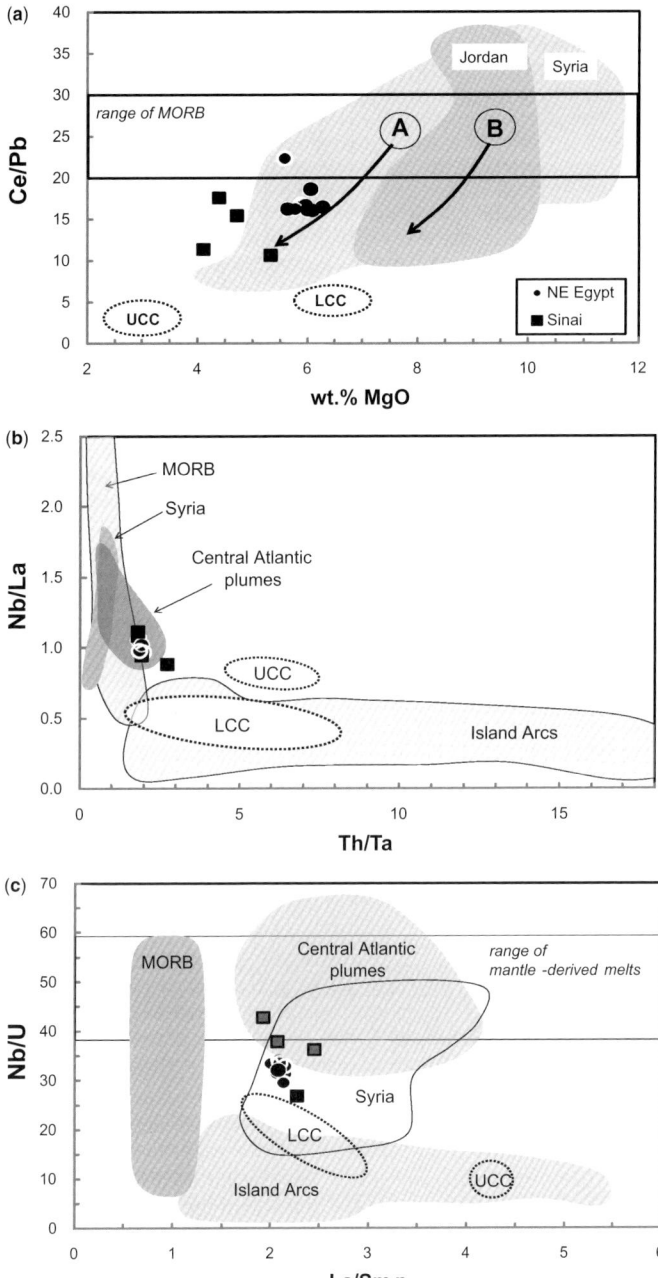

Fig. 6. (**a**) Ce/Pb values are lower than those observed in MORB and their positive correlation with MgO suggests incorporation of materials from the upper (UCC) or lower continental crust (LCC) (crustal values from Rudnick & Fountain 1995; Taylor & McLennan 1995; Wedepohl 1995). Basalts erupted across Syria (Krienitz *et al.* 2006, 2008, 2009) and in Jordan (Shaw *et al.* 2003) show similar trends, with the Jordanian samples more closely approaching lower crustal values. (**b**) Nb/La–Th/Ta values cluster tightly and fall within the range of Central Atlantic mantle plumes; samples derived from the Afar plume also plot within this field (data from GEOROC, Barrat *et al.* 1998). (**c**) La/Sm$_n$–Nb/U values in the Egyptian basalts fall outside the range of mantle-derived melts, and trend towards the average crustal compositions. Shown for comparison are tholeiites from NW Syria (Krienitz *et al.* 2006) and Red Sea MORB (data from GEOROC, Schilling *et al.* 1992); the MORB component does not appear to contribute to geochemical variations observed in Egypt.

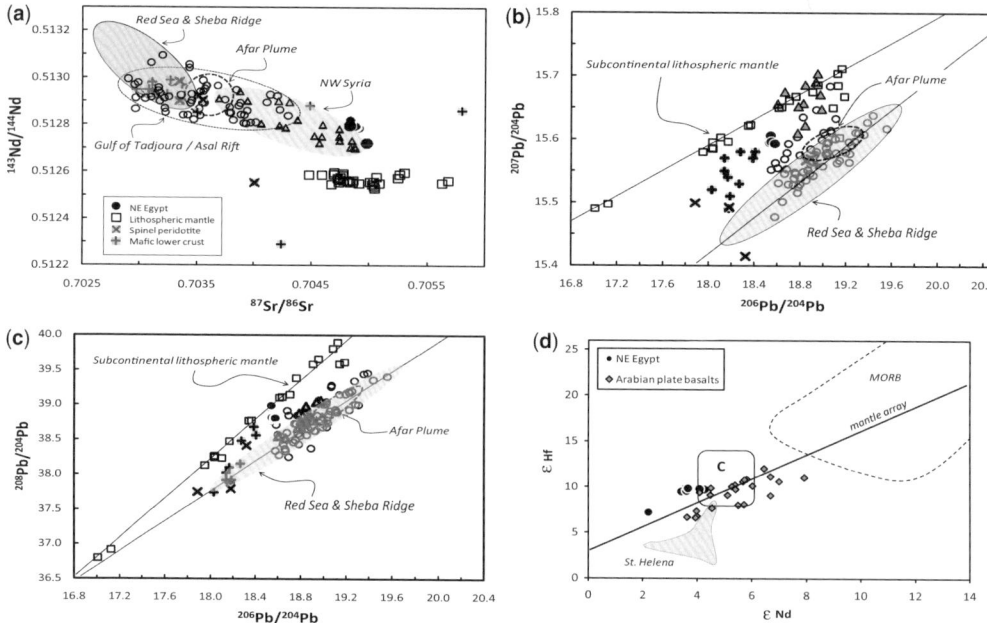

Fig. 7. (a) Sr–Nd isotopic variations. Several distinct isotopic reservoirs are apparent in this diagram: (1) the Afar plume, including basalts from Erta 'Ale in northern Ethiopia (Barrat *et al.* 1998); (2) Red Sea and Sheba Ridge (Gulf of Aden) MORB, which extends from the Afar plume field to more depleted values (Altherr *et al.* 1988, 1990; Schilling *et al.* 1992; Volker & McCullough 1993); and (3) the field of metasomatized Pan-African lithospheric xenoliths from Uganda (Davies & Lloyd 1989), indicated as 'lithospheric mantle' on the figure. The Egyptian basalts have Sr–Nd isotopic values among the most highly radiogenic observed in the East African Rift System; they plot adjacent to the field of basalts from the Gulf of Tadjoura and the Asal Rift in the southernmost Red Sea (Schilling *et al.* 1992; Barrat *et al.* 1998) that record the early stage of regional crustal rifting. Lower crustal mafic lithologies from Saudi Arabia (Hegner & Pallister 1997) and spinel peridotite xenoliths from the Arabian and Red Sea region (Baker *et al.* 1998; Shaw *et al.* 2003) have compositions that overlap the Afar plume and extend to more radiogenic compositions. **(b)** Values of $^{208}Pb/^{204}Pb-^{206}Pb/^{204}Pb$ measured in the Egyptian basalts lie above the regional mantle array defined by basalts from NE Syria and the Dead Sea Fault region, the Gulf of Aden and Red Sea MORB; they overlap basalts from NW Syria, Gulf of Tadjoura and the Asal Rift and some lower crustal and spinel peridotite xenoliths. Pb isotopic values of the Egyptian basalts trend towards the range of metasomatized subcontinental lithospheric mantle xenoliths from Uganda. **(c)** Values of $^{207}Pb/^{204}Pb$ $^{206}Pb/^{204}Pb$ define a field between the local mantle array and the suite of Pan-African upper crustal xenoliths; this relationship is also observed in lavas from NW Syria and the southernmost Red Sea. **(d)** Hf–Nd isotopic values of the Egyptian basalts plot above the mantle array (Chauvel & Blichert-Toft 2001) and include compositions within the range of the common 'C' mantle component (Hanan & Graham 1996; Hanan *et al.* 2004). Data for Arabian plate basalts from Bertrand *et al.* (2003).

those estimated for the Afar plume and observed in the Arabian mantle lithosphere (*c.* 19.0, *c.* 38.6). The Pb isotopic trends of the Egyptian, Syrian and southernmost Red Sea suites approach the array of values observed in metasomatized lithospheric mantle xenoliths from Uganda (Davies & Lloyd 1989).

Hf–Nd isotopic data are particularly valuable for investigating source compositions because both the high field strength (Hf) and rare earth elements (Sm, Nd, Lu) involved are immobile relative to the large ion lithophile elements during low-temperature and hydrothermal alteration. Taken together, the Hf–Nd isotopic data provide insights

into source compositions and contamination histories that are not available from Sr–Nd–Pb data alone. The Hf isotopes constrain the depths of previous melting events in mantle sources. This interpretation is possible because the presence of garnet during melting has a large fractionation effect on the Lu/Hf ratio. In general, a positive variation in εHf relative to the mantle array (i.e. data plot above the array) in Nd–Hf isotope diagrams indicates a garnet peridotite source whereas negative εHf values that plot below the array suggest a spinel peridotite source (Johnson & Beard 1993). All of the study samples have Hf–Nd isotopic values that plot near but slightly above the mantle

Table 2. *Radiogenic isotopic data (see appendix for analytical details)*

Sample	Location	$^{87}Sr/^{86}Sr$	$2s/\sqrt{n}$	$^{143}Nd/^{144}Nd$	$2s/\sqrt{n}$	eNd	$^{206}Pb/^{204}Pb$	$2s/\sqrt{n}$	$^{207}Pb/^{204}Pb$	$2s/\sqrt{n}$	$^{208}Pb/^{204}Pb$	$2s/\sqrt{n}$	$^{176}Hf/^{177}Hf$	$2s/\sqrt{n}$	eHf
AR05	Abu Roash	0.7	0	0.51	0	3.0	18.5682	0	15.59	0	38.8191	0.0033	0.28303	0	8.9
AZ03	Abu Zaabal	0.7	0	0.51	0	3.0	18.5607	0	15.59	0	38.8125	0.0014	0.28304	0	9.3
CS02A	Cairo Suez	0.71	0	0.51	0	1.6	18.5405	0	15.61	0	38.9861	0.0013	0.28297	0	6.7
FQ06B	Fayyum Qatrani	0.7	0	0.51	0	2.8	18.5347	0	15.59	0	38.7970	0.0013	0.28303	0	8.9
FQ10B	Qatrani Mountains	0.7	0	0.51	0	3.0	18.5858	0	15.59	0	38.8055	0.0014	0.28303	0	9.0
FQ11A	Qatrani Mountains	0.7	0	0.51	0	3.7	18.5813	0	15.59	0	38.8090	0.0009	0.28304	0	9.2
WH01	Wadi Hagul	0.7	0	0.51	0	3.1	18.5470	0	15.6	0	38.8121	0.0024	0.28304	0	9.3
WH08	Wadi Hagul	0.7	0	0.51	0	3.5	18.5790	0	15.59	0	38.8056	0.0020	0.28304	0	9.2

array (Vervoort *et al.* 1999; Chauvel & Blichert-Toft 2001) and show little variation in ε_{Hf} with increasing ε_{Nd} (Fig. 7). The trend of the Egyptian samples contrasts with that of Mio-Pliocene basalts from across the Arabian Plate (Bertrand *et al.* 2003) which define an elongate field that broadly overlaps but, for the most part, plots below the mantle array and extends towards MORB-like values. The Egyptian basalts were apparently derived from greater depth than the Arabian Plate basalts. Two of the Egyptian basalts (FQ11A and WH08) plot within the range of the common 'C' mantle component (Hanan & Graham 1996; Hanan *et al.* 2004), consistent with trace element evidence that they have experienced only minor interaction with crustal materials. Sample CS02A has a significantly lower ε_{Hf} and ε_{Nd} than the other Egyptian samples, consistent with its more evolved nature. For the most part, however, the samples plot at much higher ε_{Hf}–ε_{Nd} values than crustal materials, similar in range to OIB, indicating that contamination was limited and likely restricted to incorporation of more highly mobile elements (e.g. U, Pb, Th).

Discussion

Evolution of Egyptian basalts

The restricted range of bulk compositions makes it difficult to interpret a precise fractionation history for this suite of lavas. The mineral chemistry requires an episode of crystallization at mid-crustal depth: clinopyroxene cores with up to 5.7 wt% Al_2O_3 suggest initial growth at 0.4–0.6 GPa, that is, depths of *c.* 12–18 km based on crustal densities of 2.8–3.0 g cm^{-3} (Nimis 1999; Putirka *et al.* 2003). Plagioclase feldspar and clinopyroxene phenocrysts are generally found in proportions predicted for low-pressure crystallization (Bryan *et al.* 1981; Tormey *et al.* 1987), indicating that a protracted period of differentiation occurred at shallow crustal levels. Petrographic observations suggest initial fractionation of minor olivine followed by plagioclase feldspar, and shortly thereafter by concurrent plagioclase feldspar and clinopyroxene growth. The range of magma compositions present in this shallow chamber may have been restricted to <6.5 wt% MgO as seen in our samples, or could have included more mafic liquids. The broad spatial distribution of basalts with near-homogeneous bulk compositions suggests a large evolving magma chamber at mid-crustal levels.

Both lower and upper crustal magma chambers are favourable sites for assimilation as inferred from petrographic observations and select trace element variations. Indeed, sparse quartz xenocrysts

in the Cairo–Suez Road basalts provide compelling evidence for contamination by interaction with granitic material. It is worth emphasizing that the degree of upper crustal assimilation is quite low. In particular, the generally smooth primitive mantle-normalized ITE diagrams (Fig. 5), the diversity of ITE ratios that overlap the field of mantle-derived basalts (Fig. 6) and the observation that lavas with the highest $^{206}Pb/^{204}Pb$ values plot along the mantle array with positive $\Delta\varepsilon Hf$ (the difference between measured ε_{Hf} and that predicted by the Nd–Hf mantle array correlation line) within the field for the 'C' mantle component (Fig. 7) all restrict the amount of assimilation to a small fraction of the evolving magma by weight.

Dynamics of Egyptian magmatism

The Egyptian basalts were erupted during a time of widespread volcanic activity in northern Africa and the circum-Mediterranean. Two key factors contributing to this period of volcanism are (1) prolonged rifting, leading to localized lithospheric thinning and possible decompression melting of the lithospheric mantle, and (2) ascent of plume-derived materials beneath the region, providing heat and potential magmatic and metasomatic fluids to the overlying lithosphere. As detailed below, it appears most likely that a combination of these processes led to the generation of the NE Egyptian basalts.

The isotopic compositions of Egyptian basalts reflect contributions from two distinct source regions: one comparable to the Afar mantle plume and the other analogous to the rejuvenated Pan-African lithosphere that likely underlies much of the continent. This latter source is readily identified in the Sr–Nd–Pb isotopic plots (Fig. 7), in which the Egyptian basalts lie on consistent mixing lines between the two proposed source domains. This physical model is consistent with our understanding of the regional crustal structure. Seismic studies in Jordan and Egypt (El-Isa *et al.* 1987; Gahrib 2006; Radwan *et al.* 2006; Saleh & Badawy 2006) that suggest predominantly felsic upper crustal lithologies persist to depths of *c.* 15–20 km and are underlain by more mafic compositions that extend to the Moho at *c.* 30 km depth. Away from the Red Sea margin, the lithospheric mantle persists to depths of *c.* 100–110 km (Radwan *et al.* 2006), typical of the stable cratonic portions of the north African plate (Dugda & Nyblade 2006). The hydrous lithospheric mantle sampled in xenoliths from Uganda provides a possible representation of this source region, with melting triggered by the ascent of warm asthenospheric material with Afar-plume-like isotopic and trace element characteristics. The parental mafic liquids underwent initial fractionation in the lower crust, most likely stagnating at the boundary between upper and lower crust, and then ascended to very shallow levels where they continued to fractionate prior to eruption. The geochemistry of the Egyptian basalts is consistent with generation and emplacement at a very early stage of continental rifting.

There is good indication that rift structures also played a role in emplacement of the Egyptian basalts as well as other Miocene volcanic centres. Localized rifting that occurred between Cretaceous and Early Eocene time was accompanied by volumetrically insignificant magmatism; in northern Egypt, early east–west trending faults likely facilitated magma ascent to the surface rather than directly triggering mafic volcanism through decompression melting (Wilson & Guiraud 1992). Many (though not all) of the Miocene and younger volcanic provinces (e.g. Gharyan, Hoggar, Tibesti, Turkana) are situated at the intersections of extensional structures associated with the modern East African Rift and these older faults (Guiraud *et al.* 1992; Wilson & Guiraud 1992; Farahat *et al.* 2006). In NE Egypt, including the Sinai Peninsula, basaltic dikes and flows were emplaced along localized fault zones contemporaneous with the opening of the Gulf of Suez (Abdel-Monem & Heikel 1981; Meneisy 1990; Baldridge *et al.* 1991; Schandelmeier & Reynolds 1997; Guiraud & Bosworth 1999). In the absence of high degrees of extension, several authors (Wilson & Guiraud 1992; Wilson & Patterson 2001; Beccaluva *et al.* 2008) have proposed that individual volcanic provinces are ultimately supported by ascending asthenosphere, most likely in the form of small plumes or in isolated sites of shallow mantle upwelling.

Tomographic results (Ritsema *et al.* 1999; Castle *et al.* 2000; Kuo *et al.* 2000; Becker & Boschi 2002) consistently indicate the presence of a large region of anomalously low shear wave velocity beneath southern Africa, interpreted as a thermochemical anomaly. The South African Superplume extends from the core mantle boundary towards the 660 km discontinuity beneath the East African Rift (Ritsema *et al.* 1999). Numerical models and analogue experiments predict that secondary plumes of the order 100–200 km diameter will form above the margins of such large plumes that become stalled at 660 km depth, as the South African Superplume may have done (Wilson & Patterson 2001; Burke *et al.* 2008; Tackley 2008). It is interesting in this context to note that global large igneous provinces and hotspots all lie within 5° of the inferred boundaries of the large South African and Pacific low shear velocity provinces (Castle *et al.* 2000; Kuo *et al.* 2000; Burke *et al.* 2008). We suggest that Tertiary volcanism in northern Africa was

supported in part by secondary plumes derived directly from the margins of, or from the transition zone above, this very large mantle feature. An analogous model was proposed (Goes *et al.* 1999; Wilson & Patterson 2001) to explain widespread magmatism beginning *c.* 25 Ma in western and central Europe.

The deep mantle structure beneath Europe and Africa is not fully constrained. It seems most likely, however, that the basalts studied here derive in some measure from the South African Superplume as a source of heat and/or material. We suggest that volcanism in NE Egypt reflects contributions from a mantle source domain with 'C'-like radiogenic isotopic characteristics, with inputs from the overlying metasomatized spinel-facies lithosphere located within the continental keel. We envisage lithospheric melting to be driven by impingement of a thermal/thermochemical upwelling derived from the margins of the South African Superplume or through small-scale convection in the mantle near the 660 km discontinuity. Ascent of mafic liquids is facilitated by a regional network of extensional features, many of which were reactivated by incipient rifting in the northern Red Sea. Similar preexisting zones of lithospheric weakness throughout northern Africa and the Mediterranean enabled this process to occur across a wide region, although the deep mantle source features may not in all cases be uniform (Guiraud *et al.* 1992; Wilson & Guiraud 1992; Farahat *et al.* 2006; Beccaluva *et al.* 2008).

Conclusions

Mid-Tertiary NE Egyptian basalts were erupted during a time of widespread volcanic activity beginning at *c.* 35 Ma in northern Africa. Petrographic observations and mineral chemistry indicate a polybaric fractionation history with initial crystallization at 12–18 km, followed by more extensive mineral growth at near-surface conditions. Bulk rock major and trace element data reveal that the primary melts were generated at depths approaching *c.* 65 km, that is, within the spinel stability field of the lithospheric mantle. The presence of quartz xenocrysts in one basalt sample, as well as ubiquitous elevated Pb contents, low Nb/U values and a trend toward highly radiogenic Sr and Pb isotopic compositions, is consistent with contributions from both granitic crustal and metasomatized lithospheric materials. The basalts have isotopic compositions that lie close to or along isotopic mixing trends between the Afar Plume and this lithology, suggesting that both lithospheric and sublithospheric materials were involved in their genesis. A model for regional magmatism is proposed whereby heat and/or parental mafic magma are supplied by shallow thermal upwellings derived from the margins of the South African Superplume or along the 660 km discontinuity beneath northern and central Africa. The upwellings provide heat and local metasomatic agents to the lithosphere, with melt ascent facilitated by both ancient and young rift structures.

This contribution represents the MSc thesis research of Endress, who gratefully acknowledges financial support from the Geological Society of America, the Department of Geosciences and the Alliance for Education, Science, Engineering and Development in Africa at the Pennsylvania State University. Support was also provided from NSF grant GEO-0631377 to Furman and EAR-0738963 to Hanan. Field work was generously supported by the Geology Department, Assiut University. J. Miller provided technical assistance in the laboratory at San Diego State University. The manuscript benefited greatly from thoughtful and constructive comments by G. Bianchini, K. Haase and B. Bingen.

Appendix

Prior to dissolution, rock chips were cleaned with distilled 2N HCl, 1N HBr and ultrapure H_2O. Sr isotopes were measured on a VG Sector 54 seven-collector thermal ionization mass spectrometer in multi-dynamic mode. Hf, Nd and Pb were analysed on a Nu Plasma HR multi-collector ICP-MS using the Nu Instrument desolvating nebulizer equipped with a $100 \, \mu l \, min^{-1}$ nebulizer (DSN100). The $^{87}Sr/^{86}Sr$ isotope ratios were normalized to $^{88}Sr/^{86}Sr = 0.1194$ to correct for mass fractionation and the Sr isotopic data are reported relative to NIST SRM $987 = 0.710250$. The Nd and Hf isotope ratios were corrected for instrumental mass fractionation and machine bias by applying a discrimination factor determined by bracketing sample runs with analyses of the SDSU AMES Nd and the JMC 475 standards (every 2 or 3 samples). Hf isotope sample data were normalized to $^{179}Hf/^{177}Hf = 0.7325$ and are reported relative to JMC $475 = 0.282162$; Nd isotope sample data were normalized to $^{146}Nd/^{144}Nd = 0.7219$ and are reported relative to the lab value of $^{143}Nd/^{144}Nd = 0.512130$ for the SDSU AMES Nd standard; the measured value of La Jolla Nd at SDSU $= 0.511844$.

The Pb was separated on two anion exchange columns, a 500 ul followed by a 200 ul column in order to maximize the Pb separation for optimum ICP-MS analyses. The lead yields were determined to be c. 100%, therefore the columns do not have a fractionation effect. The in-run mass discrimination for Pb was monitored with NIST SRM 997 Tl (White *et al.* 2000). The Pb isotope ratios were corrected for instrumental mass fractionation and machine bias by applying a discrimination factor determined by bracketing sample analyses with analyses of the NIST standard SRM 981 (every two samples), using the NIST SRM 981 values determined by Todt *et al.* (1996).

Total procedural blanks were: <25 pg Hf, <90 pg Pb, <200 pg Nd, <250 pg Sr. No blank corrections were applied to the data because they are insignificant. The reported uncertainties are 2 standard errors of the mean ($2\sigma\sqrt{n}$) for the in-run precision, except for some Sr samples where the uncertainties represent $2\sigma\sqrt{n}$ for 2–3 replicate analyses, and thus provide a measure of sample reproducibility. The NIST SRM987 standard averaged $^{87}Sr/^{86}Sr = 0.712022 \pm 0.000006 \, 2\sigma\sqrt{n}$, $n = 9$, AMES Nd standard averaged $^{143}Nd/^{144}Nd = 0.512093 \pm 0.000003 \, 2\sigma\sqrt{n}$, $n = 4$, JMC475 standard averaged $0.282145 \pm 0.000006 \, 2\sigma\sqrt{n}$, $n = 21$, and NIST SRM 981 averaged $^{206}Pb/^{204}Pb = 16.9438 \pm 0.0003$; $^{207}Pb/^{204}Pb = 15.5015 \pm 0.0003$; $^{208}Pb/^{204}Pb = 36.7311 \pm 0.0010 \, 2\sigma\sqrt{n}$, $n = 17$. NIST SRM 997 averaged $^{205}Tl/^{203}Tl = 2.4304 \pm 0.0004 \, 2\sigma\sqrt{n}$, $n = 17$.

References

ABDEL-MONEM, A. A. & HEIKEL, M. A. 1981. Major element composition, magma type and tectonic environment of the Mesozoic to Recent basalts, Egypt: A Review. *Bulletin of the Faculty of Earth Science, K.A.U.*, **4**, 121–148.

ABOAZOM, A. S., ASRAN, A. S. H., ABDEL GHANI, M. S. & FARAHAT, E. S. 2006. Geologic and geochemical constraints on the origin of some Tertiary alkaline rift volcanics, the Gharyan Area, Northwestern Libya. *Assiut University Journal of Geology*, **35**, 25–47.

AÏT-HAMOU, F., DAUTRIA, J.-M., CANTAGREL, J.-M., DOSTAL, J. & BRIQUEUE, L. 2000. Nouvelles données géochronologiques et isotopiques sur le volcanisme cénozoïque de l'Ahaggar (Sahara algérien): des arguments en faveur de l'existence d'un panache. *Earth and Planetary Sciences*, **330**, 829–836.

ALTHERR, R., HENJES-KUNST, F., PUCHELT, H. & BAUMANN, A. 1988. Volcanic activity in the Red Sea axial trough – Evidence for a large mantle diapir? *Tectonophysics*, **150**, 121–133.

ALTHERR, R., HENJES-KUNST, F. & BAUMANN, A. 1990. Asthenosphere v. lithosphere as possible sources for basaltic magmas erupted during formation of the Red Sea: Constraints from Sr, Pb and Nd isotopes. *Earth and Planetary Science Letters*, **96**, 269–286.

BAKER, J., SNEE, L. & MENZIES, M. 1996. A brief Oligocene period of flood volcanism in Yemen: implications for the duration and rate of continental flood volcanism at the Afro-Arabian triple junction. *Earth and Planetary Science Letters*, **138**, 39–55.

BAKER, J., CHAZOT, G., MENZIES, M. & THIRLWALL, M. 1998. Metasomatism of the shallow mantle beneath Yemen by the Afar plume – Implications for mantle plumes, flood volcanism, and intraplate volcanism. *Geology*, **26**, 431–434.

BALDRIDGE, W. S., EYAL, Y., BARTOV, Y., STEINITZ, G. & EYAL, M. 1991. Miocene magmatism of Sinai related to the opening of the Red Sea. *Tectonophysics*, **197**, 181–201.

BARRAT, J. A., FOURCADE, S., JAHN, B. M., CHEMINEE, J. L. & CAPDEVILA, R. 1998. Isotope (Sr, Nd, Pb, O) and trace-element geochemistry of volcanics from the Erta'Ale range (Ethiopia). *Journal of Volcanology and Geothermal Research*, **80**, 85–100.

BECCALUVA, L., BIANCHINI, G., ELLAM, R. M., MARZOLA, M., OUN, K. M., SIENA, F. & STUART, F. M. 2008. The role of HIMU metasomatic components in the North African lithospheric mantle: petrological evidence from the Gharyan lherzolite xenoliths, NW Libya. *In*: COLTORTI, M. & GRÉGOIRE, M. (eds) *Metasomatism in Oceanic and Continental Lithospheric Mantle*. Geological Society, London, Special Publications, **293**, 253–277.

BECCALUVA, L., BIANCHINI, G., NATALI, C. & SIENA, F. 2009. Continental flood basalts and mantle plumes: a case study of the Northern Ethiopian Plateau. *Journal of Petrology*, **50**, 1377–1403.

BECKER, T. W. & BOSCHI, T. W. 2002. A comparison of tomographic and geodynamic models. *Geochemisty. Geophysics, Geosystems*, **3**, 1–48.

BERTRAND, H., CHAZOT, G., BLICHERT-TOFT, J. & THORAL, S. 2003. Implications of widespread high-μ volcanism on the Arabian Plate for Afar mantle plume and lithosphere composition. *Chemical Geology*, **198**, 47–61.

BRYAN, W. B., THOMPSON, G. & LUDDEN, J. N. 1981. Compositional variation in normal MORB from 22°–25°N: Mid-Atlantic Ridge and Kane Fracture Zone. *Journal of Geophysical Research*, **86**, 11 815–11 836.

BURKE, K., STEINBERGER, B., TORSVIK, T. H. & SMETHURST, M. A. 2008. Plume generation zones at the margins of Large Low Shear Velocity Provinces on the core–mantle boundary. *Earth and Planetary Science Letters*, **265**, 49–60.

CASTLE, J. C., CREAGER, K. C., WINCHESTER, J. P. & VAN DER HILST, R. D. 2000. Shear wave speeds at the base of the mantle. *Journal of Geophysical Research*, **105**, 21 543–21 558.

CHAUVEL, C. & BLICHERT-TOFT, J. 2001. A hafnium isotope and trace element perspective on melting of the depleted mantle. *Earth and Planetary Science Letters*, **190**, 137–151.

CHAZOT, G. & BERTRAND, H. 1993. Mantle sources and magma-continental crust interactions during early Red Sea-Gulf of Aden rifting in Southern Yemen: elemental and Sr, Nd, Pb isotope evidence. *Journal of Geophysical Research*, **98**, 1819–1835.

COLEMAN, R. G. & MCGUIRE, A. V. 1988. Magma systems related to the Red Sea opening. *Tectonophysics*, **150**, 77–100.

COWARD, M. P. & RIES, A. C. 2003. Tectonic development of North African basins. *In*: ARTHUR, T. J., MAC-GREGOR, D. S. & CAMERON, N. R. (eds) *Petroleum Geology of North Africa: New Themes and Developing Technologies*. Geological Society, London, Special Publications, **207**, 61–83.

DAVIES, G. R. & LLOYD, F. E. 1989. Pb–Sr–Nd isotope and trace element data bearing on the origin of the potassic subcontinental lithosphere beneath Southwest Uganda. *Geological Society of Australia, Special Publications*, **14**, 784–794.

DENIEL, C., VIDAL, P., COULON, C., VELLUTINI, P.-J. & PIGUET, P. 1994. Temporal evolution of mantle sources during continental rifting: the volcanism of Djibouti (Afar). *Journal of Geophysical Research*, **99**, 2853–2869.

DUGDA, M. T. & NYBLADE, A. A. 2006. New constraints on crustal structure in eastern Afar from the analysis of receiver functions and surface wave dispersion in Djibouti. *In*: YIRGU, G., EBINGER, C. J. & MAGUIRE, P. K. H. (eds) *The Afar Volcanic Province within the East African Rift System*. Geological Society, London, Special Publications, **259**, 239–251.

EISSEN, J.-P., JUTEAU, T., JORON, J.-L., DUPRE, B., HUMLER, E. & AL'MUKHAMEDOV, A. 1989. Petrology and geochemistry of basalts from the Red Sea axial rift at 18 degrees North. *Journal of Petrology*, **30**, 791–839.

EL-HINNAWI, E. E. 1965. Petrographical and geochemical studies on Egyptian (UAR) basalts. *Bulletin of Volcanology*, **28**, 283–293.

EL-ISA, Z. H., MECHIE, J., PRODELH, C., MAKRIS, J. & RIHM, R. 1987. A crustal structure study of Jordan derived from seismic refraction data. *Tectonophysics*, **138**, 235–253.

ENDRESS, C. 2010. *Geochemistry of 24 Ma basalts from NE Egypt: implications for widespread magmatism in northern Africa*. MS thesis, Pennsylvania State University.

FARAHAT, E. S., ABDEL GHANI, M. S., ABOAZOM, A. S. & ASRAN, A. M. H. 2006. Mineral chemistry of Al Haruj low-volcanicity rift basalts, Libya: implications for petrogenic and geotectonic evoltion. *Journal of African Earth Sciences*, **45**, 198–212.

FRANZ, G., STEINER, G., VOLKER, F., PULDO, D. & HAMMERSCHMIDT, K. 1999. Plume related alkaline magmatism in central Africa—the Meibod Hills (W Sudan). *Chemical Geology*, **157**, 27–47.

FURMAN, T., BRYCE, J. G., KARSON, J. & IOTTI, A. 2004. East African Rift System (EARS) Plume Structure: Insights from Quaternary Mafic Lavas of Turkana, Kenya. *Journal of Petrology*, **45**, 1069–1088.

FURMAN, T., BRYCE, J. G., ROONEY, T., HANAN, B. B., YIRGU, G. & AYALEW, D. 2006a. Heads and tails: 30 million years of the Afar plume. *In*: YIRGU, G., EBINGER, C. J. & MAGUIRE, P. K. H. (eds) *The Afar Volcanic Province within the East African Rift System*. Geological Society, London, Special Publications, **259**, 95–119.

FURMAN, T., KALETA, K. M., BRYCE, J. G. & HANAN, B. 2006b. Tertiary mafic lavas of Turkana, Kenya: constraints on East African plume structure and the occurrence of high-μ volcanism in Africa. *Journal of Petrology*, **47**, 1221–1244.

GAHRIB, A. A. 2006. Crustal structure of Tushka Region, Abu-Simbel, Egypt, inferred from spectral ratios of P waves of local earthquakes. *Acta Geophysica*, **54**, 361–377.

GEORGE, R., ROGERS, N. & KELLEY, S. 1998. Earliest magmatism in Ethiopian: evidence for two mantle plumes in one flood basalt province. *Geology*, **26**, 923–926.

GOES, S., SPAKMAN, W. & BIJWAARD, H. 1999. A lower mantle source for central European volcanism. *Science*, **286**, 1928–1931.

GUIRAUD, R. & BOSWORTH, W. 1999. Phanerozoic geodynamic evolution of northeastern Africa and the northwestern Arabian Platform. *Tectonophysics*, **315**, 73–108.

GUIRAUD, R., BINKS, R. M., FAIRHEAD, J. D. & WILSON, M. 1992. Chronology and geodynamic setting of

Cretaceous–Cenozoic rifting in West and Central Africa. *Tectonophysics*, **213**, 227–234.

GUIRAUD, R., BOSWORTH, W., THIERRY, J. & DELPLAN-QUE, A. 2005. Phanerozoic geological evolution of Northern and Central Africa: An overview. *Journal of African Earth Sciences*, **43**, 83–143.

HANAN, B. B. & GRAHAM, D. W. 1996. Lead and helium isotope evidence from oceanic basalts for a common deep source of mantle plumes. *Science*, **272**, 991–995.

HANAN, B. B., BILCHERT-TOFT, J., PYLE, D. G. & CHRISTIE, D. M. 2004. Contrasting origins of the upper mantle revealed by hafnium and lead isotopes from the SE Indian Ridge. *Nature*, **432**, 91–94.

HANAN, B. B., SHERVAIS, J. W. & VETTER, S. K. 2008. Plume–lithosphere interaction beneath the Snake River Plain. *Geology*, **36**, 51–54.

HEGNER, E. & PALLISTER, J. S. 1989. Pb, Sr, and Nd isotopic characteristics of Tertiary Red Sea Rift volcanics from the central Saudi Arabian coastal plain. *Journal of Geophysical Research*, **94**, 7749–7755.

HOFMANN, A. W., JOCHUM, K. P., SEUFERT, M. & WHITE, W. M. 1986. Nb and Pb in oceanic basalts: new constraints on mantle evolution. *Earth and Planetary Science Letters*, **79**, 33–45.

HOFMANN, C., COURTILLOT, V., FERAUD, G., ROUCHETT, P., YIRGU, G., KETEFO, E. & PIK, R. 1997. Timing of the Ethiopian flood basalt event and implications for plume birth and global change. *Nature*, **389**, 838–841.

JOHNSON, C. M. & BEARD, B. L. 1993. Evidence from hafnium isotopes for ancient sub-oceanic mantle beneath the Rio Grande rift. *Nature*, **362**, 441–444.

KIEFFER, B., ARNDT, N. *ET AL.* 2005. Flood and shield basalts from Ethiopia: Magmas from the African Superswell. *Journal of Petrology*, **45**, 793–834.

KLEIN, E. M., LANGMUIR, C. H. & STAUDIGEL, H. 1991. Geochemistry of basalts from the SE Indian Ridge, 115°E–138°E. *Journal of Geophysical Research*, **96**, 2089–2107, doi: 10.1029/90JB01384.

KRIENITZ, M.-S., HAASE, K. M., MEZGER, K., ECKARDT, V. & SHAIKH-MASHAIL, M. A. 2006. Magma genesis and crustal contamination of continental intraplate lavas in northwestern Syria. *Contributions to Mineralogy and Petrology*, **151**, 698–716.

KRIENITZ, M.-S., HAASE, K. M., MEZGER, K., VAN DEN BOGAARD, P., THIEMANN, V. & SHAIKH-MASHAIL, M. A. 2009. Tectonic events, continental intraplate volcanism, and mantle plume activity in northern Arabia: Constraints from geochemistry and Ar–Ar dating of Syrian lavas. *Geochemistry, Geophysics, Geosystems*, **10**, doi: 10.1029/2008GC002254.

KUO, B.-Y., GARNERO, E. J. & LAY, T. 2000. Tomographic inversion of S-SKS times for shear velocity heterogeneity in D': degree 12 and hybrid models. *Journal of Geophysical Research*, **105**, 28 139–28 157.

LEBAS, M. J., LEMAITRE, R. W., STRECKEISEN, A., ZANETTIN, B. & IUGS SUBCOMMISSION ON THE SYSTEMATICS OF IGNEOUS ROCKS 1985. A chemical classification of volcanic rocks based on the total alkali-silica diagram. *Journal of Petrology*, **27**, 745–750.

LUCASSEN, F., FRANZ, G., ROMER, R., PUDLO, D. & DULSKI, P. 2008. Nd, Pb, and Sr isotope composition of Late Mesozoic to Quaternary intra-plate magmatism in NE-Africa (Sudan, Egypt): high-μ signatures from the mantle lithosphere. *Contributions to Mineralogy and Petrology*, **156**, 765–784.

MENEISY, M. Y. 1990. Vulcanicity. *In*: SAID, R. (ed.) *The Geology of Egypt*, Balkema, Rotterdam, 157–172.

MOGHAZI, A.-K. M. 2003. Geochemistry of a Tertiary continental basalt suite, Red Sea coastal plain, Egypt: petrogenesis and characteristics of the mantle source region. *Geological Magazine*, **140**, 11–24.

MORSY, A. & ATIA, M. S. 1983. Effects of weathering on the mineralogy and chemical composition of some Egyptian basalts. *Mineralogia Polonica*, **14**, 101–112.

NIMIS, P. 1999. Clinopyroxene geobarometry of magmatic rocks. Part 2. Structural geobarometers for basic to acid, tholeiitic and mildly alkaline magmatic systems. *Contributions to Mineralogy and Petrology*, **135**, 62–74.

OCHIENG', J. O., WILKINSON, A. F., KAGASI, J. & KIMOMO, S. 1988. *Geology of the Loiyangalani area*. Report 107 (Reconnaissance) Republic of Kenya Ministry of Environment and Natural Resources, Mines and Geology Department.

PIK, R., DENIEL, C., COULON, C., YIRGU, G., HOFMANN, C. & AYALEW, D. 1998. The northwestern Ethiopian Plateau flood basalts. Classification and spatial distribution of magma types. *Journal of Volcanology and Geothermal Research*, **81**, 91–111.

PIK, R., DENIEL, C., COULON, C., YIRGU, G. & MARTY, B. 1999. Isotopic and trace element signatures of Ethiopian flood basalts: evidence for plume–lithosphere interactions. *Geochimica et Cosmochimica Acta*, **63**, 2263–2279.

PIK, R., MARTY, B. & HILTON, D. R. 2006. How many mantle plumes in Africa? The geochemical point of view. *Chemical Geology*, **226**, 100–114.

POLLOCK, M. A., KLEIN, E. M., KARSON, J. A. & COLEMAN, D. S. 2009. Compositions of dikes and lavas from the Pito Deep Rift: Implications for crustal accretion at superfast spreading centers. *Journal of Geophysical Research*, **114**, B03207, doi: 10.1029/2007JB005436.

PUTIRKA, K. D., MIKAELIAN, H., RYERSON, F. & SHAW, H. 2003. New clinopyroxene-liquid thermobarometers for mafic, evolved, and volatile-bearing lava compositions, with application to lavas from Tibet and the Snake River Plain, Idaho. *American Mineralogist*, **88**, 1542–1554.

RADWAN, A. H. A., ISSAWY, E. A., DÉREROVÁ, J., BIELIK, M. & KOHUT, I. 2006. Integrated lithospheric modeling in the Red Sea area. *Contributions to Geophysics and Geodesy*, **36**, 373–384.

RITSEMA, J., VAN HEIJST, H. J. & WOODHOUSE, J. H. 1999. Complex shear wave velocity structure imaged beneath Africa and Iceland. *Science*, **286**, 1925–1928.

ROLLINSON, H. R. 1993. *Using Geochemical Data: Evaluation, Presentation, Interpretation*. Longman Group, Essex.

ROONEY, T., HANAN, B., FURMAN, T. & GRAHAM, D. 2008. Multi-component isotopic mixing in the Ethiopian Rift: modeling plume contributions to recent magmatism. *Geochimica et Cosmochimica Acta*, **72**, A804.

RUDNICK, R. L. & FOUNTAIN, D. M. 1995. Nature and composition of the continental crust; a lower crustal perspective. *Reviews of Geophysics*, **33**, 267–309.

SAID, R. 1990. *The Geology of Egypt*. Balkema, Rotterdam.

SALEH, S. & BADAWY, A. 2006. 3D crustal modeling and seismicity of the Eastern Mediterranean and the Egyptian coastal zone. *Acta Geodaetica et Geophysica Hungarica*, **41**, 101–116.

SCHANDELMEIER, H. & REYNOLDS, P. O. 1997. *Paleogeographic-Paleotectonic Atlas of North-Eastern Africa, Arabia, and Adjacent Areas*. Balkema, Rotterdam.

SCHILLING, J. G., KINGSLEY, R. H., HANAN, B. B. & MCCULLY, B. L. 1992. Nd–Sr–Pb Isotopic variations along the Gulf of Aden – evidence for Afar mantle plume continental lithosphere interaction. *Journal of Geophysical Research*, **97**, 10 927–10 966.

SHAW, J. E., BAKER, J. A., MENZIES, M. A., THIRLWALL, M. F. & IBRAHIM, K. M. 2003. Petrogenesis of the largest intraplate volcanic field on the Arabian Plate (Jordan): a mixed lithosphere–asthenosphere source activated by lithospheric extension. *Journal of Petrology*, **44**, 1657–1679.

STERN, R. J., KRÖNER, A., BENDER, R., REISCHMANN, T. & DAWOUD, A. S. 1994. Precambrian basement around Wadi Halfa, Sudan: a new perspective on the evolution of the East Saharan Craton. *Geologische Rundschau*, **83**, 564–577.

SUN, S. S. & MCDONOUGH, W. F. 1989. Chemical and isotopic systematics oceanic basalts: implications for mantle composition and process. *In*: SAUNDERS, A. D. & NORRY, M. J. (eds) *Magmatism in the Ocean Basins*. Geological Society Special Publications, **42**, 313–345.

TACKLEY, P. J. 2008. Layer cake or plume pudding? *Nature Geoscience*, **1**, 157–158.

TAYLOR, S. R. & MCLENNAN, S. M. 1995. The geochemical evolution of the continental crust. *Reviews of Geophysics*, **35**, 241–265.

TODT, W., CLIFF, R. A., HANSER, A. & HOFMANN, A. W. 1996. Evaluation of a 202Pb–205Pb double spike for high precision lead isotope analysis. *In*: BASU, A. & HART, S. (eds) *Earth Processes: Reading the Isotopic Code*. American Geophysical Union, Washington, 429–437.

TORMEY, D. R., GROVE, T. L. & BRYAN, W. B. 1987. Experimental petrology of normal MORB near the Kane Fracture Zone: $22°–25°N$, mid-Atlantic ridge. *Contributions to Mineralogy and Petrology*, **96**, 121–139.

VERVOORT, J. D., PATCHETT, P. J., BLICHERT-TOFT, J. & ALBAREDE, F. 1999. Relationships between Lu–Hf and Sm–Nd isotopic systems in the global sedimentary system. *Earth and Planetary Science Letters*, **168**, 79–99.

VOLKER, F. & MCCULLOCH, M. T. 1993. Submarine basalts from the Red Sea: New Pb, Sr, and Nd isotopic data. *Geophysical Research Letters*, **20**, 927–930.

WANG, K., PLANK, T., WALKER, J. D. & SMITH, E. I. 2002. A mantle melting profile across the Basin and Range, SW USA. *Journal of Geophysical Research*, **107**, doi: 10.1029/2001JB000209.

WEDEPOHL, K. H. 1995. The composition of the continental crust. *Geochimica et Cosmochimica Acta*, **59**, 1217–1232.

WHITE, W. M., ALBARÈDE, F. & TÉLOUK, P. 2000. High-precision analysis of Pb isotope ratios by multi-collector ICP-MS. *Chemical Geology*, **167**, 257–270.

WILLIAMS, G. & SMALL, J. 1984. A study of the Oligo-Micoene basalts in the Western Desert. *In*: *Proceedings of the 7th Petroleum Exploration Seminar*, EGPC, Cairo, 252–268.

WILSON, M. & GUIRAUD, R. 1992. Magmatism and rifting in Western and Central Africa, from Late Jurassic to Recent times. *Tectonophysics*, **213**, 203–225.

WILSON, M. & PATTERSON, R. 2001. Intraplate magmatism related to short-wavelength convective instabilities in the upper mantle: Evidence from the Tertiary–Quaternary volcanic province of western and central Europe. *In*: ERNST, R. E. & BUCHAN, K. L. (eds) *Mantle Plumes: Their Identification Through Time*. Geological Society of America, Special Papers, **352**, 37–58.

The Mid-Miocene East African Plateau: a pre-rift topographic model inferred from the emplacement of the phonolitic Yatta lava flow, Kenya

HENRY WICHURA[1]*, ROMAIN BOUSQUET[1], ROLAND OBERHÄNSLI[1],
MANFRED R. STRECKER[1,2] & MARTIN H. TRAUTH[1]

[1]*Institut für Erd- und Umweltwissenschaften, Universität Potsdam, Karl-Liebknecht-Str. 24,
D-14476 Potsdam, Germany*

[2]*DFG Leibniz Center for Surface Process and Climate Studies, Universität Potsdam,
Karl-Liebknecht-Str. 24, D-14476 Potsdam, Germany*

*Corresponding author (e-mail: wichura@geo.uni-potsdam.de)

Abstract: High topography in the realm of the rifted East African Plateau is commonly explained by two different mechanisms: (1) rift-flank uplift resulting from mechanical and/or isostatic relaxation and (2) lithospheric uplift due to the impingement of a mantle plume. High topography in East Africa has far-reaching effects on atmospheric circulation systems and the amount and distribution of rainfall in this region. While the climatic and palaeoenvironmental influences of high topography in East Africa are widely accepted, the timing, the magnitude and this spatiotemporal characteristic of changes in topography have remained unclear. This dilemma stems from the lack of datable, geomorphically meaningful reference horizons that could unambiguously record surface uplift. Here, we report on the formation of high topography in East Africa prior to Cenozoic rifting. We infer topographic uplift of the East African Plateau based on the emplacement characteristics of the *c.* 300 km long and 13.5 Ma Yatta phonolitic lava flow along a former river valley that drained high topography, centred at the present-day rift. The lava flow followed an old riverbed that once routed runoff away from the eastern flank of the plateau. Using a compositional and temperature-dependent viscosity model with subsequent cooling and adjusting for the Yatta lava-flow dimensions and the covered palaeotopography (slope angle), we use the flow as a 'palaeo-tiltmeter'. Based on these observations and our modelling results, we determine a palaeoslope of the Kenya dome of at least 0.2° prior to rifting and deduce a minimum plateau elevation of 1400 m. We propose that this high topography was caused by thermal expansion of the lithosphere interacting with a heat source generated by a mantle plume. Interestingly, the inferred Mid-Miocene uplift coincides with fundamental palaeoecological changes including the two-step expansion of grasslands in East Africa as well as important radiation and speciation events in tropical Africa.

Thermal and dynamic processes associated with mantle plumes and active continental rifting across East Africa have resulted in long-wavelength plateaus, pronounced regional topographic contrasts, and widespread geophysical anomalies in the different branches of the East African Rift System (EARS; e.g. White & McKenzie 1989; Prodehl *et al.* 1994; Huerta *et al.* 2009). High-plateau topography exists in Ethiopia and Kenya and reaches average elevations of 2400 and 1900 m, respectively. Both plateaus are traversed by Cenozoic rifts constituting the tectonically active EARS (Fig. 1). It has been suggested that the Eastern branch of the EARS propagated from north to south, accompanied by plateau-type volcanism at *c.* 45 Ma in Ethiopia and *c.* 8 Ma in Tanzania (George *et al.* 1998), thus causing the distinct rift segments of Ethiopia, Kenya and northern Tanzania. Separated from the eastern branch by the

Tanzania Craton, the western branch of the EARS comprises a region of high topography extending from lakes Albert and Tanganyika southward to Lake Malawi, with isolated volcanism starting at *c.* 12–10 Ma (e.g. Ebinger 1989; Fig. 1). The extensional structures in both rift branches exploit pre-existing crustal anisotropies in Proterozoic mobile belts (e.g. Ebinger 1989; Hetzel & Strecker 1994; Nyblade & Brazier 2002).

Spatially, high east African topography correlates with volcanism, extensional structures, long-wavelength negative gravity Bouguer anomalies and reduced thickness of mantle lithosphere (e.g. Achauer *et al.* 1994; Class *et al.* 1994; Smith 1994; Fuchs *et al.* 1997; Simiyu & Keller 1997; Nyblade & Brazier 2002; Fig. 1). Radiometrically datable rocks and a wealth of geophysical information have made this region a premier location to investigate the relationship between potential mantle

From: VAN HINSBERGEN, D. J. J., BUITER, S. J. H., TORSVIK, T. H., GAINA, C. & WEBB, S. J. (eds) *The Formation and Evolution of Africa: A Synopsis of 3.8 Ga of Earth History.* Geological Society, London, Special Publications,
357, 285–300. DOI: 10.1144/SP357.15 0305-8719/11/$15.00 © The Geological Society of London 2011.

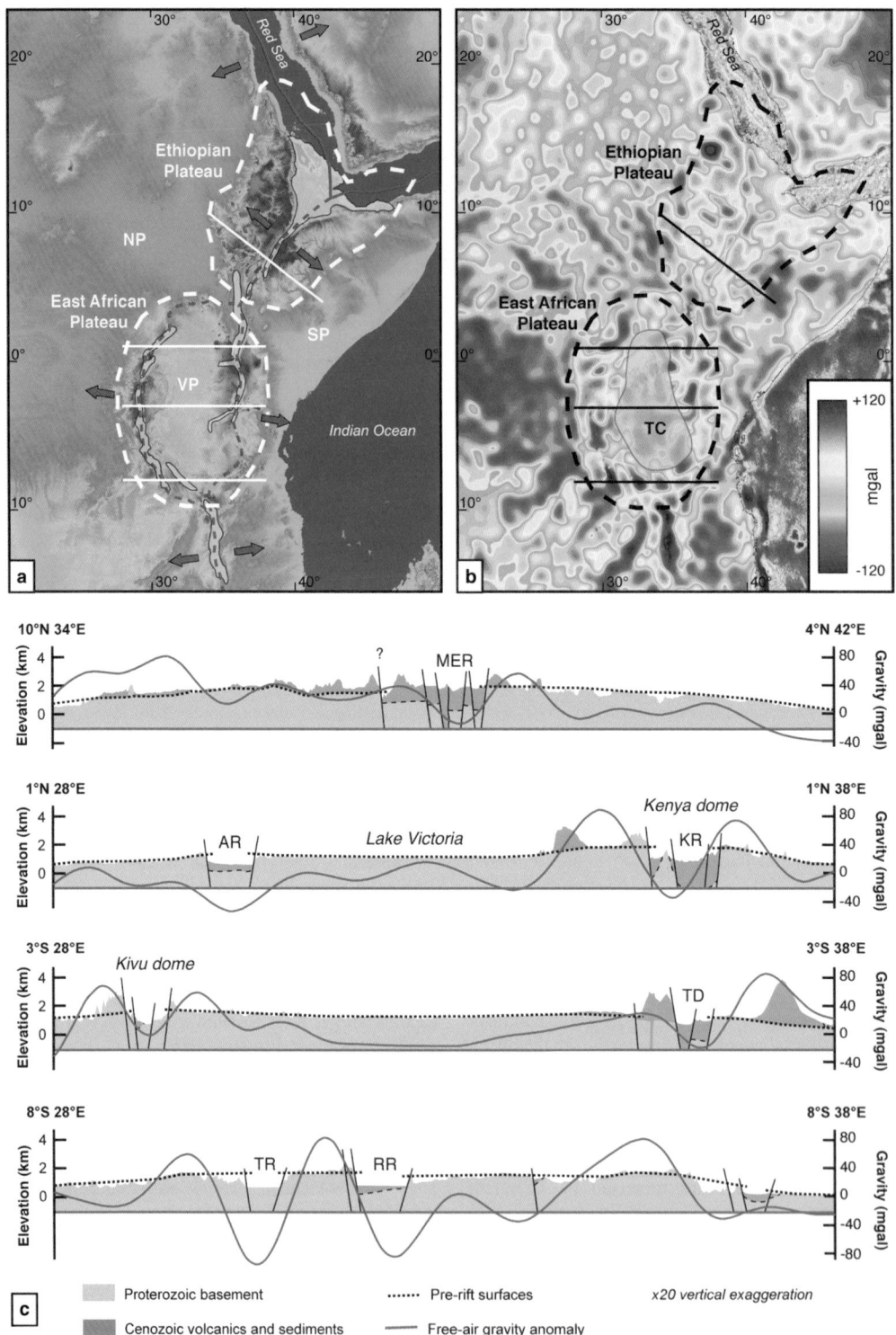

Fig. 1.

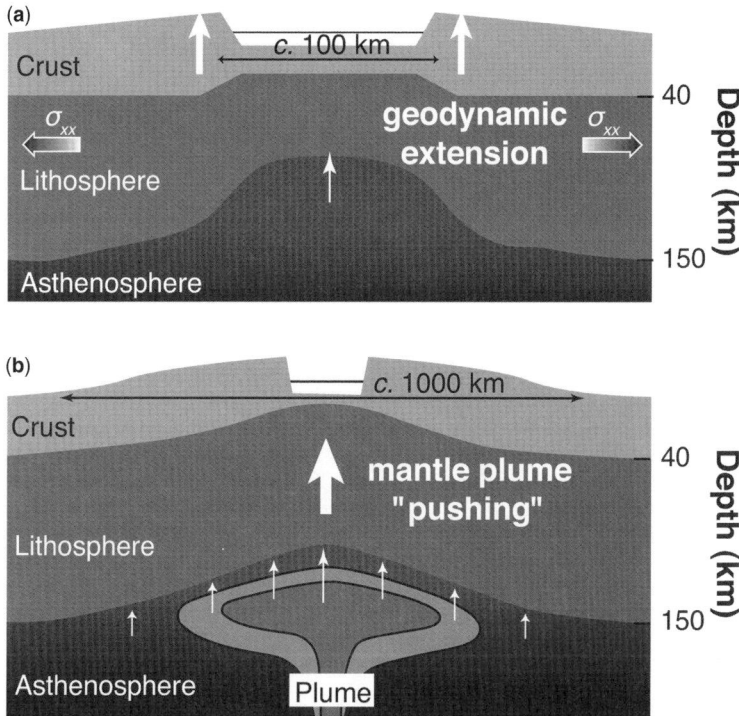

Fig. 2. Sketch of plume–lithosphere interaction in a geodynamically active rift setting. (**a**) Conventional post-rift uplift model provides only rift shoulder uplift due to mechanical relaxation and/or isostatical adjustment at regional-scale. (**b**) Syn-rift uplift model predicts large-scale uplifted dome provoked by mantle plume impingement on continental lithosphere.

plume-induced plateau uplift, tectonic rift-flank development and magmatic processes. Furthermore, Cenozoic uplift in East Africa has also formed an efficient orographic barrier to moisture-bearing winds from the Indian Ocean and the Congo Basin, thus governing the distribution and amount of rainfall in East Africa and influencing the geomorphic process regime and environmental conditions (Sepulchre *et al.* 2006; Wichura *et al.* 2010*a*).

Topographic uplift and related rift evolution in East Africa has been cast into two simplified end-member models: (1) late-stage uplift associated with major rift structures, regional-scale extensional detachments and mechanical relaxation and/or isostatic adjustment (e.g. Kenya Rift; Bosworth 1987; Ebinger *et al.* 1989; Fig. 2a) and (2) regional domal uplift followed by extensional processes associated with mantle–plume impingement on continental lithosphere (e.g. Ethiopian Plateau; Braun & Beaumont 1989; Burov & Cloetingh 2009; Fig. 2b). For the Kenya Rift and neighbouring regions it is however not well known which

Fig. 1. East African topography and free-air gravity anomalies. (**a**) Topography showing the main uplifted plateaus (white dashed lines) and the eastern and western branch of the East African Rift System (yellow areas). Red arrows display the main extension direction. Red lines mark the plate boundaries (solid) and developing plate boundaries (dashed) between the Nubian (NP), Somalian (SP) and Victoria (VP) plates. White lines denote the topographic profiles in (c). (**b**) Free-air gravity anomaly in East Africa for the same map section as in (a). Map is based on gravity anomaly grid of Sandwell & Smith (1997) and created with the gravity anomaly on-line map construction tool (http://www.serg. unicam.it/Geo.html). Grey area highlights the extent of the Tanzania Craton (TC). Black lines denote free-air gravity anomaly profiles in (c). (**c**) Four topographic and free-air gravity anomaly profiles from north to south. The topographic profiles (vertical exaggeration ×20) integrate the potential pre-rift surface and contact between the Proterozoic basement and Cenozoic cover (MER: Main Ethiopian Rift; AR: Albert Rift; KR: Kenya Rift; TD: Tanzania Divergence; TR: Tanganyika Rift; RR: Rukwa Rift).

mechanisms have taken place and in which order of events rift processes have occurred.

Due to the lack of unambiguous topographic and geomorphic reference horizons, quantitative information about a pre-rift uplift across the Tanzania Craton and the adjacent two rift branches is limited. During the Cretaceous, East Africa acquired its identity as a continent-wide topographic feature of low mean elevation due to low tectonic activity, high sea levels and a climate conducive

to chemical weathering and downwearing reducing relief contrasts (Burke & Gunnel 2008). For this reason, the Congo Basin and regions to the NE are covered by shallow-water Cretaceous marine strata that are related to the trans-Saharan seaway, supporting the notion of low topography in East Africa (e.g. Giresse 2005). Smith (1994) estimated < 1000 m of crustal uplift of an elongated pre-rift area, locally defined as the Kenya dome, in Early–Mid-Miocene time (Figs 1c & 3). This term has

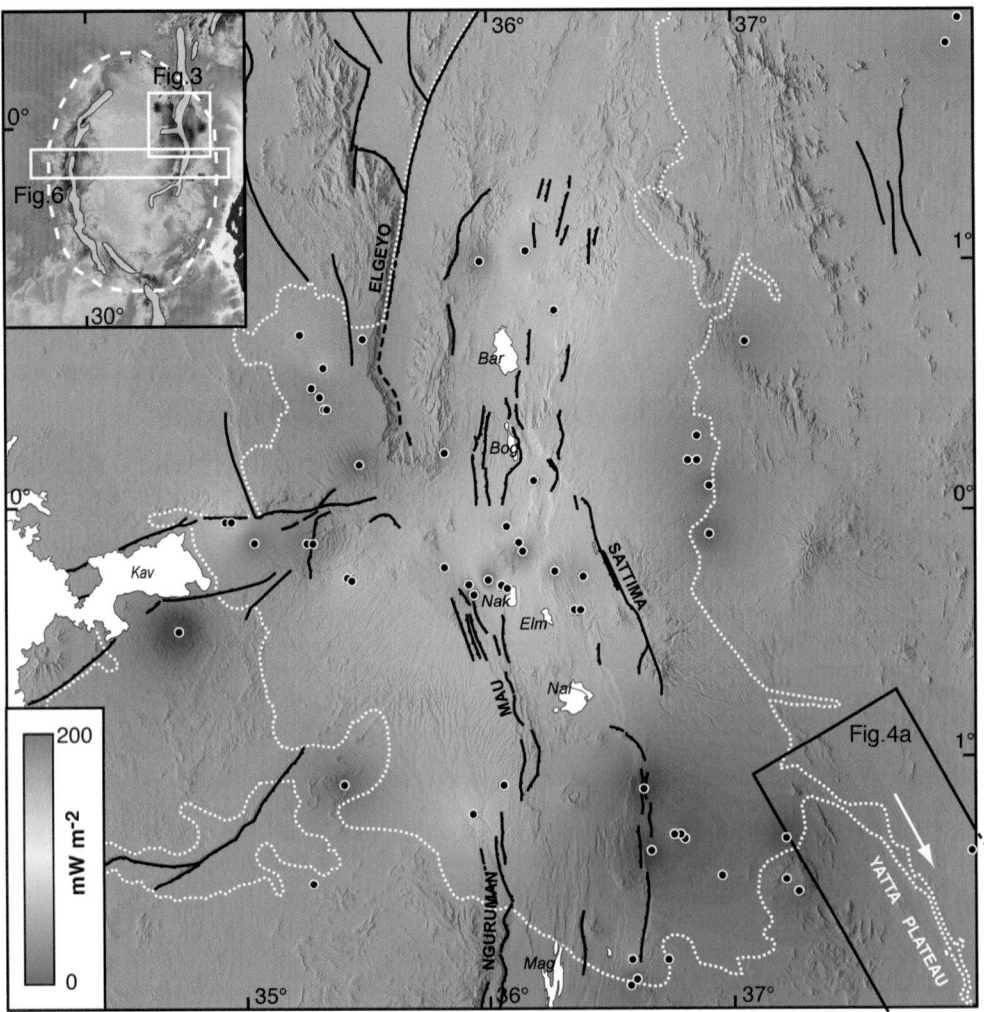

Fig. 3. Recent heat flow in the Central Kenya Rift. East African Plateau inbox shows map section of the main figure and topographic swath profile rectangle used for Figure 6. Heat-flow map is compiled from 62 heat-flow measurements (black dots) taken from the global heat-flow database (http://www.heatflow.und.edu/index2.html) and Wheildon *et al.* (1994). Data points were inverse distance weighted with a *d*-value of 12. White areas show present rift lakes in Central Kenya. Black lines display major escarpments and normal faults. Volcanic extent of the Miocene plateau volcanism is shown by white dotted line. Black outlines show the north-western part of the Yatta Plateau (Fig. 4a). Flow direction of the Yatta lava flow is from NW to SE (white arrow) (Bar: Lake Baringo; Bog: Lake Bogoria; Kav: Kavirondo Golf; Nak: Lake Nakuru; Elm: Lake Elmenteita; Nai: Lake Naivasha; Mag: Lake Magadi).

also been used to describe the topography of the overall rift region, including phonolitic and basaltic plateau volcanics accumulated within and on the flanks of the Central Kenya Rift (e.g. King 1978), thus making a rigorous assessment of topographic evolution difficult. Equally unclear is the topographic evolution of the amagmatic Kivu dome in the Western branch (Chorowicz 2005; Fig. 1c).

Apatite fission-track thermochronology has been used to attempt to define the timing of rift-related shoulder uplift from exhumation signals. However, due to the relatively high annealing temperature of apatite and slow geomorphic process rates, these studies only provided Cretaceous ages related to an earlier rifting event and thus prevented an assessment of Cenozoic uplift and exhumation of the East African Plateau (EAP) using this method (Wagner et al. 1992; Foster & Gleadow 1996). However, more temperature sensitive (U–Th)/He thermochronometry by Spiegel et al. (2007) suggests a rapid cooling episode in Kenya during the Late Neogene that has been related to uplift and exhumation along the eastern and western flanks of the Kenya Rift. In contrast, sustained exhumation since the early Miocene has been reported for southern Ethiopia (Pik et al. 2008).

Due to many ambiguities in the available geologic records (e.g. King 1978; Smith 1994) and only indirect indicators of uplift determined by thermochronology, we chose a different approach to determine the Cenozoic evolution of topography of the EAP using phonolitic flow characteristics on the eastern rift shoulder in Kenya. The Yatta lava flow may provide direct evidence for topography and relief conditions along the eastern flank of the EAP prior to rifting (Wichura et al. 2010a). The Yatta phonolites flowed away from higher topography centred in the region of the present-day rift, utilizing a river channel that formerly drained the plateau region towards the east. Based on these topographic and geomorphic conditions we posit that viscosity and slope angle of the flow may furnish important insights into the gradients for the emplacement of this lava flow, which ultimately provide information on the palaeo-elevation of the source area.

The eastern flank of the EAP during the Miocene

It is known from seismic studies that hot material exists below the uplifted East African Plateau (EAP) and that Cenozoic uplift and volcanism in that region are the consequence of significant temperature changes with depth (Nyblade et al. 2000; Kohn et al. 2008). This spatial phenomenon can be only explained by the activity of a mantle

plume beneath the Tanzania Craton that is proposed to have been active since the Eocene–Oligocene (Ebinger & Sleep 1998; George et al. 1998; Pik et al. 2008). This mantle plume caused different pronounced axial volcanism in the eastern and western branch of the EARS as mentioned earlier.

Considering the total volume of effusive rocks (2.3×10^5 km^3) involved in the volcanic evolution of the Kenya Rift (e.g. Williams 1982; Hay et al. 1995), the plateau phonolites represent the product of the largest phase of volcanism with estimates of $4.0–5.0 \times 10^4$ km^3 (Lippard 1973) erupted between the Mid- and Late Miocene (Smith 1994). The greatest amount of phonolites erupted between 13 and 11 Ma (Smith 1994), as suggested by their isotopic ages and spatial extent (e.g. Baker & Wohlenberg 1971; Fig. 3). Lippard (1973) reported on individual flows with thicknesses ranging from a few tens of metres to as much as 270 m, emplaced over relatively short periods of time. However, one of the most prominent and spectacular flows is the 13.51 Ma Yatta lava flow (Fairburn 1963; Veldkamp et al. 2007), which was emplaced prior to the formation of the oldest rift structures in northern Kenya. The Yatta phonolites constitute one of the longest lava flows on Earth, which flowed south-eastwards along a palaeovalley. Although the location of the volcanic centre of the Yatta phonolites must have been centred in the present-day rift region, the exact source has however not been unambiguously identified (e.g. Williams & Chapman 1986). Smith (1994) inferred the eruptive centre to have been in the vicinity of Sattima Fault, which bounds the Central Kenya Rift on the east.

Yatta lava-flow characteristics

The Yatta flow was channelized for a distance of at least 300 km (length L) from NE to SW into the Tsavo plains (e.g. Walsh 1963; Fujita 1977; Wichura et al. 2010b). Due to differential erosion, this lava flow now constitutes positive relief and stands tall above the Athi River (Fujita 1977) (parallel to the flow), mimicking the course of the palaeo-river (Fig. 4a, b).

The Yatta lava-flow dimensions have been derived using satellite imagery (Shuttle Radar Topography Mission or SRTM and LandSat) and field observations including GPS support and geomorphic mapping (Fig. 4a). The observable thickness of the Yatta Plateau phonolites from the basement contact to the flow-top remnants is between 12 and 25 m (Fig. 4b). The maximum width is between 8 km in the NW and 1 km in the remote SE part. A compilation of 30 (every 10 km) topographic SW–NE profiles yielded

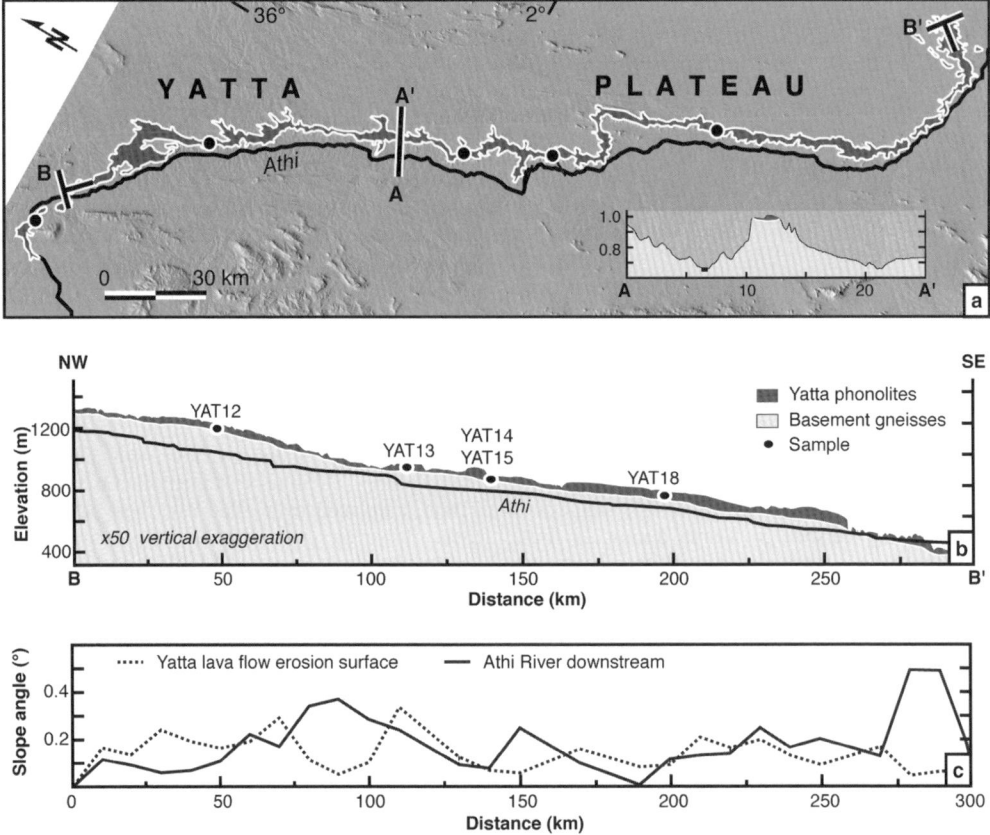

Fig. 4. The Yatta Plateau. (**a**) SRTM image highlighting the Yatta lava flow and the Athi River coursing parallel to the flow. North is 60° anticlockwise rotated. Black dots mark sample locations. A–A′ and B–B′ denote the geological cross-sections of the Yatta lava flow perpendicular to the flow direction and along the central flow top (b), respectively. Labelling of cross section A–A′ is distance in kilometres (*x*-axis) and elevation in kilometres (*y*-axis). (**b**) Geological cross-sections constructed along the Yatta lava flow and present Athi River downstream. Black circles mark the location of phonolitic samples which were used to calculate viscosities by applying the model by Hui & Zhang (2007). (**c**) Recent slope angle distribution corresponding to the erosion surface of the Yatta lava flow and the Athi River relative to the lava flow distance. Slope angles were measured every 10 km.

estimations of the average width of the flow, the altitude of the upper flow surface and the slope angle (Fig. 4c). The remnant average width is estimated to be 3 km; along the eroded top of the flow exists an average elevation difference of 2.8 m km^{-1}. This corresponds to the average slope angle of 0.16° along the bottom of the palaeo-valley (as determined from erosional cuts) which is in good agreement with the early estimate by Saggerson (1963). As the Yatta flow is very narrow compared to its length and the distinguishing ratio of maximal width to maximal length (W_m/L_m) is 0.028, it appears that this flow comprises a single major flow (Walsh 1963) namely of S-type (following the classification of Kilburn & Lopes 1991). Indeed, our field observations and assessments by

earlier workers (e.g. Saggerson 1963; Fujita 1977) clearly show that the lava flow is neither the result of multiple volcanic eruptions nor fissure eruptions along its axis.

Previous studies on the Columbia River basalt in the western US, the Deccan Volcanic Province in India and the island of Hawaii led to the conclusion that the formation of length-dominated lava flows is related to the efficient delivery of lava through well-developed tube systems (e.g. Self *et al.* 1996; Bondre *et al.* 2004; Harris & Rowland 2009). The efficiency of lava tubes in transferring turbulent-flowing lava over long distances and in minimizing (isothermal) cooling is well known, and has previously been discussed in detail by Keszthelyi (1995) for example. However, the features associated with this particular flow formation are generally easily discernible and include collapse pits, longitudinal cracks along the roof and secondary vents (Sakimoto *et al.* 1997; Calvari & Pinkerton 1999). Additionally, porphyric rock texture due to turbulent flow and coated spatter fragments generated by constantly degassing and inclined grooves on the internal wall remnants are characteristic of many lava tubes (Calvari & Pinkerton 1999). All these tube-fed characteristics do not apply to the Yatta phonolites.

For a detailed analysis of flow dynamics and the influence of chemical composition on the final dimension of the Yatta lava flow and its style of emplacement, we examined five different localities with well-preserved flow sections (Table 1). Fortunately, the Yatta Plateau encompasses several fault-related erosion cuts perpendicular to the palaeo-flow direction, which provides access to the internal part of the flow. We observed evidence for laminar flow indicated by aligned feldspars parallel to the flow direction, both along flow rims and in the interior. As turbulent flow does not allow crystallization of aligned phenocrysts (e.g. Bryan 1983; Smith 2002) and evidence for tube-fed flow is not recognized, we consequently interpret the Yatta lava flow as a giant, channelized surface flow.

Locally, pillow lavas at the contact between the phonolites and the basement rocks suggest the presence of water during emplacement. Interestingly, despite all the different features suggesting emplacement in a palaeo-drainage system, no fluvial sediments were observed in the contact zone between the flow and the basement rocks. This phenomenon may however be explained by the thermal erosion effect of flowing lava (e.g. Greeley *et al.* 1998) or the fact that the palaeo-riverbed developed on a more-or-less polished crystalline basement and did not enable large accumulations of sands or gravels. The pillow lavas may simply have resulted from local small ponds formed by rainfall in the palaeo-drainage system.

Basic assumptions

Lava flows can be either considered to behave as a Bingham or Newtonian fluid, rheologically characterized by plastic or dynamic viscosities (e.g. Griffith 2000), respectively, depending on the initial yield strength supporting a fluid to flow. The flow advance is mainly controlled by the rheological conditions in its rear frontal zone (RFZ), the part of a flow between its convex-curved outer front and its established channel structure upstream (Lopes & Kilburn 1990; Kilburn & Lopes 1991). This zone is assumed to be characterized by isotropic and nonfractionated conditions and a uniform density (e.g.

Lopes & Kilburn 1990; Tallarico & Dragoni 1999). Due to low deformation rates in the RFZ (Lopes & Kilburn 1990), we follow the approach by Tallarico & Dragoni (1999) and assume the lava to be an approximately isothermal Newtonian liquid flowing in a rectangular channel. Accordingly, this channel simulates a model-based shape of the Athi river palaeo-valley, which was occupied by the Yatta flow along an inferred constant slope angle β. The solution for viscous flow in a channel with a width W and thickness H can be obtained from the solution for a filled rectangular conduit having the same width and double thickness (e.g. White 1991). The channel dimensions are compatible with the average Athi palaeo-valley width (3 km) and channel margins (12–25 m) over the entire length (290 km). Clearly, downstream variations of width or slope can hydraulically affect the velocity field; in this case the one-dimensional (1D) model is an approximation at a single cross-section, thus assuming laminar flow in flow direction (Tallarico & Dragoni 1999). Furthermore, we regard the lava flow from its eruption temperature near the vent, where lava has not yet developed significant yield strength and can therefore be treated as a Newtonian liquid (Pinkerton & Stevenson 1992; Dragoni & Tallarico 1994). It is important to note that we keep this assumption of Newtonian flow along the entire length and with consequent cooling, because it is not yet known at which temperatures lava flows change to Bingham rheology.

Modelling lava-flow emplacement

Basaltic lava flows reach distances in excess of 150 km along slope angles $<0.1°$ (Self *et al.* 1996). Since viscosities of phonolites are higher compared to basalts, we tested the hypothesis of whether phonolitic lava could reach a distance of *c.* 300 km along a slope of 0.16°. For this simulation we developed a 1D empirical model. The model relies on the viscosity equation by Hui & Zhang (2007):

$$\log \eta = A + \frac{B}{T} + \exp\left(C + \frac{D}{T}\right) \quad (1)$$

This equation simulates viscosity η as a function of temperature T, chemical composition and water content of the lava. A, B, C and D are linear functions of X_i (mole fractions of oxide component i) (Hui & Zhang 2007). The mole fraction for water is integrated in D and combined with an internal temperature dependence of the hydrous component (Hui & Zhang 2007). The 2σ deviation of this fit is 0.61 log η units (Hui & Zhang 2007). As input data we took information from eight standard

Table 1. X-ray fluorescence major element chemistry for eight samples from the Yatta lava flow (YAT_{mean} is a synthetic composition representing the average of all displayed samples; latitude and longitude localities for these samples are also displayed and major oxides in wt%)

Sample	YAT12-4	YAT12-5	YAT13-1	YAT13-2	YAT14-2	YAT15-1	YAT15-2	YAT18-2	YAT_{mean}
Longitude	37°39.34'E	37°39.81'E	37°57.59'E	37°57.86'E	37°58.36'E	38°05.03'E	38°05.29'E	38°21.44'E	
Latitude	1°22.16'S	1°22.58'S	1°58.50'S	1°58.34'S	1°58.76'S	2°10.90'S	2°11.46'S	2°31.36'S	
SiO_2	54.60	54.80	54.30	53.50	55.20	54.30	54.30	54.80	54.48
TiO_2	0.80	0.83	0.81	0.81	0.80	0.82	0.82	0.82	0.81
Al_2O_3	19.00	19.10	19.10	19.20	19.00	18.90	19.00	18.60	18.99
FeOT	5.49	5.72	5.60	5.90	5.53	5.63	5.54	5.56	5.62
MnO	0.25	0.27	0.25	0.45	0.25	0.25	0.23	0.24	0.27
MgO	0.90	0.86	0.80	0.81	0.88	0.81	0.89	0.80	0.84
CaO	2.16	1.44	1.45	1.74	1.33	1.45	1.83	2.04	1.68
Na_2O	7.33	7.57	8.06	6.51	7.42	8.25	6.43	6.43	7.25
K_2O	4.92	5.59	5.39	5.67	5.60	5.29	6.33	5.48	5.53
P_2O_5	0.27	0.29	0.26	0.28	0.27	0.28	0.32	0.29	0.28
H_2O	3.88	2.89	3.51	4.78	3.28	3.48	3.79	4.27	3.74
Total	99.60	99.36	99.53	99.65	99.56	99.46	99.48	99.33	99.50

X-ray fluorescence analysed samples from the Yatta Plateau (Table 1). Fresh hand-specimens were collected on five different outcrops along the general flow trend (Fig. 4a, b). Furthermore, we added one synthetic sample YAT_{mean} to capture the full chemical range of phonolites related to the Yatta Plateau eruptions. This sample is an averaged mixture of oxide components of all Yatta samples (Table 1).

Starting from the eruption temperature typical for phonolitic magmas ($T_0 = 870\ °C$; Ablay et al. 1995) we defined a stepwise (n) cooling process with $\Delta T = 10\ °C$. The gradually increasing viscosity of the lava during cooling corresponds to the stepwise decreasing rear frontal velocity v_n following Jeffreys' (1925) equation:

$$v_n(T_n, X_i) = \frac{H^2\, g\rho\, \sin\beta}{a\eta_n(T_n, X_i)} \quad (2)$$

where β is the slope angle, g is the gravity constant ($9.81\ \mathrm{m\ s^{-2}}$), ρ is the rock density (phonolite $2.4 \times 10^3\ \mathrm{kg\ m^{-3}}$), H is the mean lava flow thickness, a is 3 as suggested by Booth & Self (1973) for channels that are wide relative to their depth and η_n is the viscosity computed stepwise from the model by Hui & Zhang (2007). Furthermore, we defined the simulation to finish when no flow advance can be recognized. This occurs when cooling forces the velocity to fall below $0.01\ \mathrm{m\ s^{-1}}$.

In order to evaluate the time for the RFZ of the Yatta lava flow to reach its final position we used estimates of effusion rates F, the flux of lava expressed by the volume of effused material per unit of time, approximately given by:

$$F = WH\overline{V} \quad (3)$$

where W is the mean width, H is the mean thickness and V bar is the mean velocity of the RFZ (the sum of all v_n divided by n cooling steps). Furthermore, we assume a constant production rate of material at the vent. Estimates of the emplacement time (t_{em}) were obtained using the Graetz model, an empirical conduction-limited cooling model (Pinkerton & Sparks 1976) in which lava-flow cooling is compared to the heat loss of a hot fluid moving through a cool conduit. The heat advection within the flow along its length is described by the dimensionless Graetz (G_z) number (Carrasco-Núñez 1997) and is given by:

$$G_Z = \frac{H^2\overline{V}}{kL} \quad (4)$$

where k is the thermal diffusivity (phonolite $3.5 \times 10^{-7}\ \mathrm{m^2\ s^{-1}}$; Martin & White 2002) and L is the lava-flow length. It has been empirically determined that lava flows come to rest before G_z falls to about 300 (Carrasco-Núñez 1997). Hence,

the emplacement time (t_{em}) can be considered as:

$$t_{em} = \frac{L}{\overline{V}} \quad (5)$$

so that the equation can be rewritten as:

$$t_{em} = \frac{H^2}{G_z k}. \quad (6)$$

Finally, we modelled the lava flow in two ways: (1) pre-defining a fixed thickness and calculating how far the lava can flow and (2) assuming the present lava-flow length and estimating the required mean thickness to compute the mean frontal velocity and effusion rate simultaneously. In addition, in both simulations we calculated the mean thickness and length as a function of the underlying topography (Wichura et al. 2010b).

Modelling results

During the advance of long channelized lava flows from the vent to the attainment of their final morphology, simultaneous viscosity changes are essentially temperature controlled. Therefore, viscosity equation (1) is suitable to be applied to our model, integrating a composition dependency and a validation over the full range of temperatures considered (600–870 °C). Starting from the eruption temperature, the model computed viscosities for five selected Yatta samples (YAT12-5, YAT14-2, YAT15-2, YAT18-2, YAT_{mean}; Table 1 and Fig. 5a) according to n temperature steps. As the number of cooling steps increases, the trend toward higher viscosities is maintained for all samples with varying degrees (Fig. 5a) and the model predicts the highest viscosities between 6.2×10^3 and $1.3 \times 10^7\ \mathrm{Pa\ s}$ for the composition with the lowest water content (2.89 wt%; YAT12-5). We also calculated the lowest viscosities ranging from 2.6×10^3 to $3.4 \times 10^6\ \mathrm{Pa\ s}$ for the sample with the highest water content (4.27 wt%; YAT18-2). Viscosities for sample YAT15-2 and YAT_{mean} overlap over the entire temperature range from 3.6×10^3 to $6.0 \times 10^6\ \mathrm{Pa\ s}$ (Fig. 5a).

The development of the rear frontal velocity for the sample YAT_{mean} is shown in Figure 5b, determining the constraints (mean thickness, slope angle and effusion rate) under which a theoretical lava flow reaches the morphological parameters of the Yatta lava flow, specifically the length of 290 ± 10 km. The rear frontal velocities at the eruption temperature range from 3.5 to $1.7\ \mathrm{m\ s^{-1}}$ with slope angles from 2 and 0.5°, respectively. Constant slope angles in the range of 0.1–0.2° are more realistic with respect to the slope angle of 0.16° estimated for the Yatta flow. For that range, the mean frontal velocities at the vent were calculated as

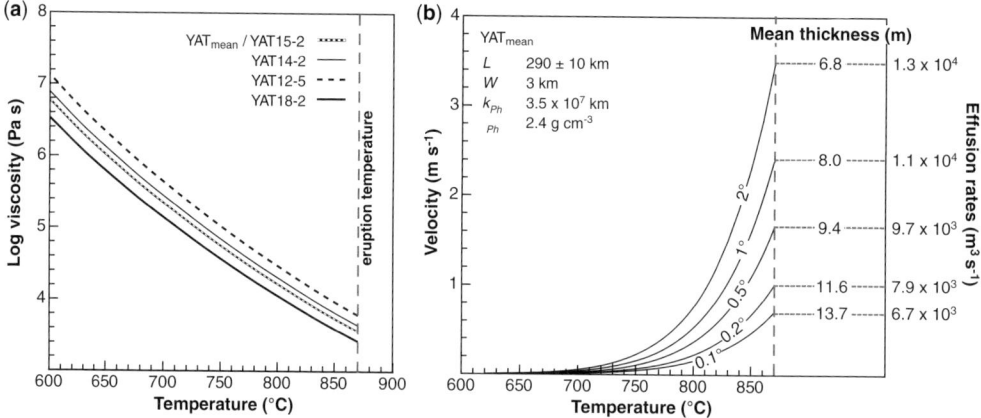

Fig. 5. Emplacement of the Yatta lava flow. (**a**) Viscosities as a function of temperature and composition of four phonolitic samples from the Yatta lava flow. Sample compilation is added by one synthetic composition considered as representative for all studied samples (YAT$_{mean}$). All viscosities were stepwise ($\Delta T = 10$ °C) calculated with the Hui & Zhang (2007) model starting from the eruption temperature $T_0 = 870$ °C (Ablay *et al.* 1995). 2σ for calculated viscosities is 0.61 log μ units. (**b**) Rear frontal velocity profiles of sample YAT$_{mean}$ as a function of temperature, mean thickness, slope angle and effusion rate. Every profile represents the parameter fixing the conditions (slope angle, mean thickness and effusion rate) under which the recent Yatta lava flow dimension can be achieved (*L*: lava flow length; *W*: lava flow width; k_{ph}: thermal diffusivity for phonolite; ρ_{ph}: density for phonolite).

0.7–1.0 m s^{-1}. The calculated stepwise flow advance in the channel comes to a halt at temperatures between 620 °C for a slope angle of 2° and 670 °C for a slope angle of 0.1°.

Modelling small active lava flows on different volcano types has shown that the shape and roughness of the channel used by the lava are of great importance (Del Negro *et al.* 2007). Sensitivity tests of our model however suggest that topographic disturbances along the course of the river valley do not significantly influence the model result, particularly when dealing with high effusion rates of 6.7×10^3 to 1.3×10^4 m^3 s^{-1} (Fig. 5b). We found that the most important parameters controlling lava-flow length are the water content of the lavas, the viscosity change with ongoing cooling and the lava-flow thickness (Wichura *et al.* 2010*b*).

Our simulation of the Yatta flow suggests that sustained flow may occur at slope angles of at least 0.18–0.20°, with an estimated uncertainty of 15% (Wichura *et al.* 2010*a*). Under these circumstances and a lava-flow thickness of 12 ± 3 m, the lava-flow front must have been able to reach the eastern Tsavo plains over a distance of c. 300 km. For comparable thicknesses and the present-day slope of 0.16° the flow simulation yields a maximal flow length of 260 km, which is 13% shorter than the Yatta flow. It is important to note that the slope angles we calculated are minimum values.

In this respect, the applicability of the model is limited because of two crucial facts observed on the Yatta flow. First, the source of the lava flow is

not well known. Hence, it is possible that the Yatta lava flow may have been even longer than the 290 ± 10 km used in our model. The simulation would therefore either predict a higher mean flow thickness or flow with a slope angle >0.2°. The latter scenario is more likely because the thickness depends not just on the slope but also on mass loss to levees, volume loss (or gain) by degassing or inflation and rheological changes during transport (e.g. Baloga *et al.* 2001). Second, the flow direction is NW–SE and therefore rotated by 60° northwards with respect to the slope of the EAP in this area. The projection of the lava flow onto an east–west profile along the same inclined plane yields a slope angle of up to 0.38° (Fig. 6).

Using the calculated slope angle of 0.18–0.20° across the EAP and assuming a virtually flat top of the uplifted region and a uniform uplift at both flanks of the plateau, the phonolites would have originated in a plateau setting with a maximum elevation of 1400 m (Fig. 6). This supports our initial hypothesis of an areally expansive, high pre-rift topography prior to 13.5 Ma that comprised a region at least 1300 km in diameter in the present-day EAP region (Wichura *et al.* 2010*a*; Fig. 6). This assessment is also in line with earlier interpretations by Le Bas (1971) and Burke (1996). Furthermore, Ebinger (1989) proposed a temporally parallel doming phase for the eastern and western branch in the Early Miocene and observed that several indirect lines of evidence indicate that uplift of the rift flanks are superimposed on the plateau

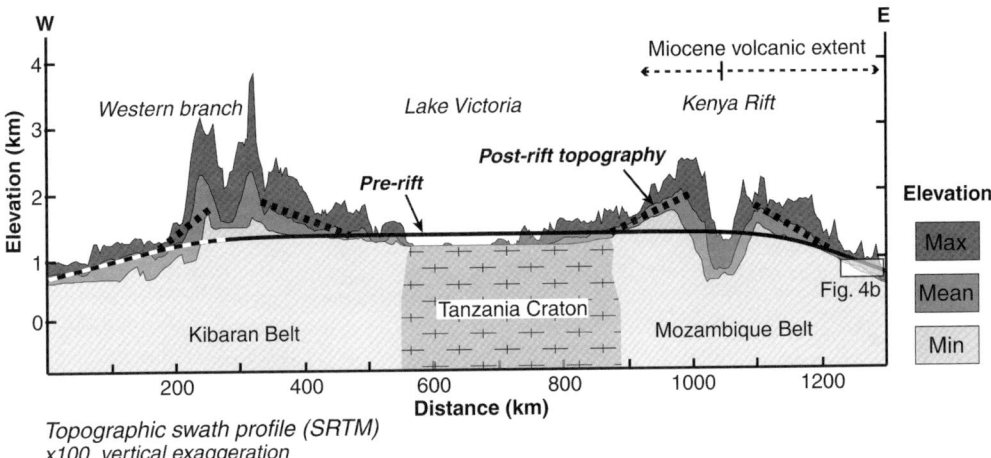

Fig. 6. Mid-Miocene East African Plateau derived from slope angle modelling on the eastern flank. The pre-rift topographic model uses a slope angle of 0.2°, which yields a flat plateau 1400 m high and 1300 km wide across the Tanzania Craton and adjacent Proterozoic belts (Kibaran and Mozambique). The dotted line represents the pre-rift topography of the western flank assuming that the East African Plateau rose uniformly at both flanks (e.g. Ebinger 1989). Different grey areas display the modern topography across the centre of the East African Plateau (1° × 12° topographic swath profile extracted from SRTM data; see also insert in Fig. 3). Thick dashed line marks the portion of syn- and post-rift uplift, respectively. White box shows the location of the profile shown in Figure 4b. East–west arrows display the Miocene volcanic extent from the central Kenya Rift axis.

topography (Fig. 6). The lower slope angle measured today on the eastern rim of the EAP may therefore be related to crustal bending during limited uplift of the rift shoulders.

Links between tectonics, topography, climate and evolution

Our results provide important new insights into east African topographic evolution prior to rifting and ensuing environmental change. Today, the EAP constitutes an important topographic feature that influences the distribution and amount of rainfall associated with the African–Indian monsoon and westerly moisture sources at a continental scale (Sepulchre et al. 2006; Roberts & White 2010). Exacerbated by the additional effects of rift-shoulder uplift and the construction of volcanoes adjacent to the rift flanks (Shakleton 1945), the uplift of the EAP has consequently impacted the drainage system in East Africa. In combination, the long-wavelength uplift, the build-up of volcanic edifices and the rifting process have helped compartmentalize the EAP into semi-independent basins. Indeed, the eastern branch of the East African Rift System comprises a patchwork of closely spaced humid and arid sedimentary environments in the vicinity of the equator (Sepulchre et al. 2006; Bergner et al. 2009) which have repeatedly

changed their characteristics during episodes of climatic change (Trauth et al. 2003, 2005; Garcin et al. 2009). Many of these basins hold lakes ranging from freshwater to alkaline in character, and these changes must have impacted palaeoenvironmental conditions and evolutionary patterns in a fundamental way. For example, the Mid-Miocene uplift of the EAP documented in our study correlates with an early expansion of C_4 plants in East Africa at c. 14 Ma (Kingston et al. 1994; Morgan et al. 1994; Fig. 7). In addition, Retallack et al. (1990) reported a change from grassy woodland to wooded grassland, based on fossil soils and grasses analysed at the Miocene mammal fossil locality at Fort Ternan, western Kenya. Jacobs (2004) studied a grass-dominated savannah biome that began expanding in the Mid = Miocene (16 Ma) and became widespread during the Late Miocene (c. 8 Ma), as documented by pollen and carbon isotopes from both West and East Africa (Fig. 7).

Taken together, the history of palaeo-vegetation development in East Africa suggests a two-step evolution towards more pronounced dry conditions at the expense of rainforests at 16–14 Ma, and the expansion of grasslands at 8–7 Ma (Jacobs 2004, Fig. 7). Based on our analysis, the first step towards increasingly arid conditions may have been linked with incipient uplift of the East African and Ethiopian plateaus prior to 13.5 and 20 Ma (Pik et al. 2008), respectively. In contrast,

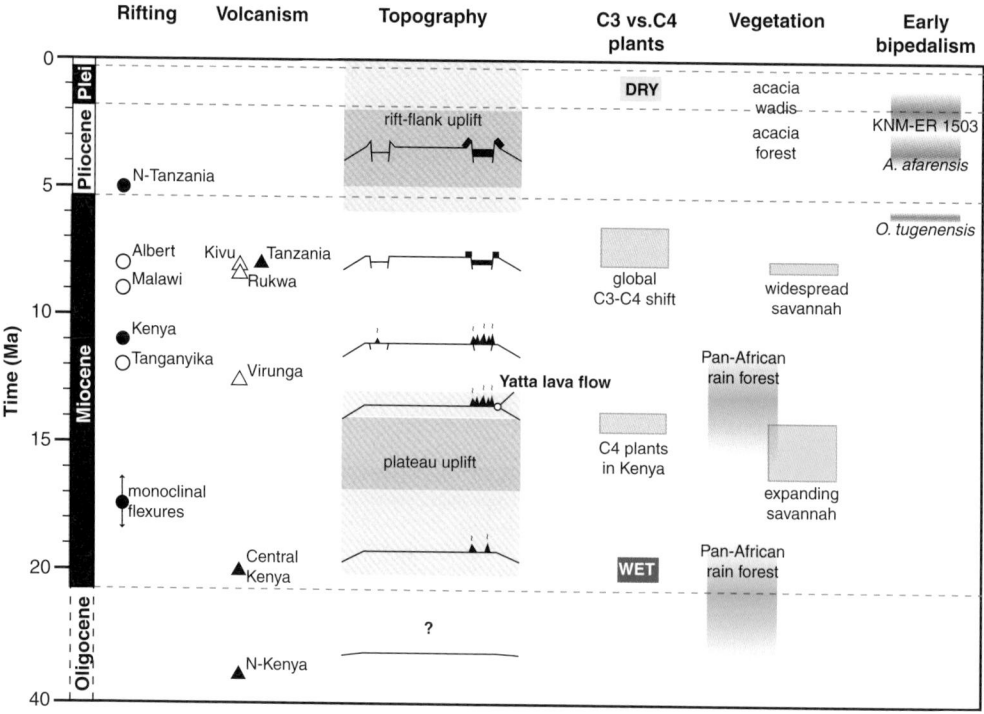

Fig. 7. Cenozoic uplift chronology, vegetation changes and early hominid evolution in East Africa. Mid-Miocene uplift correlates with major climatic and environmental shifts. Data compilation as follows: onset of volcanic activity (triangle) and rifting (circle) in the eastern (filled) and western branch (empty) of the East African Rift System (Baker & Wohlenberg 1971; Cohen *et al.* 1993; Chorowicz 2005; McDougall & Brown 2009), Pliocene rift flank uplift (Ebinger *et al.* 1993; Spiegel *et al.* 2007), K–Ar age of the Yatta lava flow (Veldkamp *et al.* 2007), pollen and carbon isotopes (Retallack *et al.* 1990; Kingston *et al.* 1994; Cerling *et al.* 1997; Jacobs 2004), Pan-African rainforest expansion–isolation phases (Vincens *et al.* 2006; Couvreur *et al.* 2008) and early hominid bipedalism (Richmond & Jungers 2008).

the second shift towards a dominance of grassland vegetation may have taken place in response to the global decrease in pCO_2 (Cerling *et al.* 1997; Ségalen *et al.* 2007), possibly severed by the additional topographic effects of rifting, rift-shoulder uplift and volcanism.

The Mid-Miocene uplift of the East African and Ethiopian plateaus and coeval environmental changes also impacted radiation and speciation. For example, molecular phylogenetic studies have been performed on the evolutionary history of Pan-African lineages of the rainforest-restricted plant family *Annonaceae*. The Pan-African clade of *Annonacae* unravels a pattern in diversification of rainforest-restricted trees occurring in west/ central and east African rain forests and provides strong evidence for diachronous origins (Couvreur *et al.* 2008). Successive connection–isolation events between the Guineo-Congolian rainforest and East Africa at *c.* 16 and *c.* 8 Ma (Couvreur *et al.* 2008; Fig. 7) temporally correlate with our interpretation of two-tiered environmental change

due to plateau uplift prior to 13.5 Ma and global pCO_2 changes in the Late Miocene (Fig. 7).

Model for EAP uplift

Although poorly constrained, the evolution of the oldest major erosion surface in East Africa falls into the age range between 53 and 24 Ma and is attributed to the Africa Surface (Burke & Gunnel 2008). This surface can be viewed as a basal reference surface which was successively uplifted and finally became an integral part of the East African Plateau (EAP). Arguments in support of such a surface in East Africa are (1) Miocene volcanic flows (e.g. Yatta Plateau) and lake deposits in the Kavirondo Gulf of Lake Victoria covering this erosional surface sculpted into basement rocks; (2) the volcanic rocks related to the regions encompassing Mount Elgon (20–22 Ma; Walker *et al.* 1969) and Lake Turkana (18–16 Ma; Joubert 1966) in western and northern Kenya, respectively, overlie an erosion surface in basement rocks (Morley *et al.* 1992); and (3) palaeo-drainage reconstructions by King *et al.* (1972) and King (1978) suggest that the Kenya Rift had developed along an ancient drainage divide, possibly uplifted during the Palaeogene (Smith 1994). Vestiges of the basement-cut erosion surface exist in the form of broad interfluves or wide plateaus, such as the surface comprising the region across the Serengeti Plains at an elevation of 1500 m (Burke & Gunnel 2008).

In an active geodynamic setting such as the East African Rift System (EARS), which has been affected by buoyant asthenospheric material and high heat flow since *c.* 45 Ma (Ebinger & Sleep 1998; Fig. 3), large temporal and spatial variations of geothermal gradients are to be expected when the disparate onset of volcanism and extensional faulting are taken into account. The quantification of lithospheric warping due to a mantle anomaly and consequent uplift and deformation of geomorphic reference horizons is therefore complex and depends on several parameters. While dimension, temperature and density differences constitute main controlling factors between the plume, the surrounding asthenosphere and the overlying lithosphere (e.g. White & McKenzie 1989; Moore *et al.* 1999), the mantle/crust strength ratio ('coupled crust–mantle' model) and the Moho temperature can also fundamentally influence the configuration of a domal uplift, rift segments and the ensuing topographic response (Gueydan *et al.* 2008). In this regard, inherited crustal anisotropies could induce an asymmetric structure as observed in the East African Rift System (e.g. Smith & Mosley 1993; Hetzel & Strecker 1994; Tommasi & Vauchez 2001).

However, based on the fact that the mantle plume below the Tanzania Craton not only gives rise to the observed volcanism in the adjacent rifts but is also significantly interacting with the lithosphere and generating lithospheric uplift across the EAP (Nyblade *et al.* 2000), Koehn *et al.* (2008) proposed that a component of thermal uplift is probably realistic and would add to the complexity of the rift system. As inferred from our study, the evolution of the EAP pre-rift topography reached an elevation of 1400 m, represents a change of 1% thermal induced lithospheric expansion and can be interpreted as a large-scale surface expression caused by lithosphere–plume interactions (Fig. 8). While coeval regional-scale lithospheric folding (Burov & Cloetingh 2009) may have triggered the reactivation of pre-existing Proterozoic structures straddling the Tanzania Craton, initial monoclinal flexures developed in the Mid-Miocene and fundamentally influenced the rifting process in East Africa.

Conclusion

Using a modelling approach involving the channelized and length-dominated phonolitic Yatta lava flow of the eastern Kenya Rift shoulder during its emplacement in the Mid-Miocene, we determine that this lava flow moved on a palaeotopography with a slope angle of 0.18–0.2°, originating at a maximum elevation of 1400 m. We conclude that this elevation corresponds to the Mid-Miocene elevation of the East African Plateau (EAP) in Kenya, which integrated a Palaeogene erosion surface sculpted into basement rocks, dynamically uplifted due to plume–lithosphere interactions. Furthermore, topographic changes (of the order 1000 m) clearly affect moisture transport and spatial rainfall patterns and amounts. For this reason it is most likely that the uplift of the EAP

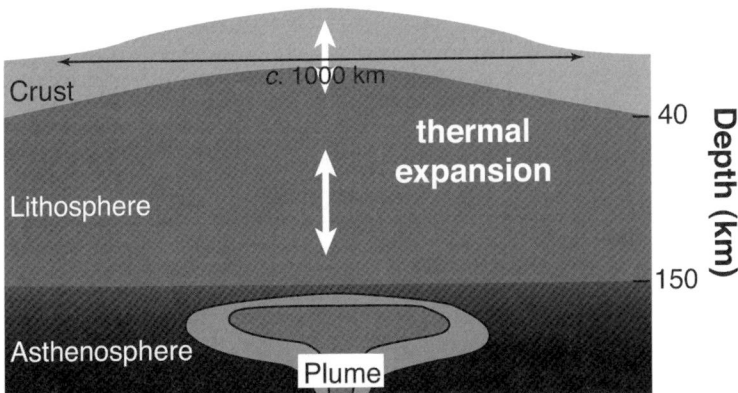

Fig. 8. Thermal dilatation model for pre-rift plateau. The heating of a mantle plume affecting the east African lithosphere is inducing a volume change of the lithosphere and allowing a large-scale uplift of the East African Plateau.

and the formation of pronounced palaeotopography by the Mid-Miocene had a first-order impact on the evolution of the east African tropical climate. The timing of uplift is consistent with hydrologically induced vegetation shifts, such as the two-step expansion of grasslands at *c.* 16 and *c.* 8 Ma. In addition, the evolution of topography coincides with important radiation and speciation events.

This work was conducted in the Graduate School GRK1364 Shaping Earth's Surface in a Variable Environment funded by the German Research Foundation (DFG), co-financed by the federal state of Brandenburg and the University of Potsdam. We thank the Government of Kenya (Research Permits MOST 13/001/30C 59/10) and the University of Nairobi for research permits and support. Finally, we thank the reviewers R. M. S. Fernandes and J. J. Tiercelin for their fruitful comments on the first manuscript version.

References

ABLAY, G. J., ERNST, G. G. J., MARTI, J. & SPARKS, R. S. J. 1995. The ~2 ka subplinian eruption of Montana Blanca, Tenerife. *Bulletin of Volcanology,* **57,** 337–355.

ACHAUER, U., GLAHN, A. *ET AL.* 1994. New ideas on the Kenya rift based on the inversion of the combined dataset of the 1985 and 1989/90 seismic tomography experiments. *Tectonophysics,* **236,** 305–329.

BAKER, B. H. & WOHLENBERG, J. 1971. Structure and evolution of the Kenya rift valley. *Nature,* **229,** 538–542.

BALOGA, S. M., GLAZE, L. S., PEITERSON, M. N. & CRISP, J. A. 2001. Influence of volatile loss on thickness and density profiles of active basaltic flow lobes. *Journal of Geophysical Research,* **106,** 13395–13405.

BERGNER, A. G. N., STRECKER, M. R., TRAUTH, M. H., DEINO, A., GASSE, F., BLISNIUK, P. & DÜHNFORTH, M. 2009. Tectonic and climatic controls on evolution of rift lakes in the Central Kenya Rift, East Africa. *Quaternary Science Reviews,* **28,** 2804–2816.

BONDRE, N. R., DURAISWAMI, R. A. & DOLE, G. 2004. Morphology and emplacement of flows from the Deccan Volcanic Province, India. *Bulletin of Volcanology,* **66,** 29–45.

BOOTH, B. & SELF, S. 1973. Rheological features of the 1971 Mount Etna lavas. *Philosophical Transactions of the Royal Society of London,* **274,** 99–106.

BOSWORTH, W. 1987. Off-axis volcanism in the Gregory Rift, east Africa: implications for models of continental rifting. *Geology,* **15,** 397–400.

BRAUN, J. & BEAUMONT, C. 1989. A physical explanation of the relation between flank uplifts and the breakup unconformity at rifted continental margins. *Geology,* **17,** 760–764.

BRYAN, W. B. 1983. Systemantics of modal phenocrysts assemblages in Submarin basalts: petrologic implications. *Contributions to Mineralogy and Petrology,* **83,** 62–74.

BURKE, K. 1996. The African Plate. *South African Journal of Geology,* **99,** 341–409.

BURKE, K. & GUNNEL, Y. 2008. *A continental-scale synthesis of geomorphology, tectonics, and environmental change over the past 180 million years.* Geological Society of America, Memoirs, **201.**

BUROV, E. & CLOETINGH, S. 2009. Controls of antle plumes and lithospheric folding on modes of intraplate continental tectonics: differences and similarities. *Geophysical Journal International,* **178,** 1691–1722.

CALVARI, S. & PINKERTON, H. 1999. Lava tube morphology on Etna and evidence for lava flow emplacement mechanism. *Journal of Volcanology and Geothermal Research,* **90,** 263–280.

CARRASCO-NÚÑEZ, G. 1997. Lava flow growth inferred from morphometric parameters: a case study of Citlaltépetl Volcano, Mexico. *Geological Magazine,* **134,** 151–162.

CERLING, T. E., HARRIS, J. M., MACFADDEN, B. J., LEAKEY, M. G., QUADE, J., EISENMANN, V. & EHLERINGER, J. R. 1997. Global vegetation change through the Miocene/Pliocene boundary. *Nature,* **389,** 153–158.

CHOROWICZ, J. 2005. The East African Rift System. *Journal of African Earth Sciences,* **43,** 379–410.

CLASS, C., ALTHERR, R., VOLKER, F., EBERZ, G. & MCCULLOCH, M. T. 1994. Geochemistry of Pliocene to Quaternary alkali basalts from the Huri Hills, northern Kenya. *Chemical Geology,* **113,** 1–22.

COHEN, A. S., SOREGHAN, M. J. & SCHOTZ, C. A. 1993. Estimating the age of formation of lakes: an example of lake Tanganyika, East African Rift. *Geology,* **21,** 511–514.

COUVREUR, T. L. P., CHATROU, L. W., SOSEF, M. S. M. & RICHARDSON, J. E. 2008. Molecular phylogenetics reveal multiple tertiary vicariance origins of the African rain forest trees. *BioMed Central Biology,* **6,** 54–63.

DEL NEGRO, C., FORTUNA, L., HERAULT, A. & VICARI, A. 2007. Simulations of the 2004 lava flow at Etna volcano using the Magflow cellular automata model. *Bulletin of Volcanology,* **70,** 805–812.

DRAGONI, M. & TALLARICO, A. 1994. The effect of crystallization on the rheology and dynamics of lava flows. *Journal of Volcanology and Geothermal Research,* **59,** 241–252.

EBINGER, C. J. 1989. Tectonic development of the western branch of the East African Rift System. *Geological Society of American Bulletin,* **101,** 885–903.

EBINGER, C. J. & SLEEP, N. H. 1998. Cenozoic magmatism throughout east Africa resulting from impact of a single plume. *Nature,* **395,** 788–791.

EBINGER, C. J., BECHTEL, T., FORSYTH, D. & BOWIN, C. 1989. Effective elastic plate thickness beneath the East African and Afar plateaus and dynamic compensation of the uplifts. *Journal of Geophysical Research,* **94,** 2883–2901.

EBINGER, C. J., YEMANE, T., WOLDEGABRIEL, G., ARONSON, J. L. & WALTER, R. C. 1993. Late eocene-recent volcanism and faulting in the southern main Ethiopian Rift System. *Journal of the Geological Society,* **150,** 99–108.

FAIRBURN, W. A. 1963. *Geology of the North Machakos-Thika area.* Geological Survey of Kenya, Report **59.**

FOSTER, A. & GLEADOW, A. J. W. 1996. Structural framework and denudation history of the flanks of the Kenya and Anza Rifts, East Africa. *Tectonics*, **15**, 258–271.

FUCHS, K., ALTHERR, R., MULLER, B. & PRODEHL, C. 1997. Structure and dynamic processes in the lithosphere of the Afro-Arabian Rift System. *Tectonophysics*, **278**, 1–352.

FUJITA, M. 1977. *Geology of the Yatta Plateau, Kenya.* 2nd Prelimary Report African Studies, Nagoya University, Earth Sciences **2**, 161–164.

GARCIN, Y., JUNGINGER, A., MELNICK, D., OLAGO, D. O., STRECKER, M. R. & TRAUTH, M. H. 2009. Late Pleistocene–Holocene rise and collapse of the Lake Suguta, northern Kenya Rift. *Quaternary Science Reviews*, **28**, 911–925.

GEORGE, R., ROGERS, N. & KELLEY, S. 1998. Earliest magmatism in Ethiopia: evidence for two mantle plumes in one flood basalt province. *Geology*, **26**, 923–926.

GIRESSE, P. 2005. Mesozoic–Cenozoic history of the Congo Basin. *Journal of African Earth Sciences*, **43**, 301–315, doi: 10.1016/j.jafrearsci.2005.07.009.

GREELEY, R., FAGENTS, S. A., HARRIS, R. S., KADEL, S. D. & WILLIAMS, D. A. 1998. Erosion by flowing lava: field evidence. *Journal of Geophysical Research*, **103**, 27325–27334.

GRIFFITH, R. W. 2000. The dynamics of lava flows. *Annual Review of Fluid Mechanics*, **32**, 477–518.

GUEYDAN, F., MORENCY, C. & BRUN, J.-P. 2008. Continental rifting as a function of lithosphere mantle strength. *Tectonophysics*, **460**, 83–93.

HARRIS, A. J. L. & ROWLAND, S. K. 2009. Effusion rate controls on lava flow length and the role of heat loss: a review. *In*: THORDARSON, T., SELF, S., LARSEN, G., ROWLAND, S. K. & HOSKULDSSON, A. (eds) *Studies in Volcanology: The Legacy of George Walker.* Special Publications of IAVCEI, **2**. Geological Society, London, 33–51.

HAY, D. E., WENDLANDT, R. F. & KELLER, G. R. 1995. Origin of Kenya Rift Plateau-type flood phonolites: integrated petrologic and geophysical constraints on the evolution of the crust and upper mantle beneath the Kenya Rift. *Journal of Geophysical Research*, **100**, 10549–10557.

HETZEL, R. & STRECKER, M. R. 1994. Late Mozambique Belt structures in western Kenya and their influence on the Evolution of the Cenozoic Kenya Rift. *Journal of Structural Geology*, **16**, 189–201.

HUERTA, A. D., NYBLADE, A. A. & REUSCH, A. M. 2009. Mantle transition zone structure beneath Kenya and Tanzania: more evidence for deep-seated thermal upwelling in the mantle. *Geophysical Journal International*, **177**, 1249–1255.

HUI, H. & ZHANG, Y. 2007. Toward a general viscosity equation for natural anhydrous and hydrous silicate melts. *Geochimica et Cosmochimica Acta*, **71**, 403–416.

JACOBS, B. F. 2004. Palaeobotanical studies from tropical Africa: relevance to the evolution of forest, woodland and savannah biomes. *Philosophical Transactions of the Royal Society of London*, **359**, 1573–1583.

JEFFREYS, H. J. 1925. The flow of water in an inclined channel of rectangular section. *Philosophical Magazine*, **48**, 793–807.

JOUBERT, P. 1966. *Geology of the Loperot area.* Geological Survey of Kenya, Report **74**.

KESZTHELYI, L. 1995. A preliminary thermal budget for lava tubes on the earth and planets. *Journal of Geophysical Research*, **100**, 20411–20420.

KILBURN, C. R. J. & LOPES, R. M. C. 1991. General patterns of flow field growth: Aa and blocky lavas. *Journal of Geophysical Research*, **96**, 19721–19732.

KING, B. C. 1978. Structural and volcanic evolution of the Gregory rift valley. *In*: BISHOP, W. W. (ed.) *Geological Background to Fossil Man.* Geological Society, London, Special Publications, **6**, 29–54.

KING, B. C., LE BAS, M. J. & SUTHERLAND, D. S. 1972. The history of the alkaline volcanoes and intrusive complexes of eastern Uganda and western Kenya. *Journal of Geological Society, London*, **128**, 173–205.

KINGSTON, J. D., MARINO, B. D. & HILL, A. 1994. Isotopic evidence for Neogene hominid paleoenvironments in the Kenya rift valley. *Science*, **264**, 955–959.

KOHN, D., AANYU, K., HAINES, S. & SACHAU, T. 2008. Rift nucleation, rift propagation and the creation of basement micro-plates within active rifts. *Tectonophysics*, **458**, 105–116.

LE BAS, M. J. 1971. Per-alkaline volcanism, crustal swelling, and rifting. *Nature: Physical Science*, **230**, 85–87.

LOPES, R. M. C. & KILBURN, C. R. J. 1990. Emplacement of lava flow fields: application of terrestrial studies to Alba Patera, Mars. *Journal of Geophysical Research*, **95**, 14383–14397.

LIPPARD, S. J. 1973. The petrology of phonolites from the Kenya Rift. *Lithos*, **6**, 217–234.

MARTIN, U. & WHITE, J. D. L. 2002. Melting and mingling of phonolitic pumice deposits with intruding dykes: an example from the Otago Peninsula, New Zealand. *Journal of Volcanology and Geothermal Research*, **114**, 129–146.

MCDOUGALL, I. & BROWN, F. H. 2009. Timing of volcanism and evolution of the northern Kenya Rift. *Geological Magazine*, **146**, 34–47.

MOORE, W. B., SCHUBERT, G. & TACKLEY, P. J. 1999. The role of rheology in lithospheric thinning by mantle plumes. *Geophysical Research Letters*, **26**, 1073–1076.

MORGAN, M. E., KINGSTON, J. D. & MARINO, B. D. 1994. Carbon isotopic evidence for the emergence of C_4 plants in the Neogene from Pakistan and Kenya. *Nature*, **367**, 162–165.

MORLEY, C. K., WESTCOTT, W. A., STONE, D. M., HARPER, R. M., WIGGER, S. T. & KARANJA, F. M. 1992. Tectonic evolution of the Northern Kenyan rift. *Journal of the Geological Society, London*, **149**, 333–348.

NYBLADE, A. A. & BRAZIER, R. A. 2002. Precambrain lithospheric controls on the development of the East African Rift System. *Geology*, **30**, 755–758.

NYBLADE, A. A., OWENS, T. J., GURROLA, H., RITSEMA, J. & LANGSTON, C. A. 2000. Seismic evidence for a deep upper mantle thermal anomaly beneath east Africa. *Geology*, **28**, 599–602.

PIK, R., MARTY, B., CARIGNAN, J., YIRGU, G. & AYALEW, T. 2008. Timing of East African Rift development in southern Ethiopia: implication for mantle plume activity and evolution of topography. *Geology*, **36**, 167–170.

PINKERTON, H. & SPARKS, R. S. J. 1976. The 1975 sub-terminal lavas, Mount Etna: a case history of the formation of a compound lava field. *Journal of Volcanology and Geothermal Research*, **1**, 167–182.

PINKERTON, H. & STEVENSON, R. J. 1992. Methods of determining the rheological properties of magmas at sub-liquidus temperature. *Journal of Volcanology and Geothermal Research*, **53**, 47–66.

PRODEHL, C., KELLER, G. R. & KHAN, M. A. 1994. Crustal and upper mantle structure of the Kenya Rift. *Tectonophysics*, **236**, 1–466.

RETALLACK, G. J., DUGAS, D. P. & BESTLAND, E. A. 1990. Fossil soils and grasses of a middle Miocene East African grassland. *Science*, **247**, 1325–1328.

RICHMOND, B. G. & JUNGERS, W. L. 2008. Orrorin tugenensis femoral morphology and the evolution of hominin bipedalism. *Science*, **319**, 1662–1665.

ROBERTS, G. G. & WHITE, N. 2010. Estimating uplift rate histories from river profiles using African examples. *Journal of Geophysical Research*, **115**, doi: 10.1029/2009JB006692.

SAGGERSON, E. P. 1963. *Geology of the Simba-Kibwezi Area*. Geological Survey of Kenya, Report **58**.

SAKIMOTO, S., CRISP, J. & BALOGA, S. M. 1997. Eruption constraints on tube-fed planetary lava flows. *Journal of Geophysical Research*, **102**, 6597–6613.

SANDWELL, D. T. & SMITH, W. H. F. 1997. Marine gravity from Geosat and ERS 1 satellite Altimetry. *Journal of Geophysical Research*, **102**, 10039–10054.

SÉGALEN, L., LEE-THORP, J. A. & CERLING, T. E. 2007. Timing of C_4 grass expansion across sub-Saharan Africa. *Journal of Human Evolution*, **53**, 549–559.

SELF, S., THORDARSON, T. *ET AL.* 1996. A new model for the emplacement of Columbia River basalts as large, inflated pahoehoe lava flow fields. *Geophysical Research Letters*, **23**, 2689–2692.

SEPULCHRE, P., RAMSTEIN, G., FLUTEAU, F., SCHUSTER, M., TIERCELIN, J. J. & BRUNET, M. 2006. Tectonic Uplift and Eastern Africa Aridification. *Science*, **313**, 1419–1423.

SHAKLETON, R. M. 1945. *Geology of the Nyeri Area*. Mineralogical and Geological Department of Kenya, Report **12**.

SIMIYU, S. M. & KELLER, G. R. 1997. An integrated analysis of lithospheric structure across the East African Plateau based on gravity anomalies and recent seismic studies. *Tectonophysics*, **278**, 291–313.

SMITH, J. V. 2002. Structural analysis of flow-related textures in lavas. *Earth-Science Reviews*, **57**, 279–297.

SMITH, M. 1994. Stratigraphic and structural constraints on meachnism of active rifting in the Gregory Rift, Kenya. *Tectonophysics*, **236**, 3–22.

SMITH, M. & MOSLEY, P. 1993. Crustal heterogeneity and basement influence on the development of the Kenya Rift, East Africa. *Tectonics*, **12**, 591–606.

SPIEGEL, C., KOHN, B. P., BELTON, D. X. & GLEADOW, A. J. W. 2007. Morphotectonic evolution of the Central Kenya Rift Flanks: implications for Late Cenozoic environmental change in East Africa. *Geology*, **35**, 427–430.

TALLARICO, A. & DRAGONI, M. 1999. Viscous Newtonian laminar flow in a rectangular channel: application to Etna lava flows. *Bulletin of Volcanology*, **61**, 40–47.

TOMMASI, A. & VAUCHEZ, A. 2001. Continental rifting parallel to ancient collisional belts: an effect of the mechanical anisotropy of the lithospheric mantle. *Earth and Planetary Science Letters*, **185**, 199–210.

TRAUTH, M. H., DEINO, A., BERGNER, A. G. N. & STRECKER, M. R. 2003. East African climate change and orbital forcing during the last 175 kyr BP. *Earth and Planetary Science Letters*, **206**, 297–313.

TRAUTH, M. H., MASLIN, M. A., DEINO, A. & STRECKER, M. R. 2005. Late Cenozoic moisture history of East Africa. *Science*, **309**, 2051–2053.

VELDKAMP, A., BUIS, E., WIJBRANS, J. R., OLAGO, D. O., BOSHOVEN, E. H., MARÉE, M. & VAN DEN BERG VAN SAPAROEA, R. M. 2007. Late Cenozoic fluvial dynamics of the River Tana, Kenya, an uplift dominated record. *Quaternary Science Reviews*, **26**, 2897–2912.

VINCENS, A., TIERCELIN, J. J. & BUCHET, G. 2006. New Oligocene–early Miocene microflora from southwestern Turkana Basin Palaeoenvironmental implication in the northern Kenya Rift. *Palaeogeography, Palaeoclimatology, Palaeoecology*, **239**, 470–486.

WAGNER, M., ALTHERR, R. & VAN DEN HAUTE, P. 1992. Apatite fission-track analysis of Kenyan basement rocks: constraints on the thermotectonic evolution of the Kenya dome. A reconnaissance study. *Tectonophysics*, **204**, 93–110.

WALKER, A., BROCK, P. W. G. & MACDONALD, R. 1969. Fossil mammal locality on Mount Elgon, eastern Uganda. *Nature*, **223**, 591–596.

WALSH, L. E. 1963. *Geology of the Ikutha Area*. Geological Survey of Kenya, Report **56**.

WHEILDON, J., MORGAN, P., WILLIAMSON, K. H., EVANS, T. R. & SWANBERG, C. A. 1994. Heat flow in the Kenya rift zone. *Tectonophysics*, **236**, 131–149.

WHITE, F. M. 1991. *Viscous Fluid Flow*. McGraw-Hill, New York.

WHITE, R. S. & MCKENZIE, D. P. 1989. Magmatism at rift zones: the Generation of continental margins and flood basalts. *Journal of Geophysical Research*, **94**, 7685–7729.

WICHURA, H., BOUSQUET, R., OBERHÄNSLI, R., STRECKER, M. R. & TRAUTH, M. H. 2010a. Evidence for middle Miocene uplift of the East African Plateau. *Geology*, **38**, 543–546.

WICHURA, H., BOUSQUET, R. & OBERHÄNSLI, R. 2010b. Emplacement of the mid-Miocene Yatta lava flow, Kenya: implications for modeling long channelled lava flows. *Journal of Volcanology and Geothermal Research*, **198**, 325–338, doi: 10.1016/j.jvolgeores.2010.09.017.

WILLIAMS, L. A. J. 1982. Physical aspects of magmatism in continental rifts. *In*: PALMÁSON, G. (ed.) *Continental and Oceanic Rifts*. American Geophysical Union, Washington D. C., 193–222.

WILLIAMS, L. A. J. & CHAPMAN, G. R. 1986. Relationships between major structures, salic volcanism and sedimentation in the Kenya rift from the equator northwards to Lake Turkana. *In*: FROSTICK, L. E, RENAUT, R. W., REID, I. & TIERCELIN, J. J. (eds) *Sedimentation in the African Rifts*. Geological Society, London, Special Publications, **25**, 59–74.

Late Neogene volcanics and interbedded palaeosols near Mount Kenya

W. C. MAHANEY[1]*, RENÉ W. BARENDREGT[2], MIKE VILLENEUVE[3],
JAROSLAV DOSTAL[4], T. S. HAMILTON[5] & MICHAEL W. MILNER[1]

[1]*Department of Geography, 4700 Keele St., N. York, Ontario, M3J 1P3, Canada and
Quaternary Surveys, 26 Thornhill Ave., Thornhill, ON L4J 1J4, Canada*

[2]*University of Lethbridge, Department of Geography, Lethbridge, Alberta, T1K 3M4, Canada*

[3]*Continental Geoscience Division, Geological Survey of Canada, Ottawa, ON, K1A 0E8, Canada*

[4]*Saint Mary's University, Department of Geology, Halifax, Nova Scotia, 83H3C3, Canada*

[5]*Camosun College, Department of Chemistry and Geosciences, Lansdowne Campus,
3100 Foul Bay Rd., Victoria, British Columbia, V8P 5J2, Canada*

*Corresponding author (e-mail: arkose@rogers.com)

Abstract: Two lava flows with interbedded palaeosols outcrop *c.* 40 km SW of Mount Kenya, near the Amboni River north of Mweiga, Kenya along the Nyeri/Thompson Falls Road, at $0°18'S$; $37°48'E$. These flows, overlain by loess, are principally trachyandesite and form the base of the Mount Kenya Volcanic Series which, in the early literature, is described as being of probable Miocene/Pliocene age. Here we report $^{39}Ar/^{40}Ar$ dates (*c.* 5.2–5.5 Ma) and reversed magnetizations which establish a Latest Miocene to Earliest Pliocene age for these flows. Weathering characteristics of palaeosols interbedded with the lavas indicate generally dry climatic conditions during the Late Miocene, punctuated with humid events during the Pliocene and Quaternary. These Late Miocene–Quaternary palaeosols depict a relatively long and complex weathering history, followed by loess deposition. The palaeosols appear to have been episodically deflated, initially in phase with the deposition of lavas when surfaces were devoid of vegetation and later during periods of climatic deterioration when wind systems intensified. Such weathering histories within palaeosol profiles are also documented on nearby Mount Kenya, where well-weathered lower palaeosol horizons developed on Matuyama-age tills are overlain by much younger less-weathered horizons developed on Brunhes-age loess. The geochronology of Late Miocene lavas reported here provides maximum ages for weathering histories of palaeosols formed in a xeric tropical highland climate.

Little is known of the volcanic pile of Mount Kenya (Mahaney 1990) apart from that outlined and discussed in Shackleton (1945) and Baker (1967), the original seminal works on the bedrock geology of the mountain. Its location almost directly on the equator, its elevation to *c.* 500 mb level in the atmosphere and its dissected lava dome combine to make it a principal region in which to reconstruct Upper Neogene stratigraphy, geomorphology and palaeoclimate, however. Because the mountain contains well-preserved ancient deposits and bedrock surfaces mantled with equally old buried and relict soils (palaeosols), it has much to offer in the way of reconstructing episodic changes of climate (Mahaney 1990). Still older bedrock and sediment outcrops in the adjoining plains and foothills, discussed here, provide an extended timescale beyond the Quaternary to the Late Miocene. Basement rocks outcrop on the southern, eastern and

northern flanks of the mountain and in several catchments. Precambrian inliers occur beneath the Simbara basalts and agglomerates of Miocene age. The distribution of basement rocks indicates that Precambrian crust nonconformably underlies the Tertiary volcanic massif. The Simbara Series of presumed Miocene age (Baker 1967) thins to the north of Nyeri (Fig. 1, cf. Kihuyo), where it is replaced by Nyeri Tuff (trachytic tuffs with minor agglomerates of Late Miocene–Early Pliocene age). The Laikipian Basalt of presumed Pliocene age, first observed and described by Gregory (1921) and later described in detail by Shackleton (1945) and Baker (1967), outcrops on the south-western slopes of the mountain between Nyeri and Naro Moru. It usually occurs as a dissected lava flow, covered with a thick red 'lithosol' (cf. Oxisol; NSSC 1995) in places and often overlain by thin Mollisols or Alfisols formed in Quaternary loess.

From: Van Hinsbergen, D. J. J., Buiter, S. J. H., Torsvik, T. H., Gaina, C. & Webb, S. J. (eds) *The Formation and Evolution of Africa: A Synopsis of 3.8 Ga of Earth History*. Geological Society, London, Special Publications, **357**, 301–318. DOI: 10.1144/SP357.16   0305-8719/11/$15.00 © The Geological Society of London 2011.

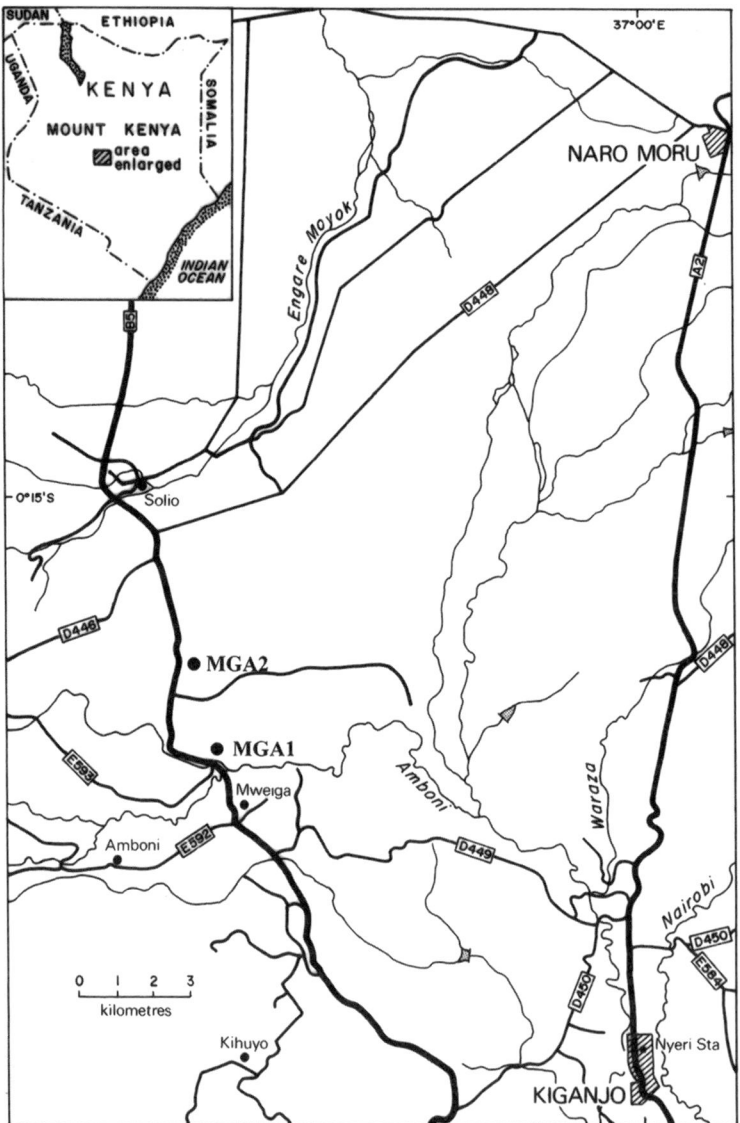

Fig. 1. Location of study sites west of Mount Kenya on the Aberdare dip-slope on the flanks of the Eastern Rift Valley.

Eruptive rocks from Mount Kenya and its parasitic cones cover an area of 4300 km^2. The area forms an elliptical shape, oriented NE–SW and has a diameter of *c.* 100 km. The complex of rock comprising the Mount Kenya Volcanic Series of Pleistocene age includes lower basalt, overlain by phonolite, rhomb porphyry, trachyte and kenyte (phonolite with distinct rhomb-shaped anorthoclase phenocrysts) from the main volcanic vent, now closed with a nepheline syenite plug. A second period of volcanism occurred late in the Pleistocene when lava and pyroclastics erupted from several

satellite cones and fissures (Baker 1967), the most prominent being Ithanguni (3894 m a.s.l.). A single sample of trachytic tuff from Ithanguni has an age of 320 ± 20 ka (Rundle 1985; Charsley 1989).

While Late Neogene palaeoclimate records are well known from recovered ocean cores (Lisieck & Raymo 2005) or regional lacustrine geological records (Trauth *et al.* 2005; Scholz *et al.* 2007), reconstructing climatic change of terrestrial environments is problematic given the poor preservation of geological archives (especially palaeosols). Palaeoclimate records obtained from near-equatorial terrestrial environments, such as those described here, provide a new and important correlation with Mount Kenya from the inception of its evolution and its resultant glacial history to the present day. In this paper we date early eruptive rocks associated with its Late Miocene–Early Pliocene history, and provide documentation of Late Neogene climatic conditions for East Africa by applying an integrated, multidisciplinary suite of techniques. The palaeomagnetic data provide spot readings of the Earth's magnetic field during this important period in East African volcanism. It is the prominent outcrops of trachytic andesite and tephrite lava and their loessic cover deposits to the north of the Amboni River (Fig. 1) on the west flank of the mountain (1825 m a.s.l.) that are described here. New $^{39}Ar/^{40}Ar$ dates (Fig. 2a, b) and palaeomagnetic measurements of the lavas provide maximum ages for the interbedded and overlying palaeosols, and provide a first estimate of weathering histories and palaeoclimate trends from this region spanning Late Miocene to present.

Regional geology

The Central Plateau of East Africa, of which Mount Kenya is a part, is an area of intense tectonic activity that has waxed and waned since the Early Miocene (Baker 1967; Schluter 1997). Following fractures as old as Cretaceous, a system of rift valleys formed in the Neogene and saw the formation of enormous lakes. Major faulting led to down-dropping of several rift structures while others remained several hundred metres above sea level. Following fractures in the crust, immense lava flows and volcanoes, such as the Aberdare Mountains (Satima: 4001 m a.s.l.) and Mount Kenya (5199 m a.s.l.) dating from the Pliocene and Pleistocene, formed in echelon aligned from west to east adjacent to the Eastern Rift Valley. Older volcanoes such as Mount Elgon (4321 m a.s.l.), a shield volcano of Miocene age, provide a prominent marker on the boundary between Kenya and Uganda. Not all mountain massifs are volcanic; some are inlier erosional remnants of Late Proterozoic age (Cherangani Hills north of Mount Kenya). In the study area these basement lithologies are typically high-grade gneisses and granitic rocks.

The succession of volcanic rock which forms the Mount Kenya Massif amounts to a thickness of c. 6200 m overlying the African Precambrian basement series (Baker 1967; Vernacombe 1983; Ries et al. 1992; Fig. 2c). Eruption of Late Miocene–Pliocene phonolite, trachyte and basaltic rocks, tentatively dated by Baker (1967), were considered to have been derived mainly from the mantle (Rogers 2006; Dawson 2008). This volcanism, accompanied by eruption of tuff and agglomerate, probably began just prior to the development of an erosion surface of Late Pliocene age (Baker 1967). The lavas described here probably erupted just prior to the Laikipian Basalts, observed further to the north, and form part of the underpinnings of the Mount Kenya Massif. The lavas are tentatively correlated with the Nyeri Tuff.

The tectonic history of the East African Rift and adjoining Aberdare Dip Slope involved several stages (Foster & Gleadow 1996; Dawson 2008). The Kenyan Dome uplift of more than 400 m began in the Late Cretaceous and initial stages of rifting began in the Late Miocene (Baker et al. 1971), culminating with a 'true' rift valley in the Pliocene (Foster et al. 1997; Ebinger et al. 2000; Veldkamp et al. 2007). Between 3.0 and 3.5 Ma initial eruptions occurred from Mount Kenya; the final eruption from the main vent was originally dated to 2.65 Ma (Everden & Curtis 1965) but has been revised to c. 2.71 Ma (Veldkamp et al. 2007). Net updoming of the Eastern Rift Valley rim is calculated at c. 1 km between 3.21 and 2.65 Ma and further indeterminate uplift is postulated from incision rates in Tana Valley occurring c. 0.9 Ma. Mount Kenya must have been close to its present c. 5000 m a.s.l. elevation to accommodate the growth of Quaternary ice estimated to have started c. 2.0 Ma (Mahaney 1990).

Estimates of erosion of the summit of Mount Kenya vary from over 2000 m (Gregory 1921) to <1000 m (Mahaney 1990). The mountain itself has been stable since the time of its formation; crystallization of the plug in the Late Pliocene (2.65 Ma; Everden & Curtis 1965; cf. Veldkamp et al. 2007) with faulting restricted displacements to minor amounts of only tens of metres.

The history of glaciations on Mount Kenya is extensive. The oldest glaciations, named Gorges, Lake Ellis and Naro Moru, are known only from the eastern and south-eastern flanks of the mountain. They are important in the context of this study because the loess in the upper beds of the succession (Fig. 2a, b) may be sourced primarily from these glaciated areas. The Gorges Glaciation, which occurred partly within the Olduvai subchron, is older than 1.9 Ma and has an indeterminate upper limit at <1.8 Ma (Mahaney 1990; Mahaney et al. 1997). The reversely magnetized sediments of the younger Lake Ellis Glaciation are older than 0.78 Ma (Barendregt & Mahaney 1988) and younger than underlying sediments, which are of Olduvai age. Weathering characteristics of glacial deposits and palaeosols of the Gorges and Lake Ellis glaciations suggest a long, punctuated weathering history in these tropical mountains (Mahaney 2004). Both the Gorges and Lake Ellis moraines comprise highly

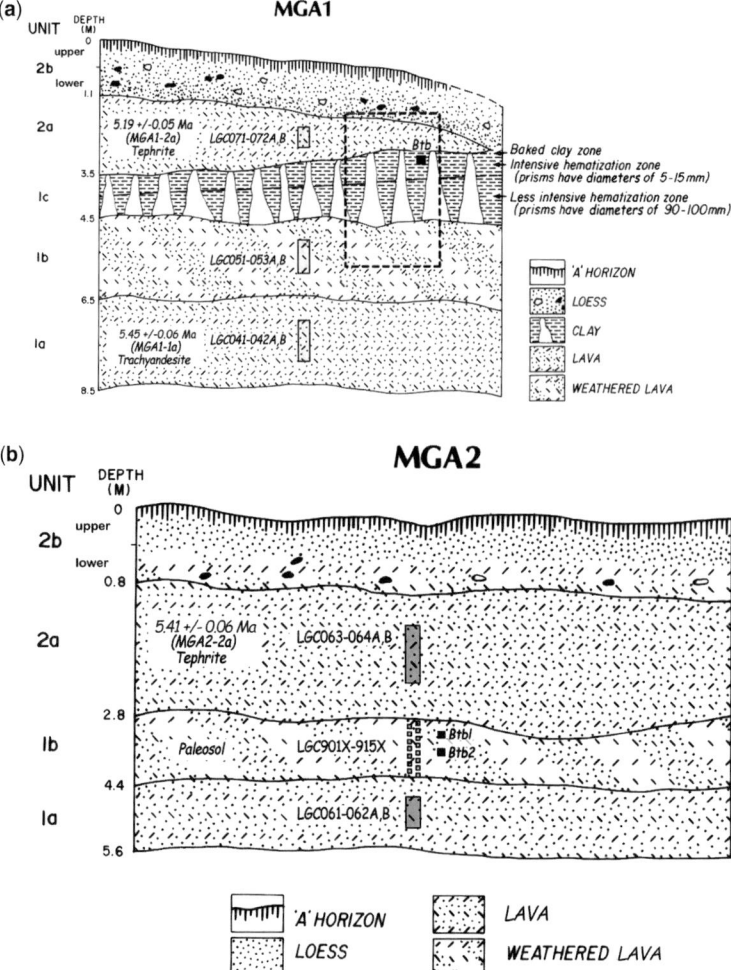

Fig. 2. Stratigraphy, Amboni River sites: (**a**) MGA1 300 m north of the Amboni River; (**b**) MGA2 3 km north of MGA1 on the Nyeri-Naro Moru Road, west flank of Mount Kenya Area; (**c**) generalized Late Neogene stratigraphy of the East African Plateau (from Baker 1967, modified from Mahaney 1990).

weathered tills, and are covered with normally magnetized loess of Brunhes age. This suggests a period or periods of loess stripping (deflation) followed by renewed aeolian input during a subsequent glaciation (Mahaney *et al.* 1997). The loess covering the lavas at Mweiga is most likely derived from these sources.

Methods and materials

The soil descriptions follow guidelines set out by the Soil Survey Staff (NSSC 1995) and Birkeland (1999). Soil colour assessments are based on Oyama and Takehara soil colour chips (1970). Samples of

c. 500 g were collected at the sites to allow for particle size, clay mineral and chemical analyses. Samples were air dried and treated with H_2O_2 to decrease organic matter content. The materials were then wet sieved and the <63 μm fraction was subjected to analysis by hydrometer (Day 1965). Samples were further subjected to various physical, chemical and mineralogical analyses.

Soil samples were subjected to additional laboratory analyses including conductivity (Bower & Wilcox 1965) and pH, determined by electrode. The clay fraction was analysed by X-ray diffraction (Whittig 1965) to identify the primary and secondary mineral composition of the <2 μm size material. The chemistry of sand size material was

(c)

Age				Unit	MAX. OBSERVED THICKNESS (M)
RECENT				SURFICIAL DEPOSITS (mostly nonglacial)	~20
PLEISTOCENE		GLACIATION		LIKI TILLS	~20
	MT. KENYA VOLC. SERIES	PARASITIC VENTS		NYAMBENI VOLCANICS / THIBA BASALTS	>150
		GLACIATION		TELEKI TILL	~10
	MT. KENYA VOLC. SERIES	PARASITIC VENTS		ITHANGUNI & OTHER TRACHYTIC VOLCANICS & NECKS	c. 365
		GLACIATION		PRE-TELEKI TILLS	>40
	MOUNT KENYA VOLCANIC SERIES	PARASITIC VENTS		AGGLOMERATES	>180
		FISSURE ERUPTIONS		RIEBECKITE TRACHYTES	25
				OLIVINE BASALTS	210
				MUGEARITES, OLIVINE TRACHYTES	245
		MAIN ERUPTIVE EPISODES		NEPHELINE SYENITE & PHONOLITE of the PLUG	—
				KENYTES, PHONOLITES & PYROCLASTICS	>550
				PHONOLITES & TRACHYTES	>490
				RHOMB PORPHYRIES	335
				PHONOLITES	0–120
				UNEXPOSED VOLCANICS	c. 2900
				LOWER BASALTS	c. 45
PLIOCENE				LAIKIPIAN BASALTS / olivine basalts, basanites, olivine nephelinites	c. 120
MIOCENE (?)			Mweiga Trachytes latest Miocene	NYERI TUFF / trachytic tuffs & agglomerates	30
MIOCENE (?)				SIMBARA SERIES / basalts & agglomerates	>75
PRE-CAMBRIAN				BASEMENT SYSTEMS / micaceous & graphitic gneisses, amphibolites, quartzo-feldspathic gneisses, & pegmatites	—

Fig. 2. *Continued.*

determined by Energy-Dispersive Spectrometer (EDS) (Mahaney 2002).

To determine the composition of the source minerals and their weathered state, it proved necessary to subject each sample to intensive investigations by light microscopy, randomly counting out 300–400 grains per sample for more detailed analysis. Out of this population, a subpopulation of grains was selected for more intensive investigation by Field Emission Scanning Electron Microscopy (FESEM) and EDS using a Link-Isis System (Mahaney 2002).

To help constrain the age of the Mweiga stratigraphic sequence, palaeomagnetic samples were collected from both lavas and sediments (including palaeosols). Blocks of oriented lava were subsampled in the lab, while sediments were collected in plastic cylinders pressed into cleaned vertical faces and oriented in the field using magnetic compass. Remanent magnetization was measured at the Rock Properties Lab (Pacific Geoscience Centre, Sidney, BC, Canada) using a JR-5A spinner magnetometer (Agico, Brno, Czech Republic) after stepwise alternating-frequency (AF) demagnetization of all sediments and lavas (0–100 mT) and thermal demagnetization of selected lavas (100–650 °C). Characteristic remanent magnetization

directions and site means were calculated using principle component analysis (Kirschvink 1980).

Results

Geology

The lavas described here outcrop prominently on the dissected and inclined Aberdare dip-slope. Lying on the Nyeri Tuff of Late Miocene age, they underlie the Mount Kenya Volcanic Suite of presumed Plio/Pleistocene age. They only occasionally outcrop at the surface, where exposures (Fig. 3a, b) reveal flows of variable thickness. The flows thicken to the south, towards Nyeri (Baker 1967). The sections discussed here contain columnar jointing in places and lack the dark-grey colour and frequent fractures observed elsewhere by Baker nearly a half century ago. Thin outcrops of weathered rock (palaeosols; Fig. 3c) are found within and at the top of both sections. A reddish-brown soil is interbedded between flows 1a and 2a (Fig. 3a; unit 1a is obscured by detritus below unit 1b) with hues in places increasing to 2.5 YR, and is weathered to a depth of 80–100 cm thickness (Table 1).

Fig. 3. (**a**) Close-up view of MGA1 profile, units 1b–2a. (**b**) Distant view of MGA1 profile unit 1a–2b. Contacts are approximate.

The petrography of the Laikipian Formation is known to range from olivine alkali basalt to olivine nephelinite (Baker 1967). At the sites described here, the lack of a contact with the underlying Simbara Series and a different petrography to that described by Baker (1967) leads to a tenuous assignment of the lower trachytic andesite outcrops at Mweiga as correlative with the Late Miocene Nyeri Tuff of Baker (1967). The MGA1 and MGA2 flows, with their trachytic andesite and tephrite compositions (discussed below) and lack of a dark grey colour, are distinct from previously studied volcanics of the Aberdare dip-slope and warrant further investigation.

Stratigraphy

The two lavas (Fig. 2a, b) emanated from a common vent on the west flank of Mount Kenya and

terminated on the drainage divide leading to the Amboni River just north of Nyeri (Fig. 1). Low local relief (of the order 2–3 m) makes it difficult to trace the flows on aerial photography. Both the MGA1 and MGA2 sections have a similar stratigraphy (Fig. 2a, b). The lower flow (units 1a, 1b) is a trachytic andesite, of which the upper half is weathered and contains aeolian sediment. Unit 1c (MGA1 only) is a hematized and baked zone, which appears to have been baked when the overlying flow was deposited. The upper flow (unit 2a, 2b (lower)) is a tephrite of which the upper half of the flow is extensively weathered and covered with loess.

At MGA1 the lava flows are interbedded with palaeosols. A lower palaeosol (units 1b & 1c) is interbedded between units 1a and 2a, and contains minor amounts of aeolian grains. These aeolian grains were transported for only short distances,

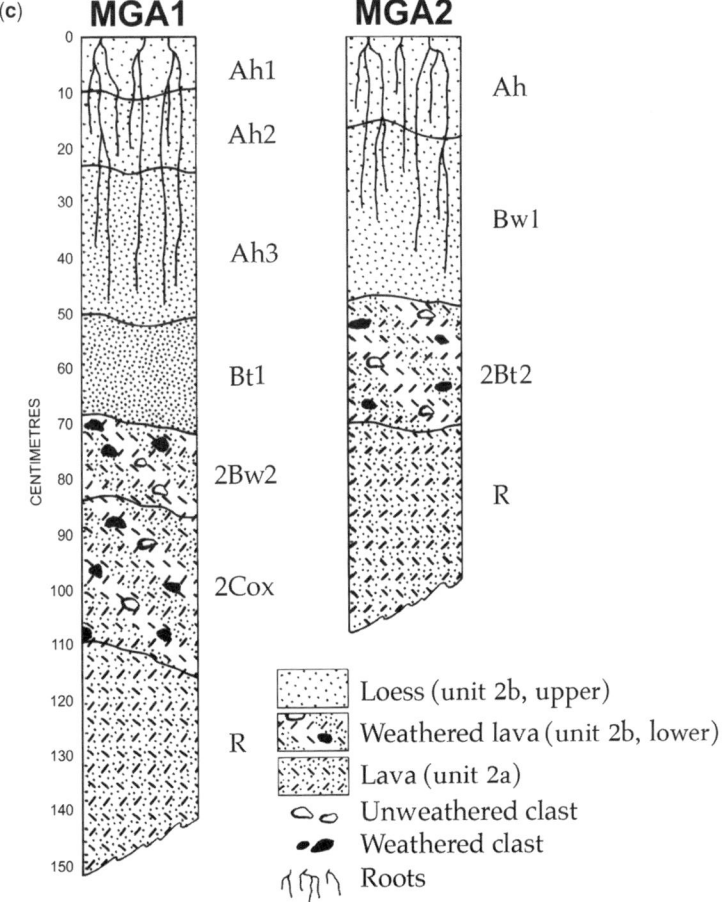

Fig. 3. (*Continued*) (**c**) Multi-storey palaeosols at MGA1 and 2, formed in the upper weathered lava and loess (Bed 2b). Pebble clasts shown at the base of unit 2b considered to result from deflation.

Table 1. *Soil reaction, electrical conductivity (EC), colour and mean phi calculations for palaeosols in the MGA1 and 2 sections, Mount Kenya area, Kenya*

Site	Horizon/Bed	Depth (cm)	pH (1:5)	EC (1:5)	Colour[1]	Mean Phi[2]
MGA1	Ah1	0–9	7.7	0.15	7.5YR2/3	5.8
	Ah2	9–23	7.7	0.10	7.5YR3/3	5.4
	Ah3	23–50	7.7	0.13	7.5YR3/2	6.4
	Bt1	50–68	7.7	0.33	10YR4/3	8.5
	2Bw2	68–84	8.1	0.12	10YR4/4	3.7
	2Cox	84–110	8.1	0.05	10YR5/2,5/3	1.0
	Btb 1c	340	7.2	0.05	2.5YR5/8	4.7
MGA2	Ah	0–18	7.9	0.14	7.5YR4/3	7.8
	Bw1	18–50	7.9	0.17	10YR4/4	7.5
	2Bt2	50–70	7.9	0.28	10YR4/2,5/2	10.7
	1b Bt1	300–310	8.1	0.32	2.5YR5/6	7.9
	1b Bt2	420–430	8.1	0.16	2.5YR4/8	8.4

[1] Oyama & Takehara (1970).
[2] Mean phi based on: (25th + 50th + 75th)%/3.

and were mixed with weathered volcanic minerals which formed *in situ* under subaerial conditions.

The contact between units 1b/1c and 2a is sharp at both sections and was formed when lava (unit 2a) truncated the palaeosol (units 1b/1c). In the case of MGA1, the upper hematized zone (Btb horizon) was baked by emplacement of the upper lava (unit 2a); unit 1b at MGA2 was also truncated by the upper lava but lacks any evidence of baking. The contact between units 2a and 2b is erosional. At its base, unit 2b exposes lower horizons of a multi-storey palaeosol. Unit 2b is a complex compound palaeosol formed chiefly from weathered volcanics (lower part) which merges with weathered loess in the upper part. At the contact between the upper and lower parts of unit 2b, deflation appears to have removed part of the palaeosol. While no dates are available for the upper palaeosol (unit 2b; Fig. 3b), sands and silts are weathered and coated throughout the unit at both sections.

The upper palaeosols shown in Figure 3c formed in the lava and loess of unit 2b in both sections (unit 2a is labeled R (rock) in Fig. 3c). Arabic numbers preceding horizon labels indicate lithologic discontinuities (2 is a mixture of loess and weathered lava) and unnumbered horizon labels indicate loess (NSSC 1995). Cox designation follows Birkeland (1999).

Palaeomagnetism

Palaeomagnetic results helped to further constrain the age range of the Mweiga stratigraphy. Both lavas and sediments show simple mono-component reversed magnetization. The predominant remanence carrier is magnetite, with lavas being demagnetized between 550–600 °C (Fig. 4). Mean

directions (Table 2) of lavas and sediments are nearly identical, and are about 10 degrees steeper than the Geocentric Axial Dipole for the sampling latitude (0.3°S). The palaeomagnetic results are from individual cooling units, and thus record spot readings of the Earth's magnetic field. The mean inclination is well within the range of palaeo-secular variation and results can therefore be regarded as reliable.

Based on the three $^{39}Ar/^{40}Ar$ dates ranging from 5.19 to 5.45 Ma and the reversed polarity, these lavas are Latest Miocene and Earliest Pliocene (Messianian) in age, and fall within the early Gilbert Reversed Chron. The age of the upper flow MGA1-2a falls just within the earliest Pliocene and since the lava is reversely magnetized it must predate the Thvera normal subchron (5.26–5.0 Ma).

Particle size

From the data available (Fig. 5a, ternary diagram), particle size distributions show four distinct groups ranging from loam with a loamy sand outlier to sandy clay loam, clay loam and clay (claystone) as defined within the soil textural classification system (NSSC 1995). The MGA1-1c sample is from the baked zone. The MGA2-1b samples contain similar amounts of clay compared to the MGA2-2b sediment above. The MGA1-1c and MGA2-1b units are both truncated by unit 2a at both sections.

Analysis of particle size curves from both sections to determine mean phi or centre-of-gravity values, using 25th plus 50th plus 75th percentiles, is shown in Table 1. In the MGA1 profile, phi values within the loessic materials (Ah1 to Bt1, unit 2b) increase gradually from top to bottom resulting

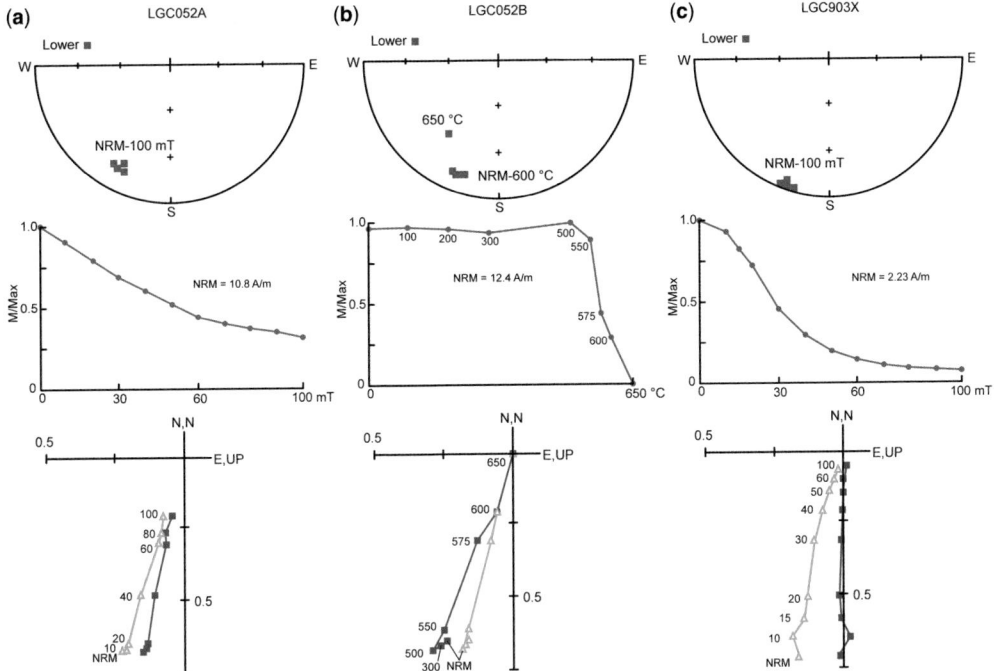

Fig. 4. Representative demagnetization characteristics for (**a, b**) lavas LGC052A and B and (**c**) palaeosol LGC903X at Mweiga, Kenya. For each sample the upper figure shows the stereographic plot of magnetization direction at each demagnetization step (squares on lower hemisphere depict reversed magnetization). The centre plots show normalized intensities during demagnetization steps. Alternating field demagnetization steps are shown in millitesla (mT) and thermal demagnetization steps in degrees Celsius. Lower figures are orthogonal plots showing the horizontal (vertical) projection at each step with an open triangle (solid square). These show simple mono-component magnetization with reverse polarity for all three samples, and are typical of all samples measured in this study. The predominant remanence carrier is magnetite (samples subject to thermal demagnetization were demagnetized between 550–600 °C.

from clay movement into the Bt1 horizon. Within the weathered lava and loess inclusions (2Bw2-2Cox horizons; lower unit 2b, MGA1) mean phi decreases as silt and clay decrease from top to bottom as weathering intensity diminishes with depth. The baked horizon (unit 1c, Fig. 2a) in MGA1 shows a mean phi value of 4.7 which places the sample in the loamy sand range, which

Table 2. *Summary of palaeomagnetic remanence directions*

Unit	Sample labels	n	D (°)	I (°)	K	α_{95} (°)	Polarity
MGA-1							
2a (Tephrite)	(LGC071-072A,B)	2	190	10	660	10	R
1b weathered lava	(LGC051-053A,B)	6	192	10	101	7	R
1a (Trachyandesite)	(LGC041-042A,B)	4	190	9	2864	2	R
Mean of MGA-1		12	191	10	203	3	R
MGA-2							
2a (Tephrite)	(LGC063-064A,B)	4	187	7	224	6	R
1b (palaeosol)	(LGC901X-915X)	14	191	12	174	3	R
1a lava	(LGC061-062A,B)	4	190	9	141	8	R
Mean of MGA-2		22	190	11	155	3	R

Abbreviations: n, number of samples; D and I, declination and inclination of mean remanence direction; k, precision parameter; $\alpha95$, circle of confidence ($P = 0.05$); R, reverse magnetization). Present Earth's Field (PEF) directions at the sampling site: $D = 0°$, $I = -22°$. The mean inclination expected for a Geocentric Axial Dipole (GAD) field for this locality is $-0.6°$. Sampling latitude = 0.3°S and longitude = 37.8°E.

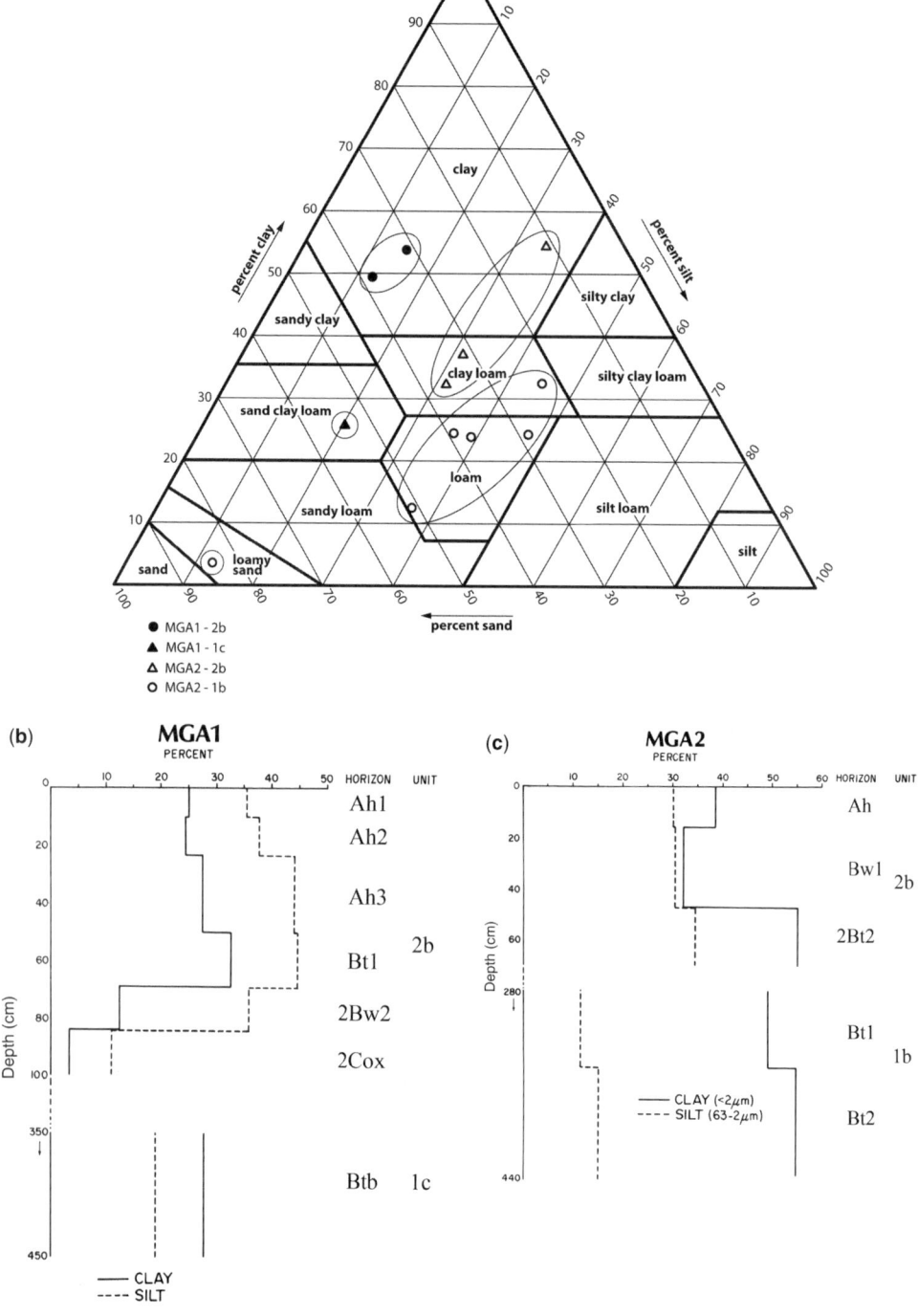

Fig. 5. (a) Ternary diagram showing particle size distributions in MGA1-1c and MGA2-1b weathered rock and MGA1 and 2 palaeosols; (b) depth distributions of silt and clay in MGA1; and (c) depth distributions of silt and clay in MGA2.

is clearly anomalous given the textures of all other weathered horizons in the two sections.

The MGA2 profile contains a higher mean phi in the surface horizons (Ah and Bw1), increasing dramatically into the 2Bt2 horizon. The lower B horizons in unit 1b (Fig. 2b) show equally high mean phi values indicating that appreciable amounts of clay were formed in the 300 ka period of weathering in the Early Pliocene. This interpretation assumes that the age of lava in unit 1a of MGA2 is equivalent in age to the dated lava (unit 1a) at the base of MGA1. However, the high clay content in unit 1b of MGA2 indicates the weathering window might be of greater duration.

Silt and clay distributions (Fig. 5b, c) with depth in MGA1 and MGA2 (unit 2b) are somewhat similar with the exception that clay and silt percentages with depth are reversed in the two profiles. There is somewhat less clay on average in MGA1 when compared with MGA2, whereas clay production in MGA2 far outstrips MGA1. The data clearly show the ultimate production of a 10–20 cm thick argillic (Bt) horizon in both loessic profiles overlying either a truncated profile of lava-plus-loess (lower unit 2b in MGA1) or a truncated lava (unit 2a, MGA2). These different clay percentages may be related to differing times of deflation or to past vegetation history.

Chemistry

The soil colours in Table 1 are used as a proxy for weathering intensity. The colours of both the palaeosols and baked zone are strong bright brown and reddish brown. The palaeosols display strong 7.5YR hues in the surface epipedon and grade to 10YR hues in the lower horizons of the surface loess (horizons 2Bw loess and weathered lava at MGA1 and 2Bt2 at MGA2). Clearly, the extreme reddish colours in the interbedded palaeosols (unit 1b, 1c) indicate intense weathering. The darker hues, stronger value and chroma in the surface horizons of MGA1 suggest a microclimate favourable to higher humus accumulation, in comparison to the hues noted in the surface epipedon of MGA2. Slope conditions and present vegetation at both sections are similar and therefore build-up of organic matter should be similar at the two sites. The age of the palaeosols is well beyond the time required for the establishment of a dynamic equilibrium between organic carbon and nitrogen; soil colours can therefore be considered as reliable indices of weathering intensity (see Mahaney 1990). Within the surface palaeosols, colour indices indicate strong pre-Last Glacial Maximum (LGM) colours, suggesting considerable antiquity (Mahaney 1990).

The pH (Table 1) of the weathered bedrock, baked clay bed and palaeosols ranges from near neutral to moderately and strongly alkaline. The lowest pH occurs in the baked clay bed. In MGA2, the weathered bedrock is strongly alkaline with a consistent pH of 8.1. The surface palaeosols are either consistent (MGA2) at 7.9 or vary slightly (MGA1) from 8.1 at base, consistent with the weathered bedrock in MGA2, lowering slightly to 7.7 in the uppermost horizons. This pH trend for the palaeosols is representative of a present-day climate of approximately c. 750 mm of precipitation, with available soil moisture at 50% or c. 375 mm. The soils are therefore xeric, taxonomically classed as paleudalfs (NSSC 1995), that is, dry but periodically moist enough to allow translocation of clay into the B horizons.

The distribution of total salts (Table 1), estimated from the electrical conductivity of the samples, correlates approximately with the pH. Concentrations of total salts range from 0.05 to 0.32 $\mu S\ cm^{-1}$ with the greatest variability within the bedrock. The palaeosols reveal increased salt concentration with depth correlating closely with the pH.

Mineralogy of clay fraction

Analysis of the <2 μm fraction (Table 3) of the loessic sediment (unit 2b, upper, at both sections), the clay bed of unit 1c at MGA1 and the weathered bedrock (unit 1b) in MGA2 was undertaken to determine the distribution of primary minerals, weathering products and clay minerals. Clay minerals in the weathered bedrock include Ca-smectite in small to moderate amounts and traces of illite-smectite, as well as traces of kaolinite, meta-halloysite and illite that appear to be detrital. Moderate amounts of Ca-smectite in MGA2-unit 1b decrease to small concentrations in MGA1-unit 1c. Small concentrations of Ca-smectite in unit 1c argue against a hydrothermal origin, as suggested by other workers (Mizota & Faure 1998).

Among secondary oxides and hydroxides, maghemite, hematite and goethite range from trace to small amounts. Within the primary minerals, quartz and plagioclase are evenly distributed in small amounts within the sections. The quartz composition here, as on nearby Mount Kenya (Mahaney 1990), is considered to originate from exposed outcrops of the granitic Precambrian basement complex. The higher quartz concentration within the upper unit 2b of MGA1 possibly reflects younger aeolian input.

In the weathered lava/loessic horizons of unit 2b, Ca-smectite dominates as the main clay mineral along with trace amounts of illite-smectite; the latter is an allochthonous airfall component or is inherited from the tephrite. As in the weathered bedrock of unit 1b/1c, secondary oxides and

Table 3. *Semi-quantitative distributions of minerals (psilomelane in trace quantities detected in MGA1, unit 1c; Bt horizon) in palaeosols of the MGA1 and 2 sections, west flank of Mount Kenya*

Sample	K	mH	I	IS	S	Go	Mic	Q	A	M	O	Ca-P	Mag	H
MGA1-Ah1	x	x	tr	tr	x	tr	X	x	x	x	–	–	–	–
MGA1-Ah2	–	–	tr	tr	x	tr	Tr	xx	tr	–	–	tr	–	tr
MGA1-Ah3	–	–	tr	tr	x	tr	–	xx	tr	–	–	tr	–	tr
MGA1-Bt	–	–	–	x	x	tr	X	x	tr	x	–	x	–	–
MGA1-2Bw1	–	–	–	–	x	tr	–	x	–	–	–	x	x	–
MGA1-2Cox	–	–	–	tr	x	–	–	x	–	tr	tr	x	x	tr
MGA2-Ah	–	–	–	–	x	tr	–	x	–	–	–	x	x	x
MGA2-Bw1	–	–	–	tr	x	tr	–	x	–	–	–	x	tr	tr
MGA2-2Bt2	–	–	–	tr	x	–	–	x	–	tr	–	x	tr	tr
MGA2-1b Bt1	–	–	–	–	x	–	–	x	–	–	–	x	x	–
MGA2-1b Bt2	–	–	–	–	x	tr	–	x	–	–	–	x	x	x

Clay minerals: kaolinite (K), metahalloysite (mH), illite (I), smectite (S) and illite-smectite (IS). Primary minerals include Quartz (Q), Albite (A), Ca-plagioclase (Ca-P), Microcline (Mic), Goethite (Go), Muscovite (M), Orthoclase (O), Maghemite (Mag) and Hematite (H). Semi-quantitative estimate of amounts of individual mineral species is based on relative peak height: nil (–), trace (tr), small amount (x) and moderate amount (xx).

hydroxides include goethite, occasional maghemite and hematite, and are mainly resident in the B horizons. Ca-plagioclase, likely derived from the weathered lava, is present in small amounts near the contact with the underlying flow. Small amounts of microcline and muscovite, mostly absent in MGA2, are common in the MGA1 loess, probably from airfall derived from exposed inliers of the Precambrian crust. Because quartz exists in small quantities throughout the palaeosols, the weathered beds in both sections must have been open to the subaerial atmosphere for some time. As on Mount Kenya, quartz in the fine fraction signals aeolian influx (Mahaney 1990).

Scanning electron microscopy

The collection of grains shown in Figure 6a–f is representative of the weathered lower lava flow (unit 1b). Most notable is the prevalence of differential weathering states for grains of similar composition, with some grains being weathered and others relatively fresh within the same horizon. This presents a conundrum; while some grains may be aeolian in origin, most are a weathering residue from an estimated 300 ka period when soil was exposed to the subaerial atmosphere. The concentration of lightly weathered grains among a population of well-weathered sands is unexpected, but could be explained if the fresh grains are the remains of mineral nuclei where the surface armour has been removed by weathering. Advanced weathering of grains in the lower palaeosol (unit 1b) of MGA1 and MGA2 over a 300 ka period under a xeric soil climate regime is indicated by a higher Si/Al ratio and the presence of abundant Ca-smectite. Despite

the dry climate, desiccated clay/Fe coatings and Fe and Mn precipitates are observed on mineral grains (Fig. 6a, c, e, f). Abrasion fatigue and parallel ridges and fractures observed on <10% of the total population of grains analysed are nonetheless representative of aeolian transport (Fig. 6e). It is therefore likely that periodic alterations in wind systems contributed to the influx of a portion of the sediments in the lower palaeosol. Further analysis of the grains derived from the lower lava reveals an assortment of quartz, mica and sanidine that is characteristically andesitic, unlike the 'basaltic' flows previously described by other workers in the region.

A similar conclusion may be reached from analysis of the palaeosol developed in the upper lava (unit 2b, Fig. 7). The vesicular features and tubules (Fig. 7a, b), while only occasionally observed in basalts, are abundant here and are typical of tephritic flows. Sheet structures developed within the grain hollows may be biogenic in nature (Mahaney *et al.* 2009), possibly explaining the high C in the EDS spectra. Due to extended exposure of unit 2b to the atmosphere and other weathering agents, such growths are not unexpected and may play an important role in chemical weathering processes (possibly including chelation). Coatings are slight in the lower and upper segments of unit 2b and somewhat fewer than in unit 1b, and consist largely of clay minerals such as smectite (Fig. 7f). This reinforces yet again the suggested dry environment that persisted through the Late Neogene. The higher frequency of abrasion fatigue or broken surface structures on the population of grains studied argues for increasing airfall of minerals in the sediments of unit 2b. Grain etching, solutional features and breakdown of phenocrysts

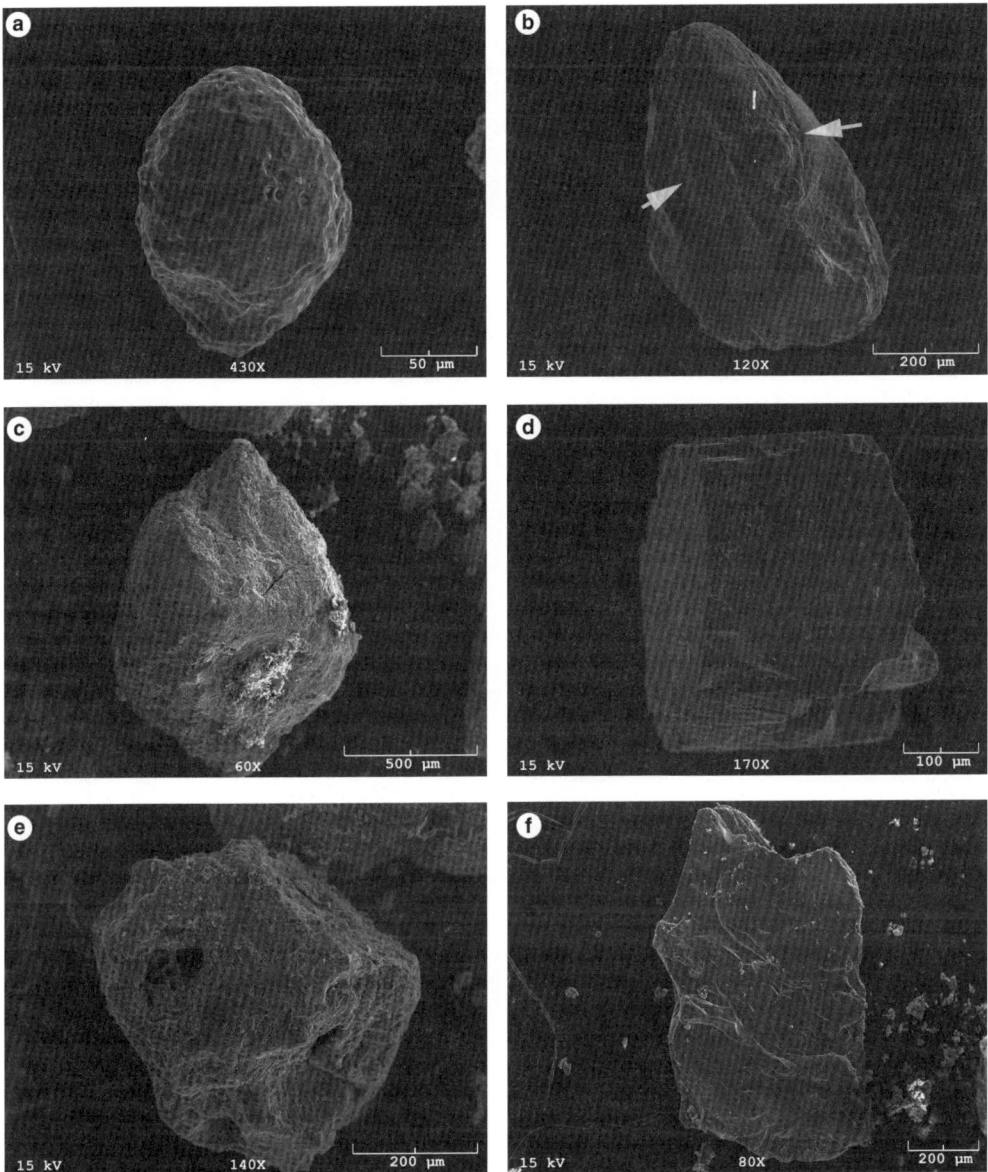

Fig. 6. Lower weathered lava (unit 1b): (**a**) weathered quartz with a slight K coating; (**b**) clear abrasion fatigue (arrows) on windblown quartz with minor etching; (**c**) Fe-coated partly etched sanidine with a Fe-precipitate; (**d**) conchoidal fracture on an uncoated crystalline sanidine; (**e**) Mn- and Fe-encrusted tubule in rhodochrosite; and (**f**) fine sand-sized biotite with etching on the edges.

(Fig. 7c) are observed on feldspar and sanidine grains (Fig. 7c–e) and indicate a somewhat accelerated rate of weathering in comparison to that seen on grains from unit 1b. This is largely a function of either a wetter, stronger climate (as deduced from lower concentrations of smectite) or a longer time period of weathering (Latest Miocene–Quaternary).

In the absence of a radiometric date for unit 2b, it is not possible to estimate weathering history or palaeoclimate.

The grains in Figure 8 are typical of loess mixed with weathered volcanics from the upper flow (see Mahaney 1990 for comparable SEM imagery of palaeosols higher up the slopes of Mount Kenya).

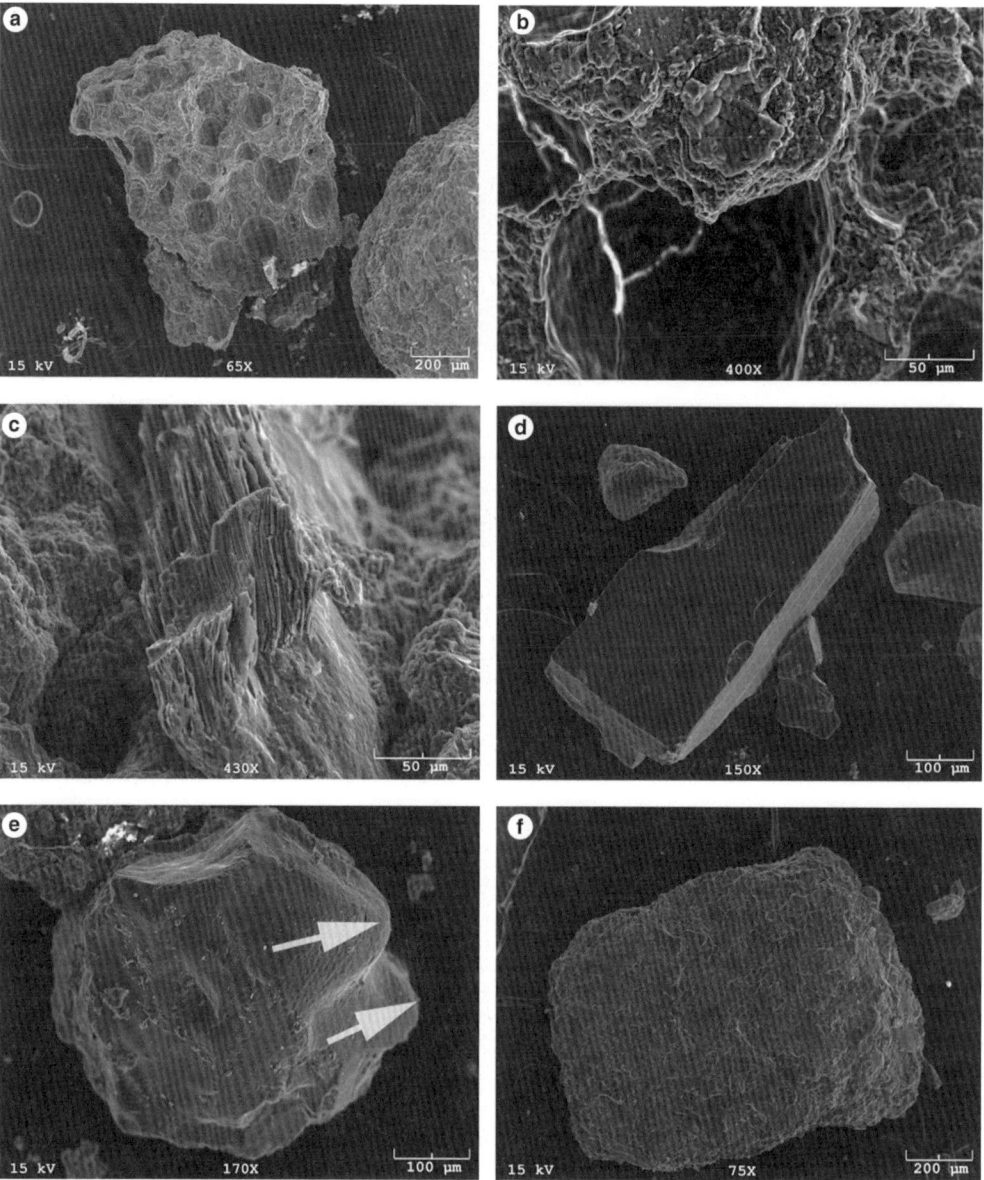

Fig. 7. Upper weathered lava (unit 2b, lower): (**a**) vesicles in a possible stilbite (zeolite) grain; (**b**) enlargement of the sheet structure growth within the vesicle hollows (further analysis revealed biogenic-like weathering features); (**c**) high Al-weathered biotite phenocryst within sanidine; (**d**) differentially edge-weathered feldspar grain; (**e**) abrasion fatigue (arrows) and solution features observed on bulbous quartz; and (**f**) smectite coating on sanidine.

As with aeolian sediment in the lower beds (unit 1b), a combination of weathered and fresh material persists (presumably from weathering of armoured grains). V-shaped percussion cracks indicate fluvial transport followed by aeolian import (Mahaney 2002); the high number of desiccation features on certain grains tells of long weathering (wetting/drying) histories (Fig. 8b, f), while others are either more recent additions to the rock cycle (Fig. 8a, c, d, e) or are fresh mineral nuclei freed after weathering of outer armoured skins. The presence of vesicular features and sanidine in the three units are further evidence of their trachytic-andesite and tephrite origin.

Fig. 8. Loess (unit 2b, upper): (**a**) minor etching and bulbous edges (arrows) indicative of aeolian transport on quartz; (**b**) highly weathered quartz grain with v-shaped percussion cracks indicating fluvial transport; (**c**) relatively fresh twinned orthoclase (etching is observed on the twin plane); (**d**) detritus in the depression of fresh crystal anorthoclase with etching along crystal planes; (**e**) abrasion fatigue on a Fe-coated sanidine grain; and (**f**) primary positive and negative octahedral crystal growth faces and etched-out exsolution lamellae that probably occurred prior to aeolian transport.

Representative EDS spectra are shown in Figure 9a (MGA1-2b-Ah1 horizon) and Figure 9b (MGA2-2b-2Bt2 horizon) with the Fe concentrations on quartz increasing slightly down-section. From the surface Ah1 horizons (Fig. 9a) down-section, secondary Fe coatings reach an apex in unit 2b-Bw and Bt horizons as shown in Figure 9b (decreasing lower down in units 1b and1c). The presence of Al in the MGA2-2b-2Bt2 horizon may represent aggressive leaching (gibbsite) at some time in the Quaternary as discussed below. However, gibbsite was not detected by X-ray diffraction (XRD) in the clay fraction.

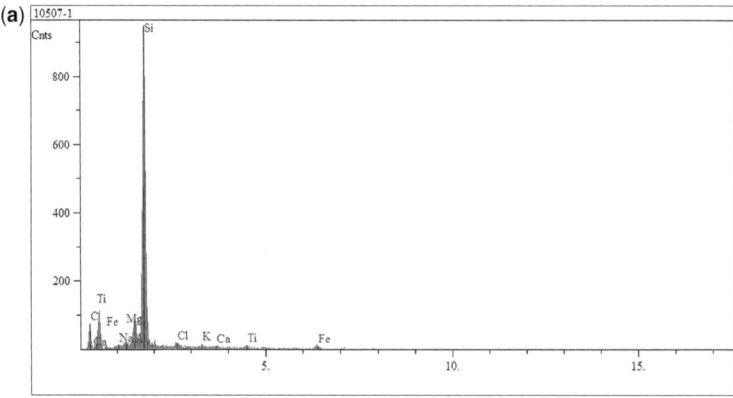

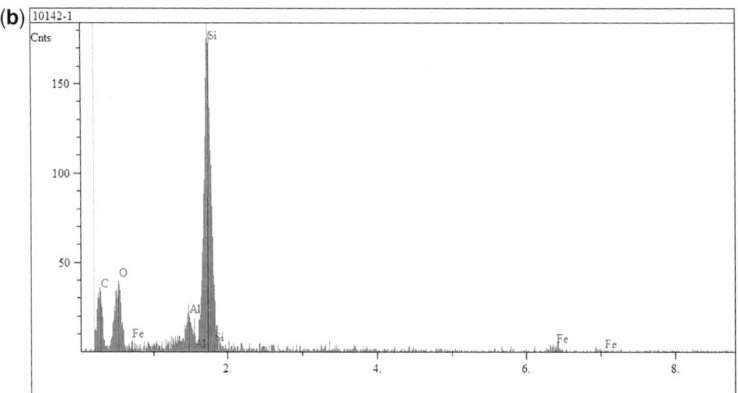

Fig. 9. EDS spectra: (**a**) MGA1-2b-Ah1 horizon with very thin Fe (1 line) coating with unexplained Mg composition; and (**b**) MGA2-2b-2Bt2 horizon with stronger Fe (2 lines) and Al coating. The Al might be gibbsite which formed along with clay illuviation.

Palaeoclimate

An estimate of Late Miocene climate from a 300 ka period of weathering (5.45–5.19 Ma) is provided from the stratigraphy described at Mweiga, Kenya. During this interval of time, weathering led to the production of reasonably high concentrations of clay and appreciable hydrolysis of anorthoclase, analbite and albite along with accessory minerals to produce Fe/clay coatings, the latter derived in large quantities from mafic minerals. The lack of appreciable mica in these sediments together with a prevailing dry climate and limited soil moisture indicates a slow conversion of plagioclases and mafic minerals into illite-smectite and smectite. Clay mineral concentrations over the entire *c.* 5 Ma period support an initial drier climate followed by alternating wetter/drier conditions, and the onset of a greater influx of loess constituting the bulk of soil parent materials in the surface soil. The mobilization of loess is, in all likelihood, related to the

glaciations on Mount Kenya as described by Mahaney (1990).

Following emplacement of the upper lava (unit 2a), weathering intensity remained the same or slowed somewhat with less Ca-smectite being produced but with similar concentrations of randomly interstratified illite-smectite being present. Concomitantly, in the lower beds of unit 2b, clay coatings and secondary Fe have about the same intensity as in the underlying (unit 1b) but formed over a much longer time period. The concentration of Ca-smectite declines in these horizons from moderate to small amounts (Table 3), dropping slightly again in overlying loessic sediments. These fluctuations, while not detectable within the semi-quantitative results presented in Table 3, are clearly shown on XRD traces and amount to ±10%. The Early–Mid-/Late Pliocene climate is therefore adduced from the clay mineral composition and the weathered state of sands and silts, following a similar pattern to Pliocene climatic reconstructions from global benthic ^{18}O records (Lisiecki & Raymo 2005). Small amplitude oscillations of warm/wet to cold/dry climate, as reconstructed from global records, might be slightly adjusted within this sequence to suggest: first a drier climate through the 300 ka window in the Early Pliocene, then a less dry climate through the Middle and Late Pliocene ending with a somewhat greater amplitude of wet and dry climate during the Quaternary. Such a sequence deduced from the clay mineral/SEM data outlined here, albeit crude, fits well with Lisiecki & Raymo's (2005) reconstruction. Moreover, argillic (Bt) horizon development may well have occurred during punctuated humid intervals of 2.7–2.5, 1.9–1.7 and 1.1–09 Ma superimposed on longer-term aridification of Africa as determined from lacustrine records described by Trauth *et al.* (2005).

The exact timing of the unconformity between units 2a and 2b at both sections is unknown but is suspected to have occurred sometime in the Early Pleistocene when temperature amplitudes are known to have widened, based on the Lisiecki & Raymo (2005) database. Ca-smectite production certainly continued, but at a somewhat reduced rate in the upper horizons of 2b compared to the lower horizons. The presence of Bt and 2Bt horizons in MGA1 and MGA2, respectively, suggest periodic accumulation of clay with depth in the two profiles, probably associated each time with minor leaching during a succession of interglacial episodes on nearby Mount Kenya (Mahaney 1990). In the case of MGA1 the Bt horizon is contained within the loess; in MGA2 the 2Bt horizon forms part of the weathered beds in the upper lava, but the clay movement in both profiles tells a mutually reinforcing story of clay formation and translocation.

While pollen stratigraphy is not available for the two pedostratigraphic sections, the clay mineralogy and soil properties are compatible with soil weathering seen in the Alfisol soil taxonomic order. This leads to a Xeralf (Birkeland 1999) designation, probably formed when the climate was more humid during parts of the Quaternary. From the lack of carbonates in these palaeosols, it is clear that the grassland climate was never dry enough to generate calcic horizons or, if they formed at some time in the past, they were completely eradicated by subsequent leaching. Despite soil pH ranging from 7.2 to 8.1 in surface horizons throughout unit 2b, there is no appreciable build-up of carbonate.

Conclusions

Late Miocene sections at Mweiga, Kenya provide a record of outpouring of lavas, weathering histories

and loess deposition. The latter is thought to be partly related to the timing of glaciations on Mount Kenya and the nearby Aberdare Mountains. The ^{39}Ar/^{40}Ar dates of lavas and the magnetostratigraphy of lavas and sediments provide an estimate of the passage of time for some of these events, in a region where few dates are available. The findings reported here provide a first estimate of weathering histories of outcrops on the Aberdare dip-slope to the west of Mount Kenya and provide new information on soil morphogenesis, palaeoclimatic trends and loess ingress and episodic deflation tentatively correlated with Pleistocene glaciations. Clay mineral analysis of the weathered bedrock and loessic cover sediments indicates the presence of sufficient Ca-smectite. Together with small concentrations of illite-smectite this argues for a dry climate from the Latest Miocene–Pleistocene, becoming more subhumid during the Pleistocene with humid phases possibly correlated to interglacial periods. Given the small sample population (two sections) and ages of the bracketing flows for only a 300 ka weathering window in MGA1 during the Latest Miocene, the evidence indicates that the early weathering phase (although dry) was sufficiently long enough to weather alkalic and alkali minerals in the trachytic andesite lava flow and to remove their armoured skins and release fresh internal cores of different minerals. Weathering of the tephrite mineral complex in unit 2b stretched over a longer period of time, initially under a less dry climate. A higher frequency of weathered grains formed over a longer duration, perhaps punctuated periodically with a more humid climate. However, both buried and relict palaeosols contain appreciable quantities of Ca-smectite that diminish in concentration upwards in the profile as quantities of Fe-oxihydroxides and clay coatings on sands and silts increase.

The combination of weathering products and their trends indicate that, while there may have been periods of more subhumid climate in the Latest Miocene–Early Pliocene, the predominate Quaternary soil climate was xeric. A shallow (at best) wetting depth fluctuated with nearly random punctuated humid periods of *c.* 200 ka, leading to the development of argillic horizons which produced soils which fall in the Alfisol order (suborder Xeralf; NSSC 1995; Birkeland 1999).

Funding from the National Geographic Society – Quaternary History of Mount Kenya (Phase III) – and from Quaternary Surveys, Toronto to WCM and NSERC funding to RWB is gratefully acknowledged. We thank B. Kapran (York University) for laboratory assistance. B. Kapran and D. Tessler (University of Lethbridge) prepared and improved some of the illustrations.

References

BAKER, B. H. 1967. *Geology of the Mt. Kenya area.* Geology Report 79, Kenya Geological Survey, Nairobi.

BAKER, B. H., WILLIAMS, L. A. J., MILLER, J. A. & FITCH, F. G. 1971. Sequence and geochronology of the Kenya rift volcanics. *Tectonophysics,* **11**, 191–215.

BARENDREGT, R. W. & MAHANEY, W. C. 1988. Paleomagnetism of selected Quaternary sediments on Mount Kenya, East Africa; a reconnaissance study. *Journal of African Earth Sciences,* **7**, 219–225.

BIRKELAND, P. W. 1999. *Soils and Geomorphology.* Oxford University Press, Oxford.

BOWER, C. A. & WILCOX, L. V. 1965. Soluble salts. *In:* BLACK, C. A. (ed.) *Methods of Soil Analysis.* American Society of Agronomy, Madison, Wisconsin, 933–951.

CHARSLEY, T. J. 1989. Composition and age of older outwash deposits along the northwestern flank of Mount Kenya. *In:* MAHANEY, W. C. (ed.) *Quaternary and Environmental Research on the East African Mountains.* Balkema, Rotterdam, 165–174.

DAWSON, J. B. 2008. *The Gregory Rift Valley and Neogene – Recent Volcanoes of Northern Tanzania.* Geological Society, London, Memoirs, **33**.

DAY, P. 1965. Particle fractionation and particle size analysis. *In:* BLACK, C. A. (ed.) *Methods of Soil Analysis.* American Society of Agronomy, Madison, Wisconsin, 545–567.

EBINGER, C. J., YEMANE, T., HARDING, D. J., TESFAYE, S., KELLEY, S. & REX, D. C. 2000. Rift deflection, migration and propagation: linkage off the Ethiopian and Eastern rifts, Africa. *Geological Society of America Bulletin,* **112**, 163–176.

EVERDEN, J. F. & CURTIS, G. H. 1965. The potassium-argon dating of late Cenozoic rocks in East Africa and Italy. *Current Anthropology,* **6**, 343–385.

FOSTER, D. A. & GLEADOW, A. J. W. 1996. Structural framework and denudation history of the flanks of the Kenya and Anza Rifts, East Africa. *Tectonics,* **15**, 258–271.

FOSTER, D. A., EBINGER, C., MBEDE, E. & REX, D. 1997. Tectonic development of the northern Tanzanian sector of the East African Rift System. *Journal of the Geological Society, London,* **154**, 689–700.

GREGORY, J. W. 1921. *The Rift Valleys and Geology of East Africa.* London.

KIRSCHVINK, J. L. 1980. The least-squares line and plane and the analysis of paleomagnetic data. *Geophysical Journal of the Royal Astronomical Society,* **62**, 699–718.

LISIECKI, L. E. & RAYMO, M. E. 2005. A Pliocene/Pleistocene stack of 57 globally distributed benthic oxygen-18 records. *Paleoceanography,* **20**, 1–17.

MAHANEY, W. C. 1990. *Ice on the Equator.* Wm Caxton Press, Ellison Bay, Wisconsin.

MAHANEY, W. C. 2002. *Atlas of Sand Grain Surface Textures and Applications.* Oxford University Press, Oxford.

MAHANEY, W. C. 2004. Quaternary glacial chronology of Mount Kenya Massif (with 1:250,000 map of Mt. Kenya). *In:* EHLERS, J. & GIBBARD, P. L. (eds) *Quaternary Glaciations: Extent and Chronology, Part III.* INQUA subcommission on glaciation, Oxford University Press, Oxford, 227–231.

MAHANEY, W. C., BARENDREGT, R. W. & VORTISCH, W. 1997. Relative ages of loess and till in two Quaternary paleosols in Gorges Valley, Mount Kenya, East Africa. *Journal of Quaternary Science,* **12**, 61–72.

MAHANEY, W. C., DOHM, J., BARENDREGT, R. W., KYEONG, K. & MILNER, M. W. 2009. Microbes in Pliocene paleosols in volcanic terrane on Earth correlated with similar exposures on Mars. Abstract, American Geophysical Union Meeting, San Francisco, CA., Dec. 14–19.

MIZOTA, C. & FAURE, K. 1998. Hydrothermal origin of smectite in volcanic ash. *Clays and Clay Minerals.* **46**, 178–182.

NATIONAL SOIL SURVEY CENTER (NSSC). 1995. *Soil Survey Laboratory Information Manual.* Soil Survey Investigations Report no. 45, version 1.00, USDA.

OYAMA, M. & TAKEHARA, H. 1970. *Standard Soil Color Charts.* Japan Research Council for Agriculture, Forestry and Fisheries.

RIES, A. C., VERNACOMBE, J. R., PRICE, R. C. & SHACKLETON, R. M. 1992. Geochronology and geochemistry of the rocks associated with a late proterozoic ophiolite in West Pokot, NW Kenya. *Journal of African Earth Sciences,* **14**, 25–35.

ROGERS, N. W. 2006. Basaltic magmatism and the geodynamics of the East African Rift system. *In:* YIRGU, G., EBINGER, C. J. & MAGUIRE, P. K. H. (eds) *The Afar Volcanic Province within the East African Rift System.* Geological Society, London, Special Publications, **259**, 77–93.

RUNDLE, C. C. 1985. K–Ar ages for Miocene to Pleistocene volcanic rocks from northern Kenya: 1. Samples from sheet 20 of the Samburu-Marsabit Mapping Project, Rept. 85/24, British Geological Survey, Isotope Geology Unit.

SCHLUTER, T. 1997. *Geology of East Africa (with contributions by Craig Hampton).* Science Publishers, Stuttgart.

SCHOLZ, C. A., JOHNSON, T. C. *ET AL.* 2007. East African megadroughts between 135 and 75 thousand years ago and bearing on early-modern human origins. *PNAS,* **104**, 16416–16421.

SHACKLETON, R. M. 1945. *Geology of the Nyeri Area.* Dept. 12, Geological Survey, Nairobi.

TRAUTH, M. H., MASLIN, M. A., DEINO, A. & STRECKER, M. R. 2005. Late Cenozoic moisture history of East Africa. *Science,* **309**, 2051–2053.

VELDKAMP, A., BUIS, E. *ET AL.* 2007. Late Cenozoic fluvial dynamics of the River Tana, Kenya, an uplift dominated record. *Quaternary Science Reviews,* **26**, 2897–2912.

VERNACOMBE, J. R. 1983. A proposed continental margin in the Precambrian of Western Kenya. *Geologische Rundschau,* **72**, 663–670.

WHITTIG, L. D. 1965. X-ray diffraction techniques for mineral identification and mineralogical composition. *In:* BLACK, C. A. (ed.) *Methods of Soil Analysis.* Amererican Society of Agronomy, Madison, Wisconsin, **1**, 671–696.

Recent tectonics of Tripolitania, Libya: an intraplate record of Mediterranean subduction

F. A. CAPITANIO[1]*, C. FACCENNA[2], R. FUNICIELLO[2] & F. SALVINI[2]

[1]*School of Geosciences, Monash University, Clayton, Victoria 3800, Australia*

[2]*Dipartimento Scienze Geologiche Università Roma TRE,*
Largo S.L. Murialdo 1 00146 Roma

*Corresponding author (e-mail: fabio.capitanio@monash.edu)

Abstract: High-energy seismicity is historically recorded in Tripolitania, Libya suggesting that this area, far from Mediterranean convergent margin, is currently deforming. How this deformation relates to surrounding tectonics of the Africa-Europe convergence is still poorly known. Here, we use remote sensing image analysis and structural survey to show the recent deformation history that affected Tripolitania and reactivated the western bordering structures of Sirte Basin. This tectonic regime onset long after the Paleocene–Oligocene deformation correlated to the Hellenic subduction evolution (Libyan tectonics have been quiescent since then) and is compatible with age and trends of the Sicily Channel rift zone, a deformational belt that developed across the Maghrebian chain. We show that the continuity of this belt reaches farther than that previously acknowledged, as far as *c.* 1400 km from the collisional front. We speculate on the causes of deformation in this remote area, suggesting that the extensional belt formed in response to the strong slab-pull gradients at the central Mediterranean subduction margin which followed the progressive closure of the oceanic basin.

The convergence of major plates results in large deformation, mostly accommodated along narrow belts around the plate margins. In the Mediterranean (Fig. 1), the current strain field (Serpelloni *et al.* 2007) is compatible with the geological record of the convergence and progressive collision of the African and Eurasian plates (Faccenna *et al.* 2004); however, strain rates are also recorded in Tripolitania, Libya at a large distance from the collisional front. There is still no explanation for the occurrence of such large deformation far from convergent margins and into continental interiors. As a consequence, although some aspects of the recent and active deformation in the area are well studied (Finetti 1984; Westaway 1990), these have not yet been acknowledged in the Mediterranean tectonic reconstruction.

We have collected structural data from the interpretation of five remotely sensed Landsat Thematic Mapper (TM) 5 images and a field survey in the Tripolitania region. We show evidence of a Pliocene event that reactivated pre-existing structures of the Hun Graben on the western flank of Sirte Basin and formed a new widespread regional fracture pattern associated with volcanism. The timing and trends of the deformation are compatible with the coeval structures and kinematics of the Sicily Channel rifts (Argnani 1993), cross-cutting the Maghrebian front, while Sirte Basin and Cyrenaica (facing the Hellenic trench) were quiescent

since the Oligocene (Bosworth *et al.* 2008). Breakout analysis (Schäfer *et al.* 1981) and rift-related volcanism in the Zalla Graben, aged 6–0.2 Ma (Ade-Hall *et al.* 1974; Farahata *et al.* 2009; Cvetković *et al.* 2010), support the idea that this regime has been continuous from the Pliocene to today and extends far deeper in the African Plate than that indicated by seismicity (Westaway 1990; Serpelloni *et al.* 2007).

To explain the dynamic context of such tectonics, we propose that the deformation in Tripolitania has accommodated the differential motions of the fragmenting African plate during its northwards subduction.

Recent tectonics of the Tripolitania region

The Tripolitania region (NW Libya) is a *c.* 500 m high plateau neighbouring the Sirte Basin. In the Jebel region (northernmost Tripolitania), the Mesozoic sequences are gently folded into a WNW–ESE-trending large anticline (Desio 1951; Desio *et al.* 1963). The northern flank of the fold is faulted by a system of EW to ENE–WSW normal faults, and it is buried below a thick sequence of Miocene marine and Pliocene–Recent deposits in the Jefara plain (Lipparini 1940; Desio 1951). The carbonates shallow upwards towards the east and are exposed in Al Khums, where the reef facies of

From: Van Hinsbergen, D. J. J., Buiter, S. J. H., Torsvik, T. H., Gaina, C. & Webb, S. J. (eds) *The Formation and Evolution of Africa: A Synopsis of 3.8 Ga of Earth History.* Geological Society, London, Special Publications, **357**, 319–328. DOI: 10.1144/SP357.17 0305-8719/11/$15.00 © The Geological Society of London 2011.

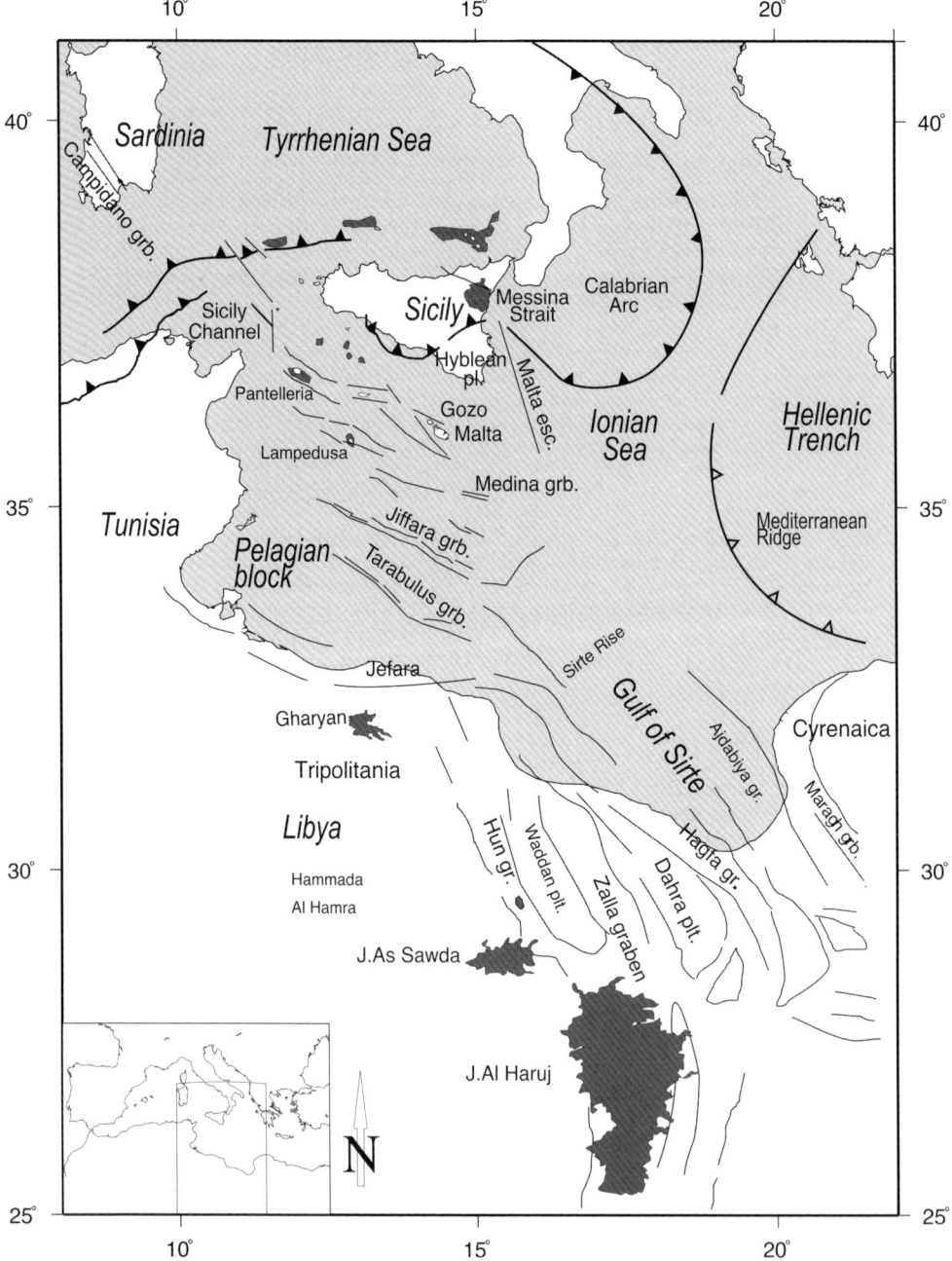

Fig. 1. Tectonic map of Central Mediterranean Sea and Tripolitania–Sirte Basin, Libya. Grey areas represent volcanism, thin lines major lineaments, thick lines collisional fronts, black triangles the Maghreb–Appenninic front and hollow triangles the Hellenic trench. Modified after Anketell (1992) and Finetti (1984).

Mid-Miocene unconformably overlay Cretaceous Dolomites (Floridia 1939; Salem & Spreng 1981). This is the only outcrop of Miocene units in the Tripolitania–Sirte domain.

The southern Jebel flank is overlain by marine ingressive Paleocene carbonates which extend south to Hammada Al Hamra (Jordi & Lonfat 1963), constraining the age of the uplift to *c.* 55–50 Ma, accompanied by minor volcanic activity (Piccoli 1970). This is the only deformation event since the Mesozoic. Large volcanic activity is recorded later in Gharyan in the Jebel region and the Jebel As Sawda volcanics on the southern end of Hun Graben by 6 Ma (Piccoli 1970; Ade-Hall *et al.* 1974; Giraudi 1995). This is accompanied by the opening of another large volcanic centre further south of the Hun Graben, the Al Haruj volcanic field (Ade-Hall *et al.* 1974) which was active between 6 and 0.2 Ma along the reactivated western flank of the Zalla Graben (Farahata *et al.* 2009; Cvetković *et al.* 2010).

Towards the east, the surface of the Tripolitania Plateau deepens into Sirte Basin. The Sirte Basin wide rift consists of a series of NW–SE-trending grabens separated by platforms. After an initial Cretaceous rifting, major subsidence was achieved in the Paleocene under shallow marine conditions that lasted until the Late Eocene (Goudarzi 1981; Anketell 1992; Baird *et al.* 1992; van der Meer & Cloetingh 1993; Guiraud & Bosworth 1997). On the eastern flank of the basin, inversion pervades the Cyrenaica region in the Paleocene–Oligocene (Bosworth *et al.* 2008); however, from this time onwards, the whole Libyan region has been tectonically quiescent.

Landsat TM 5 images

Five 185 × 185 km Landsat TM 5 scenes with a nominal pixel resolution of 30 × 30 m have been processed covering the Tripolitania–Hun Graben–Waddan platform area (Fig. 2). A survey in the northernmost eastern flank of Hun Graben, in Bu Njem and in the Jebel Msellata provided ground-truth data for reflectance clusters as well as structural data.

Profile analysis and band stretching led to the differentiation of lithotypes outcropping in the Gharyan-Msellata area: ash cones (high reflectance in bands 7 and 5), basaltic flows and dikes (low reflectance, black in Fig. 2), Mesozoic Carbonates/Dolomites and Palaeogene Carbonates (high reflectance in band 5, green colours in Fig. 2).

The Jebel region is intersected by a dense pattern of roughly NW–SE trending lineaments. Most of them are the site of fluvial deposits with marked catchments reorientation. The main tectonic trends are N120° and N145° (Fig. 2b). This pattern is widespread over Mesozoic–Palaeogene rocks. In the Jebel Gharyan, basaltic flows and ash cones are aligned within the two main trends. The N120° trending lineaments are more frequent and longer (up to 30 km) than the N145° trending lineaments (<15 km). The latter are coeval and subordinates to the first lineaments, do not cross-cut the longer lineaments and show no offset; we therefore consider these lineaments as joints achieved in one event. We infer a tensional stress regime, roughly oriented NE–SW on the basis of (1) the presence of volcanic centres requiring extension and (2) their alignment with the joint pattern, suggesting that volcanism is fissural (i.e. followed extension).

The Hun Graben is a 50 km wide NW–SE-trending depressed asymmetric structure. The central depression gently dips towards the SE where the Palaeogene Carbonates are overlain by basaltic plateau of the Jebel As Sawda. In the northern Hun Graben, structural grain is marked by N143° lineaments and elongated dark belts. The relief analysis and the fluvial patterns indicate no offset along lineaments.

The western Hun Graben flank is a gently NE-dipping monocline, ending with a narrow scarp. Along the scarp a few long and continuous lineaments are visible, trending N125° and N147°, and no offset is observed. On the eastern flank, a steeper scarp is bounded by hundreds of kilometre-long lineaments trending N140°. The offset along the eastern shoulder increases southwards where it develops into a N140° SW-dipping monocline. Secondary faulting patterns are long N125° faults and shorter N154° faults. North of Bu Njem, lineaments change into N140°-trending dark joint belts.

East of the Hun Graben, the Waddan Platform dips north-eastward into the Sirte Basin and plunges under sand dunes. Here the main trends observed are aligned with Hun Graben structures (N150° and N128°) with an additional N171° trend. No offset is shown in the northernmost Waddan Platform, whereas small a offset occurs in southern areas.

In this sector, subsidence is achieved during Palaeogene–Neogene deformation which reactivated the Hun Graben buried structures, as suggested by the close distribution around the N140° trend in Palaeogenic rocks north of Bu Njem. The reactivation can be constrained by the age of the volcanic centre of Jebel As Sawda on the southern border of the Hun Graben, coeval to the Gharyan event.

Field structural survey

Our structural survey is focused on the Miocene outcrops in the Al Khums area, the Hun Graben eastern flank in Bu Njem. In these areas we can characterize the major NW trends found in the Landsat imager, as well as the Mediterranean offshore structures (Fig. 3a).

In the Al Khums area (Fig. 3b), Mid-Miocene reef carbonates unconformably overlay Cretaceous Dolomites of Jebel Msellata (Floridia 1939; Salem & Spreng 1981). The Ras al Hammam outcrops of Miocene reef carbonates are deformed by 1–2 m spaced joints clustered around N135°. In Funduk

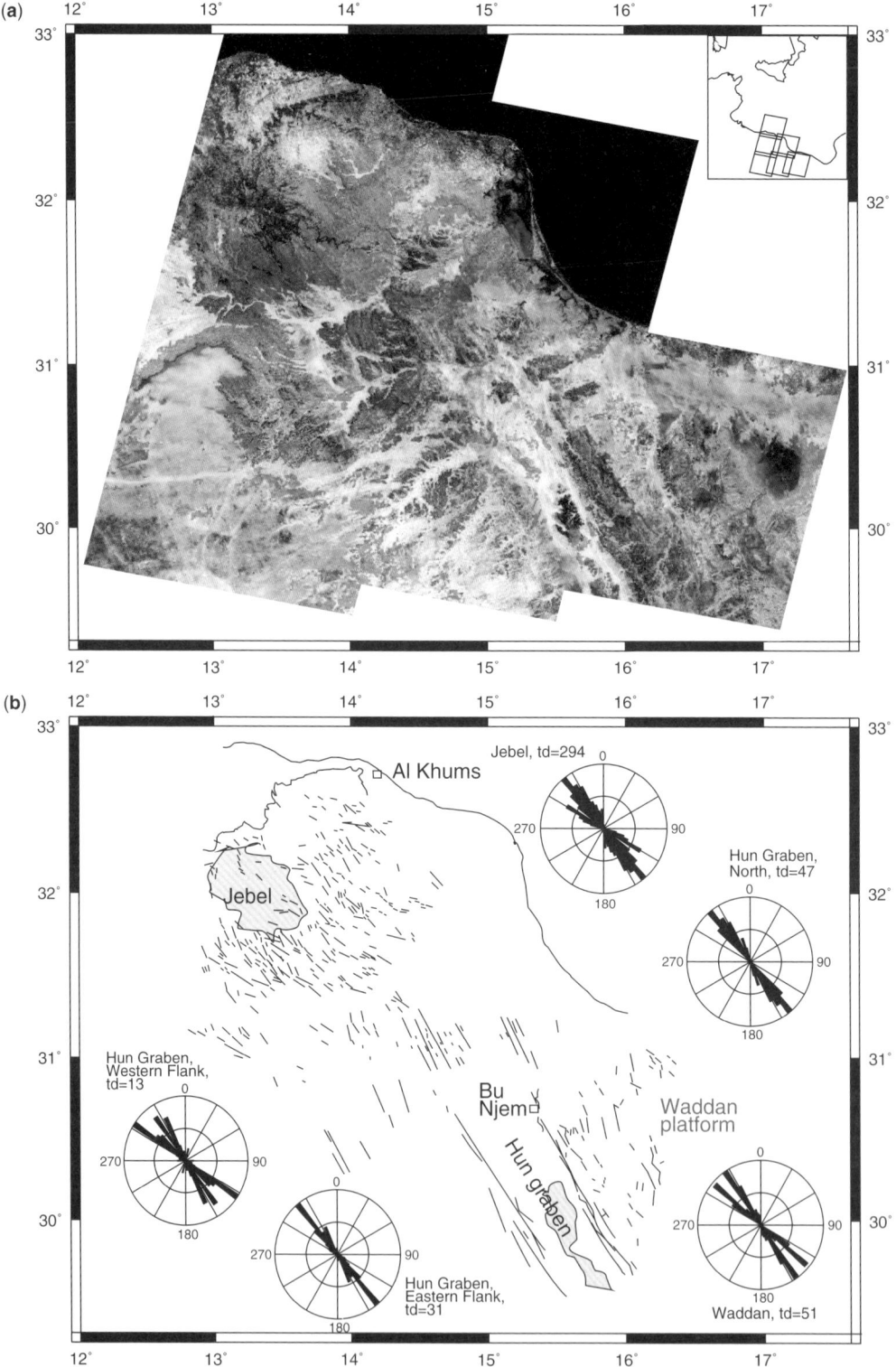

Fig. 2. (**a**) Mosaic of rectified Landsat TM 5 images (bands 7, 5, 4) with location map. Image source ESA. (**b**) Lineaments and trend analysis rose plots. Grey represents volcanics of Jebel Gharyan in the Jebel region and Jebel As Sawda, Hun Graben.

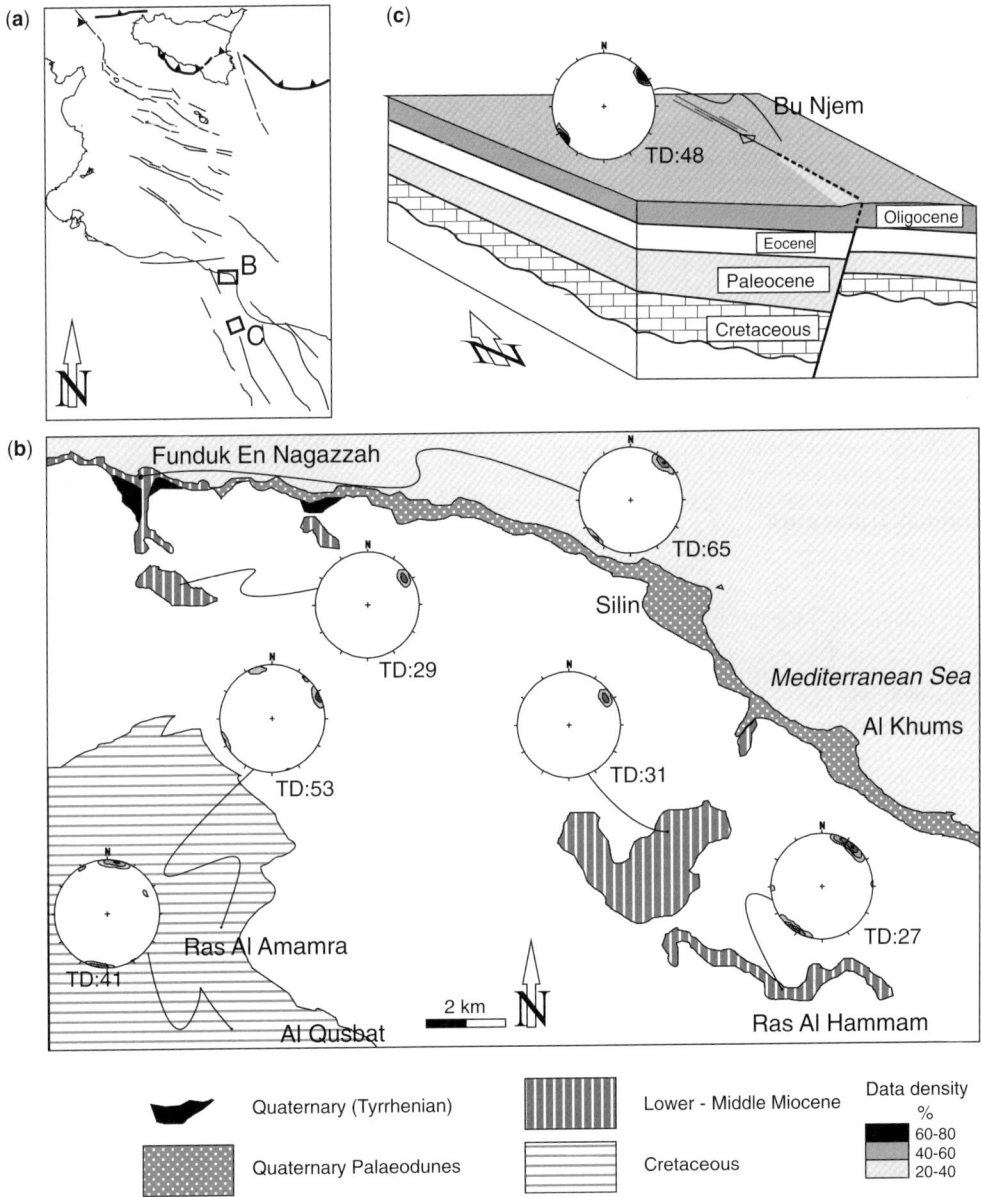

Fig. 3. (**a**) Location map of the surveyed areas with structural regional features. (**b**) Geological map of Al Khums–Jebel Msellata area (modified after Floridia 1939) and stereographic projection of surveyed fracture poles, contour plots. (**c**) Block diagram of the Bu Njem area (modified after Jordi & Lonfat 1963) with stereographic projection of surveyed fracture poles. Total structural data (TD) are plot on equal area, lower hemisphere net; contour lines are each 20%.

En Naggazzah, this trend intensifies into 10–20 cm spaced joints in the Miocene carbonates and propagates in the underlying Upper Cretaceous Dolomites. In the Jebel Msellata, the Upper Cretaceous sandstones are affected by a high-density closely spaced (20–100 cm) joint system. This trend is clustered around N150° in Ras al Hamamrah and

N100° in Al Qusbat sites in the Jebel region. Bedding in the Jebel is subhorizontal and no faulting is observed, although the Al Khums trend is slightly rotated here. The deformation in Al Khums is consistent with the Jebel pattern.

The trend observed in the Al Khums area extends southwards to Bu Njem in the Hun Graben.

Structural survey in this area reveals the presence of centimetric–metric scale N140° joints on the hinge of a N140°-trending, 20° south-westward-dipping draping fold in the outcropping Upper Eocene–Oligocene marine carbonates (Fig. 3c). This pattern characterizes the dark belt in the Landsat images. The offset between the rift shoulder and the basin floor (*c.* 30 m, increasing southwards) is indicative of Neogene reactivation of the underlying extensional structures, allowing a tensile regime compatible with the Tripolitania trend to be inferred. Although it is not possible to constrain further the age of the Neogene reactivation, we correlate it to the age of the continental rift-related volcanism in the Jebel As Sawda (i.e. 6 Ma) as this is most likely to represent the major tectonic pulse in the region.

Tectonic framework

The tectonic frame of the African Plate in the Mediterranean can be appreciated within the context of the Meso-Cenozoic convergence between Africa and Eurasia and the following formation of collisional belts along their margins (Dercourt *et al.* 1986). The North African Plate has subducted below the Eurasian continent along the Maghrebian-Apennine front in the west and below the Hellenic front in the east (Faccenna *et al.* 2003). Important heterogeneities are present in this segment of the subducting African plate. In the west, the continental Pelagian block underthrusts the Maghrebian front. The Malta escarpment separates form the deep Ionian oceanic basin (Catalano *et al.* 2001) eastwards, currently subducting past the Calabrian arc and the Hellenic trench in the east (Fig. 1).

The progressive closure of the Ionian basin and the pull of subducting Hellenic slab have conferred complex deformation along the Libyan margin during the Paleocene, with inversion in Tripolitania and Cyrenaica bordering the subsiding domain of the Sirte Basin (van der Meer & Cloetingh 1993; Abadi *et al.* 2008; Bosworth *et al.* 2008; Capitanio *et al.* 2009). The following tectonic quiescence in Libya, from the Oligocene onward, can be tentatively attributed to the closure of oceanic basin and the consequent reduction of available pull force. More recent results of numerical modelling of the Hellenic subduction system (Capitanio *et al.* 2010) show that the onset of back-arc spreading on the upper plate has a minor effect on the subducting plate strain; the evolution of the Hellenic–Aegean sea and the African subducting plate can therefore be considered dynamically distinct.

The current state of stress of the Libyan structures, revealed by seismicity in the region, further corroborates these results. In fact, two distinct styles are defined in Libya (Suleiman & Doser 1995). The first, in the Cyrenaica region and offshore which borders the Mediterranean Ridge and where low-energy earthquakes have *P*-axes *c.* NE-directed, shows the typical focal mechanisms of thrust faults and that they are aligned with the direction of tectonic convergence in the Aegean. The second, in Tripolitania where *P*-axes of high-energy seismic events are rotated to a *c.* NW direction, is characterized by strike–slip motions. This suggests that the two areas are under the control of different stress regimes that can be related to different dynamic contexts. We therefore eliminate the possibility that the recent deformation of Tripolitania can be linked to the late evolution of the Hellenic convergent margin. In the next section, we explore the relationship between the Tripolitanian recent deformation and the central Mediterranean tectonics.

Implications for the central Mediterranean tectonics

During the Africa–Europe convergence, extensional belts have developed dissecting the continuity of the Maghrebian–Apennine front in the western-central Mediterranean (Faccenna *et al.* 2004). The Sicily Channel rift zone developed in the Pliocene, cross-cutting the Miocene Maghrebian–Apennine front to the north in response to NE–SW-trending tensional regime (Tricart *et al.* 1994) (Fig. 4a). The rift system consists of a series of *c.* NW-trending basins, namely the Malta, Pantelleria and Lampedusa grabens, active from the Early Pliocene onwards (Henning Illies 1981; Jongsma *et al.* 1987; Argnani 1990; Grasso *et al.* 1990; Tricart *et al.* 1994). Tectonic frames of this area of the Mediterranean do not extend further than the Sicily Channel; however, to the south, the NW structural trend propagates to the Tarabulus and Jiffara troughs (offshore Tripolitania) which are of the same extensional trend and age (Finetti 1984). Strike of structures is NNW due to the reactivation of Cretaceous lineaments (Goudarzi 1981) in the Hun Graben, and has a general *c.* NW-trend in the joints in the Jebel region, coeval with volcanism by 6 Ma. The reactivation of faults is recorded as far south as the Zalla Graben, where continental-rifting-related volcanism in the Jebel Al Haruj is coeval to volcanism in the Hun Graben and Tripolitania (Ade-Hall *et al.* 1974).

The thrust front reached the Calabrian realm (Fig. 4b) in the Pleistocene. A second extensional belt formed at a high angle off the front in the Messina strait (Argnani & Bonazzi 2005; Argnani *et al.* 2009; Fig. 4b). By that time, extension in the Sicily Channel troughs had rotated towards an

(a)   **(b)**   **(c)**

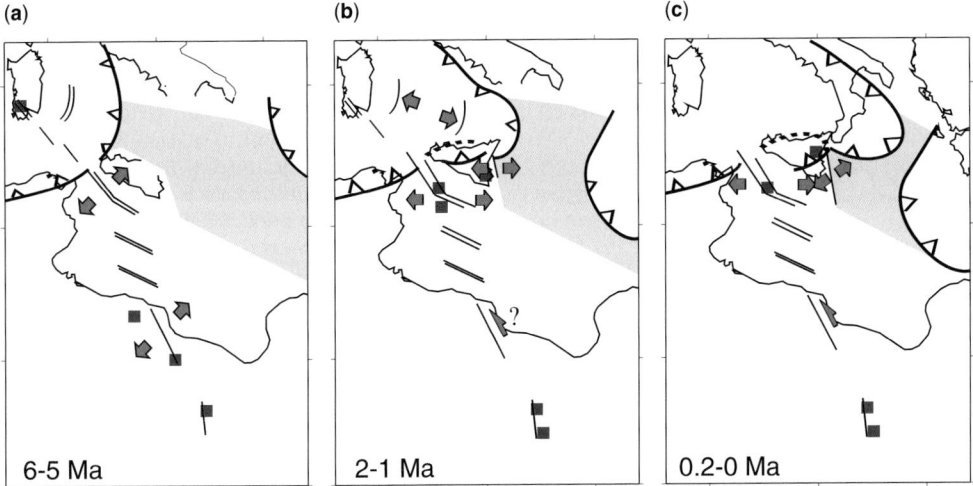

6-5 Ma   2-1 Ma   0.2-0 Ma

Fig. 4. Tectonics reconstruction of the central Mediterranean–Libyan domains in (**a**) Pliocene, (**b**) Pleistocene, and (**c**) Recent; Ionian basin in grey. Thick lines with triangles are for the convergent margins, dashed thick line for slab detachment (after Argnani 2009) and thin lines indicate only active major extensional lineaments. Red squares for anorogenic volcanism (from Faccenna *et al.* 2004). Plate kinematics rotation plots from Müller *et al.* 2008, with respect to Eurasia fixed.

EW-direction (Catalano *et al.* 2009) and localized between Pantelleria and Lampedusa troughs only (Catalano *et al.* 2009) where negligible offsets (Argnani 1990) and fissural volcanisms suggest minor extension. Volcanism, fracturing and reactivation of extensional structures have ceased in Tripolitania; however, increased volcanic activity in Jebel Al Haruj (Farahata *et al.* 2009; Cvetković *et al.* 2010), continuity of strain along the northernmost Tripolitania from stress analysis of break-outs (Schäfer *et al.* 1981) and Pleistocene faulting in the Jefara plain (Giraudi 1995) suggest that the tectonic regime did not come to a complete halt.

This tectonic regime was active until recent time (Fig. 4c). Strike–slip motions in north-eastern Sicily accompany the progressive migration of thrust front in the Calabrian arc (Argnani *et al.* 2009). The seismotectonics and slip rates along the Malta escarpment computed from the global positioning satellite (GPS) motions show that Pliocene decoupling has recently extended to the south along the lineament, and do not exclude a sinistral strike–slip component (Jenny *et al.* 2006). Whereas tectonic activity has mostly faded in the Sicily Channel, activity of Al Haruj volcanism in Libya is recorded until 0.2 Ma (Ade-Hall *et al.* 1974; Farahata *et al.* 2009).

The patterns of GPS motions and CMT (Centroid Moment Tensor) solutions from historical catalogues are compatible with the Pleistocene tectonics. The GPS Pantelleria-relative motions (Hollenstein *et al.* 2003) shown in Figure 5a are in good agreement with the Pleistocene deformation trends of the Sicily Channel rift (Catalano *et al.* 2009). Deformation has faded here and is larger around the Malta escarpment.

Centroid moment tensor solutions for historical earthquakes provide an image of present-day deformation (Suleiman & Doser 1995; Pondrelli *et al.* 2006; Serpelloni *et al.* 2007). These show a *c.* NE-directed extension consistent from the Sicily Channel rift zone to offshore Tripolitania troughs and Hun Graben (Fig. 5b, red/white balls). The focal mechanisms in the Sicily Channel are compatible with the trend of the Malta escarpment, whereas in Tripolitania these are slightly rotated. Here, large earthquakes localized on the bordering faults of the Hun Graben result in clear sinistral strike–slip motions (Suleiman & Doser 1995). The regime in Tripolitania is possibly rotated with respect to the Pliocene trend. We tentatively constrain the rotation to the Pleistocene, when the Malta escarpment activatesd and deformation migrated eastwards (Fig. 4b).

We have conferred present-day left-lateral shear along the Sicily Channel–Tripolitania belt (Fig. 4c), as constrained by the tectonics and the CMT solutions. Note that in Serpelloni *et al.* (2007), the Sicily Channel–Tripolitania belt has dextral motions on the basis of the large velocity of Nubia–Eurasia motions with respect to Pantelleria. However, these two stations are both on the western flank of the Pelagian block; their relative motion does not therefore describe the current deformation

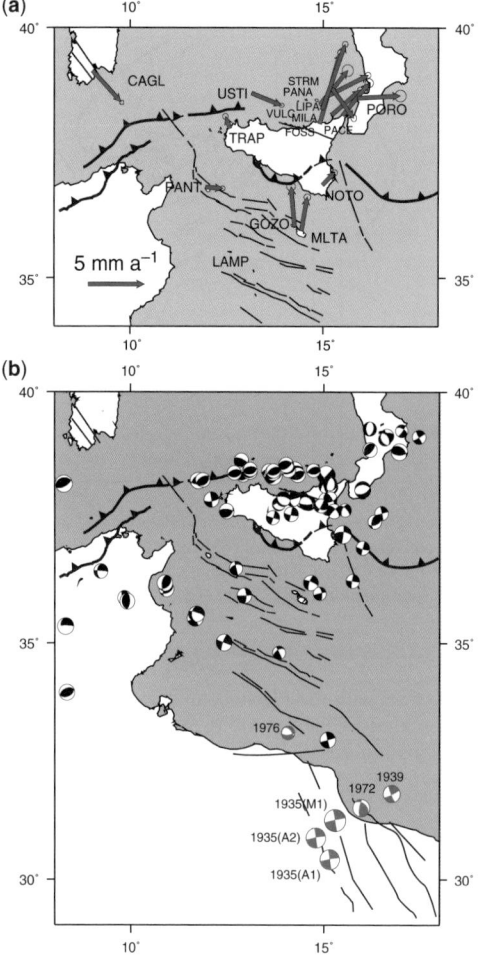

Fig. 5. (**a**) GPS velocity vectors transformed into Lapedusa (LAMP) reference frame from Hollenstein *et al.* (2003). Stations labels: Cagliari (CAGL), Pantelleria (PANT), Lampedusa (LAMP), Gozo (COZO), Malta (MLTA), Noto (NOTO), Trapani (TRAP) Ustica (USTI), Stromboli (STRM), Panarea (PANA), Lipari (LIPA), Vulcano (VULC), Milazzo (MILA), Fossazzo (FOSS), Pace (PACE) and Capo Poro (PORO). (**b**) Solution of focal mechanisms ($M_W > 4$) from seismic catalogue RCMT (Pondrelli *et al.* 2006) in black and from Suleiman & Doser (1995) in red.

in the Sicily Channel rift, and is at odds with the shear sense of the CMT in the Hun Graben.

Discussion

Previous models have considered the portion of Pelagian block bounded by the Sicily Channel, the Malta escarpment and the Medina Graben as an

independent block, completely decoupled from the African plate. In these models the motions between this block and the surrounding are accommodated by a set of faults extending from Malta and Medina grabens to the Malta escarpment (Jongsma *et al.* 1987; Catalano *et al.* 2008; Argnani 2009). This implies that the northward migration of the microplate is driven by the slab pull attached to its north margin, or by surrounding plate motions. However, the progressive slab break-off initiated in the Pliocene below the Sicily Channel and which migrated eastwards (Goes *et al.* 2003; Faccenna *et al.* 2004, 2005; Argnani 2009) removed the slab pull, thus ruling out the first hypothesis. On the other hand, the lack of northward compression along the Malta–Medina grabens suggests that the northward motions of Nubia, Malta and Gozo are possibly accommodated in the north–south compressional belt along the Tyrrhenian margin of Sicily (Serpelloni *et al.* 2007) (Fig. 5b); an additional margin along the Malta Graben is not required.

Here we propose that the Pelagian block is decoupled along the extensional belt of the Sicily Channel–Tripolitania. The delamination due to the rollback of Tyrrhenian subducting plate (Argnani 1990) or the lateral tear propagation (Faccenna *et al.* 2004, 2005) possibly contributed to plate rifting, although this does not account for the stress regime found in Tripolitania. Bellahsen *et al.* (2003) showed in their models that large slab-pull gradients can form along the trench where ongoing subduction is flanked by collided margin, where slab pull has faded. These are the most effective in inducing strain deep in plate interiors at a high angle with the trench. Similar conditions are found in the Mediterranean in Pliocene. By this time, the Pelagian block separated two domains where oceanic plate had been completely consumed to the west and where oceanic lithosphere was still available for subduction to the east (Fig. 4a). Slab-pull gradients that formed along the trench might have been large enough to drive the deformation in the Sicily Channel–Tripolitania belt. This dynamic frame is compatible with the initial lithospheric stretching and following sinistral motions of the eastern flank of the Pelagian block with respect to the western flank, following the decoupling.

Conclusions

A combination of structural survey and remotely sensed image analysis allow us to describe a tectonic model for the Tripolitania area, so far poorly known. We found evidences of Pliocene deformation along a belt stretching from the Jebel-Al Khums region to Jebel As Sawda, southern Hun Graben, where volcanism and reactivation of normal faults suggest a NE-directed extension. The age of the deformation and its orientation are compatible with the opening of the Sicily Channel rift zone, representing the southernmost culmination of a belt that propagates from the convergent margin for more than 1400 km. Seismotectonics and GPS analysis shows that this regime is still active, indicative of an important role in the tectonics of the Mediterranean.

Revised reconstructions of the central Mediterranean tectonics since the Pliocene are proposed to include these structures. We suggest that the Sicily Channel–Tripolitania belt has accommodated the differential pull along the convergent margin that arises from the northward subduction of heterogeneous lithosphere since the Pliocene. This offers an understanding of the link between the convergent margins and the enigmatic intraplate deformation.

We thank L. Jolivet for comments on a previous version of the paper and M. Seton for the rotation poles sets. G. Bertotti, editor D. van Hinsbergen and an anonymous reviewer helped to improve the paper. All figures were prepared with GMT Generic Mapping Tool (Wessel & Smith 1991). This research is supported by the Australian Research Council's Discovery Project funding DP0987374, and is dedicated to R. Funiciello and his group 'hic sunt leones'.

References

ABADI, A. M., VAN WEES, J. D., VAN DIJK, P. M. & CLOETINGH, S. A. P. L. 2008. Tectonics and subsidence evolution of the Sirt Basin, Libya. *American Association of Petroleum Geologists Bulletin*, **92**, 993–1027.

ADE-HALL, J. M., REYNOLDS, P. H., DAGLEY, P., MUSSETT, A. E., HUBBARD, T. P. & KLITZSCH, E. H. 1974. Geophysical studies of North African Cenozoic volcanic areas: I. Haruj Assuad, Libya. *Canadian Journal of Earth Sciences*, **11**, 998–1006.

ANKETELL, J. 1992. History of the Sirt Basin and its relationship to the Sabratah Basin and Cirenaica Platform. *In*: SALEM, M., HAMMUDA, O. S. & ELIAGOUBI, B. A. (eds) *The Geology of Sirte Basin*. Elsevier, Amsterdam, 57–88.

ARGNANI, A. 1990. The Strait of Sicily Rift Zone: foreland deformation related to the evolution of a back-arc basin. *Journal of Geodynamics*, **12**, 311–331.

ARGNANI, A. 1993. Neogene basins in the Strait of Sicily (central Mediterranean): tectonic setting and geodynamic implications. *In*: BOSCHI, E., MANTOVANI, E. & MORELLI, A. (eds) *Recent Evolution and Seismicity of the Mediterranean Region*. Springer, Berlin, 173–187.

ARGNANI, A. 2009. Evolution of the southern Tyrrhenian slab tear and active tectonics along the western edge of the Tyrrhenian subducted slab. *In*: VAN HINSBERGEN, D. J. J., EDWARDS, M. A. & GOVERS, R. (eds) *Collision and Collapse at the Africa–Arabia–Eurasia*

Subduction Zone. Geological Society, London, Special Publications, **311**, 193–212.

ARGNANI, A. & BONAZZI, C. 2005. Malta Escarpment fault zone offshore eastern Sicily: Pliocene-Quaternary tectonic evolution based on new multichannel seismic data. *Tectonics*, **24**, doi: 10.1029/2004TC04656.

ARGNANI, A., BRANCOLINI, G., BONAZZI, C., ROVERE, M., ACCAINO, F., ZGUR, F. & LODOLO, E. 2009. The results of the Taormina 2006 seismic survey: possible implications for active tectonics in the Messina Straits. *Tectonophysics*, **476**, 159–169.

BAIRD, D. W., ABURAWI, R. M. & BAILEY, N. J. L. 1992. Geohistory and petroleum in the central Sirt Basin. *In*: SALEM, M., HAMMUDA, O. S. & ELIAGOUBI, B. A. (eds) *The Geology of Sirt Basin*. Elsevier, Amsterdam, 3–56.

BELLAHSEN, N., FACCENNA, C., FUNICIELLO, F., DANIEL, J. M. & JOLIVET, L. 2003. Why did Arabia separate from Africa? Insights from 3-D laboratory experiments. *Earth and Planetary Science Letters*, **216**, 365–381.

BOSWORTH, W., EL-HAWAT, A. S., HELGESON, D. E. & BURKE, K. 2008. Cyrenaican 'shock absorber' and associated inversion strain shadow in the collision zone of NE Africa. *Geology*, **36**, 695–698.

CAPITANIO, F. A., FACCENNA, C. & FUNICIELLO, R. 2009. Opening of Sirte Basin: result of slab avalanche? *Earth and Planetary Science Letters*, **285**, 210–216.

CAPITANIO, F. A., ZLOTNIK, S. & FACCENNA, C. 2010. Controls on subduction reorganization in the Hellenic margin, eastern Mediterranean. *Geophysical Research Letters*, **37**, L14309, doi: 10.1029/2010GL044054.

CATALANO, R., DOGLIONI, C. & MERLINI, S. 2001. On the Mesozoic Ionian basin. *Geophysical Journal International*, **144**, 49–64.

CATALANO, S., DE GUIDI, G., ROMAGNOLI, G., TORRISI, S., TORTORICI, G. & TORTORICI, L. 2008. The migration of plate boundaries in SE Sicily: influence on the large-scale kinematic model of the African promontory in southern Italy. *Tectonophysics*, **449**, 41–62.

CATALANO, S., DE GUIDI, G., LANZAFAME, G., MONACO, C. & TORTORICI, L. 2009. Late Quaternary deformation on the island on Pantelleria: new constraints for the recent tectonic evolution of the Sicily Channel Rift (southern Italy). *Journal of Geodynamics*, **48**, 75–82.

CVETKOVIĆ, V., TOLJIĆ, M., AMMAR, N. A., RUNDIĆ, L. & TRISH, K. B. 2010. Petrogenesis of the eastern part of the Al Haruj basalts (Libya). *Journal of African Science*, **58**, 37–50, doi: 10.1016/j.jafrearsci.2010.01.006.

DERCOURT, J., ZONENSHAIN, L. P. *ET AL.* 1986. Geological evolution of the Tethys belt from the Atlantic to the Pamirs since the Lias. *Tectonophysics*, **123**, 241–315.

DESIO, A. 1951. Cenno riassuntivo sulla costituzione geologica della Libia. *Report of the 18th session of the International Geological Congress*, Great Britain.

DESIO, A., ROSSI RONCHETTI, C., POZZI, R., CLERICI, F., INVERNIZZI, C., PISONI, G. & VIGANÓ, P. L. 1963. Stratigraphical studies in the Tripolitanian Jebel (Libya). *Revista Italiana Di Paleontologia*, **IX**, 26–42.

FACCENNA, C., JOLIVET, L., PIROMALLO, C. & MORELLI, A. 2003. Subduction and the depth of convection in the Mediterranean mantle. *Journal of Geophysical Research*, **108**, doi: 10.1029/2001JB001690.

FACCENNA, C., PIROMALLO, C., CRESPO-BLANC, A., JOLIVET, L. & ROSSETTI, F. 2004. Lateral slab deformation and the origin of the western Mediterranean arcs. *Tectonics*, **23**, doi: 10.1029/2002TC001488.

FACCENNA, C., CIVETTA, L., D'ANTONIO, M., FUNICIELLO, F., MARGHERITI, L. & PIROMALLO, C. 2005. Constraints on mantle circulation around the deforming Calabrian slab. *Geophysical Research Letters*, **32**, doi: 10.1029/2004GL021874.

FARAHATA, E. S., ABDEL GHANI, M. S., ABOAZOM, A. S. & ASRANC, A. M. H. 2009. Mineral chemistry of Al Haruj low-volcanicity rift basalts, Libya: implications for petrogenetic and geotectonic evolution. *Journal of African Science*, **45**, 198–212.

FINETTI, I. 1984. Geophysical study of the Sicily Channel Rift Zone. *Bollettino di Geofisica Teorica ed Applicata*, **XXXVI**, 345–368.

FLORIDIA, G. B. 1939. Osservazioni sul Miocene dei dintorni di Homs. *Bollettino della Societa Geologica Italiana*, **LVIII**, 34–56.

GIRAUDI, C. 1995. Geomorphology and sedimentology in the Ghan and Zargha Wadis and on the Jado Plateau (Libya). *In*: BARICH, B. E., CONATI BARBARO, C., CAPEZZA, C. & GIRAUDI, C. (eds) *Geoarchaeology of the Jebel Gharbi Region, Libya Antiqua*, Volume 1. L'Erma di Bretschneider, Rome, 18–26.

GOES, S., GIARDINI, D., JENNY, S., HOLLENSTEIN, C., KAHLE, H. G. & GEIGER, A. 2003. A recent tectonic reorganization in the south-central Mediterranean. *Earth and Planetary Science Letters*, **226**, 335–345.

GOUDARZI, G. H. 1981. Structure – Libya. *In*: SALEM, N. J. & BUSREWIL, M. T. (eds) *Geology of Libya*. Al-Fateh University, Tripoli, 879–892.

GRASSO, M., DE DOMINICIS, A. & MAZZOLDI, G. 1990. Structures and tectonic setting of the western margin of the Hyblean-Malta shelf, Central Mediterranean. *Annales Tectonicae*, **4**, 140–145.

GUIRAUD, R. & BOSWORTH, W. 1997. Senonian basin inversion and rejuvenation of rifting in Africa and Arabia: synthesis and implications to plate-scale tectonics. *Tectonophysics*, **282**, 39–82.

HENNING ILLIES, J. 1981. Graben formation – The Maltese Island – A case history. *Tectonophysics*, **73**, 151–168.

HOLLENSTEIN, C., KAHLE, H. G., GEIGER, A., JENNY, S., GOES, S. & GIARDINI, D. 2003. New GPS constraints on the Africa–Eurasia plate boundary zone in southern Italy. *Geophysical Research Letters*, **30**, doi: 10.1029/2003GL017554.

JENNY, S., GOES, S., GIARDINI, D. & KAHLE, H. G. 2006. Seismic potential of southern Italy. *Tectonophysics*, **415**, 81–101.

JONGSMA, D., WOODSIDE, J. M., KING, G. C. P. & VAN HINTE, J. E. 1987. The medina wrench: a key to the kinematics of the central and eastern Mediterranean over the past 5 Ma. *Earth and Planetary Science Letters*, **82**, 87–106.

JORDI, H. A. & LONFAT, F. 1963. Stratigraphic subdivision and problems in upper Cretaceous – lower Tertiary deposits in nortwestern Libya. *Rev. Inst. Fr. Pétr.*, **XVIII**, 56–70.

LIPPARINI, T. 1940. Tettonica e geomorfologia della Tripolitania. *Bollettino della Societa Geologica Italiana*, **LIX**, 33–45.

MÜLLER, R. D., SDROLIAS, M., GAINA, C. & ROEST, W. R. 2008. Age, spreading rates, and spreading asymmetry of the world's ocean crust. *Geochemistry, Geophysics, Geosystems*, **9**, Q04006, doi: 10.1029/2007GC001743.

PICCOLI, G. 1970. Outlines of volcanism in northern Tripolitania (Libya). *Bollettino della Societa Geologica Italiana*, **89**, 449–461.

PONDRELLI, S., SALIMBENI, S., EKSTRÖM, G., MORELLI, A., GASPERINI, P. & VANNUCCI, G. 2006. The Italian CMT dataset from 1977 to the present. *Physics of the Earth and Planetary Interiors*, **159**, 286–303.

SALEM, M. & SPRENG, A. C. 1981. Middle Miocene Stratigraphy. Al Khums Area, Northwestern Libya. *In*: SALEM, M. & BUSREWIL, M. T. (eds) *The Geology of Libya*. Academic Press, London, 97–116.

SCHÄFER, K., KRAFT, K. H., HAUSLER, H. & ERDMAN, J. 1981. In situ stresses and paleostresses in Libya. *In*: SALEM, N. J & BUSREWIL, M. T. (eds) *Geology of Libya*. Al-Fateh University, Tripoli, 907–922.

SERPELLONI, E., VANNUCCI, G. *ET AL.* 2007. Kinematics of the Western Africa-Eurasia plate boundary from focal mechanisms and GPS data. *Geophysical Journal International*, **169**, 1180–1200.

SULEIMAN, A. S. & DOSER, A. I. 1995. The seismicity, seismotectonics and earthquake hazards of Libya, with detailed analysis of the 1935 April 19, M = 7.1 earthquake sequence. *Geophysical Journal International*, 312–322.

TRICART, P., TORELLI, L. & ARGNANI, A. 1994. Extensional collapse related to compressional uplift in the alpine chain off Northern Tunisia (central Mediterranean). *Tectonophysics*, **238**, 317–329.

VAN DER MEER, F. & CLOETINGH, S. 1993. Intraplate stresses and the subsidence history of the Sirte Basin (Libya). *Tectonophysics*, **226**, 37–58.

WESSEL, P. & SMITH, W. H. F. 1991. Free software helps map and display data. *EOS Transactions, American Geophysical Union*, **72**, 441.

WESTAWAY, R. 1990. The Tripoli, Libya, earthquake of September 4, 1974: implications for the active tectonics of the Central Mediterranean. *Tectonics*, **9**, 231–248.

The enigmatic Chad lineament revisited with global gravity and gravity-gradient fields

CARLA BRAITENBERG[1]*, PATRIZIA MARIANI[1], JÖRG EBBING[2,3] & MICHAL SPRLAK[1,4]

[1]*Dipartimento di Geoscienze, Università di Trieste, via Weiss 1, 34100 Trieste, Italy*

[2]*Geological Survey of Norway (NGU), Leiv Eirikssons vei 39, 7491 Trondheim*

[3]*Department of Petroleum Engineering and Applied Geophysics, NTNU, 7491 Trondheim, Norway*

[4]*Research Institute of Geodesy and Cartography, Chlumeckého 4, 826 62 Bratislava, Slovakia*

*Corresponding author (e-mail: berg@units.it)

Abstract: The crustal structure of northern Africa is puzzling, large areas being of difficult access and concealed by the Sahara. The new global gravity models are of unprecedented precision and spatial resolution and offer a new possibility to reveal the structure of the lithosphere beneath the Sahara. The gravity gradients correlate better than gravity with geological features such as rifts, fold belts and magmatic deposits and intrusions. They are an ideal tool to follow geological units (e.g. basement units) below a stratigraphic layer of varying density (e.g. sediments). We focus on the Chad lineament, a 1300 km arcuate feature located between the west and central African rift system. The gravity fields show differences between the lineament and the west and central African rift system. Along the centre of the lineament high-density rocks must be present, which relate to either magmatic or metamorphic rocks. This is very different from the lineaments of the western and central-west African rift system which are filled with sediments. Considering present models of rifting and the absence of topography, the lineament cannot be coeval to the west and central African rift system and is most likely older. We suggest that the lineament is a structural element of the Saharan Metacraton.

Geodetic satellite missions, for example CHAMP (CHAllenging Mini-Satellite Payload), GRACE (Gravity Recovery And Climate Experiment) and GOCE (Gravity field and steady-state Ocean Circulation Explorer), have been flown in recent years to map the Earth's gravity field to an unprecedented precision and spatial resolution. Preliminary data of the GOCE mission are now available, but with lower spatial resolution than global models based on the observations of satellite data combined with terrestrial data (e.g. Rummel *et al.* 2002; Förste *et al.* 2008). Terrestrial data include observations acquired by a gravimeter on land, ship and airplane; altimeter satellite observations of the sea surface also contribute to terrestrial observations, because the sea surface follows the geoid. The global satellite-terrestrial gravity field models are available in spherical harmonic expansion with maximum degree and order of $N = 360$ (EIGEN05C – European Improved Gravity model of the Earth by New techniques; Förste *et al.* 2008) or $N = 2159$ (EGM2008 – Earth Gravitational Model 2008; Pavlis *et al.* 2008). The preliminary model derived with GOCE observations is only available with $N = 210$, 224 or 250, according to different calculation methods (Bruinsma *et al.* 2010; Migliaccio *et al.* 2010; Pail *et al.* 2010). The maximum degree and order of the expansion is an important parameter, because a higher degree will show smaller-scale structures. By subtracting the potential field of the reference ellipsoid from the Earth's gravity potential, the disturbing potential is obtained. The disturbing potential represents the gravity effect of the density variations inside the Earth with respect to a homogeneous ellipsoid. The gravity anomaly is equal to the first spatial derivative, and the gravity-gradient tensor (Marussi tensor) is equal to the second derivatives of the disturbing potential (e.g. Hofmann-Wellenhof & Moritz 2005).

In the present study we calculate the gravity anomaly and the gravity gradients for north-central Africa, using the global model EGM2008 which has the highest spatial resolution. We show that the fields have sufficient spatial resolution and accuracy to detect geological boundaries. We compare the gravity signals to a schematic geological map of north-central Africa that includes those features which are presumably accompanied by density variations. The wavelengths of the geological units must

From: VAN HINSBERGEN, D. J. J., BUITER, S. J. H., TORSVIK, T. H., GAINA, C. & WEBB, S. J. (eds) *The Formation and Evolution of Africa: A Synopsis of 3.8 Ga of Earth History*. Geological Society, London, Special Publications, **357**, 329–341. DOI: 10.1144/SP357.18 0305-8719/11/$15.00 © The Geological Society of London 2011.

be larger than the spatial resolution of the gradient fields, in order to be detected. We seek the relationship between the gravity signals and the different geological structures, in order to demonstrate that the fields can be used to identify geological units. This knowledge can be transferred to areas in which the geology is not well known (e.g. Braitenberg & Ebbing 2009), in order to guide future geological campaigns and determine geological boundaries.

Another application is to investigate outcropping structures by tracing the gravity signal. A good example is the arched gravity anomaly in eastern Chad, which had been noticed in the early 1970s (e.g. Louis 1970). In our work we compare the characteristics of the linear anomaly, which we call the Chad lineament, with the gravity signal of the central African rift and the Pan-African suture. The present data are more complete with respect to the studies of the 1960s and 1970s, because the terrestrial data have been merged with satellite observations allowing the anomaly to be viewed in a regional context. Previous interpretations associated the anomaly either with a rift or a suture; we show that the lineament differs in several aspects from the nearby rifts and thus is not a coeval rift. We discuss the other possible interpretations.

Satellite and terrestrial gravity field

Nominally the smallest wavelength resolvable with the EGM2008 potential field model is equal to $\lambda_{min} = 2\pi R/N_{max} = 19$ km (Hofmann-Wellenhof & Moritz 2005), where R is Earth's radius and N_{max} is the maximum degree and order of the expansion. The spatial wavelengths of EGM08 longer than 572 km (maximal degree and order of spherical harmonic expansion $N = 70$), which correspond to features greater than 286 km (half wavelength), are based entirely on satellite observations and are therefore truly global (Förste *et al.* 2008). Wavelengths between 572 ($N = 70$) and 334 km ($N = 120$) depend increasingly on terrestrial data, whereas wavelengths smaller than 334 km ($N = 120$) depend entirely on terrestrial data. For the GOCE-derived model, wavelengths down to *c.* 191 km ($N = 210$) are resolved and are ideal for testing the validity of the EGM08 model. In terms of degree of the spherical harmonic expansion, EGM08 relies entirely on terrestrial data for degrees between $N = 120$ and $N = 2159$. The GOCE model allows us to control the EGM08 field up to degree $N = 210$, that is, we are testing the contribution of terrestrial data between $N = 120$ and $N = 210$. A good correspondence between the two fields allows us to be confident that up to degree 210 terrestrial data correctly represent the field, and gives us confidence in the correctness of the higher orders.

In Figure 1 the gravity anomaly derived from the GOCE satellite (Migliaccio *et al.* 2010) is compared to the gravity anomaly derived from the EGM08 model (Pavlis *et al.* 2008). Among the three available GOCE models we choose the space-wise solution (Migliaccio *et al.* 2010), which takes advantage of the spatial correlation of the gravity field by applying a local numerical solution method to produce intermediate grid data; for this reason, it is expected that it can better describe local or regional behaviour of the field. The fields are in good agreement and only differ slightly in a few places, which is due to the varying density of terrestrial data. The statistical parameters for the difference between the two fields are: average difference = 0.02 mGal, standard deviation = 6 mGal, maximal value of difference = 36 mGal. The only evident mismatch is found in Nigeria (longitude 7°, latitude 10°): a high-gravity value is found in EGM08 (bright red spot) which is missing in the GOCE field, and is therefore probably due to a confined problem in the terrestrial data used in EGM08.

The use of satellite data in north Africa is particularly useful, because the combination of different terrestrial gravity campaigns is hampered since a unified height system is not available everywhere; the combined fields therefore may have errors in the wavelength range equal to multiples of the individual length scale of the campaign.

Ideally, all data are linked to the reference system ISGN71 (International Gravity Standardization Net 1971) which provides absolute reference stations. ISGN71 has a bias to western countries however, where the number of available absolute gravity stations is of greater density. The gravimeter being a relative instrument, measurements must always start at an absolute gravity station in order to reduce the gravity differences to gravity values. Some datasets are therefore linked to a reference station which is far away and are only linked internally. In such cases, a long-wavelength shift might occur. This problem is relieved with the global gravity fields that merge the terrestrial data using the control on the longer wavelengths through the satellite data. A further advantage of the global availability of the potential field is to be able to calculate the gradient tensor, a quantity well-suited for highlighting upper crustal density. The terrestrial data nonetheless are important in defining wavelengths shorter than 334 km of the EGM08 field.

The gradient field or Marussi tensor

The gravity-potential field EGM2008 (Pavlis *et al.* 2008) is published in terms of the spherical harmonic expansion, for which spherical coordinates are

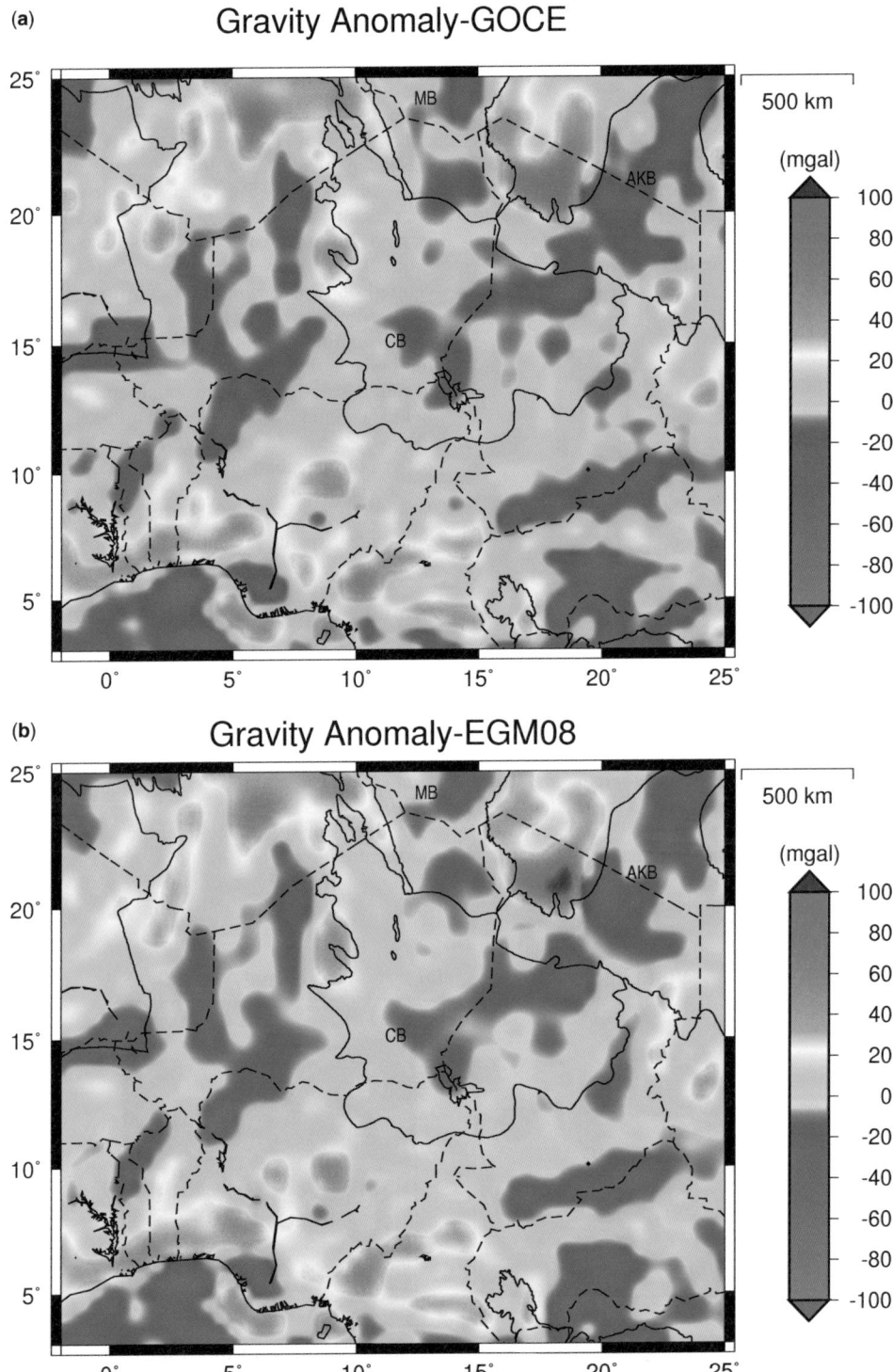

Fig. 1. Control of the combined terrestrial-satellite gravity model with a truly global satellite-derived gravity model. Maximum degree and order $N = 210$. (**a**) Gravity anomaly observed from satellite GOCE (Migliaccio *et al.* 2010) and (**b**) combined satellite-terrestrial gravity model EGM08 (Pavlis *et al.* 2008). CB, Chad basin; MB, Murzuk basin; AKB, Al Kufra basin. Outlines of major sedimentary basins depicted as continuous line; national borders depicted as dashed line.

preferable to the Cartesian coordinate system. The Marussi tensor is symmetric and is defined by six quantities which, in the north-oriented system centred on the observation point, are T_{NN}, T_{EE}, T_{ZZ}, T_{NZ}, T_{EZ}, T_{EN}; five of these are independent as the trace $(T_{NN} + T_{EE} + T_{ZZ})$ is equal to zero. The Marussi tensor and gravity field both reflect density variations in the crust, but outline very different subsurface features. Starting from the disturbing potential $T(r, \varphi, \lambda)$ we use the following conventions for defining the Marussi tensor:

$$T_{NN} = \frac{1}{r^2}\left(\frac{\partial^2 T}{\partial \varphi^2} + \frac{r\partial T}{\partial r}\right)$$

$$T_{NE} = \frac{1}{\cos\varphi r^2}\left(\frac{\partial^2 T}{\partial\varphi\partial\lambda} + \frac{\tan\varphi\partial T}{\partial\lambda}\right) = T_{EN}$$

$$T_{NZ} = \frac{1}{r}\left(\frac{\partial^2 T}{\partial\varphi\partial r} - \frac{1}{r}\frac{\partial T}{\partial\varphi}\right) = T_{ZN}$$

$$T_{EE} = \frac{1}{\cos^2\varphi r^2}\left(\frac{\partial^2 T}{\partial\lambda^2} + r\cos^2\varphi\frac{\partial T}{\partial r} - \cos\varphi\sin\varphi\frac{\partial T}{\partial\varphi}\right)$$

$$T_{EZ} = \frac{1}{r\cos\varphi}\left(\frac{\partial^2 T}{\partial\lambda\partial r} - \frac{1}{r}\frac{\partial T}{\partial\lambda}\right) = T_{ZE}$$

$$T_{ZZ} = \frac{\partial^2 T}{\partial r^2}$$

where r is radial distance and φ, λ are latitude and longitude, respectively. All components except T_{ZZ} depend on the orientation of the planar coordinates (φ, λ), that is, the above field-operators (or field quantities) are not rotationally invariant. It is therefore useful to define rotationally invariant quantities such as the horizontal gradient and other

invariant quantities such as the curvature and the total gradient (e.g. Pedersen & Rasmussen 1990). For the scope of illustration we take a spherical prism (spherical mass element bounded by two meridians, two latitudes and two spherical surfaces with radius r_1 and r_2) and map the gravity and T_{ZZ}. The prism has a side length of 100 km by 100 km, is 1 km thick, density 200 kg m^{-3} and its top is set at 4 km depth. The calculation height is 5000 m and is made in spherical coordinates. The two fields (Fig. 2) have the following characteristics: T_{ZZ} is centred on the mass, giving a positive signal over the body; along the borders of the body a small-amplitude negative stripe is observed. When the body is either deep or with slant borders, the negative stripe is reduced or absent. The gravity field is centred over the mass, but it does not show the negative values along the borders and the pattern of the anomaly is broader. We find that the T_{ZZ} component is ideal for geological mapping, as it highlights the centre of the anomalous mass with higher spatial resolution than gravity.

Main geological and tectonic features

The main geological features of interest in our study are shown in Figure 3. In north-central Africa, the West African Craton and the Sahara Metacraton have been identified (Abdelsalam *et al.* 2002). The southern part of the West African Craton consists of the Leo shield; its central part is the basement of the vast Taoudenni Basin. The Congo Craton dominates central Africa and its northern edge is seen in the south-eastern part of Figure 3. The African cratons are defined on the basis of their structural composition, age and

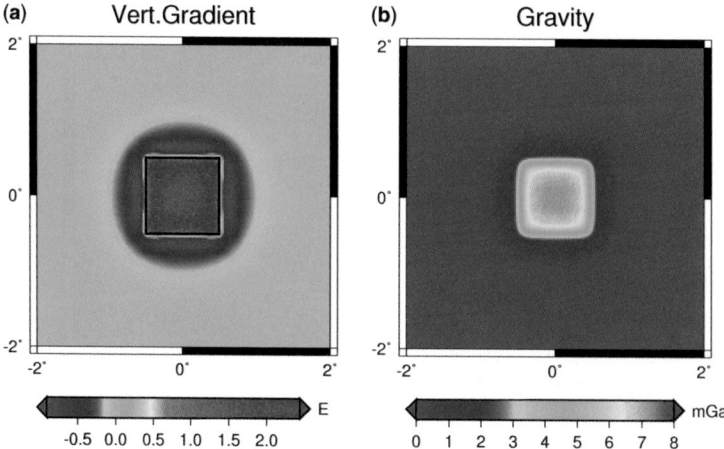

Fig. 2. (a) Vertical (T_{ZZ}) gradient and (b) gravity field over a spherical prism of size longitude $= 1°$, latitude $= 1°$, $dz = 1$ km, top at 4 km depth. Calculation height 5 km, density $= 200$ kg m^{-3}. The black box outlines the prism.

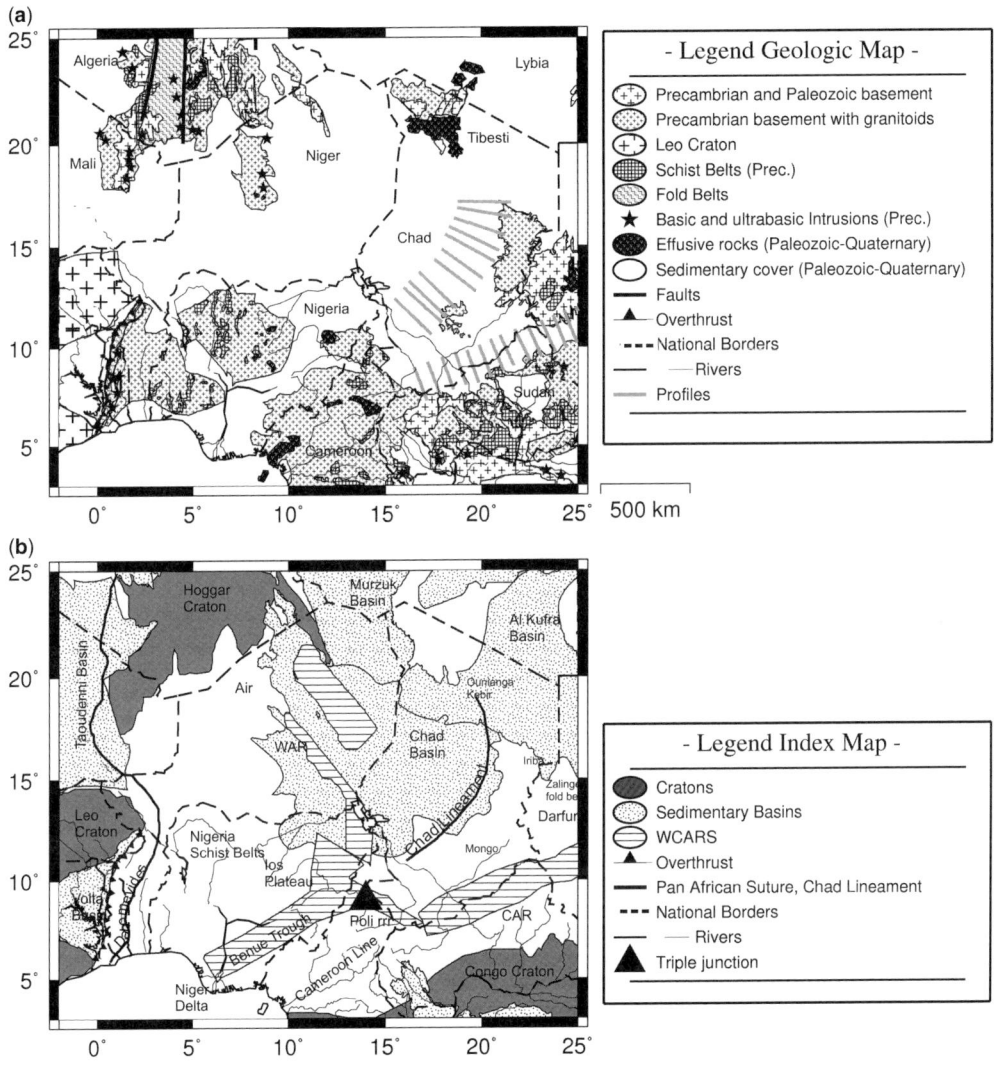

Fig. 3. Main geological features of north-central Africa. (**a**) Geologic map with the position of the profiles used in demonstrating fundamental differences in the gravity and gradient fields for the Chad lineament and the central African rift. Country names are shown. (**b**) Geographic index map with locations mentioned in the text and names of geological units shown in (a). The information stems from the Geologic map of Africa in six sheets (CGMW/UNESCO 1990). The outlines of cratons are based on numerous studies (e.g. Schlüter 2006; Priestley *et al.* 2008). The basin outlines coincide greatly with those of Persits *et al.* (2001). The principal rifts refer to Schlüter (2006), based on Guiraud *et al.* (2005). The magmatism relies on the works of Wilson & Guiraud (1992) and McHone (2000). CAR, central African rift; WAR, West African Rift; WCARS, west and central African rift system; Poli-rrr, position of the hypothetical Poli triple junction.

tectono-metamorphic evolution. They mostly comprise granite–gneiss complexes of greenstone belts associated with basic and ultra-basic rocks and of metamorphosed sediments.

On the eastern margin of the West African Craton, the Trans-Saharan Fold Belt of Pan-African age (Neoproterozoic) is located (Dauteuil *et al.*

2009). The Trans-Saharan Fold Belt marks the eastern border of the West African Craton and extends from Hoggar in the north to the Dahomeyides in the south. The fold belt is 100 to 400 km wide and 1300 km long, consisting of Pan-African meta-sediments. Along the fold belt a suture has been identified. It is marked by strings of mafic and ultra-mafic massifs in its southern and central portion. To the west of the Dahomeyides the overturned lower and middle sequences of the Volta Basin have been deformed and overthrusted, containing volcano–sedimentary complexes (Trompette 1994).

Volcanism in north-central Africa started in the Eocene (35–30 Ma) with the eruptions of the Hoggar magmatic province. The Cameroon line is a volcanic alignment oriented NNE–SSW that extends for more than 1000 km from Chad into the Atlantic Ocean. It was built between 32 Ma and the present day on a series of grabens and horsts which formed at the end of the Cretaceous (Dauteuil *et al.* 2009). The Tibesti volcanic province is emplaced on Precambrian basement; the volcanic activity seems to have started in the Lower Miocene and continued in the Quaternary (Gourgaud & Vincent 2004).

The west and central African rift system (WCARS) is a series of NNW–NW-trending extensional basins of the Early Cretaceous. The rift basins have widths greater than 150 km and depths in excess of 5 km. The WCARS lacks volcanism but is associated with a broad zone of crustal thinning (Fairhead & Green 1989). However, recent studies show that crustal thickness beneath the Cameroon volcanic line, beneath part of the WCARS and the Pan-African Oubanguides Belt, to the south is 35–39 km and the crust is even thicker under the northern margin of the Congo Craton (43–48 km) (Tokam *et al.* 2010). In our study area several large-scale intracratonic basins are found: the Chad, Taoudenni, Al Kufra and Murzuk basins. They are vast subsiding depressions, tectonically calm and developed on the Gondwana supercontinent (Dauteuil *et al.* 2009). Their relation to the crust and lithospheric structure has however not been studied in detail due to the absence of reliable data.

The modern gravity and gravity-gradient fields

Gravity and the components of the Marussi tensor are calculated for northern Africa starting from the EGM2008 (Pavlis *et al.* 2008) spherical harmonic expansion of the gravity potential field. We adopt a geocentric spherical coordinate system and calculate the fields at the height of 4000 m above a spherical Earth model of radius equal to the ellipsoidal radius at intermediate latitude (Earth radius 6375.37853 km; Janák & Šprlák 2006). The 4000 m height guarantees all values to be above topography, which is necessary for topographic reductions. The values are calculated on a regular grid of 0.05° (*c.* 6 km) grid cell size, with

a maximum degree and order equal to 2159 of the harmonic expansion.

We correct the fields for the topographic effect. The topography is modelled with the ETOPO1 DEM (1 arc-minute global relief Digital Elevation model of Earth's surface; Amante & Eakins 2009) using a resolution of 0.02° (*c.* 2 km). We perform a spherical calculation, in which the spherical mass elements are approximated with a prismatic mass element (Forsberg 1984). We use a standard density of 2670 kg m^{-3} for land and a density of 1030 kg m^{-3} for the sea. The ETOPO1 grid is given in ellipsoidal geodetic coordinates, which must be converted into geocentric spherical coordinates for the calculations. The topography-corrected gravity anomaly and the vertical gradient (T_{ZZ}) are shown in Figure 4. The topographic correction removes the obvious correlation of the fields with the topography and produces the Bouguer anomaly field, if applied to the gravity anomaly. It is an analogous correction when applied to the vertical gradient. The correction amounts to up to a few hundreds of mGal for gravity and up to a few tens of Eötvös for the vertical gradient. It is greatest over the highest topographic elevations, for example the Tibesti massif.

Comparison of the gravity and the gradient field reveals that anomalies are located in the same regions, but that the gradient field highlights more details. Where the gravity shows either a negative or positive signal, the gradient is also anomalous but reflects shorter wavelengths and does not reproduce the same pattern. In the following we demonstrate the usefulness of the gradient field for highlighting geological features. The simplest way to achieve this task is to critically inspect the mapped fields, describe the characteristics and match them to the known geology. The correspondence between the characteristics of the field and the particular geological features allows us to apply the fields in other regions to reveal unknown structures that are either concealed by sediments or which have not been mapped before.

Large sag basins (e.g. Chad Basin) have low-gradient levels between −5 and +5 Eötvös for T_{ZZ} and are limited to −50 mGal for gravity. The volcanic deposits (e.g. Cameroon line, Tibesti massif) are distinct by high T_{ZZ} values exceeding +30 Eötvös. The gravity reflects the crustal thickness variations, and has either negative values as in Tibesti and Hoggar or positive values (schist belt of western Nigeria) lacking the short-wavelength characteristic of the gradient. Examples are the Tibesti massif, the Cameroon volcanic line and the Early Jurassic magmatic deposits of western Africa (compare Figs 3 & 4). Sutures generate a positive signal in T_{ZZ} (near to 30 Eötvös or slightly greater) and a positive gravity signal; the

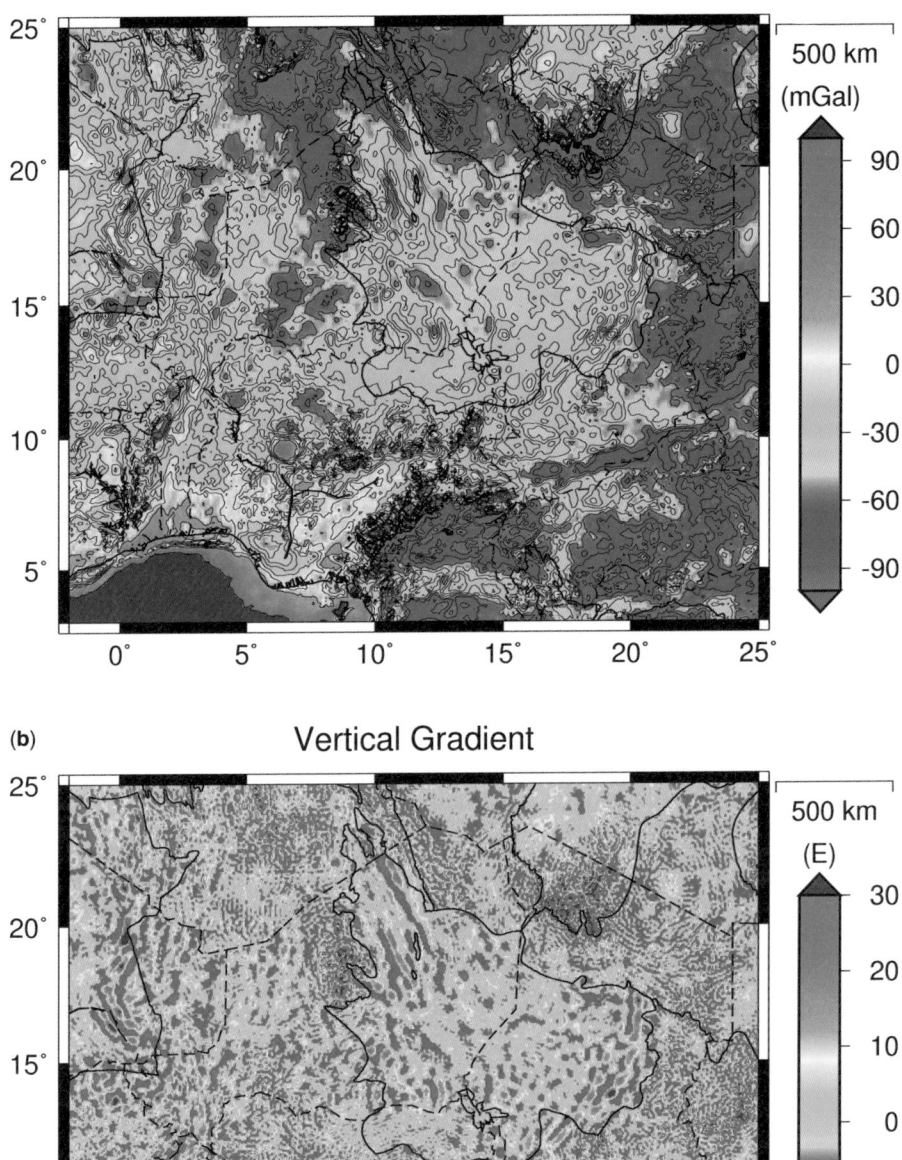

Fig. 4. (a) Bouguer gravity field in mGal and (b) vertical gradient corrected for topography in Eötvös units for the WCARS and the Chad lineament of north-central Africa. Outlines of major sedimentary basins depicted as continuous line; national borders depicted as dashed line.

Trans-Sahara or Pan-African suture (Trompette 1994) is easily identified. The high values in the southern part of the suture (also termed Daho-meyides) are explained by the presence of high-pressure granulites and eclogite (Attoh & Nude 2008).

The branches of the WCARS (Burke & Dewey 1972) generate linear gradient signals with coupled positive parallel anomalies in T_{ZZ} and a negative signal in gravity. Folded and metamor-phosed sediments or basement with magmatic intrusions generate a characteristic signal in T_{ZZ}, with prevalence at short wavelengths. To the north-west of the Benue trough the short-wavelength content in T_{ZZ} is explained by the presence of the schist belt of western Nigeria (Trompette 1994), which does not produce a clear gravity signal. The belt comprises ancient basement intruded by granitic plutons and amphibolites, both high-density rocks, which produces the short-wavelength signal in T_{ZZ} in contrast to the lower density of schists. This high-frequency content in T_{ZZ} is very characteristic and allows basement affected by granitic intrusions to be distinguished from basement without intrusions. Another good example is given by the basement of Air or the Ios Plateau.

We now focus on the WCARS, which produces a series of prominent linear signals, and the above-mentioned Chad lineament which produces the striking arched elongated anomaly. In both the gravity field (Fig. 4a) and the vertical gradient (Fig. 4b), the WCARS can be clearly identified as well as magmatic deposits such as the Tibesti massif. In between the west and central rift zone, at the eastern flank of the Chad Basin, a slightly bent gravity high is observed which corresponds to the Chad lineament. This third linear signal connects the volcanic province of Tibesti in the north to the western limit of the CAR (central African rift) over a length of 1300 km. The lineament has no expression in topography or outcrop and is entirely covered by the sediments of the Chad Basin. Note also that it separates areas of basement outcrop to the east from areas of younger sediment cover to its west.

In Figure 5 we compare the gravity, gravity gradient and topography across the Chad lineament and the central African rift zone. The series of profiles (for location of profiles see Fig. 3a) reveals the characteristic signal of the Chad lineament: the central gravity high with over 50 mGal amplitude and a central maximum in T_{ZZ} up to 40 Eötvös strong and up to 75 km wide flanked by two smaller minima (Fig. 5a); the southern half of the lineament is double and has a secondary high. The topography is flat so the lineament is completely buried by sediments. The central African Rift (Fig. 5b) has a completely different characteristic

signal of a central gravity low up to 80 mGal in amplitude with respect to the flanks. The T_{ZZ} has two lateral maxima 150 km apart with amplitude of up to 40 Eötvös strong and a central minimum. For the T_{ZZ} a relative central high is superimposed on the central minimum. The topography is high at one or the other sides of the rift. Both the west and central African rift systems are almost linear and are not bent like the Chad lineament. The differences in the characteristics of the Chad lineament and the WCARS highlight their different origin, as we will discuss in more detail in the following.

Discussion

We have mapped the gravity field and gravity gradients for central-north Africa using the recently available global potential field expansion in spherical harmonics (Pavlis *et al.* 2008). The gravity potential model blends terrestrial and satellite observations and has an increased database compared to the previous compilation of Louis (1970). To our knowledge, the latter is the most complete work on a gravity campaign in Chad and is based on a similar terrestrial dataset but contains no satellite observations. Louis (1970) explains in his pioneering work how interpolation of the measurements was carried out in two ways: (1) by manual interpolation of the points drawing iso-anomalies of 10 mGals or (2) by least-squares interpolation of the points with a polynomial surface and successive automatic drawing of the iso-anomalies at 10 mGal intervals. Available computing facilities three to four decades ago dictated that the processing and interpolation of the raw data be simplified.

The final maps of gravity anomaly and Bouguer anomaly therefore contain less information than the original observations that have a measurement error of 1–5 mGal, depending on the surveyed area (Louis 1970). The modern representation in terms of spherical harmonic expansion of the field contains locally more information than the maps presented by Louis (1970), as the iso-anomalies can be represented with a smaller interval than 10 mGal. A further improvement of the global field data is due to the geographical limitation of the maps of Louis (1970) that do not include data in Sudan, northern Chad, large parts of Nigeria or Congo.

We next consider the gravity gradient which highlights more superficial masses than basic gravity measurements.

The arcuate lineament 1300 km long in eastern Chad (Figs 3 & 4) deserves special attention, because the origin of the gravity signal is presently unknown. The lineament was first mentioned in Louis (1970), who describes the prominent gravity anomaly together with the other gravity highs and lows that can be found in Chad. Tentatively, Louis

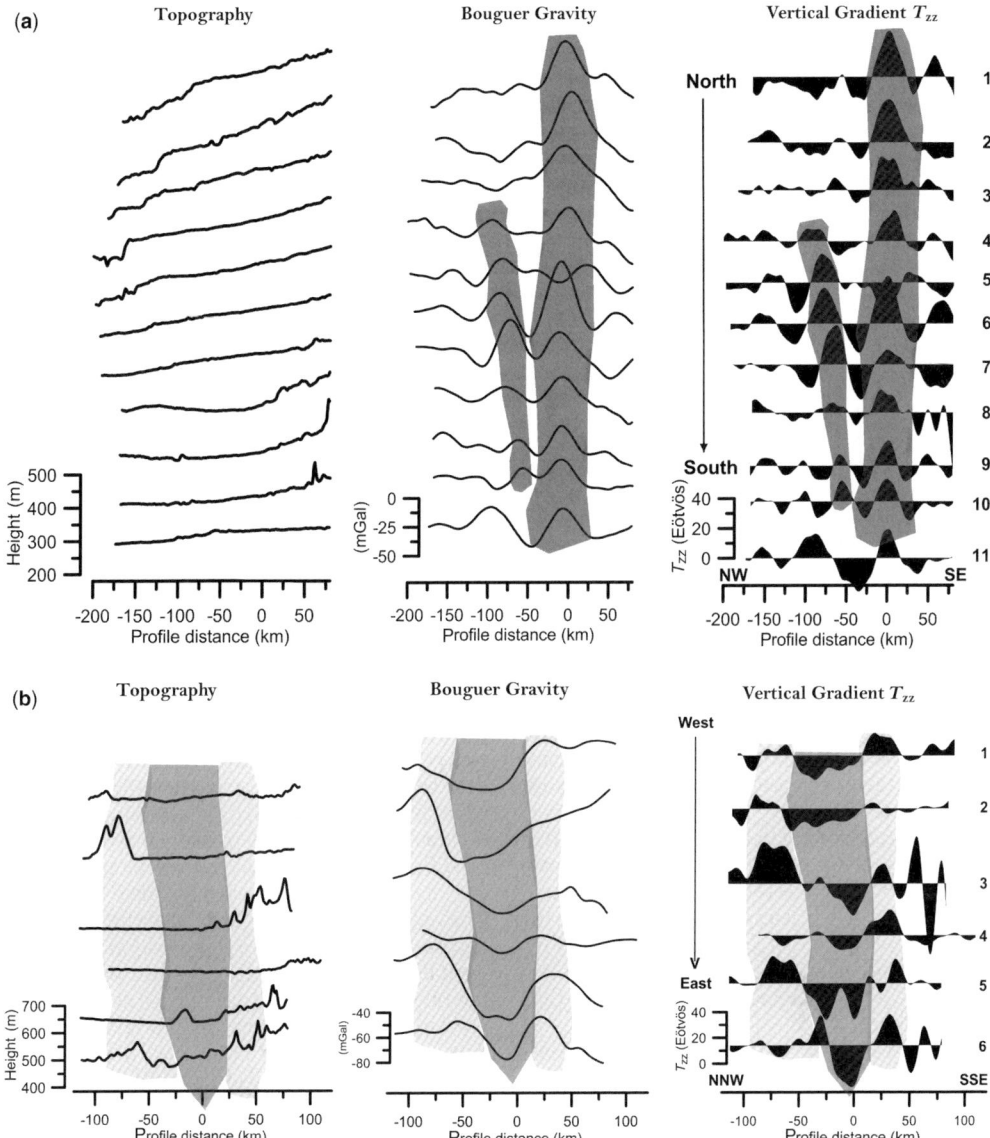

Fig. 5. Profiles comparing the central African rift with the Chad lineament. Left: topography; centre: Bouguer gravity; and right: vertical gravity gradient (dark shaded area depicts central axis; lighter shaded area depicts flanks of central axis). Profiles across (**a**) Chad lineament and (**b**) central African rift. Note the systematic differences between the two: Chad lineament has no topography and a gravity high associated with it, whereas the rift is associated with topographic highs along its flanks and is marked by a central gravity low.

(1970) interprets the anomaly as the western border of the Nile Craton, a block discussed by Rocci (1965). The lineament was interpreted to represent subcrustal material that was transferred to upper crustal levels. Louis (1970) suggests that the age of the feature is older than Cretaceous, and thus older than the Benue rift or the volcanism of Cameroon and Tibesti. The author considers it more probably related to a consolidated fracture with the presence of heavy elements of Caledonian or Precambrian age, contemporaneous with the Pan-African Trans-Saharan suture.

Burke & Whiteman (1973) review 16 uplifts and 25 rifts of Mesozoic and Tertiary age in Africa, including 11 rrr-junctions (rift–rift–rift junction produced by uplift). Among these the Poli rrr-junction (centred at 9°N, 14°E) is included, from which the Chad anomaly emanates, and which is suggested to mark a rift (termed Ati rift) forming one arm of the Poli structure. The positive gravity anomaly is interpreted (without any control from modelling) as a long series of basic intrusions. The approximate age of the three hypothesized rifts emanating from the Poli rrr-junction starts at 130 Ma and stops at 80 Ma.

The Chad gravity anomaly is roughly delineated in Fairhead & Green (1989) who model the East Niger Rift Basin with crustal thinning and successive sediment filling, which leads to a broad gravity high and an axial gravity low. Fairhead & Green (1989) explain the WCARS using the lithospheric extension model of basin formation (Mckenzie 1978). Here, stretching of the lithosphere results in passive upwelling of hotter less-dense asthenosphere and a concomitant necking of the low-density crust. The isostatic response is surface subsidence, which continues after extension ceases. The tectonic model for the WCARS is that of a strike–slip motion extending from the mid-ocean ridge along reactivated fracture zones into the Benue trough. The central African shear zone is ideally orientated to transmit strike–slip motion from the ocean ridge into the African continent. The WCARS rift basins are generated parallel or perpendicular to the shearing orientation and have widths 150 km or greater, depths in excess of 5 km and lack volcanism.

The interpretation of the Chad lineament as a rift arm does not fit the above framework for two main reasons: (1) it lacks the broad positive gravity signal due to crustal thinning and it does not have the axial gravity minimum due to the sedimentary infill of the subsiding rift; and (2) the orientation of the Chad lineament is oblique to the expected rifting direction.

An alternative interpretation of the Chad lineament was introduced by Freeth (1984) who tried to model the crustal dyke or intrusion following the hypothesis of Burke & Whiteman (1973). He concluded that an exceptionally and improbably great dimension of the dyke is needed to explain the gravity observations (over 1000 km length, 35 km width, *c.* 30 km thickness). The observed gravity signal also demonstrates abrupt changes, a feature easily explained by a superficial density contrast. Instead, the author proposed that the anomaly reflects the presence of a band of basic volcanic and/or volcanoclastic sediments within the basement, and therefore at shallower depths than the dyke or intrusions.

The gravity and gradient maps we present here highlight the fundamental differences between the Chad anomaly and the rifts in central-north Africa. The anomaly compares better to the Pan-African suture that borders the West African Craton, as it is particularly evident in the gradient field. The curvature radius of the lineament and the alignment next to reworked Precambrian basement are common features of the data. Two separate outcrops of the Saharan Metacraton to the east of the lineament fit the convex, eastern side of the Chad lineament and could belong to the rock unit limited by the lineament. We have calculated the signal produced by a body 30 km wide, 8 km thick and its top at 1 km depth, with a density of 300 kg m^{-3}, which we find resembles the observed gradient and gravity signal (Fig. 6). In analogy to the Pan-African suture, the signal could be generated by high-density rocks as eclogite, diorite or amphibolite. As we have no additional geophysical constraints, this is only one of many possible models that can explain the anomaly. However, the possibility that the anomaly is caused by crustal thinning and a superficial rift basin (the preferred model for the WCARS) can be ruled out (e.g. Browne & Fairhead 1983). Crustal thinning would clearly produce a gravity signal of greater wavelength.

Another argument is the absence of an expression in topography for the Chad lineament. If the Chad lineament had been formed by Cretaceous rifting, rift shoulder uplift would be observed analogous to the WCARS as the formation mechanism could not be entirely different. The topography over the Chad anomaly is perfectly flat (Fig. 5a), while the topography over the central African rift is over 700 m high (Fig. 5b). Similar environmental conditions imply analogous erosion, and consequently similar annual rates of erosion. Unless the Chad lineament is much older than the WCARS, traces of a rift-related topography should be observable today. Only a much older age of the Chad lineament would allow erosion to flatten the topographic elevation. A Cretaceous origin of rifting must therefore be ruled out. Freeth (1984) claims the northernmost part of the Chad lineament lies below undeformed sediments of Devonian age, which is further evidence in favour of the considerably older age than the Mesozoic rifts.

The interpretation of the Chad lineament in terms of an old structure due to the above argument is in good agreement with its geographical setting. The area we have investigated is part of the Saharan Metacraton defined by Abdelsalam *et al.* (2002): the crustal rocks occupying a rectangular-shaped region bounded by longitudes 8°30′E and 35°E and latitudes 3°N and 24°N. The metacraton relates to Neoproterozoic continental crust that has been remobilized during deformation, metamorphism, emplacement of igneous bodies and local episodes of crust formation related to rifting and oceanic basin development. The Saharan Metacraton is dominated by metamorphic rocks intruded

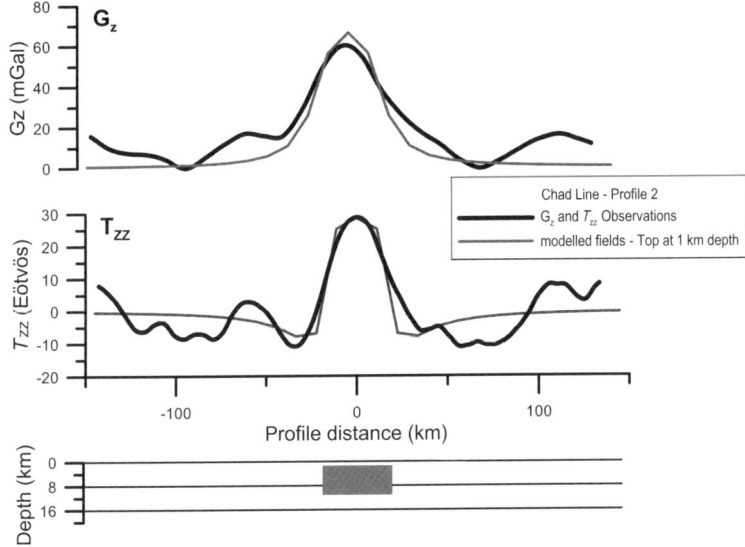

Fig. 6. Observed (black) and modelled (grey) vertical gradient (T_{ZZ}) and gravity field over a possible high density body (300 kg m^{-3} density, 30 km width, 8 km thickness, top at 1 km depth). The body represents one of many possible models. A deep origin of the observed signal in the Chad lineament can be excluded.

by Neoproterozoic granitoids ranging in age from 750 to 550 Ma. It seems that most of the gneisses were formed or metamorphosed during Neoproterozoic time, involving Palaeoproterozoic or even Archaean continental crust. Two structural trends dominate the Saharan Metacraton: an early ENE–WSW trend and a younger NS trend. The ENE–WSW trend would be compatible with the Chad lineament, and has been defined in the outcrops east of the Chad lineament as the Zalingei Fold Belt which is thought to continue SW into Cameroon. This trend has been interpreted as representing a pre-Neoproterozoic structure marked by an event of age 630–620 Ma, or was inherited from early Neoproterozoic terrane collisions. North-trending structures in the form of fold belts and faults have been found in some parts of the metacraton, as in the Tibesti massif. These north-trending upright folds and strike–slip faults were suggested to be due to east–west-aligned crustal shortening accompanying collision between the Saharan Metacraton and the Arabian–Nubian shield. In the context of crustal shortening, the Chad lineament could well be interpreted as overthrusted lower crust or high-grade compressional metamorphism that generated high-density rocks such as ultrabasics, eclogites and amphibolites.

Conclusions

The recent geodetic satellite missions CHAMP and GRACE have produced global observations of the gravity field with a resolution (half-wavelength) of 165 km. The integration with terrestrial data has resulted in a global gravity model with a nominal spatial resolution (half-wavelength) of about 10 km (Pavlis *et al.* 2008). The realistic resolution however depends on the availability of the terrestrial data and may change locally. We have mapped the vertical gravity gradient and the gravity field for north-central Africa and compared this field to existing geological maps and known tectonic structures. The gravity gradient is very well correlated to the geological map and to known lineaments such as fold belts, rifts and magmatic deposits associated with density variations. It shows that the global fields are suitable for tracing geological structures. The global availability makes direct access to any study area feasible. The advantage over regional or national databases is that possible datum shifts due to differences in reference systems are ruled out. Features extending several degrees can be traced, a task that may be difficult when combining several terrestrial databases. The use of the gradient fields is very valuable for highlighting superficial structures as deep structures are attenuated naturally by the potential field law of gravity (e.g. Blakely 1995).

A review of all major linear anomalies found in north-central Africa highlights a 1300 km long lineament which is comparable in amplitude to the well-known WCARS or the Trans-Sahara suture east of the Taoudenni Basin. It is a slightly arcuate linear feature, convex eastwards, running through Chad Basin to Tibesti. It has no topographic expression or geologic outcrop, and demonstrates a central gravity high in comparison to the WCARS rift (that demonstrates a central gravity low). The WCARS has entered the subsequent literature and geological maps (CGMW/UNESCO 1990; Guiraud *et al.* 2005; Schlüter 2006) whereas the Chad lineament has been omitted in the more recent literature. This may be due to the fact that its arcuate north–south orientation does not fit the model that explains the Cretaceous rift system as a result of strike–slip tectonics caused by the opening of the central and south Atlantic Oceans (e.g. Fairhead 1988; Fairhead & Green 1989; Guiraud *et al.* 2005; Moulin *et al.* 2010).

The lineament could mark an ancient suture and could therefore be much older than the Cretaceous WCARS rift system.

We thank I. Nagy for her assistance with the bibliographic search and retrieval of documents. We thank C. Gaina for providing the compilation on the cratons of northern Africa. We thank Statoil for financial support and in particular H. M. Bjørnseth from Statoil for discussions. We thank the Italian Space Agency (ASI) for supporting the GOCE-Italy project. The work was partially supported by PRIN contract 2008CR4455_003. We acknowledge the use of the GMT-mapping software of Wessel & Smith (1998) and the tesseroid forward calculation software developed by L. Uieda in the frame of our GOCE-projects. We acknowledge the use of the EGM2008 gravity model and software of Pavlis *et al.* (2008). We thank R. Rummel, C. Reeves and the editor S. Buiter for their meticulous reviews.

References

ABDELSALAM, M. G., LIEGEOIS, J.-P. & STERN, R. J. 2002. The Saharan metacraton. *Journal of African Earth Sciences*, **34**, 119–136.

AMANTE, C. & EAKINS, B. W. 2009. ETOPO1 1 Arc-Minute Global Relief Model: Procedures, Data Sources and Analysis. NOAA Technical Memorandum NESDIS NGDC-24, 19, March 2009.

ATTOH, K. & NUDE, P. M. 2008. The tectonic significance of carbonatite and ultra-high pressure rocks in the Pan-African Dahomeyide suture zone, southeastern Ghana. *In*: ENNIH, N. & LIEGEOIS, J.-P. (eds) *The Boundaries of the West African Craton*. Geological Society, London, Special Publications, **297**, 217–231.

BLAKELY, R. J. 1995. *Potential Theory in Gravity and Magnetic Applications*. Cambridge University Press, Cambridge.

BRAITENBERG, C. & EBBING, J. 2009. New insights into the basement structure of the west-Siberian basin from forward and inverse modelling of Grace satellite gravity data. *Journal of Geophysical Research*, **114**, B06402, doi: 10.1029/2008JB005799.

BROWNE, S. E. & FAIRHEAD, J. D. 1983. Gravity study of the Central African Rift System: a model of continental disruption. The Ngaoundere and Abu Gabra rifts. *Tectonophysics*, **94**, 187–203.

BRUINSMA, S. L., MARTY, J. C., BALMINO, G., BIANCALE, R., FÖRSTE, C., ABRIKOSOV, O. & NEUMAYER, H. 2010. GOCE Gravity Field Recovery by Means of the Direct Numerical Method. *In*: *Proceedings of ESA Living Planet Symposium 2010, Bergen, June 27–July 2*, Bergen, Norway, 2010.

BURKE, K. C. & DEWEY, J. F. 1972. Orogeny in Africa. *In*: DESSAUVAGIE, T. F. J. & WHITEMAN, A. J. (eds) *African Geology, Ibadan 1970*. University of Ibadan, Ibadan, 583–608.

BURKE, K. C. & WHITEMAN, A. J. 1973. Uplift, rifting and the break-up of Africa. *In*: TARLING, D. H. & RUNCORN, S. K. (eds) *Implications of Continental Drift to the Earth Sciences*. Academic Press, London, 735–755.

CGMW/UNESCO. 1990. *International Geological Map of Africa at 1:5 000 000*. CGMW/UNESCO, Paris.

DAUTEUIL, O., BOUFETTE, J. & TOTEU, F. 2009. *The Changing Faces of Africa*. Commission for the Geological Map of the World (CGMW), Paris, 1–48.

FAIRHEAD, J. D. 1988. Mesozoic plate tectonic reconstructions of the central South Atlantic Ocean: the role of the West and Central African rift system. *Tectonophysics*, **155**, 181–191.

FAIRHEAD, J. D. & GREEN, C. M. 1989. Controls on rifting in Africa and the regional tectonic model for the Nigeria and east Niger rift basins. *Journal of African Earth Sciences*, **8**, 231–249.

FORSBERG, R. 1984. *A study of terrain reductions, density anomalies and geophysical inversion methods in gravity field modelling*. Ohio State University, Scientific Report No. 5, Report Number AFGL-TR-84-0174, 133.

FÖRSTE, C., SCHMIDT, R. ET AL. 2008. The Geo-ForschungsZentrum Potsdam/Groupe de Recherche de Gèodésie Spatiale satellite-only and combined gravity field models: EIGEN-GL04S1 and EIGEN-GL04C. *Journal of Geodesy*, **82**, 331–346, doi: 10.1007/s00190-007-0183-8.

FREETH, S. J. 1984. How many rifts are there in West Africa? *Earth and Planetary Science Letters*, **67**, 219–227.

GOURGAUD, A. & VINCENT, P. M. 2004. Petrology of two continental alkaline intraplate series at Emi Koussi volcano, Tibesti, Chad. *Journal of Volcanology and Geothermal Research*, **129**, 261–290.

GUIRAUD, R., BOSWORTH, W., THIERRY, J. & DELPLANQUE, A. 2005. Phanerozoic geological evolution of Northern and Central Africa: an overview. *Journal of African Earth Sciences*, **43**, 83–143.

HOFMANN-WELLENHOF, B. & MORITZ, H. 2005. *Physical Geodesy*. Springer, Berlin.

JANÁK, J. & ŠPRLÁK, M. 2006. New software for gravity field modelling using spherical harmonics. *Geodetic and Cartographic Horizon*, **52**, 1–8 (in Slovak).

LOUIS, P. 1970. Contribution geophysique à la connaissance géologique du basin du lac Tschad. *Memoires Orstom*, **42**, 1–311.

MCHONE, J. G. 2000. Non-plume magmatism and rifting during the opening of the central Atlantic Ocean. *Tectonophysics*, **316**, 287–296.

MCKENZIE, D. 1978. Some remarks on the development of sedimentary basins. *Earth and Planetary Science Letters*, **40**, 25–32.

MIGLIACCIO, F., REGUZZONI, M., SANSÒ, F., TSCHERNING, C. C. & VEICHERTS, M. 2010. GOCE data analysis: the space-wise approach and the first space-wise gravity field model. *In*: *Proceedings of ESA Living Planet Symposium 2010, Bergen, June 27–July 2*, Bergen, Norway, 2010.

MOULIN, M., ASLANIAN, D. & UNTERNEHR, P. 2010. A new starting point for the South and Equatorial Atlantic Ocean. *Earth-Science Reviews*, **98**, 1–37.

PAIL, R., GOIGINGER, H. *ET AL*. 2010. GOCE gravity field model derived from orbit and gradiometry data applying the time-wise method. *In*: *Proceedings of ESA Living Planet Symposium 2010, Bergen, June 27–July 2*, Bergen, Norway, 2010.

PAVLIS, N. K., HOLMES, S. A., KENYON, S. C. & FACTOR, J. K. 2008. An Earth Gravitational Model to degree 2160: EGM2008. *Geophysical Research Abstracts*, **10**, EGU2008-A-01891.

PEDERSEN, L. B. & RASMUSSEN, T. M. 1990. The gradient tensor of potential field anomalies: some implications on data collection and data processing of maps. *Geophysics*, 1558–1566.

PERSITS, F. M., AHLBRANDT, T. S., TUTTLE, M. L., CHARPENTIER, R. R., BROWNFIELD, M. E. & TAKAHASHI, K. I. 2001. *Maps showing geology, oil and gas fields and geological provinces of Africa*, Open file report 97–470A, Version 2.0, US Department of Interior, US Geological Survey.

PRIESTLEY, K., MCKENZIE, D., DEBAYLE, E. & PILIDOU, S. 2008. The African upper mantle and its relationship to tectonics and surface geology. *Geophysical Journal International*, **175**, 1108–1126.

ROCCI, G. 1965. Essai d'interprétation de measures géochronologiques: la structure de l'Ouest africain. *Sciences de la Terre*, **10**, 462–478.

RUMMEL, R., BALMINO, G., JOHANNESSEN, J., VISSER, P. & WOODWORTH, P. 2002. Dedicated gravity field missions-principles and aims. *Journal of Geodynamics*, **22**, 3–20.

SCHLÜTER, T. 2006. *Geological Atlas of Africa*. Springer Verlag, Berlin.

TOKAM, A. P. K., TABOD, C. T., NYBLADE, A. A., JULIÀ, J., WIENS, D. A. & PASYANOS, M. E. 2010. Structure of the crust beneath Cameroon, West Africa, from the joint inversion of Rayleigh wave group velocities and receiver functions. *Geophyical Journal International*, **183**, 1061–1076.

TROMPETTE, R. 1994. *Geology of Western Gondwana (2000–500 Ma)*. Balkema A. A. Publishers, Rotterdam, 1–350.

WESSEL, P. & SMITH, W. H. F. 1998. New, improved version of generic mapping tools released. *Transactions of American Geophysical Union*, **79**, 579.

WILSON, M. & GUIRAUD, R. 1992. Magmatism and rifting in Western and Central Africa, from Late Jurassic to Recent times. *Tectonophysics*, **213**, 203–225.

Towards a better understanding of African topography: a review of passive-source seismic studies of the African crust and upper mantle

STEWART FISHWICK[1]* & IAN D. BASTOW[2]

[1]*Department of Geology, University of Leicester, University Road, Leicester, LE1 7RH, UK*

[2]*Department of Earth Sciences, University of Bristol, Wills Memorial Building, Queen's Road, Bristol BS8 1RJ, UK*

*Corresponding author (e-mail: sf130@le.ac.uk)

Abstract: Explaining the cause and support of Africa's varied topography remains a fundamental question for our understanding of the long-term evolution of the continent. As geodynamical modelling becomes more frequently used to investigate this problem, it is important to understand the seismological results that can be incorporated into these models. Crustal thickness estimates are crucial for calculating components of topography that are isostatically compensated. Variations in seismic velocity help constrain variations in subsurface temperature and density and thus buoyancy; measurements of anisotropy can also be used to determine the contribution of the mantle flow field to dynamic topography. In this light, we review the results of passive seismic studies across Africa. At the continental scale there are significant differences in crustal models, meaning large uncertainties in corrections for isostatic topography. In east Africa, multiple seismic experiments have provided firm constraints on crustal and mantle structure. Tomographic images illuminate a broad (c. 500 km wide) low-velocity region in the upper mantle, with possible connection to the African Superplume in the lower mantle. These observations, alongside the variations in radial anisotropy, strongly suggest that the mantle flow field contributes significantly to the uplift of the region. Beneath southern Africa, low velocities are observed near the base of the continental lithosphere; the depth to transition zone discontinuities however suggests that they are not linked to the superplume beneath. It is thus less clear what role the sublithospheric mantle plays in supporting the region's high topography. Many of Africa's secondary topographic features (e.g. Atlas, Hoggar, Bie Dome) are underlain by slow velocities at depths of 100–150 km and are adjacent to rapid changes in lithospheric thickness. Whether these variations in lithospheric structure promote small-scale convection or simply guide the larger-scale mantle flow field remains ambiguous.

Forty years after the advent of plate tectonic theory, while the vertical motion of the oceanic plates can be explained by relatively simple arguments concerning their age and thermal structure (McKenzie 1978) the vertical movements in continental regions often remain a matter of considerable debate. For example, for a region surrounded by mostly extensional plate boundaries there is significant topography on the African continent. The dominant first-order feature is the near bimodal topography described by Doucouré & de Wit (2003): elevated topography (c. 500–3000 m) is observed in eastern and southern Africa, while lower topography occurs in western and central Africa and towards the north-eastern (Egyptian) margin of the continent (Fig. 1). Superimposed on these long-wavelength structures are a series of basins and swells (Fig. 1), famously described by Holmes (1944). In northern Africa the swells are frequently topped by relatively young (c. 35 Ma–Recent) volcanism (e.g. Hoggar,

Tibesti, Fig. 1) suggesting a link to elevated mantle temperatures and dynamic support. In the south, swells are also observed on top of the generally high topography (e.g. Namibia, Bie; Fig. 1) but in these regions are not always associated with recent volcanism. A simple correlation between hotspot tectonism and uplift is therefore not necessarily applicable continent-wide. The timing and cause of Africa's considerable topography thus remains a matter of continued debate.

In his work *The African Plate*, Burke (1996) presented the view that the majority of African topography, including the Great Escarpment, has been formed during the last c. 30 Ma. One indication for this young continent-wide uplift has come from interpretations of the correlations of an African surface defined using geomorphological evidence. Secondly, the prominent Oligocene unconformity observed in seismic reflection data all around the continental margin is compelling evidence that uplift commenced around this time

From: Van Hinsbergen, D. J. J., Buiter, S. J. H., Torsvik, T. H., Gaina, C. & Webb, S. J. (eds) *The Formation and Evolution of Africa: A Synopsis of 3.8 Ga of Earth History*. Geological Society, London, Special Publications,
357, 343–371. DOI: 10.1144/SP357.19   0305-8719/11/$15.00 © The Geological Society of London 2011.

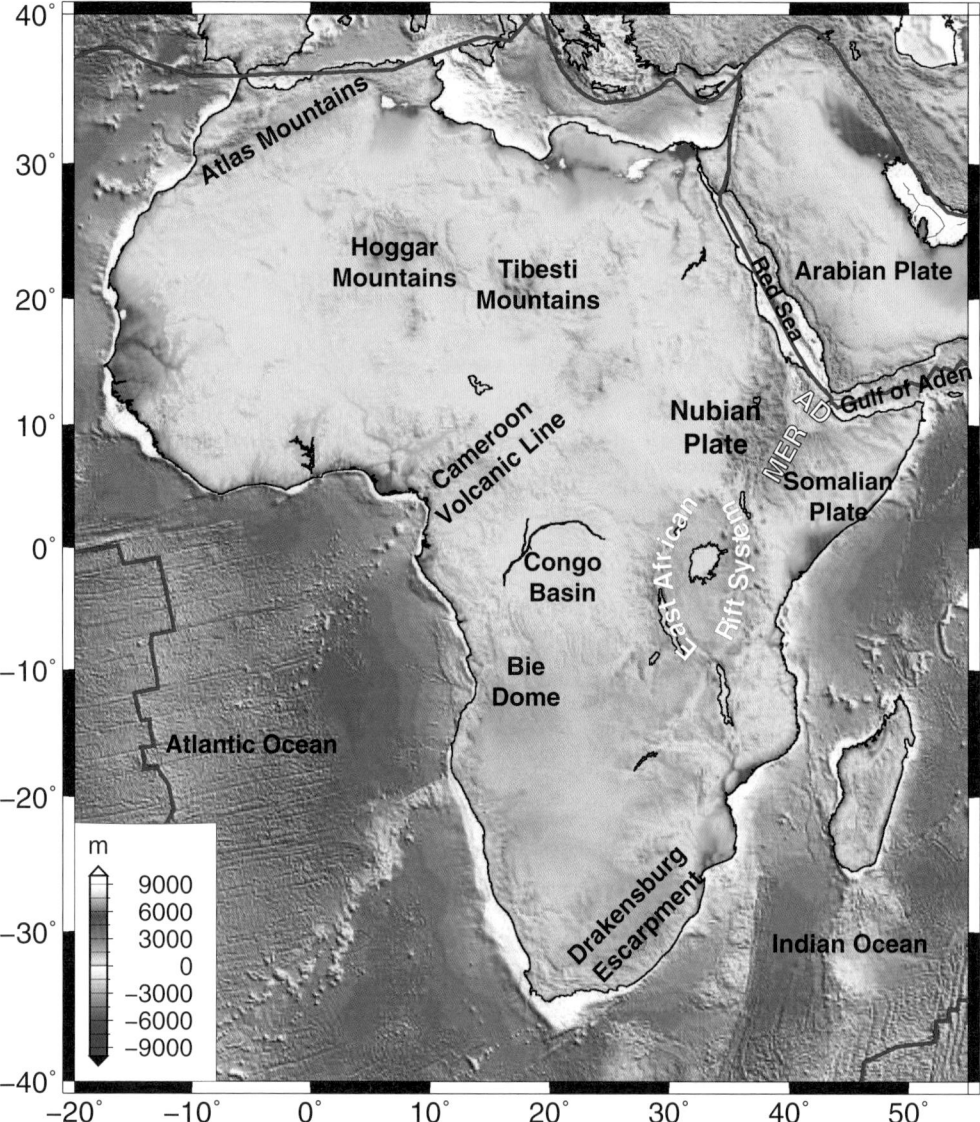

Fig. 1. Location map of Africa showing major tectonic features superimposed on regional topography (MER, Main Ethiopian Rift; AD, Afar Depression). Red lines are major plate boundaries.

(e.g. Burke 1996). Roberts & White (2010) estimate uplift histories from river profiles for a number of regions in Africa; they also suggest that significant uplift has occurred since 30–40 Ma. The timing, however, is not uniform across the continent, with recent, rapid, localized uplift occurring in regions such as the Bie Dome (Fig. 1).

In contrast, others believe that a significant part of the high African topography has a much older history. Doucouré & de Wit (2003) attempt to reconstruct the Mesozoic topography of Africa by

removing the topographic effect of features known to be of Cretaceous age or younger. They observe that the bimodality of African topography was already in place in the Early Mesozoic. Significant evidence for a Mesozoic timing of the uplift comes in the form of apatite fission-track thermo-chronology. De Wit (2007) reviews the various apatite fission track studies alongside offshore stratigraphic studies to suggest that denudation across southern Africa operated on a wide scale (>1000 km) during the Cretaceous. Are the inferred

Mesozoic and Cenozoic histories incompatible with each other, or simply a result of differing techniques being sensitive to specific time periods?

The tectonic and geodynamic processes relating to topographic variation also appear to be variable. Collisional processes, commonly associated with regions of high topography, are the predominant factor only in the Atlas mountains (Fig. 1) of northwestern Africa (e.g. Frizon de Lamotte *et al.* 2000); even in this region, however, the high topography has also been associated with convective upwelling of the underlying mantle (e.g. Teixell *et al.* 2005). Elsewhere, other factors such as the emplacement of volcanic rock at the surface during hotspot tectonism, crustal thinning and rift flank flexure during extensional tectonics (e.g. the East African Rift or EAR) are likely to contribute to the development of Africa's variable topography (Fig. 1).

Are all these parts of the African plate presently in a state of isostatic equilibrium, or do regions require dynamic components towards the support of the topographic variations? The classical view of isostasy can be separated into two end-member models: Airy and Pratt. In Airy isostasy, the excess mass of high topography is compensated by a thickened root of low-density material relative to the surroundings. In contrast, in the Pratt model there is no thickening of the root; high topography is compensated instead by a lower-density column directly beneath. In order to assess the cause of any isostatic support, it is therefore necessary to know both the crustal thickness and density. Depending on the choice of compensation depth, variations in the lithospheric mantle density must also be considered (see e.g. Crosby *et al.* 2010).

If the topography is not isostatically supported, alternatives include regional support due to the strength of the lithosphere or a dynamic mantle component to the support. Many of Africa's domal regions appear to have a thin (<40 km) effective elastic plate thickness (Pérez-Gussinyé *et al.* 2009), so flexural support appears an unlikely candidate to explain the first-order topographic observations. Dynamic support, or dynamic topography, will occur when flow (caused by density variations within the Earth's mantle) interacts with the surface layer (for a recent review see Braun 2010). If there are dynamic components to the support of Africa's high topography, are these dominated by perturbations in the deep mantle (e.g. Lithgow-Bertelloni & Silver 1999; Gurnis *et al.* 2000), more localized asthenospheric upwellings (e.g. King & Ritsema 2000; Montagner *et al.* 2007; Al-Hajri *et al.* 2009) or a combination of both?

Crustal thickness and density, mantle density and temperature and flow direction are therefore of first-order importance in understanding topography. Serendipitously, many of these are manifest as measurable seismic signatures. For example, receiver function and controlled source seismic studies provide detailed constraints on the depth of velocity discontinuities such as the Moho, the lithosphere–asthenosphere boundary and the mantle transition zone. Seismic velocities measurable using body and surface-wave tomography are strongly influenced by variations in mantle temperature and density. The alignment of olivine crystals in the flowing mantle results in seismic anisotropy, which can be quantified via analysis of shear-wave splitting and surface-wave studies of the directional dependence of seismic velocities. Many studies on the African continent in the last 20 years have performed these analyses and thus provide valuable constraints for the geodynamic community. It is the goal of this manuscript to review the seismological experiments that have been performed in Africa to date, with a view to understanding better the variable topography of the continent.

Following the approach of Burke (1996), discussion will be focused on three areas: southern Africa; east Africa; and central, north and western Africa. An overview of the topographic features and a brief discussion of the plausible tectonics and geodynamics is initially presented. We then review the recent regional seismic studies, focusing on crustal thickness, mantle velocity structure and observations of seismic anisotropy. Finally, we consider whether these studies provide direct evidence of the processes causing topographic variation and, in particular, how the seismic results may provide additional constraints for the ongoing geodynamical modelling. It is clear that there is no single factor controlling the topographic expression of the African continent.

African topography

Eastern Africa

Nyblade & Robinson (1994) proposed that southern Africa, the East African Plateau and part of the south-eastern Atlantic form a contiguous region: the African superswell. Comparisons of bathymetry and elevation with models of plate cooling and the global mean elevation of 565 m, respectively, shows that the African superwell has c. 500 m residual topography (Nyblade & Robinson 1994). Within east Africa there appear, however, two quite distinct regions with higher elevations: the Kenya/Tanzania dome and the Ethiopian Plateau (Ebinger *et al.* 1989). The topography of Kenya and Tanzania appears to be contiguous with the higher elevations throughout the south of the continent, although its appearance is accentuated by the low-lying Congo Basin to the west (Fig. 1). In contrast, the Ethiopian Plateau is clearly separated from this area by a distinct NW–SE-trending topographic low (the Turkana Depression). On top of these broad swells, the scars of the EARS can also be seen. In the north, the Afar depression marks the region of the rift–rift–rift triple junction where the Red Sea and Gulf of Aden spreading centres meet with the EAR. The rift-related topography can be followed through Ethiopia, and then into the two branches surrounding the Archaean Tanzania Craton (Fig. 1).

Many authors have invoked thermal plumes to explain Africa's hotspot tectonism and uplift, but the location and number of plumes or upper mantle convective cells is often debated. For example, Ebinger & Sleep (1998) suggested that one large plume spread beneath the African Plate near Turkana at *c.* 45 Ma, but only small amounts of melt volume were produced until lithospheric thinning commenced in the Red Sea and Gulf of Aden rifts. Topography at the base of the lithosphere channelled buoyant material up to *c.* 1000 km from Turkana to the evolving Red Sea rift and the Mesozoic rift zones of eastern and central Africa. Alternatively, two Cenozoic plumes may have existed beneath East Africa, one rising and dispersing beneath southern Ethiopia at *c.* 45 Ma and the other rising beneath the Afar depression (e.g. George *et al.* 1998). More recent geochemical studies have explained their observations based on a modified single-plume hypothesis: multiple upwellings from the African superplume (Kieffer *et al.* 2004; Furman *et al.* 2006).

The timing of Ethiopia's high topography is perhaps somewhat better constrained than that of southern Africa due to the dating of the volcanism in the region. A thick pile (500–2000 m) of continental flood basalts and rhyolites were emplaced in the central Ethiopian Plateau *c.* 30 Ma (Hofmann *et al.* 1997; Ayalew *et al.* 2002; Coulie *et al.* 2003). This is thought to have been due to the impact of the Afar mantle plume, which was coeval with marked uplift of the plateau (*c.* 20–30 Ma; Pik *et al.* 2003). Isolated shield volcanism in the period *c.* 30–10 Ma occurred across the Ethiopian Plateau (Fig. 1) and added *c.* 2 km of additional local relief (e.g. Coulie *et al.* 2003; Kieffer *et al.* 2004). The temporal evolution of rift-related topography remains somewhat controversial. In southern Ethiopia, results from low-temperature (U–Th)/He thermochronometry suggest that the rift development has been continuous since initiation in the Miocene (Pik *et al.* 2008). In contrast, Gani *et al.* (2007) suggest that significant recent (*c.* 10 Ma) uplift was due to foundering of the Plateau lithosphere following extensive heating and weakening since the onset of flood basalt volcanism at *c.* 30 Ma.

Timing of plateau formation around Kenya and Tanzania is less well constrained. The earliest volcanism in Kenya started in the Turkana region of northern Kenya at *c.* 35–40 Ma (e.g. Furman *et al.* 2006), and has also been associated with the impingement of a mantle plume (e.g. George *et al.* 1998). Following the rift system to the south through Kenya and Tanzania, the onset of Cenozoic volcanism becomes progressively younger (e.g. Morley *et al.* 1992). Locally, there is also evidence for Neogene uplift along the flanks of some rift valleys (e.g. Spiegel *et al.* 2007).

Southern Africa

Within this broad area of uplifted topography (Fig. 1) there are regions with significant local variations. For example, significant topography is observed along much of the continental margin and has been associated with the break-up of Gondwana (*c.* 130 Ma). Although this relationship is questioned by Burke (1996), both voluminous sediment accumulation offshore of the southern coast (Tinker *et al.* 2008) and apatite–fission track studies (e.g. Brown *et al.* 2002) support a hypothesis of uplift and erosion at this time. However, recent analyses of river profiles on each of the southern African subswells (Bie, Namibia, Drakensburg; Fig. 1) indicates more recent (<30 Ma) uplift. Both the Bie Dome and the Drakensburg Escarpment show significant (0.5–1 km) uplift since *c.* 10 Ma, while the river profiles on the Namibian swell have been modelled successfully by continued uplift since the Oligocene (30 Ma; Roberts & White 2010). More recent post-Pliocene uplift has also been inferred on the Angolan margin from the analysis of offshore seismic data (e.g. Al-Hajri *et al.* 2009).

The cause of the elevated topography in southern Africa is less clear than in east Africa, where models involving mantle plumes have been invoked to explain volcanism that correlates spatially with the observed domal features. Geodynamic modelling from global tomographic models suggest that in southern Africa much of the elevation could be explained by density anomalies below 1000 km depth (e.g. Lithgow-Bertelloni & Silver 1999). Gurnis *et al.* (2000) noted that if the continental lithosphere is thick and has a high effective viscosity, an increased coupling between the deep mantle and surface can lead to increased elevations. Nyblade & Sleep (2003) proposed an alternative mechanism for supporting the high topography, suggesting that the combined effects of multiple plumes in the Mesozoic could generate sufficient long-lasting uplift to account for the present elevations.

Northern, central and western Africa

The remaining portion of Africa covers a wide region from the Congo Basin and Cameroon

Volcanic Line (CVL) in central Africa, across much of the Sahara and into the Atlas Mountains in the north (Fig. 1). At the northern margin of the continent, the high relief of the Atlas Mountains has been associated with the convergence of Africa and Europe (e.g. Frizon de Lamotte et al. 2000). Although the tectonic shortening appears to be relatively modest (e.g. Teixell et al. 2003), Buiter et al. (2009) show that significant topography can be reached through the inversion of a failed rift. Additionally, recent modelling of the lithospheric structure beneath the Atlas, using a variety of geoid, gravity, topography and heat-flow data, suggests that a NE–SW-trending strip of thinned lithosphere underlies the region (e.g. Zeyen et al. 2005; Fullea et al. 2007). A variety of geodynamical models have thus been proposed for this region: one suggestion being that a shallow mantle upwelling (Teixell et al. 2005) contributes to the regional topography.

The topographic swells in the central part of north Africa (e.g. Hoggar, Tibesti; see Fig. 1) have associated volcanism dating from the Oligocene (c. 30 Ma) through to the Quaternary (e.g. Wilson & Guiraud 1992). The volcanism has been related to the time period when the African plate became relatively stationary to mantle circulation and the distribution was explained by regions of fertile lithospheric mantle (e.g. Ashwal & Burke 1989; Burke 1996). Ebinger & Sleep (1998) showed that while channelling along regions of thin lithosphere would lead flow of plume material beneath the Darfur swell and towards the CVL, it could not explain the volcanism of Hoggar and Tibesti. Similarly, geochemical data also suggest very different characteristics between the volcanism in Hoggar and that found near the Afar hotspot (Pik et al. 2006).

In addition to the volcanic swells in northern Africa are a number of intracontinental basins (e.g. Taoudeni, Ghadames, Sirte and Al-Kufrah). The mechanisms for the subsidence of these and, indeed, all intracontinental basins remain heavily debated. Low strain-rate stretching (e.g. Armitage & Allen 2010), cooling of the lithosphere (e.g. Kaminski & Jaupart 2000) and dynamic subsidence (e.g. Heine et al. 2008) have all been proposed to explain the general behaviour of this style of basin. For northern Africa, Holt et al. (2010) recently illustrated that models of lithospheric cooling (from an originally thin lithosphere) can explain the subsidence history of the Ghadames and Al-Kufrah basins. There are, however, significant differences in estimates of the present-day lithospheric thickness beneath all the north African basins (see Priestley et al. 2008; Pasyanos 2010; Global and continental studies). For example, the Taoudeni Basin sits atop the thick lithosphere of the West African Craton, thus

indicating that varied mechanisms may be required for the different basins.

Further to the south, another intracontinental basin (the Congo Basin) provides the most noticeable region of low topography on the African continent (Fig. 1). Flanked by the CVL, Kenyan Dome and Bie Dome, the Congo Basin has accumulated 4–9 km of sediments beginning with extension in the Late Precambrian time and with subsidence continuing until the present day (see e.g. Daly et al. 1992; Giresse 2005). The long sedimentary history, in contrast to other extensional basins such as the North Sea, is difficult to explain and has important implications for the low topography observed today. Hartley & Allen (1994) suggested recent subsidence may relate to convective downwelling beneath the region, a hypothesis supported in the recent geodynamic modelling of Forte et al. (2010). Analysis of geophysical data over the basin has given varied results. From analysis of the gravity signature, Downey & Gurnis (2009) inferred a high-density anomaly within the mantle lithosphere. In contrast, the modelling of Crosby et al. (2010) suggested that the extended period of subsidence could be explained by extension and subsequent cooling of initially thick lithosphere, and interpreted the gravity anomaly in terms of recent convective drawdown.

Passive seismic experiments in Africa

Our understanding of the structure of the crust and upper mantle beneath Africa has advanced significantly over the last 15 years, primarily due to the deployment of broadband seismometers either as permanent stations or as temporary networks designed to investigate specific geological and tectonic problems. Over the last 20 years, networks such as Geoscope, Geofon, US Geological Survey (USGS), Incorporated Research Institutions for Seismology/ International Deployment of Accelerometers, IRIS/ USGS and the Mediterranean Broadband Seismographic Network have provided publicly available broadband data from permanent stations in many countries, including: Algeria, Botswana, Central African Republic, Djibouti, Egypt, Ethiopia, Gabon, Ivory Coast, Kenya, Madagascar, Mali, Morocco, Namibia, Senegal, South Africa, Tunisia, Uganda and Zambia (Fig. 2).

In addition to the data available from the global networks, the ongoing AfricaArray project (Nyblade et al. 2008) is rapidly increasing the number of permanent broadband seismometers within the African continent (Fig. 2). Supplementing the permanent networks, a number of large active and passive land-based regional experiments have been carried out and significantly increase the

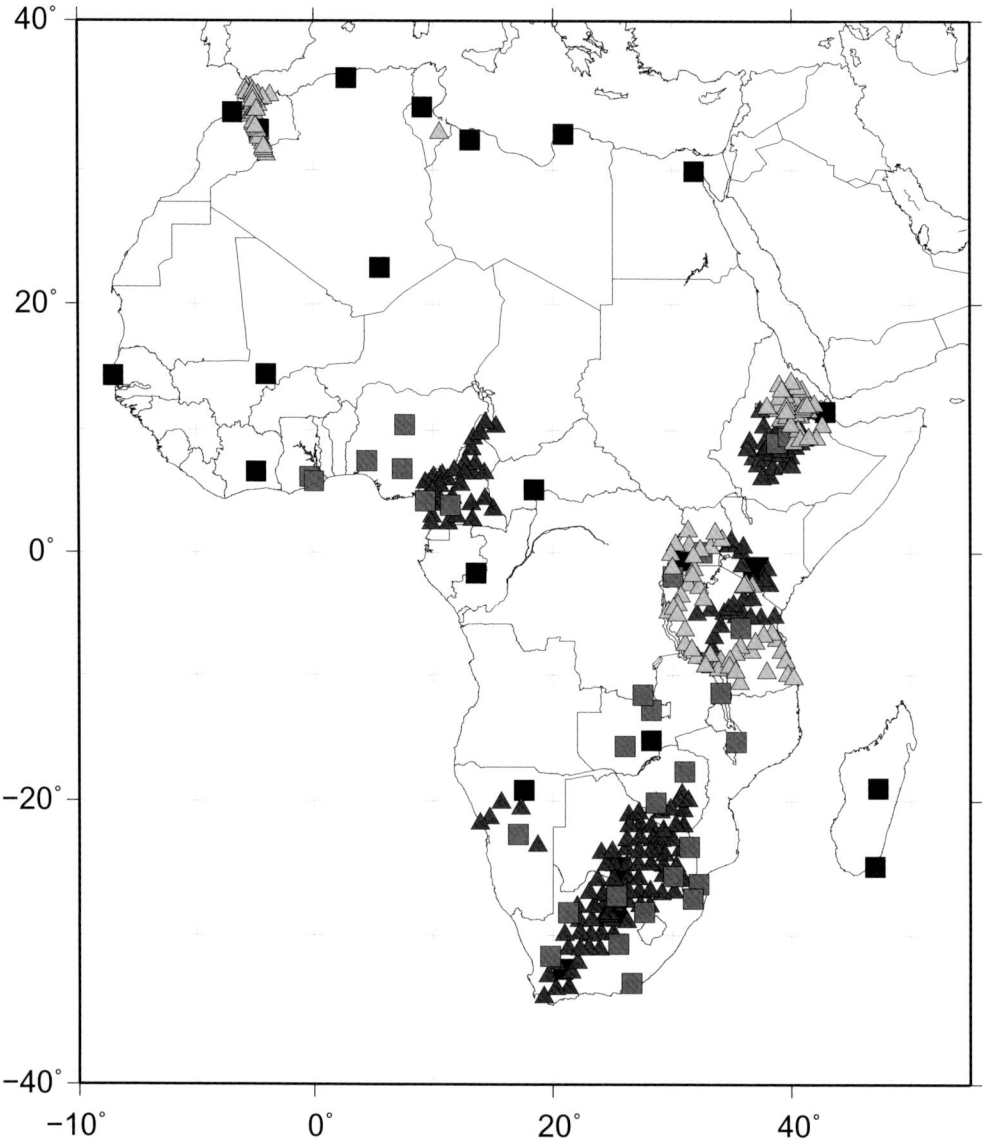

Fig. 2. Map of broadband seismic stations (1990–2010) with data available through IRIS-Data Management Centre or Geofon-GFZ Potsdam. Black squares show permanent seismograph stations from global networks; grey squares show stations as part of Africa Array (Nyblade *et al.* 2008); dark grey triangles show stations from temporary networks operating between 1995–2008 (e.g. Carlson *et al.* 1996; Nyblade *et al.* 1996; Hanka *et al.* 2000; Nyblade & Langston 2002; Maguire *et al.* 2003; Tibi *et al.* 2005; Wölbern *et al.* 2010); light grey triangles show temporary networks operating from 2008 to present.

available data (Fig. 2). The purpose of Africa's portable seismic experiments has been broad-ranging. In east Africa, the principal goal has been to understand hotspot tectonism and the development of various stages of rifting from embryonic continental rifting in the south to incipient sea-floor spreading in Afar (e.g. Nyblade & Langston 2002; Maguire *et al.*

2003). The Precambrian cratons of Tanzania and southern Africa have also been studied seismically, with the goal of understanding the processes that shaped the early Earth (e.g. Carlson *et al.* 1996; Nyblade *et al.* 1996). In Cameroon, seismic data have been collected in order to investigate the origins of the CVL (Tibi *et al.* 2005) which, unusually for an intra-plate volcanic line, has no age progression and has been attributed variously to plume flow and small-scale convection models. Each of these tectonic and geodynamic processes have seismological signatures with important implications for the development of topography. Further projects have either recently begun (e.g. Morocco) or are scheduled to begin soon (e.g. Zambia, Mozambique, Morocco), so the African seismic database will continue to grow in the coming years.

Review of seismological observations

In the following sections we follow the approach of Burke (1996) by summarizing seismological constraints from three areas: east Africa, southern Africa and central/north/west Africa. In doing so, we focus on the seismological results that have first-order importance for understanding topography. We first consider variations in crustal thickness, which have significant implications for isostacy. Mantle seismic velocity structure is then reviewed in the light of the causes and support of topography. Finally, because of its potential to place constraints on the mantle flow field beneath the continent, we summarize constraints from studies of mantle anisotropy.

Estimates of crustal thickness

Global and continental studies. Many geodynamic studies of the African continent proceed on the assumption that the crust can be adequately described by global models such as 3SMAC

(Nataf & Ricard 1996) or Crust 2.0 (Bassin *et al.* 2000; Fig. 3a, b). These models are however relatively simple and lack the fine-scale details revealed by regional seismic experiments. 3SMAC, for example, is defined by only three layers: sediments and upper and lower crust. Thicknesses of the layers are constrained from a variety of geophysical compilations, inherently having non-uniform coverage (Nataf & Ricard 1996). In contrast, Crust 2.0 is more complex with five layers to allow for variable sediments, and an additional middle crustal layer (Bassin *et al.* 2000). Each point on the 2 × 2 degree grid is given a key 1D profile to assign various types of crustal structure. The velocities and densities are then dependent on the type of crustal structure; these key profiles and associated density variations are thus important when estimating crustal isostacy.

Although lacking the resolution of local studies, short-period surface-wave studies using broadband seismic stations deployed across the African continent provide an improved spatial resolution compared to the global crustal models. Using an analysis of group velocity dispersion, Pasyanos & Nyblade (2007) present a crustal thickness map for the African continent (Fig. 3c) as well as estimates of crustal and mantle velocities. Although the path coverage is uneven, at short periods the uncertainties on the group velocity measurements should be small; uncertainties in crustal thickness will be due principally to the model parameterization and the trade-off between crustal thickness and velocity.

In contrast to the global models, a thinner crust (c. 30 km) is observed beneath much of northern

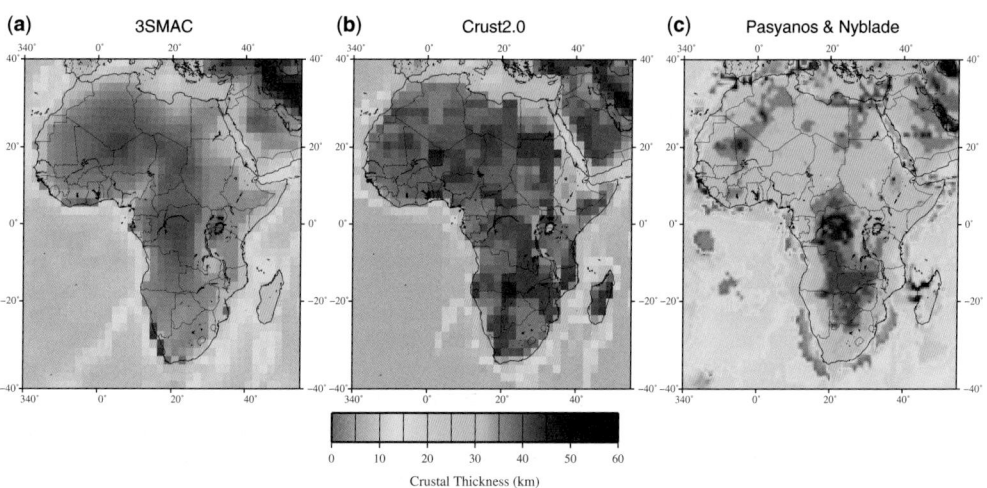

Fig. 3. Crustal thickness variation across Africa. Estimates taken from: (**a**) the global model 3SMAC (Nataf & Ricard 1996); (**b**) the global model Crust2.0 (Bassin *et al.* 2000); and (**c**) the surface-wave study of Pasyanos & Nyblade (2007).

Africa and a thicker crust (c. 40 km) is seen beneath many of the cratonic keels of the African land mass (Kaapvaal, Congo, West African Craton; Fig. 3). The general increase in crustal thickness in the south of the continent is consistent with the higher elevations observed in the region. However, perhaps surprisingly given their low elevation, the thickest crust in the cratonic regions is observed beneath the Congo Basin (c. 45–50 km) and the Taudeni Basin (c. 40–45 km) in NW Africa. Other features absent from the global crustal models but identified in the Pasyanos & Nyblade (2007) study include the thinner crust between the Ethiopian and East African plateaus, and thin crust beneath the Benue Trough in west Africa (Fig. 3c).

Eastern Africa. East Africa has been host to some of the largest seismic experiments in the world whose focus has been the improved understanding of the transition from embryonic continental rifting in the south to incipient sea-floor spreading in Afar (e.g. Maguire *et al.* 2003; Bastow *et al.* 2011). Although detailed constraints on crustal structure come from wide-angle profiles in select parts of the EAR system (e.g. KRISP Working Group 1987; Prodehl & Mechie 1991; Maguire *et al.* 2006), the best spatial coverage of crustal thickness comes from analysis of P-to-S converted phases from the Moho captured during receiver function analysis. A common approach is the so-called H–K stacking procedure of Zhu & Kanamori (2000), which uses P-to-S converted phases from teleseismic earthquakes to constrain bulk-crustal properties (crustal thickness and VpVs ratio) beneath a seismograph station.

In Ethiopia, receiver function studies (Dugda *et al.* 2005; Stuart *et al.* 2006; Cornwell *et al.* 2010) corroborate the wide-angle results (Mackenzie *et al.* 2005; Maguire *et al.* 2006), indicating that extension in the Ethiopian rift is achieved without the marked crustal thinning that would be predicted from many traditional models of break-up. Moho depths are c. 40 km on the Ethiopian and Somalian plates, and 32–36 km within the rift (Fig. 4b). These observations, coupled with evidence of crustal melt zones imaged by magnetotelluric study (Whaler & Hautot 2006) and the spatial coincidence of seismicity in and around the rift associated with these zones of partial melt (Keir *et al.* 2009), indicate that extension is accommodated principally by magma intrusion into the extending plate. Only further north in Afar does the crust appreciably thin to c. 25 km (Dugda *et al.* 2005), coincident with a marked reduction in elevation towards and ultimately (in the Danakil Depression) below sea level.

Further south adjacent to the Kenya rift, receiver functions suggest a crustal thickness of 39–42 km

to the east of the rift and c. 38 km to the west (Dugda *et al.* 2005). While there is little information from passive seismic studies on the crustal thickness along the rift itself, analysis of seismic refraction and wide-angle reflection data from the Kenya Rift International Seismic Project (KRISP) experiments suggest that the crustal thickness varies from 35 km in the south beneath the Kenyan dome to as little as 20 km beneath the low-lying Turkana depression (Mechie *et al.* 1997). Regional Rayleigh wave dispersion data (Benoit *et al.* 2006b) and the continent-wide study of Pasyanos & Nyblade (2007) corroborate these controlled source results and show that Mesozoic crustal thinning can explain the low-elevation region that separates the Ethiopian and East African plateaus.

Beneath the western branch of the rift system, recent analysis of teleseismic data has suggested crustal thickness of c. 30 km on the rift shoulder and significantly thinner crust (c. 20–28 km) beneath the Rwenzori block, where the mountains reach elevations of more than 5000 m (Wölbern *et al.* 2010; Fig. 4d). These recent observations of thin crust beneath regions of very high elevation are in marked contrast to normal Airy isostasy, suggesting that alternative mechanisms such as delamination of a lithospheric block may have caused the uplift (Wölbern *et al.* 2010). Further to the SE below the cratonic region of Tanzania, crustal thickness estimates are c. 40 km with slightly thicker crust beneath the Ubendian Belt on the SW margin of the craton (Last *et al.* 1997; Julià *et al.* 2005; Fig. 4d).

Southern Africa. Within southern Africa a number of studies have used receiver functions to determine the crustal structure (e.g. Nguuri *et al.* 2001; Niu & James 2002; Stankiewicz *et al.* 2002; Nair *et al.* 2006). These studies have been in broad agreement and show crust of c. 38 km beneath much of the Kaapvaal Craton, with the exception of a c. 44 km thick region adjacent to the Limpopo Belt (see Appendix A). Further to the south, the Namaqua Natal Mobile Belt and Cape Fold Belt is underlain by variable, thick crust (c. 36–50 km), and there is significantly thinner crust (c. 26 km) on the oceanic side of the escarpment at the very southwestern margin of the continent (see also Harvey *et al.* 2001). One of the most recent studies of the crust beneath the Southern Africa Seismic Experiment (SASE) array (Kgaswane *et al.* 2009) also incorporates group-velocity surface-wave data from Pasyanos & Nyblade (2007) and Larson & Ekström (2001) alongside the receiver functions, and finds similar estimates of Moho depth (see Fig. 4c) beneath the cratonic region. However, beneath the Namaqua Natal Mobile Belt, the results from the joint inversion generally suggest

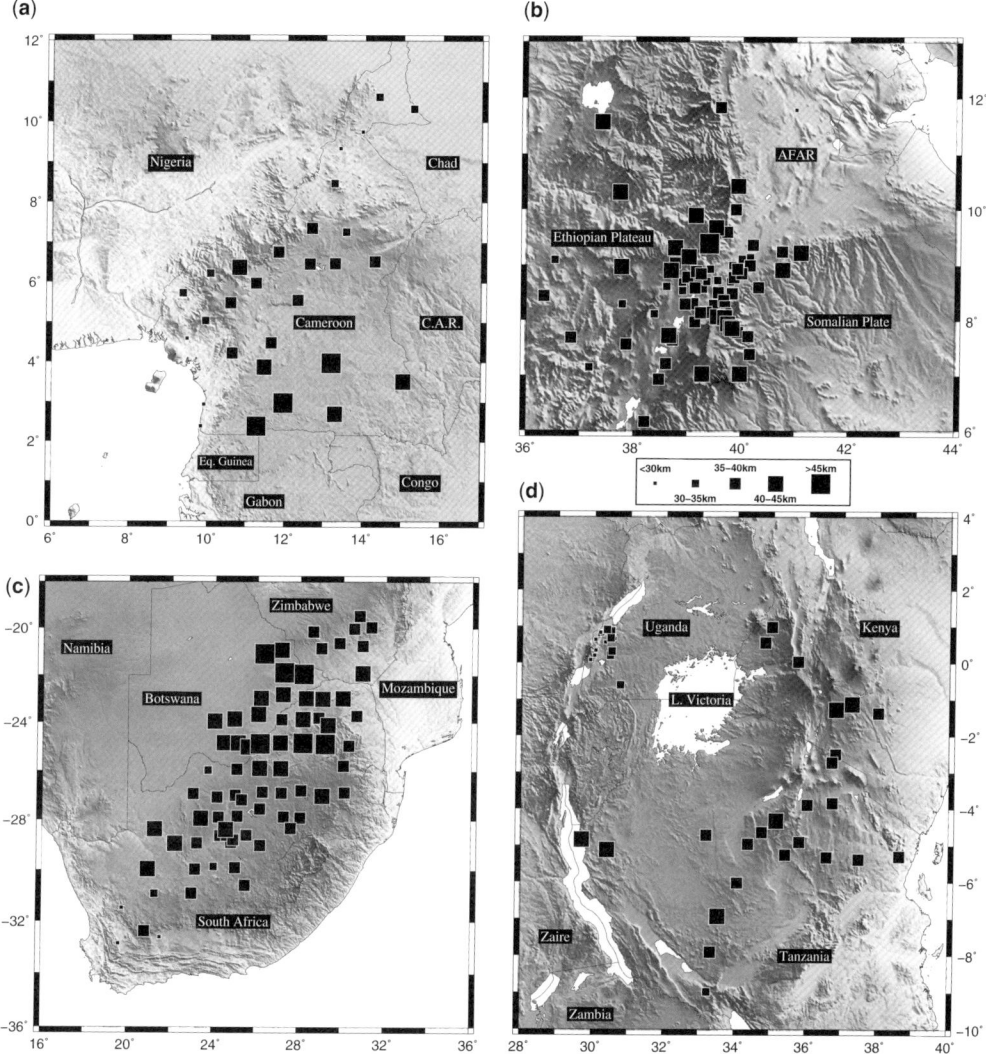

Fig. 4. Crustal thickness constraints for the African continent constrained using receiver function analysis. Data are sourced from (**a**) West Africa – Tokam *et al.* (2010); (**b**) Ethiopia – Dugda *et al.* (2005); Stuart *et al.* (2006); highest signal-to-noise ratio crustal constraints from the study of Cornwell *et al.* (2010) are also used; (**c**) Southern Africa – Kgaswane *et al.* (2009); (**d**) Eastern Africa – Last *et al.* (1997); Dugda *et al.* (2005); and Wölbern *et al.* (2010). CAR, Central African Republic. Station locations and crustal thicknesses are listed in Appendix A.

thinner crust than indicated by the receiver functions alone (Fig. 4c; Appendix A).

Further constraints on crustal thickness come from the ambient noise tomography of Yang *et al.* (2008), with additional information from the teleseismic two-plane-wave tomography study of Li & Burke (2006). Encouragingly, beneath much of the cratonic region the results are in good agreement with the study of Nair *et al.* (2006) and Kgaswane *et al.* (2009), indicating thickened crust through the northern Kaapvaal Craton and Limpopo Belt.

Because the results from different seismic techniques show good agreement, the confidence in the crustal thickness estimates is enhanced. It is noticeable, however, that the abrupt changes in crustal thickness within the Kaapvaal Craton are not always mirrored by topographic variations expected from simple Airy isostasy (Nair *et al.* 2006; Yang *et al.* 2008). This observation has been confirmed by modelling of the Bouguer gravity anomaly over the Bushveld complex, where high-density mafic rocks are required to provide the mass within the crustal section to balance the thickened root (Webb *et al.* 2004).

Northern, central and western Africa. Seber *et al.* (2001) developed a 3D crustal model for the middle-east and north Africa, combining data from crustal-scale refraction and reflection profiles, receiver functions, gravity modelling and surface-wave data. The limited seismic data through much of north Africa means that this model is predominately based on inferences from the geology of the region with significant interpolation between few data points. The results indicated thickest crust (>40 km) towards the SW margin of the West African Craton (Seber *et al.* 2001). Elsewhere in northern Africa, detailed knowledge is absent due to the lack of data. However, the recent study by Liu & Gao (2010) analysed the crustal structure beneath the GEOSCOPE seismograph station at Tamnarasset, immediately adjacent to the Hoggar swell. Almost 20 years of data were available for the station, leading to an exceptionally high-resolution study. Crustal thickness is observed to be *c.* 34 km, although there is significant contrast in the observed receiver functions between adjacent geological terranes (Liu & Gao 2010). There is no obvious increase in crustal thickness detected by raypaths that have Moho conversions close to the Hoggar swell. However, the lack of seismic stations deployed directly on the swell precludes any unambiguous testing of isostatic crustal thickness variations.

In western Africa, the recent seismic experiment investigating the CVL provides detailed information on crustal thickness. Using a joint inversion of receiver functions with Rayleigh wave-group velocities, Tokam *et al.* (2010) find significant variation in Moho depths (Fig. 4a). Beneath the CVL the Moho is observed at 35–39 km depth, similar to the depths observed in the adjacent Pan-African belt. Thicker crust is observed under the margin of the Congo Craton (*c.* 43–48 km), associated with the continent–continent collision during the formation of Gondwana; thinner crust to the north (*c.* 26–31 km) is likely to be related to the formation of the Benue Trough (Tokam *et al.* 2010). These recent estimates of crustal thickness from receiver functions are also in agreement with earlier seismic refraction studies (Stuart *et al.* 1985). Across the CVL there is also a good correlation between the crustal thickness estimates and the Bouguer gravity anomaly (Tokam *et al.* 2010).

While topographic variations on the African continent are often mirrored by isostatically predictable variations in crustal thickness (e.g. thinned crust beneath the rifts of western and eastern Africa), the lack of correlation between elevation and Moho depth in other areas implies that other mechanisms must be contributing to the observed topography. In the following section we review

present-day understanding of Africa's mantle seismic structure in an effort to examine the role of the mantle in controlling topography.

Velocity variations in the upper mantle

While seismic velocities are affected by factors such as melt, water, composition and grain size, within the upper mantle temperature is generally considered the dominant control on velocity (e.g. Goes *et al.* 2000; Faul & Jackson 2005). Studies that provide information on the lateral and vertical variation in seismic velocity are therefore crucial in providing information on likely temperature variations. In turn, knowledge of mantle velocity structure helps determine mantle contributions to the development and maintenance of topography. The structure of the mantle transition zone (MTZ) is also sensitive to variations in temperature. The opposite Clapeyron slopes of the 410 and 660 km discontinuities should lead to a thickened MTZ in cold regions and a thinned MTZ in hot regions (Bina & Helffrich 1994). The depth of the discontinuities can be determined, for example, by receiver function analysis, adding seismological constraints on thermal structure to those derived from tomographic studies.

Global and continental studies. The deep seismic structure of Africa is presented by numerous global tomographic inversions using data from the global seismic networks (e.g. Grand 2002; Ritsema & Allen 2003; Montelli *et al.* 2006; Li *et al.* 2008; Simmons *et al.* 2010). One dominant feature common to all models is the African Superplume, a broad (*c.* 500 km wide), *c.* 3% S-wave slow-velocity anomaly that originates at the core–mantle boundary beneath southern Africa, and rises towards the base of the lithosphere somewhere in the region of Ethiopia and the Red Sea/Gulf of Aden. These global models, and the influence of the African Superplume, have been the focus of much of the geodynamic modelling beneath Africa and the adjacent regions (e.g. Lithgow-Bertelloni & Silver 1999; Gurnis *et al.* 2000; Daradich *et al.* 2003; Simmons *et al.* 2007; Forte *et al.* 2010).

The majority of global studies provide lower-resolution images of upper mantle structure than regional studies. Body wave tomographic inversions generally require relatively close station spacing ($\leq$200 km) in a region of interest to place accurate constraints on structure in the upper *c.* 300 km. Given the uneven distribution of seismic stations across Africa, surface-wave studies that constrain seismic structure along the great circle path between the source and the receiver are therefore a vital technique to image the velocity variations in the upper mantle on a continental-scale. Since the propagation of surface waves is predominately within the crust and upper mantle (dependent on the frequency and mode of surface wave studied), it is therefore possible to achieve good path coverage and resolution in regions with no seismic stations (see e.g. Fishwick 2010), given a good distribution of seismic sources.

A number of global and regional surface-wave studies including Africa have been performed during the last fifteen years (e.g. Ritsema & van Heijst 2000; Debayle et al. 2001; Sebai et al. 2006; Pasyanos & Nyblade 2007; Lebedev & van der Hilst 2008; Priestley et al. 2008; Fishwick 2010). Within the upper mantle the dominant feature of these models is the contrast between the fast velocities associated with Precambrian lithosphere and much slower velocities (5–10%) beneath the rest of the continent. Figure 5 shows the results from the studies of Fishwick (2010) and Lebedev & van der Hilst (2008). While both studies use surface-wave data, different inversions codes and datasets (regional and global, respectively) are used. The similarity in results, which show the lateral extent of fast velocities and clearly highlight the different cratonic regions, is encouraging and suggests they are a reliable representation of the velocity structure. If these ancient cores, depicted by regions of fast velocity, have thermal anomalies that are completely balanced by compositional density differences (the isopycnal hypothesis proposed by Jordan 1975), there should be no residual topographic expression. In Africa, the cratonic regions of the NW and the south have contrasting topographic signatures, implying that the physical properties of the cratonic lithosphere are, indeed, not the principal cause of the broad-scale topographic variation.

A secondary geodynamic feature that may be associated with changes in lithospheric structure is upwelling caused by edge-driven convection (e.g. King & Anderson 1998). This process has been invoked as a possible cause for the intraplate volcanoes on the African continent (King & Ritsema 2000), and could therefore contribute to the observed topographic swells. In east Africa, seismic results show a clear link between melt, upper mantle slow velocities and slow velocities in the deep mantle (see below), indicating that the volcanism and high topography is not primarily caused by edge-driven convection. However, in other African regions the results are less clear. Montagner et al. (2007) suggest that many of the hotspots in northern Africa are only associated with shallow slow-velocity anomalies, compatible with an edge-driven origin. In SW Africa, Al-Hajri et al. (2009) associated the Bie Dome with slow-velocity anomalies in the upper mantle.

More recently, attempts have been made to quantify lithospheric thickness variations beneath parts of Africa from surface-wave results (e.g.

Priestley et al. 2008; Fishwick 2010; Pasyanos 2010). Figure 6 shows the lithospheric thickness estimates for the whole of Africa from the recent tomographic study of Fishwick (2010), which is broadly similar to that of Priestley et al. (2008) and Pasyanos (2010). For example, all the models show thick lithosphere (c. 200 km) beneath NW Africa, although whether or not this extends to the northern continental margin beneath the Atlas mountains remains ambiguous. Within central Africa, there remains significant debate as to the lateral extent of thick lithosphere (and fast seismic velocities) beneath the Congo Basin. It appears that the lithosphere thins towards the eastern margin of the basin; however, the thickness estimates differ significantly (c. 180 km – Fig. 6 and c. 120 km – Pasyanos 2010). Improved knowledge of lithospheric thickness and structure beneath the basin has important consequences in understanding the subsidence history and present-day topography of the region (Downey & Gurnis 2009; Crosby 2010; Forte 2010).

The filtered free-air gravity anomaly (e.g. GRACE data, Tapley et al. 2005) provides an additional source of information to make inferences on the cause of the topographic variation. Many regions with high topography (e.g. Atlas, Hoggar, Tibesti and Bie) correlate with positive gravity anomalies, are close to the edge of the regions of thick lithosphere and/or are underlain by thin lithosphere (Fig. 6). These results suggest that the surface expression of mantle upwelling may be controlled by the edge of the cratonic lithosphere. Without further geodynamical modelling however, it is not possible to say whether this is edge-driven convection (e.g. the model proposed by King & Ritsema 2000) or simply rapid changes in lithospheric structure influencing the larger-scale mantle flow field.

Eastern Africa. P- and S-wave tomographic images (Fig. 7) presented by Bastow et al. (2008) corroborate a growing body of evidence that Ethiopia is underlain by a broad (c. 500 km wide) low-velocity zone (Debayle et al. 2001; Grand 2002; Benoit et al. 2006a; Li et al. 2008; Simmons et al. 2009) that connects to the African Superplume in the lower mantle. There is no evidence for a narrow (c. 100–200 km diameter) plume tail beneath Ethiopia, as would be expected if a starting plume existed today. Absolute delay times at permanent station AAE in Addis Ababa, which indicate the mantle beneath Ethiopia is among the slowest worldwide (Poupinet 1979; Bastow et al. 2008), strongly suggest that high temperatures and partial melt are necessary to explain the seismic observations. The inference from these results is that it is reasonable to suggest that the upper mantle thermal anomaly is likely to have a significant role in the formation and support of the Ethiopian Plateau.

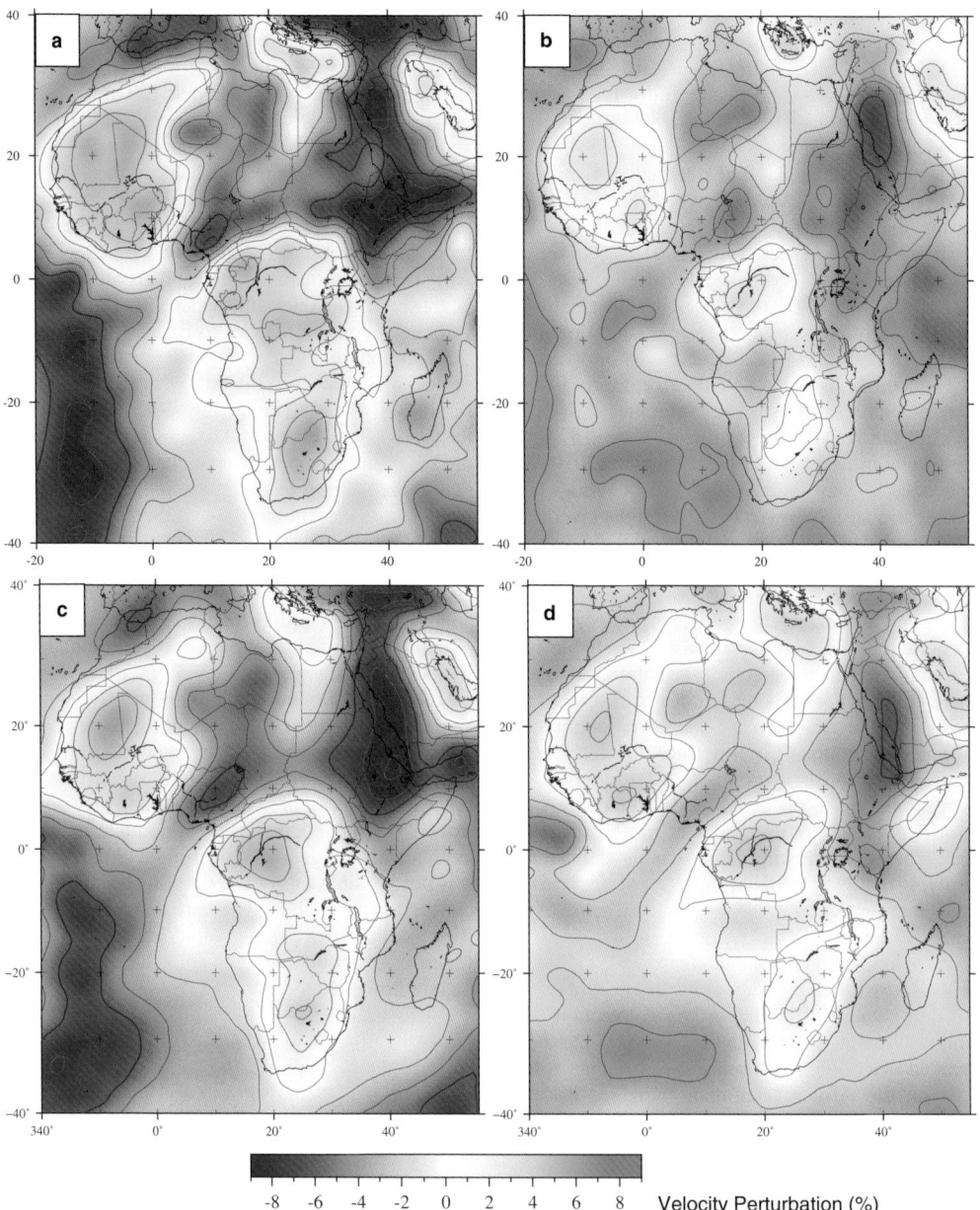

Fig. 5. Comparison of surface-wave models for the African continent. All data are plotted with the same colour scale and relative to the reference model ak135 (Kennett *et al.* 1995). Grey lines delineate 2% contour intervals in velocity perturbation. (**a**, **b**) Shear-wave speeds at 100 km and 200 km depth from the regional tomographic study of Fishwick (2010). (**c**, **d**) Shear-wave speeds at 110 km and 200 km depth extracted from the global tomographic study of Lebedev & van der Hilst (2008).

Further south in Kenya, low-velocity zones beneath the EAR are also imaged using body-wave seismic tomography (e.g. Achauer & Masson 2002; Park & Nyblade 2006). In Tanzania, tomographic images illuminate the *c.* 200–300 km deep high-

velocity lithosphere, which overlies low-velocity material that extends to at least 400 km depth (Ritsema *et al.* 1998). Surface-wave studies suggest a slightly thinner lithosphere than observed in other cratonic regions (e.g. Weeraratne *et al.*

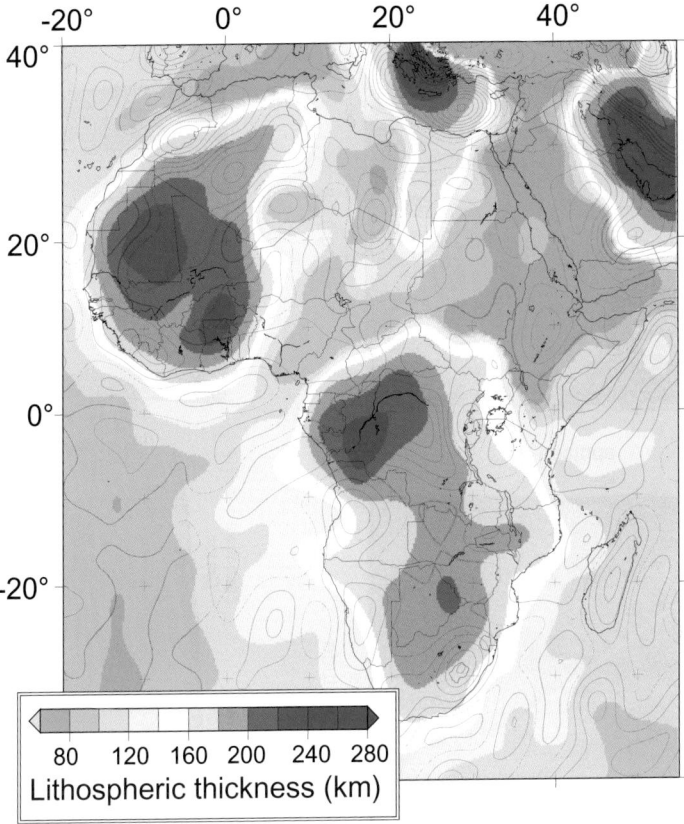

Fig. 6. Estimates of lithospheric thickness from the tomographic model of Fishwick (2010); velocities are converted to temperatures using the empirical parameterization of Priestley & McKenzie (2006). Red (positive) and blue (negative) contours show the long-wavelength free-air gravity anomalies from the GRACE data (Tapley *et al.* 2005).

2003; Fishwick 2010); shear velocities beneath the Tanzanian Craton are higher than a global average to a depth of 150 $\pm$ 20 km. Beneath Kenya and Tanzania there is significant variation in the depth of the 410 discontinuity, indicating potential variation in mantle temperatures at these depths; for large regions the estimated depth is 20–40 km deeper than normal (Nyblade 2011).

The joint inversion of receiver functions and surface-wave dispersion data also suggests that there is significant difference in seismic velocities within the lithospheric mantle beneath Ethiopia and east Africa. The much slower velocities beneath Ethiopia suggest that this region has been affected to a much greater extent by a mantle thermal anomaly (e.g. Julià *et al.* 2005; Dugda *et al.* 2009). Given the similar domed topography seen in each region, the effects at lithospheric depths may be controlled by differences in the pre-existing lithospheric structure rather than significantly different thermal anomalies in the mantle

upwellings. A fundamental question that remains in east Africa is whether the Turkana Depression is low simply due to crustal isostasy (e.g. Benoit *et al.* 2006*b*) or whether there are also differences in mantle upwellings in the region.

Southern Africa. A significant focus of the work on southern Africa has been to obtain high-resolution models of the lithospheric structure beneath the Kaapvaal Craton. For example, array-based techniques using surface waves have been applied to map fine-scale variation in velocity structure (Li & Burke 2006; Chevrot & Zhao 2007). Detailed body-wave tomography has also been performed using the SASE data (James *et al.* 2001; Fouch *et al.* 2004), indicating the continuation of fast wave speeds (relative to an unknown background model) to depths of 300–400 km (Fig. 7). All these models show heterogeneity within the upper mantle, but provide little conclusive evidence as to the possible cause of the elevated topography of southern Africa.

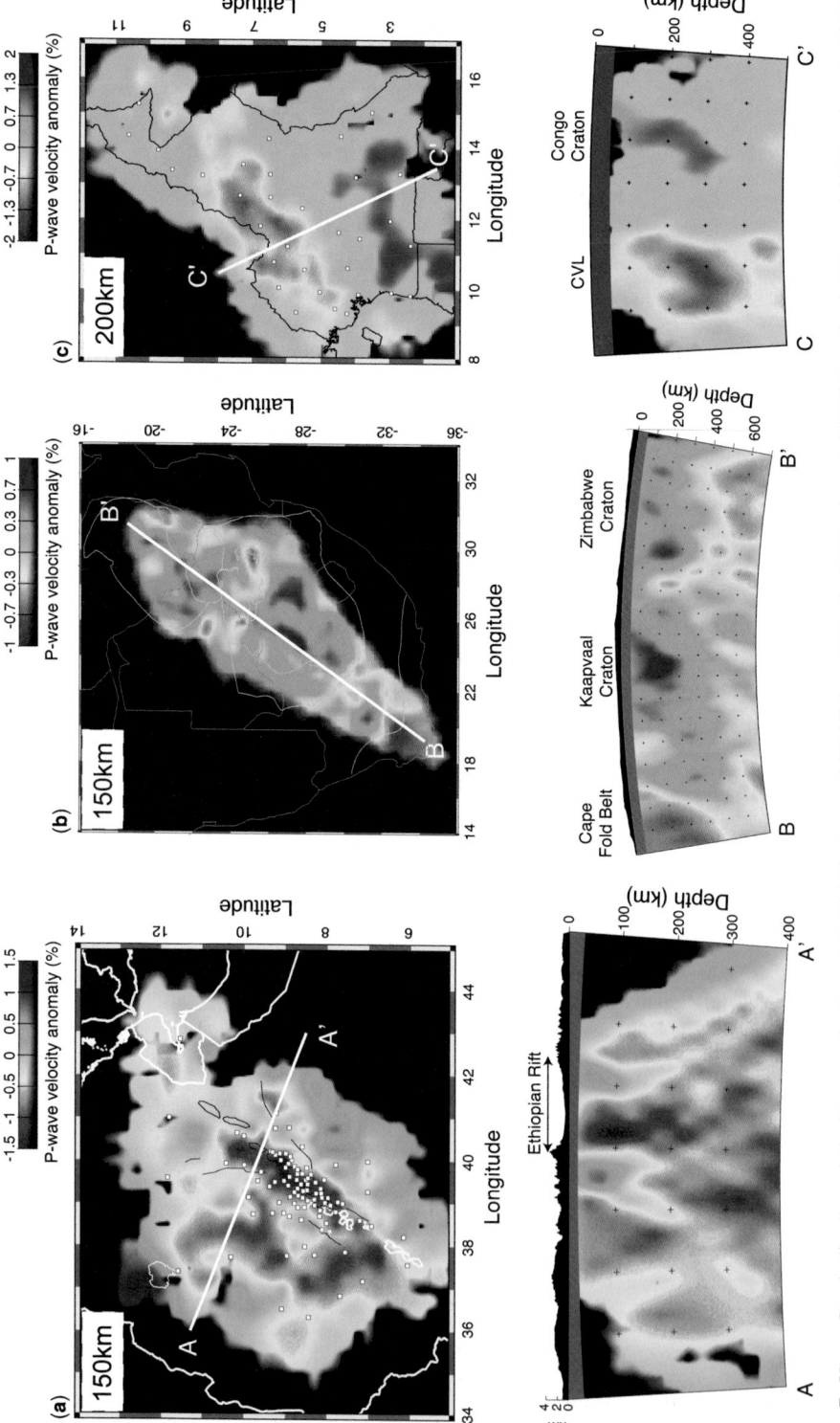

Fig. 7. Upper: Depth slice through the P-wave velocity model of (**a**) Bastow *et al.* (2008) from Ethiopia, (**b**) Fouch *et al.* (2004) from South Africa and (**c**) Reusch *et al.* (2010) from Cameroon. White squares are seismic station locations in (a) and (c). In (a) black lines in the depth slices show locations of Quaternary magmatic centres and Mid-Miocene border faults. Areas of low ray density are black. Lower: Cross-sections through each of the models (orientation of the cross-section is shown on the depth slice by a white line. The grey bands at the top of these preclude the view of shallow structure unresolved in the inversions. In each model, velocity structure is retrieved using regularized non-linear least-squares inversion of relative arrival-time residuals as described by VanDecar *et al.* (1995). The use of relative arrival-time residuals and not absolute delay times means that the zero anomaly level in each model is the background mean value of the region, which is not necessarily the global average.

However, one area of significant debate has been the existence of any upper mantle low-velocity zone beneath the Kaapvaal Craton. While the body-wave tomography suggests fast velocities to >300 km depth, modelling of triplicated body-wave phases suggests a distinct low-velocity zone beneath 150 km depth (Wang *et al.* 2008). Results from surface-wave analysis show similar discrepancies; some studies indicate a low-velocity zone is required to fit dispersion data (e.g. Priestley 1999; Pedersen *et al.* 2009); other studies show no indication of the low-velocity zone but instead have a change in anisotropy (Freybourger *et al.* 2001). Recently, Priestley & Tillmann (2009) have shown that the surface-wave tomography and body-wave tomography results are not necessarily incompatible, due to the difficulty of estimating the absolute velocity variation with depth for the body-wave models.

Imaging of mantle discontinuities using receiver functions has also indicated the existence of a low-velocity zone beneath southern Africa. A number of studies have placed the depth to this transition of *c.* 150 km (e.g. Savage & Silver 2008; Hansen *et al.* 2009*a*, *b*; Vinnik *et al.* 2009; Moorkamp *et al.* 2010). Whether this discontinuity is the lithosphere–asthenosphere boundary or is instead related to a change in physical properties within the subcontinental lithospheric mantle remains an area of debate (see e.g. Fishwick 2010; Rychert *et al.* 2010 for further discussion).

If the low-velocity zone beneath the Kaapvaal is caused by thermal anomalies and thus provides support to the high topography, then discontinuities within the MTZ may provide further indications of the depth extent of any anomaly and links to whole mantle convection. Using P- and S-receiver functions, Vinnik *et al.* (2009) suggest the presence of a distinct low-S-velocity layer above the 410 km discontinuity (also observed by Wittlinger & Farra 2007), high attenuation in the upper mantle and a discontinuity at 450 km depth. Although these features were attributed to the effects of plume-like phenomena in the upper mantle, there was no apparent thinning of the transition zone; the different arrival times from P660s and P410s phases were close to those from the IASP91 global reference model (Vinnik *et al.* 2009). Stankiewicz *et al.* (2002) also investigated the transition zone discontinuities for southern Africa. Beneath Kimberley the discontinuities were similar to the global average, while for other sections of the Kaapvaal Craton the 410 was slightly elevated and the 660 depressed (indicating a slightly cooler mantle temperature than normal; Stankiewicz *et al.* 2002). It therefore remains unclear as to whether there is any present-day thermal support for the high topography beneath southern Africa.

Northern, central and western Africa. The limited number of dense seismic networks makes it difficult to ascertain the extent of any anomalously high-temperature mantle underlying the volcanic swells in northern Africa. Hadiouche & Jobert (1988) indicated an east–west band of low velocities in the north of the continent; the slowest velocities were however observed further north than the Hoggar swell, beneath the Saharan basins and correlating to the location of high heat-flow measurements (Lesquer *et al.* 1990). Recent studies (e.g. Montagner *et al.* 2007; Priestley *et al.* 2008; Sicilia *et al.* 2008; Fishwick 2010) indicate low velocities within the upper mantle (Fig. 5), but do not provide a consistent view on either the exact location or depth extent. Due to the resolution limits of fundamental-mode-dominated surface-wave tomography, body-wave studies may eventually offer greater insight into the structure below 300 km depth; however, the lack of stations is presently a problem. Ayadi *et al.* (2000) showed slow upper mantle velocities beneath the Hoggar swell extending to depths of 300 km in some places. However, this was interpreted as modification of the lithospheric mantle rather than evidence of asthenospheric upwelling (Ayadi *et al.* 2000); the length of the profile also limits the maximum depth to where there is resolution. Studies using multiple-bounce body waves have the potential for good resolution in the upper mantle and have indicated slow wave speeds beneath Hoggar throughout the upper mantle (Begg *et al.* 2009) or, alternatively, very slow wave speeds in the transition zone (Simmons *et al.* 2009).

In western Africa, the velocity structure beneath the Cameroon Volcanic Line was investigated in the 1980s (Dorbath *et al.* 1986; Plomerová *et al.* 1993) and again in a recent seismic experiment (Reusch *et al.* 2010). All studies have shown low velocities in the uppermost mantle beneath the region (e.g. Fig. 7), although Dorbath *et al.* (1986) found no simple correlation with topography. Both Plomerová *et al.* (1993) and Reusch *et al.* (2010) attribute the velocity anomalies in the upper mantle to asthenospheric upwelling. The most recent results (Fig. 7) show a tabular low-velocity zone extending to at least 300 km depth which, when combined with the timing of volcanism in the region, corroborates the edge-driven convection hypothesis for the origins of volcanism in the region (Reusch *et al.* 2010). However, the depth extent of the low-velocity anomaly has yet to be resolved.

Seismic anisotropy

Seismic anisotropy (the directional dependence of seismic wave speed) can result from the alignment of minerals in the crust and mantle, the preferential alignment of fluid or melt or some combination thereof (e.g. Blackman & Kendall 1997). This has the implication that studies of seismic anisotropy can place fundamental constraints on the present-day strain field that characterizes a tectonically active region; they are also capable of elucidating the strain field that acted in the past in regions now seismically and volcanically inactive (fossil anisotropy). Beneath the deforming plates the mantle flow field also imparts a seismic anisotropic fabric, principally due to the lattice preferred orientation (LPO) of anisotropic olivine crystals.

When a shear wave encounters an anisotropic medium, it splits into two orthogonal shear waves with one travelling faster than the other. Splitting can be quantified by the time delay (δt) between the two shear waves and the orientation (φ) of the fast shear wave (Silver & Chan 1991). Body-wave studies (e.g. Kendall *et al.* 2005) generally have good lateral resolution, but are limited to regions with seismic stations. In contrast, surface waves can be used to investigate seismic anisotropy along the great circle path between stations. Surface-wave particle motions decay with depth, dependent on their wavelength. Their velocity is therefore period dependent, a characteristic that can be exploited to resolve velocity as a function of depth. In tandem, therefore, combined study of body and surface waves can be used to place detailed constraints on strain and flow with depth (e.g. Bastow *et al.* 2010).

Geodynamic studies that predict mantle flow patterns (e.g. Forte *et al.* 2010) inherently make testable predictions about seismic anisotropy. For example, a vertically oriented olivine fabric resulting from a buoyant mantle plume would not be expected to produce significant shear-wave splitting in vertically propagating SKS phases. On the other hand, lateral flow of material along the lithosphere–asthenosphere boundary or aligned olivine at the base of a moving plate ('basal drag') are likely to be characterized by φ measurements parallel to the direction of flow (although complications do exist in the relationship between LPO and flow, e.g. Kaminski & Ribe 2002) and observations of large δt.

Global and continental studies. Given the number of proposed mantle upwellings beneath the African plate, the relatively slow horizontal plate motion and variations in lithospheric thickness, the mantle flow field at the lithosphere–asthenosphere boundary is likely to be particularly complicated. In a global model of azimuthal anisotropy, Debayle *et al.* (2005) indicate weak (<1%) anisotropy beneath much of central and southern Africa at depths of 200 km. Slightly larger values are observed towards the northern and eastern margin of the continent. Using the same inversion technique but on a continent scale, Priestley *et al.* (2008) also

found similarly low magnitudes of anisotropy at depths below 150 km. The low magnitude of azimuthal anisotropy suggests either weak horizontal flow or spatially variable anisotropy that shows no consistent direction when averaged over the spatial resolution of the surface-wave studies. However, using surface waves once more but applying a different methodology, Sebai *et al.* (2006) indicate that a general north–south trend in the fast direction of anisotropy is observed beneath the continent at depths of 200–300 km away from Afar.

The uncertainties in magnitude and direction of anisotropy beneath Africa in surface -ave studies means that comparisons of geodynamical models with seismic anisotropy have presently focused on results from shear-wave splitting analyses. Behn *et al.* (2004) compare shear-wave splitting measurements at thirteen ocean island stations surrounding Africa with varying geodynamic models. For stations away from mid-ocean ridges, the anisotropy was best explained by a model which incorporated large-scale upwelling originating in the lower mantle alongside plate motions (Behn *et al.* 2004). Forte *et al.* (2010) compare both flow direction and the horizontal component of maximum stretching from their geodynamic model with SKS results. While the relationship between either flow or stretching and the SKS results appears complicated, along the East African Rift there is a very strong subhorizontal flow field that likely also contributes to observed seismic anisotropy (Forte *et al.* 2010).

Eastern Africa. A number of workers have presented studies of SKS shear-wave splitting along the East African Rift System (e.g. Kendall *et al.* 2006; Fig. 8). In Tanzania and Kenya, fast polarization directions parallel the eastern and western branches of the rift system that surround the Tanzania Craton (e.g. Gao *et al.* 1997; Walker *et al.* 2004). The strength of seismic anisotropy is much lower in the south where the rift is least evolved. Measurements away from the rift, from the craton and the SE part of the EAR extending into the Mozambique Belt, are somewhat different however (Walker *et al.* 2004). Beneath the Tanzanian Craton, splitting is much weaker and oriented in a more east–west direction parallel to Precambrian structural trends (e.g. Shackleton 1986). Walker *et al.* (2004) interpret the splitting patterns in Tanzania in terms of a number of mechanisms including asthenospheric flow around a cratonic keel, plume–lithosphere interactions, pre-existing lithospheric fabric and melt-induced anisotropy.

Ayele *et al.* (2004) interpreted splitting measurements in Kenya, Ethiopia and Djibouti in terms of melt inclusions noting that the magnitude of splitting increases from the south to the north, consistent

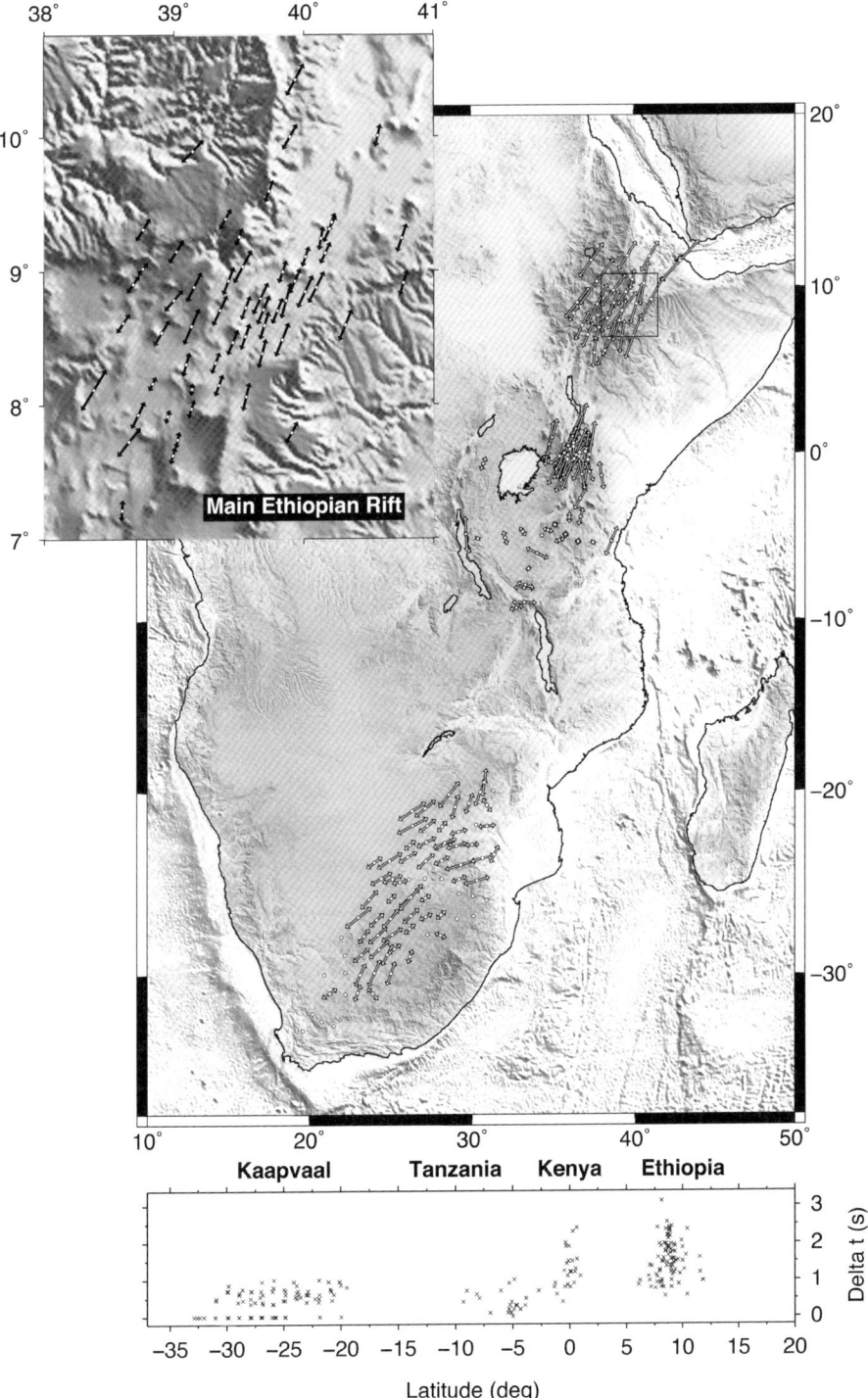

Fig. 8. Compilation of SKS shear-wave splitting measurements from the East African Rift (Gao *et al.* 1997; Ayele *et al.* 2004; Gashawbeza *et al.* 2004; Walker *et al.* 2004) and southern Africa (Silver *et al.* 2001). In the inset figure, detailed splitting observations from the Ethiopia Afar Geoscientific Lithospheric Experiment (EAGLE) experiment are from Kendall *et al.* (2005). In both maps, arrows show the orientation of fast shear wave and the length of arrow is proportional to magnitude of splitting. The bottom figure shows δt as a function of latitude.

with the expectation that more melt is expected towards the mature Afar Depression. Building on this study and later SKS splitting analyses in Ethiopia (e.g. Gashawbeza *et al.* 2004; Kendall *et al.* 2005, 2006), Bastow *et al.* (2010) combined analysis of surface and body waves to show that three mechanisms for anisotropy act beneath the Ethiopian rift: periodic thin layering of seismically fast and slow material in the uppermost *c.* 10 km; oriented melt pockets (with aspect ratio *c.* 0.02) at *c.* 20–75 km depth; and olivine LPO in the upper mantle beneath. The results are explained best by a model in which low-aspect-ratio melt inclusions (dykes and veins) are being intruded into an extending plate during late stage break-up (Bastow *et al.* 2010). In conjunction with the results on crustal structure (see 'Eastern Africa' section) these observations from Ethiopia again show that magma plays an important role in accommodating strain during the late stages of continental break-up. In northern Afar, where the strain field is characterized by the subaerial Red Sea and Gulf of Aden rifts, two anisotropic layers are identified in SKS shear-wave splitting analysis (Gao *et al.* 2010): the upper layer, interpreted as WNW-trending dyke intrusions; and the lower layer with a stronger NE-trending anisotropic fabric, interpreted as LPO due to horizontal flow in the mantle beneath the extending plate.

The extent to which the mantle beneath east Africa has a vertical flow component is unclear from the body-wave studies, because the method delivers only null measurements when anisotropic fabrics are perpendicular to the shear-wave particle motion. However, using data from both Rayleigh and Love waves, surface-wave studies can estimate the variation in radial anisotropy (the difference in velocity between vertically and horizontally propagating shear waves). Montagner *et al.* (2007) and Sicilia *et al.* (2008) find low values of radial anisotropy beneath Afar which, combined with the slow wave speeds, indicates an upwelling consistent with geodynamic models that also points towards a vertical and thus buoyant component of the flow field. These surface-wave studies also show a complex pattern of azimuthal anisotropy with significant variations of fast direction with depth, not completely consistent with the SKS results.

Southern Africa. Shear-wave splitting observations from the SASE experiment in southern Africa were some of the first to show that Archaean mantle deformation could be preserved as fossil mantle anisotropy (Silver *et al.* 2001): fast polarization directions systematically follow the trend of Archaean structures. The most anisotropic regions are Late Archaean in age (Zimbabwe Craton, Limpopo Belt, western Kaapvaal Craton), with δt reducing dramatically in off-craton regions to the

SW and Early Archaean regions to the SE (Fig. 8). Silver *et al.* (2001) also proposed that small or vertically incoherent seismic anisotropy was a likely explanation for the Early Archaean regions of southern Africa. In contrast, Freybourger *et al.* (2001) modelled the structure beneath the array using surface-wave data; azimuthal anisotropy compatible with the SKS results was not observed at periods sensitive to lithospheric mantle depths. Radial anisotropy, with SH faster than SV in the top 100 km, was however required to fit the data.

Below the thick lithosphere of southern Africa, one estimate of seismic anisotropy comes from the detailed analysis of converted phases (Vinnik *et al.* 2009); interpretation of these results indicated a NE fast direction at *c.* 300 km depth. Given the inferred depth to this anisotropy and correlation with plate motion, it has been related to deformation of the base of the lithosphere or an asthenospheric fabric. In the continental-scale study by Sebai *et al.* (2006) radial anisotropy is positive beneath southern Africa, although the anisotropy extends to greater depths than in the model of Freybourger *et al.* (2001), and the Love wave path coverage is limited. There is therefore no strong evidence for vertical upwelling acting as a support for the southern Africa topographic high.

Northern, central and western Africa. In the rest of Africa there have been no local-to-regional studies of seismic anisotropy. Given the evidence for slow velocities and possible mantle upwelling beneath the Cameroon Volcanic Line (see 'Northern, central and western Africa' section), detailed estimates of radial anisotropy could provide fundamental constraints on the extent of vertical flow. Similarly, the geodynamic modelling of Forte *et al.* (2010) indicates significant upwelling from a region termed the West African Superplume. Further measurements of anisotropy in this region would improve the understanding of the contribution to topography of the mantle flow fields.

Discussion and outlook

Seismic data can help constrain a number of factors that are fundamental to the cause and support of Africa's topography. Receiver function analysis, for example, can help constrain crustal thickness and density variations. These in turn are essential to appreciate what proportion of the observed topography is isostatically compensated. Mantle velocities, revealed using seismic tomography, are readily interpreted in terms of temperature variations and buoyancy. Coupled with measurements of seismic anisotropy, which place constraints on the direction of mantle flow, these observations help to identify regions where topography is dynamically supported. With the inclusion of receiver function study of the mantle transition zone, seismological observations can also be used to link some regions of high topography and recent volcanism to upwellings that originate in the lower mantle. Given constraints on topography on the LAB, the geodynamicist can also constrain more subtle effects on topography from edge-driven mantle convection; this level of investigation is however only possible via a suite of seismological observations.

We now know from global-to-regional seismic tomographic inversions that the mantle beneath eastern and southern Africa is characterized by a broad low-velocity zone that extends from the core mantle boundary beneath southern Africa and impinges on the LAB somewhere in the region of Ethiopia. Geodynamic and seismic studies (e.g. Lithgow-Bertelloni & Silver 1999; Gurnis et al. 2000; Daradich et al. 2003; Simmons et al. 2007; Forte et al. 2010) each show that the mantle flow field is important in explaining the first-order superswell topography. Detailed correlations of dynamic topography, however, require a calculation of the isostatically compensated topography. Nyblade & Robinson (1994) used average elevation on the continents and the expected bathymetry to estimate the residual topography for the African superswell, and this residual was matched in the studies of Lithgow-Bertelloni & Silver (1999) and Gurnis et al. (2000). The recent seismic tomography of Simmons et al. (2009) and geodynamic modelling of Forte et al. (2010) use estimates of dynamic topography as a constraint on their models. In these cases the isostatic topography is estimated from the Crust2.0 model (Bassin et al. 2000).

Figure 3 shows that when compared to the relatively smooth continent-scale crustal model of Pasyanos & Nyblade (2007) derived using surface-wave data, Crust2.0 provides very different a priori constraints on crustal structure. The model of Pasyanos & Nyblade (2007) for example shows generally thicker crust in the southern half of the African continent, as might be expected due to the high topography. This bimodality in crustal thickness is not, however, observed in Crust2.0 (Bassin et al. 2000). Fortunately, the ever-growing body of literature reporting on regional variations in crustal thickness in Africa means that future geodynamic models should be able to place tighter a priori constraints on the crust.

Despite these uncertainties at a continental scale, dense seismic networks have successfully improved our understanding of regional topographic variation. In Ethiopia, for example, where controlled-source (e.g. Keranen et al. 2004; Maguire et al. 2006) and passive-source seismic study (e.g. Dugda et al. 2005; Daly et al. 2008) provides detailed

constraints on the structure across the EAR, modelling of the gravity field has been successful in resolving fine-scale variations in crustal density (e.g. Cornwell et al. 2006; Mickus et al. 2007). A well-resolved crust has led to a better-understood mantle: the observed low-density buoyant upper mantle (e.g. Bastow et al. 2008) is required beneath the under-plated uplifted Ethiopian Plateau and the Main Ethiopian Rift (MER) in order to account for the long-wavelength features of the gravity field (Cornwell et al. 2006). Consistent with these observations, thermochronological analysis of incision rates on the Ethiopian Plateau point strongly towards a mantle contribution to the observed plateau uplift (e.g. Pik et al. 2008).

While significant progress has been made in eastern Africa in understanding topography (see also Nyblade 2011), several challenges remain. Many tomographic studies provide only relative measures of seismic velocity because they invert relative and not absolute arrival-time data (e.g. Bastow et al. 2005, 2008; Benoit et al. 2006c; Park & Nyblade 2006). Linkage of these low-velocity zones to the lower mantle is also unclear from receiver function analyses of 410 and 660 km discontinuities that do not, as yet, confidently identify broad regions where the transition zone is thinned due to elevated temperatures. Despite these and other remaining questions in east Africa, given the number of seismic experiments and interdisciplinary studies it is perhaps unsurprising that eastern Africa is now the best understood region of the continent.

Elsewhere, correlations between seismic results and topography are often unclear. In southern Africa, for example, local crustal thickness variations within the Kaapvaal Craton are not matched by changes in topography (Webb et al. 2004; Yang et al. 2008). Furthermore, the cause and extent of the observed upper mantle low-velocity zone remains ambiguous (e.g. Savage & Silver 2008; Hansen et al. 2009b), and the observations within the MTZ do not indicate a present-day link to the underlying superplume (Stankiewicz et al. 2002). Recent geodynamic models that incorporate chemical heterogeneity to the mantle density field also find no present-day dynamic uplift beneath much of southern Africa (Forte et al. 2010). There remains no single, unambiguous explanation for the broad region of high topography in southern Africa.

Many of the smaller-scale topographic features (Fig. 1) such as the volcanic swells of north Africa, the Atlas Mountains, the Cameroon Volcanic Line and the swells around the margin of southern Africa are underlain by low velocities in the uppermost mantle (e.g. King & Ritsema 2000; Montagner et al. 2007; Al-Hajri et al. 2009;

Reusch *et al.* 2010; Fig. 5). These topographic features are often located close to abrupt changes in lithospheric thickness (e.g. Priestley *et al.* 2008; Fishwick 2010; Fig. 6) and edge-driven convection may therefore influence topography in these regions (King & Ritsema 2000; Reusch *et al.* 2010). Alternatively, these rapid variations in lithospheric thickness could act as a guide to the larger-scale mantle flow, focusing upwellings at the margin of cratonic regions.

As the volume of seismic data from Africa continues to increase many of these issues will be addressed, adding fundamental new constraints to future generations of geodynamic models. To conclude, we highlight a number of key seismic observations and methodologies that will be particularly important in establishing improved links between seismological and topographic variations.

(1) Continent-wide high-resolution measurements of crustal thickness and density are required for accurate quantification of Africa's isostatically compensated topography. This goal will be achieved best via receiver function analysis at future dense deployments across the continent, although improved surface-wave studies will also help. Residual topography, in the absence of compressional tectonics, can then be attributed more confidently to mantle effects.

(2) In order to understand the effects of the mantle in dynamically supporting topography, improved tomographic images of mantle structure will be required. The development of techniques to estimate absolute velocities from regional body-wave studies to replace the present generation of relative arrival-time models will help constrain physical properties (density, viscosity, etc.) of the mantle more accurately.

(3) Ray tracing through these updated 3D tomographic models, rather than a 1D Earth model, will help remove the effects of upper mantle heterogeneity and thus sharpen migrated images of the transition zone revealed by receiver function analysis. The

resulting maps of P660s-P410s will subsequently help identify uplifted and volcanically active regions where upper-mantle low velocities are sourced from the lower mantle.

(4) Understanding small-scale convection patterns in the asthenosphere can be appreciated only when variations in lithospheric thickness are constrained better from combined surface wave and receiver function study. These lateral variations in lithospheric structure then need to be incorporated in future generations of geodynamic models. The related lateral changes in temperature, density and, particularly, viscosity may have profound implications for the flow regime in the uppermost mantle.

(5) Estimates of seismic anisotropy become increasingly important to validate the geodynamical modelling. Improved resolution of Love wave velocities is needed to determine shorter-wavelength features of radial anisotropy. Shear-wave splitting measurements provide short-length-scale constraints on the direction of the flow field, and concurrent analyses of these data with estimates of azimuthal anisotropy from surface waves will help separate lithospheric and asthenospheric contributions to the observations. The latter will provide benchmark constraints for future geodynamic models that predict the flow field.

It is clear that while seismic studies can provide a number of important constraints, it is only through the integration of many areas of Earth sciences that a full understanding of Africa's varied topography will emerge.

All those involved in African seismic deployments must be acknowledged, as without them there would be no data and no results to review. We are grateful to A. Reusch for providing images from her Cameroon tomographic study. M. Pasyanos and S. Lebedev are also thanked for providing datasets from their tomographic studies for inclusion in the manuscript. Two anonymous reviewers and editor S. Buiter provided comments that helped focus the contribution. IB is funded by a Leverhulme Trust Early Career Fellowship.

Appendix

Table A1. *Compilation of Moho depth estimates from receiver functions and joint inversion of receiver functions and surface-wave dispersion data (see Fig. 4)*

Southern Africa				Southern Africa			
Station	Latitude	Longitude	Depth (km)	Station	Latitude	Longitude	Depth (km)
SA01	−34.29	19.25	26[b], 30.0[c]	SA53	−24.11	29.33	43[a]
SA02	−33.74	20.27	44[b], 38.5[c]	SA54	−23.72	30.67	35.5[a], 38.0[c]

(Continued)

Table A1. *Continued*

Southern Africa				Southern Africa			
Station	Latitude	Longitude	Depth (km)	Station	Latitude	Longitude	Depth (km)
SA03	−33.66	21.34	36–44[b], 48.7[c]	SA55	−22.98	28.3	43[a], 42.7[c]
SA04	−32.85	19.62	25.5[a], 36–44[b]	SA56	−23.01	29.07	43[a], 41.7[c]
SA05	−32.61	21.54	25.5[a], 40–45[b], 44.1[c]	SA57	−22.98	30.02	40.5[a], 40.3[c]
SA07	−31.98	20.23	41–46[b], 46.2[c]	SA58	−23.52	31.4	43.7[c]
SA08	−31.91	22.07	50[b], 49.3[c]	SA59	−24.84	24.46	40.5[a], 41.0[c]
SA09	−30.92	22.99	38[a], 46[b], 47.6[c]	SA60	−23.85	24.96	40.5[a], 41.1[c]
SA10	−30.97	23.91	33.0[a], 45.5[a], 42–46[b], 46.8[c]	SA61	−23.95	24.02	43[a], 43.2[c]
SA11	−29.97	20.95	40.5[a], 41.4[c]	SA62	−24.85	25.14	40.5[a], 41.3[c]
SA12	−29.85	22.25	43.9[c]	SA63	−23.66	26.08	43[a], 42.3[c]
SA13	−29.98	23.14	35.5[a], 35.9[c]	SA64	−22.97	26.2	40.5[a], 41.2[c]
SA14	−29.87	24.02	33[a], 34.2[c]	SA65	−22.82	27.22	40.5[a], 43.0[c]
SA15	−29.9	25.03	35.5[a], 36.4[c]	SA66	−21.9	26.37	38.0[a], 48.0[a], 46.9[c]
SA16	−28.95	22.2	40.5[a], 34.7[c]	SA67	−21.89	27.27	45.5[a]
SA17	−28.93	23.23	35.5[a], 38.0[c]	SA68	−21.95	28.19	45.5[a], 50.3[c]
SA18	−28.63	24.31	35.5[a], 36.5[c]	SA69	−22.31	29.27	52.6[c]
SA19	−28.91	24.83	35.5[a], 36.6[c]	SA70	−21.09	26.34	50.5[a], 51.6[c]
SA20	−29.02	26.2	35.5[a], 36.4[c]	SA71	−20.93	27.14	43[a], 43.6[c]
SA22	−27.97	22.01	35.5[a], 48.0[a]	SA72	−20.14	28.61	35.5[a], 37.7[c]
SA23	−27.93	23.41	40.5[a], 40.4[c]	SA73	−21.85	30.28	35.5[a], 45.5[a], 49.6[c]
SA24	−27.88	24.24	38[a], 38.4[c]	SA74	−21.92	30.93	40.5[a], 42.2[c]
SA25	−27.85	25.13	38[a], 37.8[c]	SA75	−20.86	29	37.5[a], 39.0[c]
SA26	−27.55	26.18	37.5[a], 39.1[c]	SA76	−20.64	29.85	35.5[a], 36.5[c]
SA27	−27.86	27.29	37.5[a], 39.1[c]	SA77	−20.76	30.92	38[a], 39.0[c]
SA28	−27.9	28.07	37.5[a], 37.4[c]	SA78	−19.47	30.77	35.5[a], 37.2[c]
SA29	−26.93	23.04	35.5[a], 35.8[c]	SA79	−20.02	30.52	37.5[a], 37.7[c]
SA30	−27.07	24.17	35.5[a], 36.6[c]	SA80	−19.96	31.32	35.5[a], 37.5[c]
SA31	−27	25.02	38[a], 38.5[c]	SA81	−30.93	21.27	33[a], 42–46[b], 46.6[c]
SA32	−26.87	26.28	37.5[a], 38.9[c]	SA82	−30.98	22.25	42–48[b], 49.1[c]
SA33	−26.9	27.18	37.5[a], 37.6[c]	POGA	−27.35	31.71	33.0[a], 40.5[a]
SA34	−26.8	28.1	35.5[a], 37.5[c]	HVD	−30.61	25.5	35.5[a]
SA35	−27.02	29.09	40.5[a], 39.6[c]	SEK	−28.32	27.63	37.5[a]
SA36	−26.88	30.13	38[a], 36.5[c]	MOPA	−23.52	31.4	35.5[a], 43.0[a]
SA37	−25.97	23.72	33[a], 34.6[c]	SWZ	−27.18	25.33	38[a]
SA38	−25.93	25.09	38[a], 39.2[c]	UPI	−28.36	21.25	40.5[a]
SA39	−25.9	26.15	40.5[a], 41.7[c]	CVNA	−31.48	19.76	28[a]
SA40	−25.9	27.15	43[a], 44.5[c]	GRM	−33.31	26.57	28.0[a], 35.5[a]
SA42	−25.67	29.22	38[a], 42.0[c]	BB02	−28.38	24.59	40.5[a]
SA43	−25.79	30.07	35.5[a], 43.0[a], 43.3[c]	BB08	−28.43	24.63	35.5[a]
SA44	−26.03	30.9	25.5[a], 40.5[a], 41.2[c]	BB14	−28.54	24.68	38[a]
SA45	−24.88	26.16	45.5[a], 43.8[c]	BB24	−28.62	24.63	35.5[a]
SA46	−24.84	27.11	40.5[a], 39.4[c]	BB31	−28.79	24.93	35.5[a]
SA47	−24.85	28.16	45.5[a], 48.9[c]	BOSA	−28.61	25.56	35.5[a], 35.8[c]
SA48	−24.9	29.22	45.5[a], 45.2[c]	LBTB	−25.02	25.6	43[a], 41.4[c]
SA49	−24.96	30.31	38[a], 53.5[c]	SUR	−32.38	20.81	35.5[a], 42–46[b], 49.2[c]
SA50	−23.87	27.17	38[a], 39.7[c]				
SA51	−23.86	28.16	43[a], 48.9[c]				
SA52	−23.8	28.9	38[a], 39.7[c]				

Western Africa				Eastern Africa			
Station	Latitude	Longitude	Depth (km)	Station	Latitude	Longitude	Depth (km)
CM01	2.39	9.83	28[d]	NYAN	0.21	30.45	32[e]
CM02	2.7	13.29	43[d]	RUBO	0.34	30.04	21[e]
CM03	3.52	15.03	43[d]	RUGA	−0.26	30.1	38[e]
CM04	2.98	11.96	45.5[d]	RWEB	0.32	30.49	32[e]
CM05	2.94	9.91	28[d]	SEML	0.91	30.36	34[e]

(Continued)

Table A1. *Continued*

Western Africa				Eastern Africa			
Station	Latitude	Longitude	Depth (km)	Station	Latitude	Longitude	Depth (km)
CM06	2.39	11.27	45.5^d	SEMP	0.84	30.17	27^e
CM07	3.87	11.46	43^d	ANGA	−2.5	36.8	39^f
CM09	4.23	9.33	25.5^d, 40.5^d	KAKA	0.56	34.8	37^f
CM10	4.22	10.62	38^d	KITU	−1.37	38	40^f
CM11	3.98	13.19	48^d	KMBO	−1.13	37.25	41^f
CM12	4.48	11.63	38^d	KR42	0.04	35.73	38^f
CM13	4.59	9.46	28^d	NAI	−2.37	36.8	42^f
CM15	5.03	9.93	33^d	TALE	0.98	34.98	38^f
CM16	5.48	10.57	35.5^d	BASO	−4.32	35.14	41^g
CM17	5.55	12.31	35.5^d	MBWE	−4.96	34.35	37^g
CM18	5.72	9.36	30.5^d	MITU	−6.02	34.06	38^g
CM19	5.98	11.23	35.5^d	MTAN	−7.91	33.32	37^g
CM20	6.23	10.05	33^d	MTOR	−5.25	35.4	38^g
CM21	6.47	12.62	35.5^d	PUGE	−4.72	33.18	37^g
CM22	6.48	13.27	35.5^d	RUNG	−6.94	33.52	42^g
CM23	6.37	10.79	40.5^d	SING	−4.64	34.73	37^g
CM24	6.52	14.29	35.5^d	HALE	−5.3	38.62	39^g
CM25	6.76	11.81	38^d	KIBA	−5.32	36.57	36^g
CM26	7.27	13.55	33^d	KIBE	−5.38	37.48	37^g
CM27	7.36	12.67	35.5^d	KOMO	−3.84	36.72	36^g
CM28	8.47	13.24	30.5^d	KOND	−4.9	35.8	37^g
CM29	9.35	13.39	25.5^d	LONG	−2.73	36.7	37^g
CM30	9.76	13.95	28^d	TARA	−3.89	36.02	37^g
CM31	10.33	15.26	30.5^d	GOMA	−4.84	29.69	44^g
CM32	10.62	14.37	33^d	INZA	−5.12	30.4	42^g
				PAND	−8.98	33.24	35^g
				BURO	0.86	30.17	24^e
				ITOJ	0.84	30.23	21^e
				KABA	0.78	30.13	24^e
				KABE	0.87	30.47	34^e
				KABG	0.63	30.65	30^e
				KAGO	0.68	30.46	32^e
				KARA	0.09	29.9	28^e
				KARU	0.79	30.22	21^e
				KASE	−0.03	30.15	30^e
				KASS	0.57	30.31	24^e
				KILE	0.21	30.01	22^e
				KISA	0.59	30.74	30^e
				KMTW	0.74	30.38	30^e
				MBAR	−0.6	30.74	30^e, 33^f
				MIRA	0.66	30.57	29^e
				MWEY	−0.19	29.9	28^e
				NGIT	0.64	30.03	25^e

Ethiopia				Ethiopia			
Station	Latitude	Longitude	Depth (km)	Station	Latitude	Longitude	Depth (km)
ADEE	7.79	39.91	37.6^h	AAUS	9.04	38.77	37^f
ADUE	8.54	38.9	33.9^h	ARBA	6.07	37.56	30^f
AMME	8.3	39.09	37.6^h	BELA	6.93	38.47	38^f
ANKE	9.59	39.73	37.6^h	BIRH	9.67	39.53	41^f
AREE	8.93	39.42	33.6^h	DMRK	10.31	37.73	41^f
ASEE	7.97	39.13	38.2^h	GOBA	7.03	39.98	42^f
AWAE	8.99	40.17	36.0^h	GUDE	8.97	37.77	36.5^f
BEDE	8.91	40.77	41.9^h	HERO	7.03	39.28	42^f

(Continued)

Table A1. *Continued*

Ethiopia				Ethiopia			
Station	Latitude	Longitude	Depth (km)	Station	Latitude	Longitude	Depth (km)
BORE	8.73	39.55	32.0[h]	HIRN	9.22	41.11	41[f]
BUTE	8.12	38.38	32.0[h]	HOSA	7.56	37.86	37[f]
CHAE	9.31	38.76	40.7[h]	JIMA	7.68	36.83	36[f]
DIKE	8.06	39.56	42.6[h]	KARA	10.42	39.94	44[f]
DZEE	8.78	39	38.4[h]	NAZA	8.57	39.29	27[f]
FURI	8.9	36.69	37.4[f], 37.4[h]	NEKE	9.09	36.52	34[f]
GTFE	8.99	39.84	31.4[h]	SELA	7.97	39.13	27[f]
INEE	9.9	39.14	40.7[h]	WANE	10.17	40.65	30[f]
HIRE	9.22	41.11	39.4[h]	WASH	8.99	40.17	35[f]
KARE	10.42	39.94	43.8[h]	WELK	8.3	37.78	33[f]
KOTE	9.39	39.4	45.7[h]	BAHI	11.57	37.39	44[f]
LEME	8.61	38.61	33.3[h]	CHEF	6.16	38.21	37[f]
MECE	8.59	40.32	37.6[h]	DELE	8.44	36.33	36[f]
MELE	9.31	40.2	35.0[h]	DIYA	11.83	39.6	37[f]
MIEE	9.24	40.76	35.0[h]	TEND	11.79	41	25[f]
NURE	8.7	39.8	32.6[h]	TERC	7.15	37.18	34[f]
SENE	9.15	39.02	41.9[h]	1179	8.85	39.24	35.8[i]
SHEE	10	39.9	35.9[h]	1246	8.5	39.57	38.2[i]
E31	8.78	39.86	35.4[h]	1266	8.36	39.67	35.5[i]
E32	8.85	40.01	35.2[h]	1281	8.25	39.7	36.4[i]
E33	8.93	39.93	35.5[h]	1296	8.13	39.72	38.6[i]
E34	7.21	38.6	38.9[h]	1306	8.05	39.69	40.1[i]
E35	9.13	40.17	34.3[h]	1324	7.93	39.75	41.1[i]
E36	9.11	40.01	35.0[h]	1333	7.87	39.81	41.0[i]
E40	9.36	40.22	35.0[h]	1337	7.85	39.84	40.4[i]
E50	8.27	39.5	36.9[h]	1351	7.78	39.94	38.9[i]
E51	8.15	39.35	38.6[h]	1373	7.71	40.14	37.5[i]
E52	8.14	39.24	38.6[h]	1400	7.38	40.17	38.8[i]
E55	8.3	38.95	38.3[h]	E59	8.71	39.35	32.3[i]
E57	8.59	39.13	37.3[h]				
E64	8.57	39.29	33.6[h]				
E67	8.38	39.68	37.0[h]				
E70	8.88	39.15	35.7[h]				
E71	8.69	38.9	33.9[h]				
E72	8.49	39.83	38.4[h]				
E76	7.73	38.65	40.2[h]				
E79	7.64	38.72	39.2[h]				

[a]Kgaswane *et al.* (2009); [b]Harvey *et al.* (2001); [c]Nair *et al.* (2006); [d]Tokam *et al.* (2010); [e]Wölbern *et al.* (2010); [f]Dugda *et al.* (2005); [g]Last *et al.* (1997); [h]Stuart *et al.* (2006); [i]Cornwell *et al.* 2010.

References

ACHAUER, U. & MASSON, F. 2002. Seismic tomography of continental rifts revisited: from relative to absolute heterogeneities. *Tectonophysics*, **358**, 17–37.

AL-HAJRI, Y., WHITE, N. & FISHWICK, S. 2009. Scales of transient convective support beneath Africa. *Geology*, **37**, 883–886.

ARMITAGE, J. J. & ALLEN, P. A. 2010. Cratonic basins and the long-term subsidence history of continental interiors. *Journal of the Geological Society, London*, **167**, 61–70.

ASHWAL, L. D. & BURKE, K. 1989. African lithospheric structure, volcanism, and topography. *Earth and Planetary Science Letters*, **96**, 8–14.

AYADI, A., DORBATH, C., LESQUER, A. & BEZZEGHOUD, M. 2000. Crustal and upper mantle velocity structure of the Hoggar swell (Central Sahara), Algeria. *Physics of the Earth and Planetary Interiors*, **118**, 111–123.

AYALEW, D., BARBEY, P., MARTY, B., REISBERG, L., YIRGU, G. & PIK, R. 2002. Source, genesis, and timing of giant ignimbrite deposits associated with Ethiopian continental flood basalts. *Geochimica et Cosmochimica Acta*, **66**, 1429–1448.

AYELE, A., STUART, G. & KENDALL, J.-M. 2004. Insights into rifting from shear wave splitting and receiver functions; an example from Ethiopia. *Geophysical Journal International*, **157**, 354–362.

BASSIN, C., LASKE, G. & MASTERS, G. 2000. The current limits of resolution for surface wave tomography in North America. *Eos, Transactions, American Geophysical Union*, **81**, F897.

BASTOW, I. D., STUART, G. W., KENDALL, J. M. & EBINGER, C. J. 2005. Upper-mantle seismic structure in a region of incipient continental breakup: northern Ethiopian rift. *Geophysical Journal International*, **162**, 479–493.

BASTOW, I. D., NYBLADE, A. A., STUART, G. W., ROONEY, T. O. & BENOIT, M. H. 2008. Upper mantle seismic structure beneath the Ethiopian hotspot: rifting at the edge of the African low velocity anomaly. *Geochemistry, Geophysics, Geosystems*, **9**, Q12022, doi: 10.1029/2008GC002107.

BASTOW, I. D., PILIDOU, S., KENDALL, J. M. & STUART, G. W. 2010. Melt-induced seismic anisotropy and magma assisted rifting in Ethiopia: evidence from surface waves. *Geochemistry, Geophysics, Geosystems*, **11**, Q0AB05, doi: 10.1029/2010GC003036.

BASTOW, I. D., KEIR, D. & DALY, E. 2011. The Ethiopia Afar Geoscientific Lithospheric Experiment (EAGLE): probing the transition from continental rifting to incipient sea floor spreading. *In*: BECCALUVA, L., BIANCHINI, G. & WILSON, M. (eds) *Volcanism and Evolution of the African Lithosphere*. Geological Society of America, Special Papers, **478**, 51–76.

BEGG, G. C., GRIFFIN, W. L. *ET AL.* 2009. The lithospheric architecture of Africa: seismic tomography, mantle petrology and tectonic evolution. *Geosphere*, **5**, 23–50.

BEHN, M. D., CONRAD, C. P. & SILVER, P. G. 2004. Detection of upper mantle flow associated with the African Superplume. *Earth and Planetary Science Letters*, **224**, 259–274.

BENOIT, M., NYBLADE, A., OWENS, T. & STUART, G. 2006*a*. Mantle transition zone structure and upper mantle S velocity variations beneath Ethiopia: evidence for a broad, deep-seated thermal anomaly. *Geochemistry, Geophysics, Geosystems*, **7**, Q10113, doi: 10.1029/2006GC001398.

BENOIT, M., NYBLADE, A. & PASYANOS, M. 2006*b*. Crustal thinning between the Ethiopian and East African Plateaus from modeling Rayleigh wave dispersion. *Geophysical Research Letters*, **33**, L13301, doi: 10.1029/2006GL025687.

BENOIT, M., NYBLADE, A. & VANDECAR, J. 2006*c*. Upper mantle P wavespeed variations beneath Ethiopia and the origin of the Afar hotspot. *Geology*, **34**, 329–332.

BINA, C. R. & HELFFRICH, G. 1994. Phase transition Clapeyron slopes and transition zone seismic discontinuity topography. *Journal of Geophysical Research*, **99**, 15853–15860.

BLACKMAN, D. & KENDALL, J. M. 1997. Sensitivity of teleseismic body waves to mineral texture and melt in the mantle beneath a mid-ocean ridge. *Philosophical Transactions of the Royal Society, London*, **355**, 217–231.

BRAUN, J. 2010. The many surface expressions of mantle dynamics. *Nature Geoscience*, **3**, 825–833.

BROWN, R. W., SUMMERFIELD, M. A. & GLEADOW, A. J. W. 2002. Denudational history along a transect across the Drakensberg Escarpment of southern Africa derived from apatite fission-track thermochronology. *Journal of Geophysical Research*, **107**, doi: 10.1029/2001JB000745.

BUITER, S. J. H., PFIFFNER, O. A. & BEAUMONT, C. 2009. Inversion of extensional sedimentary basins: a numerical evaluation of the localisation of shortening. *Earth and Planetary Science Letters*, **288**, 492–504.

BURKE, K. 1996. The African plate. *South African Journal of Geology*, **99**, 341–409.

CARLSON, R. W., GROVE, T. L., DE WIT, M. J. & GURNEY, J. J. 1996. Anatomy of an Archean craton: a program for interdisciplinary studies of the Kaapvaal Craton, southern Africa. *Eos, Transactions, American Geophysical Union*, **77**, 273–277.

CHEVROT, S. & ZHAO, L. 2007. Multiscale seismic tomography finite-frequency Rayleigh wave tomography of the Kaapvaal craton. *Geophysical Journal International*, **169**, 201–215.

CORNWELL, D. G., MACKENZIE, G. D., ENGLAND, R. W., MAGUIRE, P. K. H., ASFAW, L. M. & OLUMA, B. 2006. Northern Main Ethiopian Rift crustal structure from new high precision gravity data. *In*: YIRGU, G., EBINGER, C. J. & MAGUIRE, P. K. H. (eds) *The Afar Volcanic Province within the East African Rift System*. Geological Society, London, Special Publications, **256**, 307–321.

CORNWELL, D. G., MAGUIRE, P. K. H., ENGLAND, R. W. & STUART, G. W. 2010. Imaging detailed crustal structure and magmatic intrusion across the Ethiopian rift using a dense linear broadband array. *Geochemistry, Geophysics, Geosystems*, **11**, Q0AB03, doi: 10.1029/2009GC002637.

COULIE, E., QUIDELLEUR, P. Y., GILLOT, P. Y., COURTILLOT, V., LEFÈVRE, J. C. & CHIESA, S. 2003. Comparative K–Ar and Ar–Ar dating of Ethiopian and Yemenite Oligocene volcanism: implications for timing and duration of the Ethiopian traps. *Earth and Planetary Science Letters*, **206**, 477–492.

CROSBY, A., FISHWICK, S. & WHITE, N. 2010. Structure and evolution of the intracratonic Congo Basin. *Geochemistry, Geophysics, Geosystems*, **11**, Q06010, doi: 10.1029/2009GC003014.

DALY, E., KEIR, D., EBINGER, C. J., STUART, G. W., BASTOW, I. D. & AYELE, A. 2008. Crustal tomographic imaging of a transitional continental rift: the Ethiopian rift. *Geophysical Journal International*, **172**, 1033–1048.

DALY, M. C., LAWRENCE, S. R., DIEMU-TSHIBAND, K. & MATOUANA, B. 1992. Tectonic evolution of the Cuvette Centrale, Zaire. *Journal of the Geological Society, London*, **149**, 539–546.

DARADICH, A., MITROVICA, J., PYSKLYWEC, R., WILLETT, S. & FORTE, A. 2003. Mantle flow, dynamic topography, and rift-flank uplift of Arabia. *Geology*, **31**, 901–904.

DE WIT, M. 2007. The Kalahari Epeirogeny and climate change: dfferentiating cause and effect from core to space. *South African Journal of Geology*, **110**, 367–392.

DEBAYLE, E., LÉVÊQUE, J.-J. & CARA, M. 2001. Seismic evidence for a deeply rooted low-velocity anomaly in the upper mantle beneath the northeastern Afro/Arabian continent. *Earth and Planetary Science Letters*, **193**, 423–436.

DEBAYLE, E., KENNETT, B. L. N. & PRIESTLEY, K. 2005. Global azimuthal seismic anisotropy: the unique plate-motion of Australia. *Nature*, **433**, 509–512.

DORBATH, C., DORBATH, L., FAIRHEAD, J. D. & STUART, G. W. 1986. A teleseismic delay time study across the

Central African Shear Zone in the Adamawa region of Cameroon, West Africa. *Geophysical Journal of the Royal Astronomical Society*, **86**, 751–766.

DOUCOURÉ, C. M. & DE WIT, M. J. 2003. Old inherited origin for the present near bimodal topography of Africa. *Journal of African Earth Sciences*, **36**, 371–388.

DOWNEY, N. J. & GURNIS, M. 2009. Instantaneous dynamics of the cratonic Congo basin. *Journal of Geophysical Research*, **114**, B06401, doi: 10.1029/2008JB006066.

DUGDA, M. T., NYBLADE, A. A., JULIÀ, J., LANGSTON, C. A., AMMON, C. A. & SIMIYU, S. 2005. Crustal structure in Ethiopia and Kenya from receiver function analysis: implications for rift development in eastern Africa. *Journal of Geophysical Research*, **110**, B01303, doi: 10.1029/2004JB003065.

DUGDA, M. T., NYBLADE, A. A. & JULIÀ, J. 2009. S-wave velocity structure of the crust and upper mantle beneath Kenya in comparison to Tanzania and Ethiopia: implications for the formation of the East African and Ethiopian plateaus. *South African Journal of Geology*, **112**, 241–250.

EBINGER, C. J. & SLEEP, N. H. 1998. Cenozoic magmatism throughout east Africa resulting from impact of a single plume. *Nature*, **395**, 788–791.

EBINGER, C., BECHTEL, T., FORSYTH, D. & BOWIN, C. 1989. Effective elastic plate thickness beneath the East African and Afar plateaus and dynamic compensation of the uplifts. *Journal of Geophysical Research*, **94**, 2883–2901.

FAUL, U. H. & JACKSON, I. 2005. Seismic signatures of temperature variations in the upper mantle. *Earth and Planetary Science Letters*, **234**, 119–134.

FISHWICK, S. 2010. Surface wave tomography: imaging of the lithosphere asthenosphere boundary beneath central and southern Africa. *Lithos*, **120**, 63–73.

FORTE, A., QUÉRÉ, S., MOUCHA, R., SIMMONS, N. A., GRAND, S. P., MITROVICA, J. X. & ROWLEY, D. B. 2010. Joint seismic-geodynamic-mineral physical modelling of African geodynamics: a reconciliation of deep-mantle convection with surface geophysical constraints. *Earth and Planetary Science Letters*, **295**, 329–341.

FOUCH, M. J., JAMES, D. E., VANDECAR, J. C., VAN DER LEE, S. & KAAPVAAL SEISMIC GROUP. 2004. Mantle seismic structure beneath the Kaapvaal and Zimbabwe Cratons. *South African Journal of Geology*, **107**, 33–44.

FREYBOURGER, M., GAHERTY, J. B., JORDAN, T. H. & THE KAAPVAAL SEISMIC GROUP. 2001. Structure of the Kaapvaal craton from surface waves. *Geophysical Research Letters*, **28**, 2489–2492.

FRIZON DE LAMOTTE, D., SAINT BEZAR, B., BRACÈNE, R. & MERCIER, E. 2000. The two main steps of the Atlas building and geodynamics of the western Mediterranean. *Tectonics*, **19**, 740–761.

FULLEA, J., FERNÁNDEZ, M., ZEYEN, H. & VERGÈS, J. 2007. A rapid method to map the crustal and lithospheric thickness using elevation, geoid anomaly and thermal analysis. Application to the Gibraltar Arc System, Atlas Mountains and adjacent zones. *Tectonophysics*, **430**, 97–117.

FURMAN, T., BRYCE, J., HANAN, B., YIRGU, G. & AYALEW, D. 2006. Heads and tails: 30 Million years

of the Afar plume. *In*: YIRGU, G., EBINGER, C. J. & MAGUIRE, P. K. H. (eds) *The Afar Volcanic Province within the East African Rift System*. Geological Society, London, Special Publications, **259**, 95–119.

GANI, N., GANI, M. & ABDELSALAM, M. 2007. Blue Nile incision on the Ethiopian Plateau: pulsed plateau growth, Pliocene uplift, and hominin evolution. *GSA Today*, **17**, 4–11.

GAO, S., DAVIS, P. *ET AL.* 1997. SKS splitting beneath the continental rift zones. *Journal of Geophysical Research*, **102**, 22781–22797.

GAO, S. S., LIU, K. H. & ABDELSALEM, M. G. 2010. Seismic anisotropy beneath the Afar Depression and adjacent areas: implications for mantle flow. *Journal of Geophysical Research*, **115**, B12330, doi: 10.1029/2009JB007141.

GASHAWBEZA, E., KLEMPERER, S., NYBLADE, A., WALKER, K. & KERANEN, K. 2004. Shear wave splitting in Ethiopia: Precambrian mantle anisotropy locally modified by Neogene rifting. *Geophysical Research Letters*, **31**, L18602, doi: 10.1029/2004GL020471.

GEORGE, R., ROGERS, N. & KELLEY, S. 1998. Earliest magmatism in Ethiopia: evidence for two mantle plumes in one continental flood basalt province. *Geology*, **26**, 923–926.

GIRESSE, P. 2005. Mesozoic–Cenozoic history of the Congo Basin. *Journal of African Earth Sciences*, **43**, 301–315.

GOES, S., CAMMARANO, F. & HANSEN, U. 2000. Shallow mantle temperatures under Europe from P and S wave tomography. *Journal of Geophysical Research*, **105**, 11153–11169.

GRAND, S. P. 2002. Mantle shear-wave tomography and the fate of subducted slabs. *Philosophical Transactions of the Royal Society*, **360**, 2475–2491.

GURNIS, M., MITROVICA, J. X., RITSEMA, J. & VAN HEIJST, H. J. 2000. Constraining mantle density structure using geological evidence of surface uplift rates: the case of the African superplume. *Geochemistry, Geophysics, Geosystems*, **1**, 1020, doi: 10.1029/1999GC000035.

HADIOUCHE, O. & JOBERT, N. 1988. Geographical distribution of surface-wave velocities and 3D upper mantle structure in Africa. *Geophysical Journal International*, **95**, 87–109.

HANKA, W., YUAN, X., KIND, R., WACKERLE, R., WYLEGALLA, K., BOCK, G. & TRUMBULL, R. 2000. First insights to upper mantle structure under the Damara Belt Namibia, from receiver function study. *Eos, Transactions, American Geophysical Union Supplement*, Fall Meeting Abstracts, F831.

HANSEN, S. E., NYBLADE, A. A. & JULIÀ, J. 2009a. Estimates of crustal and lithospheric thickness in subsaharan Africa from S-wave receiver functions. *South African Journal of Geology*, **112**, 229–240.

HANSEN, S. E., NYBLADE, A. A., JULIÀ, J., DIRKS, P. H. G. M. & DURRHEIM, R. J. 2009b. Upper-mantle low-velocity zone structure beneath the Kaapvaal craton from S-wave receiver functions. *Geophysical Journal International*, **178**, 1021–1027.

HARTLEY, R. & ALLEN, P. A. 1994. Interior cratonic basins of Africa: relation to continental breakup and role of mantle convection. *Basin Research*, **6**, 95–113.

HARVEY, J. D., DE WIT, M. J., STANKIEWICZ, J. & DOUCOURÉ, C. M. 2001. Structural variations of the crust in the Southwestern Cape, deduced from seismic receiver functions. *South African Journal of Geology*, **104**, 231–242.

HEINE, C., MÜLLER, R. D., STEINBERGER, B. & TORSVIK, T. H. 2008. Subsidence in intracontinental basins due to dynamic topography. *Physics of the Earth and Planetary*, **171**, 252–264.

HOFMANN, C., COURTILLOT, V., FERAUD, G., ROCHETTE, P., YIRGU, G., KETEFO, E. & PIK, R. 1997. Timing of the Ethiopian flood basalt event and implications for plume birth and global change. *Nature*, **389**, 838–841.

HOLMES, A. 1944. *Principles of Physical Geology*. Thomas Nelson and Sons Limited, Edinburgh.

HOLT, P., ALLEN, M. B., VAN HUSEN, J. & BJØRNSETH, H. M. 2010. Lithospheric cooling and thickening as a basin forming mechanism. *Tectonophysics*, **495**, 184–194.

JAMES, D. E., FOUCH, M. J., VANDECAR, J. C., VAN DER LEE, S. & KAAPVAAL SEISMIC GROUP. 2001. Tectospheric structure beneath southern Africa. *Geophysical Research Letters*, **28**, 2485–2488.

JORDAN, T. H. 1975. The continental tectosphere. *Reviews of Geophysics and Space Physics*, **13**, 1–12.

JULIÀ, J., AMMON, C. & NYBLADE, A. 2005. Evidence for mafic lower crust in Tanzania, East Africa, from joint inversion of receiver functions and Rayleigh wave dispersion velocities. *Geophysical Journal International*, **162**, 555–569.

KAMINSKI, E. & JAUPART, C. 2000. Lithosphere structure beneath the Phanerozoic intracratonic basins of North America. *Earth and Planetary Science Letters*, **178**, 139–149.

KAMINSKI, E. & RIBE, N. M. 2002. Timesclaes for the evolution of seismic anisoptropy in mantle flow. *Geochemistry, Geophysics, Geosystems*, **3**, 1051, doi: 10.1029/2001GC000222.

KEIR, D., BASTOW, I. D., WHALER, K. A., DALY, E., CORNWELL, D. G. & HAUTOT, S. 2009. Lower crustal earthquakes near the Ethiopian rift induced by magmatic processes. *Geochemistry, Geophysics, Geosystems*, **10**, Q0AB02, doi: 10.1029/2009GC002382.

KENDALL, J., STUART, G. W., EBINGER, C. J., BASTOW, I. D. & KEIR, D. 2005. Magma assisted rifting in Ethiopia. *Nature*, **433**, 146–148.

KENDALL, J., PILIDOU, S., KEIR, D., BASTOW, I. D., STUART, G. W. & AYELE, A. 2006. Mantle upwellings, melt migration and the rifting of Africa: insights from seismic anisotropy. *In*: YIRGU, G., EBINGER, C. J. & MAGUIRE, P. K. H. (eds) *The Afar Volcanic Province within the East African Rift System*. Geological Society, London, Special Publications, **259**, 55–72.

KENNETT, B. L. N., ENGDAHL, E. R. & BULAND, R. 1995. Constraints on seismic velocities in the Earth from travel times. *Geophysical Journal International*, **122**, 108–124.

KERANEN, K., KLEMPERER, S., GLOAGUEN, R. & EAGLE WORKING GROUP. 2004. Three dimensional seismic imaging of a protoridge axis in the main Ethiopian rift. *Geology*, **32**, 949–952.

KGASWANE, E. M., NYBLADE, A. A., JULIÀ, J., DIRKS, P. H. G. M., DURRHEIM, R. J. & PASYANOS, M. E. 2009. Shear wave velocity structure of the lower crust in southern Africa: evidence for compositional heterogeneity within Archaean and Proterozoic

terrains. *Journal of Geophysical Research*, **114**, B12304, doi: 10.1029/2008JB006217.

KIEFFER, B., ARNDT, N. *ET AL.* 2004. Flood and shield basalts from Ethiopia: magmas from the African superswell. *Journal of Petrology*, **45**, 793–834.

KING, S. D. & ANDERSON, D. L. 1998. Edge-driven convection. *Earth and Planetary Science Letters*, **160**, 289–296.

KING, S. D. & RITSEMA, J. 2000. African hot spot volcanism: small-scale convection in the upper mantle beneath cratons. *Science*, **290**, 1137–1140.

KRISP WORKING GROUP. 1987. Structure of the Kenya rift from seismic refraction. *Nature*, **325**, 239–242.

LARSON, E. W. F. & EKSTRÖM, G. 2001. Global models of surface wave group velocity. *Pure and Applied Geophysics*, **158**, 1377–1399.

LAST, R. J., NYBLADE, A. A., LANGSTON, C. A. & OWENS, T. J. 1997. Crustal structure of the East African Plateau from receiver functions and Rayleigh wave phase velocities. *Journal of Geophysical Research*, **102**, 24469–24483.

LEBEDEV, S. & VAN DER HILST, R. D. 2008. Global upper-mantle tomography with the automated multimode inversion of surface and S-wave forms. *Geophysical Journal International*, **173**, 505–518.

LESQUER, A., TAKHERIST, D., DAUTRIA, J. M. & HADIOUCHE, O. 1990. Geophysical and petrological evidence for the presence of an 'anomalous' upper mantle beneath the Sahara basins (Algeria). *Earth and Planetary Science Letters*, **96**, 407–418.

LI, A. & BURKE, K. 2006. Upper mantle structure of southern Africa from Rayleigh wave tomography. *Journal of Geophysical Research*, **111**, B10303, doi: 10.1029/2006JB004321.

LI, C., VAN DER HILST, R. D., ENGDAHL, R. & BURDICK, S. 2008. A new global model for P wave speed variations in Earth's mantle. *Geochemistry, Geophysics, Geosystems*, **9**, Q05018, doi: 10.1029/2007GC001806.

LITHGOW-BERTELLONI, C. & SILVER, P. 1999. Dynamic topography, plate driving forces and the African superswell. *Nature*, **395**, 269–272.

LIU, K. H. & GAO, S. S. 2010. Spatial variations of crustal characteristics beneath the Hoggar swell, Algeria, revealed by systematic analyses of receiver functions from a single seismic station. *Geochemistry, Geophysics, Geosystems*, **11**, Q08011, doi: 10.1029/2010GC003091.

MACKENZIE, G. D., THYBO, H. & MAGUIRE, P. K. H. 2005. Crustal velocity structure across the Main Ethiopian Rift: Results from two-dimensional wide-angle seismic modelling. *Geophysical Journal International*, **162**, 994–1006.

MAGUIRE, P. K. H., EBINGER, C. J. *ET AL.* 2003. Geophysics project in Ethiopia studies continental breakup. *Eos, Transactions, American Geophysical Union*, **84**, 342–343.

MAGUIRE, P. K. H., KELLER, G. R. *ET AL.* 2006. Crustal structure of the northern Main Ethiopian Rift from the EAGLE controlled-source survey; a snapshot of incipient lithospheric break-up. *In*: YIRGU, G., EBINGER, C. J. & MAGUIRE, P. K. H. (eds) *The Afar Volcanic Province within the East African Rift System*. Geological Society, London, Special Publications, **259**, 269–292.

McKENZIE, D. P. 1978. Some remarks on the development of sedimentary basins. *Earth and Planetary Science Letters*, **40**, 23–32.

MECHIE, J., KELLER, G. R., PRODEHL, C., KHAN, M. A. & GACIRI, S. J. 1997. A model for the structure, composition and evolution of the Kenya rift. *Tectonophysics*, **278**, 95–119.

MICKUS, K., TADESSE, K., KELLER, G. & OLUMA, B. 2007. Gravity analysis of the main Ethiopian rift. *Journal of African Earth Sciences*, **48**, 59–69.

MONTAGNER, J.-P., MARTY, B. *ET AL*. 2007. Mantle upwellings and convective instabilities revealed by seismic tomography and helium isotope geochemistry beneath eastern Africa. *Geophysics Research Letters*, **34**, L21303, doi: 10.1029/2007GL031098.

MONTELLI, R., NOLET, G., DAHLEN, F. & MASTERS, G. 2006. A catalogue of deep mantle plumes: new results from finite-frequency tomography. *Geochemistry, Geophysics, Geosystems*, **7**, Q110007, doi: 10.1029/2006GC001248.

MOORKAMP, M., JONES, A. G. & FISHWICK, S. 2010. Joint inversion of receiver functions, surface wave dispersion and magnetotelluric data. *Journal of Geophysical Research*, **115**, B04318, doi: 10.1029/2009JB006369.

MORLEY, C., WESCOTT, W., STONE, D., HARPER, R., WIGGER, S. & KARANJA, F. 1992. Tectonic evolution of the northern Kenyan Rift. *Journal of the Geological Society*, **149**, 333.

NAIR, S. K., GAO, S. S., LIU, K. H. & SILVER, P. G. 2006. Southern African crustal evolution and composition: constraints from receiver function studies. *Journal of Geophysical Research*, **111**, B02304, doi: 10.1029/2005JB003802.

NATAF, H.-C. & RICARD, Y. 1996. 3SMAC: an a priori tomographic model of the upper mantle based on geophysical modeling. *Physics of the Earth and Planetary Interiors*, **95**, 101–122.

NGUURI, T. K., GORE, J. *ET AL*. 2001. Crustal structure beneath southern Africa and its implications for the formation and evolution of the Kaapvaal and Zimbabwe cratons. *Geophysics Research Letters*, **28**, 2501–2504.

NIU, F. & JAMES, D. E. 2002. Fine structure of the lowermost crust beneath the Kaapvaal Craton and its implications for crustal formation and evolution. *Earth and Planetary Science Letters*, **200**, 121–130.

NYBLADE, A. A. 2011. The upper mantle low velocity anomaly beneath Ethiopia, Kenya and Tanzania: Constraints on the origin of the African Superswell in eastern Africa and plate v. plume models of mantle dynamics. *In*: BECCALUVA, L., BIANCHINI, G. & WILSON, M. (eds) *Volcanism and Evolution of the African Lithosphere*. Geological Society of America, Special Papers, **478**, 37–50.

NYBLADE, A. & ROBINSON, S. 1994. The African superswell. *Geophysics Research Letters*, **21**, 765–768.

NYBLADE, A. & LANGSTON, C. A. 2002. Broadband seismic experiments probe the East African Rift. *Eos, Transactions, American Geophysical Union*, **83**, 405–408.

NYBLADE, A. A. & SLEEP, N. H. 2003. Long lasting epeirogenic uplift from mantle plumes and the origin of the Southern African Plateau. *Geochemistry, Geophysics, Geosystems*, **4**, 1105, doi: 10.1029/2003GC000573.

NYBLADE, A., BIRT, C., LANGSTON, C. A., OWENS, T. J. & LAST, R. J. 1996. Seismic experiment reveals rifting of craton in Tanzania. *Eos, Transactions, American Geophysical Union*, **77**, 520–521.

NYBLADE, A. A., DIRKS, P., DURRHEIM, R., WEBB, S., JONES, M., COOPER, G. & GRAHAM, G. 2008. Africaarray: developing a geosciences workforce for africa's natural resource sector. *Leading Edge*, **27**, 1358–1361.

PARK, Y. & NYBLADE, A. 2006. P-wave tomography reveals a westward dipping low velocity zone beneath the Kenya Rift. *Geophysics Research Letters*, **33**, L07311, doi: 10.1029/2005GL025605.

PASYANOS, M. 2010. Lithospheric thickness modelled from long-period surface wave dispersion. *Tectonophysics*, **481**, 38–50.

PASYANOS, M. & NYBLADE, A. A. 2007. A top to bottom lithospheric study of Africa and Arabia. *Tectonophysics*, **444**, 27–44.

PEDERSEN, H. A., FISHWICK, S. & SNYDER, D. B. 2009. A comparison of cratonic roots through consistent analysis of seismic surface waves. *Lithos*, **109**, 81–95.

PÉREZ-GUSSINYÉ, M., METOIS, M., FERNÀNDEZ, M., VERGÉS, J., FULLEA, J. & LOWRY, A. R. 2009. Effective elastic thickness of Africa and its relationship to other proxies for lithospheric structure and surface tectonics. *Earth and Planetary Science Letters*, **287**, 152–167.

PIK, R., MARTY, B., CARIGNAN, J. & LAVÉ, J. 2003. Stability of the Upper Nile drainage network (Ethiopia) deduced from (U–Th)/He thermochronometry: implications for uplift and erosion of the Afar plume dome. *Earth and Planetary Science Letters*, **215**, 73–88.

PIK, R., MARTY, B. & HILTON, D. R. 2006. How many mantle plumes in Africa? The geochemical point of view. *Chemical Geolology*, **226**, 100–114.

PIK, R., MARTY, B., CARIGNAN, J., YIRGU, G. & AYALEW, T. 2008. Timing of East African Rift development in southern Ethiopia: implication for mantle plume activity and evolution of topography. *Geology*, **36**, 167–170.

PLOMEROVÁ, J., BABUŠKA, V., DORBATH, C., DORBATH, L. & LILLIE, R. J. 1993. Deep lithospheric structure across the Central African Shear Zone in Cameroon. *Geophysical Journal International*, **115**, 381–390.

POUPINET, G. 1979. On the relation between P-wave travel time residuals and the age of the continental plates. *Earth and Planetary Science Letters*, **43**, 149–161.

PRIESTLEY, K. 1999. Velocity structure of the continental upper mantle: evidence from southern Africa. *Lithos*, **48**, 45–56.

PRIESTLEY, K. & McKENZIE, D. 2006. The thermal structure of the lithosphere from shear wave velocities. *Earth and Planetary Science Letters*, **244**, 285–301.

PRIESTLEY, K. & TILLMANN, F. 2009. Relationship between the upper mantle high velocity seismic lid and the continental lithosphere. *Lithos*, **109**, 112–124.

PRIESTLEY, K., MCKENZIE, D., DEBAYLE, E. & PILIDOU, S. 2008. The African upper mantle and its relationship to tectonics and surface geology. *Geophysical Journal International*, **175**, 1108–1126.

PRODEHL, C. & MECHIE, J. 1991. Crustal thinning in relationship to the evolution of the Afro-Arabian rift system: a review of seismic refraction data. *Tectonophysics*, **198**, 311–327.

REUSCH, A. M., NYBLADE, A. A., WIENS, D. A., SHORE, P. J., ATEBA, B., TABOD, C. T. & NNANGE, J. M. 2010. Upper mantle structure beneath Cameroon from body wave tomography and the origin of the Cameroon Volcanic Line. *Geochemistry, Geophysics, Geosystems*, **11**, Q10W07, doi: 10.1029/2010GC003200.

RITSEMA, J. & VAN HEIJST, H. 2000. New seismic model of the upper mantle beneath Africa. *Geology*, **28**, 63–66.

RITSEMA, J. & ALLEN, R. M. 2003. The elusive mantle plume. *Earth and Planetary Science Letters*, **207**, 1–12.

RITSEMA, J., NYBLADE, A. A., OWENS, T. J., LANGSTON, C. A. & VANDECAR, J. C. 1998. Upper mantle seismic velocity structure beneath Tanzania, East Africa: implications for the stability of cratonic lithosphere. *Journal of Geophysical Research*, **103**, 21201–21213.

ROBERTS, G. G. & WHITE, N. 2010. Estimating uplift rate histories from river profiles using Adaman examples. *Journal of Geophysical Research*, **115**, B02406, doi: 10.1029/2009JB006692.

RYCHERT, C. A., SHEARER, P. M. & FISCHER, K. M. 2010. Scattered wave imaging of the lithosphere–asthenosphere boundary. *Lithos*, **120**, 173–185.

SAVAGE, B. & SILVER, P. G. 2008. Evidence for a compositional boundary within the lithospheric mantle beneath the Kalahari craton from S receiver functions. *Earth and Planetary Science Letters*, **272**, 600–609.

SEBAI, A., STUTZMANN, E., MONTAGNER, J.-P., SICILIA, D. & BEUCLER, E. 2006. Anisotropic structure of the African upper mantle from Rayleigh and Love wave tomography. *Physics of the Earth and Planetary Interiors*, **155**, 48–62.

SEBER, D., SANDVOL, E., SANDVOL, C., BRINDISI, C. & BARAZANGI, M. 2001. Crustal model for the Middle East and North Africa region: implications for the isostatic compensation mechanism. *Geophysical Journal International*, **147**, 630–638.

SHACKLETON, R. 1986. Precambrian collision tectonics in Africa. *In*: COWARD, M. P. & RIES, A. C. (eds) *Collision Tectonics*. Geological Society, London, Special Publications, **19**, 329–349.

SICILIA, D., MONTAGNER, J. P. *ET AL*. 2008. Upper mantle structure of shear-waves velocities and stratification of anisotropy in the Afar Hotspot region. *Tectonophysics*, **462**, 164–177.

SILVER, P. & CHAN, G. 1991. Shear wave splitting and subcontinental mantle deformation. *Journal of Geophysical Research*, **96**, 16429–16454.

SILVER, P. G., GAO, S. S. & LIU, K. H. (THE KAAPVAAL SEISMIC GROUP). 2001. Mantle deformation beneath Southern Africa. *Geophysics Research Letters*, **28**, 2493–2496.

SIMMONS, N. A., FORTE, A. M. & GRAND, S. P. 2007. Thermochemical structure and dynamics of the African superplume. *Geophysics Research Letters*, **34**, L02301, doi: 10.1029/2006GL028009.

SIMMONS, N. A., FORTE, A. M. & GRAND, S. P. 2009. Joint seismic, geodynamic and mineral physical constraints on three-dimensional mantle heterogeneity: implications for the relative importance of thermal v. compositional heterogeneity. *Geophysical Journal International*, **177**, 1284–1304.

SIMMONS, N. A., FORTE, A. M., BOSCHI, L. & GRAND, S. P. 2010. GyPSuM: A joint tomographic model of mantle density and seismic wave speeds. *Journal of Geophysical Research*, **115**, B12310, doi: 10.1029/2010JB007631.

SPIEGEL, C., KOHN, B. P., BELTON, D. X. & GLEADOW, A. J. W. 2007. Morphotectonic evolution of the central Kenya rift flanks: implications for late Cenozoic environmental change in East Africa. *Geology*, **35**, 427–430.

STANKIEWICZ, J., CHEVROT, S., VAN DER HILST, R. D. & DE WIT, M. J. 2002. Crustal thickness, discontinuity depth, and upper mantle structure beneath southern Africa: constraints from body wave conversions. *Physics of the Earth and Planetary Interiors*, **130**, 235–251.

STUART, G. W., FAIRHEAD, J. D., DORBATH, L. & DORBATH, C. 1985. A seismic refraction study of the crustal structure associated with the Adamawa Plateau and Garoua Rift, Cameroon, West Africa. *Geophysical Journal of the Royal Astronomical Society*, **81**, 1–12.

STUART, G. W., BASTOW, I. D. & EBINGER, C. J. 2006. Crustal structure of the northern Main Ethiopian rift from receiver function studies. *In*: YIRGU, G., EBINGER, C. J. & MAGUIRE, P. K. H. (eds) *The Afar Volcanic Province within the East African Rift System*. Geological Society, London, Special Publications, **253**, 271–293.

TAPLEY, B., RIES, J. *ET AL*. 2005. GGM02 – an improved Earth gravity field model. *Journal of Geodynamics*, **79**, 467–478.

TEIXELL, A., ARBOLEYA, M.-L., JULIVERT, M. & CHARROUD, M. 2003. Tectonic shortening and topography in the central High Atlas (Morocco). *Tectonics*, **22**, 1051, doi: 10.1029/2002TC001460.

TEIXELL, A., AZARZA, P., ZEYEN, H., FERNÀNDEZ, M. & ARBOLEYA, M.-L. 2005. Effects of mantle upwelling in a compressional setting: the Atlas Mountains of Morocco. *Terra Nova*, **17**, 456–461.

TIBI, R., LARSON, A. M. *ET AL*. 2005. A broadband seismological investigation of the Cameroon Volcanic Line. *Eos, Transactions, American Geophysical Union*, **86** (Abstract S11B-0170).

TINKER, J., DE WIT, M. & BROWN, R. 2008. Linking source and sink: evaluating the balance between onshore erosion and offshore sediment accumulation since Gondwana break-up, South Africa. *Tectonophysics*, **455**, 94–103.

TOKAM, A.-P. K., TABOD, C. T., NYBLADE, A. A., JULIÀ, J., WIENS, D. A. & PASYANOS, M. E. 2010. Structure of the crust beneath Cameroon, West Africa, from the joint inversion of Rayleigh wave group velocities and receiver functions. *Geophysical Journal International*, **183**, 1061–1076.

VANDECAR, J. C., JAMES, D. E. & ASSUMPÇÃO, M. 1995. Seismic evidence for a fossil mantle plume beneath

South America and implications for plate driving forces. *Nature*, **378**, 25–31.

VINNIK, L., ORESHIN, S., KOSAREV, G., KISELEV, S. & MAKEYEVA, L. 2009. Mantle anomalies beneath southern Africa: evidence from seismic S and P receiver functions. *Geophysical Journal International*, **179**, 279–298.

WALKER, K., NYBLADE, A., KLEMPERER, S., BOKELMANN, G. & OWENS, T. 2004. On the relationship between extension and anisotropy: constraints from shear wave splitting across the East African Plateau. *Journal of Geophysical Research*, **109**, B083202, doi: 10.1029/2003JB002866.

WANG, Y., WEN, L. & WEIDNER, D. 2008. Upper mantle SH- and P- velocity structures and compositional models beneath southern Africa. *Earth and Planetary Science Letters*, **267**, 596–608.

WEBB, S. J., CAWTHORN, R. G., NGUURI, T. & JAMES, D. 2004. Gravity modeling of Bushveld Complex connectivity supported by Southern African Seismic Experiment results. *South African Journal of Geology*, **107**, 207–218.

WEERARATNE, D. S., FORSYTH, D. W., FISCHER, K. M. & NYBLADE, A. A. 2003. Evidence for an upper mantle plume beneath the Tanzanian craton from Rayleigh wave tomography. *Journal of Geophysical Research*, **108**, 2427, doi: 10.1029/2001JB001225.

WHALER, K. A. & HAUTOT, S. 2006. The electrical resistivity structure of the crust beneath the northern Ethiopian rift. *In*: YIRGU, G., EBINGER, C. J. & MAGUIRE, P. K. H. (eds) *The Afar Volcanic Province within the East African Rift System*. Geological Society, London, Special Publications, **256**, 293–305.

WILSON, M. & GUIRAUD, R. 1992. Magmatism and rifting in Western and Central Africa, from Late Jurassic to Recent times. *Tectonophysics*, **213**, 203–225.

WITTLINGER, G. & FARRA, V. 2007. Converted waves reveal a thick and layered tectosphere beneath the Kalahari super-craton. *Earth and Planetary Science Letters*, **254**, 404–415.

WÖLBERN, I., RÜMPKER, G., SCHUMANN, A. & MUWANGA, A. 2010. Crustal thinning beneath the Rwenzori region, Albertine rift, Uganda, from receiver-function analysis. *International Journal of Earth Sciences*, **99**, 1545–1557.

YANG, Y., LI, A. & RITZWOLLER, M. H. 2008. Crustal and uppermost mantle structure in southern Africa revealed from ambient noise and teleseismic tomography. *Geophysical Journal International*, **174**, 235–248.

ZEYEN, H., AYARZA, P., FERNÀNDEZ, M. & RIMI, A. 2005. Lithospheric structure under the western African–European plate boundary: a transect across the Atlas Mountains and the Gulf of Cadiz. *Tectonics*, **24**, doi: 10.1029/2004TC001639.

ZHU & KANAMORI 2000. Moho depth variation in southern California from teleseismic receiver functions. *Journal of Geophysical Research*, **105**, 2969–2980.

Index

Page numbers in *italics* refer to Figures. Page numbers in **bold** refer to Tables.